Proceedings of the 12th International Zeolite Conference IV

Conference held July 5-10, 1998
Baltimore, Maryland, U.S.A.

MATERIALS RESEARCH SOCIETY CONFERENCE PROCEEDINGS

Proceedings of the 12th International Zeolite Conference IV

Conference held July 5-10, 1998
Baltimore, Maryland, U.S.A.

Editors:

M.M.J. Treacy
NEC Research Institute, Inc.
Princeton, New Jersey, U.S.A.

B.K. Marcus
Zeolyst International
Conshohocken, Pennsylvania, U.S.A.

M.E. Bisher
NEC Research Institute, Inc.
Princeton, New Jersey, U.S.A.

J.B. Higgins
Air Products & Chemicals, Inc.
Allentown, Pennsylvania, U.S.A.

Materials Research Society
Warrendale, Pennsylvania

Volume IV cover: Principle of particle fractionation in a CHDF capillary.
(K.T. Jung and Y.G. Shul, Volume III, page 1523)

Single article reprints from this publication are available through
University Microfilms Inc., 300 North Zeeb Road, Ann Arbor, Michigan 48106

CODEN: MRSPDH

Copyright 1999 by Materials Research Society.
All rights reserved.

This book has been registered with Copyright Clearance Center, Inc. For further
information, please contact the Copyright Clearance Center, Salem, Massachusetts.

Published by:

Materials Research Society
506 Keystone Drive
Warrendale, PA 15086
Telephone (724) 779-3003
Fax (724) 779-8313
Website: http://www.mrs.org/

Library of Congress Cataloging in Publication Data

International Zeolite Conference (12th : 1998 : Baltimore, Maryland)
Proceedings of the 12th International Zeolite Conference / editors,
    M.M.J. Treacy, B.K. Marcus, M.E. Bisher, J.B. Higgins
        p.cm.—(Materials Research Society conference proceedings)
        Includes bibliographical references and indexes.
        ISBN 1-55899-463-7 (set)
        1. Zeolites—Congresses. I. Treacy, M.M.J. (Michael M.J.) II. Marcus, B.K.
    III. Bisher, M.E. IV. Higgins, J.B. V. Title VI. Series
TP245.S5163 1998   98-46348
666'.86—dc21   CIP

Manufactured in the United States of America

TABLE OF CONTENTS

VOLUME I

VOLUME 1 3
VOLUME 2 733
VOLUME 3 1479
VOLUME 4 2259

Prefaces
   Chairs of the 12th International Zeolite Conference xlv
   Publications Chair xlvii

Executive Committee of the 12th IZC xlviii

Program Committee of the 12th IZC xlix

Local Arrangements Committee xlix

Pre-Conference Zeolite School l

Publications l

International Advisory Board to the 12th IZC li

Session Chairs of the 12th IZC lii

Financial Support liii

PART I: BARRER SYMPOSIUM

R.M. Barrer Symposium: Contributions to Zeolite Synthesis
and Modification 3
   D.E.W. Vaughan

Richard Barrer Symposium: His Contributions to Sorption and
Ion Exchange in Zeolites 13
   L.V.C. Rees

Three Score Years with Membranes: The Legacy of R.M. Barrer 25
   W.J. Koros and A.S. Michaels

PART II: FUNDAMENTALS OF DIFFUSION

PLENARY LECTURE
Molecular Transport in Zeolites—Miracles, Insights and
Practical Issues 35
   J. Kärger

Site-Hopping Dynamics of Benzene Adsorbed on
Calcium-Exchanged Faujasite-Type Zeolites 43
    D.E. Favre, D.J. Schaefer, and B.F. Chmelka

Different Translational Mobility Types of Confined
Molecules in the One-Dimensional Micropores of Model
Zeolites: AlPO$_4$-5 and AlPO$_4$-11 51
    J.P. Coulomb, C. Martin, Y. Grillet, and R. Kahn

Diffusivities of Xylenes in X-Type Zeolites: A Quasi-Elastic
Neutron Scattering Study 59
    H. Jobic, M. Bee, B. Frick, and A. Methivier

Frequency Response Studies of the Adsorption and the
Transport of C_1 to C_6 n-Alkanes in Silicalite-1 67
    L. Song and L.V.C. Rees

PART III: ADSORPTION

Diffusion of Hydrogen in Various Zeolites Studied by
Pulsed-Field Gradient NMR and Quasi-Elastic Neutron
Scattering Techniques 77
    N-K. Bär, H. Jobic, and J. Kärger

Accessibility of Cation (Li, Na) Sites in Li,Na-LSX Zeolite Using
Paramagnetic O_2 as a Chemical Shift Agent—A ^{6}Li and ^{23}Na
MAS NMR Spectroscopic Study 85
    L.C. de Ménorval, J. Plévert, R. Dutartre, F. Di Renzo, and F. Fajula

Experimental and Computational Studies of the Adsorption
Properties of n-Alkanes in Ferrierite 91
    W.J.M. van Well, X. Cottin, J.W. de Haan, B. Smit,
    J.H.C. van Hooff, and R.A. van Santen

The Influence of Microwave Energy on Adsorption in Zeolites 97
    M.D. Turner, W.C. Conner, R.L. Laurence, and S. Yngvesson

Experimental Studies of Adsorption of Mixtures in Zeolites
Using Microcalorimetry 103
    F. Siperstein, R.J. Gorte, and A.L. Myers

Sorption Thermodynamics of Oxygen and Argon on a
Faujasite Type Zeolite 111
    D. Shen and M. Bülow

Adsorption Sites in Zeolites A and X Probed by Competitive
Adsorption of H_2 with N_2 or O_2: Implications for N_2/O_2
Separation 119
    J. Eckert, F. Trouw, A.L.R. Bug, and R. Lobo

Sorption of $C_{6,7,8}$ Alkanes in Aluminophosphate Molecular
Sieve, AlPO$_4$-5 127
    B.L. Newalkar, R.V. Jasra, V. Kamath, and S.G.T. Bhat

Cation Positions in Dehydrated Zeolites Li-LSX and Li,Na-LSX 135
    J. Plévert, F. Di Renzo, F. Fajula, and G. Chiari

The Diffusion Property, External Acidity and Catalytic
Performances of Nano-Crystalline HZSM-5 Zeolite 141
    G. Hongchen, W. Xiangsheng, and W. Guiru

Theoretical and Experimental Results for Diffusion Coefficients
of Mixtures in Zeolites 149
    S. Jost, N-K. Bär, S. Fritzsche, R. Haberlandt, and J. Kärger

Theoretical Analysis and Simulation of Diffusion and Catalysis
in Single-File Systems 153
    K. Hahn, J. Kärger, and C. Rödenbeck

A Frequency-Response Study on the Sorption of CO_2 in Zeolite
A, X and Y 159
    G. Onyestyák, J. Valyon, and L.V.C. Rees

Toward a Better Understanding on the Benzene Location in
Sodium-Exchanged Beta Zeolite: An *In Situ* Infrared Study of
Molecular Recognition 167
    V. Norberg and B-L. Su

Modification of Zeolites by Spontaneous Thermal Dispersion
of Salts and Metal Oxides and Its Applications 175
    C-B. Wang, Y. Xie, and Y. Tang

FTIR Microscopy on Single Crystals: Diffusion of Hydrocarbons
in Silicalite 183
    V.L. Zholobenko and J. Dwyer

An Illustrative Example of One-Dimensional Phase:
Neopentane Confined in the Model $AlPO_4$-5 Zeolite 189
    C. Martin, J.P. Coulomb, P.L. Llewellyn, and Y. Grillet

Structural Analysis of Methane Sorbed Phases on Silicalite I,
Mordenite, $AlPO_4$-5 and $AlPO_4$-11 197
    J.P. Coulomb, C. Martin, N. Floquet, P.L. Llewellyn, and Y. Grillet

Solvent Recovery by Pressure Swing Adsorption with High
Silica Zeolite 203
    K. Chihara, H. Kimbara, and Y. Takeuchi

Role of Immobilization and Mobilization Processes in the
Kinetics of Transformation of p-Ethyltoluene Over HZSM-5 209
    A. Zikánová, M. Derewinski, J. Krysciak, and M. Kocirik

Sorbate Immobilization in MFI and FAU Types of Zeolites 215
    M. Jama, M. Kocirik, M. Eic, J. Dubsky, A. Zikánová, M. Mello,
    and A. Micke

An Attempt to Correlate the Non-Isothermal Desorption
Behavior of Different Molecules on a NaX Zeolite with
Their Softness 223
    B. Hunger, O. Klepel, C. Engler, M. Heuchel, and E. Geidel

Direct Observation of Intracrystalline Transport Diffusion
by Interference Microscopy 231
    U. Schemmert and J. Kärger

Determination of Adsorption Constants and Diffusion
Coefficients of Aromatics in Zeolites by Transient
Experiments in a Recycle Reactor 235
    E. Klemm and G. Emig

Sorption and Diffusion of Propane, Propylene, Pentane and
Hexane in Pellets of 13X and 5A Zeolites Studied by the
Gravimetric and ZLC Techniques 243
    J.A.C. Silva, F.A. da Silva, and A.E. Rodrigues

Molecular Segregation Effects on Adsorption in Zeolites 251
    L.A. Clark, A. Gupta, and R.Q. Snurr

Prediction of the Binary Adsorption Equilibria of CO_2 and CH_4
on SAPO-5 and SAPO-11 257
    A. Sirkecioglu, L.F. Petrik, and A. Erdem-Senatalar

Inverse Gas Chromatography Studies of Alkali Cation
Exchanged X Zeolites 263
    J. Xie, M. Bousmina, G. Xu, and S. Kaliaguine

Low Temperature Dihydrogen Adsorption as a Probe for the
Study of Alkaline Cationic Forms of Faujasites 269
    V.B. Kazansky, V.Yu. Borovkov, and H.G. Karge

Application of Alanine Adsorbed on the Molecular Sieves for
the Dosimetric Measurements 277
    S. Kowalak, Z. Stuglik, K. Kruszona, and A. Jankowska

Sorption of Water on Molecular Sieves of AFI Structure Type—
An Explanation of Mechanism and Sorption Capacity 285
    J. Kornatowski and M. Rozwadowski

Water Adsorption and Dealumination Behavior of HZSM-12
Zeolite 293
    Z.B. Wang, H. Ikeya, T. Sano, and K. Soga

Reversible and Irreversible Adsorption of Water on HZSM-5
Zeolite 301
    T. Sano, S. Fukuya, Z.B. Wang, and K. Soga

Adsorption of Nitrogen and Carbon Dioxide on ZSM-5 309
    F.H. Tezel, P. Harlick, and A. Sirkecioglu

Adsorption of H_2, N_2, CO, NO and O_2 on Sodium and
Potassium Titanosilicate (Na-ETS10 and K-ETS10):
A FTIR Study 317
    A. Zecchina, C. Otero Areán, C. Lamberti, G. Turnes Palomino,
    G. Spoto, D. Scarano, S. Bordiga, and F. Geobaldo

PART IV: THEORY AND MODELLING

Adsorption of Linear and Branched Alkanes in Ferrierite:
A Computational Study 325
    T.J.H. Vlugt, B. Smit, and R. Krishna

A Computational Study of Ethane Cracking in Cluster Models
of Zeolite H-ZSM-5 333
    S.A. Zygmunt, L.A. Curtiss, and L.E. Iton

Study of Short and Long Range Effects on Brønsted Acidity in
AFI and CHA Zeotypes 341
    G. Sastre and D.W. Lewis

Structure and Reaction Pathways for Methylamine/
Zeolite System 349
    J. Limtrakul and M. Kuno

Modelling Benzene Diffusion in Na-X and Na-Y Zeolites at
Finite Loadings 357
    S.M. Auerbach, C. Saravanan, and F. Jousse

Grand Canonical Monte Carlo Simulation of Single
Component and Binary Mixture Adsorption in Zeolites 363
    M.D. Macedonia and E.J. Maginn

Transport and Self-Diffusion Coefficients in Zeolites—
An MD Study 371
    M. Gaub, S. Fritzsche, R. Haberlandt, and D.N. Theodorou

Theoretical Studies of the Structure and Properties of
Cobalt-Substituted Aluminophosphates 379
    N.J. Henson, P.J. Hay, and A. Redondo

Semi-Empirical and Ab Initio Calculations of the Spectroscopic
Properties of Co(II) Coordinated to Six-Rings in Zeolites 387
    A.A. Verberckmoes, R.A. Schoonheydt, A. Ceulemans, A. Delabie,
    and K. Pierloot

Applications of Density Functional Theory to Study the
Physical and Chemical Properties of Transition Metal
Cation-Exchanged Zeolites 393
    M.J. Rice, N.O. Gonzales, A.K. Chakraborty, and A.T. Bell

Coordination and Electronic Structure of Extraframework
Transition Metal-Ions by High-Resolution X-ray Photoelectron
and Optical Spectroscopy 401
    K. Klier, B. Wichterlová, and J. Dedecek

How Big Are the Pores of Zeolites? 409
  M. Cook and W.C. Conner

Ab Initio and Density Functional Studies of Hydrocarbon
Interaction with Zeolitic Clusters 415
  L.A. Curtiss, S.A. Zygmunt, and L.E. Iton

Vibrational Spectra of Highly Aluminated Faujasites:
MD Simulation Study 423
  M. Koudriachova and E. Geidel

Theoretical Studies on the Effect of Site Proximity on the
Zeolite Y Acid Strength 429
  P.M. Esteves, M.A.C. Nascimento, and C.J.A. Mota

A Computational Study of NOx Adsorption on Co-ZSM-5
Zeolite 437
  J.S. Kim, S-E. Park, J-S. Chang, Y.K. Park, and C.W. Lee

A Molecular Dynamic Approach on the Determination of
HMI(Hexamethleneimine) Configurations in MCM-22 445
  M. Sato, K. Ando, H. Uehara, and M. Miyake

A New Electron Spin Density Equation for Alkene Cation
Radicals on Zeolitic Surface 451
  S. Shih

Characterization of Defect Centers in Zeolites Using Computer
Modelling Techniques 457
  A.A. Sokol, C.R.A. Catlow, J.M. Garces, and A. Kuperman

An Explanation for the Enhanced Hydrocarbon Cracking
Activity of Steamed Y Zeolites 465
  B.A. Williams, S.M. Babitz, J.T. Miller, R.Q. Snurr, and H.H. Kung

Interaction of NO Molecules with Highly Dispersed Titanium
Oxides Incorporated into Silicalite and Zeolite Cavities:
A Theoretical Ab Initio Study 473
  N.U. Zhanpeisov, H. Yamashita, H. Mishima, M. Matsuoka, and M. Anpo

Statistical Thermodynamics of Ammonia-Alkali Cation
Complexes in Zeolite ZSM-5 481
  B. Boddenberg, G.U. Rakhmatkariev, J. Viets, and K.N. Bakhranov

Activity of C-H and C-C Bond in the Cracking Reaction Over
Isomorphously Substituted ZSM-5—A Density Functional Study 489
  A. Chatterjee, T. Iwasaki, T. Ebina, H. Tsuruya, T. Kanougi,
  Y. Oumi, M. Kubo, and A. Miyamoto

Rotational Dynamics of CH_4 Molecules Sorbed in Zeolites:
A Computer Modelling Study 497
  M.V. Koudriachova

Chabazite: Enthalpy of Formation and Energetics of Hydration 505
  A. Navrotsky, S-H. Shim, T.R. Gaffney, and J. MacDougall

An Ising Model for Benzene Diffusion in Na-Y Zeolite 509
  C. Saravanan, F. Jousse, and S.M. Auerbach

PART V: STRUCTURE

PLENARY LECTURE
The Search for New Zeolite Frameworks 517
  M.M.J. Treacy, K.H. Randall, and S. Rao

Combining a Structure Envelope with Chemical Information to
Solve Complex Zeolite Structures from Powder Data 533
  S. Brenner, L.B. McCusker, and C. Baerlocher

Using Textured Samples to Facilitate Zeolite Structure Analysis
with Powder Diffraction Data 537
  T. Wessels, C. Baerlocher, and L.B. McCusker

The Synthesis of Zeolite ERS-7 and Its Structure Determination
Using Simulated Annealing and Synchrotron X-ray Powder
Diffraction 541
  R. Millini, G. Perego, L. Carluccio, G. Bellussi, D.E. Cox,
  B.J. Campbell, and A.K. Cheetham

PART VI: INDUSTRIAL APPLICATIONS

PLENARY LECTURE
Observations on Zeolite Applications 551
  J.M. Garces

Converting Natural Gas to Ethylene and Propylene by the
UOP/HYDRO MTO Process 567
  P.T. Barger and S.T. Wilson

Development and Industrial Application of a New β Zeolite
Catalyst for the Production of Cumene 575
  C. Perego, S. Amarilli, G. Bellussi, O. Cappellazzo, and G. Girotti

Optimization of the Thickness of Zeolite 4A Coatings Grown
on Metal Surfaces for Heat Pump Applications 583
  M. Tatlier and A. Erdem-Senatalar

Activity, Selectivity, Deactivation and Crystal Size Behavior
of MTO Conversion Over ZSM-5 in a Jetloop Recycle Reactor 591
  K.P. Möller, W. Böhringer, A.E. Schnitzler, E. van Steen, and
  C.T. O'Connor

PART VII: MEMBRANES

PLENARY LECTURE
Zeolitic Coatings—Potential Use in Catalysis and Separations 603
  J.C. Jansen, J.M.v.d. Graff, N.v.d. Puil, S.B.G. Seijger, and S.P.J. Smith

Effect of Pinholes on Permselectivities of MFI Membranes 613
    M. Matsukata, J.D. Wang, N. Nishiyama, M. Nomura, and K. Ueyama

On the Role of Renucleation and Crystal Incorporation During
Secondary Growth of Precursor (Seed) Zeolite Layers 621
    A. Gouzinis, L.C. Boudreau, G. Xomeritakis, and M. Tsapatsis

Bilayered Hollow Zeolite Tubes 629
    L. Tosheva, V. Valtchev, and J. Sterte

Growth of Zeolites Firmly Attached to Metal Surfaces for Use
in Catalysis 637
    M.J. van der Eerden, D.C. Koningsberger, and J.W. Geus

PART VIII: MESOPOROUS MATERIALS—CHARACTERIZATION

Heterogenization of a Novel Epoxidation Catalyst: Phase
Immobilization of a Titanium Silsesquioxane in an MCM-41
Molecular Sieve 645
    S. Krijnen, H.C.L. Abbenhuis, R.W.J.M. Hanssen, J.H.C. van Hooff,
    and R.A. van Santen

Selective Hydrogenation of Cinnamaldehyde Over Novel
Mesoporous (Cu, Zn, Al) Mixed Oxide Catalysts 651
    S. Valange, Z. Gabelica, J.M. Clacens, A. Derouault, and J. Barrault

Neutron Diffraction Study of Phase Transitions Observed
During the Sorption of D_2O on MCM-41 (40 Å and 25 Å) 659
    N. Floquet, J.P. Coulomb, C. Martin, Y. Grillet, P.L. Llewellyn,
    and G. André

Photoluminescence Properties and Photocatalytic Reactivities
of Cr- and V-HMS Mesoporous Molecular Sieve Catalysts 667
    H. Yamashita, M. Ariyuki, S. Higashimoto, T. Ono, S.G. Zhang,
    J-S. Chang, S-E. Park, J.M. Lee, Y. Ichihashi, Y. Matsumura,
    and M. Anpo

PART IX: MESOPOROUS MATERIALS—SYNTHESIS

Design of Well-Defined Catalysts Supported on
Micelle-Templated Silicas 675
    P. Sutra, N. Bellocq, D. Brunel, F. Di Renzo, F. Fajula,
    A. Galarneau, M. Lasperas, and P. Moreau

New Insights into the MCM-41 Hydrothermal Restructuring
Process 681
    A. Sayari, M. Kruk, and M. Jaroniec

Crystallization of Mesoporous Silica MCM-48 689
    R. Ryoo and J.M. Kim

Structure and Synthesis of MCM-48 697
    M.W. Anderson, C.L. Jackson, and D.P. Luigi

PART X: MESOPOROUS MATERIALS—SYNTHESIS AND MODIFICATION

Synthesis and Characterization of Composite MCM-41/ZSM-5 Material 707
    L. Huang and Q. Li

Attempts of Structuring the Pore Walls of Mesoporous MCM-41 Materials 713
    A. Karlsson, M. Stöcker, and R. Schmidt

Preparation of Fine Aggregate and Rod-Like Ordered Mesoporous Silicas by the Use of Behenyltrimethylammonium Chloride with an Aqueous Sodium Metasilicate Solution 719
    S. Shio, A. Kimura, M. Yamaguchi, K. Yoshida, and K. Kuroda

Surface Modification of Mesoporous Ordered Silicates with Organosilanes 725
    M. Jaroniec, C.P. Jaroniec, M. Kruk, and A. Sayari

Author Index lv

Subject Index lxv

TABLE OF CONTENTS

VOLUME II

VOLUME 1 3
VOLUME 2 733
VOLUME 3 1479
VOLUME 4 2259

Prefaces
    Chairs of the 12th International Zeolite Conference .... Volume I - xlv
    Publications Chair .... Volume I - xlvii

Executive Committee of the 12th IZC .... Volume I - xlviii

Program Committee of the 12th IZC .... Volume I - xlix

Local Arrangements Committee .... Volume I - xlix

Pre-Conference Zeolite School .... Volume I - l

Publications .... Volume I - l

International Advisory Board to the 12th IZC .... Volume I - li

Session Chairs of the 12th IZC .... Volume I - lii

Financial Support .... Volume I - liii

PART I: MESOPOROUS MATERIALS

NMR as a Mechanistic Tool for the Study of Mesoporous M41S Syntheses 733
    B.E. Gore, M.W. Anderson, A. Steel, G.J.T. Tiddy, and S.W. Carr

Mesoporous (M)-MSU-X Metallo-Silicates by Nonionic Polyethylene Oxide Surfactant Templating: Acid $((N^0N^+)X^-I^+)$ and Base $(N^0M^+I^-)$ Catalyzed Pathways 741
    S.A. Bagshaw, T. Kemmitt, and N.B. Milestone

Immobilization of Chiral Manganese-Salen-Complexes on Mesoporous MCM-41-Type Molecular Sieves 749
    S. Ernst and M. Selle

A Novel Pathway Towards MCM-41 Related Mesoporous Monodisperse Silica Spheres in the Submicrometer and Micrometer Range 757
    M. Grün and K.K. Unger

Critical Appraisal of Different Methods of Mesopore
Analysis for MCM-41 Silicates 761
  M. Kruk, M. Jaroniec, and A. Sayari

Synthesis and Characterization of Mesoporous Molecular
Sieves Based on Silica and Some Organosiloxanes 767
  A.S. Kovalenko, J.V. Chernenko, V.G. Il'in, and A.P. Filippov

Synthesis of Mesoporous Aluminophosphates Using
Surfactants with Long Alkyl Chains 771
  T. Kimura, Y. Sugahara, and K. Kuroda

Controlled Thermal Extraction of Template from Zeolite
Type Materials—Part II: MCM41 Mesopores 779
  M.T.J. Keene, P.L. Llewellyn, R. Denoyel, R.D.M. Gougeon, R.K. Harris,
  and J. Rouquerol

Investigation into the Engineering of Molybdenum(VI)
Polyoxoanions Under Hydrothermal Conditions 787
  Y. Xu, J.J. Lu, N.K. Goh, and L.S. Chia

A Comparison of the Coordination Geometry of the
Tetrapyridine Copper(II) Complex in MCM-41, NaX and
NaY: A Two-Dimensional Electron Spin Echo Envelope
Modulation Study 795
  A. Pöppl, M. Hartmann, M. Gutjahr, W. Böhlmann, and R. Böttcher

Characterization of the Acidity and Catalytic Activity of
MCM-48 and Amorphous Silica-Alumina Materials 801
  J. Carrazza, F. González, R. Adrián, D. Djaouadi, J.G. Moore,
  D.Y. Shahriari, C.C. Landry, and J. Lujano

Synthesis and Characterization of Ruthenium-Containing
MCM-41 and MCM-48 Mesoporous Materials 809
  C. Bischof and M. Hartmann

Synthesis, Characterization and Catalytic Activity of
Ti-MCM-41 Materials Obtained Under Highly Acidic Media 817
  A. Corma, J.L. Jorda, M.T. Navarro, J. Pérez-Pariente, and F. Rey

Structure and Redox Properties of Vanadium Species in
MCM-41 825
  G. Grubert, W. Grünert, J. Rathousky, A. Zukal, G. Schulz-Ekloff,
  and M. Wark

Adsorption Properties of Niobium- and Aluminum-Containing
MCM-41 Molecular Sieves—FTIR Studies 833
  M. Ziolek, I. Nowak, P. Decyk, O. Saur, and J.C. Lavalley

Computer Modelling of Mesoporous Frameworks 839
  R.G. Bell

Porous Materials Based on Layered Silicate Kanemite Via
Pillaring Process 847
  U. Brenn, W. Schwieger, and H.G. Karge

Redox Behavior of V-MCM-41 Studied by *In Situ* X-ray
Absorption and EPR 857
    A.C. Wei, H.C. Wu, J.F. Lee, and K.J. Chao

ESR Study on Fe-Silicate with ZSM-48 Structure: Influence
of Template Amount and Treatment Process on the
Distribution of Iron 863
    H. Du, E. Roduner, W. Fan, R. Li, and J. Cao

Synthesis and Characterization of Fe- and Cr-Containing
Silicates Having ZSM-48 Structure Prepared with a Low
Amount of 1,6-Hexanediamine 869
    W. Fan, R. Li, B. Zhong, H. Du, and E. Roduner

Catalytic Behavior of Vanadium Containing Molecular
Sieves for Selective Oxidation of Ethanol 877
    S. Kannan and S. Sivasanker

Methanol and β-Naphthol Coupling to β-Naphthyl Methyl
Ether Over Mesoporous Molecular Sieve Al-MCM-41 885
    A-N. Ko, L-W. Chen, and C-Y. Chou

Liquid Phase Oxidation of 2-Methyl Naphthalene to
2-Methyl-1,4-Naphthaquinone Over Encapsulated
Transition Metal Complexes in Y and MCM-41 Molecular Sieves 893
    K.V.V.S.B.S.R. Murthy, N. Srinivas, S.J. Kulkarni, and K.V. Raghavan

The Shape Selectivity in the Alkylation of Naphthalene
with n-Butanol, Isobutanol or Tert-Butanol Over Modified Y
Zeolites, Mordenite, SAPO-5 and MCM-41 Molecular Sieves 897
    G. Kamalakar, M. Ramakrishna Prasad, S.J. Kulkarni, and K.V. Raghavan

Catalytic Performances of Metallosilicates with ZSM-48
Structure 901
    W. Fan, H. Yang, B. Fan, R. Li, and B. Zhong

The Adsorption on Modified Mesoporous Molecular Sieve
MCM-41 909
    Y-C. Long, T-M. Xu, Y-J. Sun, and W-Y. Dong

Preparation of Carbon Nanotubes on the Surface of
Magadiite Impregnated with Cobalt (II) or Iron (III) Ions 917
    K. Mukhopadhyay, A. Fudala, K. Hernadi, I. Kiricsi, A. Fonseca,
    and J.B. Nagy

Preparation and Characterization of Oriented MCM-41 Films 923
    Q. Cai, W. Pang, Y. Xu, W. Lin, and B. Liu

Optimum Synthesis Conditions for Tubular Shape MCM-41 931
    T-R. Ling, B-Z. Wan, H-P. Lin, and C-Y. Mou

PART II: CATALYSIS

PLENARY LECTURE
Redox Catalysis Over Molecular Sieves: Structure and
Function of Active Sites 941
    B. Wichterlová, J. Dedecek, and Z. Sobalík

Cracking of n-Heptane Over USY Zeolites: Effect of the
Extra-Framework Aluminum Content and the Temperature
in the Deactivation and Selectivity 975
    R.F. Santos and E.A. Urquieta-González

Effect of the Allocation of Alkali Cations on the Basicity
of Zeolites 981
    M.L. Valcheva-Traykova and N.P. Davidova

Progress in Preparation of Strong Basic Zeolites:
Application of Redox Method 989
    Y. Chun, J.H. Zhu, Y. Wang, D.K. Sun, and Q.H. Xu

Modification of Y-Type Zeolites with Niobium Compounds 997
    M. Ziolek and I. Nowak

Coking During Ethene Conversion in Zeolite Beta 1003
    B. Paweewan, P.J. Barrie, S. Ashtekar, and L.F. Gladden

Influence of the Extra-Framework Aluminum Species on
the Activity and Selectivity of Y Zeolites in n-Decane
Hydrocracking 1011
    E. Benazzi, J. Lynch, A. Gola, and C. Marcilly

Disproportionation of Ethylbenzene Over SAPO-11 1019
    G.G. de Medeiros and D. Cardoso

Acid-Catalyzed Benzene Hydroconversion Using Mordenite
and Beta Zeolites; Role of Molecular Hydrogen 1025
    A. Chambellan, O. Cairon, and T. Chevreau

Mode of Deactivation by Coke of a H-MWW Zeolite During
n-Heptane Cracking 1033
    D. Meloni, D. Madier, D. Martin, M. Guisnet, and H. Kessler

Effect of Dealumination of Zeolite Omega on the Activity
for Transalkylation of Toluene and Trimethylbenzene 1041
    S.H. Park, J.H. Lee, and H-K. Rhee

A New Modification of Low-Alumina Y Zeolite Prepared by
Direct Synthesis Without Templates; Physical, Chemical
and Catalytic Properties in Vacuum Gas Oil Cracking and
Isomerization of C_7-C_{10} Paraffins 1049
    M.I. Levinbuk, L.M. Kustov, M.L. Pavlov, T.V. Vasina, J.P. Fraissard,
    A.V. Kazakov, and V.K. Smirnov

Effect of Co-Feeding Ethene or Propene on the
Conversion of Methanol Over H-ZSM-5. Use of Isotopic
Labeling for Mechanistic Studies 1057
    P.O. Rønning, Ø. Mikkelsen, and S. Kolboe

The Effect of Hydrothermal Treatment of Zeolites on the
Methanol Ammination Reaction 1065
    L.H. Callanan, C.T. O'Connor, and E. van Steen

The Role of Mesoporosity, Acidity and Extra-Framework
Aluminum in the Synthesis of Cumene Over Zeolite Beta 1073
    S. Hurgobin, L.F. Petrik, J.C. Jansen, and C.T. O'Connor

Acid Catalyzed Degradation of Polyolefins 1081
    J. Dwyer, A.A. Garforth, R.S. Holsenburger, Y-H. Lin,
    and P.N. Sharratt

Effects of the Modification of Zeolite BEA on Its Catalytic
Behavior in the Acylation of Aromatics 1085
    H.K. Heinichen and W.F. Hölderich

Cresol Transformation on USHY and HZSM5 1093
    F.E. Imbert, M. Guisnet, and S. Gnep

Selective Cyclization of C_{4+} Alcohols/Carbonyls,
Formaldehyde and Ammonia to Heterocyclics Over
Modified Zeolites 1101
    N. Srinivas, S.J. Kulkarni, and K.V. Raghavan

Properties of Zeolitizised Fly-Ash as a Cracking Catalyst 1105
    E. López-Salinas, M. Hernández-Del Angel, M.L. Guzmán, N. Nava,
    J.A. Toledo, M.A. Cortés-Jácome, E. Mogica, F. Hernández,
    and I. Schifter

Synthesis of 1,3-Dioxolan-4-One by Carbonylation of
Formaldehyde on Zeolite Catalysts 1113
    T. Sano, T. Sekine, T. Shimada, Z.B. Wang, K. Soga, I. Takahashi,
    and T. Masuda

In Situ FTIR Studies of the Alkylation of Benzene by
Propene on a Series of H-Beta Zeolites 1121
    S. Siffert, L. Gaillard, and B-L. Su

Rearrangement of Styrene Oxides Under Zeolite Catalysis 1129
    K. Smith and M. Al-Shamali

Effect of Dealumination of Mordenite Catalysts on
Shape-Selective Isopropylation of Naphthalene 1133
    C. Song, A.D. Schmitz, and K.M. Reddy

Dehydroisomerization of *n*-Butane into *iso*-Butene Through
n-Butene Over Platinum-Loaded MFI-Type Metallosilicate
Catalysts 1141
    H. Nagata, Y. Takiyama, S. Tashiro, M. Kishida, and K. Wakabayashi

Hydrothermal Stability of Zeolite Catalysts for the
Reduction of NO by Hydrocarbons 1149
    S-H. Oh, S.Y. Chung, M.H. Kim, I-S. Nam, and Y.G. Kim

The Role of Cobalt Oxide Species in the SCR of NO with
Propylene on Co-Loaded ZSM-5 Catalysts 1157
    Y-K. Park, S.S. Goryashenko, D.S. Kim, and S-E. Park

Influence of Incorporated Silver Clusters on Properties of
Lewis and Brønsted Acid Sites in Mordenites 1165
    V. Petranovskii, N. Bogdanchikova, E. Paukshtis, S. Fuentes,
    Y. Sugi, T. Hanaoka, and A. Licea-Claverie

Selective Catalytic Reduction of Nitrogen Oxides Using
Hydrocarbons on Cobalt Ion-Exchanged Beta Zeolite 1169
    T. Tabata, H. Ohtsuka, G. Bellussi, and L.M.F. Sabatino

Zeolite HZSM-5-Encapsulated Cu(I) Complexes as Butanol
Carbonylation Catalysts 1177
    Q. Shi, X. Zhao, C. Li, Y. Xu, and K. Guo

Use of Zeolites to Inhibit the Transition Metal Catalyzed
Decomposition of H_2O_2 in the Bleaching of Pulp 1185
    K. Dyhr and J. Sterte

Siting and Coordination of Co(II) Ions in High Silica Zeolites 1193
    J. Dedecek, D. Kaucky, and B. Wichterlová

Hydrothermally Stable Diesel de-NOx Catalyst 1199
    J.E. Yie, J.W. Kim, Y.S. Oh, and R. Ryoo

Mechanistic Study of Propylbenzenes Formation Over
Zeolite Catalysts 1207
    A.V. Smirnov, A.A. Shashkov, I.I. Ivanova, B.V. Romanovsky,
    and Z. Gabelica

Is Singlet Molecular Oxygen Involved in Oxidations
Catalyzed by Ti Molecular Sieves? 1213
    F.M.P.R. van Laar, D.E. De Vos, D.L. Vanoppen, F. Pierard,
    A. Brodkorb, A. Kirsch-De Mesmaeker, and P.A. Jacobs

Oxidation of Hydrocarbons and Alcohols by Hydrogen
Peroxide Over TAPO-5, -11, -31, and -36 1221
    M.H. Zahedi-Niaki, M.P. Kapoor, and S. Kaliaguine

Chemo-Selective Oxidation of Organic Halides
Catalyzed by TS-1 Under Solvent-Free Triphase Conditions
Using Dilute H_2O_2 1227
    R. Kumar, A. Bhaumik, and P. Mukherjee

VPI-5 Containing Framework Chromium as Catalyst for
the Liquid- and Gas-Phase Oxidation of Cyclohexane 1233
    F.J. Luna, M. Wallau, P.A. Jacobs, and U. Schuchardt

Synthesis, Characterization and Catalytic Properties of
Bi-Tetravalent Metals Substituted AlPO-11 1239
    P. Mériaudeau, V.A. Tuan, V.T. Nghiem, J. Fraissard,
    and C. Naccache

Zeolite Modification in Relation to Catalysis of Benzene to
Phenol Oxidation by Nitrous Oxide 1245
    L.V. Pirutko, D.P. Ivanov, K.A. Dubkov, V.V. Terskikh,
    A.S. Kharitonov, and G.I. Panov

Direct Hydroxylation of Benzene with H_2 and O_2 Over
2-Ethylanthraquinone Encapsulated Pd(0)-Y Zeolite 1253
    S-E. Park, J.W. Yoo, W.J. Lee, J-S. Chang, and C.W. Lee

Synthesis, Characterization and Catalytic Activity of
Saponite-Like Materials Isomorphously Substituted with
Cobalt, Chromium and Copper 1261
    M. Sychev

Alkali-Containing Ti-Zeolites: New Catalysts for the
Oxidation of Diethylamine with H_2O_2 1269
    E. Jorda, A. Tuel, R. Teissier, and J. Kervennal

Catalytic Activity in Phenol Conversion Over Novel
Catalysts of Microporous Lead Titanate (JPT-1) and
Layered Titanosilicate (JDF-L1) 1277
    F-S. Xiao, R. Yu, H. Du, S. Qiu, W. Pang, and R. Xu

Characterization and Catalytic Properties of TS-1 from the
System of TPABr and TEAOH and the System of TPABr and
n-Butylamine 1283
    X. Wang, X. Guo, G. Li, Q. Zhao, X. Bao, X. Han, and L. Lin

Friedel-Crafts Reactions of Furan Derivatives Over Strong
Acid Zeolites 1293
    W.M. Van Rhijn, W. Verrelst, and P.A. Jacobs

Post-Mortem Analysis of FCC Model Catalysts 1299
    P.P.H.J.M. Knops-Gerrits, I.V. Mishin, and P.A. Jacobs

Alkylation of Naphthalene with Propylene Over
Dealuminated Mordenites 1307
    T. Li and X. Liu

Effect of Reactor Type on the Activity Obtained During the
Hydration of Propylene Over Zeolite Beta Catalysts 1313
    W. Li, J. Wang, N. Guan, S. Zheng, and K. Tao

Selective Propionylation of Toluene Over Solid Catalysts 1317
    A.P. Singh and B. Jacob

Mesoporous Molecular Sieve as Novel Adsorbent for GSC 1323
    L. Li and W. Jialiang

PART III: CATALYSIS BY ION-EXCHANGED METALS

Production of Hydrogen Peroxide from Hydrogen and
Oxygen Over Pd-Zeolites 1331
    C. Montes de Correa, P. Jacobs, and J. Garcés

Low Temperature Catalytic Oxidation of VOC's Over
Zeolite Supported Metal Catalysts 1337
    S.A. Bagshaw and N.B. Milestone

Significant Roles of Substituted Iron-Group Metal Ions in
SAPOs for Exhibiting High Catalytic Performance of
Selective Olefin Synthesis from Methanol 1345
    T. Inui, M. Kang, and Y. Nomura

Synthesis of Propylene Oxide from Propylene, Oxygen
and Hydrogen Catalyzed by Palladium-Platinum-
Containing Titanium Silicalite 1351
    W. Laufer, R. Meiers, and W.F. Hölderich

Mono and Polynuclear Cu Species in ZSM-5. Nature and
Reactivity in the Reduction of NO in the Presence of
Various Reductants 1359
    G. Centi and A. Galli

Control of the Structure of Iron/ZSM-5 Catalysts 1367
    R.W. Joyner and M. Stockenhuber

In Situ ESR of Gd^{3+}/H-ZSM-5: An Approach to Study of
Distribution and Properties of Lanthanide Promoters in
High-Silica Zeolites 1373
    A.V. Kucherov, A.A. Slinkin, and M. Shelef

Direct Conversion of Methane to Aromatics Over Transition
Metal Ion-Loaded H-ZSM-5 Zeolites 1381
    B.M. Weckhuysen, D. Wang, M.P. Rosynek, and J.H. Lunsford

PART IV: OXIDATION CATALYSIS

Characteristics of Liquid Phase Oxidation Over Ti-Beta
Zeolite Synthesized by a Dry Gel Conversion Method 1391
    T. Tatsumi, Q. Xia, and N. Jappar

n-Hexane Oxyfunctionalization by Hydrogen Peroxide
Over Titanium Silicalite Containing Catalytic Membranes 1399
    S.Q. Wu, C. Bouchard, and S. Kaliaguine

Titanium Modified UTD-1 as a Catalyst for Oxidation
Reactions 1403
    K.J. Balkus, Jr., A.K. Khanmamedova, A. Scott, and J. Hoefelmeyer

Steam Reforming of Methanol for the Production of
Hydrogen on Molecular Sieve Catalysts 1409
    S. Wellach, M. Hartmann, S. Ernst, and J. Weitkamp

PART V: ACID CATALYSIS

The Effect of Molecular Sieve Structure on Benzene
Propylation Selectivity to iso-/n-Propylbenzenes.
Catalytic and Theoretical Study 1419
    J. Cejka, J.E. Sponer, N. Zilková, and B. Wichterlová

Alkylation of Isobutane with Light Olefins Over Zeolite BEA 1425
    G.S. Nivarthy, K. Seshan, and J.A. Lercher

Selective Skeletal Butene Isomerization Through a
Bimolecular Mechanism 1433
    M. Guisnet, P. Andy, N.S. Gnep, E. Benazzi, and C. Travers

Effect of the ZSM-5 Crystal Size on the Cracking Activity of
Lube Oils and Polyolefins 1441
    J.L. Sotelo, R. van Grieken, J. Aguado, D.P. Serrano, J.M. Escola,
    and J.M. Menéndez

Zeolite Catalyzed Isopropylation of Naphthalene at
Supercritical Reaction Conditions 1447
    R. Gläser and J. Weitkamp

The Role of Higher Alcohols in the Highly Selective
Beckmann Rearrangement of Cyclohexanone Oxime
Catalyzed by Zeolites 1455
    T. Tatsumi and L-X. Dai

Acylation of Phenol with Acetic Acid Over Beta Zeolite 1463
    E.V. Sobrinho, E. Falabella, S. Aguiar, D. Cardoso, F. Jayat,
    and M. Guisnet

Zeolites as Catalysts for the Selective *Para*-Nitration of
Toluene 1471
    D. Vassena, D. Malossa, A. Kogelbauer, and R. Prins

Author Index lv

Subject Index lxv

TABLE OF CONTENTS

VOLUME III

VOLUME 1 3
VOLUME 2 733
VOLUME 3 1479
VOLUME 4 2259

Prefaces
    Chairs of the 12th International Zeolite Conference .... Volume I - xlv
    Publications Chair .... Volume I - xlvii

Executive Committee of the 12th IZC .... Volume I - xlviii

Program Committee of the 12th IZC .... Volume I - xlix

Local Arrangements Committee .... Volume I - xlix

Pre-Conference Zeolite School .... Volume I - l

Publications .... Volume I - l

International Advisory Board to the 12th IZC .... Volume I - li

Session Chairs of the 12th IZC .... Volume I - lii

Financial Support .... Volume I - liii

PART I: SYNTHESIS

Organic-Functionalized Molecular Sieves: A New Class of Shape-Selective Catalysts 1479
    C.W. Jones, K. Tsuji, and M.E. Davis

Zeolite Crystallization and Transformation Determined by Atomic Force Microscopy 1487
    M.W. Anderson, J.R. Agger, N. Pervaiz, S.J. Weigel, and A.K. Cheetham

Synthesis of Defect-Free Pure Silica Polymorphs of Low Framework Density in Aqueous Fluoride Media 1495
    P.A. Barrett, E.T. Boix, M.A. Camblor, A. Corma, M.J. Díaz-Cabañas, S. Valencia, and L.A. Villaescusa

Guest/Host Interactions and Dynamics Studies of Organic Molecules in High-Silica Zeolites 1503
    D.F. Shantz and R.F. Lobo

Iron Substitution in the Microporous Titanosilicate ETS10 1507
    A. Eldewik, V. Luca, N.K. Singh, and R.F. Howe

Synthesis of Pure Silica Beta and Al-Free Ti-Beta Using
TEAOH and Their Characterization 1515
 P.R. Hari Prasad Rao, K. Ueyama, E. Kikuchi, and M. Matsukata

Characterization of Particle Growth Mechanism of TS-1
Zeolite by Capillary Hydrodynamic Fractionation (CHDF) 1523
 K.T. Jung and Y.G. Shul

Nucleation and Growth Mechanism of Zeolites:
A Small-Angle X-ray Scattering Study on Si-MFI 1529
 P-P.E.A. de Moor, T.P.M. Beelen, and R.A. van Santen

Synthesis of Crystalline Porous Solids in Ammonia 1535
 D.M. Millar and J.M. Garces

Thermally Stable ZSM-5 Zeolite Materials with New
Microporosities 1543
 R. Le Van Mao and D. Ohayon

Continuous Synthesis of Zeolites Using a Tubular Reactor 1553
 P.M. Slangen, J.C. Jansen, H. van Bekkum, G.W. Hofland,
 F. van der Ham, and G.J. Witkamp

Structure Formation at Ambient Temperature in the
Al/P-System 1561
 W. Schwieger, T. König, H. Toufar, G. Fu,
 H. Meyer zu Altenschildesche, G.T. Kokotailo, and C.A. Fyfe

TAPO-34 and TAPSO-34 Synthesized by Using Morpholine
as Templating Agent: Spectroscopic Studies 1569
 L. Marchese, A. Frache, S. Coluccia, and J.M. Thomas

News About CrAPO-5—A Framework Incorporation of Cr 1577
 J. Kornatowski and G. Zadrozna

Characterization of ZnAPO-50 1585
 A. Ristic, N. Novak Tusar, N. Zabukovec Logar, G. Mali,
 A. Meden, and V. Kaucic

The Potential of Microwave Heating in (Al)ZSM-5 Synthesis 1591
 J.P. Zhao, C.S. Cundy, R.J. Plaisted, and J. Dwyer

Role of Solubility in Zeolite Synthesis 1595
 J. Sefcík and A.V. McCormick

Experimental and Computational Studies of Magnesio-
Aluminophosphates Synthesized Using Polymeric and
Oligomeric Templates 1603
 P.A. Wright, P.A. Cox, G.W. Noble, and V. Patinec

Promoter Assisted Low-Temperature Synthesis of High
Silica Molecular Sieves with MFI, ZSM-48 and MTW
Topologies 1611
 R. Kumar, R.K. Pandey, P. Mukherjee, and A. Bhaumik

Influence of Alkali Cations on the Incorporation of
Aluminum, Boron and Gallium into the MFI Framework
in Fluoride Containing Media 1619
    J.B. Nagy, R. Aiello, F. Crea, and F. Testa

Synthesis of Platelet-Like Faujasitic Zincophosphate X
Crystals 1627
    M.J. Castagnola and P.K. Dutta

Crystallization Mechanisms of Novel Aluminophosphate
Materials in Quasi Non-Aqueous Mono- and
Dialkylformamide Media 1633
    L. Vidal and Z. Gabelica

Solvothermal Synthesis and Characterization of
Silica-Pillared Layered Stannic Phosphate 1641
    X. Jiao, W. Pang, F. Zhou, and R. Xu

Composite Compounds with Open Frameworks (MIL-n)
Hydrothermal Synthesis and Structure Determinations of
Some New Vanadodiphosphonates 1649
    D. Riou and G. Férey

Synthesis of Zeolite ZSM-12 and Its Borosilicate Analogue
Using Benzyltrimethylammonium Chloride as a Template 1655
    Z-Y. Yuan, T-H. Chen, Z-B. Long, J-Z. Wang, and H-X. Li

Clays as Raw Materials for Zeolite Synthesis: Effect of
Impact Grinding on Kaolinite Structure and Reactivity 1663
    E.I. Basaldella, R. Torres Sánchez, S.L. Pérez de Vargas, D. Caputo,
    and C. Colella

Following the Crystallization of Microporous Solids Using
EDXRD Techniques 1671
    A.T. Davies, G. Sankar, C.R.A. Catlow, and S.M. Clark

Comparative Investigation of Zeolite Formation in Silicate
and Phosphate Systems 1679
    P.S. Yaremov, N.V. Turutina, and V.G. Ilin

Transformation of Layer Silicates into Three Dimensional
Structure by Pillaring of Aluminium Containing Magadiite
with Inorganic Polymer Ions 1685
    A. Fudala, Y. Kiyozumi, S-I. Niwa, M. Toba, F. Mizukami,
    J.B. Nagy, and I. Kiricsi

Effect of Organic Structure-Directing Agents on the Fine
Structure of ZSM-12 1693
    Y. Kubota, A. Seriu, Y. Moriyama, Y. Sugi, S. Ritsch, K. Hiraga,
    and O. Terasaki

Synthesis of Large Single Crystals of Silica-Sodalite and
Silicalite-I in the Presence of Pyrocatechol 1701
    C. Shao, S. Qiu, F. Xiao, X. Li, Q. Zhai, S. Zheng, and Z. Zhang

Controlled Thermal Extraction of Template from Zeolite
Type Materials—Part I: MFI and BEA Micropores 1707
    C. Sauerland, P. Llewellyn, Y. Grillet, J. Patarin, and F. Rouquerol

Synthesis and Structures of As-Synthesized and Calcined
AlPO$_4$-14 Revealing a Three-Dimensional Channel System
with 8-Ring Pores 1715
    R.W. Broach, S.T. Wilson, and R.M. Kirchner

Synthesis and Characterization of Chabazitic
Aluminophosphates 1723
    J.K. Wyles, G. Sankar, D.W. Lewis, C.R.A. Catlow, and
    J.M. Thomas

Synthesis of Microporous Aluminophosphates in the
Presence of Diaza-Polyoxa-Macrocycles: Co-Structuring
Role of F$^-$ and/or (CH$_3$)$_4$N+ Ions 1731
    L. Schreyeck, P. Caullet, J.C. Mougenel, and B. Marler

Stability of the ULM-n Microporous Gallophosphates in the
System GaPO$_4$-HF-Amine-H$_2$O 1737
    C. Gérardin, A. Navrotsky, T. Loiseau, and G. Férey

Highly Selective Template for Synthesis of ALPO-31 (ATO) 1743
    O.V. Kikhtyanin, R.F. Vogel, C.L. Kibby, T.V. Harris, K.G. Ione,
    and D.J. O'Rear

Preparation of LTA Type Gallophosphate Large Single
Crystals 1751
    Y. Yao, W. Pang, Y. Xu, and R. Xu

Room-Temperature Synthesis of Zincophosphates in the
Presence of Diaminoalkanes and Their Characterization 1757
    P. Reinert, A. Khatyr, J. Patarin, and B. Marler

Encapsulation of a Chelate Ni(II) Complex into AlPO$_4$-5
Molecular Sieve 1765
    N. Rajic, D. Stojakovic, A. Meden, and V. Kaucic

Synthesis and Acidity Characterization of CoAPO-37 1771
    C.S. Costa, J.P. Lourenço, C. Henriques, A.P. Antunes, F.R. Ribeiro,
    M.F. Ribeiro, and Z. Gabelica

Pulsed Laser Deposition of Molecular Sieve Films and
Membranes 1779
    M.E. Gimon-Kinsel, T. Munoz, Jr., A. Ayala, L. Washmon,
    and K.J. Balkus, Jr.

Nanoscale Films of AlPO$_4$-5 Prepared by Microwave
Synthesis 1787
    S. Mintova, S. Mo, and T. Bein

The Preparation of Zeolite 4A Membrane Reactor and Its
Use in Dehydration of Diethylene Glycol 1795
    W. Guiru, G. Hongchen, and L. Yushan

Morphological Evolution of Mordenite Crystals 1803
  F. Hamidi, R. Dutartre, F. Di Renzo, A. Bengueddach,
  and F. Fajula

Preferred Orientation in Thin Silicalite-1 Films Synthesized
by Seeding 1809
  J. Hedlund, S. Mintova, and J. Sterte

Control of Morphology of ZSM-5 Zeolite Crystal 1817
  A. Iwasaki and T. Sano

Synthesis, Isolation and Characterization of Nano-Powder
of Silicalite-1 Type Molecular Sieves 1825
  R. Ravishankar, C. Kirschhock, B.J. Schoeman, D. De Vos,
  P.J. Grobet, P.A. Jacobs, and J.A. Martens

Preparation of Thin Films of ZSM-5 on Ceramic Supports 1833
  N.B. Milestone, F. Mizukami, Y. Kiyozumi, K. Maeda, and S. Niwa

Control of Intercrystalline Regions of Silicalite Membrane
by Pressurized Sol-Gel Technique 1841
  T. Sano, A. Hayashi, K. Yamada, Z.B. Wang, K. Soga,
  and H. Yanagishita

Biphasic Silicate Materials Based on Porous Glasses—
Preparation and Properties 1849
  W. Schwieger, M. Rauscher, F. Scheffler, D. Freude, U. Pingel,
  and F. Janowski

Thin Zeolite NA Films by the Seed Film Method 1857
  J. Hedlund, E. Babouchkina, and J. Sterte

Synthesis of Zeolite LTA Single Crystals of Macro- to
Nanometer Size 1863
  G. Zhu, S. Qiu, J. Yu, F. Gao, F. Xiao, R. Xu, T. Sakamoto,
  and O. Terasaki

Extended Insertion of Gallium in EMT Zeolite Framework 1871
  S. Iwamoto, T. Inui, and Z. Gabelica

Incorporation Level and Nature of the Framework Metal
Sites Versus Crystallization Time and Temperature in TS-1
and FES-1 1877
  B. Echchahed, D.T. On, F. Béland, and L. Bonneviot

Synthesis of (Ti,Al)-Beta Using Hexafluorotitanate as
Titanium Source 1885
  S.L. Jahn and D. Cardoso

Isomorphous Substituted Early Transition Metal Containing
BEA Via Post-Synthesis Modification of H-(B)-BEA 1893
  J.P.M. Niederer and W.F. Hölderich

Synthesis of Microporous Metallosilicates Using
Ammonium Hexafluoro Complexes of Metals in
a Fluoride Medium 1901
    N.B. Milestone and N.S. Sahasrabudhe

Zirconium Containing Alumino-Silicate with BEA Structure 1909
    B. Rakshe, V. Ramaswamy, S.G. Hegde, and A.V. Ramaswamy

Fast Synthesis of TS-1 Zeolite by Microwave Heating 1917
    M.A. Uguina, D.P. Serrano, R. Sanz, and E. Castillo

Synthesis of Boro-Titanosilicate MFI Zeolites from
Alkali-Free Media: The Limitations of the Ammonium
Fluoride Neutral Route 1925
    Z. Gabelica and M. Shibata

Synthesis and Characterization of Zeolites Prepared Using
Metallocene Templates 1931
    K.J. Balkus, Jr., A. Ramsaran, R. Szostak, and M. Mitchell

Synthesis and Characterization of a New Microporous
Aluminophosphate $Al_{16}P_{20}O_{80}H_4 \cdot 4C_6H_{18}N_2$ with Intersecting
12- and 8-Membered Channels 1937
    J. Yu, K. Sugiyama, N. Togashi, S. Zheng, S. Qiu, J. Chen, R. Xu,
    Y. Sakamoto, O. Terasaki, K. Hiraga, Y. Tanaka, S. Nakata, M. Light,
    M.B. Hursthouse, and J.M. Thomas

Factors Influencing the Synthesis of Novel Large Pore
Zeolite SSZ-42 and Its Subsequent Characterization 1945
    C.Y. Chen, S.I. Zones, L.T. Yuen, T.V. Harris, and S.A. Elomari

Kinetics of HZSM-5 Dealumination in Steam 1953
    C.D. Hughes, A. Labouriau, S. Neugebauer Crawford, R. Romero,
    J. Quirin, and W.L. Earl

Synthesis and Characterization of Low-Silica
Erionite-Offretite 1961
    S.S. Khvoshchev, M.A. Shubaeva, I.V. Karetina, and Yu.V. Shapoval

Controlling Acidity and Shape Selectivity of Acidic Zeolites
by Silanation 1969
    M. Seitz, E. Klemm, and G. Emig

Characterization and Passivation of the External Surface
of the Zeolites Mordenite and Beta 1975
    P.J. Kunkeler, J.A. Elings, R.A. Sheldon, and H. van Bekkum

Crystallization Kinetics, Formation, and Characterization of
Nanocrystalline Particle Agglomerate of Silicalite-1 1983
    Y. Yan, Y-C. Long, A-M. Wu, Y-J. Sun, H-W. Jiang, and H-Y. He

Structural Characterization, Adsorption and Catalysis of
FER-Type Zeolite Synthesized in $TMEDA-Na_2O-Al_2O_3-SiO_2-H_2O$
System 1991
    Y. Yan, M-H. Ma, H-W. Jiang, Y-C. Long, Y-J. Sun, and L. Zhao

The Initial Stage in the Synthesis of Zeolite X from Silica
Xerogel 2001
    X.M. Luo

Synthesis of OU-1, A Large-Pore, High Silica Zeolite,
by Dry Gel Conversion 2009
    M. Matsukata, M. Kato, T. Suzuki, Y. Sasaki, E. Kikuchi, K. Ueyama,
    and P.R. Hari Prasad Rao

Effect of Different Silanization Procedures on the External
Surface Activity and Shape Selectivity of HZSM-5 2015
    R.W. Weber, H.P. Röger, K.P. Möller, and C.T. O'Connor

Studies on ZSM-25 Zeolite 2023
    A.V. Totktarev, T.V. Harris, C.L. Kibby, K.G. Ione, and D.J. O'Rear

Possibilities of Compositional Engineering in Zeolite Based
Novel Precursors for Electronic Ceramics 2033
    K. Selvaraj, V. Ramaswamy, and A.V. Ramaswamy

Synthesis of TON Type Zeolite in Presence of 1-Butanol 2041
    T. Sano, A. Suzuki, S. Fukuya, F. Matsuoka, Z.B. Wang, K. Soga,
    and Y. Kohtoku

Influence of the Concentrations of Aluminum and Silicon
in the Liquid Phase on the Kinetics of Crystal Growth of
Zeolite A 2049
    T. Antonic and B. Subotic

On the Real Significance of the "Induction Period" of
Zeolite Crystallization 2057
    B. Subotic, J. Bronic, and T. Antonic

Further Investigations into the Synthesis of Pure-Silica
Molecular Sieves Using N-Methyl, $N-R_1$-Piperidinium
Moieties 2065
    K. Tsuji and M.E. Davis

PART II: SOLID STATE

Electrochemistry of Methyl Viologen-Exchanged Zeolite Y
Modified-Electrodes 2073
    M.D. Baker and T.W. Hui

Influence of the Acidity of the Support and of the Nature
of the Ligand on the Formation of Supported Gold- and
Palladium-Based Nanoparticles 2079
    D. Guillemot, M. Polisset-Thfoin, D. Bonnin, V.Yu. Borovkov,
    and J. Fraissard

New Zeolite Pigments and Inclusion Complexes 2087
    E.M. Hughes, D.M. Kurten, and M.T. Weller

Studies on Chiral Induction Within Zeolites:
Photoelectrocyclization of Tropolone Alkyl Ethers 2095
    A. Joy, D.R. Corbin, and V. Ramamurthy

Metal-Insulator Transition of Potassium Clusters in KX 2103
    Y. Ikemoto, T. Nakano, and Y. Nozue

Laser Action from a Zeolite Based Host-Guest Composite 2111
    G. Ihlein, F. Schüth, O. Krauss, U. Vietze, F. Laeri, B. Limburg,
    and M. Abraham

Scanning Pyroelectric Microscopy of Zeolites Loaded with
Polar Molecules 2117
    G.J. Klap, S.M. van Klooster, M. Wübbenhorst, J.C. Jansen,
    H. van Bekkum, and J. van Turnhout

New Zeolite/Dye Composites with Second-Order
Nonlinear and Switchable Optical Properties 2121
    F. Marlow and K. Hoffmann

Incorporation of Cobalt and Copper Cations in Cationic
Positions of Clinoptilolite Via Solid-State Reaction 2129
    I.M. Astrelin, T. Enhbold, and M. Sychev

Silver Nucleation and Growth at Zeolite Modified
Electrodes 2137
    M.D. Baker, M. McBrien, C. Liu, and D.H. Brouwer

Structure of Silver Clusters Stabilized in Mordenite and
Erionite Channels 2143
    N.E. Bogdanchikova, J.S. Ogden, J.M. Corker, S. Fuentes,
    and V.P. Petranovskii

Encapsulation of Chalcogens in $AlPO_4$-5 and SAPO-44 2147
    G. Li, J. Chen, and R. Xu

Preparation and Catalytic Properties of Zeolite-
Encapsulated Palladium-Salen-Complexes in the
Hydrogenation of Selected Unsaturated Compounds 2155
    S. Ernst, S. Sauerbeck, and X. Yang

Biomimetic Vanadium Oxidation Catalysis 2163
    P.P.H.J.M. Knops-Gerrits, P. Rouxhet, and P.A. Jacobs

Reactivity Toward Vanadium Species of the Defect Sites
Generated by Dealuminating a β Zeolite 2171
    E.M. El Malki, S. Dzwigaj, P. Massiani, A. Davidson, and M. Che

Roles of Adsorbed Water and Exchangeable Cations in
Microwave Heating of 'A' Type Zeolite 2179
    T. Ohgushi, S. Komarneni, and A.S. Bhalla

Photoinduced Activation of CO_2 by Rhenium (I)
Tricarbonyl Bipyridyl Chloride Encapsulated in Zeolite Y 2187
    H.M. Sung-Suh, W.Y. Kim, J.S. Chang, C.W. Lee, and S-E. Park

Zeolite-Encapsulated Copper and Manganese (X_2-Salen) Complexes with SOD, Catalase and Peroxidase Activities 2195
    C.R. Jacob, S.R. Varkey, and P. Ratnasamy

Environmentally Benign Electrophilic Bromination Reactions of Aromatics Catalyzed by Zeolites 2203
    A.P. Singh, S.P. Mirajkar, and S. Sharma

Methacrylate Polymers in Zeolites and Mesoporous Hosts. In Situ Incorporation of Acrylate Groups in MCM-Materials 2209
    K. Moller, T. Bein, and R.X. Fischer

Raman and X-ray Absorption Spectra of Selenium Species Stabilized in the Channels of $AlPO_4$-5 Single Crystals 2217
    V.V. Poborchi, A.V. Kolobov, H. Oyanagi, J. Caro, V.V. Zhuravlev, and K. Tanaka

Electron Microscopy Study of PbI_2 Clusters in Zeolite LTA 2225
    Y. Sakamoto, N. Togashi, T. Ohsuna, Y. Nozue, and O. Terasaki

Novel Pigments Via Microwave-Assisted Crystallization Inclusion of Chromophores in $AlPO_4$-5 or Ship-in-the-Bottle Synthesis of Dyes in HY 2233
    I. Braun, C. Schomburg, M. Bockstette, G. Schulz-Ekloff, and D. Wöhrle

The Development of New Luminescent Materials from Zeolite X 2241
    C. Borgmann, J. Sauer, T. Jüstel, U. Kynast, and F. Schüth

Titanium and Tin Oxide-Loaded Zeolites as Optical Sensor Materials for Reductive Atmospheres 2249
    M. Warnken, G. Grubert, N.I. Jaeger, and M. Wark

Author Index lv

Subject Index lxv

TABLE OF CONTENTS

VOLUME IV

VOLUME 1 3
VOLUME 2 733
VOLUME 3 1479
VOLUME 4 2259

Prefaces
    Chairs of the 12th International Zeolite Conference .... Volume I - xlv
    Publications Chair .... Volume I - xlvii

Executive Committee of the 12th IZC .... Volume I - xlviii

Program Committee of the 12th IZC .... Volume I - xlix

Local Arrangements Committee .... Volume I - xlix

Pre-Conference Zeolite School .... Volume I - l

Publications .... Volume I - l

International Advisory Board to the 12th IZC .... Volume I - li

Session Chairs of the 12th IZC .... Volume I - lii

Financial Support .... Volume I - liii

PART I: CHARACTERIZATION

Investigation of Acid Sites in Zeolites by 1-H NOESY NMR of
Probe Molecule 2259
    S-I. Lee and H. Chon

Probing Zeolite Internal Structures Using Very Low-Temperature
^{129}Xe NMR 2265
    A. Labouriau, T. Pietrass, S. Neugebauer Crawford, W.A. Weber,
    G. Panjabi, B.C. Gates, and W.L. Earl

Activation of Light Alkanes in the Presence of Benzene Over
Acidic Zeolites: A ^{13}C MAS NMR Study 2273
    I.I. Ivanova and F. Fajula

Interaction of CD_3OH with HY and HZSM-5 Zeolites Studied by
^{1}H NMR: Broad-Line at 4 K and High Resolution MAS at 298 K:
Comparison with Superacidic Compounds 2279
    P. Batamack and J. Fraissard

A ^{13}C NMR and FTIR Investigation of Acetonitrile Adsorption in
H-MFI Between 290 and 523K 2287
    J. Sepa, R.J. Gorte, D. White, B.H. Suits, and V.S. Swaminathan

Hydrochlorofluorocarbon Reactivity and Structural
Characterization of Zinc Exchanged NaX 2295
    M.F. Ciraolo, P. Norby, J.C. Hanson, D.R. Corbin, and C.P. Grey

Hydrofluorocarbon Zeolite Interactions: NMR and X-ray
Diffraction Studies 2301
    C.P. Grey, M.F. Ciraolo, and K.H. Lim

Thermal Transformations of Zeolite LiA(BW), (Li,Na)LTA and
Their Derivatives Obtained by Mechanochemical Treatment 2309
    C. Kosanovic, B. Subotic, P. Norby, and M. Soufek

NMR Studies of Oxygen-Zeolite Interactions at Low Temperatures 2317
    H. Liu, H-M. Kao, and C.P. Grey

The ^{13}C MAS NMR Detection of Organic Templates in Zeolites 2325
    M. Kovalakova, B.H. Wouters, and P.J. Grobet

Characterization and Quantitation of Lewis-Acid Sites in Solid
Acids by ^{31}P Solid-State NMR of the TMPO Complex 2331
    A.W. Peters, K.T. Mueller, K.J. Sutovich, E.F. Rakiewicz,
    and R.F. Wormsbecher

T-O-T Framework and Ligand Vibrations for Characterization
of Co(II) Ion Complexation in High Silica Zeolites 2339
    Z. Sobalík, Z. Tvaruzková, and B. Wichterlová

Crystal Structure of Hydrated Partially and Completely
NH_4-Exchanged Forms of Stilbite 2345
    A. Alberti, A. Martucci, M. Sacerdoti, S. Quartieri, G. Vezzalini,
    P. Ciambelli, and M. Rapacciuolo

Determination of the Location of Template Molecules in
Zeolite EU-1 Via a Combined Molecular Modelling and X-ray
Diffraction Approach 2355
    S.J. Andrews, J.L. Casci, P.A. Cox, and M.D. Shannon

Crystal Structure of Zeolite Ferrierite in As-Synthesized,
NH_4- and H-Forms 2361
    G. Cruciani, A. Alberti, A. Martucci, K.D. Knudsen, P. Ciambelli,
    and M. Rapacciuolo

Ion Exchange Between Cd^{2+} Solution and Clinoptilolite Mineral 2371
    J.C. Torres

Ion Exchange in Zeolite P, Zeolite JBW and Frameworks Containing
Low Valent Cations 2379
    A.M. Healey, S.E. Dann, and M.T. Weller

Non-Aqueous Synthesis and Structural Characterization of
Microporous Cobalt (II) Phosphates: $(NH_3CH_2CH_2NH_3)_{0.5}(CoPO_4)$
and $H_2Co_{3.5}P_3O_{12}$ 2387
    Y. Xu, X.L. Jiao, and W.Q. Pang

Ruherford Backscattering Spectroscopy, An Easy Method
to Vizualize and Quantify Metal Concentration Gradients
Through Metallosilicate Zeolite Crystals: The Case of
MFI Gallosilicates 2395
    Z. Gabelica, S. Valange, M. Jacobs, and G. Demortier

Characterization of K+ Ion Exchange into Na-LSX Using
Time-Resolved Synchrotron X-ray Powder Diffraction and
Rietveld Refinement 2401
    Y. Lee, C.L. Cahill, J.C. Hanson, J.B. Parise, S.W. Carr, M.L. Myrick,
    U.W. Preckwinkel, and J.C. Phillips

Joint X-ray Diffraction/NMR Structure Elucidation of Microporous
Fluorinated Alumino-Phosphates: ULM-3 Al and ULM-4 Al 2409
    F. Taulelle, V. Munch, C. Huguenard, A. Samoson, T. Loiseau, N. Simon,
    J. Renaudin, and G. Férey

An Attempt to Locate Protons in the ZSM-5 Structure by
Combined Synchrotron and Neutron Diffraction 2413
    B. Toby, S. Purnell, R. Hu, A. Peters, and D.H. Olson

Single Crystal Structure Analysis of a Microcrystal of ZSM-11
Using Synchrotron X-ray Data 2419
    H. van Koningsveld, M.J. den Exter, J.H. Koegler, C.D. Laman,
    S.L. Njo, and H. Graafsma

Laser-Induced Luminescence Investigation of Y Zeolites
Catalysts Surfaces 2425
    C. Lalo, J. Deson, A. Gedeon, and J. Fraissard

Structural Investigation of Sorbate-Induced Phase Transitions
in ZSM-5 by FT-Raman Spectroscopy 2431
    Y. Huang and P. Qiu

ESCA Study of Zeolites 2439
    I. Jirka

Faulting Effects in the CHA-GME Group of ABC-6 Materials 2445
    J. Plévert, R.M. Kirchner, and R.W. Broach

Hydrothermal Synthesis and Structural Study of a New
Fluorinated Gallophosphate $Ga_4(PO_4)_3(HPO_4)F_3 \cdot T$ (T = amine) 2453
    S.J. Weigel, T. Loiseau, G. Férey, V. Munch, F. Taulelle,
    R.E. Morris, G.D. Stucky, and A.K. Cheetham

Cation Siting in Microporous Materials: A 2-D Triple-Quantum
MAS NMR Study 2457
    J.R. Agger, M.W. Anderson, J. Rocha, D.P. Luigi, M. Naderi,
    and A.K. Baggaley

Magic-Angle-Turning NMR and Theoretical Studies of Chemical
Shift Tensors on Microporous Catalysts 2465
    A. Philippou, F. Salehirad, D.P. Luigi, and M.W. Anderson

Magnetic Resonance Studies on VAPO-5 and MgVAPO-5
Microporous Materials 2473
    T. Blasco, P. Concepcion, L. Fernandez, J.M. Lopez Nieto,
    and A. Martinez-Arias

^{31}P and ^{27}Al MAS NMR of MAPO-36 and MAPO-5 with
High Mg Content 2481
    M.V. Giotto, M. da S. Machado, S.P.O. Rios, J. Pérez-Pariente,
    and D. Cardoso

Probing the Structure of Metal-Substituted Molecular Sieves
by Solid-State NMR 2489
    A. Labouriau, S. Neugebauer Crawford, K.C. Ott, and W.L. Earl

The Reversible Coordination of Framework Aluminum in
Zeolites 2497
    B.H. Wouters, T-H. Chen, and P.J. Grobet

Influence of Guest Compounds on the Base Strength of
Zeolites Y and X Investigated by NMR Spectroscopy 2503
    M. Hunger, U. Schenk, B. Burger, and J. Weitkamp

The Use of Tertiary Amines in the Elemental Characterization
of Zeolites and Catalysts 2511
    M.E. Tatro

A Detailed NMR Study to the Polarization of Non-Framework
La^{3+} Cations with the Framework Y Zeolite: Application of
^{29}Si, ^{27}Al MAS and ^{27}Al MQ MAS NMR 2515
    J.A. van Bokhoven, A.L. Roest, A.P.M. Kentgens, and
    D.C. Koningsberger

As-Synthesized ITQ-1, The All-Silica Analog of MCM-22(P):
Ordered, Disordered, or Something in Between? 2519
    S.L. Njo, H. van Koningsveld, B. van de Graaf, Ch. Baerlocher,
    and L.B. McCusker

Structure Analysis of the Ion Exchanged Forms of the New
Microporous Titanosilicate $M_2TiSi_3O_9 \cdot 2.5H_2O$
(M = NH_4^+, K+ and Li+) 2525
    J-L. Paillaud, V. Valtchev, S. Mintova, and H. Kessler

PART II: CATALYSIS AND CHARACTERIZATION

Adsorption Complexes of Metal Halides at the Brønsted
Acid Sites in Zeolites 2535
    A.I. Biaglow

Skeletal Isomerization of C_4-C_8 Paraffins: Comparison of
Zeolites and Sulfated Oxides 2541
    L.M. Kustov, T.V. Vasina, O.V. Masloboishchikova, A.V. Ivanov,
    E.G. Khelkovskaya-Sergeeva, and P. Zeuthen

Strong Acid Sites Formed by a Combination of Framework
Acid Site and Extra-Framework Cation in Porous Materials
as Measured by Temperature Programmed Desorption
of Ammonia 2549
    T. Kunieda, N. Katada, and M. Niwa

The Effect of Oxygenates on the n-Hexane Aromatization
Activity of Pt/KL 2557
    M.E. Dry, R.J. Nash, and C.T. O'Connor

Zeolite Beta: The Relationship Between Calcination Procedure,
Aluminum Configuration and Catalytic Activity in the MPV
Reduction of Ketones 2565
    P.J. Kunkeler, B.J. Zuurdeeg, and H. van Bekkum

The Vibrational Spectroscopy of Acid-Base Interactions in
Zeolite Cavities as a Tool for Acid-Strength Investigation 2571
    A. Zecchina, D. Scarano, C. Lamberti, G. Spoto, C. Pazé,
    and S. Bordiga

FTIR Study of the Reactivity of the Surface Methoxy Species
Formed by the Reaction of Methanol on H-ZSM-5 2577
    F. Wakabayashi, J.N. Kondo, C. Hirose, and K. Domen

Comparative Studies of Catalytic Effect for the Oxidative
Methylation of Toluene with Methane Over Basic Zeolite
Catalysts 2585
    L. Zhou, W. Li, Q. Fu, N. Guan, S. Zheng, and K. Tao

Lewis Acid Sites in Zeolite Y Studied by Adsorption, EPR and
NMR Techniques 2589
    A. Seidel, A. Gutsze, and B. Boddenberg

The Distribution of Acid Strength of OH Groups in Steamed
HY Zeolites Studied by IR Spectroscopy 2595
    J. Datka, B. Gil, J. Fraissard, P. Massiani, and P. Batamack

The Properties of Alkoxyl Groups in Zeolites Studied by
IR Spectroscopy 2601
    J. Datka, J. Rakoczy, and G. Zadrozna

Vibrational Spectroscopic Investigations of Pyrrole
Adsorption in Faujasites: Studies by Infrared, Raman
and Neutron Spectroscopy 2609
    E. Geidel, H. Jobic, and S.F. Parker

Novel Frequency Response Techniques for the Study of the
Kinetics of Heterogeneous Catalysis 2615
    I.R. Harkness, M. Cavers, L.V.C. Rees, J.M. Davidson, and
    G.S. McDougall

Entropically Determined Adsorption Peculiarities Studied on KFI by TPD, Microcalorimetry, ^{13}C CP MAS NMR and FTIR 2623
J. Jänchen, W.J.M. van Well, J.H.M.C. van Wolput, and H. Stach

Catalytic Significance of Strong Acid Sites in Dealuminated Faujasites and Mordenites 2629
I.V. Mishin, T.R. Brueva, and G.I. Kapustin

Distribution of Acid Sites in Zeolites of Different Types Based on Microcalorimetric Measurements 2637
G.I. Kapustin, T.R. Brueva, and I.V. Mishin

Principle for Generation of Acidity in Y Zeolite Found by Ammonia Temperature-Programmed Desorption: Stoichiometric Generation of Acid Sites with a Constant Strength by Isolated Framework Al Atoms 2643
H. Igi, N. Katada, and M. Niwa

Adsorption, Acidic and Catalytic Properties of Decationized Low-Alumina Zeolites Obtained Through Direct Synthesis 2651
J.Ya. Smorodinskaya, Yu.I. Azimova, M.I. Levinbuk, and M.Ya. Melnikov

Study of the Adsorption State of Phenol on HY Zeolite by Infrared Spectroscopy 2659
X-W. Li, X. Su, and X-Y. Liu

ESR Study of NaY Supported Pd and Pt Ions and Clusters 2665
H. Du, R. Klemt, F. Schell, J. Weitkamp, and E. Roduner

Study of Pyridine and Ammonia Sorption in Faujasite and Mordenite Zeolites by the Frequency-Response Technique 2673
D. Shen, Gy. Onyestyák, and L.V.C. Rees

Effect of Acido-Basicity of Beta Zeolites on the Conversion of Chloromethane to Hydrocarbons as Studied by FTIR and TPD-MS 2681
B-L. Su, D. Jaumain, K. Ngalula, and M. Briend

Chloromethane as Probe Molecule to Characterize the Brønsted Acidity of Zeolites: An *In Situ* FTIR Study 2689
B-L. Su and D. Jaumain

Peculiarities of Brønsted Acid Sites in FER-Type Zeolites 2697
J. Weitkamp, M. Breuninger, H.G. Karge, and M. Hunger

Basic and Acidic Sites in Cs/Na Faujasites: An IR Study 2705
E. Garrone, P. Marturano, B. Onida, M. Laspéras, and F. Di Renzo

Nature of Zn Sites in Zn-MFI: An FTIR Investigation 2711
S. Valange, Z. Gabelica, B. Onida, and E. Garrone

Comparative Study of n-Heptane Hydrocracking Over
Pt HEMT and Pt HFAU Catalysts 2719
    A. Berreghis, P. Magnoux, and M. Guisnet

Aromatization of 1,3-Butadiene on Basic Zeolites in the
Vapor Phase 2727
    J. Ackermann, E. Klemm, and G. Emig

Controlled Removal of Extra-Framework Aluminum Species
in USY Zeolite 2735
    E. Benazzi, J. Lynch, A. Gola, S. Lacombe, and C. Marcilly

Reinsertion of Nonframework Aluminums in Dealuminated
HZSM-5 Zeolite by Acid Treatment 2743
    T. Sano, R. Tadenuma, Y. Uno, Z.B. Wang, and K. Soga

CO and NO Adsorption on the Copper Exchanged SAPO-34
Molecular Sieve Catalyst 2751
    D.B. Akolekar and S.K. Bhargava

Aromatization of n-Heptane on the Modified MFI Zeolites 2759
    N. Bilba, I. Asaftei, Gh. Iofcea, and N. Naum

Predicting Extra-Framework Cation Positions in Zeolites:
Energy Minimization and (N,V,T) Monte Carlo Simulations
in LiLSX(X) 2767
    C.F. Mellot and A.K. Cheetham

Reactivity of NO on Co^{2+}/Co^{3+} Redox Sites in CoAPO-18:
FTIR and UV-Vis-NIR Studies 2775
    E. Gianotti, L. Marchese, G. Martra, and S. Coluccia

N_2O Decomposition Over (Fe)-ZSM-22 Zeolites 2781
    L. Matachowski, M. Kasture, T. Machej, and M. Derewinski

Structure and Activity of Cerium-Promoted Ag-ZSM-5
for the Selective Catalytic Reduction of Nitric Oxide
with Methane 2787
    Z. Li and M. Flytzani-Stephanopoulos

CeO_2-H-ZSM-5 Composites—A Bifunctional System for
the Selective Catalytic Reduction of NO by Methane 2795
    T. Liese, D. Rutenbeck, and W. Grünert

In Situ Synthesis of Zeolites on Cordierite and Their Catalytic
Behavior in Decomposition of NO 2803
    N.J. Guan, X.L. Shan, K. Zhang, D.S. Wang, and S.H. Xiang

Preparation, Characterization and $DeNO_x$ Activity of
(Pt-Co)ZSM-5 and (Pt-Cu)ZSM-5 Zeolite Catalysts 2809
    A. Tamási, I. Kiricsi, Z. Kónya, Z. Schay, J. Halász, and
    L. Guczi

Hydroconversion of Heptane and Octane Over Bifunctional
Zeolites; Influence of Structure and Metal Distribution on
Activity and Selectivity 2817
    A. Jentys, A. Lugstein, G. Kinger, and H. Vinek

Coordination Chemistry of Titanium in Titanosilicate
Molecular Sieves Studied by Electron Spin Resonance
and Electron Spin Echo Modulation Spectroscopy 2825
    A.M. Prakash and L. Kevan

Palladium Species in Co/Pd/H-ZSM-5 Catalysts for
CH_4-SCR of NOx 2833
    M. Ogura, M. Hayashi, and E. Kikuchi

Crystal Structure of a Benzene Sorption Complex of
Dehydrated Fully Mn(II)-Exchanged Zeolite X 2839
    Y. Kim, A.N. Kim, Y.W. Han, and K. Seff

Cation Exchanged Zeolites ZSM-5 for the Hydroxylation
of Benzene with Nitrous Oxide 2847
    S. Kowalak, K. Nowinska, M. Swiecicka, M. Sopa, A. Jankowska,
    G. Emig, E. Klemm, and A. Reitzmann

Modification of Zeolites by Mo with the Use of Chemical
Transport Reaction 2855
    A.V. Kucherov and A.A. Slinkin

Catalytic Decomposition of N_2O Over NaY-Supported and
USY-Supported Rh Catalysts 2863
    K. Yuzaki, T. Yarimizu, K. Aoyagi, S. Ito, T. Sato, S. Hayashi,
    and K. Kunimori

Total Catalytic Oxidation of Acetic Acid by H_2O_2 Over
Transition Metal-Exchanged NaY Zeolites 2869
    S. Lévesque, Y. Yang, F. Larachi, and A. Sayari

Reduction of Iron Ions to the Metallic State in X and Y
Zeolites by Sodium Azide 2875
    H.K. Beyer, G. Onyestyák, B.J. Jönsson, K. Matusek, and K. Lázár

Dehydroisomerization of n-Butane Over Bifunctional Catalysts 2881
    G.D. Pirngruber, K. Seshan, and J.A. Lercher

Water and Sulfur Resistant Pt-Based Zeolite Catalyst for
NOx Reduction 2889
    S.E. Maisuls, S. Feast, K. Seshan, J.G. van Ommen, and J.A. Lercher

Faujasite Y Confined Ni(II)-Tetrakis(N-methyl-4-pyridyl)-
Porphyrin as Hydrogenation Catalyst 2897
    B-Z. Zhan, P.A. Jacobs, and X-Y. Li

Nitrile-to-Amide Hydrolysis Catalyzed by Faujasite-Y
Confined Chromium(III) Porphyrin 2905
    B-Z. Zhan, P.A. Jacobs, and X-Y. Li

Restricted Transition State at Pore Mouth Catalysis in the
Selective Hydroisomerization of Normal and Methyl
Branched C_8 Paraffins Over Monodimensional 10-Ring
Molecular Sieves 2913
    P. Mériaudeau, Vu.A. Tuan, G. Sapaly, Vu.T. Nghiem,
    and C. Naccache

Catalytic Cracking of Palm Oil to Hydrocarbon Liquid Fuels
Over Various Zeolite Catalysts: Optimization Studies 2921
    S. Bhatia, N.A.M. Zabidi, and F. Twaiq

Carcinogenicity of Mineral Erionite Fibers: Measurements
and Hypothesis of Activity 2927
    B.D. Hogg, P.K. Dutta, J.F. Long, and A. Vaidyalingam

Influence of the Rare Earth Content on the Amount and
on the Nature of the Coke Formed from n-Heptane Over
Y Zeolites 2935
    C.A. Henriques and J.L.F. Monteiro

PART III: NMR CHARACTERIZATION

Enhanced Surface NMR of Zeolites and Related Materials
Using Laser-Polarized Xenon 2943
    E. Brunner, M. Haake, A. Pines, J. Reimer, and R. Seydoux

^{19}F and ^{29}Si Solid-State NMR Spectroscopy on Five-Coordinate
Silicon Sites, $(SiO)_4SiF^-$, in Zeolites 2951
    H. Koller, A. Wölker, S. Valencia, L.A. Villaescusa, M.J. Díaz-Cabañas,
    and M.A. Camblor

High Temperature ^{1}H MAS NMR Studies of the Proton Mobility
in Zeolites 2955
    H. Ernst, D. Freude, T. Mildner, and H. Pfeifer

Interaction of Chlorfluorocarbons with Zeolites Studied by
In Situ IR and Multinuclear NMR Spectroscopy 2963
    I. Hannus, Z. Kónya, P. Lentz, J.B. Nagy, and I. Kiricsi

In Situ and *Ex Situ* NMR Methodology to Study Microporous
Phase Crystallization 2971
    C. Gerardin, M. Haouas, F. Taulelle, C. Estournes, T. Loiseau,
    and G. Ferey

Solid-State NMR Investigation of Cation Siting in LiX Zeolites 2979
    M. Feuerstein, A. Burton, and R.F. Lobo

Characterization of Extra-Framework Cations in ETS-10
Studied Through MQ-MAS NMR 2985
    L. Delevoye, S. Ganapathy, T. Kumar, C. Fernandez,
    and J-P. Amoureux

Applications of ^{1}H NMR Imaging and ^{129}Xe NMR to the Study of Hydrocarbon Diffusion in Zeolites and Coke Distribution 2991
    T. Domeniconi, P. N'Gokoli-Kekele, J-L. Bonardet,
    M-A. Springuel-Huet, and J. Fraissard

Faulted Zeolite Framework Structures 2999
    H. Gies, R. Kirchner, H. van Koningsveld, and M.M.J. Treacy

Author Index lv

Subject Index lxv

VOLUME IV

Part I
Characterization

INVESTIGATION OF ACID SITES IN ZEOLITES BY 1-H NOESY NMR OF PROBE MOLECULE

SANG-ICK LEE[†] and HAKZE CHON

Department of Chemistry, Korea Advanced Institute of Science and Technology, 373-1, Kusong-dong, Yusong-ku, Taejon 305-701, Korea; hzchon@sorak.kaist.ac.kr

ABSTRACT

The acid sites in silicoaluminophosphate-5, silicoaluminophosphate-11 and HZSM-5 were investigated using magic angle spinning (MAS) nuclear magnetic resonance (NMR) and nuclear Overhauser effect (NOE) spectroscopy. MAS NMR and NOESY experiments have permitted the measurement of different types (or different locations) of hydroxyl groups. The specificity of distance between the adsorbate (benzene as a probe molecule) and different types of hydroxyl groups has been verified by 1-H NOESY techniques. Measurements of through space distance with 1-H NOESY techniques between hydroxyl groups and the probe molecule provides a way for characterizing their locations in zeolites.

INTRODUCTION

The Magic angle spinning (MAS) NMR techniques are used routinely to average the anisotropic interactions that broaden the resonance lines from magnetically dilute nuclei in polycrystalline solids. In recent years, additional information about the structure of zeolites has been obtained from MAS NMR studies [1-3]. Particularly, the 2-D MAS NMR techniques have proven to be a reliable method for the direct observation and characterization of nuclear environments in solids. The 2-D MAS NMR have been demonstrated recently in zeolites as useful for tracing the connectivity's between tetrahedral ^{29}Si atoms in three-dimensional framework [2,3]. One important development is the ability to determine the distances between nuclei [4] which are coupled *via* dipolar interactions. This technique seems to become the solid-state equivalent of the NOESY experiment, since it provides through space distance information.

In zeolites, the bridging hydroxyl groups are well documented, sufficient information is not

---

[†] Present address: Process Research Dept. 4, Semiconductor Research Division, San 136-1, Ami-ri, Bubal-eup, Ichon-si, Kyoungki-do, 467-701 Korea; silee@sr.hei.co.kr

available on the location and acidic properties. There is considerable interest in obtaining information on the strength and the location of the bridging hydroxyl groups.

In the present study, the acid sites of zeolites are investigated by means of ^{1}H and ^{29}Si MAS NMR and ^{1}H NOESY. To determine their geometry and location of the different environments of hydroxyl groups on zeolites, benzene has been applied as a probe molecule. The specificity of distance between benzene and hydroxyl groups in zeolites is investigated by 2-D ^{1}H NOESY NMR.

EXPERIMENTAL

High-resolution solid state MAS ^{29}Si and ^{1}H NMR spectra were obtained with a Bruker MSL-300 spectrometer. The experimental conditions were: for ^{29}Si, pulse length of 2.8 μs, flip angle $\pi/4$, recycle delay 5 s, spinning rate 4 kHz; for ^{1}H, these parameters were 3 μs, $\pi/4$, 5 s, 4.5 kHz, respectively.

The 2-D NOESY experiments were carried out using conventional pulse sequences and benzene as a probe molecule at room temperature. A schematic diagram of the NOESY NMR pulse sequence is shown in below:

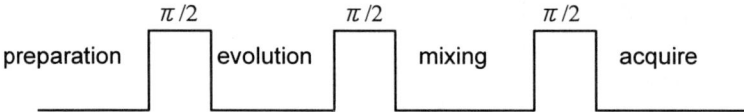

Three 2-D NOESY spectra were acquired with variable mixing times. The relaxation delay between acquisitions was 2 s, the data size in t_2 was 512 complex and the size in t_1 was 256. For each t_1, 16 or 64 scans were averaged to improve the signal-to-noise ratio. As standard in multi-dimensional NMR experiments, nuclear frequencies during the evolution period are monitored indirectly by incrementing t_1, which modulates the free-induction-decay signal detected directly during t_2. The length of mixing time is usually allowed to vary randomly by ±10% to suppress scalar-coupling effects [5]. The standard pulse sequence and TPPI phase cycling were used for phase sensitive NOESY. Not having exchange, off-diagonal cross-peak intensites measured by the nuclear Overhauser effect should be directly proportional to $1/r^6$ [6], where r is the distance between the pair of protons that cross-relax each other.

The crystallinity of all the samples was checked by X-ray powder diffraction (XRD) analysis. Before ^{1}H NMR measurements, the calcined zeolites were heated at 5 K/min to 623 K and held at 623 K for 3 hours under the vacuum and subsequently lowered to room temperature. All the NMR samples were kept in the sealed Pyrex tubes and transferred into ZrO_2 rotors in a dry glove box, and sealed off.

RESULTS AND DISCUSSION

Figure 1 shows the ^{1}H MAS NMR spectra of SAPO-5, SAPO-11 and ZSM-5 samples, which contain several types of hydroxyl groups, measured at room temperature. SAPO-5 and SAPO-11 consist of a maximum of three lines at 1.9, 3.8 and 4.8 ppm, and 1.5, 3.4 and 4.4 ppm, respectively. While the line at 1.9 ppm (SAPO-5) and 1.5 ppm (SAPO-11) is due to SiOH and POH groups at the lattice defects or at the outer surface. Bridging hydroxyl groups have caused the latter two signals at each sample [7]. The ^{1}H NMR spectrum of ZSM-5 exhibits the well known two signals at 2.0 ppm (SiOH groups at the outer surface or on lattice defects) and 4.2 ppm (bridging hydroxyl groups). ^{1}H MAS NMR chemical shifts are considered to be related to the proton donor ability of the corresponding sites and can thus provide information on their acid

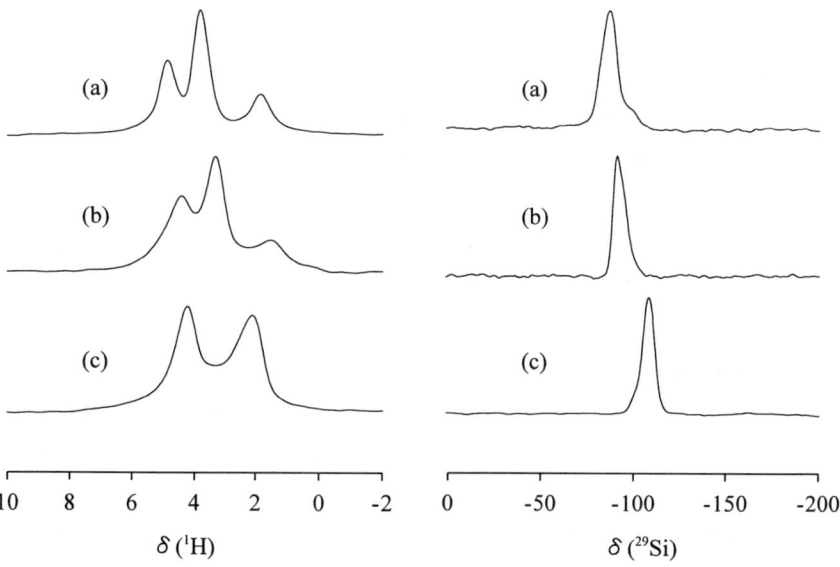

Figure 1. ^{1}H MAS NMR spectra of the calcined and dehydrated SAPO-5 (a), SAPO-11 (b) and ZSM-5 (c)

Figure 2. ^{29}Si MAS NMR spectra of the calcined and dehydrated SAPO-5 (a), SAPO-11 (b) and ZSM-5 (c)

strength [8]. For comparable hydroxyl groups, higher values of chemical shift may be taken as resulting from higher acid strength.

Figure 2 shows the ^{29}Si MAS NMR spectra of the calcined SAPO-5, SAPO-11 and ZSM-5 samples. The major peak of SAPO-5 and SAPO11 are located at about -90 ppm~-95 ppm, which is attributed to the Si (4Al) configuration belonging to SAPO domains [9]. The ^{29}Si NMR spectra show that Si was incorporated into the framework. The incorporation of Si usually contributes the charge-balancing protons, which act as Brønsted acid sites. The ^{29}Si NMR spectrum of ZSM-5 under study is shown in Figure 1 (c). Signal with a maximum at about -112 ppm and a weak signal at -104 ppm for Si (1Al) are observed.

Contour plots of 2-D ^{1}H NOESY spectra with various mixing times of the SAPO-5, SAPO-11 and ZSM-5 samples adsorbed by benzene were presented in Figure 3. In the NOESY spectra of 100 ms mixing time on SAPO-5, one medium NOE cross peak from benzene protons to 3.8 ppm was shown in Figure 3 (a). No NOE connectivity between benzene and 4.8 ppm was observed at this mixing time, suggesting that 4.8 ppm is more distant from benzene than 3.8 ppm [10]. Even at 300 ms mixing time, we observed two off-diagonal cross peaks between the pair of protons that cross-relax each other: a medium and a weak NOE for benzene↔3.8 ppm and benzene↔4.8 ppm, respectively. The more the mixing time was increased, the lower the intensity of diagonal and off-diagonal peaks was observed.

Figure 3 (b) shows the contour plots of NOESY spectra of the SAPO-11. One medium NOE cross peak for benzene↔3.4 ppm was observed at 60 ms mixing time but not for benzene↔4.4 ppm. Whereas two off-diagonal cross-peaks were observed at 150 ms mixing times, suggesting that 3.4 ppm is closer from benzene than 4.4 ppm.

A correlation between benzene and the bridging hydroxyl group at 12-ring O-atom (SAPO-5 corresponds to 3.8 ppm and SAPO-11 to 3.4 ppm) leads to the medium NOE cross peak. The weak NOE cross peak for benzene↔4.8 ppm (SAPO-5) and 4.4 ppm (SAPO-11) then was also observed at 300 and 150 ms mixing time, respectively. The location of bridging hydroxyl groups in SAPO-5 and –11 can be directly monitored by ^{1}H NOESY technique. Off-diagonal cross-peak intensities measured by NOE's should be directly proportional to $1/r^6$ [6], so that the 4.8 ppm (SAPO-5) and 4.4 ppm (SAPO-11) is further from the benzene proton than the 3.8 ppm (SAPO-5) and 3.4 ppm (SAPO-11). None of the NOE connectivity related to 1.9 ppm was observed for the variation of mixing times. From 2-D ^{1}H NOESY studies one can assign the position of 3.8 (SAPO-5) and 3.4 (SAPO-11) ppm to the proton at the 12-ring O-atom, and also the position

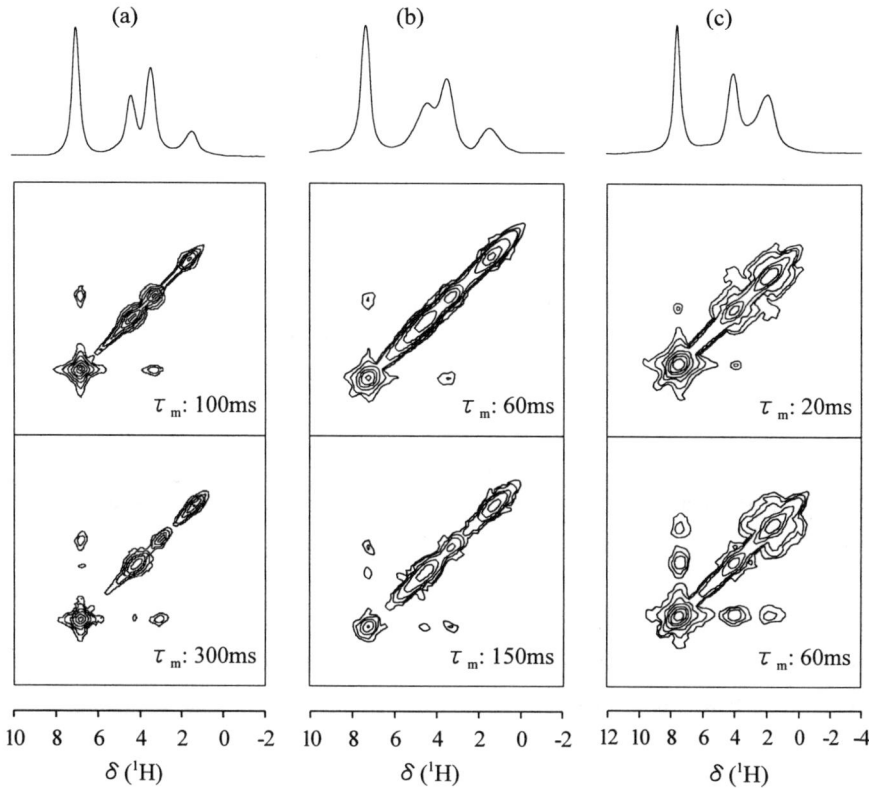

Figure 3. 2D ^{1}H NOESY spectra of the SAPO-5 (a), SAPO-11 (b) and ZSM-5 (c) after adsorption of benzene as a probe molecule at room temperature with different mixing times.

of 4.8 (SAPO-5) and 4.4 (SAPO-11) ppm to the proton at the 6-ring O-atom. This result agrees with that of previous papers [7,10,11].

In the NOESY spectra on ZSM-5 at 20 ms mixing time (Figure 3 (c)), a NOE for benzene↔4.2 ppm was observed but not for benzene↔2.0 ppm at this mixing time. The two intense off-diagonal cross-peaks were observed even at 60 ms mixing time. NOE connectivity between benzene and 2.0 ppm was observed at this mixing time, suggesting that the proton at benzene correlate with 2.0 ppm which is caused by hydroxyl group on lattice defects.

2-D ^{1}H NOESY studies with benzene as a probe molecule show that hydroxyl groups in SAPO-5, SAPO-11 and ZSM-5 framework are not identical in location. The characteristics of their presence had to be associated with a different location, *i.e.* the proton at the 6-ring and the 12-ring O-atom in SAPO-5 and SAPO-11 framework. In the case of ZSM-5, we observed NOE connectivity between the hydroxyl group on lattice defects and probe molecule. The investigation has presented us with information about differences in distance between the bridging hydroxyl groups and the probe molecule in zeolites.

REFERENCES

1. M. Hunger, M.W. Anderson, A. Ojo and H. Pfeifer, Microporous Mater. **1**, 17 (1993)
2. C.A. Fyfe, H. Gies, Y. Feng and G.T. Kokotailo, Nature **341**, 223 (1989)
3. C.A. Fyfe, H. Grondey, G.T. Kokotailo, S. Ernst and J. Weitkamp, Zeolites **12**, 50 (1992)
4. L.W. Jelinski, Anal. Chem. **62**, 212R (1990)
5. S. Macura, K. Wthrich and R.R. Ernest, J. Magn. Reson. 46, 269 (1982)
6. J.K.M. Sanders and B.K. Hunter, in Modern NMR Spectroscopy, (Oxford, New York, 1988)
7. B. Zibrowius, E. Löffler and M. Hunger, Zeolites **12**, 167 (1992)
8. G. Engelhardt, Trends Anal. Chem **8**, 343 (1989)
9. J.A. Martens, P.J. Grobet and P.A. Jacobs, J. Catal. **126**, 299 (1990)
10. S.-I. Lee and H. Chon, J. Chem. Soc., Faraday Trans. **93**(9), 1855 (1997)
11. S.G. Hedge, P. Ratnasamy, L.M. Kustov and V.B. Kazansky, Zeolites **8**, 137 (1988)

PROBING ZEOLITE INTERNAL STRUCTURES USING VERY LOW TEMPERATURE ^{129}XE NMR

ANDREA LABOURIAU[†], TANJA PIETRASS[‡], SUSAN NEUGEBAUER CRAWFORD[†], WILLIAM A. WEBER[+], GHANSHAM PANJABI[+], BRUCE C. GATES[+], and WILLIAM L. EARL[†].

[†] Chemical Science and Technology Division, Los Alamos National Laboratory, Los Alamos, New Mexico 87545.
[‡] Department of Chemistry, New Mexico Institute of Mining and Technology, Socorro, New Mexico 87801.
[+] Department of Chemical Engineering and Materials Science, University of California, Davis, California 95616.

ABSTRACT

We have measured the ^{129}Xe chemical shift of xenon in very well characterized zeolite Y and in a sample of zeolite Y containing small rhodium clusters. By measuring the chemical shifts over an extremely wide temperature range, the data are interpretable in terms of very simple models that yield effective van der Waals attraction energies.

INTRODUCTION

In 1981 and 82 [1-4] ^{129}Xe NMR was proposed as a method for measuring pore sizes and pore structures in porous materials in general and zeolites in particular. Unfortunately the technique has not proven to be as generally applicable or as useful as originally promulgated. Ripmeester and coworkers have pointed out some of the problems with ^{129}Xe NMR [5-7]. The result of these problems is a general belief that ^{129}Xe NMR is not a useful structural tool. Our experience is that this method is not useful for *routine* pore characterization but it can be effectively used to understand more complex physical chemical interactions in porous materials.

In our work, we rely on variable temperature studies over a wide temperature range to extract information about the energetics of adsorption.

Our eventual goal is to use xenon as a probe of cations and metal particles in micropores, especially zeolites. In this case, the problem is to relate ^{129}Xe chemical shifts or relaxation times as a function of temperature to energies of adsorption or electronic properties. Cheung published a little known (or little appreciated) paper that discusses a model for xenon-pore wall interactions [8]. This model uses square well potentials for van der Waals attraction in cylindrical pores. In spite of the simplicity of the model, Cheung is able to fit low temperature ^{129}Xe NMR data for a few zeolites. However, he only has experimental data at two temperatures, room temperature and 144 K. We have made ^{129}Xe chemical shift measurements in zeolite Y as a function of temperature from room temperature to 40 K at several different xenon loadings and applied Cheung's interpretation to our data.

Another model, due to Raftery, et al, [9] uses a standard NMR two site chemical exchange model with the sites being: xenon in the gas (center of the pore) and in contact with the wall. They assign different chemical shifts to these two sites and are able to describe the variable temperature shifts for xenon adsorbed on poly(acrylic acid). We have modified this model also and applied it to our variable temperature ^{129}Xe shift data.

The key differences between our work and prior literature is the ability to make measurements of chemical shifts as a function of temperature to very low temperatures, where xenon atoms become essentially immobile. The other requirement is to have very well characterized and understood zeolites and metal clusters in the zeolites to make it possible to interpret the model data.

EXPERIMENTAL

The starting zeolite used for these experiments is a commercial zeolite Y obtained from Grace Chemical Company. We have characterized it extensively with powder x-ray diffraction (XRD), ^{29}Si MAS NMR, elemental analysis, and scanning electron microscopy (SEM). The XRD pattern is characteristic of zeolite Y with no crystalline impurities and no indications of significant amorphous material. The ^{29}Si NMR spectrum has very little intensity in the silanol region, indicating high crystallinity. From the ratios of the Si(1Al), Si(2Al), etc. peaks we calculate the Si:Al ratio to be 2.6. The SEM photographs show very regular crystallites with dimensions approximately 0.5 μm on a side. The elemental analysis agrees with the spectroscopic data but with an iron impurity level of about 200 ppm (wt %). Overall this is a very good zeolite Y although the iron level is slightly higher than desired for some NMR measurements.

Samples were prepared for ^{129}Xe NMR by carefully weighing the zeolite into an 8 mm (outside diameter) glass tube that was mounted on a vacuum system. The zeolite was carefully dried by slow heating to 723 K and held at that temperature for 12 hours. The zeolite was then cooled and natural abundance xenon gas was measured into the tube, volumetrically, and the tube was flame sealed. We have prepared many samples in this fashion, the sample reported in this communication contained approximately 1 xenon atom for every 3 α cages of the zeolite Y. The sample was also prepared with approximately 20 torr of He gas to facilitate thermal equilibrium in the variable temperature experiments. We also report preliminary data on a sample that contains small rhodium metal clusters. These are prepared by synthesizing the rhodium-6 hexadecacarbonyl in situ, in the α cages. The carbonyl is then carefully decomposed, leaving rhodium metal clusters that are six atoms in size. Many preparations of this type of sample have been synthesized at The University of California at Davis in recent years and characterized by a variety of techniques, especially by extended X-ray fine structure (XAFS). The sample reported here contained approximately one Rh-6 cluster every 12 α cages and one xenon every 2 α cages. It also contained 20 torr of helium for temperature equilibration.

NMR measurements were made on a Varian Unity-400 NMR spectrometer with a 9.4 T magnet and a nominal resonance frequency of 110.6 MHz for ^{129}Xe. Variable temperature was accomplished with an Oxford model CF 1200 cryostat and a homebuilt transmission line probe [10]. Sample temperature was measured with a Lakeshore Cryogenics calibrated carbon-glass resistor close to the NMR sample. Chemical shifts were measured relative to an external standard and corrected to the chemical shift for gaseous xenon at zero pressure.

RESULTS AND DISCUSSION

Variable temperature, ^{129}Xe chemical shifts can be related to van der Waals energies through a model that describes the interaction. The models chosen here are those due to Cheung and Raftery, et al. We recently submitted a detailed discussion of the models for publication so the details are not presented here [11]. Cheung's model uses a square well potential at the zeolite wall to approximate Lennard-Jones potentials. The chemical shift equation is:

$$\delta(T) = \frac{c\varepsilon}{\left(1 + F\exp\left(\frac{-\varepsilon}{k_B T}\right)\right)} \quad (1)$$

where ε is the depth of the square well potential, c is a phenomenological constant, and F is a geometrical term given by :

$$F = (L - 2a_{xe})/2lm - 1.$$

a_{xe} is the van der Waals radius of the xenon atom and m is 1, 2, or 3 for one-, two-, or three-dimensional pores. The potential well has a width of l and a value of $-\varepsilon$ inside the well and zero on the outside. It becomes infinite when the distance between Xe and O atoms is smaller than the sum of their Van der Walls radii. The quantity, L, expresses the mean 'free' pore size which is related to the xenon mean-free path (λ), i.e., $L = a_{xe} + 2\lambda$.

The model used by Raftery, et al, starts with the standard "two-site chemical exchange" equation:

$$\delta = \delta_s P_s + \delta_g P_g$$

where δ_g is the shift associated with xenon in the gas phase and δ_s on the inner surface and P_s and P_g are the probabilities of finding a xenon at each site. The probability of xenon on the surface can be defined in terms of a sticking time, τ_s. This results in a chemical shift equation,

$$\delta = \frac{\delta_s \tau_0 \exp\left(\varepsilon/k_B T\right)}{\tau_0 \exp\left(\varepsilon/k_B T\right) + \lambda/\sqrt{3k_B T/m}} \quad (2)$$

where ε is the van der Waals potential energy for sticking to the surface, τ_0 is the shortest residence time for xenon on the pore wall, δ_S is the chemical shift of xenon on the pore wall (the chemical shift in the center of the pore is defined as zero), and m is the mass of a xenon atom. The average sticking time at the surface can be given by the Arrhenius relationship.

Figure 1 is a plot of the measured ^{129}Xe chemical shift versus temperature with curve fits to equations 1 and 2. It is clear that there are enough fitting parameters in these equations that an excellent fit will be obtained in any case so the quality of fit is not a test of the "goodness" of the models.

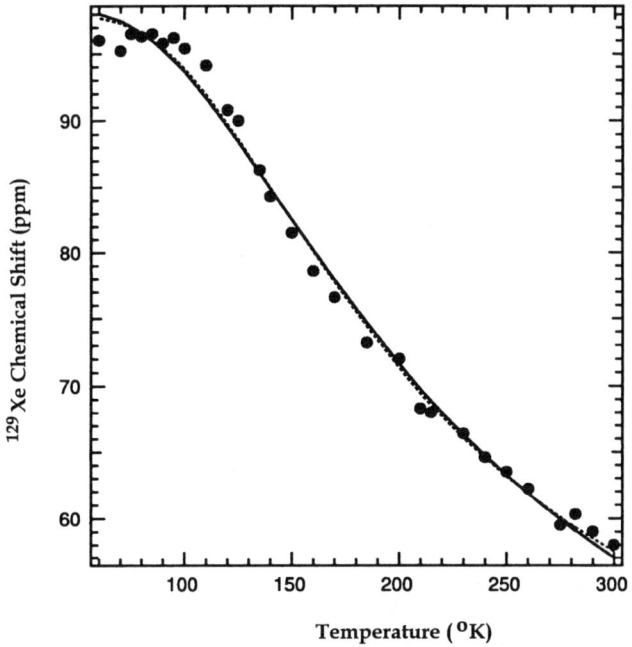

FIGURE 1: A plot of the measured ^{129}Xe chemical shift vs temperature for xenon on Na-Y. The solid curve is a fit to equation 1 and the dashed one is the fit to equation 2.

However, it is reassuring that the experimental data reproduces the low temperature flattening in chemical shift predicted by both models. Qualitatively this obtains because at very low temperature xenon motion becomes slow or non-existent and the asymptotic shift is that for xenon on the pore wall. The parameters obtained from the fits are:

Model	ε	L	τ_0	δ_S
Cheung	3.3 kJ/mol	15.0 Å		
Raftery, et al	4.1 kJ/mol		7×10^{-11} s	100 ppm

The van der Waals potentials are self consistent and are close to the 1.5 kJ/mol calculated by Kiselev and Du [12]. Their calculation is for an isolated Xe-O two body potential. The ^{129}Xe chemical shift measurements are for xenon in a pore where it certainly interacts with more than one oxygen in the zeolite wall at a time. Additionally, we have not taken into consideration the electrostatic interactions with the Na$^+$ cations in the zeolite. Overall, these results are particularly satisfying given the simplicity of the models used. Although the models start from somewhat different points, basically they both result in a similar description of the xenon atom in a pore: there is a van der Walls interaction between the xenon atom and oxygen atoms in the zeolite structure. Xenon in the "center" of the pore behaves like a gas. Xenon atoms jump between positions where they are stuck to the wall through this gas phase state. Neither model takes into account the detailed shape of the pore nor the presence of cations or other species in the pores.

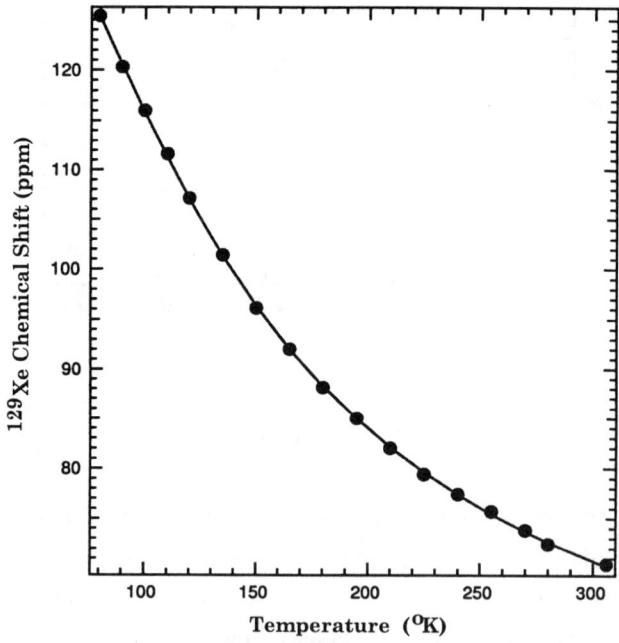

FIGURE 2: A plot of the measured ^{129}Xe chemical shift vs temperature for xenon on Na-Y containing Rh-6 clusters.

The data in Figure 2 are preliminary measurements of the ^{129}Xe chemical shift versus temperature for a sample of the same zeolite as above except for the addition of Rh-6 clusters. The solid line represents a fit to equation 1. The van der Waals potential obtained for this experiment is virtually the same as that obtained in the pure zeolite. The ^{129}Xe spectral linewidths obtained in this experiment are quite broad, ranging from about 900 Hz at 300 K to 12000 Hz at 65 K. The breadth of the NMR lines makes it difficult to obtain accurate shift data at the very low temperatures. We also note that the shifts in Figure 2 are only slightly larger than those in Figure 1, which indicates that there is some interaction between the xenon and the metal cluster. Ryoo and coworkers have made many measurements of the ^{129}Xe chemical shift in zeolites with platinum and other metal clusters [13-15]. They believe that the shift due to xenon-platinum interactions is on the order of several hundred Hz. If we assume that the xenon-rhodium interactions are similar to xenon-platinum interactions then we would expect to see significantly larger shifts than are evident in our data. There is a very significant difference in their samples. The metal clusters in Ryoo's work were made by impregnating the zeolite with a metal salt then reducing to the zerovalent metal. This produces clusters consisting of many more than 50 metal atoms per cluster. These relatively large clusters could have very different properties than the clusters in our work. In spite of this, we have no completely satisfactory explanation for the small shifts seen here.

Our goals now are to continue xenon adsorption on these well defined rhodium clusters and also on similarly prepared iridium clusters. We are also working on improving the models to allow us to explicitly include xenon-cation, e.g., Xe-Na$^+$ and xenon-metal cluster interactions in the model so we can derive more meaningful van der Waals attraction energies for the complex zeolite systems containing cations and clusters.

CONCLUSIONS

The early, simplified models of ^{129}Xe NMR chemical shifts in pores are inadequate to accurately describe the interactions. However, the large size and polarizability of xenon make it a sensitive chemical shift probe for studying porous materials. By exploiting variable temperature ^{129}Xe NMR over an extended temperature range we are able to use simple van der Waals attraction models to extract information about the energy of attraction between xenon and pore walls and potentially between xenon and metals (cations and clusters) in the pores. Successful modeling of ^{129}Xe chemical shifts requires the use of very well characterized and monodisperse zeolite samples. Part of the reason for the inconsistent ^{129}Xe NMR results in the literature is that xenon shifts are <u>extremely</u> sensitive to sample parameters and these are often not as well controlled or documented as they might be.

REFERENCES

1. T. Ito and J. Fraissard, Proc. Int. Conf. Zeolites, 5th 510; L. V. C. Rees (Ed), Heyden (London) 1980.
2. J. A. Ripmeester, J. Am. Chem. Soc. **104**, 289 (1982).
3. L. C. DeMenorval and J. Fraissard, J. Chem. Soc. ,Faraday Trans. 1 **78**, 403 (1982).
4. T. Ito and J. Fraissard, J. Chem. Phys. **76**, 5225 (1982).
5. J. A. Ripmeester, J. Magn. Reson. **56**, 247 (1984).
6. J. A. Ripmeester, and C. I. Ratcliffe, Anal. Chim. Acta **283**, 1103 (1993).
7. J. A. Ripmeester, and C. I. Ratcliffe, Anal. Chem. **283**, 1103 (1993).
8. T. T. P. Cheung, J. Phys. Chem. **99**, 7089 (1995).
9. D. Raftery, L. Reven, H. Long, A. Pines, P. Tang, and J. A. Reimer, J. Phys. Chem. **97**, 1649 (1993).
10. Y.-W. Kim, W. L. Earl, and R. E. Norberg, J. Magn. Reson. ,A **116**, 139 (1995).
11. A. Labouriau, T. Pietrass, W. A. Weber, B. C. Gates, and W. L. Earl, submitted to J. Phys. Chem.
12. A. V. Kiselev, and P. Q. Du, J. Chem. Soc. , Faraday Trans. **77**, 1 (1981).
13. R. Ryoo, S. J. Cho, C. Pak, J.-G. Kim, S.-K. Ihm, andJ. Y. Lee, J. Am. Chem. Soc. **114**, 76 (1992).
14. D. H. Ahn, J. S. Lee, M. Nomura, W. M. H. Sachtler, G. Moretti, S. I. Woo and R. Ryoo, J. Catal. **133**, 191 (1992).
15. J.-G. Kim, S.-K. Ihm, J. Y. Lee, and R. Ryoo, J. Phys. Chem. **95**, 8564 (1991).

ACTIVATION OF LIGHT ALKANES IN THE PRESENCE OF BENZENE OVER ACIDIC ZEOLITES. A ^{13}C MAS NMR STUDY

I.I. IVANOVA[+] and F. FAJULA[§]

[+] Chemistry Department, Moscow State University, Moscow 117234, Russia

[§] UMR 5618 CNRS, Ecole Nationale Supérieure de Chimie, 8, rue de l'Ecole Normale, 34296 Montpellier Cedex 5, France. Fax: ++33 4 67 14 43 91, e-mail: fajula@cit.enscm.fr

ABSTRACT

^{13}C MAS NMR has been performed *in situ* to investigate the mechanism of alkane activation in the presence of benzene over H-ZSM-5 catalyst. Propane 2-^{13}C and 2-methylpropane 2-^{13}C were used as a labelled reactants. In order to clarify the main reaction pathways, conversion of the individual starting materials was also studied under similar conditions. NMR results suggest that two reaction routes of alkane activation are realized over H-ZSM-5 catalyst in the presence of benzene: i) low temperature carbenium ion type activation, initiated via hydride abstraction by protonated benzene molecule and leading to ^{13}C label scrambling and isobutane isomerization into n-butane and ii) high temperature carbonium ion type activation leading to benzene alkylation with the fragments formed upon splitting of proponium and butonium ions.

INTRODUCTION

The mechanism of light alkanes transformations over acidic zeolites has been widely discussed during the last two decades. However, the detailed mechanism of various steps, especially the initial stages of alkanes activation, still remains a matter of controversy. Carrying out the reaction in the presence of benzene may throw some light on the mechanism of alkanes activation as benzene can act as a trap for the active species formed upon alkane activation. On the other hand, alkylation of benzene with light alkanes such as propane and isobutane can be considered as a prospective way for the synthesis of alkylaromatic products and as a future challenge for light alkanes upgrading.

A few reports on the subject reveal that the reaction pathway leading to alkylaromatics in the presence of zeolite catalysts appear as much more complex than with Friedel-Crafts and superacidic catalysts. Actually, no products of direct addition were detected. Instead, the alkylaromatics consist of mixtures of methyl- and ethylbenzenes [1]. In the case of Ga-ZSM-5 catalysts, alkylation of benzene with propane could be achieved via bifunctional propane activation on Brönsted and Ga-sites [2]. A bifunctional mechanism can hardly operate, however, over a purely acidic catalysts.

The aim of this work is to clarify the mechanism of propane and isobutane reactions in the presence of benzene over a H-ZSM-5 catalyst. An in situ ^{13}C MAS NMR technique is used to investigate the main reaction steps: adsorption, activation, primary and secondary reactions.

EXPERIMENTAL

The study was carried out on H-ZSM-5 zeolite with a Si/Al ratio of 30. The powdered samples (0.09±0.01 g) were packed into the NMR tubes (Wilmad, 5.6 mm o.d. with constrictions), evacuated to a final pressure of $6 \cdot 10^{-6}$ Torr after heating for 8 h at 723 K, and cooled down to 298 K before adsorption. In different experiments, individual propane 2-^{13}C or 2-methylpropane 2-^{13}C (99% enriched) or their mixtures with benzene were adsorbed. The molar ratio benzene/alkane was varied from 0.25 to 8, the total amount of adsorbed reactants being 4-8 molec./u.c. After reactants loading, the NMR cells maintained at 77 K to ensure a quantitative adsorption were carefully sealed to achieve proper rotor balance and high spinning rates in the MAS NMR probe.

^{13}C MAS NMR measurements were performed on MSL-400 and ASX-200 Bruker spectrometers operating at 100.6 and 50.3 MHz, respectively. Quantitative conditions were achieved using high-power gated proton-decoupling with suppressed NOE effect (90° pulse, recycling delay = 6 s). Spinning rates were within 3.8 - 5 kHz.

In typical experiments, the NMR cells were heated outside of the spectrometer to a selected temperature and maintained at this temperature for a given period of time. MAS NMR spectra were recorded after quenching sample cells to 298 K. After collection of the

NMR data, NMR cells were returned to reaction conditions and heated for progressively longer periods of time.

RESULTS AND DISCUSSION

Reaction of labelled alkanes

The initial ^{13}C MAS NMR spectrum of propane contains the only resonance at about 17 ppm, corresponding to the initially labelled methylene group of propane. The reaction of propane begins at 548 K, the main reaction being ^{13}C scrambling in the propane molecule, evidenced by the appearance of the resonance at ca. 16 ppm, corresponding to the initially unlabelled methyl group. Under these conditions, scrambling is evidenced after 5 min heating and is complete after 160 min. In parallel signals attributable to butane, isobutane and trace amounts of ethane and methane develop. The mechanism of this reaction has been discussed in detail [3] and involves protonation on strong Brönsted sites followed by rearrangement and cracking of the resulting proponium ions.

In the case of i-butane, the reaction starts at 523 K by the formation of propane and pentanes as major products and lesser amounts of n-butane. Scrambling of the ^{13}C label could not be evidenced as definitely as in the case of propane, due to the overlap between the signals of the methyne (23.6 ppm) and methyl (24 ppm) carbon atoms, but resulted in a broadening of the ^{13}C resonance. The presence of C_3-C_5 hydrocarbons and the absence of methane and ethane is typical of a multimolecular process involving transient formation of isobutene, its addition to isobutyl cation followed by rearrangement and cracking of the intermediate octyl intermediate.

Co-adsorption of benzene and labelled alkane. Low temperature activation

In the presence of co-adsorbed benzene the reaction of both hydrocarbons was slowed down and the distribution of the products formed changed significantly. Moreover the initial signals of the unreacted propane (at 17 ppm) and isobutane (at 23.6 ppm) were significantly broader indicating a reduced mobility of the molecules in the zeolite pores.

In the case of propane, the only reaction was scrambling, which started only after heating for more than 100 min at 548 K (Fig. 1). No other products could be detected, even after prolonged heating of the NMR cell.

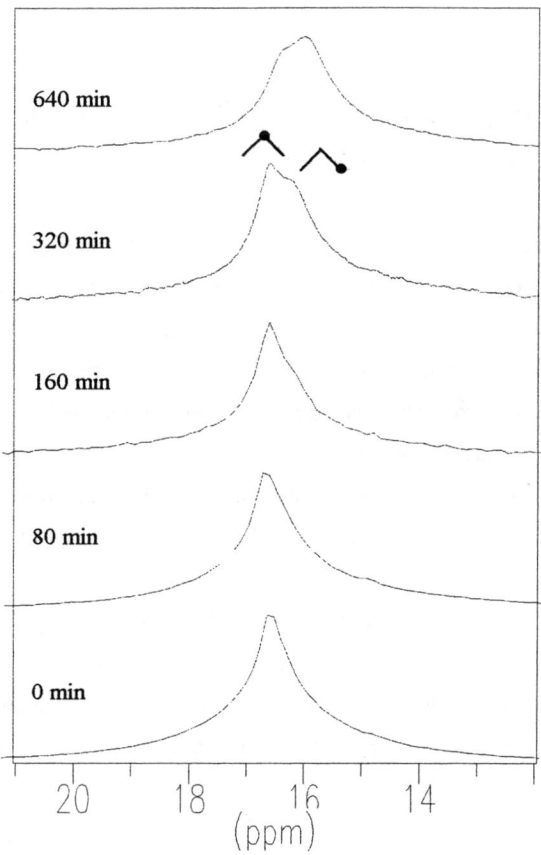

Fig. 1. Aliphatic regions of the ^{13}C MAS NMR spectra observed in course of propane 2-^{13}C activation in the presence of benzene (propane/ benzene = 1) over H-ZSM-5 zeolite at 548 K. 3000 scans were collected for each spectrum.

The results obtained for i-butane in the presence of benzene were also quite different from those observed for individual i-butane. The sole reactions here are ^{13}C label scrambling and isomerization into n-butane. No C_3 nor C_5 hydrocarbon were detected in the temperature range 523 - 573 K.

It is noteworthy that under certain conditions (excess of benzene) small amounts of the products of direct addition of $C_3H_7^+$ and $C_4H_9^+$ ions to benzene (propylbenzenes and butylbenzenes) were detected. It should be mentioned however that these products were observed only at low alkane conversions because of their low stability and preferable dealkylation at the reaction temperatures used.

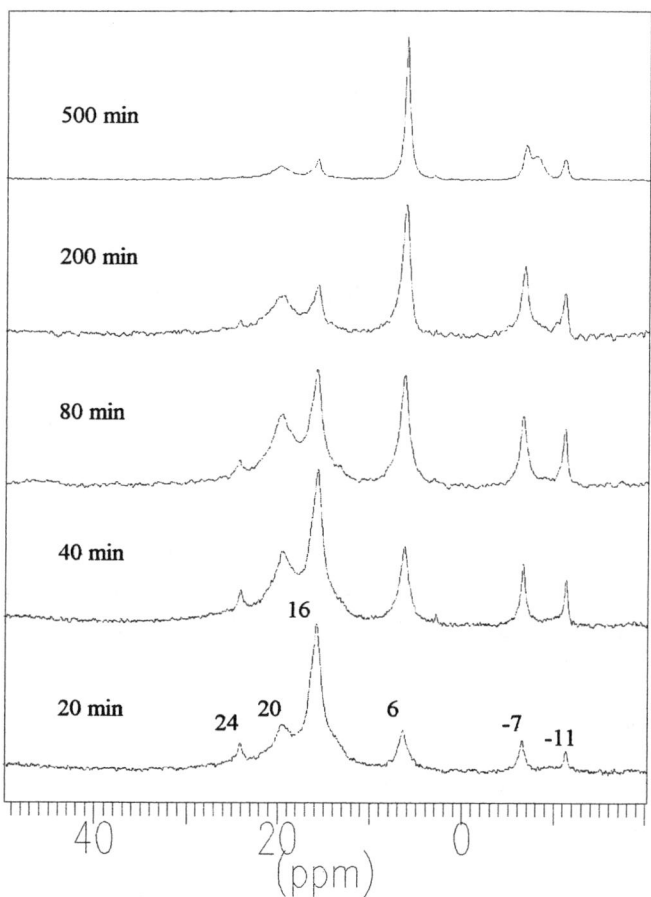

Fig. 2. Aliphatic regions of the ^{13}C MAS NMR spectra observed in course of propane 2-^{13}C reaction with benzene (propane/ benzene = 1) over H-ZSM-5 zeolite at 623 K. 2000 scans were collected for each spectrum.

High temperature activation

Alkylation of benzene begins only at 623 K. The typical ^{13}C MAS NMR spectra for the case of propane 2-^{13}C reaction with benzene are shown in Fig. 2. The major products of high temperature pathway are methylsubstituted benzenes, in particular, toluene and xylenes, evidenced by the lines at ca. 19 - 21 ppm. The formation of methylbenzenes is accompained by the appearance of methane (-7, -11 ppm) and ethane (6 ppm), pointing to carbonium ion character of alkane activation in this reaction step [3]. The probable mechanism leading to the alkylaromatics includes alkane protolysis followed by splitting of carbonium ions and alkylation of benzene with the fragments formed.

CONCLUSION

Activation of propane and isobutane in the presence of co-adsorbed benzene over H-ZSM-5 leads to product distributions different from those obtained in the reaction of the neat alkanes. At low temperature, co-adsorbed benzene suppresses the occurrence of multimolecular processes responsible for the formation of alkane products with lower and higher carbon numbers than in the starting reagent. The carbenium ions generated undergo ^{13}C label scrambling and, in the case of isobutane, isomerization into n-butane, via probably protonated cyclopropane intermediates. At higher reaction temperature methane, ethane and methylsubstituted benzenes are formed, suggesting an activation route involving carbonium ions formation by alkane protolysis.

REFERENCES

1. S.A. Isaev, T.V. Vasina, and O.V. Bragin, Izv. Akad. Nauk USSR, Ser. Khim., 10, 2228 (1991).
2. I.I. Ivanova, N. Blom, and E.G. Derouane, J. Mol. Catal., 109, 157 (1996)
3. I.I. Ivanova, E.B. Pomakhina, A.I. Rebrov, and E.G. Derouane, Topics in Catal., 6, 49 (1998)

INTERACTION OF CD3OH WITH HY AND HZSM-5 ZEOLITES STUDIED BY ^{1}H NMR: BROAD-LINE AT 4 K AND HIGH RESOLUTION MAS AT 298 K. COMPARISON WITH SUPERACIDIC COMPOUNDS.

P. BATAMACK and J. FRAISSARD

Laboratoire de Chimie des Surfaces, associé au CNRS-UPESA 7069, Université Pierre et Marie Curie, Tour 55, 4 Place Jussieu, 75252 Paris Cédex 05, France.

ABSTRACT

The adsorption of small quantities of CD_3OH on HY and HZSM-5 zeolites, Nafion-H and CF_3SO_3H is studies by means of broad-line ^{1}H NMR at 4 K and high resolution MAS at 298 K. Simulation of the broad-line spectra shows that for the strong acids (Nafion-H and CF_3SO_3H) methoxonium ion (ion-pair complex) is formed. On the contrary, for zeolites the protonation of CD_3OH is not observed. Instead, a "neutral" hydrogen bonded complex is formed. The geometry of the species is described and the proton chemical shifts from high resolution MAS are given.

INTRODUCTION

The initial step of the adsorption of methanol on zeolites has received much interest as a means to better understand and identify the intermediate species formed on the surface of microporous catalysts. The first studies on the topic claimed that an ion-pair complex was formed [1,2]. Recent theoretical calculations [3,4], IR and NMR [5] studies tend to favor "neutral" complex formation. Buzzoni et al. [6] studied by IR spectroscopy the adsorption of methanol on a film of Nafion-H heat-treated at 393 K in vacuo. At low loadings, the authors observed the gradual disappearance of the bands due to undissociated -SO_3H groups and the parallel increase in the peaks associated with -SO_3^- groups. They deduced the formation of $CH_3OH_2^+ \cdots ^-O_3S$- species. Recently, Olah et al. showed the efficiency of Nafion-H in the dehydration of alcohols [7]. In the case of species involving protons, broad-line NMR under "rigid lattice" conditions has proved to be more powerful than other spectroscopic techniques for identifying the species in which protons interact at short range [8]. In the case of the adsorption of water on zeolites, the simulation of broad-line spectra shows the coexistence of both the hydronium ion (ion-pair) and the hydrogen-bonded species (neutral complex), the latter being favored. This result is a subject of controversy with theoreticians [9]. In this study, we show that broad-line NMR is a useful technique capable of providing quantitative and structural information on the oxyprotonated species formed on the surface of a compound when methanol is adsorbed, particularly on solid catalysts.

EXPERIMENTAL

HY zeolite (Si/Al = 2.4) from U.O.P. and a commercial HZSM-5 (Si/Al = 17) were dehydrated under "shallow bed" conditions by heating at 12 K.hr^{-1} to 675 K, at which temperature they were kept for 16 h. The powder form of Nafion-H (0.95 meq/g) was provided by E. I. Du Pont de Nemours & Co., Inc. All acid sites were activated by treatment with hot 3 M HCl solution. About 0.6 g of the sample in a glass ampule were evacuated at room temperature to 10^{-2} Pa, then heated at 24 K.hr^{-1} to 463 K and held at this temperature for a week. Methanol vapor was introduced into the solids at a constant temperature of 300 K in several steps, the amount of methanol being determined gravimetrically. The samples were shaken manually and held several days at 300 K, to ensure homogeneous distribution of the adsorbed methanol, and then sealed in NMR tubes.

Acid solutions, n CD_3OH/CF_3SO_3H, with n = 0.51 and 1.0 were prepared gravimetrically by combining appropriate amounts of CD_3OH (99.8 % D) and CF_3SO_3H (99 % purity) and sealed in NMR tubes.

^{1}H MAS NMR experiments at room temperature were performed on a Bruker MSL-400 spectrometer with a home-made 5-mm probe. The rotation frequency of sealed tubes was 3.5-4.5 kHz. For triflic acid solutions MAS was not needed. Chemical shifts are expressed relative to liquid TMS as external reference using the usual conventions.

For broad-line experiments, the samples were quenched in liquid helium, and the ^{1}H NMR spectra recorded at 4 K on a home-made continuous wave 60 MHz spectrometer with phase detection and signal accumulation. The spectra are absorption derivatives. They are theoretically symmetrical with respect to the center and, in practice, the two parts of each experimental spectrum are averaged; for this reason we show only half of each spectrum. In both cases, MAS and broad-line NMR, the weak residual signal of the probe is subtracted from the total signal.

The simulated broad-line ^{1}H NMR spectra correspond to the weighted sum of the oxyprotonated species involved and for which magnetic configurations are calculated [8]: (i) H_2O or $CD_3OH_2^+$, a **r**-distant 2-spin configurations; (ii) H_3O^+, a magnetic configuration with three **r**-distant spins at the vertices of an equilateral triangle; (iii) $H_2O \cdots HO$ or distorted H_3O^+, a magnetic configuration with 3 spins at the vertices of an isoceles triangle, where **r** is the base and **r'** the equal sides; (iv) OH or two hydrogen-bonded methanol (Figure 1), a 2-spin configuration or a pure Gaussian and/or a pure Lorentzian function. Each of the corresponding functions (except the Gaussian and the Lorentzian functions) is convoluted by a Gaussian which takes into account the interaction between the protons of the configuration and those belonging to neighboring configurations and also those of the non-zero spin nuclei in the environment (^{19}F, ^{27}Al, ^{28}Si). When the effect of these non-zero spins is small the parameter of each Gaussian is related to a distance **X** which is close to the shortest distance between a proton of the configuration considered and a proton outside it.

RESULTS

MAS ^{1}H NMR at 298 K

The spectrum of the dehydrated HY zeolite presents the usual signals of HY zeolites at 4.6 ppm, a shoulder at 5.5 ppm and a small signal at 2.0 ppm. When this sample is loaded with 1.01 CD$_3$OH/Brønsted acid site a large signal appears at 9.8 ppm, attributed to methanol interacting with the acidic Brønsted sites, and two small signals, one at 3.6 ppm not assigned and another at 2.0 ppm attributed to silanol groups.

The spectrum of the anhydrous sample of HZSM-5 zeolite consists of 4 signals: 0.7 ppm (small), 1.9 ppm, 4 ppm (large), and a broad shoulder at ca. 6 ppm attributed to organic debris, SiOH groups, bridged OH groups (ZOH) and either SiOH or ZOH H-bonded to framework oxygen atoms [10,11]. In the presence of 0.99 CD$_3$OH/Brønsted acid site (Bas), the spectrum presents a signal at 8.6 ppm due to methanol exchanging with Brønsted acid sites, two shoulders at ca. 6.5 and 3.5 ppm not assigned and SiOH groups at 1.9 ppm.

The spectrum of dehydrated Nafion-H shows a signal at 10.6 ppm and that of 0.98 CD$_3$OH/Nafion-H exhibits a strong signal at 12.5 ppm due the interaction of methanol with the Brønsted site.

The exchange signals in the spectra of the samples of n CD$_3$OH/CF$_3$SO$_3$H with n = 0, 0.51 and 1.0 are at 10.6, 12.0 and 12.2 ppm, respectively.

Broad-line ^{1}H NMR at 4 K

Simulation of the spectra of 1.0 CD$_3$OH/Bas HY (Figure 2) and 0.99CD$_3$OH/Bas HZSM-5 requires two signals: that of OH groups from silanols and that of a 2-spin species with a H-H distance of 190 pm in HY zeolite and 199 pm in HZSM-5. Simulation of 0.98 CD$_3$OH/Nafion-H shows the formation of a well resolved 2-spin species with a H-H distance of 169 pm and a small signal of OH groups. The simulated spectrum of triflic acid (Figure 3) also shows a well resolved water-like 2-spin species with a H-H distance of 171 pm and a weak signal of hydroxonium ions. Table 1 summarizes the concentration of the oxyprotonated species and the distances derived from the simulations.

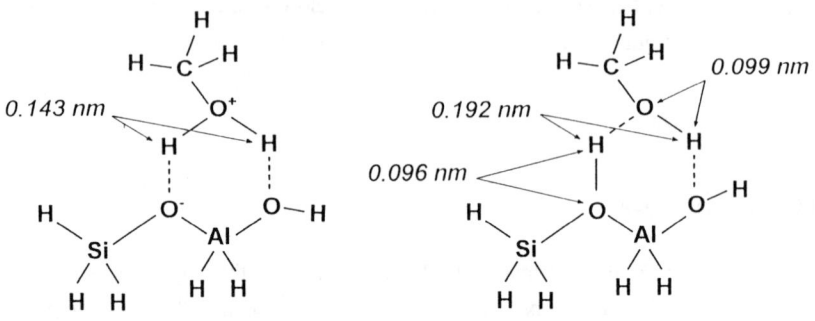

Figure 1: Geometries of methanol complexes with zeolite fragment (Ref.[5]).

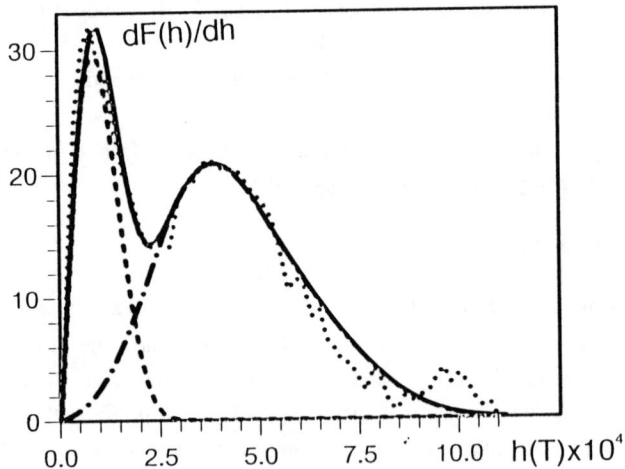

Figure 2: Half derivative broad-line ^{1}H NMR spectrum of HY/1.0CD$_3$OH. (...) Experimental; (___) fitted; (_._) 2-spin contribution and (_ _) OH groups.

sample	$CD_3OH_2^+$ ions			H-bonded CD_3OH			SiOH groups		
	n	r	X	n	r	X	n	r	X
1.0 CD_3OH/CF_3SO_3H	0.96[a]	171	251	0			-		
0.98 CD_3OH/Nafion-H	0.96[b]	169	272	0			-		
1.0 CD_3OH/Bas HY	0			0.93	190	235	0.17	326	333
1.0 CD_3OH/Bas HZSM-5	0			1.09	199	253	0.19	Lorentz.	0.9

Table 1: number of oxyprotonated species (n) per acid site of the stated samples and distances (in pm) used for simulation of the spectra. [a] 0.02 hydroxonium ion is formed; [b] 0.07 OH group is observed. Lorentz.: Lorentzian function, the parameter is in 10^{-4} T.

DISCUSSION

The relatively high proton chemical shift of methanol adsorbed on zeolites, triflic acid and Nafion-H (8.6 - 12.5 ppm) reveals a strong interaction of methanol with the catalysts. These results are comparable to those obtained with other zeolites [12-14]. Using sulfur dioxide as solvent, the interaction of methanol with superacids gave a chemical shift of 9.3 - 9.5 ppm [15]. MAS spectra and broad-line results with HY and HZSM-5 zeolites show that silanol groups do not interact with methanol, in agreement with Hunger et al. [16].

A superacid such as triflic acid is capable of protonating methanol to yield the methoxonium ion. In fact, the well resolved 2-spin species observed in the broad-line spectrum of 1.0 CD_3OH/CF_3SO_3H is that of $CD_3OH_2^+$ ions. The H-H distance of the ions in this concentrated solution is 171 ± 2 pm, greater than the calculated values (143-158 pm) [5,17]. This means that the ions in the solution are involved in H-bonds with neighboring oxygen atoms. Proton transfer from Nafion-H to methanol is also observed in 0.98 CD_3OH/Nafion-H. The H-H distance of the ions in this system (169 ± 2 pm) is comparable to that of 1.0 CD_3OH/CF_3SO_3H solution. These values are greater than those observed for water molecules (145 - 165 pm) [18,19].

For zeolites, the spectra are poorly resolved. The signal at around 5×10^{-4} T is that of OH groups interacting with each other. In the hydrogen-bonded species formed the protons are 190 ± 5 pm apart for 1.01 CD_3OH/Bas HY and 199 ± 5 pm for 0.99 CD_3OH/Bas HZSM-5. The study of 1.0 CD_3OH/Bas H-mordenite using the same technique gave 193 ± 3 pm for the H-H distance of the "neutral" complex [12]. These results are in agreement with calculations [5,17].

CONCLUSION

^{1}H broad-line NMR results show that methoxonium ions are formed when methanol interacts with triflic acid or with Nafion-H but not with zeolites. The H-H distance of these

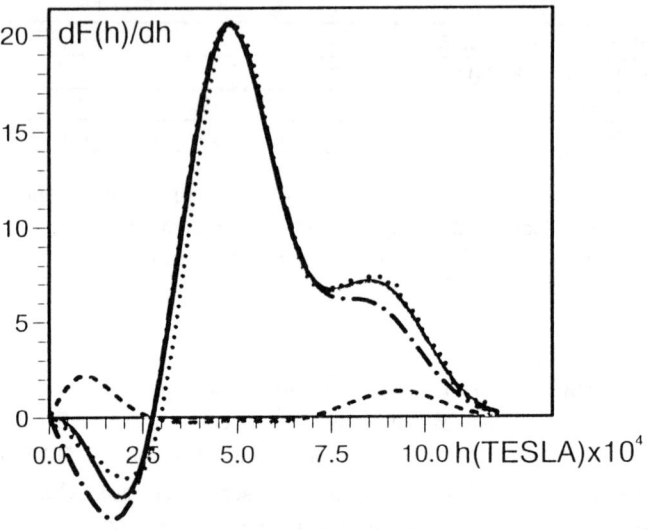

Figure 3: Half-derivative broad-line ^{1}H NMR spectrum of $CF_3SO_3H/1.0CD_3OH$. (...) Experimental; (___) fitted; (_._) 2-spin $CD_3OH_2^+$ contribution and (_ _) H_3O^+ contribution.

ions is 170 ± 2 pm, greater than the H-H distance in water molecules. The H-H distance of the hydrogen-bonded species formed in the case of zeolites is 190 ± 5 pm in HY and 199 ± 5 pm in HZSM-5.

REFERENCES

1. G. Mirth, J.A. Lercher, M. W. Anderson and J. Klinowski, J. Chem. Soc. Faraday Trans. **86**, 3039 (1990).
2. L. Kubelkova, J. Cejka and J. Novakova, Zeolites **11**, 48 (1991).
3. F. Haase and J. Sauer, J. Am. Chem. Soc. **117**, 3780 (1995).
4. C.M. Zicovich-Wilson, P. Viruela and A. Corma, J. Phys. Chem. **99**, 13224 (1995).
5. L. Kubelkova, J. Kotrla, J. Florian, T. Bolom, J. Fraissard, L. Heeribout and C. Dorémieux-Morin, Stud. Surf. Sci. and Catal. **101**, 761 (1996).
6. R. Buzzoni, S. Bordiga, G. Ricchiardi, G. Spoto and A. Zecchina, J. Phys. Chem. **99**, 11937 (1995).

7. G.A. Olah, T. Shamma and G.K.S. Prakash, Catal. Lett. **46**, 1 (1997).
8. P. Batamack, C. Dorémieux-Morin, R. Vincent and J. Fraissard, J. Phys. Chem. **97**, 9779 (1993) and references therein.
9. H. Jobic, A. Tuel, M. Krossner and J. Sauer, J. Phys. Chem. **100**, 19545 (1996).
10. L. W. Beck, J. L. White and J. F. Haw, J. Am. Chem. Soc. **116**, 9657 (1994).
11. D. Freude, Chem. Phys. Lett. 235, **69** (1995).
12. L. Heeribout, C. Dorémieux-Morin, L. Kubelkova, R. Vincent and J. Fraissard, Catal. Lett. **43**, 143 (1997).
13. M. Hunger and T. Horvath, Catal. Lett. **49**, 95 (1997).
14. V.M. Mastikhin, I.L. Mudrakovsky and A.V. Nosov, Progress in NMR Spectroscopy 23, 259 (1991).
15. G.A. Olah, G.K.S. Prakash and J. Sommer, Superacids, (Wiley, New York, 1976) p. 178.
16. M. Hunger and T. Horvath, J. Am. Chem. Soc. **118**, 12302 (1996).
17. C. Minot (private communication).
18. P. Batamack, C. Dorémieux-Morin, J. Fraissard and D. Freude, J. Phys. Chem. **95**, 3790 (1991).
19. V. Semmer, P. Batamack, C. Dorémieux-Morin, R. Vincent and J. Fraissard, J. Catal. **161**, 186 (1996).

A ^{13}C NMR AND FTIR INVESTIGATION OF ACETONITRILE ADSORPTION IN H-MFI BETWEEN 290 AND 523K

J. ŠEPA*, R.J. GORTE*, DAVID WHITE[§], B.H. SUITS[†] and V. S. SWAMINATHAN[†]

*Department of Chemical Engineering, University of Pennsylvania, Philadelphia, PA, USA
[§]Department of Chemistry, University of Pennsylvania, Philadelphia, PA, USA
[†]Department of Physics, Michigan Technological University, Houghton, MI, USA.

ABSTRACT

We have studied the structure of the adsorption complex of CH_3CN in H-MFI at temperatures up to 523K in order to examine claims for progressive, reversible proton transfer. While our results confirm that there is a reversible increase in the ^{13}C NMR isotropic shift for $CH_3{}^{13}CN$, IR results provide no evidence for a change in the order of the C-N bond. An alternative explanation for the ^{13}C NMR results is offered.

INTRODUCTION

While techniques are available for characterization of Brønsted-acid site concentrations in solid acids (1), characterization of site strengths remains a problem. For example, while site strengths are often defined by the heats of adsorption of strong bases like ammonia and pyridine, it does not appear that there is a correlation between these heats and catalytic activities (2). Spectroscopic characterization of weaker bases, ones which form only hydrogen-bonded complexes with Brønsted sites, appears to be promising for characterization of site strengths. However, because many hydrogen-bonded complexes are highly reactive (3), it is usually not possible to study thermally induced excitations of these complexes or the dynamics of accompanying structural changes in such complexes. These dynamical processes may be able to differentiate one zeolite from another and provide insights into how reactions occur at the site.

Acetonitrile (AN) is a particularly interesting probe for studying acidity. AN has a proton affinity similar to that of typical reactant molecules (PA=795 kJ/mol), readily forms a 1:1 stoichiometric adsorption complex with the Brønsted sites (4), but does not form bimolecular products nor undergo irreversible structural rearrangements upon heating to above 500K. In a recent ^{13}C NMR study of AN in H-MFI, Haw and collaborators (5) observed reversible changes in the isotropic chemical shift of the nitrile carbon with temperature and provided a theoretical model for this based on the existence of new, long-lived AN structures. The structures were predicted from an "excised", *ab-initio*, cluster calculation involving proton transfer to form imine-like species. Large experimental downfield shifts were attributed to a progressive, reversible change in the structure of the adsorption complex, from a hydrogen-bonded form in its ground rotational and vibrational state to a protonated form of higher potential energy. In this model, thermal excitations presumably drive the proton transfer as the complex moves along a one-dimensional reaction coordinate. Newly formed structures of higher potential energy (no minima) are assumed sufficiently long-lived to be observed experimentally in the NMR time scale.

In this study, we examined AN adsorption in H-MFI and the model of Haw, et al (5), using FTIR under conditions similar to those used in the NMR measurements. Unlike NMR, IR spectroscopy (6) provides a snapshot of the structure of the adsorption complex and can detect changes upon thermal excitation. Changes in the C-N bond order, from the triple-bonded nitrile of the hydrogen-bonded complex to the double-bonded imine, should be observable from the stretching frequency, ν(C-N), and intensities should be related to the average concentrations at any temperature. Since changes in the FTIR spectra should parallel the changes in the ^{13}C isotropic chemical shift, we have repeated the ^{13}C NMR studies as well.

EXPERIMENTAL

The H-MFI used in this study, prepared from Chemie Uetikon AG (Zeocat-Pentasil-PZ-2/54Na) and described elsewhere (4), had a Brønsted-acid site concentration of 0.50 mmol/g and a negligible concentration of Lewis-acid sites. The silicalite (MFI) was a commercial material (Linde S115) used in the Na form. The alumina was purchased from Fischer Scientific and had a negligible Brønsted-site density and a Lewis-site density of 0.40 mmol/g, based on irreversible AN adsorption at 295K.

The IR measurements were performed on a Bruker IFS-113, on both self-supported wafers and wafers supported on a stainless-steel mesh. The mesh did not affect the spectra, but it provided uniform temperatures in the wafer and allowed attachment of a thermocouple for accurate temperature readings. Because the peak desorption temperature for AN from H-MFI *in vacuo* is 420K (7), it was necessary to have gas-phase AN above the sample to maintain surface coverage. To minimize the amount of gas-phase AN, the free volume in the IR cell was minimized by simply using a pyrex tube, with windows glued to the ends. After admitting the required dose of CD_3CN to the evacuated sample, the tube was sealed with a torch.

In NMR, the samples were degassed in a glass sample tube overnight at 750K and 10^{-6} torr, then exposed to controlled amounts of $CH_3{}^{13}CN$, after which the sample was sealed in vacuo. NMR experiments were performed at two different magnetic field strengths, with ^{13}C resonance frequencies of 37.84 and 90.47 MHz. Two different probes were used for the high-temperature studies reported here. For the probe operating at 37.84 MHz, the sample was heated by hot N_2 gas. For the probe operating at 90.47 MHz, the sample was heated using a ceramic, wire-wound furnace. Except for spectral linewidths, the ^{13}C chemical shifts obtained from both probes were the same within experimental error. All frequencies are referenced to liquid TMS at 298K.

RESULTS AND DISCUSSION

^{13}C NMR results for AN adsorption in H-MFI below 298K have been reported previously (4). The ^{13}C NMR lineshapes of $CH_3{}^{13}CN$ in H-MFI at 298K show a partially narrowed, axially symmetric chemical-shift tensor at loadings below one per Brønsted site. At higher temperatures and loadings, the linewidth decreases dramatically due to molecular reorientation at the acid sites and exchange with physically adsorbed molecules (8), giving rise to a Lorentzian lineshape at a frequency corresponding to the isotropic chemical shift. At temperatures below 140K, one obtains the rigid-lattice, chemical-shielding tensor, which exhibits C-N dipolar fine structure. The trace of the tensor, or isotropic shift, is 115±1 ppm at 140K, compared to values of 108 ppm for the gas phase and as high as 128 ppm for the pure solid (9).

In the present study, we examined samples with adsorbed AN at temperatures above 298K. Figure 1 shows the static ^{13}C spectra for a loading of 0.40 mmol/g (0.8/site) on H-MFI as a function of temperature. For comparison we show the proton-decoupled spectrum at 78K (4) and the spectrum at 298K which displays a partially narrowed powder pattern of the shielding tensor, as described previously. Upon heating to 345K and higher, the lineshape becomes Lorentzian in the absence of both proton decoupling and magic angle spinning (MAS), due to rapid reorientations of the molecular axis. Since neither proton decoupling nor MAS affect the linewidth, the frequency of molecular reorientation is fast compared to the dipolar and chemical shielding anisotropies. This was verified in proton studies reported elsewhere (8). At 345K, the isotropic shift is 118 ppm. Further heating results in additional narrowing of the spectrum and an increase in the isotropic shift, to 120 ppm at 400K and 133 ppm at 450K. Results taken at 37.84 and 90.47 MHz were identical, although the probe used at 90.47 MHz allowed measurements at even higher temperatures. The isotropic shift was 140 ppm at 480K and 148 ppm at 520K. These isotropic shifts are somewhat smaller than those reported by Haw and coworkers (155 ppm at 473K and 176 ppm at 533K) but occur in the same temperature range.

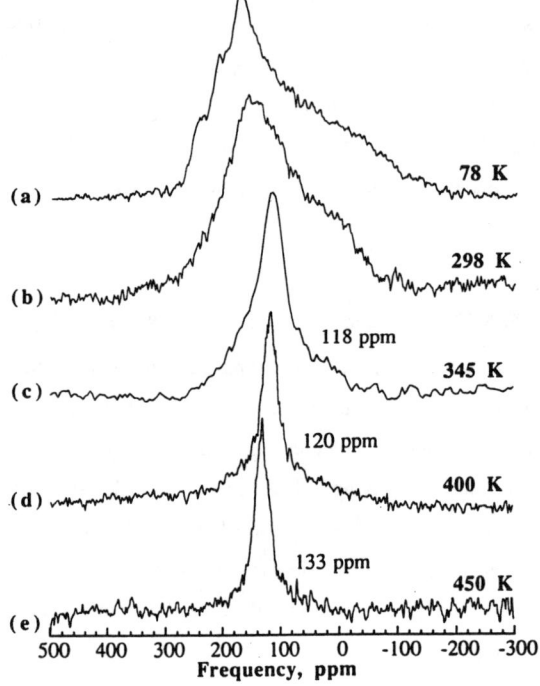

Figure 1. ^{13}C NMR spectra for $CH_3{}^{13}CN$ in H-MFI at a loading of 0.8/site, as a function of temperature.

As reported previously by Haw and coworkers (5), the isotropic shifts were, for the most part, reversible upon cooling. However, spectral features at 181 ppm and 169 ppm, which are also evident in the spectra of Haw, et al, formed irreversibly above ~500K. These features result from reaction of AN with water to produce adsorbed acetamide (181 ppm) and secondary condensation products of acetamide (169 ppm) (10). Because reaction of AN with water occurs readily at lower temperatures in H-MFI, the water in this reaction likely came from the zeolite.

The ^{13}C isotropic chemical shifts are highly dependent on the loading of AN, as shown in Figure 2. For a coverage of 1.5/site, a significant fraction of AN is physisorbed and the linewidth of the static spectrum is already narrow at room temperature due to exchange processes. The isotropic shift again increases with temperature, to 123 ppm at 400K and 128 ppm at 450K, but the changes are not as large as observed for the sample at lower coverages. (The decreased intensity at higher temperatures, more obvious in Figure 2 than in Figure 1, results from Curie's

Law and not from changes in adsorbate concentration.) The obvious conclusion is that only those molecules associated with Brønsted sites exhibit a significant change in the isotropic shift with temperature. This was confirmed in measurements on silicalite. For a loading of 0.40 mmol/g in the silicalite sample, the isotropic shift remained at ~119 ppm over the temperature range from 298K to 450K.

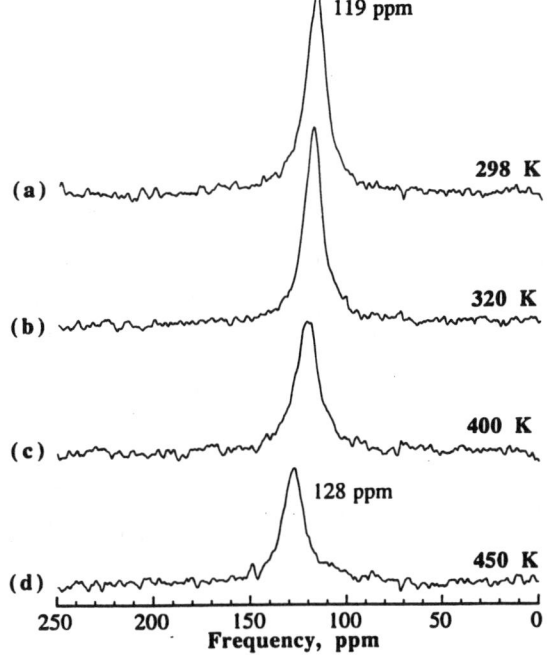

Figure 2. ^{13}C NMR spectra for $CH_3{}^{13}CN$ in H-MFI at a loading of 1.5/site, as a function of temperature.

Because some Lewis acidity is unavoidable in zeolites, we also examined AN adsorption on alumina. For a loading of 0.36 mmol/g (0.8/site), the isotropic shift at 298K is again close to the value for liquid AN, 120 ppm. The only change observed upon heating to 430K is a narrowing of the linewidth.

FTIR spectra were measured for AN on H-MFI to determine whether structural changes in the adsorbed species could be observed in the vibrational spectrum. Figure 3 shows FTIR difference spectra, with clean H-MFI used as reference, for samples loaded in the same way as the NMR samples. At 298K, the primary species is the hydrogen-bonded complex, with a ν(C-N) stretch at 2298 cm^{-1} and an ABC triplet at 2770, 2405 and 1695 cm^{-1} (11). The intense, inverse band at

3610 cm^{-1} demonstrates that the molecules are interacting primarily with Brønsted sites associated with framework Al. The small peak at 3750 cm^{-1} may indicate a slight excess of AN on the sample or, more likely, an over-subtraction of the reference. A small band at 2325 cm^{-1} can be assigned to AN adsorbed on Lewis sites (6).

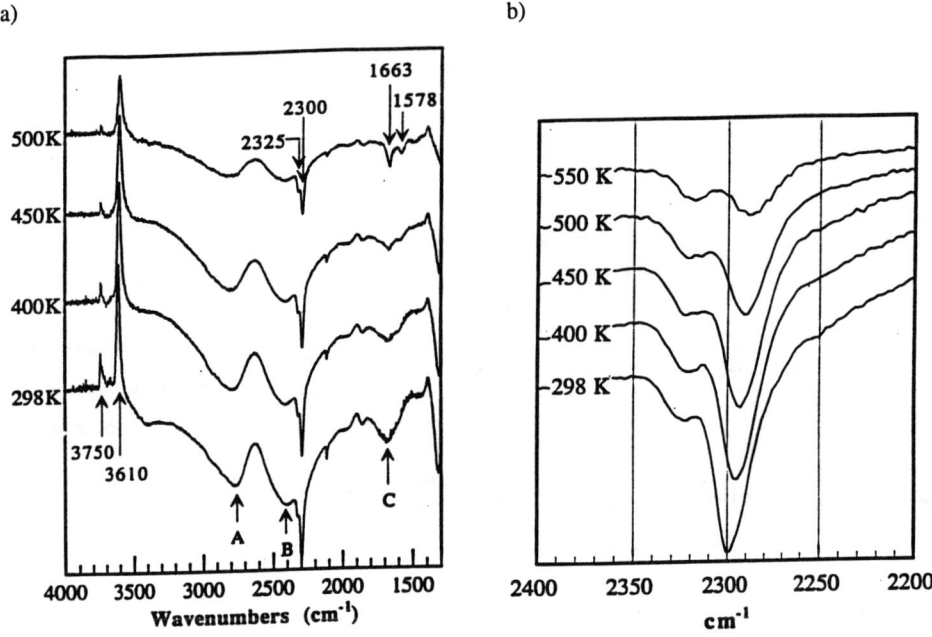

Figure 3. FTIR difference spectra for CD$_3$CN in H-MFI as a function of temperature. The frequency range from 1300 to 4000 cm^{-1} is shown in a), while b) shows the same spectra, focussing on the frequency range for ν(C-N).

Changes in the spectra as the sample is heated are shown in Figures 3a) and b). At 400 and 450K, the most significant changes occur in the ν(C-N) frequency, which decreases by 8 and 10 cm^{-1} respectively. The ABC triplet bands remain unchanged. At 450K, one also begins to see evidence for desorption of AN from the Brønsted sites from the decreased intensity of bands associated with the hydrogen-bonded complex; however, the spectrum remains essentially unchanged and there is no evidence for the formation of new species. At 500K, new peaks at

1580 and 1660 cm^{-1} appear in the spectrum. These grow in intensity with time and remain when the sample is cooled back to room temperature. They indicate the presence of a small concentration of partially deuterated acetamide, also observed in the NMR measurements. At even longer times and higher temperatures, deuteration of the zeolite hydroxyls occurs, probably due to secondary chemistry with the amides.

The FTIR results lead to several important conclusions. First, AN remains complexed to the Brønsted sites to at least 500K, after which the equilibrium shifts to the gas phase. While both the ^{13}C and proton NMR results indicate that there is rapid reorientation of the molecular axis at 500K, AN remains complexed at the Brønsted site. Second, except for the acetamide bands above 500K, there are no new features to suggest a change in C-N bond order in the temperature range at which changes occur in the ^{13}C isotropic chemical shift. Third, the ν(C-N) band shifts to lower frequencies, by as much as 12 cm^{-1} at 500K. The shift in frequency with temperature can be described by an exchange model (12) in which the ν(C-N) vibrational frequency is coupled to low-frequency, anharmonic modes of the adsorption complex (e.g. bending of the hydrogen-bond, zeolite lattice vibrations, acoustic phonons, etc.). The frequency shift caused by these dynamical processes depends on the magnitude and sign of the couplings. However, for CO adsorbed on Cu, which should have similar vibrational modes to AN on H-MFI, the change in ν(C-O) is similar in magnitude and direction to that observed for ν(C-N) in our case (13).

All of this would suggest that the changes in the ^{13}C isotropic chemical shifts with temperature are associated with dynamical processes of the AN molecule, hydrogen-bonded to the Brønsted site. Jameson (14) has shown that the temperature dependence for the chemical shielding, σ, can be accounted by a Taylor series expansion:

$$\langle\sigma\rangle_T = \sigma_e + \sum_i \left(\frac{\partial\sigma}{\partial q_i}\right)_e \langle q_i\rangle_T + \frac{1}{2}\sum_{i,j} \frac{\partial^2\sigma}{\partial q_i \partial q_j}\langle q_i q_j\rangle_T \cdots$$

where $\langle q_i\rangle_T, \langle q_i, q_j\rangle_T$ are thermal averages of the normal coordinates. According to this, the important factors in determining the magnitude of changes in the isotropic shift are distortions of the molecule due to anharmonic vibrational modes and centrifugal distortions. In the case of AN in H-MFI, the charge density is apparently affected by excitation of low-frequency modes, which change both ν(C-N) and the ^{13}C NMR shifts, without any proton transfer. While vibrational excitations can lead to changes in the structure of the complex, the fundamental physics lies in the dynamical processes, not changes in the chemical structure caused by partial or complete proton transfer. This model is currently being investigated by taking into account the coupling of the molecular modes with those of a model zeolite framework (15).

CONCLUSIONS

The large changes observed in the ^{13}C NMR isotropic chemical shifts for $CH_3^{13}CN$ in H-MFI with temperature are the result of dynamical processes associated with the excitation of low-frequency, vibrational modes. We find no evidence for changes in the chemical structure of CH_3CN resulting from either partial or complete proton transfer.

ACKNOWLEDGEMENTS

This work was supported in part by the National Science Foundation Grant #CTS9713023. Helpful discussions with Dr. M. Allavena, H.L. Dai, and R.M. Hochstrasser are gratefully acknowledged.

REFERENCES

1. W.E. Farneth and R.J. Gorte, Chemical Reviews **95**, 615 (1995).
2. D.J. Parrillo, C. Lee, R.J. Gorte, D. White, and W.E. Farneth, Journal of Physical Chemistry **99**, 8745 (1995).
3. R.J. Gorte and David White, Topics in Catalysis **4**, 57 (1997).
4. J. Šepa, R.J. Gorte, David White, E. Kassab, and M. Allavena, Chemical Physics Letters **262**, 321 (1996).
5. J.F. Haw, M.B. Hall, A.E. Alvarado-Swaisgood, E.J. Munson, Z. Lin, L.W. Beck, and T. Howard, Journal of the American Chemical Society, **116**, 7308 (1994).
6. A.G. Pelmenschikov, R.A. van Santen, J. Jänchen, and E. Meijer, Journal of Physical Chemistry **1993**, 97, 11071.
7. C.-C. Lee, R.J. Gorte, and W.E. Farneth, Journal of Physical Chemistry B **101**, 3811 (1997).
8. J. Šepa, R.J. Gorte, B.H. Suits, and David White, Chemical Physics Letters **289**, 281 (1998).
9. J. Šepa, R.J. Gorte, B.H. Suits, and David White, Chemical Physics Letters **252**, 281 (1996).
10. J. Šepa, PhD Thesis, University of Pennsylvania (1998).
11. L. Kubelkova, J. Kotrla, J. Florian, Journal of Physical Chemistry **99**, 10285 (1995).
12. R.M. Shelby, C.B. Harris and P.A. Cornelius, Journal of Chemical Physics **70**, 34 (1979).
13. J.P. Culver, M. Li, L.G. Jahn, R.M. Hochstrasser and A.G. Yodh, <u>Laser Spectroscopy and Photochemistry on Metal Surfaces, Part 1</u>, Edited by H.L. Dai and W. Ho, (World Scientific. Singapore, New Jersey, London, Hong Kong, 1995). pg. 542.
14. C.J. Jameson, Chemical Reviews, **91**, 1375 (1991).
15. M. Allavena, private communication.

HYDROCHLOROFLUOROCARBON REACTIVITY AND STRUCTURAL CHARACTERIZATION OF ZINC EXCHANGED NAX

M.F. CIRAOLO*, P. NORBY*, J.C. HANSON[†], D.R. CORBIN[§] and C.P. GREY*

*Chemistry Department, SUNY Stony Brook, Stony Brook, NY 11794-3400, USA; cgrey@sbchem.sunysb.edu
[†]Chemistry Department, Brookhaven National Laboratory, Upton, NY 11973, USA;
[§]DuPont CR&D, Wilmington, DE 19880-0262, USA;

ABSTRACT

Solid-state MAS NMR and synchrotron X-ray powder diffraction have been used to study the reactivity of fluorocarbons and to study the cation positions of Zn^{2+}-exchanged NaX. The structure of dehydrated ZnX was refined in the space group Fd3m and zinc cations were located in four different positions, all lying along the [111] direction. The residual sodium cations were located in the SII position in the supercages. Tetrahedral extra-framework aluminum species were found (by ^{27}Al MAS NMR and diffraction) in the center of the sodalite cage. The reactions of HCFC-124a (CF_2HCF_2Cl) over ZnX were studied with NMR and by temperature programmed desorption/mass spectrometry (TPD/MS). The unsaturated products of both the dehydrofluorination and dehydrochlorination reactions (CF_2CFCl and CF_2CF_2) were detected in the TPD experiments, while saturated products such as HFC-125 (CF_3CF_2H) were the major products as observed by ^{19}F NMR.

INTRODUCTION

Basic zeolites have been proposed as materials for separating different mixtures of hydrofluorocarbons (HFCs) and hydrochlorofluorocarbons (HCFCs) produced in the synthesis of the environmentally-friendly replacements to the CFCs.[1] However, HCFCs are readily decomposed in basic zeolites. For example, the conversion of $CHClF_2$ (HCFC-22) to CO_2 at ambient temperatures, over the molecular sieve 5A, was reported a number of years previously.[2] HFCs will also undergo dehydrofluorination reactions at higher temperatures.[3] We have been studying the adsorption and reactivity of HCFC-124a (CF_2HCF_2Cl) on cation exchanged X zeolites, with solid state NMR, powder diffraction and mass spectrometry, to probe the interactions of the HCFCs with the extra-framework cations and framework oxygen sites, and to relate this to the HCFC reactivity. We have chosen to study HCFC-124a since this molecule, can in principle, undergo both dehydrofluorination and dehydrochlorination reactions. Dehydrofluorinations typically occur via a carbanion mechanism. In contrast, dehydrochlorinations may occur via (concerted) E2 or E1 mechanisms, the latter being favored as the electrostatic field of the cation in the zeolite increases.[4] Thus, in theory, the preference for dehydrofluorination versus dehydrochlorination may be related to the relative importance of the interactions with the basic oxygen atoms versus with the cations.

EXPERIMENTAL

Zeolite ZnX was prepared by ion-exchanging zeolite NaX (Aldrich Chemicals) with 0.1 M $Zn(NO_3)_2$ at a temperature of 60 °C over a period of 48 hours. Dehydration of the exchanged sample was carried out by ramping the temperature under vacuum to 450 °C over 12 hours then holding at 450 °C for an additional 24 hours. HCFC-124a (Dupont) loading levels were established by monitoring the drop in pressure, on exposure of the dehydrated sample to an HCFC atmosphere, with a calibrated vacuum line and an absolute-pressure gauge. ICP elemental analysis, (Galbraith) of the exchanged Zn-X sample gave a composition (weight %) of Na 1.13, Zn 11.76, Si 14.36, and Al 11.56, giving a composition for the unit cell of $Na_{10.1}Zn_{36.8}Si_{104.3}Al_{87.7}O_{384}$.

Variable-temperature ^{27}Al and ^{19}F MAS NMR experiments were performed with a double resonance Chemagnetics probe, on a CMX-360 spectrometer. Chemical shifts for ^{23}Na, ^{27}Al and ^{19}F are quoted relative to aq. sodium chloride, aq. aluminum sulfate and CCl_3F, respectively, as external standards.

X-ray synchrotron powder diffraction data were collected at the beamline X7B at the National Synchrotron Light Source (NSLS), Brookhaven National Laboratory (BNL), with a FUJI imaging plate (200 x 400 mm, spatial resolution 100 x 100 μm) mounted perpendicular to the incoming beam. Reactions were performed with a RXM-100 Multifunctional Catalyst Testing and Characterization Machine equipped with a MKS 100C Precision Quadrupole Mass Analyzer.

RESULTS

The room temperature structure of dehydrated ZnX has been refined in the cubic space group Fd3m and agreement factors of R_{wp} = 2.22% and χ^2 = 5.34 were obtained. A total of 41.4 zinc and 8.6 sodium cations per unit cell were found (Table 1). Sodium cations were only located in the SII position in the supercage and ^{23}Na NMR demonstrated the lack of any SI sodium cations. The number of zinc cations is slightly higher than expected from the chemical analysis (37), while the number of sodium cations is lower. This is, most likely, due to the correlation between the occupancies of the Zn and Na SII positions, due to the overlap of the electron density, and is reflected in the larger e.s.d.s for these sites. It is, therefore, difficult to refine the occupancies of these two sites independently. Zinc cations were located on a split, lower symmetry (32e), SI position, consistent with the previous report for ZnY.[5] Attempts to model the electron density with a true SI position (16c), resulted in a very low occupancy for this site, and considerable residual electron density on the split position. A difference Fourier map revealed residual electron density at the center of the sodalite cage. This was attributed to extra-framework AlO_4 species, which have been previously reported in Zeolite A and in dealuminated X.[6] A resonance at 92 ppm is observed in the ^{27}Al MAS NMR of dehydrated ZnX, consistent with this species.[7] The positions of the different refined cation positions are shown in Figure 1.

Atom	Position	Occupancy
Zn	SI (32e)	5.7(2)
Zn	SI' (32e)	16.8(3)
Zn	SII' (32e)	12.7(4)
Zn	SII (32e)	6.2(6)
Na	SII (32e)	8.6(16)
Al	(8a)	1.62(2)
O	(32e)	5.41(35)

Table 1. Occupancies, per unit cell, for the refined cation positions and extra-framework AlO_4^{5-} ion. Standard deviations for the occupancies are given in parentheses.

Reactions of HCFC-124a (CF_2HCF_2Cl) adsorbed on ZnX were studied with ^{19}F and ^{27}Al variable temperature MAS NMR. Reactivity of the HCFC-124a was observed with ^{19}F MAS NMR after heating the sample for only 3 minutes at 150 °C. Two additional resonances were detected at -141 and -89 ppm which were assigned to HFC-125 (CF_3CF_2H). After longer treatment of the sample at 150 °C (up to 1 hour), a number of smaller resonances were observed between -80 and -95 ppm, which are assigned to the CF_3 groups in HCFC-124 (CF_3CFHCl), HFC-123 (CF_3CHCl_2) and HFC-116 (CF_3CF_3). A much broader, less intense resonance at ca. -126 ppm is observed. The linewidth, and the sidebands observed for this resonance, indicate that it does not come from a gaseous species, and it is assigned to either a longer chain saturated fluoro- or chlorofluorocarbon, or a silicon fluoride species. When the sample is ramped to 225 °C (for 20 mins), this resonance disappears and small broad peak at −168 ppm with associated spinning sidebands is observed. Simultaneously, a noticeable change in the ^{27}Al NMR spectrum is observed: the resonance from the extra-framework AlO_4^{5-} species at 92 ppm disappears, and a broad peak from an octahedral aluminum species at –0.8 ppm begins to appear. After 45 minutes at this temperature, the intensity of this resonance has increased, presumably as the framework aluminum atoms are attacked and more Al-F species are produced.

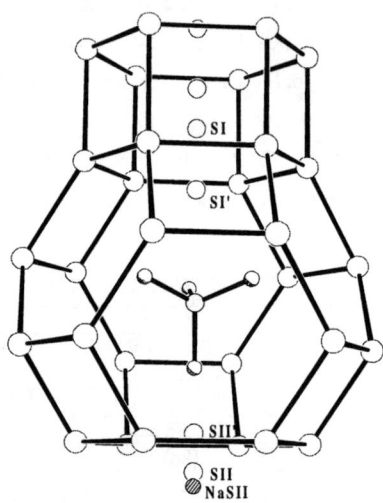

Figure 1. A view of a sodalite cage and connected double six-ring, showing the zinc (SI, SI', SII' and SII) and sodium (NaSII) positions, and the AlO_4^{5-} extra-framework species in the center of the sodalite cage.

Products from TPD studies of samples of excess HFC-124a-loaded ZnX were studied with mass spectroscopy. As the temperature was ramped above 50 °C, the dehydrochlorination and dehydrochlorination products (HFC-1114 (CF_2CF_2) and HCFC-1113 (CF_2CFCl), respectively) were observed in approximately equal amounts as the most abundant products. HCFC-124a was not observed and thus must completely react before desorption. Some HCl was observed, but no HF was seen, the latter presumably attacking the zeolite framework, at the temperatures required for desorption. Only very small concentrations of saturated products were detected.

DISCUSSION

Rietveld refinement of bare partially zinc-exchanged NaX has shown that there are both zinc and sodium cations in the supercages available for gas binding after the sample has been dehydrated. The zinc position lies only 0.53 Å above the plane formed by the 3 O2 atoms in the supercage, so will be less exposed to gas sorbates than the sodium cations which lie 1.1 Å above the O2 plane. Clearly, further ion-exchange is required to remove these sodium cations. The products in an open reaction system (TPD/MS) differed considerably from those detected in the closed reaction vessel (NMR). No unsaturated products were observed with NMR, the major product being HFC-125. Both dehydrofluorination and dehydrochlorination reactions are clearly occurring, as the products from these reactions were observed by MS. The saturated products are proposed to result from the addition of HF or HCl to the unsaturated products. Thus, HFC-125 is formed from the addition of HF to CF_2CF_2, the product of the dehydrochlorination reaction, the driving force presumably being the formation of the thermodynamically-stable CF_3 group. However, fluorine and chlorine substitution reactions cannot be ruled out. Coking is also observed in the sealed systems, which may account for the loss of some of the products of the dehydrohalogenation reactions. Although HCFC-124a dehydrohalogenation reactions were observed at 150 °C, attack of the extra-framework AlO_4^{5-} species by HF did not occur until higher temperatures. Rapid attack of the framework and further dealumination was observed at 225 °C.

ACKNOWLEDGMENTS

Support from the following grants is gratefully acknowledged: DE-FG02-96ER14681, CHE-9405436 and DE-AC02-76CH00016. CPG is a DuPont Young Professor and a Cotrell Scholar of the Research Corporation.

REFERENCES

(1) D.R. Corbin and B.A. Mahler, World Patent, W.O. 94/02440(3 February 1994).

(2) P. Cannon, J. Am. Chem. Soc. **80**, 1766 (1958).

(3) C.P. Grey and D.R. Corbin, J. Phys. Chem. **99**, 16821 (1995).

(4) W. Klading and H. Noller, J. Catal. **29**, 385 (1997).

(5) A.P. Wilkinson, A.K. Cheetham, S.C. Tang and W.J. Reppart, J. Chem. Soc. Chem. Commun, 1485 (1992).

(6) J.J. Pluth and J.V. Smith, J. Am. Chem. Soc. **105**, 51 (1983); J.J. Pluth and J.V. Smith, J. Am. Chem. Soc. **105**, 1192 (1983); J.B. Parise, D.R. Corbin and L. Abrams Acta. Cryst. **C40**, 1493 (1984).

(7) D.R. Corbin, R.D. Farlee and G.D. Stucky, Inorg. Chem. **23**, 2920 (1984).

HYDROFLUOROCARBON ZEOLITE INTERACTIONS: NMR AND X-RAY DIFFRACTION STUDIES

CLARE P. GREY, MICHAEL F. CIRAOLO and KWANG HUN LIM

Chemistry Department, SUNY Stony Brook, Stony Brook, NY 11794-3400;
cgrey@sbchem.sunysb.edu

ABSTRACT

A combination of X-ray diffraction and MAS NMR studies are used to probe cation mobilities in bare and hydrofluorocarbon (HFC)-loaded faujasites. Sodium-cation intercage motion, involving a jump though a six-ring window (SI to SI'), has been proposed to explain the high temperature ^{23}Na MAS NMR spectra of NaY. SI to SI' jumps occur with a rate of > 6 kHz at 200 °C for the loaded material and 250 °C for the bare material. Gas sorption of HFC-134 (CF_2HCF_2H) on NaY was monitored with *in-situ* methods, by using synchrotron X-ray diffraction. Structural changes, involving long range cation migrations between cages (SII to SII'/SI'), occur in seconds in these systems, on gas sorption. ^{133}Cs MAS NMR shows that the cesium-cation mobility within the supercages of CsY also increases significantly on gas sorption.

INTRODUCTION

We have been studying the sorption and reactivities of hydrofluorocarbons such as HFC-134 (CF_2HCF_2H) and HFC-134a on basic molecular sieves [1,2]. The complete phaseout of chlorofluorocarbon (CFC) production in developed countries by the year 2000 has resulted in the development and production of a variety of environmentally-friendly alternatives (the HFCs) for different applications [3]. The syntheses of the HFCs is more complex than the syntheses of the CFCs, involving many more steps, and unwanted hydrofluorocarbon (HFC) and hydrochlorofluorocarbon (HCFC) isomers are often produced. For example, HFC-134 is a common byproduct in the synthesis of HFC-134a, the replacement for the refrigerant CFC-12 (CF_2Cl_2) [3]. Basic molecular sieves have been proposed as a method of separating some of these isomer mixtures [4]. We have, therefore, been applying a variety of NMR and X-ray powder diffraction methods to determine the importance of different interactions in controlling the sorption properties of these gases in faujasite zeolites, and to rationalize trends in separations behavior and HFC reactivity.

Our previous work on the binding of HFC-134 on NaY [1] demonstrated that Na-F interactions are very important in these systems, and are strong enough to cause migration of SI' cations originally located in the sodalite cages of the faujasite structure into the supercage, where they can then bind to both ends of the HFC molecules. These ion migrations involve sodium jumps through six ring windows (from the sodalite into the supercages). This work raised a number of questions which we have systematically been trying to answer: For example, is the activation energy associated with this process low enough so that cation migrations can readily occur at room temperature? How is it affected by the presence of the HFCs in the supercages? In order to address these issues, we have designed a set-up to study sorption of HFCs in real time with the synchrotron X-ray diffraction. Sodium mobility has also been probed with variable temperature ^{23}Na MAS NMR. ^{23}Na is a non-integer spin nucleus, and thus the lineshapes of the resonances are sensitive not only to local environment, [5] but will also be sensitive to the dynamics. ^{133}Cs MAS NMR is used to study intercage motion. Reports from these studies are presented in this paper.

EXPERIMENTAL

Zeolites NaY or Cs(Na)Y were dehydrated under vacuum by slowly ramping the temperature of approximately 0.5 g of sample to 450 °C over a period of 48 hours. HFC-134 (DuPont) was adsorbed at room temperature at a loading level of 4 molecules per supercage. Loading levels were established by monitoring the drop of pressure, on exposure of the dehydrated sample to an atmosphere of the HFC, with a carefully calibrated vacuum line. The results of the characterization of the Cs(Na)Y sample have already been reported [6] and the sample contains Cs and Na in a 68:32 ratio.

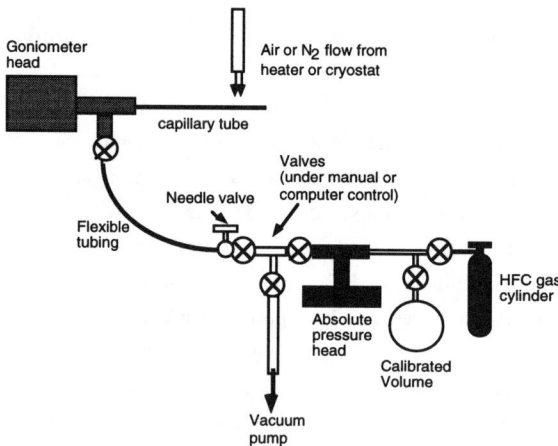

Figure 1. The design used to monitor changes on gas sorption, by *in-situ* by X-ray diffraction.

Variable temperature ^{23}Na and ^{133}Cs MAS NMR experiments were performed with a double-resonance Chemagnetics probe, on a CMX-360 spectrometer. Small flip angles were used to ensure uniform excitation of the sodium/cesium nuclei with small and large quadrupole coupling constants. Chemical shifts are quoted relative to solid sodium chloride, and 1M CsCl for ^{23}Na and ^{133}Cs, respectively as external references.

X-ray synchrotron powder diffraction data were collected with the X7B instrument at the National Synchrotron Light Source at Brookhaven National Laboratory, by using a FUJI imaging plate (IP) (200 x 400 mm, spatial resolution 100 x 100 µm) mounted perpendicular to the incoming beam. Data were collected with a translating image plate set-up as described elsewhere [6]. The set-up used for gas sorption is shown in Figure 1.

RESULTS

MAS NMR: At temperatures above ≈ 150 °C, the resonance assigned to SI cations, in the ^{23}Na MAS NMR spectra of NaY broadens, and by 250 °C is barely visible (Figure 2). The resonance from the SI cations appears to have coalesced with the broader resonances due to SI' and SII cations. For the HFC-134 loaded sample, this happens at a much lower temperature, and by 250 °C, only one broad resonance is observed.

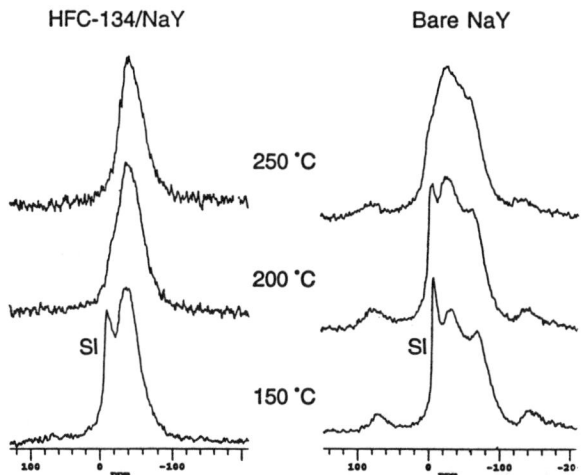

Figure 2. The ^{23}Na MAS NMR variable temperature spectra of bare and HFC-134 loaded dehydrated NaY.

The variable temperature spectra of CsY loaded with HFC-134 are shown in Figure 3. The room temperature spectrum resembles the ^{133}Cs spectra of bare anhydrous CsY at elevated temperatures (≈250 °C) [6]: only one resonance from the supercage cations is observed at -94 ppm. The two resonances from the sodalite cations at -136 ppm (SII' [6]) and -153 ppm (SI' [6]) are also clearly observed. As the temperature is lowered, no significant change is observed for the sodalite cage resonances. In contrast, the resonance at -94 ppm shifts to lower frequency. At the same time, a shoulder at higher frequency, with associated spinning sidebands, is seen, which grows in intensity and is clearly discerned at -50 °C as a separate resonance at -68 ppm. This resonance continues to shift to higher frequency and increase in intensity as the temperature is lowered further, reaching -50 ppm at -150 °C. Meanwhile, the lower frequency resonance decreases in intensity and continues to shift to lower frequency reaching -111 ppm at -100 °C; spinning sidebands are also observed for the first time at this temperature. At lower temperatures, however, the resonance splits into two, resonances being observed at -101 and -126 ppm, each with associated spinning sidebands. The resonance at -50 ppm is assigned to cations in the SII position, while the resonances at -101 and -126 ppm are assigned to SIII cations.

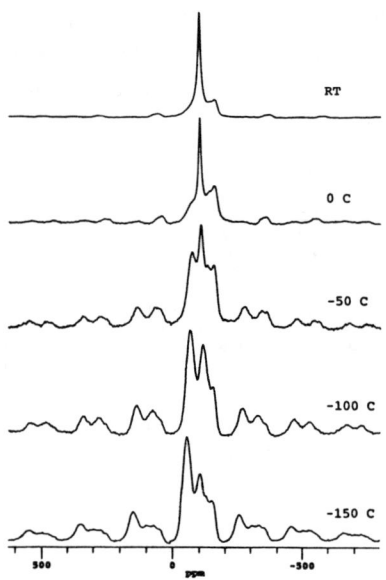

Figure 3. The ^{133}Cs MAS NMR of HFC-134 loaded Cs(Na)Y as a function of temperature.

In-situ diffraction Studies: Changes in the diffraction pattern of anhydrous NaY on sorption of HFC-134 were followed by using the translating IP. A dramatic change in the intensity of, in particular, the 111 reflection is observed on gas sorption. An decrease in the unit cell parameter is also observed. The system takes less than 1 minute to reach equilibrium at gas pressures of, for example, 90 Torr. At lower pressures (≥10 Torr), equilibrium takes longer to reach, but here, the rate limiting step is the diffusion of the gases through the zeolite particles.

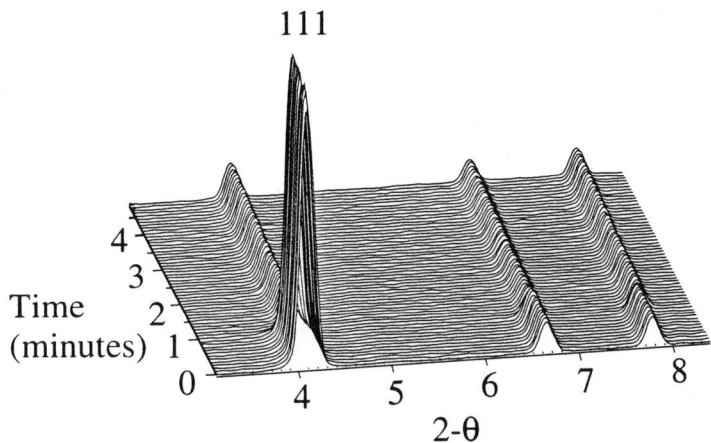

Figure 4. The X-ray powder diffraction pattern of dehydrated NaY during gas sorption. The valve was opened after 1 minute, to expose the sample to 90 Torr of HFC-134. ($\lambda = 1.003$ Å).

DISCUSSION

The ^{23}Na NMR experiment appears to be extremely sensitive to any mobility of the SI cations. These cations are located in the double-six rings of the faujasite structure. The adjacent SI' site in the sodalite cage an be reached by a jump through a six-ring window. We propose that the coalescence of the SI and broader resonances is due to rapid SI to SI' jumps, with a rate greater than the frequency separation between resonances ($\approx$ 6 kHz) [7]. Since the two SI' sites adjacent to an occupied SI site will be vacant, no concerted jump out of the SI' site is required to allow this jump process. The lowering of the onset temperature of this motion, on sorption of HFC-134, may be related to the decrease in the number of SI' cations in the sodalite cages, since the O3-O3 distances do not change significantly on gas sorption [1].

The *in-situ* diffraction studies suggest that any structural rearrangements that occur on gas sorption are very rapid. Thus, the cation migrations involving SI' to SII jumps (presumably via SII') through the six-ring window connecting the super- and sodalite cages, which have been shown to occur for a sample prepared by loading HFC-134 on NaY prior to the diffraction experiment, occur very rapidly on gas sorption.

Our previous ^{133}Cs MAS NMR results for bare CsY have shown that rapid jumps between supercage cation positions occur at temperatures above room temperature [6]. Changes in SII and SIII occupancies, as determined by *in-situ* X-ray diffraction, as the sample temperature was raised to 500 °C, allowed an estimate for the difference in energy between SII and SIII sites of approximately 12 - 18 kJ mol^{-1} to be made. Thus the differences in energy between the different cation distributions are small. This suggests that even weakly sorbed gases may alter, or perturb, cation distributions within the zeolite cages. For example, these differences are significantly smaller than the isosteric heats of adsorption obtained for these materials on sorption of HFC-134 ($\approx$-70 kJmol^{-1}) [8], and thus, the cation rearrangements are not very energetically costly, and may be likely on gas sorption, if the new cation arrangement will favor the binding of the sorbate. The ^{133}Cs NMR spectra of the HFC-134 loaded sample is consistent with considerable cesium motion at room temperature: The resonance at -94 ppm is assigned to mobile SII and SIII cations undergoing exchange. The SII cations are less mobile than the SIII cations: Mobility of the SIII cations, on the timescale of the quadrupole coupling constant, is not frozen out until -100 °C. This is consistent with higher SIII, in comparison to SII, mobility observed in the absence of the HFC molecules [6]. Thus, the activation energy for intercage jumps between different cation positions decreases significantly on HFC-134 sorption.

CONCLUSIONS

Both cation positions and cation dynamics are strongly affected by gas sorption. Activation energies for cation motion between sites in the same cages and between cages are reduced. The results demonstrate the need to study the structures of zeolitic materials in the actual environment where the sorbent is present.

ACKNOWLEDGMENTS

Support from the following grants is gratefully acknowledged: DE-FG02-96ER14681, CHE-9405436 and DE-AC02-76CH00016. CPG is a DuPont Young Professor and a Cottrell Scholar of the Research Corporation. J. C. Hanson, A.F. Gualtieri, and P. Norby are thanked for their help with the X-ray diffraction experiments.

REFERENCES

1. C.P. Grey, F.I. Poshni, A. Gualtieri, P. Norby, J.C. Hanson, D.R. Corbin, J. Am. Chem. Soc., **119**, 1981 (1997).
2. C.P. Grey, D.R. Corbin, J. Phys. Chem., **99**, 16821 (1995).
3. L.E. Manzer, Science, **249**, 31 (1990). L.E. Manzer, V.N.M Rao, Advances in Catalysis, **39**, 329 (1993). G. Webb, J. Winfield, Chemistry in Britain, **28**, 996 (1992).
4. D.R. Corbin, B.A. Mahler, U.S. Patent, US 5600040 (July 17 1995).

5. G. Engelhardt, M. Hunger, H. Koller, J. Weitkamp, Stud. Surf. Sci. Catal., **84**, 421 (1994).
6. P. Norby, F.I. Poshni, A.F. Gualtieri, J.C. Hanson, and C.P. Grey, J. Phys. Chem. B, **102**, 839 (1998).
7. A. Abragam, The Principles of Nuclear Magnetism (Oxford University Press, Oxford, 1961).
8. S. Savitz, F. Siperstein, R. Huber, S.M. Tieri, R.J. Gorte, A.L. Myers, C.P. Grey, D.R. Corbin, D.R., in preparation.

THERMAL TRANSFORMATIONS OF ZEOLITE LiA(BW), (Li,Na)LTA AND THEIR DERIVATIVES OBTAINED BY MECHANOCHEMICAL TREATMENT

C.KOSANOVIĆ*, B. SUBOTIĆ*, P.NORBY[+], M. ŠOUFEK[§]

*Ruđer Bošković Institute, Bijenička 54, 10000 Zagreb, Croatia; cleo/subotic@rudjer.irb.hr
[+]Chemistry Department, Brookhaven National Laboratory, Upton, 11973 New York, USA; norby@x7b.chm.bnl.gov or norby@kemi.aau.dk
[§]Ministry of interior, Forensic Centre, Runjaninova 2, 10000 Zagreb, Croatia

ABSTRACT

Zeolites LiA(BW) and (Li,Na)-LTA, as well as their amorphous derivatives obtained by high-energy ball milling were heated at different temperatures (760-1150°C) for 0.5 and 3h. The heating of both original (crystalline) and mechanochemically amorphized zeolite (Li,Na)-LTA resulted in the formation of a mixture of carnegieite (ca.50%) and γ-eucryptite (ca. 50%). at lower temperatures (760-822°C) and mixtures of β-eucryptite and nepheline at higher temperatures (822-1150°C).The heating of zeolite LiA(BW) resulted in the formation of γ-eucryptite at lower temperatures (760-822°C). Mechanochemically amorphized zeolite LiA(BW) transformed to mixtures of β- and γ-eucryptite at 760-822°C and to β-eucryptite at prolonged time of heating (822°C). The results obtained are discussed in terms of the influence of various factors (type of cation, present framework structure, stability of different phases) on the transformation pathways.

INTRODUCTION

Thermal transformation of zeolite precursors is a novel technique for the synthesis of aluminosilicate based ceramics [1]. By varying the ratio SiO_2/Al_2O_3 in a zeolite framework as well as the type, charge and fraction of the extra-framework cations in the channel/void system of a zeolite, different amorphous and crystalline products may be obtained by thermal treatment of the modified zeolites [2].

Mechanochemical treatment (e.g. high-energy ball milling) of the cation-exchanged zeolites may change the rate, or even the pathway of their thermal transformations [3]. In this work, the thermal transformations of the zeolites LiA(BW) and (Li,Na)-LTA and their amorphous derivatives obtained by high-energy ball milling are studied.

EXPERIMENTAL

Zeolite (Na,Li)-LTA ($0.64Li_2O:0.36Na_2O:Al_2O_3:2SiO_2:3.82H_2O$) was prepared by partial exchange of the original Na^+ ions from zeolite 4A (Union Carbide Corp.) with the Li^+ ions from 0.5 molar LiCl solution, by the procedure already described [4]. Zeolite LiA(BW) ($Li_2O:Al_2O_3:2SiO_2:2H_2O$) was synthesized hydrothermally, by the procedure described earlier [5,6].

Portions of both the zeolites were milled in a planetary ball mill until the original crystalline phases (CP) have completely been transformed to the X-ray amorphous ones (AP). The zeolites (Na,Li)-LTA and LiA(BW) and their X-ray amorphous derivatives obtained by high-energy ball milling were heated at appropriate temperatures (760-1150°C) for predetermined times (0.5 and 3h, respectively) in a temperature controlled chamber furnace (ELPA-2 Elektrosanitarij). The temperatures of heating correspond to the temperatures of the exothermic peaks in the DSC curves of zeolite (Li,Na)-LTA and its amorphous derivative obtained by ball milling, respectively. The rate of heating was 10°C min^{-1}. The samples were cooled at the room temperature in a desiccator, before analyses.

All samples (original, milled and heated) were characterized by powder X-ray diffractometry. Integral intensities of corresponding non-overlapping sharp peaks and broad "amorphous maxima" of the X-ray diffraction patterns were used for calculation of the weight fractions of different crystalline and amorphous phases in the analyzed samples [1,4]. The Hermans-Weidinger method [7] was used for the determination of fractions of the crystalline phases in two-phase systems and the external standard method and/or mixing method, respectively [8], was used for determination of the weight fractions of different amorphous and crystalline phases in the multi-phase systems.

RESULTS AND DISCUSSION

The phase composition of the samples obtained by heating of crystalline powders (CP) of zeolites (Li,Na)-LTA and LiA(BW) and their amorphous derivatives (AP) obtained by ball milling, are shown in Table 1. The presence of a high fraction (47.5 wt.%) of amorphous phase in the sample obtained by heating of zeolite (Li,Na)LTA at 760°C for 0.5h indicates that the first step of the transformation of the crystalline precursor (Li,Na)LTA is destruction of the zeolite framework and formation of an X-ray amorphous aluminosilicate [9,10]. Hence, it may be assumed that the crystalline phases (γ–eucryptite, β–eucryptite, carnegieite, nepheline), are formed by nucleation and crystal growth from the amorphous aluminosilicate precursor. The formation of the same phases by heating of ball milled zeolite (Li,Na)-LTA (AP) corroborates such an assumption. The constant phase composition (ca.50wt % of γ-eucryptite and ca. 50wt. % carnegieite) obtained by heating at 822°C is established in <0.5h (see Fig. 1). Heating for a prolonged time (e.g., 3h) at the same temperature does not markedly influence the phase

Temperature	Time (h)	CP	AP	LiA(BW)	(Na,Li)-LTA
760°C	0.5	x	-	59.6% anhydrous LiA(BW) 40.4% γ-eucryptite	52.5% γ-eucryptite 47.5% amorphous phase
760°C	0.5	-	x	21.8% (anhydrous LiA(BW) + γ-eucryptite) 78.2% β-eucryptite	39.5% carnegieite 34.5% γ-eucryptite 26% β-eucryptite
760°C	3	x	-	60.0% anhydrous LiA(BW)+ 40.0% γ-eucryptite	51.9% carnegieite 48.1% γ-eucryptite
764°C	3	-	x	18.1% (anhydrous LiA(BW) + γ-eucryptite)+ 81.9% β-eucryptite	
822°C	0.5	x	-	100% γ-eucryptite	49.9% γ-eucryptite 50.1% carnegieite
822°C	3	x	-	100% γ-eucryptite	51.8% carnegieite 48.2% γ-eucryptite +traces β-eucryptite +nepheline
822°C	0.5	-	X	90.8% β-eucryptite 9.2% γ-eucryptite	
822°C	3	-	X	100% β-eucryptite	
1150°C	0.5	x	-		87.4% β-eucryptite 12.6% nepheline
1150°C	3	x	-		86.4% β-eucryptite 13.6% nepheline
1150°C	3	-	X		83% β-eucryptite 17% nepheline

Table 1: Phase composition of the solids obtained by heating of crystalline (CP) and amorphized (AP) zeolites LiA(BW) and (Li,Na)-LTA, at different temperatures for 0.5 and 3h.

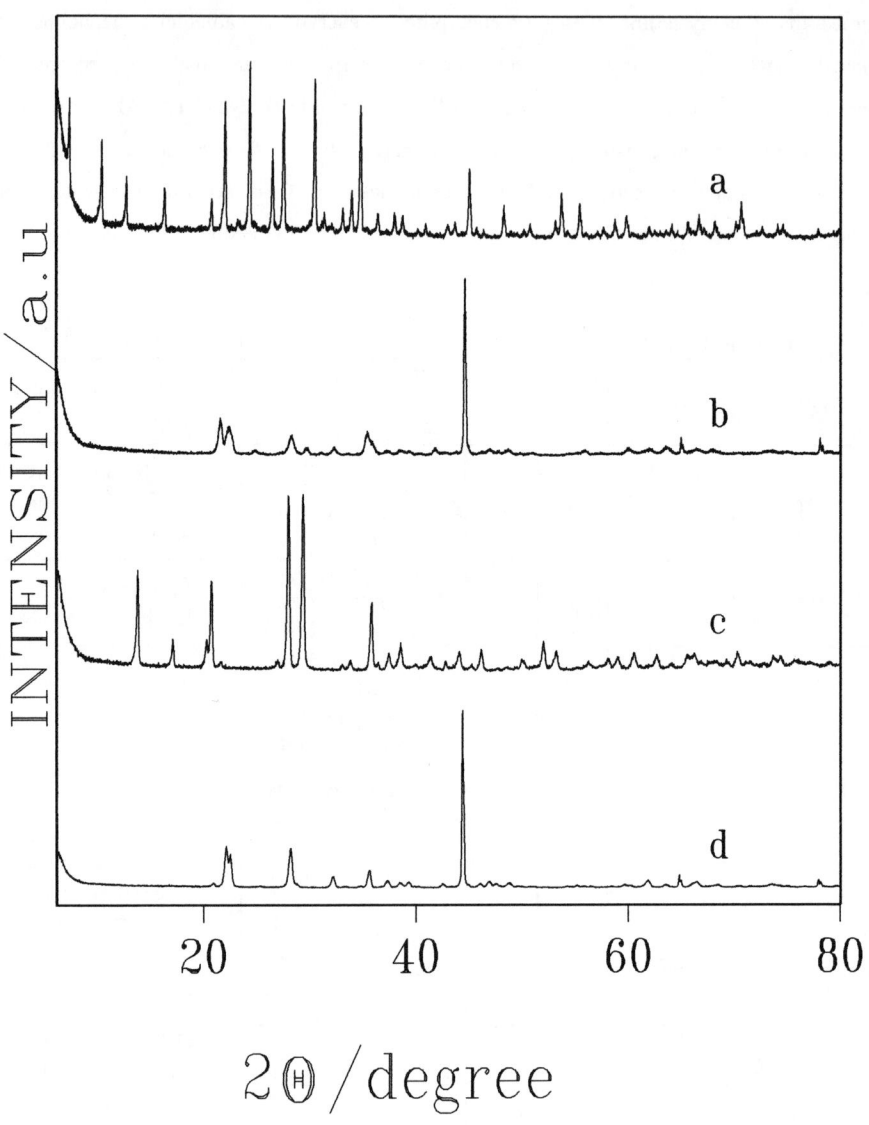

Figure 1. X-ray diffractograms of (a) the starting powder of (Na,Li)-LTA, (b) the mixture of carnegieite and γ-eucryptite obtained by heating (Na,Li)-LTA at 822°C for 0.5 h, (c) the starting powder LiA(BW) and (d) γ-eucryptite obtained by heating LiA(BW) at 822°C for 0.5 h.

composition of the products of heating (see Table 1). The presence of Li^+ ions favors the formation of eucryptite and the presence of Na^+ ions favors the formation of carnegieite. The percentages of carnegieite and eucryptite formed by heating (0.5 h at 760° C) of the amorphized (ball milled) (Li,Na)-LTA correspond to the percentages of $NaAlSiO_4$ and $LiAlSiO_4$ in the precursor. This indicates that: (i) the presence of Li^+ ions favors the formation of eucryptite and the presence of Na^+ ions favors the formation of carnegieite and (ii) all the Li^+ ions in the amorphous phase obtained by ball milling of zeolite are available for the crystallization of eucryptite(s), (β and γ). Increasing the temperature of heating above 822°C causes the transformation of γ-eucryptite into β-eucryptite and the transformation of carnegieite into nepheline, respectively. Calculations based on the chemical composition of the precursors, the composition of the products obtained after thermal treatment and on the assumption that eucriptite may be (re)-crystallized in Na-form ($NaAlSiO_4$), indicate that the β-eucryptite obtained by heating of both the crystalline and the amorphous precursors at 1150°C contains not only Li, but a mixture of Li and Na; chemical composition of the mixed (Li,Na)β-eucryptite is $Li_xNa_yAlSiO_4$ with x =0.725-0.76 and y = 0.24-0.275. Figure 2 shows that particulate properties (size and shape) of original zeolite (Li,Na)-LTA (Fig.2C) were preserved during thermal transformation (Fig. 2D), as it was earlier observed during the thermal transformation in the sequence: (Na,K)-LTA $\Rightarrow$ amorphous (Na,K)-aluminosilicate $\Rightarrow$ kalsilite + kaliophilite [9-10].

Based on the assumption that the framework structure of anhydrous LiA(BW) is that of the ABW-type, it can be concluded that anhydrous LiA(BW) zeolite may easily be transformed into γ-eucryptite with a cristobalite type structure, by moving half of the aluminium atoms from one tetrahedral hole to another while keeping the oxygen distribution almost intact [5]. Hence, a spontaneous high-temperature transformation of the anhydrous zeolite LiA(BW) into γ-eucryptite was expected. It is interesting that the transformation process at 760° C stops when ca. 40 wt. % of the LiA(BW) has been transformed into γ-eucryptite (see Table 1). Since there are two positions for the lithium ions in the LiA(BW) structure [11], it may be assumed that lithium ions from only one position, i.e., these which have lower activation energy, are available for the transformation process at 760° C; the lithium ions from another position which have higher activation energy may take part in the transformation process at higher temperatures. A complete transformation at 822° C, even in a short time (30 min), corroborates such an assumption. Again, the particulate properties (size and shape) of original zeolite LiA(BW) (Fig.2A) were preserved during its thermal transformation into γ-eucryptite (Fig. 2D).

Figure 2. Scanning electron micrographs of (a) the starting powder of LiA(BW) and (b) γ–eucryptite obtained by heating of LiA(BW) at 822°C for 3h, (c) the starting powder of (Na,Li)LTA and (d) γ–eucryptite and carnegieite obtained by heating of (Na,Li)LTA at 822°C for 3h.

Based on an analysis of the products which are formed during transformation of the ball milled (amorphized) zeolite LiA(BW) at 760° C, it may be concluded that the amorphized LiA(BW) recrystallizes into a mixture of crystalline anhydrous LiA(BW), γ-eucryptite and β-eucryptite. Phase composition of the product formed in 0.5 h does not differ considerably from the phase composition of the product formed in 3 h. This leads to an assumption that some fraction of lithium ions is "blocked" in the matrix of the amorphized LiA(BW), and that these lithium ions cannot take a part in the recrystallization/transformation processes. At an increased temperature (e.g. 822° C), the previously formed crystalline anhydrous LiA(BW) and γ-eucryptite, respectively, completely transform into more stable β-eucryptite, even in 0.5 h (see Table 1 and Figs 1). Although it is easy to assume that the relative proportions of crystalline anhydrous LiA(BW), γ-eucryptite and β-eucryptite in the products depend on the relative rates of their formation and transformation (LiA(BW) ⇒ γ-eucryptite, LiA(BW) ⇒ β-eucryptite, γ-eucryptite ⇒ β-eucryptite), the real pathways of the transformation processes cannot be defined at present.

CONCLUSION

Heating of zeolite (Na,Li)LTA in the temperature range from 760 - 822º C causes a phase transformation in the sequence: crystalline (Na,Li)-LTA ⇒ amorphous (Na,Li)-aluminosilicate ⇒ carnegieite (ca. 50 wt. %) and γ-eucryptite (ca. 50 wt. %). Hence, it can be concluded that the crystalline phases (carnegieite, eucryptite) are formed by nucleation and crystal growth from the amorphous aluminosilicate precursor. Heating of the amorphized (ball milled) zeolite (Na,Li)LTA in the same temperature range results in formation of a mixture of carnegieite (ca. 40 wt. %), γ-eucryptite (ca. 35 wt. %) and β-eucryptite (ca. 25 wt. %). The contents of carnegieite ($NaAlSiO_4$) and eucryptite ($LiAlSiO_4$) are determined by the molar ratio Na/Li in the precursor (crystalline or amorphized (Na,Li)LTA) and by the fractions of Li^+ and Na^+ ions which can take part in the transformation process at given temperature. At temperatures > 822º C, γ-eucryptite transforms into β-eucryptite and carnegieite transforms into nepheline. Part of $NaAlSiO_4$ can be crystallized in the eucryptite-type structure at 1150º C.

The degree of transformation of zeolite LiA(BW) into γ-eucryptite at 760º C (ca 40 % of the zeolite LiA(BW) can be transformed to γ-eucryptite) is determined by the number of Li^+ ions (site I) which have activation energy low enough to participate in the transformation process at 760º C. At an increased temperature (822º C), all of zeolite LiA(BW) is transformed to γ-eucryptite, thus indicating that at this temperature all Li^+ ions from zeolite LiA(BW) take part in the transformation process. Ball milled (amorphized) zeolite LiA(BW) recrystallizes into a mixture of crystalline LiA(BW), γ-eucryptite and β-eucryptite. At an increased temperature (e.g. 822º C), the previously formed crystalline anhydrous LiA(BW) and γ-eucryptite, respectively, completely transform into more stable β-eucryptite, even in 0.5 h. The pathways of the transformation processes will be studied in our further work.

REFERENCES

1. M.A. Subramanian, D.R. Corbin and R.D. Farlee, Mat. Res. Bull. **21**, 1525 (1986).
2. H.Mimura and T. Kano, Sci. Rep. RITU, **29A**, (1980) 102.
3. C. Kosanović, B. Subotić, I. Šmit, A. Čižmek, M. Stubičar and A. Tonejc, J. Mat. Sci. **32**, 73 (1977).

4. C. Kosanović, A. Čižmek, B. Subotić, I. Šmit, M. Stubičar and A. Tonejc, Zeolites **15**, 632 (1995).
5. P. Norby, Zeolites **10**, 193 (1990).
6. P. Norby, A. N. Christensen and I. E. K. Andersen, Acta Chem. Scand. **A40**, 500 (1986).
7. P. H. Hermans and A. Weidinger, Makromol. Chem. 44/46 (1961)24
8. L. S. Zevin and L. L. Zavyalova, Kolichestvenniy Rentgenographicheskiy Prazoviy Analiz, Nedra, (Moscow, 1974) p.37.
9. C. Kosanović and B. Subotić, Microporous Mater. **12**, 261 (1997).
10. C. Kosanović, B. Subotić and I. Šmit, Thermochim. Acta, in press.
11. I.S. Kerr, Z. Kristallogr. **139**, 186 (1974).

NMR STUDIES OF OXYGEN-ZEOLITE INTERACTIONS AT LOW TEMPERATURES

HAIMING LIU, HSIEN-MING KAO and CLARE P. GREY

Department of Chemistry, SUNY Stony Brook, Stony Brook, NY, USA;
cgrey@sbchem.sunysb.edu

ABSTRACT

Sorption of O_2 in the pores of zeolites has been used to discriminate between protons in the super and sodalite cages of HY. Dramatic increases in the sidebands are observed, for supercage protons that interact with the O_2; a much smaller increase in sideband intensity is seen for the sodalite protons. The ^1H spins in the O_2-containing samples show large decreases in the T_1 of approximately two orders of magnitude at -150 °C, in comparison to T_1 measurements made in air at room temperature, or at -150 °C in N_2. The T_1s for the supercage protons are significantly shorter than those for the sodalite protons. Both the decrease in the T_1s, and the large sideband manifolds, are due to interactions with the unpaired electrons present on the O_2 molecule.

Large shifts at low temperatures are observed for the ^{133}Cs NMR resonances in CsX and CsY due to Cs^+ cations in the supercages of CsY, but not for cations in the sodalite cages. The shifts result from O_2-Cs^+ interactions, and are consistent with the formation of short-lived O_2-Cs^+ complexes.

INTRODUCTION

The reduction of the ^{29}Si spin-lattice relaxation times (T_1s) of zeolites from minutes to $\approx 0.2 - 2$ s in the presence of oxygen is well established [1]. Oxygen is a paramagnetic molecule containing two unpaired electrons in the two degenerate highest occupied π^* (anti-bonding) orbitals. The decreases in the T_1s result from the interaction between the nuclear (^{29}Si) and electronic moments due to the unpaired electrons. We have been studying the interactions with O_2 at low temperatures, to explore whether MAS NMR can be used to probe O_2 sorption and binding in zeolites. The results from two systems are reported, one involving hydrogen atom - O_2 interactions and the second involving cation - O_2 interactions.

EXPERIMENTAL

Sample preparation: HZSM-5 and HY were prepared under shallow-bed conditions by deammoniation of NH_4ZSM-5(Si/Al=25) and NH_4Y(Si/Al=2.6), respectively, by first evacuating the samples at room temperature, dehydrating them at 120 °C for 12 hours and then slowly heated them to 400 °C at a rate of 12 °C/h. The temperature was then maintained at 400 °C for a further 24 h. Zeolite CsX containing 32 Cs^+ per unit cell (u.c.) and 49.5 Na^+/u.c. was provided by David Corbin from DuPont. Zeolite CsY was prepared as described by Norby et al [2]. Dehydration was carried out by ramping the temperature under vacuum to 450 °C over a period of 12 h, and holding for a further 24 h or longer. The dehydrated samples were packed into inserts in a glove box under inert N_2 atmosphere. The inserts, which help prevent rehydration, do not prevent O_2/N_2 from entering. The samples were kept sealed until immediately before the experiments.

NMR: Variable temperature ^{1}H and ^{133}Cs MAS NMR experiments were performed with a double-tuned Chemagnetics probe, on a CMX-360 spectrometer. Chemical shifts are quoted relative to TMS (^{1}H) and 1M CsCl (^{133}Cs). An ^{27}Al r.f. field strength of 63 kHz was used for the TRAPDOR experiment. Further details of this experiment can be found elsewhere [3]. Spinning speeds of 4.0 kHz and 10.0 kHz are used for ^{1}H and ^{133}Cs MAS NMR, respectively.

RESULTS

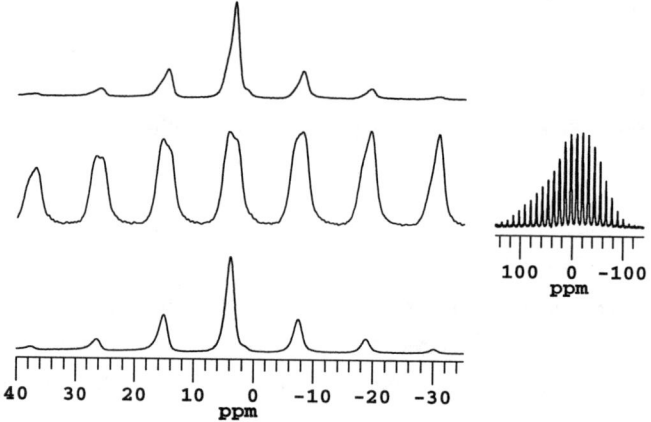

Figure 1. The ^{1}H MAS NMR spectra of HY at room temperature in air (top), -150 °C (middle) and in N_2 (bottom). An inset showing the full spectrum is shown for the middle spectrum.

Figure 1 shows the effect of temperature on the ^{1}H MAS NMR spectra of HY. As the temperature is decreased, the sideband manifolds gradually increase, and by -150 °C, a dramatic increase in the spinning sideband manifolds is observed for samples run in dry air. In contrast, when N_2 is used, very little difference is observed between the room and low temperature spectra. On close inspection, two distinct spinning sideband manifolds are observed in air at low temperatures. One sideband manifold contains spinning sidebands that cover more than 200 ppm and an isotropic resonance of 3.3 ppm. This should be compared to the room temperature value of 3.4 ppm for this resonance. This resonance is assigned to protons in the supercages of the faujasite structure. The second spinning sideband manifold is considerably narrower and has an isotropic shift of 4.4 ppm. This resonance is assigned to protons in the sodalite cages. Large differences in the T_1s were measured for the protons in N_2 and in air, and are ≈1.7 s for both protons in N_2 and 0.005 s and 0.015 s for the supercage and sodalite protons, respectively, in air. The two resonances could, therefore, be separated by acquiring the spectra in air with different recycle delays. ^{1}H MAS NMR spectra were also acquired in sealed capsules with 100 and 630 Torr of O_2, and a smaller increase in width of the sideband manifold, with decreasing temperature, was observed.

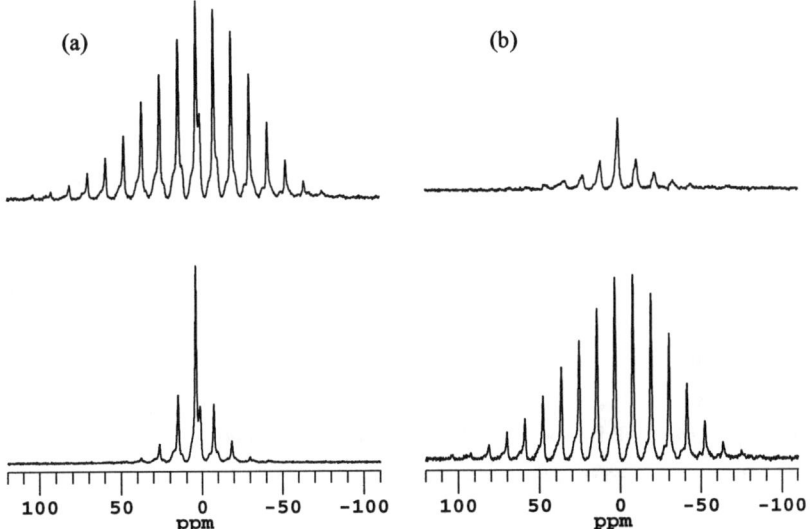

Figure 2. (a) The ^{1}H MAS NMR spectra of HZSM-5 at -150 °C in air (top), and in N_2 (bottom). (b) The ^{1}H/^{27}Al TRAPDOR NMR spectrum of HZSM-5 at -150 °C in air acquired with ^{27}Al irradiation (top) and the the difference spectrum (botttom).

Experiments were repeated for HZSM-5 (Figure 2a) and similar results were obtained. Again a large increase in the width of the sideband manifold is observed for the Brønsted-acid (4.0 ppm) protons in air. A reduction in the T₁ from 0.22 s in N₂ to 0.008 s in air is also measured. [...] Al TRAPDOR NMR was used to separate the Brønsted acid sites and silanol groups because the latter show up clearly in the experiment performed with ^{27}Al irradiation (Figure 2b). The T_1 for the silanol groups decreases from 0.36 s at room teperature to 0.014 s at -150 °C, but a smaller increase in the second moment is seen, in comparison to the results observed for the 4.0 ppm Brønsted acid site. The resonance at 7 ppm is resolved in the TRAPDOR difference spectrum, indicating that it results from protons with short H-Al internuclear distances, similar to those for the Brønsted acid site at 4.0 ppm, consistent with earlier studies [4]. A shortening of the T_1 of the resonance at 7 ppm to 0.009 s is also observed at -150 °C .

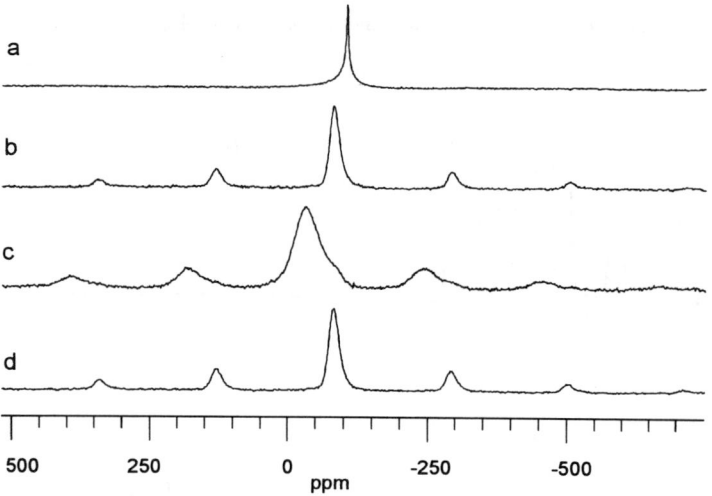

Figure 3. ^{133}Cs MAS NMR spectra of dehydrated CsX acquired in air at (a) room temperature and (b) -100 °C; (c) -150 °C. (d) was acquired in N₂ at -150 °C.

^{133}Cs MAS NMR spectra of dehydrated CsX were acquired in air and N₂, as a function of temperature (Figure 3). A sharp resonance at -101 ppm is observed at room temperature, with a Lorentzian line shape. A T_1 of 0.33 s was measured for this resonance. No spinning sidebands are present. In air, at lower temperatures this resonance shifts dramatically to higher frequencies. T_1 decreases to 0.05 s at -150 °C. In addition, a significant broadening of the isotropic resonance, which has now a Gaussian line shape, and a growth of the spinning sideband manifolds are seen.

However the sidebands are not as pronounced as in the ¹H NMR spectra. With prolonged acquisition at -150 °C this resonance continues to shift by more than 300 ppm from the room temperature value. The linewidth of the resonance also increases and the signal-to-noise ratio decreases significantly. Spinning in N_2 at -150 °C, in constrast, leads to a slightly broadened resonance at -77 ppm and the T_1 increases to 2.0 s.

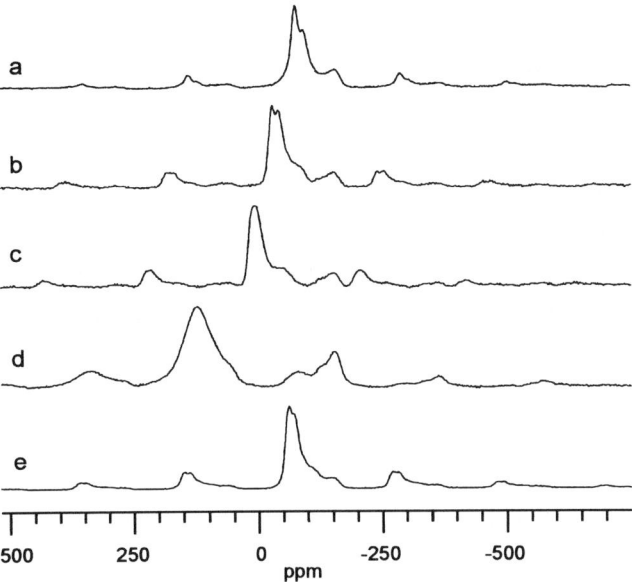

Figure 4. ¹³³Cs MAS NMR of dehydrated CsY acquired in air at (a) room temperature, (b) -70 °C, (c) -100 °C and (d) -150 °C; (e) in N_2 at -150 °C.

Figure 4 shows the ¹³³Cs MAS NMR of dehydrated CsY. At room temperature in air, resonances due to Cs⁺ in the supercages (-74 ppm and -90 ppm) and in the sodalite cages (-150 ppm), as previously assigned [2, 5, 6], are resolved. At low temperatures, the resonance due to sodalite-cage Cs⁺ remains at its original position, with a new feature being discerned at -122 ppm as a shoulder. The resonances due to supercage cations, however, shift drastically to higher frequency and at -150 °C collapse into a much broader resonance at 125 ppm (the shoulder to the lower frequency of this resonance is a spinning sideband of the resonances due to the sodalite-cage Cs^+). A T_1 of 0.07 s was measured for this resonance. On the other hand, when N_2 was used as the spinning gas, no significant changes are observed at temperatures as low as -150 °C.

DISCUSSION

The large increase in the width of the sideband manifolds, and decrease in the T_1s, as the temperature is lowered, will be caused by two major factors. Firstly, the uptake of O_2 in the pores will increase as the temperature decreases. Secondly, the susceptibility, or time averaged magnetic moment, is inversely proportional to the temperature (for a paramagnet obeying Curie's Law). The increase in second moment, with decreasing temperature, was significantly smaller for sealed samples containing O_2 with partial pressures close to 1 atm (100 - 630 Torr), than for samples acquired in air. Thus a major cause of the increase in the sidebands arises from the increased O_2 sorption. Note that large increases in sideband intensities were also observed in the ^{1}H MAS NMR spectra of HZSM-5 in sealed ampules with loading levels of 0.25 - 1.0 O_2, [7], but the loading levels of these samples are significantly higher than the sealed samples used in our experiments.

The large sidebands observed in the presence of paramagnets arise primarily from the dipolar coupling between the nuclear and electronic moments, although susceptibility effects may also cause significant increases in sidebands [8]. The dipolar coupling falls off rapidly as $1/r^3$ where r is the paramagnet-nuclear spin distance. Very rapid isotropic O_2 motion is expected to average this interaction to zero. Very little change is observed between spinning in N_2 and air at room temperature, consistent with this. The observation of such large sideband envelopes at low temperatures indicates that the O_2 motion can no longer be isotropic, in the time scale of the paramagnet-^{1}H spin interaction. The O_2 must be moving sufficiently slowly, or spend sufficient time complexed to the protons, to result in the increase in sidebands. The $1/r^3$ dependence of the dipolar interaction is responsible for the significantly smaller sideband manifolds observed for the protons in the sodalite cages, which are not accessible to the oxygen atoms.

The large sidebands for both Brønsted acid sites in H-ZSM-5 (4.0 and 7 ppm) at low temperatures indicate that both sites are accessible to the oxygen atoms. Smaller sidebands are observed for the silanol groups. The T_1s for these protons are of the same order of magnitude as the sodalite protons and are consistent with internal silanol groups that do not interact directly with O_2.

Shepelev et al [9], using X-ray single-crystal data of dehydrated partially Cs-exchanged NaX, located 36.6 Cs^+/u.c. out of a total of 43.8 cations per unit cell in SIII in the supercages. The rest (7.2 Cs^+/u.c.) were located in SI' and SII'. Our observation at room temperature of only one ^{133}Cs resonance at -101 ppm in the CsX sample, which is assigned to SIII cations, is consistent with their results. Our sample of CsX contains only 32 Cs^+/u.c. Most, if not all, Cs^+ cations must lie in the supercage and therefore, no sodalite-cage Cs^+ are seen by NMR. The Lorentzian line shape at room temperature is characteristic of motion of the cations, which is consistent with the study by Norby et al [2], where cation mobility involving the supercage Cs^+ in CsY, between SIII sites at room temperature, and between SII and SIII sites at higher temperatures, was observed by ^{133}Cs MAS NMR.

In the ^{133}Cs MAS NMR variable tempeature spectra of CsY only the resonances due to Cs^+ in the supercages are affected by O_2. Cs^+ in sodalite cages are not readily accessible to oxygen molecules, hence the corresponding resonances remain unchanged. Use of O_2 to study cation accessibility in LiLSX with ^{6}Li NMR has been recently reported in the ^{6}Li MAS NMR study of LiLSX [10], where a similar observation was made: the resonance due to Li^+ at SIII shifts to higher frequency with higher O_2 concentration and lower temperature and the chemical shift of those cations at SI' and SII are not affected by O_2. The larger shift of supercage Cs^+ in CsX (more than 300 ppm with prolonged acquisition) than that of CsY (about 200 ppm) is consistent with increased O_2 sorption in zeolite X.

Shifts in the MAS NMR of paramagnetic materials result from two different shift mechanisms. The dipolar shift mechanism, a through-space interaction, can give rise to sizable shifts if the susceptibility of the paramagnet is anisotropic [11]. The through-bond, Fermi-contact shift mechanism is a measure of the unpaired electron spin density at the site of the nucleus [12], and in this system, requires an overlap between the O_2 (anti-bonding) orbitals and the proton or cesium s-orbitals. The very large ^{133}Cs shifts are too large to originate from a dipolar mechanism only and must contain a contribution from the Fermi-contact shift mechanism. The observation of Fermi-contact provide evidence for the formation of, presumably, very short-lived Cs^+-O_2 complexes at low temperatures.

CONCLUSIONS

Large changes in the variable temperature NMR of O_2-sorbed porous materials are seen for sites that are directly accessible to O_2. These results suggest novel methods for probing O_2

binding in zeolites and molecular sieves, and for probing accessibility of pores to small gases in a wide range of porous materials. Experiments are underway to apply the method to some of these systems.

ACKNOWLEDGEMENTS
Support from the NSF is acknowledged (DMR 9458017 and CHE 9405436).

REFERENCES
1. D.J. Cookson, B.E. Smith, J. Magn. Reson., **63**, 217 (1985). J. Klinowski, T.A. Carpenter, J.M. Thomas, J. Chem. Soc., Chem. Commun., 956 (1986).
2. P. Norby, F.I. Poshni, A.F. Gualtieri, J.H. Hanson and C.P. Grey, J. Phys. Chem. B, **102**, 839 (1998).
3. C.P. Grey, A.J. Vega, J. Am. Chem. Soc., **117**, 8232 (1995).
4. E. Brunner, et al. *Stud. Surf. Sci. Cat.*, **84**, 357 (1994). L. Beck, J.L. White and J.F. Haw, J. Am. Chem. Soc., **116**, 9657 (1994). D. Freude, Chem. Phys. Lett., **235**, 69 (1995).
5. H. Koller, B. Burger, A.M. Schneider, G. Engelhardt and J. Weitkamp, Microporous. Mater., **5**, 219 (1995).
6. A. Malek, G.A. Ozin and P.M. Macdonald, J. Phys. Chem., **100**, 16662 (1996).
7. D. Zscherpel, E. Brunner and M. Koch, Zeitschrift für Physikalische Chemie, **190**, 123 (1995).
8. L.E. Drain, *Proc. Phys. Soc.*, **80**, 1389 (1962). C.P. Grey, D.Phil Thesis, University of Oxford, (1991).
9. Y.F. Shepelev, I.K. Butikova and Y.I. Smolin, Zeolites, **11**, 287 (1991).
10. J. Plévert, , L.C. de Ménorval, F.D. Renzo and F. Fajula, J. Phys. Chem. B, **102**, 3412 (1998).
11. B. Bleaney, J. Magn. Reson., **8**, 91 (1972).
12. H.M. McConnell and R.E. Roberson, J. Chem. Phys., **29**, 1361 (1958). R.J. Kurland and B.R. McGarvey, J. Magn. Reson., **2**, 286 (1970).

THE ^{13}C MAS NMR DETECTION OF ORGANIC TEMPLATES IN ZEOLITES

M. KOVALAKOVA, B. H. WOUTERS and P. J. GROBET*

Center for Surface Chemistry and Catalysis, Catholic University of Leuven, Leuven, Belgium; piet.grobet@agr.kuleuven.ac.be

ABSTRACT

Cross-polarization (CP) combined with magic-angle spinning (MAS) is the most popular technique used for the high-resolution solid ^{13}C study of occluded templates in zeolitic materials.

Depending on the mobility of the incorporated templates the high-power decoupling (DEC) technique is more suitable to detect and to quantify templates in zeolites. By this technique high-resolution MAS NMR spectra, displaying quantitative reliable ^{13}C signal intensities of the occluded templates, are obtained in ZSM-5 and SAPO-37.

INTRODUCTION

It was soon realized [1] that the presence of organic additives can facilitate the zeolite synthesis. They can act as templating agents, which implies that the zeolite cavities and channels conform closely the shape and contour of the immobilized organic species and the enclosed organic cations influence the Si:Al ratio of the zeolite framework. Two methods are mainly available to characterize the occluded templates, the temperature programmed techniques and the high-resolution solid-state ^{13}C NMR. The latter has the great potential to characterize in-situ the templates. Mostly cross-polarization (CP) combined with MAS NMR is used in the study of templates [2]. An important disadvantage of the CP technique is that the intensities and the detection of the lines in the ^{13}C NMR spectra largely depend on the cross-polarization rate of the individual carbon sites of the occluded species. The enhancement of ^{13}C NMR signals by cross-polarization is based on the magnetization transfer from a proton spin reservoir to the carbon spins of the template molecules and can be influenced by the mobility and the interaction of the templates within the zeolitic environment.

In this work the application of two high-resolution ^{13}C NMR techniques, the CP MAS and the high-power decoupling (DEC) MAS, will be compared in the study of the templates tetrapropylammonium (TPA) in ZSM-5 and SAPO-37, and of tetramethylammonium (TMA) in SAPO-37. We will prove that, depending on the mobility of the incorporated templates the high-power DEC technique is more suitable to detect and to quantify templates in zeolites.

EXPERIMENTAL

The SAPO-37 sample was synthesized from the gel containing TPAOH and TMAOH as templates, according to a recipe described in reference [3]. The ZSM-5 sample was prepared using TPABr as organic according to the synthesis recipe described in reference [4].

The ^{13}C MAS NMR spectra were recorded on a Bruker AMX-300 spectrometer. The ^{13}C CP MAS NMR spectra were acquired with a 6 µs proton 90° pulse, a repetition time of 3 s and contact times between 0.01 and 35 ms; the spinning rate was 4 kHz. The ^{13}C MAS spectra recorded with high-power decoupling (DEC) were carried out with 45° ^{13}C pulses and a recycle time of 10 s, which is sufficient to provide quantitatively reliable spectra. In case of ^{13}C MAS NMR without decoupling the same parameters were used.

RESULTS and DISCUSSIONS

^{13}C CP MAS NMR of TPA$^+$ in ZSM-5

The ^{13}C MAS NMR signals of the carbon sites of the template TPA$^+$ (Fig. 1) are enhanced by CP, which consists of a transfer of magnetization from ^{1}H to ^{13}C in a double resonance experiment during a contact time t_c under the so-called Hartmann-Hahn condition [5]. The signal enhancement by the CP method implies that the signal intensities are influenced by the kinetics of the transfer of magnetization and therefore are not a priori quantitative. Figure 1 shows the growth of the ^{13}C magnetization for the various positions in TPA$^+$ occluded in ZSM-5 as function of the contact time t_c. The splitting of the methyl resonance is attributed to the two types of methyl groups in the different environments of the linear and sinusoidal channels of the ZSM-5 structure [2]. The signal intensities in Figure 1 pass through a maximum and then fall off exponentially with increasing contact time t_c. To allow a quantitative comparison of the signal intensities in the ^{13}C CP MAS spectra, the carbon sites

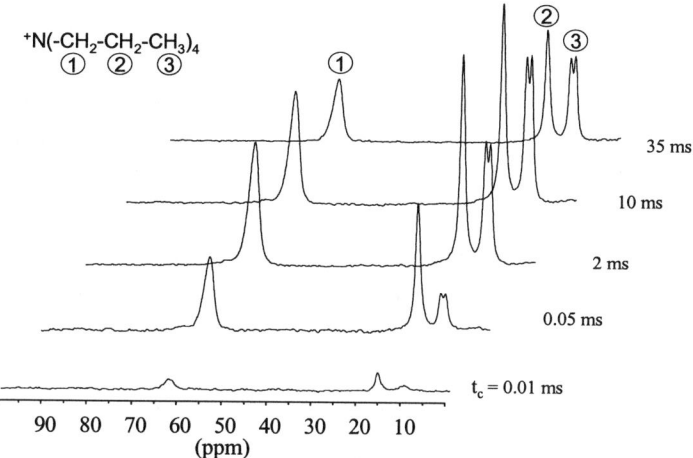

Figure 1. CP MAS ^{13}C NMR spectra of TPA$^+$ in zeolite ZSM-5, at 295 K as function of t_c.

with the slowest cross-polarization rate should be allowed sufficient time to be fully magnetized, via the spin-locked protons, before significant proton relaxation takes place. This seems to be the case for the ^{13}C CP signals of TPA$^+$ in ZSM-5.

^{13}C CP MAS NMR of TPA$^+$ and TMA$^+$ in SAPO-37

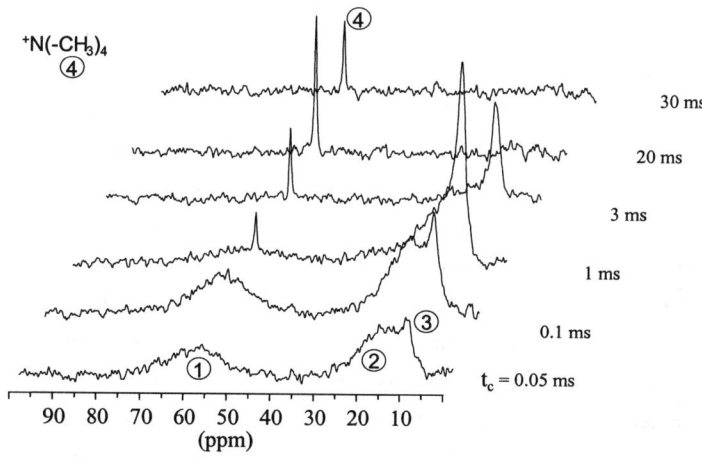

Figure 2. CP MAS ^{13}C NMR spectra of TPA$^+$/TMA$^+$ in SAPO-37 at 295 K with variable t_c.

In the as-synthesized faujasite structure of SAPO-37 there is maximum 1 TMA^+ per sodalite cage and there are between 1 and 2 TPA^+ located in the supercage [6]. The ^{13}C CP MAS NMR spectra of both templates and the signal intensities of the different ^{13}C sites are displayed in Figure 2. Only at short contact t_c the occluded TPA ions are observed; the TMA ions are seen at much longer t_c. The quasi-free rotation of the TMA protons in the sodalite cages, which reduces the ^{13}C-^{1}H dipolar coupling, is at the origin of the slow cross-polarization of the carbon sites in the TMA ions and their reduced line widths. The broadening on sites 1 and 2 of the TPA ions (Fig. 2) is probably due to the failing of the decoupling effect; the mobility of the associated protons induces a broadening on these carbon sites which is of comparable magnitude to the ^{1}H decoupling field. The fast equilibration of the proton spins, surrounding the carbon sites seems to be disturbed by the mobility of the TPA and the TMA ions within the SAPO-37 structure. So the efficiency of the CP technique used on TPA is sharply reduced compared to the CP of the same template in ZSM-5. From the study of templates on SAPO-37 we can conclude that the CP enhancement for occluded but mobile templates inside zeolites is strongly disturbed and reduced, and results in quantitative unreliable signal intensities (Fig. 2).

^{13}C DEC MAS NMR of TPA^+ and TMA^+ in SAPO-37

In order to obtain quantitative NMR data and to detect maximally both templates in one NMR spectrum we apply ^{13}C MAS NMR with and without high-power proton decoupling (DEC) on SAPO-37. In Figure 3 we compared these two techniques with the formerly used CP MAS method (with contact time t_c = 3 ms) on a SAPO-37 sample at room temperature. From the spectra it is clear that the CP technique is less efficient compared to MAS with and without DEC. Even under DEC condition it is not possible to detect the signals of carbon site 1 and 2 of the occluded TPA. But, if one performs the same set of experiments on the same sample at higher temperature, e.g. 383 K, only under DEC MAS condition all carbon sites of the two templates are observable and even quantitative NMR data are obtained. The ^{13}C signals of the three sites of TPA are exactly in a 1/1/1 ratio and the TPA/TMA ratio is very close to the value obtained from thermogravimetric analysis on that same sample. From 383 K on and higher - we went up to a temperature of 403 K, which is still far below the decomposition temperature range of TPA and TMA- the ^{13}C DEC MAS spectrum does not change appreciable.

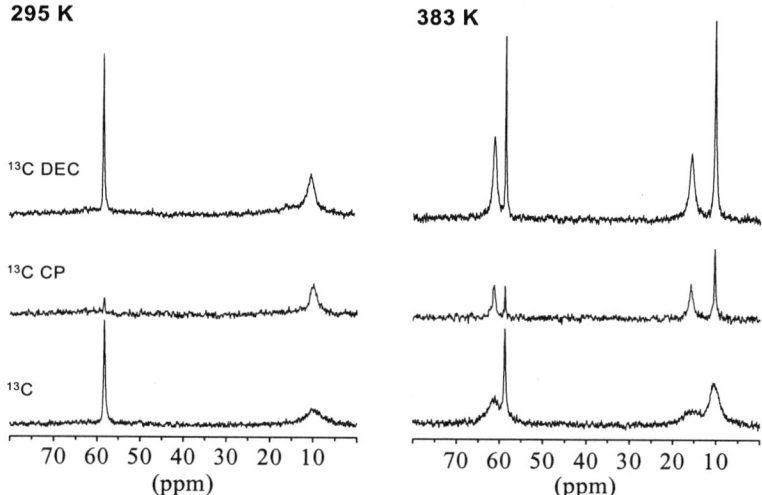

Figure 3. ^{13}C MAS NMR spectra of TPA$^+$ and TMA$^+$ in SAPO-37 taken at 295 K and 383 K, respectively, and under different high-resolution conditions.

At 383 K we suppose that most of the water molecules are desorbed from the supercages and that the occluded TPA template molecules had a higher mobility at this temperature. On the contrary, at room temperature the TPA templates are in close contact with adsorbed water molecules inside the supercages of the SAPO-37 structure and show a more restricted mobility at this lower temperature. The restricted mobility of the TPA ions at 295 K results in line widths larger than those originally from the ^{13}C-^{1}H couplings and so the decoupling is less efficient.

CONCLUSIONS

High-resolution ^{13}C CP MAS NMR was extensively used in the past for the characterization of templates occluded in the voids and channel structures of zeolites. As shown in this study the CP method is not always the most suitable technique. In cases were the templates fit tightly into the zeolitic structure (e.g. ZSM-5) the ^{13}C CP MAS NMR technique is working properly and the highly resolved spectra are quantitatively reliable. As soon as the templates show some mobility inside the molecular sieve the CP is affected and the CP enhancement is sharply reduced; templates become unobservable under certain contact time conditions (e.g. SAPO-37).

To circumvent this problem of the ^{13}C MAS NMR characterization of mobile template species in porous materials, we apply the ^{13}C DEC MAS NMR technique. By performing the experiments at high temperatures, highly resolved and quantitatively reliable spectra are obtained, and all templates are detected. The higher mobility of the templates facilitates the reduction of the ^{13}C -^{1}H dipolar coupling during high-power decoupling.

REFERENCES

1. R.M. Barrer and P.J. Denny, J. Chem. Soc., 971 (1961).
2. G. Engelhardt and D. Michel, High-Resolution Solid-State NMR of Silicates and Zeolites, (John Wiley & Sons, Chichester, 1987), pp. 348-355.
3. J.A. Martens, C. Janssens, P.J. Grobet, H.K. Beyer and P.A. Jacobs, Stud. Surf. Sci. Catal., **49A**, 215 (1989).
4. D.H. Olson, W.O. Haag and R.M. Lago, J. Catal., **61**, 390 (1980).
5. a) A. Pines, M.G. Gibby and J.S. Waugh, J. Chem. Phys., **59,** 569 (1973); b) S.R. Hartmann and E.L. Hahn, Phys. Rev., **128**, 2042 (1962).
6. L. Sierra de Saldarriaga, C. Saldarriaga and M.E. Davis, J. Am. Chem. Soc., **109**, 2686 (1987).

CHARACTERIZATION AND QUANTITATION OF LEWIS-ACID SITES IN SOLID ACIDS BY ^{31}P SOLID-STATE NMR OF THE TMPO COMPLEX

ALAN W. PETERS*, KARL T. MUELLER†, KEVIN J. SUTOVICH*, EDWARD F. RAKIEWICZ*, AND RICHARD F. WORMSBECHER*

*W.R. Grace and Company, Columbia, MD, USA; Alan.Peters@grace.com
†Department of Chemistry, Penn State University, University Park, PA, USA; ktm2@psu.edu

ABSTRACT

Characterization and quantification of both the Brønsted- and Lewis-acid sites of solid acid catalysts are accomplished using trimethylphosphine oxide (TMPO) as a probe molecule for ^{31}P MAS NMR spectroscopy. Earlier ^{31}P MAS NMR results are used as references for the ^{31}P chemical shift values of TMPO probe molecules associated with Lewis- and Brønsted-acid sites. Direct quantification of Lewis site concentrations are made for a sample of dehydroxylated alumina, while in a USY zeolite both Lewis and Brønsted sites are detected and quantified concurrently.

INTRODUCTION

The ability to determine the number of acid sites in zeolites and other solid acids is important for an understanding of the relationship of atomic-level structure to catalytic activity and selectivity. The quantification of the number of Lewis sites has been an especially difficult problem. Thermal methods, temperature-programmed desorption (TPD), and microcalorimetry are either selective for Brønsted sites [1] or do not distinguish clearly between Lewis and Brønsted sites. Infrared methods, depending on the probe used, can have the same difficulty. Even if a band is identifiable as due to an interaction with a Lewis site, relative quantification is often difficult.

In studies utilizing solid-state NMR spectroscopy, probe molecules containing ^{31}P nuclei such as trimethylphosphine (TMP) interact with both Lewis and Brønsted sites and display a range of chemical shifts [2–6]. Since ^{31}P is a spin-1/2 isotope, these results are in principle quantifiable. However, TMP is a dangerous liquid at room temperature, and there are difficulties in handling TMP and in preparing solid acid standards for the quantitation step. Studies of the more stable trimethylphosphine oxide (TMPO) have been reported on amorphous silica-alumina surfaces [3,7], and recently we reported the successful completion of ^{31}P MAS NMR studies of TMPO complexed with acid sites in γ-alumina, a number of Y-type zeolites, and a silica-alumina catalyst system [8]. For these samples, comprehensive and consistent assignments to particular types of sites are made for all resonance lines in the ^{31}P MAS NMR signals from TMPO. Based on results from dehydroxylated γ-alumina, a new chemical shift assignment (37 ppm with respect to aqueous phosphoric acid) was made for a TMPO/Lewis acid complex. Brønsted sites occur with a shift of 50 ppm and higher, and other forms of TMPO are also detected and characterized in the spectra. The assignments of ^{31}P resonances from molecules not directly associated with nearby ^{27}Al nuclei (such as crystalline or physisorbed TMPO species) were supported using ^{1}H/^{31}P/^{27}Al triple-resonance NMR methods. The concentrations of Brønsted acid sites as calculated from the NMR results were compared with concentrations obtained from isopropylamine/temperature-programmed-desorption (IPA/TPD) measurements, and substantial agreement between the methods was found.

Here we report further studies of acid sites in two Lewis-acid systems, γ-alumina and an ultrastable Y zeolite (USY). In both samples, quantitative calculations of Lewis-acid site concentrations are accomplished. For the USY sample, the concentrations of two different Brønsted-acid sites are also calculated from the NMR results and compared with results obtained from IPA/TPD measurements. The differences in the values measured by these two complimentary methods are briefly discussed.

EXPERIMENTAL

A commercial sample of Catapal A alumina obtained from Condea (Houston, TX), was calcined for 3 hours at 705°C to prepare a γ-alumina sample with 201 m^2/gram surface area. A sample of framework-dealuminated Y (USY) was prepared by hydrothermal dealumination of a sample of NaNH$_4$Y for 2 hours at 700°C. The USY had a surface area of 744 m^2/gram and a unit cell size of 2.452 nm. The elemental composition was 4.4% sodium oxide, 21.8% alumina, and the balance was silica. Prior to characterization by solid-state NMR, samples were prepared according to the following procedure for loading with TMPO. Approximately 0.8 g of solid was calcined at 600°C - 800°C for 4 hrs in a round bottom flask. After lowering the oven temperature to 150°C, the flask was capped with a rubber septum. A measured amount of a dry TMPO/CH$_2$Cl$_2$ solution of known concentration was added to the flask via a gas-tight syringe, and the mixture was agitated overnight under N$_2$ on a mechanical shaker. The solvent was removed under vacuum and the flask transferred to a glove box where the product was placed into a vial and sealed. Prior to analysis, the prepared solid samples were packed into dry MAS rotors of the Chemagnetics Pencil design. Either 5.0 mm or 7.5 mm rotors with sealed endcaps were used to hold the samples within the NMR probes. The ^{31}P chemical shifts were referenced to an external sample of solid phosphomolybdic acid (Fisher Scientific Company) which was assigned a shift value of –6 ppm with respect to aqueous 85% H$_3$PO$_4$.

For quantitative determination of the number of acid sites using NMR measurements, it is necessary to first measure the amount of phosphorus in a known mass of a sample. Bulk phosphorus quantification was carried out by sample digestion followed by standard ICP analysis. A 0.1 gram sample of the TMPO-loaded solid acid was dissolved in 10 mls of a 1:1:1 mixture of

HF, HNO$_3$, and HCl until the solution became clear. A further 60. mls of a 5.3% boric acid solution was added to the cool solution, and the ICP analysis was carried out using matched phosphorus standards.

To compare the NMR results with another well-accepted method for analyzing the concentration of Brønsted acid sites, a slightly modified version [9] of the IPA/TPD method described by Kofke et al. [1] was used.

RESULTS AND DISCUSSION

The ^{31}P MAS NMR spectra of crystalline TMPO reveal signals with an isotropic chemical shift of 39 ppm and a wide sideband pattern corresponding to a large ^{31}P chemical shift anisotropy (see Figure 1). The sideband intensity pattern at a given spinning speed allows positive identification of this phase. Earlier ^{31}P MAS NMR results [8], obtained after loading TMPO onto a range of solid acids, are used as references for determining the ^{31}P chemical shift values of probe molecules associated with Lewis- and Brønsted-acid sites. Lewis sites occur at a shift of 37 ppm in these materials, Brønsted sites occur with a shift of 50 ppm and higher, and other forms of TMPO (crystalline and physisorbed) are also detected and characterized in the spectra.

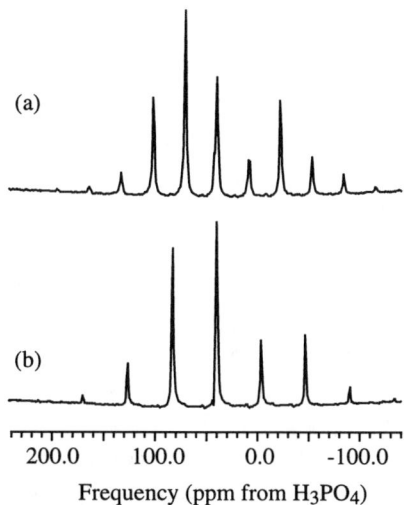

Figure 1. (a) ^{31}P MAS NMR spectrum of TMPO recrystallized from CH$_2$Cl$_2$, obtained with a MAS spinning speed of 5.0 kHz. (b) ^{31}P MAS NMR spectrum of TMPO recrystallized from CH$_2$Cl$_2$, obtained with a MAS spinning speed of 7.0 kHz.

It was also found previously that addition of a small amount of water to solid TMPO produces a new ^{31}P resonance at 41 ppm lacking a broad sideband pattern. The characteristic pattern of crystalline TMPO also remains, representing the fraction of the sample unaffected by the water. It is also important to note that with an excess of deionized distilled water (pH 7), the ^{31}P resonance from TMPO appears at 53 ppm. The shift is independent of solution pH between slightly basic to a pH of 1. In a concentrated (12M) HCl solution the resonance moves downfield to 83 ppm. In an excess of water, the resonance at 53 ppm is characteristic of the complex between the TMPO and a protonated oxygen (water), while the complex at 83 ppm is interpreted as the protonated TMPOH$^+$ species with essentially complete transfer of the proton to the TMPO. The subtle change in the spectra after the hydration of a number of solid acid/TMPO systems has also been reported [8], and it appears essential to carefully avoid the contamination of any samples with water for proper analysis of acid-site concentrations using TMPO as a probe molecule.

Catapal Alumina

A dehydroxylated calcined Catapal A alumina sample was prepared with two different loadings of TMPO. This form of alumina (γ-alumina) is a representative source of Lewis acid sites, as a large proportion of surface hydroxyl groups are removed during high-temperature treatment [10]. The solid-state ^{31}P MAS NMR spectra of samples loaded with 0.78 mmol and 2.3 mmol of TMPO/g of solid are shown in Figure 2. At low TMPO loading (Figure 2a), there is a major signal in the spectrum at 37 ppm. At higher loading (Figure 2b) a second peak appears with an isotropic chemical shift of 39 ppm and a wide sideband pattern matching that of crystalline TMPO. The close correspondence of the wide pattern to that observed from crystalline TMPO demonstrates that crystalline TMPO readily forms in these systems in a bulk phase after the titration of the Lewis acid sites. The ^{31}P NMR spectra of both TMPO/alumina samples also show the presence of a broad minor signal at approximately 53 ppm, while all other peaks in the spectrum correspond to spinning sidebands of the resonances at 37 and 53 ppm.

Analysis by ICP of the sample initially loaded with 2.3 mmol of TMPO/g yields an actual final loading of 1.56 mmol total P per gram of sample. The Lewis site concentration is obtained by measuring the integrated intensity of the peak at 37 ppm (and any identifiable spinning sidebands), and the total intensity from all peaks including sidebands is normalized to the total measured phosphorus content. From this analysis, it is calculated that there are 400. μmol of Lewis sites/gram on the measured surface area of 201 m^2/gram of the alumina sample. To compare this result to other measurements, previous determinations by Sang and coworkers [5] reported concentrations of 116 – 231 μmol of Lewis sites/gram of catalyst in chlorided γ-alumina treated with TMP and degassed over a range of different temperatures.

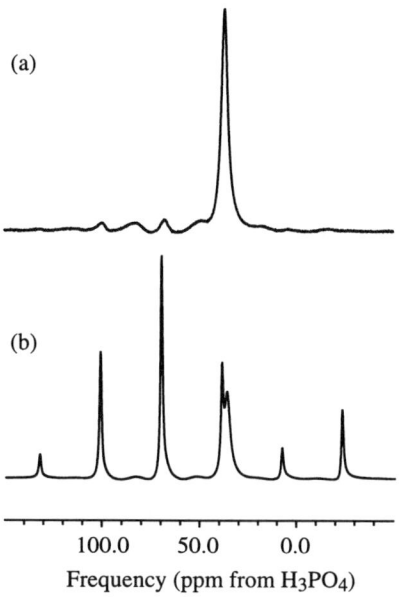

Figure 2. (a) ^{31}P MAS NMR spectrum of γ-alumina loaded with 0.78 mmol TMPO/g solid. (b) ^{31}P MAS NMR spectrum of γ-alumina loaded with 2.3 mmol TMPO/g solid. Spinning speeds for both spectra are 5.0 kHz.

USY Zeolite

The ^{31}P NMR spectrum of a sample of USY loaded initially with 3.00 mmol TMPO/g solid (1.9 mmol/g of solid by ICP analysis) is shown in Figure 3a. A broad resonance at 63 ppm is detected, along with a resonance at 53 ppm, and both are assigned to Brønsted acid sites. Confidence in the assignment of the peak at 53 ppm depends upon total exclusion of water from this system both during the loading of TMPO and the analysis by ^{31}P MAS NMR. The strong peak at 43 ppm is due to physisorbed TMPO, and the peak at 39 ppm with a strong sideband pattern is due to crystalline TMPO. The peak at 37 ppm (more easily seen in the expanded view of Figure 3b), which does not display a strong sideband pattern, occurs at the same chemical shift value as the Lewis-acid site in γ-alumina, and is therefore assigned to a TMPO/Lewis acid complex. Deconvolution and subsequent integration of this complex line shape gives 400. μmol/g of Brønsted acidity contributing to the resonance at 63 ppm, 270. μmol/g of Brønsted acidity at 53 ppm, and 90. μmol/g of Lewis acidity from the resonance at 37 ppm.

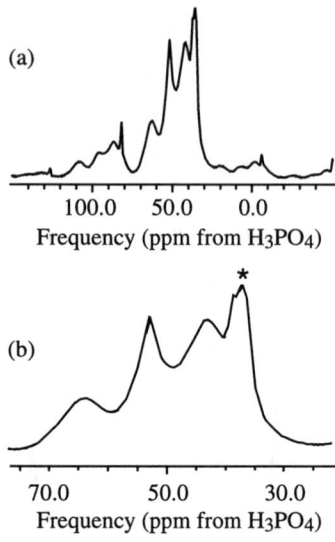

Figure 3. (a) Full ^{31}P MAS NMR spectrum of USY zeolite loaded with 3.00 mmol TMPO/g solid. (b) Expanded view of the above spectrum showing the upfield resonance (marked with an asterisk at 37 ppm) assigned to a TMPO/Lewis-acid complex.

The amount of Brønsted acidity measured by IPA/TPD analysis for the USY sample was 560. μmol/g, and this should be compared to the total Brønsted acidity of 670. μmol/g calculated using TMPO in the above ^{31}P MAS NMR analyses. One possibility for the differences in these numbers is that the sites counted by the resonance at 53 ppm in the TMPO analysis may not all be active for cracking IPA during the TPD experiment. As discussed above, the position of the ^{31}P resonances from protonated TMPO species tend to move downfield as the degree of proton transfer increases, and therefore the resonances at higher frequency are most likely due to stronger protonic acid species. The weaker protonic species, giving rise to the peak at 53 ppm, may have reduced acidic strength or behavior in these materials. Any water present in this system will also cause an increase in the intensity of signal at 53 ppm.

CONCLUSIONS

It has been shown that TMPO is a useful acidity probe in solid acid systems such as zeolites. TMPO forms several types of complexes characterized by different chemical shifts from association with protons and proton donors as well as with Lewis acid sites. The amount of each type of complex, including the complex with Lewis acid sites, can be quantified based on deconvolution and integrated ratios from ^{31}P MAS NMR spectra. In a γ-alumina system with a surface area of 201 m^2/g, the measured concentration of Lewis-acid sites was 400. μmol/g of sample. In a USY zeolite system, both Lewis- and Brønsted-acid site populations were quantified concurrently yielding concentrations of 90. μmol/g and 670. μmol/g respectively.

ACKNOWLEDGMENTS

The authors wish to thank J. Swain for conducting the IPA/TPD experiments, A. Nadjadi for assistance in sample preparation, and J. Burnham for ICP analyses. This report is based in part upon work supported by the National Science Foundation under Grant No. DMR-9458053 and an Arnold and Mabel Beckman Foundation Young Investigator Award to KTM.

REFERENCES

1. T. J. G. Kofke, T. J. Gorte, G. T. Kokotailo, and W. E. Farneth, J. Catal. **115**, 265 (1989).
2. J. H. Lunsford, W. P. Rothwell, and W. Shen, J. Am. Chem. Soc. **107**, 1540 (1985).
3. L. Baltusis, J. S. Frye, and G. E. Maciel, J. Am. Chem. Soc. **109**, 40 (1987).
4. T.-C. Sheng and I. D. Gay, J. Catal. **145**, 10 (1994).
5. H. Sang, H. Y. Chu, and J. H. Lunsford, Catal. Lett. **26**, 235 (1994).
6. L. Baltusis, J. S. Frye, and G. E. Maciel, J. Am. Chem. Soc. **108**, 7119 (1986).
7. G. E. Maciel and P. D. Ellis in NMR Techniques in Catalysis. Edited by A. T. Bell and A. Pines (Marcel Dekker, Inc., New York, 1994) pp. 231-309.
8. E. F. Rakiewicz, A. W. Peters, R. F. Wormsbecher, K. J. Sutovich, and K. T. Mueller, J. Phys. Chem. B **102**, 2890 (1998).
9. M. Juskelis, A. W. Peters, J. P. Slanga, and T. G. Roberie, J. Catal. **138**, 391 (1992).
10. J. B. Peri, J. Phys. Chem. **69**, 211 (1965).

T–O–T FRAMEWORK AND LIGAND VIBRATIONS FOR CHARACTERIZATION OF Co(II) ION COMPLEXATION IN HIGH SILICA ZEOLITES

Z. SOBALÍK, Z. TVARŮŽKOVÁ and B. WICHTERLOVÁ

J. Heyrovský Institute of Physical Chemistry, Academy of Sciences of the Czech Republic, Dolejškova 3, 182 23 Prague 8, Czech Republic; sobalik@jh-inst.cas.cz

ABSTRACT

Dehydration of several cobalt containing zeolites (Co-mordenite, Co-ZSM-5, Co-ferrierite, and Co-beta zeolite) has been shown to give formation of a new IR band at the infrared transmission window between 980 and 850 cm^{-1}. This new band has been assigned to a shifted antisymmetric mode of the T–O–T stretching vibrations due to a local deformation of the zeolite framework adjacent to the Co ion (*deformation shift*). Formation of a complex of cobalt with an extraframework ligand decreased the local lattice deformation accompanied by a reverse change of the position of this band (*relaxation shift*). Interaction of NO, NO_2, NH_3, H_2O and CO with dehydrated CoNa- and CoH-ferrierites (Co/Al from 0.05 to 0.42) was studied in detail by FTIR in the range of 4000 to 400 cm^{-1}. The value of the "relaxation shift" for various Co-extraframework ligand complexes in ferrierite has been established and the shift has been found to correlate well with the strength of the cobalt-ligand bonding.

Three Co ion which varied with Co concentration, with the main band at 918 cm^{-1}, and a lower intensity bands at 942 and 905 cm^{-1} were identified as Co ions of the types α, β and γ in H-ferrierite. A difference in the relaxation shift for α, β sites and preference for dinitrosyl complexes formation was observed upon NO adsorption.

By an *in situ* experiment, this approach could provide a description of the state of the mixed framework polyoxygen ligand–cobalt–NO_x complex under conditions relevant for NO selective reduction by methane.

INTRODUCTION

Siting-coordination-bonding of metal ions and cation-ligand complexes in high silica zeolites is generally accepted to play a decisive role in their function in deNOx processes. Importance of such data for elucidation of the SCR NO_x–CH_4 reaction mechanism over cobalt exchanged zeolites is of crucial importance. Experimental problems accompanying the XRD analysis of such low-metal cation containing samples are difficult to overcome and there is a general lack of structure sensitive methods with a potential for in-situ experiments.

Recently, it has been shown that the narrow infrared "transmission window" region at 1000 to 850 cm^{-1}, carries plentiful information on the identity of a cation interacting with the zeolite framework, and even on interaction with external ligands. The spectral features in this region were assigned to a split of the antisymmetric internal vibration of the T–O–T bonds into several components due to presence of a "bare" cation or cation-ligand complexes[1-5]. It has been shown that the position of the infrared band formed upon dehydration of several metal exchanged ferrierites (Ni, Co, Mn, Mg, Cu) correlates well with the heat of hydration of the cation and is regularly shifted with the changing of ammonia coverage of the cations[5].

The aim of this presentation is to evaluate the potential of this experimental approach for description of the bonding of the "bare" cobalt cation in various zeolites and formation of cobalt complexes with catalytically relevant extraframework ligands (e.g. NO, NO_2, NH_3, CO, H_2O) as well as for description of interaction complexes under conditions of NO selective reduction by methane.

EXPERIMENTAL

Co-Beta (Si/Al 11 , Co/Al 0.3), Co-mordenite (Si/Al 9 , Co/Al 0.21), CoH-ZSM-5 (Si/Al 11, Co/Al 0.22), and Co-ferrierites (Si/Al 8.4, Co/Al 0.1 - 0.3) were prepared by Co^{2+}ion exchange from diluted Co acetate or nitrate solutions with NH_4 or Na form of zeolites. Prior to spectra monitoring the zeolites were heated in vacuum at 480^{0}C and/or with subsequent adsorption of gases (NO, CO, NO_2, NH_3 and H_2O). To indicate the stability of the Co cations in individual cationic sites of the CoH-ferrierite some samples were prior to measurements calcined at about 550 ^{o}C for 3 hours and rehydrated on air after cooling to room temperature. FTIR spectra of samples in the form of self-supported pellets were recorded on a Magna-IR System 550 FTIR Nicolet using heatable cell connected to a vacuum/gas manifold.

RESULTS AND DISCUSSION

Increasing intensity of the new bands at the "transmission window" region, accompanying dehydration of Co-zeolite and creating an increasing amount of the bare cobalt cations, is depicted in Figure 1 for a CoNa-ferrierite.

The absorbance of the new band amounts up to 5 percent of the total absorbance T–O–T antisymmetric band, which agrees well with a low proportion of the T-atoms influenced by the adjacent Co cation at low Si/Al and Co/Al values. Position of the dominating band at the window region observed after high dehydration reflects a "deformation shift" of the T–O–T antisymmetric band of the parent zeolite due to framework perturbation.

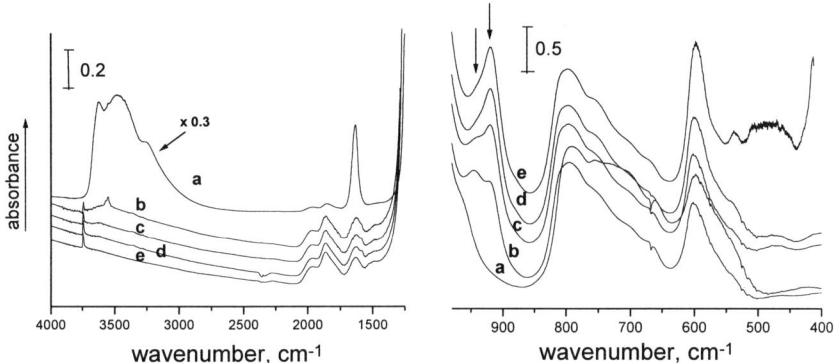

Figure 1. Infrared spectra of the fresh CoNa -ferrierite (Co/Al 0.22) (a) and after its evacuation at 150 (b), 250 (c), 350 (d), 480 (e).

Positions of the new band for various Co-zeolites (with the position of the unperturbed band given in brackets) have been found as follows: Co-ferrierite at 918 cm^{-1} (1070 cm^{-1}), Co-ZSM-5 at 942 cm^{-1} (1105 cm^{-1}), Co-beta 915 cm^{-1} (1093 cm^{-1}), Co-mordenite at 932 cm^{-1} (1075 cm^{-1}). This represents shift of about 140-180 cm^{-1} relative to the position of the main T–O–T band component of the zeolite unperturbed by a bare cobalt cation.

On dehydrated Co-ferrierites depending on Co concentration, three different Co ion bonding to framework oxygens were indicated, with the main band at 918 cm^{-1}, and additional two bands at 942 and 905 cm^{-1} of a lower intensity, which varied with Co concentration. Occupation of the three different Co ions could be estimated from the intensity of the individual T-O-T bands, and the relative concentrations of the individual Co ions before and after high temperature pretreatment has been established (see Figure 2). A sound agreement between the relative concentrations of the individual Co ions obtained by IR and by the UV-VIS technique [6] encourage us to identify the Co ions reflected at the IR bands at 942, 918 and 905 cm^{-1}of with the Co ions of the types α, β and γ in H-ferrierite obtained by analysis of the UV-VIS spectra.

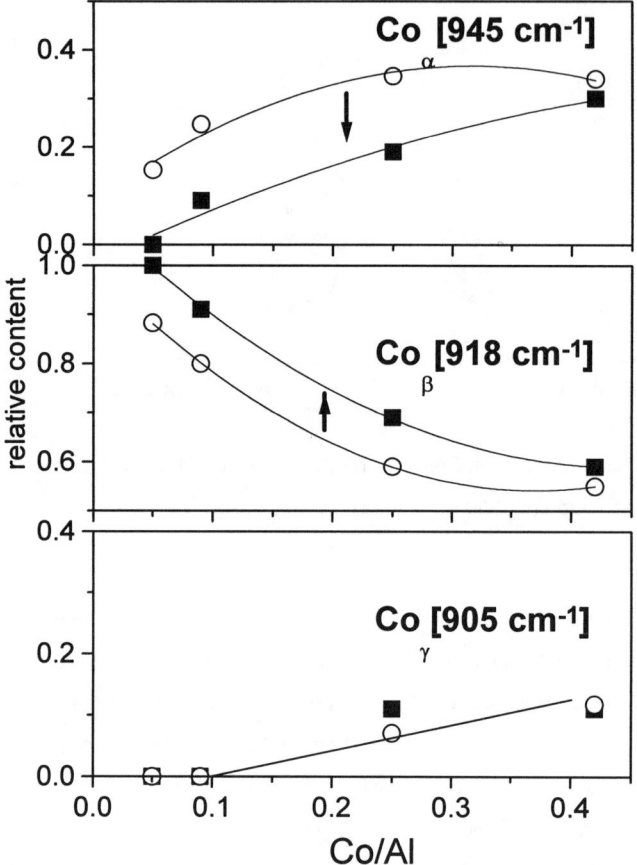

Figure 2. Relative content of the Co ions of the types α, β and γ in H-ferrierite prior (○) and after (■) high temperature pretreatment.

As analysed in detail on CoH- and CoNa-ferrierite the position of the new band is shifted in the reverse direction, i.e. towards the main T–O–T antisymmetric band, after cobalt ion interaction with an extraframework ligand. As shown in Figure 3, the process could be followed in the time scale. Formation of a single isosbestic point (here shown for NO and NO_2 adsorption) reflects transformation of a bare Co ion into Co-ligand complexes and is accompanied by a strictly parallel increase of absorbance of infrared bands of cobalt-cation complexes.

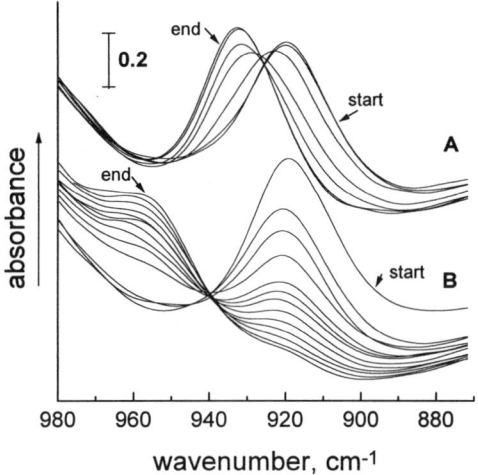

Figure 3. Time-resolved FTIR spectra of NO (20 Torr) (A) and NO_2 (0.5 Torr) (B) adsorption on CoH-ferrierite (Si/Al 8.4, Co/Al 0.05) at the transmission window region. Time scale of the experiment was 1.5 and 15 second for NO and NO_2 adsorption, resp.

Position of skeletal bands and relaxation shifts of various ligands for $Co_\beta Na$- and $Co_\beta H$-ferrierite are summarized in Table I. The relaxation shift for strongly bonded ligands, like H_2O, NO_2 and NH_3, displays a higher value while the slightly bonded NO and CO amount to about 1/2 of their value.

Table I

Ferrierite type	skeletal band (cm^{-1})	relaxation shift for ligands (cm^{-1})				
		NO	CO	NO$_2$	H$_2$O	NH$_3$
CoH-	918	14	12	32	28	26
CoNa-	918	14	16	35	25	26

The change of the cocation to Co^{2+} (H^+ and Na^+) has only small effect and the relaxation shift depends more on the zeolite structure. A high preference of the α site for Co dinitrosyl complexes formation was observed, while under the same conditions NO was bound mostly as mononitrosyls on the Co (β) sites. The deformation shift observed upon NO adsorption on the site α and β differs, with the value for α being about half of the value found for β. Thus a possibility to obtain a site-specific information on the polyoxygen - cation - ligand complexes using this experimental approach in catalytically relevant systems is demonstrated.

ACKNOWLEDGEMENTS

The authors gratefully acknowledge the financial support of the Grant Agency of the Czech Republic (project no. 203/96/1089) and of the Air Products & Chemicals, Inc., Pennsylvania.

REFERENCES

1. W.P.J.H. Jacobs, J.H.M.C. van Wolput, and R.A. van Santen, Zeolites **13**, 170 (1993).
2. J. Sárkány and W.M.H. Sachtler, Zeolites **14**, 7 (1994).
3. G.D. Lei, B.J. Adelman, J. Sárkány, W.M.H. Sachtler, Appl. Catal B: Environmental **5**, 245 (1995).
4. J. Sárkány and W.M.H. Sachtler, Stud. Surf. Sci. Catal. **94**, 649 (1995).
5. Z. Sobalík, Z. Tvarůžková and B. Wichterlová, J. Phys. Chem. B, **102**, 1077 (1998).
6. J. Dědeček and B. Wichterlová, *this meeting*.

CRYSTAL STRUCTURE OF HYDRATED PARTIALLY AND COMPLETELY NH4-EXCHANGED FORMS OF STILBITE

A. ALBERTI [*], A. MARTUCCI [*], M. SACERDOTI [*], S. QUARTIERI [†], G. VEZZALINI [†], P. CIAMBELLI [§], M. RAPACCIUOLO [§]

[*] Istituto di Mineralogia, University of Ferrara, Ferrara, Italy; alb@dns.unife.it; Fax: +39-532-293760
[†] Dipartimento di Scienze della Terra, University of Modena, Modena, Italy
[§] Dipartimento di Ingegneria Chimica e Alimentare, University of Salerno, Fisciano (Sa), Italy

ABSTRACT

Crystal structure refinements of a hydrated partially NH_4-exchanged stilbite and of a hydrated completely NH_4-exchanged stilbite were carried out on X-ray diffraction data of single crystals. Both NH_4-exchanged phases show a strong tendency from the monoclinic real F2/m symmetry towards the Fmmm topological symmetry. Unit cell volume increases by about 1% when compared with the untreated phase. An analogous behaviour has been reported in literature also for dehydrated NH_4-exchanged stilbite and hydrated NH_4-exchanged barrerite. Residual cations occupy the same sites as in the untreated phase. Many extraframework sites, all with partial occupancy, have been located. The correspondence of many of these sites with those occupied by extraframework ions in untreated stilbite suggests that, to some extent, a memory of the initial location of the extraframework atoms remains. A reasonable attribution of these sites to NH_4 and H_2O molecules has been reported.

INTRODUCTION

This work is part of a project whose purpose is to localize Brønsted and Lewis acid sites in natural or synthetic zeolites by diffractometric methods (X-ray and neutrons, on single crystals and powders). The aim of this project is to study not only H-zeolites, but also their precursors obtained by cationic exchange with NH_4.

This paper is to study the crystal structure of hydrated forms of stilbite, partially and completely exchanged with NH_4. Structural studies of the dehydrated Na/NH_4 exchanged form and of the dehydrated NH_4-exchanged form of stilbite were carried out by Mortier [1] and Pearce et al. [2], respectively.

Stilbite-type minerals are common zeolites with a two-dimensional interconnected channel system. Ten-membered rings of tetrahedra intersect smaller channels with eight-ring apertures. The channel system is readily accessible to small molecules, so that these minerals could have a potential use as molecular sieves or catalysts. The fundamental unit is the 4-4-1-1, which consists of four 5-membered rings of tetrahedra and two 4-rings. These units share a tetrahedral vertex to form chains parallel to the c direction. These chains are joined laterally by T-O-T bridges to form dense silicate layers parallel to (010).

The framework topology [STI] is orthorhombic Fmmm. This is also the real symmetry in stellerite [3], which is the Ca member of stilbite type minerals, whereas in barrerite, the Na-member, the real symmetry is orthorhombic Amma [4]. In stilbite, the (Ca, Na)-member, the real symmetry is monoclinic F12/m1. This non-standard space group is commonly used for stilbite to facilitate comparison between the related structures: stellerite, barrerite and stilbite. These different space groups are therefore related to differences in chemistry which result in different distributions of the extraframework cations, which in turn impose rotation displacements within the framework, in barrerite and stilbite, with consequent lowering of the symmetry [3,4].

EXPERIMENTAL

This study was carried out on single crystals of stilbite from Poona (India). Their NH_4-forms were prepared by exchanging natural zeolite with aqueous solution of NH_4NO_3. Typically, 30 mg of zeolite were exchanged with 150 ml of 0.1 M NH_4NO_3 solution at 70° for about 24 hours to obtain a partially exchanged stilbite (h-CaNH$_4$-S from now on) and for 125 hours to obtain a completely exchanged stilbite (h-NH$_4$-S). The samples were then washed with about 2 lt of bidistilled water, filtered, and dried overnight at 120° C. Electron microprobe analyses made on some crystals of the two batches showed a partial and an almost complete NH_4 exchange, respectively.

The composition of NH_4-exchanged forms of stilbite, determined on the same crystals used for the X-ray data collection by an ARL-SEMQ electron microprobe (wavelength dispersive mode), is reported in Table 1. Diffraction data were collected on a Siemens XP-18RA four circle diffractometer using a rotating anode X-ray generator (MoKα radiation) in the ω-2θ scan mode (2°<θ<35°). Atomic scattering factors for neutral atoms were used; curves obtained by interpolating the Si and Al percentages from chemical analysis were used for T framework sites. Water molecules and extraframework ion occupancies were refined using a combination of three-dimensional electron density synthesis and full matrix least squares techniques. An absorption correction was performed using the DIFABS method proposed by Walker and Stuart [5]. Experimental details for the data collection and structure refinements are given in Table 1. Table

Table 1 Crystal data, data collection, and refinement parameters for h-CaNH$_4$-S and h-NH$_4$-S.

	h-CaNH$_4$-S	h-NH$_4$-S
Crystal composition	(NH$_4$)$_{13.9}$(Na$_{0.79}$K$_{0.18}$Ca$_{2.39}$) (Al$_{19.70}$Si$_{52.28}$)O$_{144}$·42.3H$_2$O	(NH$_4$)$_{17.6}$(Na$_{0.07}$K$_{0.02}$Ca$_{0.92}$) (Al$_{19.58}$Si$_{52.42}$)O$_{144}$·36.6H$_2$O
Space group	F2/m	F2/m
a (Å)	13.669(5)	13.632(5)
b (Å)	18.269(5)	18.235(5)
c (Å)	17.892(3)	17.896(3)
β	89.87(2)	89.56(2)
V (Å^3)	4468	4448
Measured refl.	9313	7069
Independent refl.	5056	4587
Refl. with. Fo> 3σFo	3801	3578
No of parameters	216	216
Final R %	6.2	6.0
Final Rw^2 %	13.6	14.4
No of electrons of extrafr. cations		
From structure refinement	46	19
From chemical analysis	60	19
No of NH$_4$ and H$_2$O molecules		
From structure refinement	50.9	43.0
From chemical analysis	56.2	54.2

2 reports the coordinates of atoms, their occupancies and equivalent temperature factors; interatomic distances are given in Table 3.

DISCUSSION

Pearce et al.[2] and Mortier [1], who respectively studied a dehydrated NH$_4$-exchanged stilbite and a dehydrated Na/NH$_4$ exchanged forms of stilbite from Faeroe Islands found that these exchanged forms maintain the original F2/m symmetry, and show unit cell parameters parameters (a=13.583 Å, b=18.156, c=18.007, β=89.68° [2]; a=13.571 Å, b=18.264, c=18.095, β=89.86° [1]), very similar to those of the natural one (a=13.625 Å, b=18.253, c=17.805, β=89.10° [6]). This implies that the interaction of residual Na and NH$_4$ molecules with the framework is quite weak, and does not cause the collapse of the structure as, on the contrary, occurs in non-exchanged zeolites with stilbite type-framework when dehydrated [7-9]. In hydrated NH$_4$-exchanged forms of stilbites, as well as in hydrated NH$_4$-exchanged barrerite [10], the unit cell parameters are slightly larger than those of the natural ones, with an increase of the unit cell volume of about 1%.

Table 2 Atomic coordinates, occupancy, and thermal parameters for h-CaNH$_4$-S and h-NH$_4$-S.

Atom	h-CaNH$_4$-S x/a	y/b	z/c	Occ.	Ueq or Uiso*	h-NH$_4$-S x/a	y/b	z/c	Occ.	Ueq or Uiso*
Si1	0.3853(1)	0.3075(1)	0.3772(1)	1.0	0.0147(1)	0.3850(1)	0.3077(1)	0.3774(1)	1.0	0.0121(1)
Si11	0.6133(1)	0.3071(1)	0.3756(1)	1.0	0.0153(1)	0.6130(1)	0.3075(1)	0.3756(1)	1.0	0.0131(1)
Si3	0.1988(1)	0.0892(1)	0.4987(1)	1.0	0.0155(1)	0.1986(1)	0.0893(1)	0.4986(1)	1.0	0.0133(1)
Si4	0.1112(1)	0.3162(1)	0.5002(1)	1.0	0.0141(1)	0.1115(1)	0.3156(1)	0.5005(1)	1.0	0.0123(1)
Si5	0.2500(0)	0.2515(1)	0.2500(0)	1.0	0.0187(2)	0.2500(0)	0.2515(1)	0.2500(0)	1.0	0.0156(2)
O1	0.3150(2)	0.3058(1)	0.3031(1)	1.0	0.0350(2)	0.3153(2)	0.3066(1)	0.3041(1)	1.0	0.0303(4)
O11	0.6802(2)	0.3018(1)	0.2993(1)	1.0	0.0345(5)	0.6781(2)	0.3020(1)	0.2992(1)	1.0	0.0314(5)
O3	0.3725(2)	0.2316(1)	0.4242(1)	1.0	0.0338(5)	0.3721(2)	0.2316 1)	0.4243(1)	1.0	0.0306(4)
O31	0.6270(2)	0.2328(1)	0.4249(1)	1.0	0.0354(5)	0.6275(2)	0.2337(1)	0.4259(1)	1.0	0.0334(5)
O4	0.3577(2)	0.3790(1)	0.4268(1)	1.0	0.0396(6)	0.3587(2)	0.3786(1)	0.4265(1)	1.0	0.0381(6)
O41	0.6448(2)	0.3815(1)	0.4217(1)	1.0	0.0381(5)	0.6437(2)	0.3829(1)	0.4210(1)	1.0	0.0362(5)
O7	0.4979(2)	0.3164(1)	0.3480(1)	1.0	0.0296(4)	0.4972(2)	0.3166(1)	0.3484(1)	1.0	0.0272(4)
O8	0.3159(2)	0.1136(1)	0.4995(1)	1.0	0.0327(5)	0.3159(2)	0.1142(1)	0.4992(1)	1.0	0.0298(4)
O9	0.1917(2)	0.0000	0.5006(2)	1.0	0.0254(5)	0.1914(2)	0.5000	0.4979(2)	1.0	0.0314(6)
O10	0.0000	0.3504(2)	0.50000	1.0	0.0559(1)	0.5000	0.3496(2)	0.50000	1.0	0.0258(5)
Ca	0.4971(10)	0.0000	0.2913(3)	0.24	0.0559(1)	0.499(2)	0.0000	0.2908(7)	0.07	0.0559(1)
Na1	0.715(3)	0.066(2)	0.284(2)	0.05	0.068(8)	0.716(2)	0.044(6)	0.305(2)	0.04	0.09(2)
W1*	0.431(2)	0.111(2)	0.307(2)	0.18	0.103(8)	0.381(4)	0.096(2)	0.311(2)	0.14	0.075(9)
W1p	0.604(6)	0.100(2)	0.316(2)	0.13	0.077(8)	0.617(3)	0.100(3)	0.316(3)	0.08	0.048(9)
W2*	0.497(1)	0.1330(6)	0.3060(5)	0.43	0.064(2)	0.495(1)	0.1347(6)	0.3053(6)	0.30	0.043(2)
W3*	0.424(3)	0.0000	0.415(2)	0.33	0.126(9)	0.417(6)	0.0000	0.406(4)	0.28	0.14(2)
W3p	0.560(3)	0.0000	0.401(2)	0.32	0.15(1)	0.574(5)	0.0000	0.413(3)	0.25	0.11(1)
W5*	0.4985(7)	0.5000	0.3826(6)	0.45	0.065(2)	0.4978(6)	0.5000	0.3817(5)	0.45	0.036(1)
W6*	0.360(2)	0.5000	0.303(2)	0.28	0.126(9)	0.394(3)	0.5000	0.325(2)	0.18	0.064(7)
W6p	0.613(2)	0.5000	0.319(2)	0.26	0.066(5)	0.560(3)	0.5000	0.319(2)	0.18	0.055(7)
W8*	0.545(3)	0.5000	0.337(2)	0.14	0.060(7)	0.546(4)	0.5000	0.343(3)	0.16	0.069(9)
W10	0.511(3)	0.0000	0.424(2)	0.28	0.079(6)	0.478(5)	0.0000	0.426(3)	0.28	0.087(8)
W11	0.419(9)	0.0000	0.513(8)	0.12	0.20(4)	0.526(5)	0.0000	0.440(3)	0.29	0.10(1)
W11p	0.527(9)	0.0000	0.528(7)	0.14	0.16(3)					
W12	0.543(2)	0.124(1)	0.3053(9)	0.19	0.044(3)	0.530(2)	0.1297(9)	0.3087(9)	0.27	0.027(3)
W13	0.414(3)	0.5000	0.328(2)	0.14	0.053(7)					
N1	0.340(1)	0.075(1)	0.324(1)	0.40	0.100(4)	0.336(2)	0.075(1)	0.326(2)	0.38	0.085(5)
N1p	0.664(2)	0.067(1)	0.332(1)	0.33	0.098(5)	0.667(2)	0.070(1)	0.331(1)	0.39	0.102(5)
N2	0.505(4)	0.083(4)	0.270(3)	0.07	0.064(12)	0.504(4)	0.072(7)	0.262(3)	0.13	0.07(1)
N3	0.297(3)	0.062(2)	0.293(2)	0.15	0.095(9)	0.303(5)	0.066(3)	0.302(4)	0.11	0.06(1)

*same notation as [13]

Table 3 Selected bond distances for h-CaNH$_4$-S and h-NH$_4$-S.

h-CaNH$_4$-S				h-NH$_4$-S			
T1-O1	1.639(3)	T4-O3	1.626(3)	T1-O1	1.627(3)	T4-O3	1.631(3)
-O3	1.631(3)	-O31	1.617(3)	-O3	1.616(3)	-O31	1.614(3)
-O4	1.623(3)	-O8	1.624(3)	-O4	1.610(3)	-O8	1.618(3)
-O7	1.632(3)	-O10	1.643(2)	-O7	1.620(2)	-O10	1.642(2)
mean	1.631	mean	1.627	mean	1.618	mean	1.626
T11-O11	1.644(3)	T5-O1 [x2]	1.637(2)	T11-O11	1.629(3)	T5-O1 [x2]	1.658(2)
-O31	1.640(3)	-O11[x2]	1.623(2)	-O31	1.638(3)	-O11[x2]	1.633(2)
-O41	1.648(3)	mean	1.630	-O41	1.651(3)	mean	1.645
-O7	1.663(3)			-O7	1.664(2)		
mean	1.649	Ca-W1 [x2]	2.25(3)	mean	1.645	Ca-W1 [x2]	2.40(5)
		-W3p	2.34(3)			-W3p	2.41(6)
T3-O4	1.645(3)	-W1p[x2]	2.38(3)	T3-O4	1.645(4)	-W1p[x2]	2.48(4)
-O41	1.653(3)	-W8	2.37(3)	-O41	1.662(3)	-W8	2.47(4)
-O8	1.662(3)	-W12[x2]	2.36(2)	-O8	1.662(3)	-W12[x2]	2.43(1)
-O9	1.633(2)	-W10	2.39(3)	-O9	1.633(2)	-W10	2.43(4)
mean	1.648	-W3	2.43(3)	mean	1.650	-W3	2.34(7)
		-W2[x2]	2.44(1)			-W2[x2]	2.47(1)
		-W6p	2.48(2)			-W6p	2.39(4)
		-W6	2.57(3)			-W6	2.52(4)
		-W13	2.45(3)				

It is known that stilbite-type minerals in their cationic exchanged forms can change their real space group, moving towards higher symmetry. Barrerite in its Ca-exchanged form [11] and in its hydrated NH_4 form [10] changes its real symmetry from Amma to Fmmm, whereas stellerite in its Na-exchanged form remains Fmmm [12].

The study of hydrated NH_4-exchanged forms of stilbite allows us to tackle many problems of stilbite-type minerals, i. e.

1. does stilbite maintain its real symmetry F2/m even in its hydrated -partial or complete- NH_4-exchanged form?

2. does the location of extraframework sites in these exchanged phases have a "memory" of the location in the original ones? In other words, does the distribution of exchangeable cations in natural stilbite-type zeolites influence, or otherwise, the distribution of NH_4 and H_2O molecules in the NH_4-exchanged form?

3. are there any remarkable differences in the extraframework ions between NH_4-exchanged barrerite and NH_4-exchanged stilbites, considering that the almost complete cation NH_4 exchange makes the extraframework content very similar in both these zeolites?

4. is the distribution of residual cations, of NH_4, and H_2O recognizable in the exchanged phases?

5. is there evidence of N-H⋯O bridges that could be precursors of Brønsted acid sites in the stilbite calcinated H-form?

In answer to these questions, we can make the following observations:

1. Our results confirm the space group F2/m for both h-CaNH$_4$-S and h-NH$_4$-S. However, the crystal structures are strongly pseudosymmetric Fmmm. Many structural features support this result:

a) the β angle, which is 89.24° for natural stilbite from Poona [6], becomes 89.87° in h-CaNH$_4$-S and 89.56° in h-NH$_4$-S.

b) the T5-T5-T5 angle (where the T5-T5 vector is the length of the 4-4-1-1 fundamental unit in the c direction) is very close to 180° (179.3° both in h-CaNH$_4$-S and in h-NH$_4$-S), whereas in natural stilbite it is around 175°. This angle must be 180° in orthorhombic Fmmm symmetry.

c) the Ca atom shifts in the [100] direction towards the orthorhombic pseudo-mirror plane at x=1/2. In fact, x=0.4975 in h-CaNH$_4$-S and x=0.4986 in h-NH$_4$-S, to be compared with x=0.4840 in stilbite from Iceland [11].

d) in both h-CaNH$_4$-S and h-NH$_4$-S each extraframework site has its nearly symmetric site with respect to the orthorhombic pseudo-mirror plane at x=1/2, whereas this pseudo-symmetry does not occur in stilbite.

This result can be explained as follows: in stilbite there are two extraframework cation sites, one on a mirror plane, fully occupied by Ca (or by Ca and Na), the other partially occupied by Na; electrostatic forces between them are strong enough to push the calcium out of the mirror plane, so allowing only F2/m symmetry; in stellerite there is no sodium, and this site is partially occupied by water, so calcium can easily remain on the mirror plane, and the symmetry is Fmmm. In hydrated NH$_4$-exchanged stilbites both cation sites are weakly occupied, so that a Fmmm symmetry is now possible. The Na site, however, differs markedly from its pseudo-symmetric - by mirror plane - site, occupied by water, so that a slight deviation from the topological symmetry is justified. An analogous result was found for NH$_4$-exchanged barrerite [10]: in this case the NH$_4$-exchanged form changes its real symmetry from Amma to Fmmm. This result can be explained considering that the cation site (C3 in the notation of [4]), which imposes rotational displacement within the framework around the screw axes parallel to a, is empty in NH$_4$-barrerite, justifying the restoration of the topological symmetry.

A strong pseudo-symmetry has been found also for dehydrated Na/NH$_4$ [1] and NH$_4$-exchanged forms [2] of stilbite from the Faeroe Islands. The β angle varies from 89.10° of natural stilbite [6] to 89.86° and 89.68° respectively, whereas the T5-T5-T5 angle is 179.9° and 179.7°, respectively. Therefore there is a clear tendency of the structures of both hydrated and dehydrated NH$_4$-exchanged forms of stilbite towards a Fmmm symmetry.

2. In natural stilbite quite a large number of extraframework sites have been located (see Figure 1). Two of these are cation sites, as described above. The other sites (W1...W9 in the notation of [13]), are attributed to water molecules. All these sites, with the exception of W4, have been found in both h-CaNH$_4$-S and h-NH$_4$-S, but are now all partially occupied. It is to be noted that the W4 site is also empty in NH$_4$-exchanged barrerite [10]. Moreover, all sites found in stilbite have their pseudo-symmetric site in both hydrated NH$_4$-exchanged stilbites, and six new sites have been found. If we consider this good correspondence between the extraframework sites found in natural and hydrated NH$_4$-exchanged stilbites, we can conclude that, to some extent, a memory of the initial location of the extraframework atoms remains.

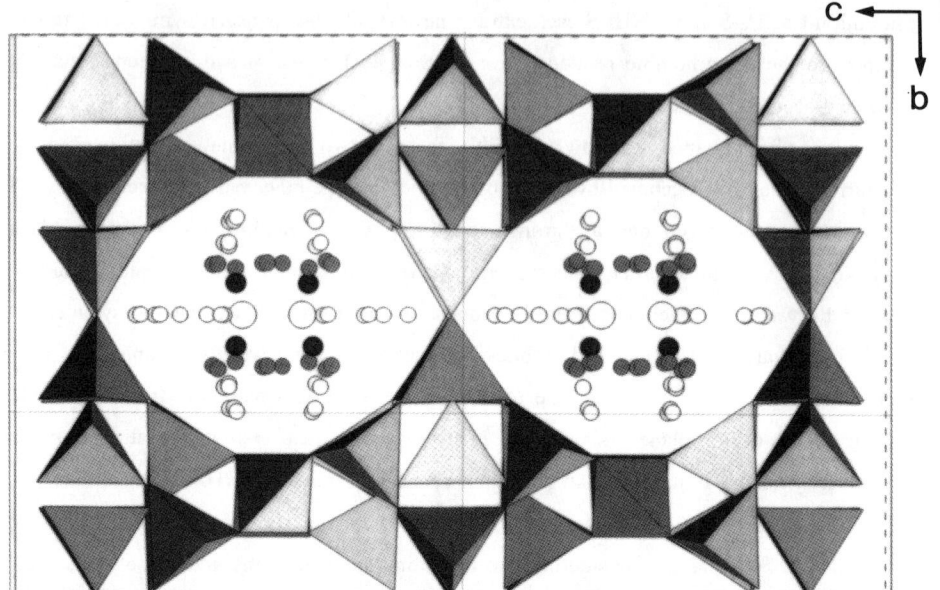

Figure 1. Projection along [100] of the crystal structure of h-NH_4-S. Large empty circles represent Ca sites, small empty circles W sites, small light gray circles N sites, and medium bright gray circles Na sites.

3. A comparison of extraframework sites in NH_4-barrerite and NH_4-stilbites shows that there are strong similarities between the two structures. In both cases, the W4 site is empty, as pointed out before, and new, weakly occupied sites have been located. There is normally a good correspondence between the coordinates of the framework sites, whereas major differences are present in their occupancies.

4. As said before, in both hydrated NH_4-exchanged stilbites, extraframework sites have been found in Ca and Na sites of untreated stilbite with, as expected, low occupancies. The bond distances and coordination indicate clearly that these sites are occupied by the residual extraframework cations. They account for 0.97 Ca atoms for h-NH_4-S and 1.90 Ca and 0.80 Na atoms for h-CaNH_4-S, respectively. It is virtually impossible to distinguish between NH_4 and H_2O molecules, because all sites have partial occupancy and large distances (> 2.8Å) from the framework oxygens. However, it is reasonable to assume that the sites with bond distances less than 2.5 Å from Ca are occupied by water molecules. The quite similar

occupancy of Ca and these sites could be seen as supporting this assumption. It is impossible to determine the true coordination of Ca, owing to the large number of H_2O sites at coordination distance and the very short distances among many of them which prevent their simultaneous occupancy. Following the above assumption, 16.1 NH_4 molecules and 26.9 H_2O molecules were located in h-NH_4-S, and 19.5 NH_4 and 31.4 H_2O molecules in h-CaNH_4-S, which reasonably compare with values given by the chemical analysis reported above. A low value of water molecules found in the structure refinement when compared with that given by the chemical analysis is usual in zeolites characterized by many partially occupied sites in the channels, as occurs in NH_4-exchanged forms of stilbite, and also in NH_4-exchanged forms of barrerite [10].

5. From the structure refinement it was not possible to localize H sites bonded to N atoms. This result was expect considering the low occupancies of N sites (never greater than 40%). We cold only hypothesize that oxygen atoms at short distance from N atoms and possible acceptor of hydrogen in a N-H···O bond could be the precursors of Brønsted acid sites in the calcinated H-form of stilbite. Such bonds involve the oxygens of Si5 tetrahedron (O1 and O11) and O7 (N-O11=3.15Å; N3-O11=2.96; N1p-O1=3.06; N2-O7=2.84). O1 and O11 head toward the channels parallel to [100] delimited by 10-membered ring of tetrahedra, whereas O7 heads toward the 8-ring parallel to [001]; therefore this oxygens could become acid sites. Unfortunately in both dehydrated Na/NH_4 stilbite [1] and NH_4 stilbite [2] any attempt to locate Brønsted sites was unsuccessful.

ACKNOWLEDGEMENTS

The Ministero della Università e della Ricerca Scientifica e Tecnologica is thanked for the financial support to the research program "Relations bewteen structure and properties in minerals: analysis and applications". The Consiglio Nazionale delle Ricerche of Italy is acknowledged for financial support. Thanks are due to "Centro Interdipartimentale Grandi Strumenti" of the University of Modena for the experimental facilities.

REFERENCES CITED

1. W.J. Mortier, Am. Mineral., **68**, 441 (1983).
2. J.R. Pearce, W.J. Mortier, G.S.D. King, J.J. Pluth, I.M. Steel, and J.V. Smith, In Proceedings

of the 5th International Zeolite Conference. Edited by L.V.C. Rees, p.261 (1980).
3. E. Galli, and A. Alberti, Bull. Soc. fr. Mineral. Cristallogr., **98**, 11 (1975a).
4. E. Galli, and A. Alberti, Bull. Soc. fr. Mineral. Cristallogr., **98**, 331 (1975b)
5. N.Walker and D. Stuart, Acta Crystallogr., **A39**, 158 (1983).
6. E. Passaglia, E. Galli, L. Leoni, and G. Rossi, Bull. Mineral., **101**, 368. (1978).
7. G. Cruciani, G. Artioli, A. Gualtieri, K. Stahl, and J.C. Hanson, Am. Mineral., **82**, 729 (1997)
8. Alberti, R. Rinaldi, and G. Vezzalini, Phys. Chem. Minerals, 82, 365 (1978).
9. Alberti, and G. Vezzalini, in Natural Zeolites. Occurrence, Properties, Use. Edited by L.B. Sand and F.A. Mumpton (Pergamon Press, Oxford) pp. 85-98. (1978)
10. Martucci, A. Alberti, M. Sacerdoti, G. Vezzalini, P. Ciambelli, and M. Rapacciuolo, submitted to Natural Zeolites. Occurrence, Properties, Use. Edited by F.A. Mumpton (1998).
11. M. Sacerdoti and I. Gomedi, Bull. Mineral., **107**, 799. (1984).
12. E. Passaglia and M. Sacerdoti, Bull. Mineral., **105**, 338 (1982).
13. S. Quartieri and G. Vezzalini, Zeolites, **7**, 163. (1987).

DETERMINATION OF THE LOCATION OF TEMPLATE MOLECULES IN ZEOLITE EU-1 VIA A COMBINED MOLECULAR MODELLING AND X-RAY DIFFRACTION APPROACH

S. J. ANDREWS[1], J. L. CASCI[2], P. A. COX[3] and M. D. SHANNON[1]

1 ICI Technology, PO Box 8, Runcorn, WA7 4QD, UK.
2 ICI Katalco, RT&E, PO Box 1, Billingham, Cleveland, TS23 1LB, UK.
3 Division of Chemistry, University of Portsmouth, St Michael's Building, White Swan Road, Portsmouth, PO1 2DT, UK; paul.cox@port.ac.uk

ABSTRACT

Molecular mechanics calculations have been used to predict the location and conformation for both hexamethonium and dibenzyldimethylammonium (DBDM) template ions in zeolite EU-1 (EUO). The validity of these predictions has been assessed experimentally using high resolution powder X-ray diffraction. The results show that the modelling study accurately predicts that both types of template molecules occupy the side-pockets in the EU-1 structure. The conformation predicted for DBDM is shown to be in excellent agreement with that obtained from Rietveld refinement of the powder pattern. For the more flexible hexamethonium ion additional rotational disorder is observed suggesting an envelope of cylindrical electron density is present in the side-pockets.

INTRODUCTION

In order to aid the design of novel zeolite frameworks with specific pore sizes and shapes via templating approaches it is important to have an understanding of the relationship between the known zeolite structures and the organic template molecules that synthesise them. Two crucial facts must be known regarding the template itself: its location within the framework and the conformation that it adopts. Precise determination of the location of these template molecules is difficult. For X-ray diffraction the reasons for this are two-fold:-

- Most zeolites occur as micro-crystalline powders precluding single crystal studies. High quality powder diffraction data are essential and conventional in-house facilities cannot normally achieve this.

- The location of small, light molecules using data which are dominated by diffraction from the framework can pose serious problems in the analysis; such problems are compounded by the existence of any disorder or partial occupancy of the extra-framework atoms.

In this study, we report the use of a combination of molecular modelling and X-ray diffraction methods to determine the location of two templates in zeolite EU-1 (EUO). The templates investigated here are:-
(i) dibenzyldimethylammonium (DBDM) $[((C_6H_6)\text{-}CH_2)_2N(CH_3)_2]^+$
(ii) hexamethonium $[(CH_3)_3N(CH_2)_6N(CH_3)_3]^{2+}$

EXPERIMENTAL

X-ray Diffraction

Samples of 'as made' EU-1 were ground, placed in a 1 mm glass capillary and then mounted on the high resolution powder diffractometer in Debye-Scherrer mode on the wiggler beam line station 9.1 at the Synchrotron Radiation Source, Daresbury Laboratory, Cheshire, UK. A Si (111) channel cut monochromator was used to give a wavelength of 1.1017 Å. Data were collected from 5° to 70° 2θ in 0.01° steps and counting for 4.25 s per step. Total data collection time was 10 hours. These data were then corrected for beam decay.

Molecular modelling

Molecular mechanics calculations were performed to predict the location and conformation for a single template ion for DBDM and hexamethonium within EU-1. The energy of the template was minimised with respect to the rigid, fully siliceous form of the zeolite framework. Calculations used the Dreiding force field and parameters within the POLYGRAF modelling package [1].

RESULTS AND DISCUSSSION

EU-1 is characterised by 10-T ring apertures in the main channels and 12-T ring apertures leading to side-pockets. Figure 1 shows the structure of EU-1. The van der Waals map is superimposed to highlight the pore system of the zeolite. The energy minimised position for hexamethonium in EU-1 is found located in the side-pocket of the zeolite. The modelling predicts that the template adopts an 'S-shaped' conformation within the pocket (Figure 2). DBDM is also predicted to reside in the side-pocket (Figure 3).

For the X-ray refinement, the space group for EU-1 and the initial values for the unit cell and atomic co-ordinates were obtained from a previous study [2]. This initial information was used as input into the Rietveld analysis program implemented at Daresbury [3] and the various scan parameters refined together with the atomic positions of the framework atoms. The latter were subject to soft constraints such that all inter-bond angles around the T-atoms were 109.5° and all T-O bond lengths were 1.60 Å.

Convergence resulted in the plot shown in Figure 4, where the observed data (dotted) for EU-1/hexamethonium is shown overlaid by the diffraction pattern calculated for the EU-1 framework (solid). The vertical tick marks below the traces show the positions of the reflections calculated for this cell and space-group. The bottom trace is the difference profile, observed-calculated. It can be seen that there is a considerable mismatch between the two patterns especially in the region between 7° and 14° 2θ, where the calculated pattern is more intense than that observed, and between 14° and 18° 2θ where the converse is true. It is these differences that are directly attributable to the presence of the template molecule in the EU-1 framework.

To test the predictions from molecular modelling, the pattern for the framework with the two types of template molecules docked in the side-pocket have been calculated. An examination of the symmetry present in the side-pocket reveals a potential four-fold degeneracy such that the single theoretical conformation is but a representation since symmetry dictates that (at least) four conformers will exist in superposition to give rise to rotational disorder along the pocket axis ([001]). Consequently, the atomic positions of the hexamethonium moiety were input to the pattern-generating program with each atom having only one quarter occupancy (except for atoms in special positions). The patterns calculated for the theoretical models for hexamethonium and DBDM in EU-1 are shown in Figures 5 and 6. In both cases, agreement between the observed and calculated profiles is very good, supporting the predictions from the modelling study.

Rietveld refinement was performed using the modelling predictions as a starting point. Different behaviour for the two systems was observed. In the case of the EU-1/ hexamethonium system, convergence was not achieved, with the template appearing to 'rotate' around the pocket axis during refinement in an attempt to represent the apparently cylindrical electron density present. The match between the observed and calculated

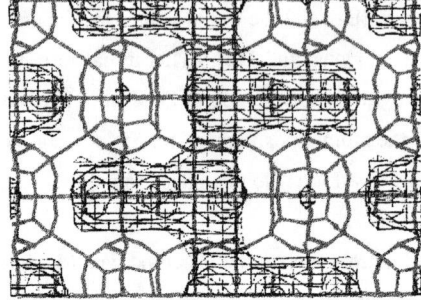

Figure 1. Structure and pore map for EU-1.

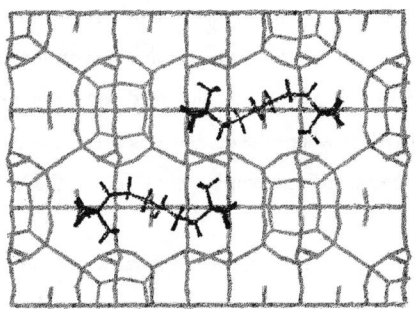

Figure 2. Hexamethonium energy minimised in the EU-1 framework.

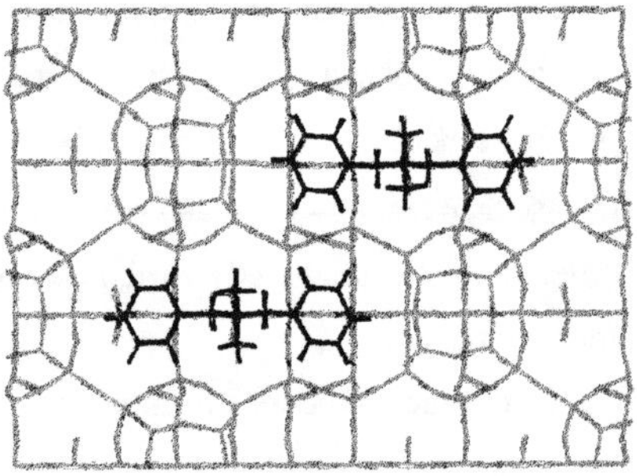

Figure 3. DBDM energy minimised in the EU-1 framework

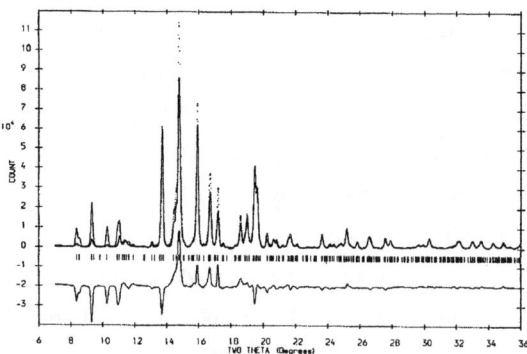

Figure 4. Observed and calculated profiles for the EU-1/hexamethonium system and the EU-1 framework with no template present respectively.

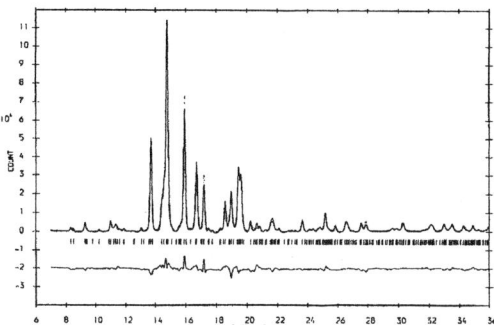

Figure 5. Observed and calculated profiles for the EU-1/hexamethonium system with the hexamethonium located in the position predicted via modelling.

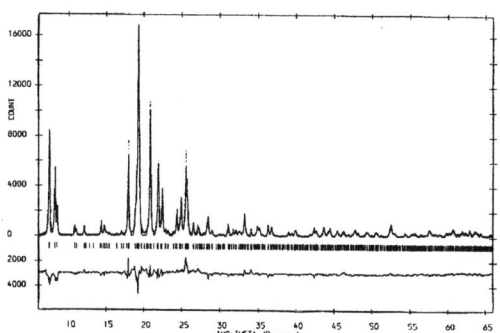

Figure 6. Observed and calculated profiles for the EU-1/DBDM system with the DBDM located in the position predicted via modelling.

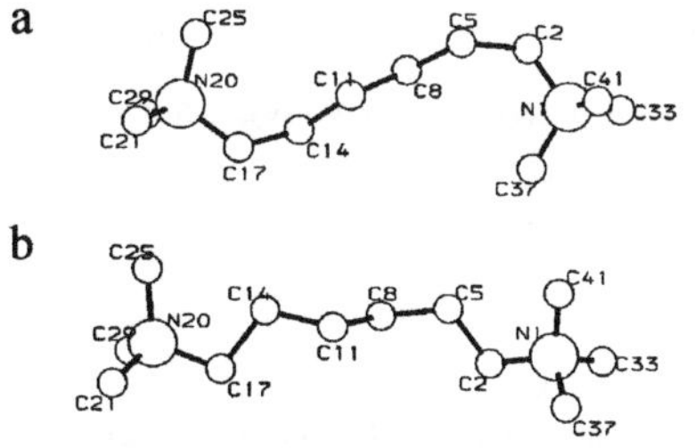

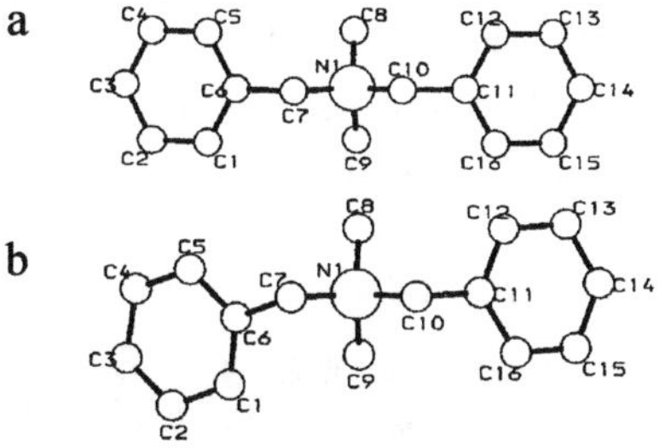

Figure 7. (a) Theoretical conformer for hexamethonium in EU-1 from modelling and (b) representation of refined conformer.

Figure 8. (a) Theoretical conformer for DBDM in EU-1 from modelling and (b) representation of refined conformer.

diffraction patterns is found to be excellent, despite the inherent disorder of the template molecule. However, little improvement is achieved compared to the unrefined model. The refined conformer is shown in Figure 7. This is merely a representation of the molecule for the reasons given above.

In contrast, for the EU-1/DBDM system convergence is achieved in this case with the template immobilised in its calculated position. The match between the observed and calculated diffraction patterns is found to be excellent, despite the inherent disorder of the template molecule, confirming the proposed model. It is emphasised that the refined conformer shown in Figure 8 merely represents one-quarter of the molecule for the reasons given above.

CONCLUSION

The good match between the observed X-ray powder patterns and those calculated using the modelling predictions show that the templates reside in the pockets along the [001] direction in EU-1, confirming previous theoretical predictions [4-6]. Further refinement of the template position against the powder data shows that the single position calculated is but one representation of the inherent, symmetry-required, disorder present along the pocket axis. In the case of hexamethonium further rotational disorder is observed along this axis giving rise to a molecular 'envelope' along the pocket. For DBDM convergence is observed confirming the observation that there is a unique stable site for this template in EU-1.

This work shows that the results of molecular mechanics calculations may be used with a high degree of confidence in the prediction of template location in zeolites. Furthermore, modelling is shown to yield excellent starting points for detailed Rietveld refinements of X-ray powder data.

REFERENCES

1. POLYGRAF from Molecular Simulations Inc., San Diego, USA.
2. N. A. Briscoe, D. W. Johnson, M. D. Shannon, G. T. Kokotailo and L. B. McCusker, Zeolites **8**, 74 (1988).
3. R. J. Cernik, P. K. Murray, P. Patterson and A.N. Fitch, J. Appl. Cryst. **23**, 292 (1990).
4. A. Moini, K. D. Schmitt, E. W. Valyocsik and R. F. Polomski, Zeolites **14(7)**, 504 (1994).
5. A. P. Stevens, A. M. Gorman, C. M. Freeman and P. A Cox, J. Chem. Soc. Faraday Trans. **92(12),** 2065 (1996).
6. D. W. Lewis, C. M. Freeman and C. R. A. Catlow, J. Phys. Chem. **99(28)**, 11194 (1995).

CRYSTAL STRUCTURE OF ZEOLITE FERRIERITE IN AS-SYNTHESIZED, NH_4- AND H-FORMS

G. CRUCIANI*, A. ALBERTI*, A. MARTUCCI*, K. D. KNUDSEN[†], P. CIAMBELLI[§] M., RAPACCIUOLO[§]

*Istituto di Mineralogia, Università di Ferrara, Ferrara, Italy; e-mail: cru@dns.unife.it; fax: +39 532 293760

[†]Swiss-Norwegian Beam Line, ESRF, Grenoble, France

[§]Dipartimento di Ingegneria Chimica, Università di Salerno, Salerno, Italy

ABSTRACT

Synchrotron X-ray Rietveld refinements of low silica ferrierite ($K_{2.7}Na_{1.1}Si_{32.2}Al_{3.8}O_{32} \cdot 12H_2O$) in its as-synthesized, NH_4-exchanged, and rehydrated H-forms were performed in the $P2_1/n$, *Immm*, and *Immm* space groups, respectively. The monoclinic distortion in the as-synthesized form is mainly caused by the strong interaction of the (K,Na) atoms, located near the center of the 8-ring, with the framework oxygens. A residual content of (K, Na) atoms is probably located near the center of the ferrierite cage. The replacement of the (K, Na) cations by NH_4^+ groups at the center of the 8-ring is likely responsible for restoring the (pseudo-) *Immm* average symmetry in the NH_4-exchanged form. This is accompanied by a significant expansion of the cell volume. The removal of NH_3 by calcination and the subsequent rehydration processes cause only slight changes in the crystal structure of the rehydrated H-form of ferrierite; this is consistent with the low contraction of the cell volume and the persistence of the (psuedo-)*Immm* symmetry. The occupancy refinement suggests that water molecules in H-ferrierite are distributed in the same extraframework positions as in the NH_4-form, with only a minimal rearrangement of populations.

INTRODUCTION

Ferrierite is known to be a natural as well as a synthetic zeolite. It is a medium-pore material described as an excellent catalyst for the isomerization of *n*-butane to isobutene[1]. The latter is an important feedstock for the production of methyl *tert*-butyl ether (MTBE), which is a commercial oxygenate additive in unleaded motor fuel.

The ferrierite framework contains two systems of mutually perpendicular one-dimensional channels of 10- and 8-membered rings. The crystal structure of a natural Mg-rich ferrierite has been solved in the orthorhombic *Immm* space group[2], which is also the maximum topological symmetry of the framework. Mg-rich ferrierites are characterized by the presence of a

Mg(H$_2$O)$_6^{2+}$ octahedron at the center of the so-called 'ferrierite-cage'[3,4], namely a [8^{2}6^{2}6^{4}5^8] cage. A monoclinic symmetry (with a metrically orthorhombic unit cell), space group $P2_1/n$, has been reported for a natural Mg-poor, Na-, K-rich ferrierite[5]. The symmetry reduction has mainly been ascribed to the movement of one bridging oxygen away from the inversion center at ¼,¼,¼ site in order to avoid 180° T-O-T bond angles[5]. Although the *Immm* space group has been successfully used in structure refinements of Mg-rich ferrierites, a lowering of the real symmetry to *Pnnm* (unit cell choice: $a > b > c$) has been suggested as due to static disorder of the Mg(H$_2$O)$_6^{2+}$ octahedron[4]. Concerning synthetic zeolites, Rietveld refinements in the *Immm* space group of K-exchanged[6] and Kr-containing[7] ferrierites, using respectively neutron and X-ray resonant powder diffraction, have been reported. The reduction of symmetry from *Immm* to *Pmnn* (the same unit cell choice as above) has been suggested in template-containing all-silica ferrierite[8], and *Pmnn* structure refinements of synthetic all-silica ferrierites have been published using single crystal and Rietveld methods[9-11]. In catalysis the flexibility of zeolite frameworks can play a role in processes like diffusion and adsorption. It is for this reason that so much attention is turned to the definition of the space group of zeolitic material. It is evident that the space group in ferrierite, as well as in other zeolites, varies both within framework and extraframework content.

The technological significance of synthetic ferrierite as shape-selective catalyst is restricted to materials whose Si/Al ratio ranges from 5 to 10[1]. This allows the presence of extraframework cations, namely Na$^+$ or K$^+$, which can be easily replaced by NH$_4^+$. The highly acidic hydrogen form[12], H-ferrierite, is thus obtained by calcinating the intermediate NH$_4$-exchanged form. The potential activity of H-zeolites strongly depends on the location and accessibility of protons with respect to the size of molecules involved in the catalytic reaction.

This study is part of a project aimed at locating proton positions in H-ferrierite by neutron powder diffraction. The present paper is devoted to clarifying the structural modifications that ferrierite underwent during the NH$_4$-exchange and calcination processes. Our particular interest is focussed on the following topics:

1- Determination of the space group of the synthetic low-silica, normally Mg-poor, ferrierite (*Immm*, $P2_1/n$, *Pnnm*, *Pmnn* or others) in the as-synthesized, NH$_4$-exchanged and rehydrated H-forms.

2- A detailed analysis of the distortions undergone by the framework of the as-synthesized ferrierite when it transforms into its NH$_4$- and H-forms.

3- Location of: a) the extraframework cations (K, Na) and water molecules in the as-synthesized form; b) the NH_4^+ groups and water molecules in the NH_4-form; c) the water molecules in the rehydrated H-form.

EXPERIMENTAL SECTION

Room temperature synchrotron X-ray powder diffraction data were collected on three samples of a low-silica ferrierite (Engelhard - ferrierite EZ^{TM}-500) in the following forms: a) 'A-S-FER': the as-synthesized form ($K_{2.7}Na_{1.1}Si_{32.2}Al_{3.8} \cdot 12H_2O$; Si/Al=8.5); b) 'NH$_4$-FER': the NH$_4$-form, as exchanged in a 1M solution of NH_4NO_3 for 139 h at room temperature (TPD and TG measurements indicate that the cation exchange is virtually completed); c) 'R-H-FER': the rehydrated H-form, as obtained by calcinating the NH$_4$-form at 550°C for 2 h, after it has been kept at room conditions for some months. Powder patterns were measured at the Swiss-Norwegian Beam Line (ESRF, Grenoble) on a triple axis diffractometer equipped with a Si(111) analyser crystal. The samples were loaded into glass capillaries which were axially spun during data collection in the Debye-Scherrer geometry.

Rietveld structure refinements were performed by the GSAS package[13]. It was necessary to account for the anisotropic Lorentzian Scherrer broadening about the [100] axis using anisotropy coefficents. For all structure refinements soft constraints were applied at the initial stage, and were then released in the final cycles. Figure 1 shows the final Rietveld profile fits for samples A-S-FER and R-H-FER. Crystallographic data and refinement details are reported in Table 1. Table 2 containing the refined positional and thermal parameters for all three samples is available upon request from the corresponding author (G.C.)

RESULTS AND DISCUSSION

The 'as-synthesized' form. Close inspection of the diffraction pattern of A-S-FER revealed the presence of weak reflections of type $h+k+l = 2n + 1$ (e.g. 021, see Figure 1a), which are forbidden in the *Immm* space group. After a preliminary refinement in *Pmnn*, as reported for all-silica synthetic ferrierites, the structure analysis was subsequently carried out in $P2_1/n$, starting from the positional parameters (same labeling was employed) of the natural Mg-poor, Na-, K-rich ferrierite[5], thus with a composition more similar to our samples. The Rietveld cell parameter

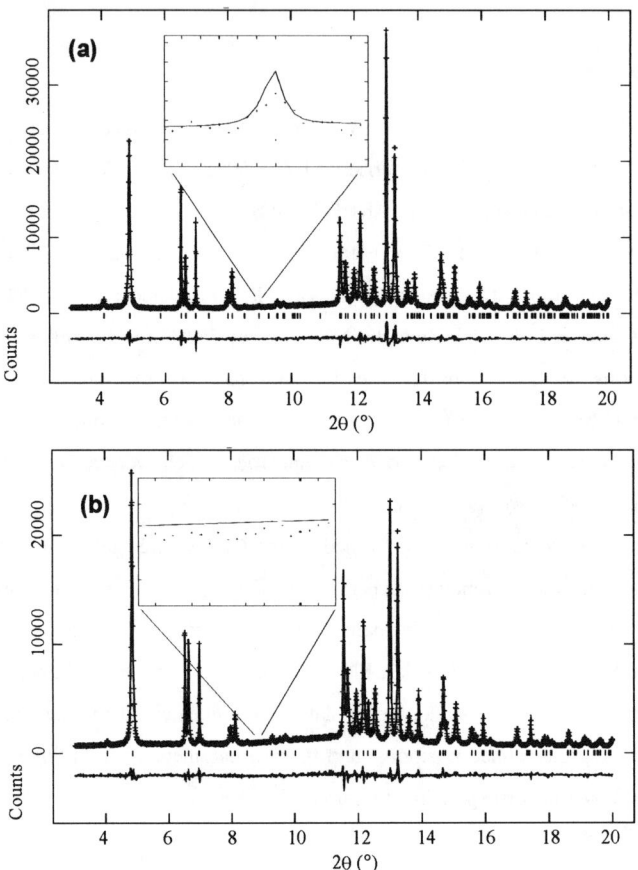

Figure 1. The observed (crosses), calculated (continuous line), and difference (bottom line) profiles of A-S-FER (**a**) and R-H-FER (**b**). Diffraction patterns are plotted in the 2θ range 3-20°. The enlarged 8.85-9.05° 2θ region (insets) shows the presence (**a**) and the absence (**b**) of the 021 reflection in A-S-FER and R-H-FER, respectively.

Table 1. Lattice parameters and refinement details for as-synthesized ferrierite (A-S-FER), NH$_4$-ferrierite (NH$_4$- FER), and rehydrated R-H-ferrierite (H-FER)

	A-S-FER	NH$_4$- FER	R-H-FER
Space Group	$P2_1/n$	$Immm$	$Immm$
a (Å)	18.5000(4)	18.9443(4)	18.9606(3)
b (Å)	14.1361(1)	14.1519(2)	14.1122(1)
c (Å)	7.4439(1)	7.4622(1)	7.4495(1)
β	90.097(3)°	90°	90°
V (Å)^3	1983.54(4)	2000.60(4)	1993.32(4)
R_{wp} (%)	6.6	6.7	5.7
R_p (%)	5.0	5.3	4.4
red-χ^2	5.1	4.1	4.2
R_F^2 (%)	5.8	7.5	7.4
N_{var}	134	64	62
N_{obs}	4836	4715	4816
N_{ref}	3318	918	962

Notes: Synchrotron X-ray radiation, λ = 0.79974(1) Å.
$R_p = \Sigma[Y_{io}-Y_{ic}]/\Sigma Y_{io}$; $R_{wp} = [\Sigma w_i(Y_{io}-Y_{ic})^2/\Sigma w_i Y_{io}^2]^{0.5}$;
red-$\chi^2 = \Sigma w_i(Y_{io}-Y_{ic})^2/(N_{obs}-N_{var})$;. $R_F^2 = \Sigma|F_o^2-F_c^2|/\Sigma|F_o^2|$

refinement gave a clear indication for the monoclinic symmetry (see Table 1). The final structure refinement showed framework distortion leading to a significant displacement of tetrahedral atoms as compared to their expected positions in *Immm*. The shift of the O51 and O52 oxygens away from the inversion center, on which they are constrained in *Immm*, is equal to 0.46 Å along [001]. It has been previously suggested that these movements could be responsible for the symmetry reduction from *Immm* to $Pnnm^4$ or $Pmnn^8$. O61 and O62 oxygens were also remarkably displaced along [100] of 0.25 Å, indicating a significant distortion of the 8-membered rings. This movement is consistent with the disappeance of the mirror plane perpendicular to [100], lowering the symmetry to $P2_1/n$. The distribution of extraframework ions closely resembles that found in natural Mg-poor ferrierite[5]. The largest electron density maximum is located inside the 8-membered rings (site I). The large occupancy fraction and bond distances from the first neighboring framework oxygens (I-O61 = 3.03 Å; I-O62 = 3.19 Å; I-O74 = 3.06 Å; I-O73 = 3.13 Å; I-O71 = 3.19 Å) suggest that mostly potassium and sodium atoms are located in the I site. This is in agreement with what was also observed in a K-exchanged ferrierite in which no lowering from the *Immm* symmetry was recognized[6]. In our case, the I site (see Figure 2a) is significantly displaced away from the center of the 8-ring along [001] to allow a better coordination environment for (K, Na) atoms, and the ring itself is distorted along [100] to accomodate (K, Na). Both of these distortions imply lowering of symmetry to $P2_1/n$. Thus it can be argued that the interaction of K and Na with the framework is mainly responsible for orthorhombic to monoclinic symmetry reduction in K- and Na-rich ferrierites. It has been suggested that, in *Immm* Mg-rich ferrierite, the presence of the $Mg(H_2O)_6^{2+}$ octahedron maintains the topological symmetry because it fits so tightly into the cage that it sterically forces the framework. As a complementary explanation, it can be observed that the Mg-octahedron exhibits the greatest affinity for the ferrierite cage and, when this octahedron is present, the I site must be vacant owing to the too short distance from one of the water molecules of the octahedron. In this way K and Na are excluded from the I site in Mg-rich ferrierite, and are therefore located within the 10-membered ring channels. In the absence of Mg, K and Na clearly show the greatest affinity for the I site. According to the refined occupancy fractions, residual amounts of K and Na atoms are likely distributed in the II' and II sites near the center of the ferrierite cage (see Figure 2a), where partial substitution by water molecules may also occur.

The NH_4-exchanged form. As commonly recognized, NH_4-exchange causes a significant enlargement of the unit cell volume (see Table 1). Since the *I*-centering forbidden peaks were absent in the diffraction pattern of NH_4-FER (see Figure 1b), the *Immm* space group was adopted in this refinement. As previously reported, the *Immm* space group provides a satisfactory description of the average crystal structure of ferrierite when the static disorder of atoms is within the range of the ordinary thermal motion. Starting parameters were those of the Mg-rich natural ferrierite[4]. The extraframework Mg site was found to be vacant, whereas a large electron density maximum in a position near the I site in A-S-FER was located by the Fourier map and refined at the center of the 8-ring (see Figure 2b; distances from framework oxygens: 2 x I-O6 = 3.03 Å; 4 x I-O7 = 3.14 Å). The occupancy fraction refined for this site exhibited the greatest decrease, compared with that found in A-S-FER. Taking into account the crystal chemical similarity between K and NH_4^+ ions[14], we might suggest that a complete substitution of (K,Na) atoms by NH_4^+ groups occurred, and that almost 80% of the I site is populated by NH_4^+. Due to the smaller ionic strength of NH_4^+ compared to K, the decreased interaction between extraframework cations and framework oxygens, related to the removal of (K, Na) atoms, is also reflected in the restoring of the (pseudo-) *Immm* symmetry. A similar effect may be attributed to the presence of the large, neutral Kr molecule at the center of the 8-ring in the *Immm* Kr-exchanged ferrierite[7]. The position of the II site in NH_4-FER is remarkably displaced from its equivalent in A-S-FER (see Figure 2); its refined occupancy suggests a partial substitution of NH_4^+. A similar substitution is also possible in the II' site (see Figure 2b), refined at the same position as in A-S-FER. Unfortunately, the clear distinction between H_2O molecules and NH_4^+ ions within statistically populated sites is a difficult task, owing to the almost identical X-ray scattering powers and their similar sizes. A detailed analysis by Fourier methods, to locate possible protons around the I site in NH_4-FER, using both synchrotron X-ray and neutron data, is in progress, and will be discussed in a future paper.

Rehydrated H-form. The unit cell volume of R-H-FER is larger than that of A-S-FER, but is remarkably smaller than that of its NH_4-exchanged form (see Table 1). This difference is mainly due to the decrease of the *a* axis length. Similarly to what is observed in NH_4-FER, the diffraction pattern of R-H-FER shows no *I*-centering forbidden peaks. The loss of NH_3 does not cause noticeable modification in the framework. The crystal structure, refined in the *Immm* space group, is very similar to that of NH_4-FER (see Figure 2c). The only differences concern the populations of the I, II and II' sites: it appears that a minimal rearrangement of the

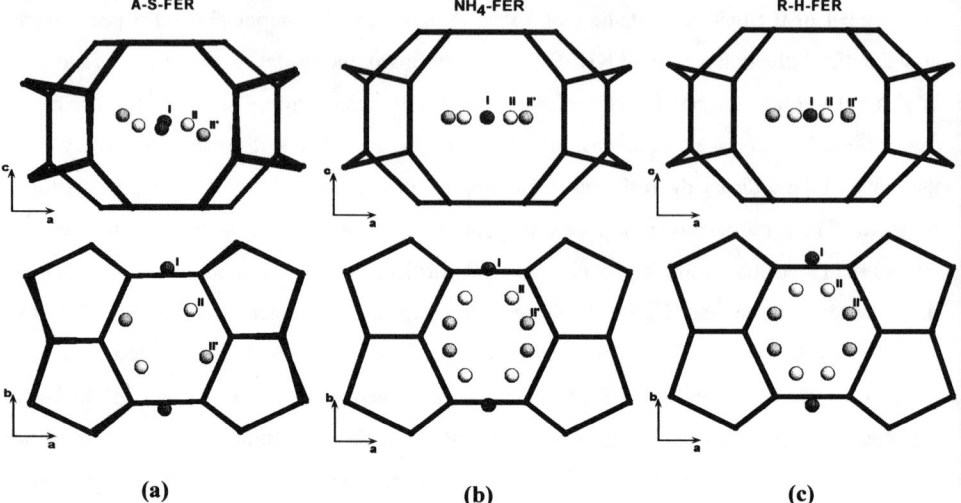

Figure 2. Stick-and-ball projections of the ferrierite cage down to [010] (top) and [001] (bottom) for the as-synthesized form (left), NH$_4$-exchanged form (middle), and rehydrated H-form (right). Tetrahedral positions are represented by threefold corners.

extraframework content has occurred in such a way that the previous NH$_4$ positions are replaced by water molecules with statistical occupancy.

ACKNOWLEDGMENTS

The Ministero dell'Università e Ricerca Scientifica e Tecnologica is thanked for the financial support to the research program "relations between structure and properties in minerals: analysis and applications". The Consiglio Nazionale delle Ricerche is also acknowledged for financial support. The access to the Swiss-Norwegian beamline and its associated facilities at ESRF was provided under the ESRF public user beamtime program.

REFERENCES CITED

1. Xu, W.-Q., Yin, Y.-G., Suib, S.C., Edwards, J.C., O'Young, C.L. (1995) J. Phys. Chem., 99, 9443.
2. Vaughan, P.A. (1966) Acta Cryst., 21, 983.
3. Gramlich-Meier, R., Meier, W.M., Smith, B.K.(1984) Z. Kristallogr., 169, 201.
4. Alberti, A., Sabelli, C. (1987) Z. Kristallogr., 178, 249.
5. Gramlich-Meier, R., Gramlich, V., Meier, W.M. (1985) Am. Mineral., 70, 619 and references cited therein.
6. Pickering, I.J., Maddox, P.J., Thomas, J.M., Cheetham, A.K. (1989) J. Catalysys, 119, 261.
7. Jones, R.H, Lightfoot, P., Ormerod, R.M. (1995) J. Chem. Soc., Chem. Commun., 783.
8. Kuperman, A., Nadimi, S., Oliver, S., Ozin, G.A., Garces, J.M., Olken, M. M. (1993) Nature, 365, 239.
9. Lewis, J.E., Freyhardt, C.C., Davis, M.E. (1996) J. Phys. Chem., 100, 5039.
10. Weigel, S.J., Gabriel, J-C., Gutierrez Puebla E., Monge Bravo, A., Henson, N.J., Bull, L.M., Cheetham, A.K. (1996) J. Am. Chem. Soc., 118, 2427.
11. Morris, R.E., Weigel, S.J., Henson, N.J., Bull, L.M., Janicke, M.T., Chmelka, B.F., Cheetham, A.K. (1994) J. Am. Chem. Soc., 116, 11849.
12. Ming, J., Karge, H. (1996) J. Chem. Soc. Faraday Trans., 92, 2641.
13. Larson, A.C., Von Dreele, R.B. (1994) Report LAUR 86-784.
14. Stuckenschmidt, E., Kassner, D., Joswig, W., Baur, W.H. (1992) Eur. J. Mineral., 4, 1229.

ION EXCHANGE BETWEEN Cd^{2+} SOLUTION AND CLINOPTILOLITE MINERAL

J. C. TORRES

Department of Catalysis of Federal University of San Carlos, San Paulo; Brazil.pjcf@iris.ufscar.br
Department of Materials Science, University of Havana, Havana; Cuba

ABSTRACT

The study of the system Na^+ / Cd^{2+} at 25° C was conducted according the ion-exchange reaction between the zeolite mineral clinoptilolite and an aqueous solution of Cd^{2+}. The experiments were designed to investigate the selectivity of clinoptilolite for Cd^{2+} in solution at low concentration.

Samples of different weights of Na-clinoptilolite were equilibrated with 0.005 N Cd^{2+} solutions. The experimental data were obtained measuring Cd^{2+} and Na^+ in both phases at equilibrium and interpreted by Bronsted and Guggenheim's classic model for calculation the Vanselow corrected selectivity coefficient and Pitzer's model for the activity coefficients of the ions in solution. The derived equilibrium constant, K and Gibbs energy of ion-exchange reaction, ΔG^0, were calculated from the isotherm data. Margules solid solution model were used to predict the behavior of the ion-exchange system studied.

INTRODUCTION

Because of their ion-exchange, adsorption, and molecular sieve properties, as well as their geographically widespread abundance, zeolite minerals have generated worldwide interest for use in a broad range of applications such as nuclear and municipal wastewater treatment, natural gas purification, petroleum production and in agriculture and aquaculture [1].

This work is based on two facts: one, the selectivity of the clinoptilolite for heavy metals in solution and two, the increasing preference of this mineral for cations of higher charge with increasing dilution [2].

The experimental data obtained were processed to obtain the ion-exchange isotherm to derive the equilibrium constant and Gibbs free energy of the reactions calculated using empirical parameters obtained from the Margules solid solution model due to compare relative selectivity of clinoptilolite for the metals studied.

Theoretical

For an exchange reaction involving a metal M^{2+} in solution and Na-clinoptilolite, the equilibrium reaction may be written as:

$$M^{2+} + 2NaZ \Leftrightarrow MZ_2 + 2Na^+, \qquad (1)$$

where **Z** is defined as a portion of zeolite framework holding unit negative charge, and **M**, the heavy metal (Cd^{2+}). The zeolite may be considered as a solid solution of two components, MZ_2 and **NaZ** [3]. The number of moles of MZ_2 and **NaZ** are respectively equal to the total number of moles $n_M(z)$ and $n_{Na}(z)$, of the ions in the zeolite, and solid phase compositions can be expressed in terms of the cationic mole fractions of Cd^{2+} and Na^+ in the zeolite:

$$X_M(z) = n_M(z) / [n_M(z) + n_{Na}(z)] \quad \text{and} \qquad (2)$$

$$X_{Na}(z) = n_{Na}(z) / [n_{Na}(z) + n_M(z)]. \qquad (3)$$

A convenient standard state of each solid phase component is the appropriate homoionic zeolite in equilibrium with an infinitely dilute solution of the same cation [4, 5]. Then the criterion for the ideal behavior in the zeolite solid solution is that $a_i(z) = X_i(z)$ for all $X_i(z)$. For the aqueous electrolyte solution, the usual standard state is defined as one molar solution referenced to infinite dilution.

The thermodynamic equilibrium constant, K, for Eqn. (1) is given by:

$$K = \{[(z) * m^2_{Na^+}]/[X^2_{NaZ}(z) * m_{M^{2+}}]\}/[(f_{MZ2})/(f_{NaZ})^2] * [(\gamma_{Na^+})^2/(\gamma_{M^{2+}})], \qquad (4)$$

or by:

$$K = K_v * [(f_{MZ2}) / (f_{NaZ})^2], \qquad (5)$$

where K_v is the Vanselow corrected selectivity coefficient [6, 7] and **m** is the molality of the aqueous ion. The quantities γ and **f** are, respectively, the single-ion activity coefficients for the aqueous ions and the rational activity coefficients for the zeolite components. The standard Gibbs free energy of the exchange reaction represented by Eqn. (1) is:

$$\Delta G^0 = - RT \ln K. \qquad (6)$$

The above thermodynamic formulation is valid if imbibition of neutral electrolyte is negligible, which for zeolites is at solution concentrations approximately < 1.0 molal [8]. An additional condition is that the effect of water activity changes in the zeolite is insignificant. Barrer and Klinowski [9] demonstrated that water activity changes are not significant for most cases of ion exchange equilibrium.

Equilibrium constants and Gibbs free energies of the exchange reaction evaluated from isotherm data depend on the values used for aqueous activity coefficients. In principle γ's which account for nonideal behavior in the aqueous solution, can be calculated using the well-established electrolyte solution theories [10-12]. Because it would be useful to investigate multicomponent exchange reactions over wide ranges of concentration and temperature, the ion-interaction model developed by Pitzer [12] was used here to calculate activity coefficients of the aqueous species.

There is no generally accepted model for activity coefficients of exchangeable ions or solid solutions [13]. Most models are based on equations for the molar excess Gibbs energy (g^{ex}) of mixing, which may be defined as:

$$g^{ex}/RT = \sum_i^n X_i \ln f_i \qquad (7)$$

where **R** is the gas constant and **T** is temperature (K). A thermodynamic model that has been found useful for both solid and liquid solutions is Margules formulation [13, 14]. For the two component system, CdZ_2 NaZ, the Margules model describes g^{ex} as:

$$g^{ex}/RT = X_{MZ2} X_{NaZ} [X_{NaZ} W_{MZ2} + X_{MZ2} W_{NaZ}], \qquad (8)$$

where W_{MZ2} and W_{NaZ} are empirical parameters which are functions only of temperature and pressure. Representation of solid phase nonidealy in terms of g^{ex} such as Eqn. (7) makes convenient the calculation of zeolite solid solution and ion-exchange properties over the whole range of zeolite composition and the prediction of ion-exchange equilibrium. Using the Gibbs-Duhem equation and Eqns. (7) and (8), $\ln f_{MZ2}$ and $\ln f_{NaZ}$ can be expressed in terms of W_{MZ2} and W_{NaZ} by:

$$\ln f_{MZ2} = X^2_{NaZ}(z) [W_{MZ2} + 2X_{MZ2}(z) *(W_{NaZ} - W_{MZ2})] \quad \text{and} \qquad (9)$$

$$\ln f_{NaZ} = X^2_{MZ2} [W_{NaZ} + 2X_{NaZ} (W_{MZ2} - W_{NaZ})]. \qquad (10)$$

The Vanselow selectivity coefficient, K_v from Eqn. [1] can be then calculated from:

$$\ln K_v = \ln K + 2X^2_{MZ2}[W_{NaZ} 2X_{NaZ}(W_{MZ2} - W_{NaZ})] - X^2_{NaZ}[W_{MZ2} + 2X_{MZ2}(W_{NaZ} - W_{MZ2})] \qquad (11)$$

K, W_{MZ} and W_{NaZ} can be derived by nonlinear regression of Eqn. [9] using the isotherm data. If the zeolite phase behaves ideally, $f_{CdZ2} = f_{NaZ} = 1$ and $g^{ex} = 0$ for all X_{CdZ2} (and X_{NaZ}), and $K = K_v$.

EXPERIMENTAL

Samples for this study were obtained from 400g fraction with particle size of 200-500 µm by a purification process with a mixture of high density liquids [2]. The fraction of clinoptilolite was washed with acetone and dried in an oven at about 60°C.

40 grams of purified were treated with 400 ml of 3M NaCl solution in a flask with condenser at 90°C, stirring vigorously for 2 weeks replacing the NaCl every two days. The clinoptilolite powder was washed several times until negative Cl⁻ qualitative analysis with 0.005N $AgNO_3$ solution. The sample was dried in an oven at about 65°C during 6 hours and then equilibrated with water vapor over saturated NaCl solution in a dissicator until constant mass. Water content of Na-clinoptilolite had been previously determined by other authors in alike purified samples of this mineral using Thermogravimetric Analysis (TGA) [15]. The average ion-exchange capacity of the samples was 1.60 meq/g.

For the ion-exchange study, the amount of clinoptilolite used in the experiments ranged from 0.05 to 1.45 +/- 0.0002 g. Samples were equilibrated with 50 ml of $CdSO_4$ solutions at

0.005N contained in polyethylene bottles and shacked at 160 cycles/min. During the reactions (for about 72 hours), temperature were kept at 25°C.

Solutions were analyzed for Na^+ and heavy cations by Flame Photometry (FP) and Atomic Absorption Spectrophotometry (AAE), respectively. Solid phase were dissolved with 10-15 ml of HF : $HClO_4$ mixture at 80°C and HCl 1:2 and analyzed for cations as aqueous phase, being measured the composition of both phases before and after equilibration [16]. The K^+, Ca^{2+} and Mg^{2+} present in the Na-clinoptilolite were assumed in inaccessible exchange sites and do not participate in the exchange process at the working temperature [2].

RESULTS

Solid purified and Na-form of the samples were analyzed by X-ray Powder Diffraction Analysis and cation-exchange capacity (CEC) given in equivalents of the metal per grams of the exchanger phase (Na-clinoptilolite) was derived from the solid analysis given in table 1. Based in the assumptions that negligible impurities are present in the samples and all Al^{3+} and Fe^{3+} substitute Si^{4+} in tetrahedral site resulting in a negative charge structure and such sites balanced by alkaline and alkaline earth ions (Na^+, K^+, Ca^{2+} and Mg^{2+}), CEC calculated was 1,92 meq/g. Nevertheless, the value taken for all calculations was the practical above indicated.

	Weight percent									
	CaO	Fe_2O_3	K_2O	Na_2O	MgO	SiO_2	Al_2O_3	ZnO	W.L.I.*	total
purified	3.32	1.82	1.12	1.45	0.71	63.27	11.78	-	10.00	83.47
Na-form	1.36	1.79	1.04	4.25	0.66	66.33	11.78	-	6.00	87.26

* Weight Loss by Ignition

Table 1. Results of X-ray Diffraction Analysis of clinoptilolite samples.

Results of binary ion-exchange study are represented in tablae 2 where the composition (equivalent cationic fraction) in both solid and liquid phase, the ratio of solution-phase single-ion activity coefficients Γ and the distribution Vanselow's constant are listed.

No. of sample	m[z] (g)	X_{Cd}^{2+}(z)	X_{Cd}^{2+}(s)	X_{Na}^{+}(z)	$\Gamma*$	ln K_v
1	0.0500	0.6390	0.9133	0.3610	1.313	-8.3018
2	0.1500	0.6117	0.6473	0.3883	1.299	-5.3112
3	0.2500	0.5873	0.3916	0.4127	1.287	-3.8427
4	0.3500	0.5582	0.1432	0.4418	1.294	-2.5751
5	0.4500	0.5087	0.0585	0.4913	1.276	-1.3105
6	0.5500	0.4402	0.0509	0.5598	1.282	-1.3414
7	0.6500	0.3787	0.0468	0.6213	1.283	-1.4512
8	0.7500	0.3475	0.0295	0.6525	1.289	-0.9779
9	0.8500	0.2911	0.0263	0.7089	1.279	-1.2810
10	0.9500	0.2623	0.0214	0.7377	1.279	-1.1969
11	1.0500	0.2479	0.0137	0.7521	1.287	-0.6647
12	1.1500	0.2255	0.0108	0.7746	1.287	-0.5528
13	1.2500	0.2085	0.0095	0.7915	1.287	-0.4995
14	1.3500	0.1949	0.0084	0.8052	1.289	-0.4332
15	1.4500	0.1791	0.0066	0.8209	1.287	-0.3256

$\Gamma = (\gamma_{Na}^+)^2 / \gamma_{Cd}^{2+}$

Table 2. Experimental data for the ion-exchange between Na-clinoptilolite and 0.005 $CdSO_4$ solution.

Values in the ion-interaction parameters used in this study were taken from Pitzer [12] and are listed as follows: $\beta^0(Na_2SO_4) = 0.01958$, $\beta^1(Na_2SO_4) = 1.113$, $C^\phi(Na_2SO_4) = 0.002487$, $\beta^0(CdSO_4) = 0.2053$, $\beta^1(CdSO_4) = 2.617$, $C^\phi(CdSO_4) = 0.0114$, $\theta_{Na,Cd} = -0.0003$ and $\Psi_{Na, Cd, SO4} = -0.0152$, $\beta^{(0)}(CoSO_4) = 0.1631$, $\beta^{(1)}(CoSO_4) = 3.346$, $C(CoSO_4) = C^\phi / 2[z_M{}^*z_a]^{1/2} = 0.00926$, $\beta^{(0)}(Na_2SO_4) = 0.0187$, $\beta^{(1)}(Na_2SO_4) = 1.0995$, $C(Na_2SO_4) = 0.00176$, $\beta^{(0)}(1\ NiCl_2) = 0.3499$, $\beta^{(1)}(NiCl_2) = 0.530$, $C(NiCl_2) = -0.00167$, $\beta^{(0)}(NaCl) = 0.0765$, $\beta^{(1)}(NaCl) = 0.2664$, $C(NaCl) = 0.000635$ and mixing parameters $\theta_{Na,Co} = -0.0194$, $\Psi_{Na,Co,SO4} = -0.0174$, $\theta_{Na,Ni} = 0.0591$, $\Psi_{Na,Ni,Cl} = 0.0115$.

DISCUSSION

Results of the ion-exchange study for Cd^{2+} is presented in figure 1, where are plotted the equivalent cationic fraction of Cd^{2+} in the solid phase vs. its equivalent cationic fraction in solution calculated using Eqn. (2). The ion-exchange isotherm shows the distribution of Cd^{2+} in both phases at equilibrium. The almost rectangular shape of the isotherm surface demonstrate the higher selectivity of clinoptilolite for heterovalent cations at low concentration [9]. Aqueous activity coefficients γ, were calculated from Pitzer's equations for aqueous electrolytes. Values of **ln K** and $X_{Cd}{}^{2+}(z)$ were fitted to the experimental data from table 2 to Margules solid solution model using Eqn. (9) in the range of compositions of the isotherm surface studied as shown in figure 2, where dashed lines curve represents the best fit of the experimental data to model used. It is important to indicate that in this study experiments were not carried out throughout the whole range of compositions of Cd^{2+} in the zeolite phase (it is, from $X_{Cd}{}^{2+}(z) = 0$ to 1), thus, to obtain the best evaluation using the Margules model to describe and predict ion-exchange equilibrium between clinoptilolite and aqueous Cd^{2+} in the zeolite phase, Eqn. (9) must be fitted to values of $X_{Cd}{}^{2+}(z)$ in the range above marked, coupled with an appropriate evaluation of aqueous activity coefficients (with sulfate as supporting anion) in order to obtain more accurate values of equilibrium constant **K** and ΔG^0. In this paper, recalculated values of **ln K_v** for modeling were used, taking its values beyond the ones corresponding to $X_{Cd}{}^{2+}(z)$ calculated from the experimental data.

Although the use of monoanionic electrolyte solution is adequate for thermodynamic characterization of ion-exchange phenomena, ion-exchange reaction that occur in geochemical systems involve aqueous solutions with variable anionic species [2]. In such cases, failure to use correct mixed electrolyte solution activity coefficients would led to a serious error in derived **K** and ΔG^0 for the ion exchange reactions, particularly for ternary or more complicated ion-exchange reactions.

Values for the empirical parameters W_{NaZ} and W_{MZ2} and for the equilibrium constant **K** obtained using Eqn. (11) were -4.797, -5.158 and 1.474, respectively. Values of K calculated for Co^{2+} and Ni^{2+} (sulfates) obtained at the same total normality were 0.723 and 0.013, respectively (paper in preparation). Results shown in figure 3 and 4 indicate the excess Gibbs energy (g^{ex}) and the activity coefficients for zeolite components calculated using Margules parameters and Eqns. (8), (9) and (10).

The energetic profile plotted in figure 3 indicates simply in what sense ion-exchange reaction has mainly occurred, being showed graphically that the exchange reaction conducted is direct-sense favored and then, the greater selectivity of natural clinoptilolite for Cd^{2+} ions than for

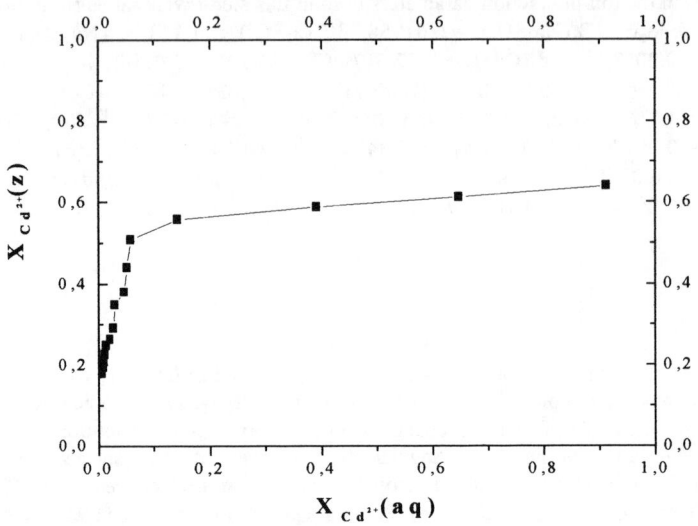

Figure 1. Isotherm points from forward experiments at 0.005 N solution calculated from Cd^{2+} and Na^+ data.

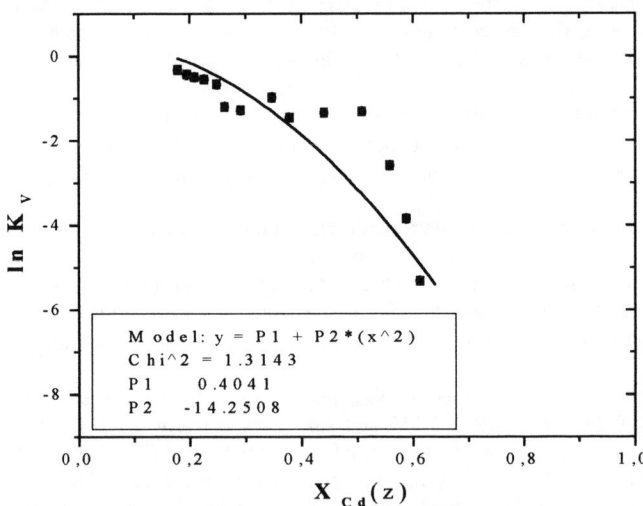

Figure 2. Vanselow selectivity coefficients vs. zeolite composition for 0.005 N isotherm calculated from Cd^{2+} and Na^+ data. The dashes line curve represents the best fit of Margules model.

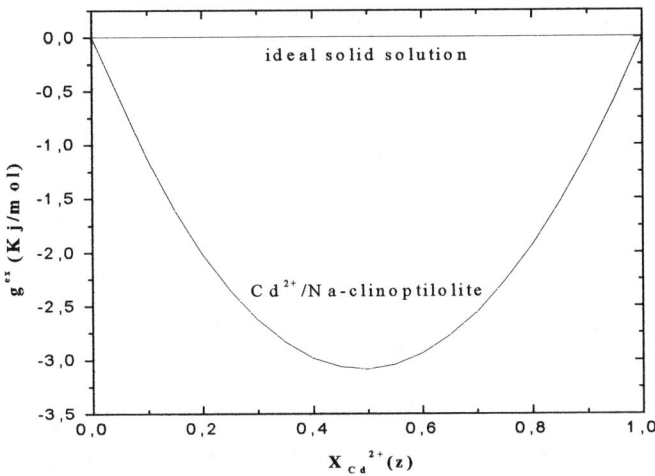

Figure 3. Excess Gibbs energy of mixing for (Cd^{2+}, Na^+)-clinoptilolite solid solutions calculated using Margules parameters fitted to 0.005 N data.

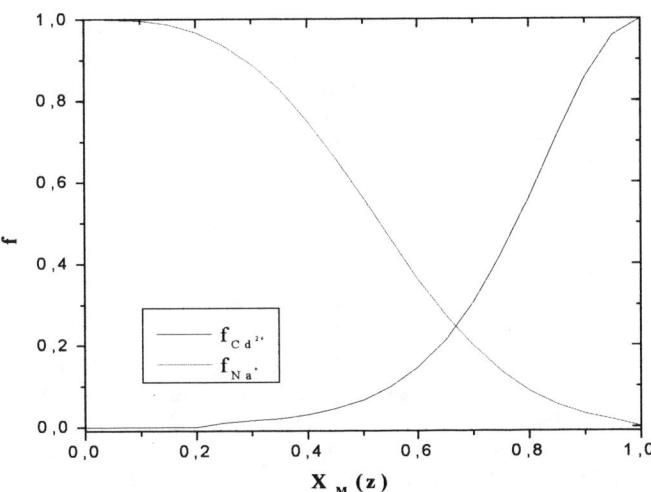

Figure 4. Activity coefficients for endmember components of (Cd^{2+}, Na^+)-clinoptilolite calculated using Margules parameters fitted to 0.005N data.

Na$^+$ which are better stabilized in the aqueous phase at the concentration and working temperature.

The estimated uncertainties [2] in analytical data for Na$^+$ and Cd^{2+} were +/- 3%, relative errors in the solution volumes and γ_i were +/-1% and +/-3%, respectively and the absolute errors in weight and exchange capacities were 2•10^{-4} g and 10^{-5} eq / g, respectively (taking the average value obtained experimentally of 1.60•10^{-3} eq / g for all calculations). The value of uncertainty of normality taken was +/-3% based on the concentration deviations caculated for the Na$^+$ and Cd^{2+} solutions measured.

CONCLUSION

Throughout this paper the greater selectivity of clinoptilolite for Cd^{2+} in solution at the working concentration has been energetically studied, predicting a probable behavior of the exchange system at 25°C using a Margules formulation for solid solution coupled with Pitzer's model for the aqueous ions involved in the exchange reaction studied.

It is important to point out the need of achieving an evenly spaced distribution of points along the ion-exchange isotherm due to give rise a better fit of the ln K$_v$ vs. X$_{Cd}^{2+}$(z) dependence resulting more accurate values for energies and activity coefficients in the zeolite phase. Nevertheless, graphic representations and quantitative data obtained give good scheme predicting relative selectivities of any pure zeolite mineral for metal cations in binary systems at equilibrium.

REFERENCES

[1]. Mumpton F. A. et al. Rev. in Min. And Geol. of Nat. Zeol. ed. by F. A. Mumpton. **4**. *Min. Soc. of Amer*. Dept. Geol. Sci. Virginia Pol. Inst. and State Univ. Blacksburg . Vrg. 24061. 1981. pgs. 177- 195.
[2]. Pabalan R. T., Geochimica et Cosmochimica Acta, Vol **58**, No. 21. 1994, pgs. 4573- 4590.
[3]. Barrer R. M. and Klinowski J.,1977. Phil. Trans. Roy. Soc. London A **285**, p. 637.
[4]. Gaines G. L. and Thomas H. G., J. Chem. Phys., 1953. **21**, p. 714
[5]. Sposito G., Oxford University Press, 1981
[6]. R. P Townsend, Pure Appl. Chem., 1986. **58**, p. 1359
[7]. Vanselow A. P., Soil Sci., 1932. **33**, p. 95.
[8]. Dyer A., Enamy H. and Townsend R. P., Sep. Sci. Technol. 1981. **16**, p. 173.
[9]. Barrer R. M. and Klinowski J. J. Chem. Soc., Faraday Trans. 1974 **I 70**, p. 2080.
[10]. J. N. Bronsted. *J. Amer. Chem. Soc.*, 1922. **44**, pgs. 827, 938.
[11]. E. A. Guggenheim. *Philos. Mag.*, 1935. **19**, p. 588.
[12]. Pitzer K. S. CRS Press. 1991. p. 75-153.
[13]. S. A. Grant D. L. Sparks, *J. Phys Chem.*, 1989. **93**, p. 6265.
[14]. Pabalan R.T. and Pitzer K. S. CRC Press, 1991. pgs. 435-490.
[15]. Mir, M., Rietveld Analysis of Natural Zeolites. M. Sc. Thesis. 1996.
[16]. Townsend R. P. 1986. *Pure Appl. Chem.* 1987. **58**, pgs. 1359-1366.

ION EXCHANGE IN ZEOLITE P, ZEOLITE JBW AND FRAMEWORKS CONTAINING LOW VALENT CATIONS

A.M. HEALEY, S.E. DANN[+] and M.T. WELLER*

Department of Chemistry, University of Southampton, Southampton, SO17 1BJ, UK.
*E-mail: mtw@soton.ac.uk. FAX: + 44 (0)1703 593592.
[+]Department of Chemistry, University of Loughborough, Leics, LE11 3TU, UK.

ABSTRACT

The ion exchange properties of zeolite P (GIS) synthesised with various framework silicon : aluminium ratios have been investigated and are compared with those of zeolite A. Denser (in terms of tetrahedral units per unit volume) frameworks have also been studied. The structures of $Na_2[ZnTO_4]$, T = Si,Ge, have been refined from powder neutron diffraction data and the ion exchange behaviours of these compounds in aqueous and melt conditions studied. The structure of the zeolite JBW, $Na_3[Al_3Si_3O_{12}].H_2O$, has been refined from powder neutron diffraction data and the ion exchange of this material with lithium and potassium ions in aqueous solution has been investigated. The ion exchange behaviours of all these systems are compared in terms of capacity, pore dimensions and the framework densities.

INTRODUCTION

The major use of zeolites is as ion exchange media and includes industrial applications such as radioactive waste management e.g. clinoptilolite for removal of strontium. However, the most widespread application is as water softening agents, especially in detergents and approximately 1 Mtonne per annum of zeolite 4A is manufactured for this purpose. Recently a new zeolite, MAP (maximum aluminium P), has been specially designed for this purpose by the Crosfield group. Its improved water softening properties are attributed to its calcium exchange selectivity, which is a result of the unusual flexibility of the zeolite P framework. The ion exchange capacity is similar to that of zeolite A due to the identical Si : Al ratio, unity. One method to further improve the properties of ion exchange materials would be the incorporation of higher levels of trivalent, or better still, divalent cations into the framework, as the capacity of framework materials to exchange non-framework ions is related to the overall charge on the framework. For example, replacement of SiO_4^{4-} with AlO_4^{5-} or ZnO_4^{6-} tetrahedra increases the negative charge on the framework and enhances the potential for ion exchange.

The majority of zeolite and zeotype materials consist of structures based on TO_4 tetrahedra, where T is trivalent (e.g. Al, B, Ga), tetravalent (e.g. Si, Ge, Ti) or pentavalent (e.g. P, As). There are, however, some examples where divalent cations have been incorporated into zeotype frameworks, these include Be [1], Zn [2], Mg [3] and Co [4]. There are also many examples where trace quantities of such cations have been successfully inserted into frameworks.

Frameworks containing MgO_4 tetrahedra are uncommon with the few examples including the recently reported sodalite $Na_8[Si_3MgO_8]_3[Cl,OH]_2$ [3]. This rarity probably stems from the relatively large size of four co-ordinate Mg^{2+} and its tendency to prefer six fold co-ordination. Structures containing tetrahedral zinc are also uncommon for the same reasons and probably also because oxygen co-ordinated solely to two zinc ions, at typical Zn-O distances 1.95Å, is rather underbonded. Tetrahedrally co-ordinated Be^{2+}, however, is of a similar size to Si^{4+} and is therefore readily incorporated into a wide range of zeotype frameworks including lovdarite and beryllosilicate-GIS. This paper describes some recent work on the ion exchange properties of zeolite P and attempts to prepare and investigate the ion exchange properties of frameworks containing zinc.

EXPERIMENTAL

Zeolite P

Zeolite P was prepared by slowly adding a preheated solution of sodium aluminate to a solution of sodium metasilicate and silica at 90 °C. The masses of the solutes depended on the target Si : Al mixture, while the ratio of solid to liquid was kept constant at 1 : 6. The mixture was stirred at constant temperature for between 1 and 6 days. The products were filtered, washed and dried at 90 °C and analysed using powder x-ray diffraction (PXD), MASNMR and thermogravimetric analysis (TGA).

Zeolite JBW

Zeolite JBW was prepared by the hydrothermal reaction of Kaolin and KOH in a Teflon lined steel autoclave. Reactions were carried out at temperatures in excess of 200 °C for up to 5 days. The products were filtered, washed and dried at 90 °C and analysed using powder x-ray diffraction (PXD), and MASNMR.

Ion exchange reactions on zeolites A, P and JBW

The ion exchange behaviour of zeolites P and JBW in solution was studied and compared with that of zeolite A under the same conditions. 1 g of zeolite was stirred in 50 cm^3 concentrated metal nitrate solution at 40 °C overnight. The products were filtered, washed with distilled water and dried at 90 °C.

Sodium zinc silicates and germanates

$Na_2[ZnSiO_4]$ and $Na_2[ZnGeO_4]$ have been synthesised using a low temperature solution technique. Zinc oxide was dissolved in a concentrated sodium hydroxide solution and aqueous alkali metal silicate added slowly with stirring. The resulting mixture was stirred at between 40 and 120 °C for 12 - 96 hours. The products were filtered, washed in distilled water and dried at 90 °C overnight. Products were characterised using PXD, powder neutron diffraction (PND), MASNMR, EXAFS and TGA.

RESULTS

Structure and ion exchange properties of zeolite P

^{29}Si MASNMR was used to obtain the silicon : aluminium ratio of the various products and showed values in the range 1:1 to 4:1 reflecting the composition of the mother liquor. The structures of the stoichiometrically different zeolites were partially refined using the Rietveld method employing the GSAS package; the starting model was that proposed by Håkanson [5]. The lattice parameters for different silicon : aluminium ratios show a trend towards smaller cell volumes for higher aluminium content products. This is not as expected with Al – O being longer than Si – O, but occurs due to a decrease in the T – O – T angle as has been seen previously for the sodalite structure [6]. These results are shown in Figure 1. The degree of ion exchange of Na-zeolite P with Ca, Li and K has been shown to be comparable with that of zeolite A in all cases. The unit cell volumes of the ion exchanged materials show a decrease in size for all cations investigated. This may be attributable to the different degrees of hydration of the materials.

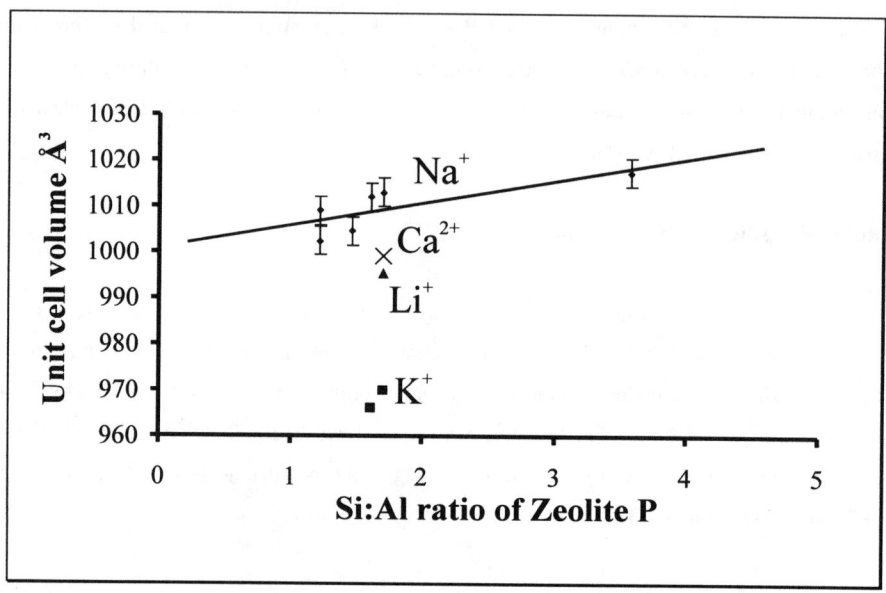

Figure 1: Lattice parameters in the zeolite P system as a function of Si:Al ratio and cation.

Structure and ion exchange properties of $Na_2[ZnTO_4]$

PND data were collected on POLARIS and HRPD at RAL. Structural analysis was carried out by the Rietveld refinement method using the GSAS package. The starting models used in the refinements were those obtained from PXD. $Na_2[ZnGeO_4]$ refined in the Pn space group, the structure agreeing with that proposed by Iljukhin et al.[7], for the sodium zinc silicate, with only small atomic shifts and larger cell dimensions. The refinement of $Na_2[ZnSiO_4]$ proved more challenging as the PND data did not show good agreement with that calculated from the proposed structure. Further investigation showed that the zinc and one of the sodium atoms required transposing. Close examination showed that the two structures were similar but with slightly different orientations of the SiO_4 tetrahedra.

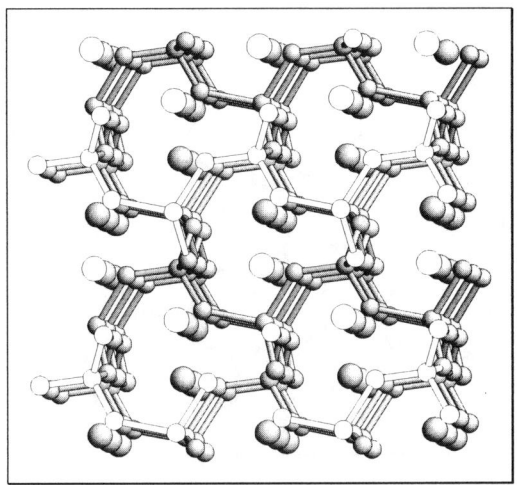

Figure 2: Structure of $Na_2[ZnTO_4]$

The high sodium ion concentration, caused by the charged framework combined with the open channels in the structure, suggest a high ion exchange potential. The framework has, however, proved to be unstable in aqueous media and all aqueous solution, ion-exchange reactions attempted have resulted in amorphous products. The sodium content of the resulting materials had fallen by up to 95 % in the case of magnesium ion exchange. IR spectra show that the Si - O bonds are still present, but no further analysis of the products has been attempted.

Ion exchange has been achieved in the molten phase by heating sodium zinc silicate with metal nitrates. The products showed good crystallinity, PXD data indicated an increase in lattice parameter with increasing cation size, and ^{29}Si MASNMR showed a more negative chemical shift with more open frameworks, as would be expected.

Structure and ion exchange properties of zeolite JBW

PND data have been collected on POLARIS at RAL and refined using the GSAS computer package. The starting model used was that proposed by Hansen and Fälth [8] in the orthorhombic space group $Pna2_1$.

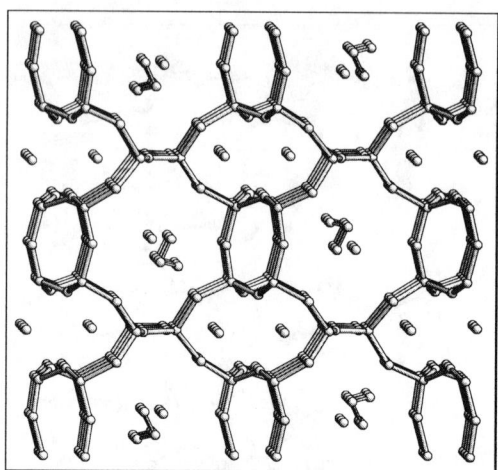

Figure 3: Structure of zeolite JBW. The non-framework species are water and sodium

The structure, shown in figure 3, shows a combination of the features studied in the previous two systems. Firstly layers of eight rings which in zeolites A and P are seen to yield good ion exchange properties, and secondly dehydrated layers of six rings similar to the cristobalite structure of the sodium zinc silicate compounds.

The solution ion exchange characteristics of zeolite JBW have been studied, and show exchange of the hydrated sodium cations for the larger K^+ ions, while the smaller Li^+ exchanges predominantly with the dehydrated sodium cations within the six ring layers of the material. The changes in lattice parameters reflect this behaviour with the Li^+ exchanged material being notably smaller. The K^+ exchanged zeolite JBW shows very little change in lattice parameter, with the non-framework species shifting within the larger eight ring channels of the hydrated region of the structure to occupy more suitable sites.

DISCUSSION

In contrast to the use of zeolites in catalytic applications, where large two and three dimensional channels are desirable, the pore dimension requirements of framework materials for ion exchange uses are generally much smaller. Indeed the use of large pore systems leads to the undesirable uptake of excessive water. The ideal ion exchange material has a three dimensional channel system, for rapid ion diffusion throughout the solid, moderately large channels formed from six and eight rings and the ability for the framework to co-ordinate strongly with the incorporated ion. These criteria are exhibited by zeolite MAP leading to its commercialisation in detergents. Zeolite P with other silicon : aluminium ratios has been shown in this work to exhibit similar ion exchange properties with facile replacement of sodium by potassium, lithium and calcium. The framework has been shown to be very flexible in response to ion exchange, with notably different cell volumes for each exchange material. This is one area where zeolite P differs from the rigid framework of zeolite A, which barely changes in size with cation exchange.

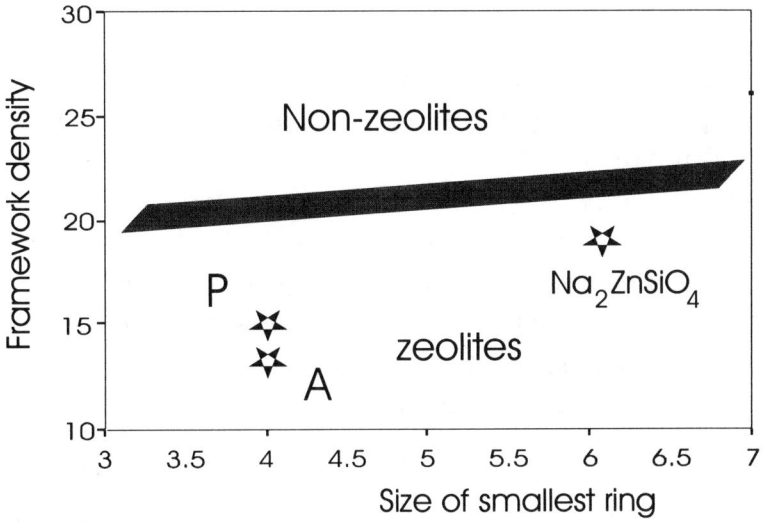

Figure 4: Framework densities of zeolites A, P and of Na_2ZnSiO_4

Stuffed cristobolites are not usually classified as zeolites as they are normally considered as semicondensed framework materials with high densities of tetrahedral units. Indeed framework density values for materials such as carnegieite, $NaSiAlO_4$ are typically near 25. However, with open T-O-T bond angles and/or larger non-framework cations the framework density of cristabolite frameworks drop well into the region associated with zeolites. If such structures contain defects, as described by Thompson [9], then the framework density falls still further. These systems therefore have structural features desirable of ion exchange media. However in non-basic media protonation of the weakly co-ordinated framework oxygens leads to rapid decomposition of the framework. If this type of structure can be stabilised in aqueous media they offer ion exchange capacities, in terms of sodium ions per unit volume, three times that of zeolites A and P.

Zeolite JBW provides a case intermediate between those of open frameworks such as zeolite P and the cristabolite structures and this may produce a system with the individual advantages of the other systems. For example high ion exchange capacities may derive from a reasonably dense structure and a potential to incorporate non hydrolysable zinc while the larger channels may allow facile diffusion of exchanging cations.

REFERENCES

1. S.E. Dann & M.T. Weller, Inorg. Chem., **35**, 555 (1996).
2. J. Grins, Solid State Ionics, **7**, 157-164 (1983).
3. J.G. Thompson, J. Dougherty, A. Melnitchenko, C. Lobo & R.L. Withers, J. Mater.Chem., **6(12)**, 1933-1937 (1996).
4. G.F. Plakhov & N.V. Belov, Sov. Phys. Crystallogr., **24(6)**, 674 (1979).
5. O. Håkanson & L. Fälth, Acta Cryst., **46**, 1363 (1990).
6. M.A. Camblor, S. Bong Hong, M. E. Davis, Chem. Commun., 425 (1996).
7. V.V. Iljukhin, A.V. Nikitin & N.V. Belov, Structure Reports, **31A**, 218 (1966).
8. S. Hansen and L Fälth, Zeolites, **vol 2**, 162 (1982).
9. J.G.Thomson, A. Melnitchenko, S..R.Palethorpe. R.L. Withers. J. Mater. Chem. **7(4)**, 673-679 (1997).

NON-AQUEOUS SYNTHESIS AND STRUCTURAL CHARACTERIZATION OF MICROPOROUS COBALT (II) PHOSPHATES: $(NH_3CH_2CH_2NH_3)_{0.5}[CoPO_4]$ AND $H_2Co_{3.5}P_3O_{12}$

Y. XU[†*], X.L. JIAO[‡] and W.Q. PANG[‡]

[†]: Division of Chemistry, School of Science, Nanyang Technological University, Singapore 259756, Republic of Singapore; xuy@nievax.nie.ac.sg
[‡]: Department of Chemistry, Jilin University, Changchun 130023, P.R.China

ABSTRACT

Two open-framework cobalt (II) phosphates, $(NH_3CH_2CH_2NH_3)_{0.5}[CoPO_4]$ (1) and $H_2Co_{3.5}P_3O_{12}$ (2), are synthesized by solvothermal crystallization and structurally characterized using the X-ray diffraction methods. (1) and (2) crystallize from reaction mixtures of $1.0Co(Ac)_2$: (1-4)H_3PO_4 : 3.0 ethylenediamine: 44 ethylene glycol and $1.0Co(Ac)_2$: (1-4)H_3PO_4 : 2.0 diethylamine: 15 triethylene glycol respectively after 72 hours at 180 °C under autogenous pressure conditions. Accurate control of the initial P/Co ratio, choice of templates and appropriate use of non-aqueous solvents are essential for the successful synthesis of these two compounds. The intensive blue color of (1) is indicative of tetrahedral Co^{2+} ions and the pink color of (2) is indicative of dominant octahedral Co^{2+} ions. *Crystal data:* (1) has tetragonal symmetry, space group $P4_2/n$, with unit cell parameters of $a = b = 10.4060(10)$, $c = 8.9389(12)$ Å, $V = 967.9(2)$ Å^3, $D_{cal} = 1.904$ g·cm^{-3}, $Z = 6$; (2) has triclinic symmetry, space group $P(-1)$, with $a = 6.4727(14)$, $b = 7.880(2)$, $c = 9.485(2)$ Å, $\alpha = 104.290(4)$, $\beta = 109.132(6)$, $\gamma = 101.336(9)°$, $V = 422.1(2)$ Å^3, $D_{cal} = 1.940$ g·cm^{-3}, $Z = 1$. The three-dimensional framework of (1) is built from corner-sharing CoO_4 and PO_4 tetrahedra arranged in a strict alteration and contains alternating 4- and 8-ring channels intersected by 6-ring windows. The three-dimensional framework of (2) is constructed from corner- and edge-sharing CoO_6 octahedra, CoO_5 trigonal bipyramids and PO_4 tetrahedra.

INTRODUCTION

The rational synthesis of crystalline microporous materials is of recent interest of much attention due to their accessible intracrystalline space essential for many commercial applications [1-4]. Engineering desired framework constituents and rendering them with microporosity is one of the adopted strategies. This has led to many efforts aimed at incorporating transition elements into three-dimensional solid architectures and the results appear to be thrilling as demonstrated by the synthesis of a number of new microporous zinco phosphates [2] and vanadium phosphates [5]. The recent success in applying CoAPO-5 and CoAPO-11 catalysts for autoxidation of cyclohexane [6] and *p*-cresol [7] drives systematic in-depth investigation towards synthesizing microporous structures in the pure cobalt phosphate system. This has resulted in the rational design of chiral tetrahedral cobalt phosphates with known zeolite topologies [8]. Being aware of the effectiveness of non-aqueous solvents for assembling microporous solids [9], we seek to explore if the same impact exists during the cooperative assembly of cobalt phosphates in the presence of organic templates. The results would serve as a useful guide on the rational synthesis of cobalt phosphate framework. In this report, we describe the non-aqueous synthesis and structural characterization of two microporous cobalt (II) phosphates, $(NH_3CH_2CH_2NH_3)_{0.5}[CoPO_4]$ **(1)** with a strict alternation of Co^{2+} and P^{5+} tetrahedra and $H_2Co_{3.5}P_3O_{12}$ **(2)** consisting of Co^{2+} in octahedral and trigonal bipyramidal configurations and P^{5+} in tetrahedral configurations.

EXPERIMENTAL

Synthesis

Compounds **(1)** and **(2)** were crystallized under solvothermal autogenous pressure conditions from specifically prepared reaction mixtures in sealed Teflon-lined stainless steel autoclaves at 180 °C for 3 days. The products were recovered by filtration and washed with deionized water. The reagents used included cobalt acetate $[Co(Ac)_2]$, 85% phosphoric acid (H_3PO_4), ethylenediamine, diethylamine, ethylene glycol and triethylene glycol.

Preparation of the reaction mixture of (1). $Co(Ac)_2$ was mixed with ethylene glycol in stoichiometry forming a homogenous solution. Ethylenediamine was then added to the above solution, and followed by the dropwise addition of H_3PO_4 with stirring. The reaction mixture was composed of $Co(Ac)_2$, H_3PO_4, ethylenediamine and ethylene glycol in the mole ratio 1 : (1-4) : 3 : 44. The optimized P/Co ratio was found to be 2.2. Application of the mixed ethylenediamine/di(*iso*-propylamine) template resulted in a change of crystal morphology from polyhedral prism to rode-shape. Attempts to use ethylamine and diethylamine templates, and triethylene glycol solvent failed to induce the crystallization of the desired phase.

Preparation of the reaction mixture of (2). Diethylamine was added to the clear solution of $Co(Ac)_2$ and triethylene glycol forming a gel mixture. H_3PO_4 was then introduced dropwise into the above mixture with stirring. The final mixture was composed of $Co(Ac)_2$,

H_3PO_4, diethylamine and triethylene glycol in the mole ratio 1 : 1-4 : 2 : 15. The best crystals of **(2)** were produced at the initial P/Co ratio of 3.8. The use of ethylene glycol and buthylene glycol solvents instead of triethylene glycol resulted in a lengthened crystallization process of **(2)**.

Characterization

Elemental analysis was performed on a Perkin-Elmer 240C element analyser. Powder XRD patterns were collected on a Rigaku D/MAX III A diffractomer with nickel-filtered Cu Kα radiation (λ = 1.5418 Å). TG and DT analyses were conducted separately using Perkin-Elmer DTA 1700 and TGA-7 at the heating rate of 10 °C min^{-1} from RT to 1000 °C. IR spectra were recorded on a Nicolet 5DX FTIR spectrometer using KBr pellets.

Crystal structure determination. X-ray crystallographic data of **(1)** and **(2)** were collected by θ/2θ scan on a Siemens P4 single crystal X-ray diffractometer with graphite monochromated Mo Kα radiation (λ = 0.71073 Å). The intensity data were corrected for absorption effect using psi-scan data at the data reduction stage along with the correction of Lorentz and polarization factors. The structures were solved by direct methods and refined using the least-square methods on F^2. All non-hydrogen atoms were refined anisotropically. The positions of the H atoms attached to O atoms were located from difference maps and those attached to C and N atoms were placed at geometrically idealized positions. All H atoms were refined isotropically by allowing them to ride on the attached atoms. All calculations were performed using the SHELXTL-PLUS crystallographic software package. The final refinement was converged to R_1 = 0.0230, wR_2 = 0.0483 {data [I > 2.0σ(I)] : parameters ratio *ca.* 7 : 1} for **(1)** and R_1 = 0.0211, wR_2 = 0.0478 {data [I > 2.0σ(I)] : parameters ratio *ca.* 10 : 1} for **(2)**.

RESULTS AND DISCUSSION

The preparation of **(1)** involves searching for the appropriate reaction conditions such as time, temperature, initial P/Co ratio, and choices of organic templates and solvents. Monophasic product of **(1)** crystallizes readily from reaction mixtures containing the P : Co ratio of 1-4 (optimized ratio = 2.2) and ethylenediamine template in predominant ethylene glycol environment. Attempts to use other organic solvents including primary alcohols and triethylene glycol fail to produce monophasic **(1)**. The crystallization of **(1)** is very sensitive to the type of templates involved as it is discovered that the single application of ethylenediamine produces polyhedra-shaped crystals, combined application of ethylenediamine/di(*iso*-propylamine) templates results in rod-shaped crystals, however, the application of simple primary (ethylamine) and secondary (diethylamine) amines is unable to give rise to any crystalline products. Similar phenomena are also observed during the crystallization of (**2**). The successful crystallization of (**2**) is ensured by the use of triethylene glycol solvent and diethylamine template, and appropriately tuned P/Co ratio. The crystallization conducted in ethylene glycol and butylene glycol solvent respectively gives a crystalline mixture of (**2**) and unidentified phases. Diethylamine is decomposed during the crystallization process leaving the micropores of (**2**) unoccupied as confirmed by the elemental analysis, TG/DT and the single crystal X-ray studies. Attempts to synthesize (**2**) in the absence of diethylamine template fails to give any crystalline product.

The DTA and TGA results of (**1**) show two endothermic peaks in the temperature ranges 340-390 and 400-600°C accompanied by weight losses of *ca.* 25.1 and 10.5 % respectively and an exothermic peak at 640 °C with a one-step weight loss of *ca.* 3.1 %. Ethylenediamine molecules interact strongly with the cobalt phosphate framework forming an integral part of (**1**). The DTA and TGA results of (**2**) show an exothermic peak at 439 °C accompanied by a weight loss of *ca.* 2.7 % and an endothermic peak at 524 °C with a one-step weight loss of *ca.* 3.9 %. Heating (**2**) at 500 °C for a prolonged period reduces the product crystallinity as indicated by the *in-situ* XRD studies.

The three-dimensional framework of (**1**) is constructed from corner-sharing tetrahedral Co^{2+} and P^{5+} ions arranged in a strict alternation. This results in a P/Co ratio of unity. The SBU of (**1**) can be viewed as 'a twin boat' *i.e.* two 4_2-screw-axis-related 4-membered rings and two generated 6-membered rings (4.6.6.4) as shown in Figure 1(a). The SBUs of (**1**) multiply through the symmetry operation of the 4_2 screw axis of *P4$_2$/n* forming the $Co_4P_4O_{16}^{4-}$ chains stretched along the crystallographic *c* axis. The adjacent $Co_4P_4O_{16}^{4-}$ chains are linked up by corner-shared CoO_4 and PO_4 tetrahedra resulting in 8-membered rings of (**1**), a double SBU, as shown in Figure 1(b). The structure of (**1**) can also be envisioned as being built up from a 4.6.8 net in space group *P4$_2$/n*. Intersecting channels of 4-, 6- and 8-membered rings are formed in the cobalt phosphate framework as shown in Figure 1(c). The 8-ring channels are two-dimensional running along the [110] and [001] directions, the 6-ring channels are two-dimensional and the 4-ring channels are three-dimensional.

The negative charge of the framework is compensated by ethylenediammoniun counterions positioned in the 8-ring channels. These ethylenediammoniun ions are disordered equally into two sets, each of SOF = 0.5, and they are packed through the symmetry operation of the 4_2 screw axis along the crystallographic c axis. An extensive hydrogen network of the type (Co)O...N(H) = 2.740-2.929 Å is observed between the anionic framework of cobalt phosphate and ethylenediammonium ions. This may explain the difficulty encountered in trying to remove the ethylenediammonium ions and the poor thermal stability of the compound (1).

The geometric parameters of the cobalt phosphate framework are unexceptional. The Co-O distances vary from 1.925 to 1.962 Å with the average of 1.945 Å. It is in consistence with those found in zeolitic cobalt phosphates [8, 10] and layered cobalt phosphates [11]. The P-O distances vary from 1.515 to 1.530 Å with the average of 1.521 Å. It is similar to that in the AlPO$_4$-n structures (1.520 Å), however, shorter than that in GaPO$_4$-n (1.526 Å). These geometric parameter deviate significantly from a reported microporous cobalt phosphate [12] [monoclinic $I_{2/b}$, a = 14.719(6), b = 14.734(5), c = 17.891(6) Å γ = 90.02(2)°] consisting of CoO$_4$ and PO$_4$ units with Co-O = 1.906-1.992 and P-O = 1.480-1.576 Å.

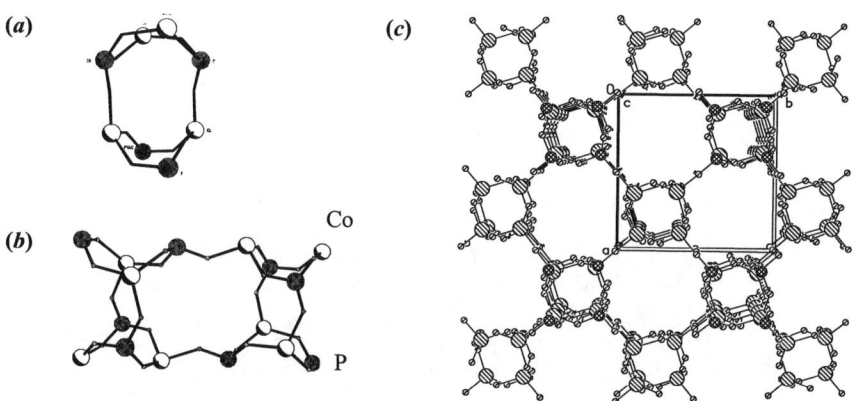

Figure 1. Ball-and-stick representation of (1): (*a*) 'twin boat' SBU; (*b*) double SBUs; (*c*) a packing view along the [001] direction showing 4- and 8-ring channels.

The structure of (2) is built up from a new Co$_{3.5}$P$_3$ SBU consisting of octahedral and trigonal bipyramidal Co^{2+} ions in a 2.5:1 ratio and tetrahedral P^{5+} ions. The CoO$_6$ octahedra and CoO$_5$ trigonal bipyramides are connected through bonding to the bridging oxygen atoms of adjacent phosphorous tetrahedra. The Co-O-Co and Co-O-P linkages are found where the adjacent cobalt polyhedra share corners and edges, and the adjacent cobalt polyhedra and phosphorous tetrahedra share vertices. Hence, the 2-, 3- and 4-membered rings of the $Co_{3.5}P_3$ SBU are formed as shown in Figure 2(a) (dashed box). Through the centre of symmetry, the SBUs of the structure of **(2)** grow to generate chains parallel to the [010] direction as shown in Figure 2(b). The portion highlighted by dashed lines can be viewed as the building unit of the chain. There are no P=O bonds found in the structure of **(2)**.

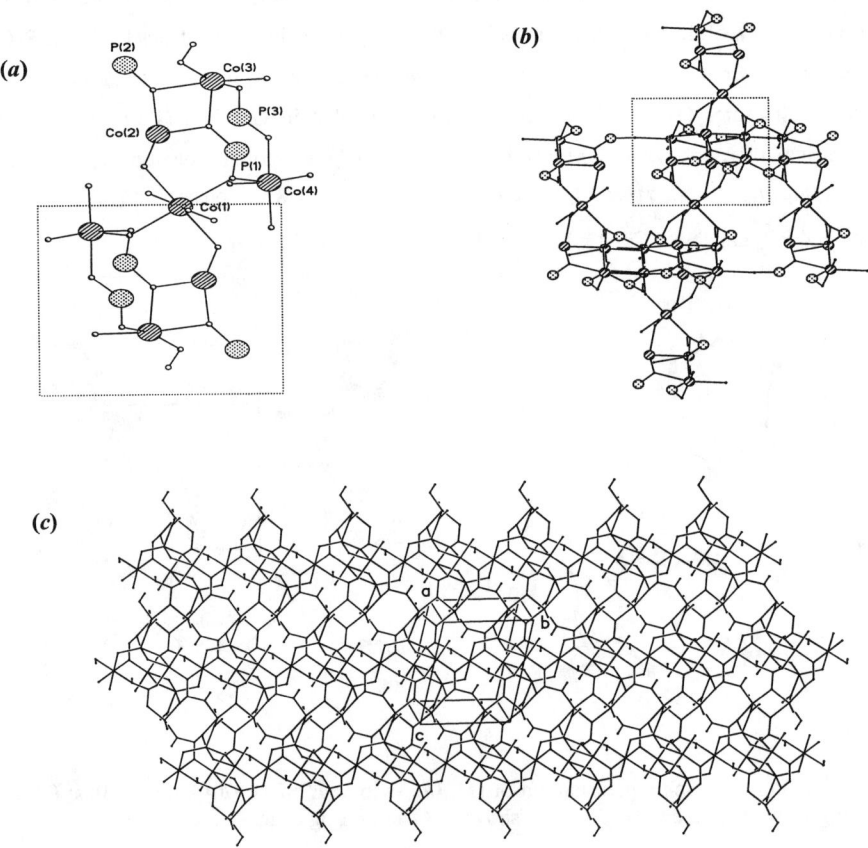

Figure 2. Ball-and-stick representation of **(2)**: (*a*) two *center-of-symmetry-related* SBUs; (*b*) connectivity of the SBUs; (*c*) a packing view along the [100] direction.

The noteworthy structural feature of **(2)** is the presence of 8-membered rings which are connected through smaller ring openings as shown in Figure 2(b) and 2(c). The structure of **(2)** can be described as microporous-like containing mixed octahedral and trigonal bipyramidal Co^{2+} ion and tetrahedral P^{5+} ions. The bond lengths in **(2)** are within expected range for the respective coordination configurations as shown by the following averaged values, $Co(O_6)$-O = 2.114 Å (2.125 Å in CoO), $Co(O_5)$-O = 2.078 Å and P-O = 1.540 Å, similar to those of a reported layered phosphate [13] and longer than those in **(1)**.

Diethylamine molecules are not found in the structure of **(2)**. The cobalt phosphate framework of **(2)**, $Co_{3.5}P_3O_{12}^{2-}$, carries two negative charges which are compensated by protons attached to bridging oxygen atoms. Therefore, it is only logical to think that **(2)** possesses Brönsted acid sites and is able to demonstrate Brönsted-acid related catalytic activity. This is yet to be investigated. The redox catalytic applications of cobalt (II) centres in microporous materials have led to great efforts aiming to introduce them into the structure frameworks. Recent reports on the synthesis of zeolitic and layer-structured cobalt (II) phosphates where Co^{2+} occupies tetrahedral sites demonstrate the power of hydrothermal technique in the assembly of transition metal phosphates. The current success in the non-aqueous synthesis of two framework cobalt phosphates serves as a useful guide on the design of microporous cobalt phosphates from another perspective.

CONCLUSIONS

In this report, we have discussed the non-aqueous synthesis and structural characterization of two three-dimensional cobalt (II) phosphates. They are crystallized under solvothermal conditions by appropriate use of template molecules and solvents. X-ray structural analysis shows that **(1)** consists of purely tetrahedral Co^{2+} ions while **(2)** of both octahedral and trigonal bipyramidal Co^{2+} ions. The success in synthesizing **(1)** and **(2)** provides further evidence towards the synthesis of $CoPO_4$ open framework materials currently observed predominantly in aluminophosphate materials.

REFERENCES

1. R.R. Xu, J.S. Chen and S.H. Feng, <u>Chemistry of Microporous Crystals</u> (Kodansha-Elsevier, 1990) p. 63 and references therein.

2. T.E. Gier and G.D. Stucky, Nature **349**, 508 (1991).
3. R. Haushalter and L. Mundi, Chem. Mater. **4**, 31 (1992) and references therein.
4. C. Bowes and G. Ozin, Adv. Mater. **8(1)**, 13 (1996) and references therein.
5. V. Soghomonian, Q. Chen, R. Haushalter, J. Zubieta and C.J. O'Connor, Science **259**, 1596 (1993).
6. S.S. Lin and H.S. Weng, Appl. Catal. **A105**, 289 (1993).
7. J. Dakka and R.A. Sheldon, The Netherlands Patent 9,200,968 (1992).
8. P.Y. Feng, X.H. Bu, S.H. Tolbert and G.D. Stucky, J. Am. Chem. Soc. **119**, 2497 (1997).
9. A. Kuperman, S. Nadimi, S. Oliver, G.A. Ozin, J.M. Garcés and M.M. Olken, Nature **365**, 239 (1993).
10. R.P. Bontchev and S.C. Sevov, Chem. Mater. **9**, 3155 (1997).
11. J.R.D. Debord, R.C. Haushalter and J. Zubieta, J. Solid State Chem. **125**, 270 (1996).
12. J.S. Chen, R.H. Jones, S. Natarajan, M.B. Hursthouse and J.M. Thomas, Angew. Chem. Int. Ed. Engl. **33**, 639 (1994).
13. P. Lightfoot, A.K. Cheetham and A.W. Sleight, J. Solid State Chem. **85**, 275 (1990).

RUTHERFORD BACKSCATTERING SPECTROSCOPY, AN EASY METHOD TO VIZUALIZE AND QUANTIFY METAL CONCENTRATION GRADIENTS THROUGH METALLOSILICATE ZEOLITE CRYSTALS: THE CASE OF MFI GALLOSILICATES

Z. GABELICA[1], S. VALANGE[1], M. JACOBS[2] and G. DEMORTIER[2]

[1] ENSCMuhouse, Université de Haute Alsace, Laboratoire de Matériaux Minéraux, UPESA 7016, 3 rue Alfred Werner, F-68093 Mulhouse-Cedex, France
[2] Facultés Universitaires de Namur, LARN, 22, Rue Muzet, B-5000 Namur, Belgium

ABSTRACT

Gallium concentration depth profiling in Ga-MFI synthesized by various routes was quantitatively evaluated by Rutherford Backscattering Spectroscopy (RBS). Gallosilicates synthesized in the presence of alkali-free methylamine involve a homogeneous Ga framework repartition. In fluoride medium, residual Ga fluoro complexes readily overcoat the gallosilicate outer surface at the end of the crystallization. Post synthesis thermal treatments of Ga-MFI/methylamine result in a partial degalliation of the framework that could be quantified by RBS. A rapid calcination in dry conditions leads to the formation of extra framework Ga oxides that migrate towards the crystal core. Under milder heating, these species stay homogeneously partitioned within the crystal channels. Under humid non oxidative atmosphere, the extra framework Ga species migrate towards the crystal surface, the migration being enhanced by a partial reduction of Ga.

INTRODUCTION

ZSM-5 type gallo- or gallo-aluminosilicates have proved very active and selective catalysts in the transformation of lower alkanes to aromatics [1-4]. Highly dispersed oxo-hydroxy Ga species, currently generated by heating Ga-exchanged, Ga-impregnated or Ga_2O_3-loaded (H,Al)-MFI [1, 3, 5] but also through degalliation of as-synthesized Ga- or (Ga,Al)-MFI zeolites under various calcination conditions and atmospheres [6, 7], were recognized to intervene in the mechanistic path of this conversion [1-3]. Besides the real nature, chemical state and amount of these extra framework Ga species, other factors such as their location, space distribution and migration upon various treatments were shown to greatly affect the activity and selectivity of the catalyst [7]. Parameters such as (hydro)thermal temperature and time or the water vapor partial pressure [7-9] can influence the (in)stability of the framework gallium in MFI gallo(alumino)silicates and, consequently, the amount, dispersion and (re)location of extra framework Ga species either as tiny aggregates within the zeolite channels [10], or stabilized as $Ga(OH)^+$ ions at the cationic exchangeable positions [7, 10] or also as more bulky isolated

Ga$_2$O$_3$ entities onto the surface of the crystallites [8]. Rutherford Back Scattering (RBS) proved very sensitive in detecting heavy elements dispersed in matrixes composed in majority of light elements [11]. Preliminary attempts to use RBS of alpha particles in combination with PIXE and XPS, allowed one to visualize the spatial distribution of (heavy) gallium species in classical Ga-loaded (H,Al)-ZSM-5 catalysts [5]. In the present approach, we demonstrate the wide potentialities of RBS to be used as a simple and complementary technique for a straightforward and quantitative evaluation of gallium concentration profiles throughout as-synthesized MFI gallosilicate crystals, and their variation upon selected post synthesis treatments simulating catalytic reactor conditions.

EXPERIMENTAL

Two MFI gallosilicates samples were synthesized from hydrogels involving Ga/96 T (T = Si + Ga) molar ratios of 4, using both the alkali-free methylamine [6] and fluoride [12] routes. The as-synthesized sample 1, supposed to involve a quasi homogeneous framework Ga distribution [6], was also calcined using variable heating programs under different atmospheres and water partial pressures within a thermobalance (Stanton Redcroft ST-780) so as to carefully control the heating conditions and weight losses/gains. Simulating the conditions currently achieved during alkane dehydrogenation steps, sample 1 was heated under mixed or successive nitrogen-hydrogen-air flows in dry and humid conditions. The various heating programmes were selected as follows:

Treatment	Atmosphere	Calcination scheme
1	Dry N$_2$/air	RT-550°C (10°C/min)/N$_2$; isotherm at 550°C/air
2	Dry N$_2$/air	RT-350°C (2°C/min)/N$_2$; isotherm at 350°C (1h); 350-550°C (10°C/min); isotherm at 550°C for 0.5 h under N$_2$ (0.5 h), then in air (1.5 h)
3	Humid N$_2$/air	As above, in humid conditions
4	Dry N$_2$/H$_2$/air	RT-120°C (5°C/min)/N$_2$; isotherm at 120°C/N$_2$+H$_2$ (1h); 120-550°C (5°C/min/N$_2$+H$_2$); isotherm at 550°C/ N$_2$+H$_2$ (0.5h), then under N$_2$ (0.5h) then in air (1h)
5	Humid N$_2$/H$_2$/air	As above, in humid conditions

After each calcination, all samples were rapidly cooled to room temperature under the appropriate atmosphere and rapidly pelleted prior to RBS measurements. The total gallium contents were determined by complexometric titration method adapted in the case of zeolitic materials, after their pre-dissolution in diluted HF [6]. High resolution solid state MAS ^{71}Ga-NMR was used under conditions appropriate for quantitative measurements of zeolitic framework Ga^{3+} contents, as previously described [13]. The experimental set-up for RBS measurements was detailed elsewhere [11]. Uniformly spherical zeolite particles were gently compressed into pellets and irradiated with an incident broad beam of 956 keV alpha particles detected at a scattering angle of 178° (Fig. 1). The incident energy has been selected so as to allow interaction in a region extending to less than the mean grain radius

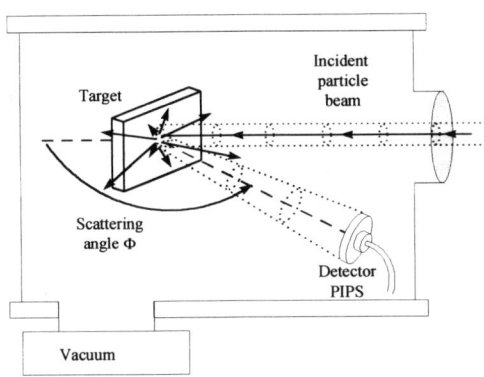

Figure 1. Schematic experimental set-up for RBS experiments

The deconvoluted depth profiles obtained were then used in the RUMP simulation program to fit the RBS spectra.

RESULTS AND DISCUSSION

As-synthesized samples

Samples 1 and 2 show the following characteristics:

Sample [Ga/96T in gel]	Nature (XRD)	% Cryst. (n-C6 ads.)	Mean cryst. size(SEM)	Ga/96T Chem.	NMR	RBS
1 [4Ga (MA)]	MFI	97	4 µm	3.71	3.37	3.63±0.2
2 [4 Ga (NH4F)]	MFI	(103)	2 µm	4.0	3.77	gradient

Sample 1 exhibits a very homogeneous Ga concentration at least through a depth of 400 nm (Fig 2a). Because of the marked intensity increase at about 0.6 MeV due to the presence of silicon, the Ga concentration could not be evaluated at higher depths so we arbitrary considered that the amount of Ga calculated from the RBS plateau ending at a depth of about 400 nm, further extrapolates through the bulk of the crystallites. The actual amount of Ga, as

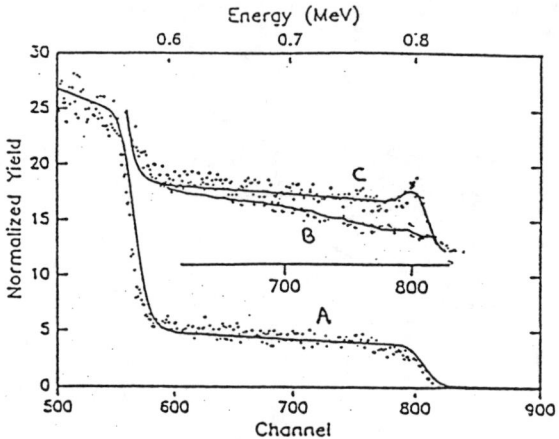

Figure 2. RBS profiles for sample 1; (A): as synthesized; (B): after treatment 1; (C) after treatment 3; "channel 800" represents the crystal surface and "channel 600" corresponds to an analysis depth of about 400 nm

calculated for a matrix supposed to involve theoretically 192 oxygens, 92 Si and 4 Ga atoms per unit cell of MFI, the target density being estimated to 2, was perfectly constant along the first 400 nanometers starting from the surface and was equal to 3.63 ± 0.2 atoms per unit cell. This value fairly well fits the one obtained by both chemical analysis and NMR, this latter technique only detecting framework Ga. RBS therefore brings for the first time a major proof that the methylamine route leads to a homogeneous Ga insertion into the framework of Ga-MFI during its crystallization (Figure 2A, horizontal plateau).

Sample 2, synthesized in the presence of fluoride mobilizing agent, does show a Ga gradient: 5.8, 5.0, 4.2 and 3.4 Ga/96 T were detected respectively within "slices" of 25 nm of depth starting from the surface towards the core. This indicates that residual Ga fluoro complexes may overcoat the outer surface, as also suggested by the (slight) differences between chemical and NMR analyses. The "bulk" value (3.4 Ga/u.c.) suggests a fairly homogeneous Ga incorporation starting from the core up to a depth of about 100 nm from the outer surface.

Sample 1 calcined in various conditions

Treatment 1 corresponds to the protocol currently used to decompose the tetrapropylammonium (TPA) template. In our case, the synthesis was achieved in the total absence of alkali cations so that the acidic (Ga,H)-ZSM-5 zeolite was straightforwardly generated [6]. Such a treatment results in a partial degalliation as suggested by the evaluation of the amount of residual Brönsted sites by ammonia TPD of NH_4^+-exchanged sample 1. The amount of such sites per unit cell was 2.9, to be compared with the initial amount of framework Ga (3.4) measured by NMR, indicating that about 0.5 Ga have left the T lattice positions. RBS shows for the first time the space profiling of these species within the calcined crystals, namely a marked gradient of Ga, increasing from the surface to the core of the crystals (Fig. 2 b). The calculated unit cell Ga amounts were respectively of 1.84 and 2.42 for the first two 100 nm layers, while the bulk (roughly corresponding to the next 200 nm) stabilizes at 2.9 Ga/u.c. As the total Ga must still be around 3.7 (chemical analysis), one must assume that the extra framework Ga species, probably as tiny aggregates, have moved from the surface to the interior of the crystals, probably by creeping through the zeolite channels, as earlier speculated [10]. It is assumed that these species readily migrate under the quite strong steaming conditions that occur during treatment 1. The presence of bulky Ga_2O_3 type surface clusters is excluded at least in this case, while the formation of $Ga(OH)_2^+$ cationic species located at the exchange positions is very unlikely in the absence of framework Al.

Treatments 2 (N_2-air) and 4, (N_2-H_2-N_2-air) achieved under milder temperature conditions than in treatment 1, thus in the absence of steaming during the low temperature isotherms (the TPA templates are still stabilizing the zeolite while the water is early released), maintain the homogeneous Ga gradients despite the decrease of the total number of Brönsted sites (respectively 2.9 and 2.5/96T). In such cases, even though similar amounts of framework Ga were extracted, these extra framework Ga species still remain homogeneously distributed through the crystallites. Under the reducing treatment 4, more Ga was extracted from the lattice positions than under nitrogen. These reduced Ga species also stay homogeneously distributed, probably within the zeolite channels. We conclude that even though (partly) reduced Ga_xO_y species are more mobile than Ga_2O_3 [1, 5], another driving force such as water vapor may be needed to induce their migration.

Calcinations achieved under humid atmospheres do confirm this hypothesis. Both under slightly reducing conditions achieved under nitrogen flow (treatment 3) or upon a more drastic reduction by hydrogen (treatment 5), the sample readily undergoes degalliation (respectively 2.6 and 2.2 Brönsted sites as measured by combined TPD) while the RBS profiling reveals marked concentration gradients. This further confirms that both steaming and reduction are needed for a steady migration of the extra framework Ga species. In our case, they appear to migrate towards the crystal surface (Fig. 2 c, treatment 3), where they readily re-oxidise into (probably) bulky Ga_2O_3 oxidic species upon the final calcination in air.

Our RBS data suggest that if mobility, redispersion and specific re-location of extra framework Ga species is desired to achieve an efficient catalytic phase, the calcination treatment must involve a combined steaming-reducing step.

REFERENCES

1. V. Kanazirev, V. Mavrodinova, L. Kosova and G.L. Price, Catal. Letters, **9,** 35 (1991) 2. G. Giannetto, G, Leon, J. Papa, R. Monque, R. Galiasso and Z. Gabelica, Catal. Letters **22,** 273 (1993)
3. G. Giannetto, R. Monque and R. Galiasso, Catal. Rev.-Sci. Eng. **36,** 271 (1994)
4. V.R. Choudhary, P. Devadas, A.K. Kinage and M. Guisnet, Zeolites, **18,** 188 (1997)
5. E.G. Derouane, S.B. Abdul Hamid, I.I. Ivanova, N. Blom and P.E. Höjlund-Nielsen, J. Mol. Catal., **86,** 371 (1994),
6. Z. Gabelica, G. Giannetto, F. Dos Santos, R. Monque and R. Galiasso, in: Proceedings of the 9th International Zeolite Conference, Edited by R. von Ballmoss, J.B. Higgins and M.M.J. Treacy (Butterworth and Heinemann, Boston, Massachussetts, 1993), pp 231-238
7. G.L. Price, V.I. Kanazirev and K. Dooley, Zeolites, **15,** 725 (1995)
8. A.V. Kucherov, A.A. Slinkin, H.K. Beyer and G. Borbely, J. Chem. Soc. Faraday Trans, 1, **85,** 2737 (1989)
9. A. Montes, Z. Gabelica, A. Rodriguez and G. Giannetto, Appl. Catal. A, **161,** L1 (1997)
10. K.J. Chao, S.P. Sheu, H.L. Lin, M.J. Genet and M.J. Feng, Zeolites, **18,** 18 (1997)
11. M. Jacobs and F. Bodart, Nucl. Instr. and Meth. in Phys. Res. B, **118,** 714 (1996)
12. Z. Gabelica and S. Valange, Res. Chem. Intermediates **24,** 227 (1998)
13. Z. Gabelica, C. Mayenez, R. Monque, R. Galiasso and G. Giannetto, in: Synthesis of Microporous Materials, Vol I: Molecular Sieves, Edited by M. L. Occelli and H. E. Robson (Van Nostrand Reinhold, New-York, 1992), pp 190-221.

CHARACTERIZATION OF K+ ION EXCHANGE INTO Na-LSX USING TIME RESOLVED SYNCHROTRON X-RAY POWDER DIFFRACTION AND RIETVELD REFINEMENT

Y. LEE[a], C. L. CAHILL[b], J. C. HANSON[c], and J. B. PARISE[a,b]
S. W. CARR[d], M. L. MYRICK[e], U. W. PRECKWINKEL[f], and J. C. PHILLIPS[f]

[a] Geosciences Department, State University of New York, Stony Brook, NY11794, USA; yollee@ic.sunysb.edu
[b] Chemistry Department, State University of New York, Stony Brook, NY11794, USA
[c] Chemistry Department, Brookhaven National Laboratory, Upton, NY11973, USA
[d] ANSTO, Private Mail Bag 1, Menai 2234, AUSTRAILIA
[e] Department of Chemistry and Biochemistry, University of South Carolina, SC 29208, USA
[f] BRUKER AXS, INC., 6300 Enterprise Lane, Madison, WI 53719-1173 USA

ABSTRACT

The pathway along which K^+ replaces Na^+ in Low-Silica X (LSX) has been determined using a combination of time-resolved synchrotron X-ray powder diffraction data collected with a translating imaging plate and Rietveld structure refinement. In agreement with the *ex situ* data collected on this same system, the exchange proceeds through a two-phase region. The two phases are distinguished by K^+ occupancy of site I in the second phase to appear during continuous ion exchange. To obtain kinetic information and to automatically monitor the experiment, Iterative Target Transformation Factor Analysis (ITTFA) was used to analyze the data. To investigate the exchange mechanism in detail, higher time resolution was provided by using a CCD detector. A preliminary modeling of the diffraction data collected using the CCD indicates that only the wall sites, II and III´, are replaced at the early stage of the exchange.

INTRODUCTION

Of the faujasite-type zeolites [1], Low-Silica X (LSX) [2-3] has improved performance characteristics in ion exchange and adsorption applications [4]. While most ion exchange studies have been performed on bulk samples using *ex situ* analysis techniques, time-resolved studies have the potential to provide information on transient phenomena masked in *ex situ* studies.

We have previously reported a combined NMR and *ex situ* X-ray powder diffraction study [5] on materials in the Na-K LSX solid solution series. Samples were equilibrated at the 20%, 42%, and 80% K^+-exchange levels following the ion exchange isotherm by Sherry [6], as well as at the end members of the solid solution series. At low levels of K^+ exchange, samples had single ^{29}Si MAS NMR bands, and potassium ions preferentially replaced sodium ions at sites I´ and II as determined from Rietveld refinement. For the sample equilibrated at the 80% K^+-exchange level,

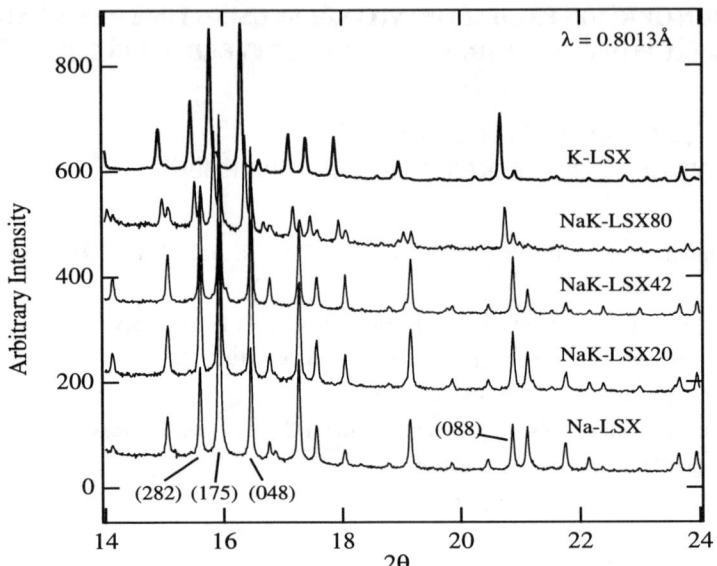

Figure 1. Five synchrotron X-ray powder diffraction patterns (from the *ex situ* study [5]) showing the changes in the relative intensities and peak positions as K^+-exchange percentage increases in the Na-K LSX solid solution series. Note the two phases in NaK-LSX80 which was equilibrated at the 80% K^+-exchange level [6].

two phases were observed in both the X-ray powder diffraction pattern (Figure 1) and as evidenced by the two ^{29}Si resonances in the NMR [5]. The phase to appear at the higher K^+-exchange levels has a larger unit cell volume. Rietveld structure refinement confirmed the abrupt increase in volume coincided with the occupancy by K^+ ions of site I, the double 6-ring site (Figure 1). In the present study, *in situ* synchrotron X-ray powder diffraction was utilized to examine the pathway along which K^+ ion exchange occurs in Na-LSX. By using imaging plates and monochromatic synchrotron X-radiation, time-resolved data of sufficient quality to allow full Rietveld refinement [7-8] were obtained. We have also performed a preliminary study using a CCD detector in an effort to monitor this reaction with higher time resolution, especially at the beginning of the exchange process.

EXPERIMENTAL

Na-LSX was prepared by ion exchange of NaK-LSX ($Na_{74}K_{22}Al_{96}Si_{96}O_{384} \cdot nH_2O$) with 1.0 M NaCl solution. The ^{29}Si and ^{27}Al MAS NMR measurements indicated a Si to Al ratio of unity, and ICP elemental analysis confirmed complete exchange after four 24-hour treatments at 80 °C. This material, Na-LSX ($Na_{96}Al_{96}Si_{96}O_{384} \cdot nH_2O$), was used for *in situ* ion exchange.

Time-resolved powder diffraction studies were performed at beamline X7B of the National Synchrotron Light Source (NSLS), using a Small Environmental Cell for Real Time Studies (SECReTS) [9] and a Translating Image Plate (TIP) [10] detector system (Figure 2). The Na-LSX powder was loaded into a 0.5 mm glass capillary that was plugged with glass fiber at both ends.

Figure 2. The Translating Imaging Plate system, as used at beamline X7B of the NSLS along with a schematic representation of the SECReTS setup for *in situ* ion exchange. The capillary is held in place with a Swagelock tee, which is then mounted on a standard goniometer head (not shown), where centering adjustments and translations can be made. The exchange solution is passed over the sample through the inlet port using N_2 over pressure and then exits through the outlet port. Diffracted X-rays are collected on an imaging plate, a portion of which is covered by shields. Translation of the plate (arrow) behind a slit allows a fresh region of the plate to receive X-rays, while covering the exposed portion.

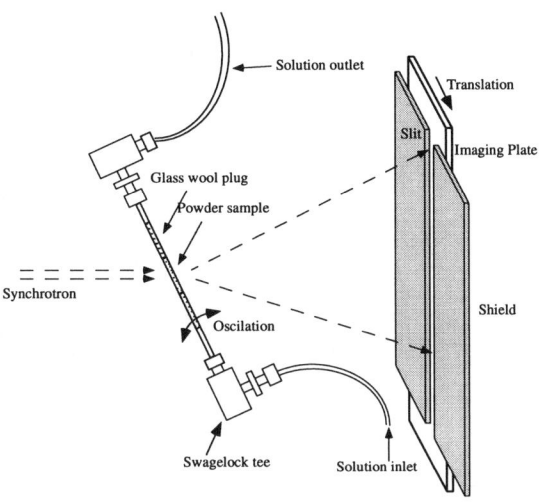

This capillary was mounted on SECReTS (Figure 2) and a 0.01 M KCl exchange solution was passed over the sample using an over pressure of 30 psi of N_2 gas at room temperature. The TIP detector recorded the diffraction pattern over 6 hours ($\lambda = 0.9949$Å) behind a slit of width 3 mm with a step counting time of 24 seconds and a step size of 0.2 mm.

A similar setup with a CCD detector was utilized at beamline X7A of the NSLS. The detector was set at $2\theta = 23°$ with a sample to detector distance of 30.9 cm, covering a 2θ range from 17.7° to 28.2° ($\lambda = 0.7000$Å). Diffraction patterns were recorded for every 60 second on a small area CCD chip (6.25 × 6.25 cm) without the slits (Figure 2).

RESULTS AND DISCUSSION

Only one set of peaks were observed during the first 30 minutes of the ion exchange (Figure 3). Two sets of peaks, which could be indexed as two separate face-centered cubic phases, were present after this initial stage. The second set of peaks were shifted to a lower 2θ value relative to the peaks from the original phase, indicating a larger unit cell. This was consistent with the *ex situ* observation where the sample of 80% K^+-exchanged LSX consisted of two different phases [5]. These two phases were designated as **A** and **B** (Figure 3) for the discussion below.

Rietveld refinement of structural models for phases **A** and **B** was performed using data integrated at different times during the exchange. The structural parameters refined from the previous *ex situ* study [5] were used to construct and constrain starting models. Since knowledge of the cation distribution in each phase was important to understand the ion exchange process, only occupancies of cations and water molecules were refined in those cases where two phases were present

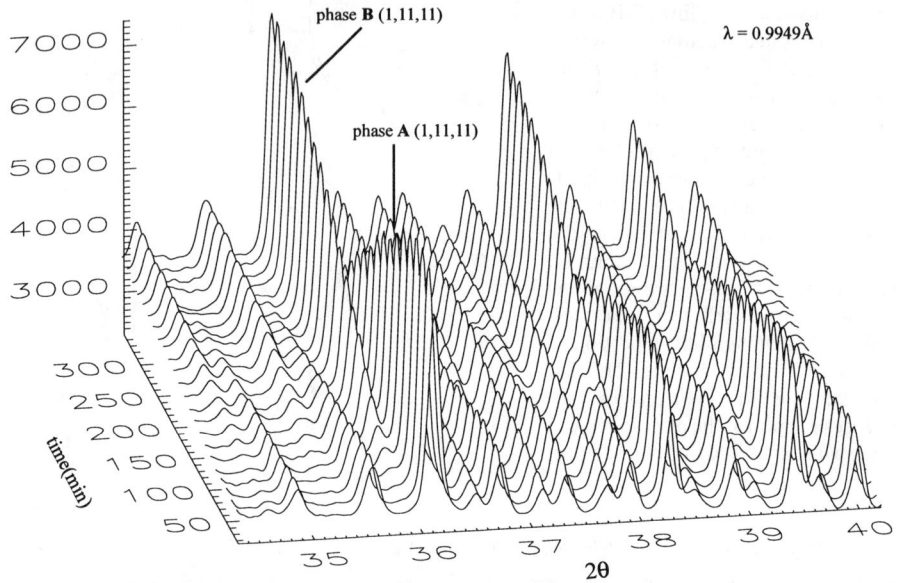

Figure 3. Plot of the X-ray powder diffraction profiles as a function of time during the 6 hours of K^+ exchange into Na-LSX. The patterns are obtained by integrating the imaging plate vertically with an integration width of 3 mm, about the size of the slit (Figure 2).

in the patterns (Figure 3). The positions of all the atoms were held fixed according to the result from the *ex situ* study (Figure 4) [5].

The results of the refinements using data taken at the initial and final portions of the *in situ* scan (Figure 3) were similar to the structural models of the Na-LSX and K-LSX from the previous *ex situ* study [5]. In the two-phase region, phase **A** was characterized by having K^+ ions at sites I′ and II while phase **B** was distinguished from phase **A** by having K^+ ions at site I as well as at sites I′ and II. It is this occupancy of site I by K^+ ions that is responsible for the lattice parameter expansion in agreement with the *ex situ* observation [5]. In both phases, sites I′ and II exhibited mixed occupancies by Na^+ and K^+ ions (Figure 4). Site III′ was modeled using an oxygen scattering factor due to the possible disordering at this weakly bound site [11].

The changes in potassium occupancy at sites I, I′ and II (Figure 5) indicated a time-dependent and site-specific ion exchange process. In phase **A**, potassium ions first replaced sodium ions at site II. The occupancy of K^+ at site II increased until sodium ions at site I′ were exchanged, at which stage phase **B** appeared. In phase **B**, potassium occupancy at sites I and I′ as a function of time trended in opposite directions, suggesting the diffusion of K^+ ions from site I′ into site I (Figure 5). There was a continuous increase of potassium ions at site II during the *in situ* scan.

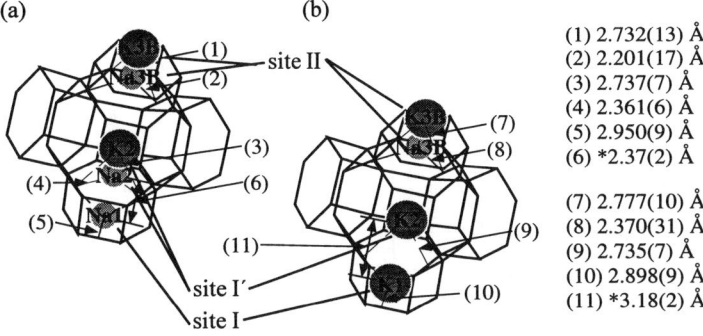

Figure 4. A schematic illustration of cation distribution in (a) phase **A** and (b) phase **B**. The coordination distances shown here were determined from the *ex situ* study on the NaK-LSX80 [5] and were used as fixed parameters in the present work (*simultaneous occupancy is not allowed at these closely separated sites).

(1) 2.732(13) Å
(2) 2.201(17) Å
(3) 2.737(7) Å
(4) 2.361(6) Å
(5) 2.950(9) Å
(6) *2.37(2) Å
(7) 2.777(10) Å
(8) 2.370(31) Å
(9) 2.735(7) Å
(10) 2.898(9) Å
(11) *3.18(2) Å

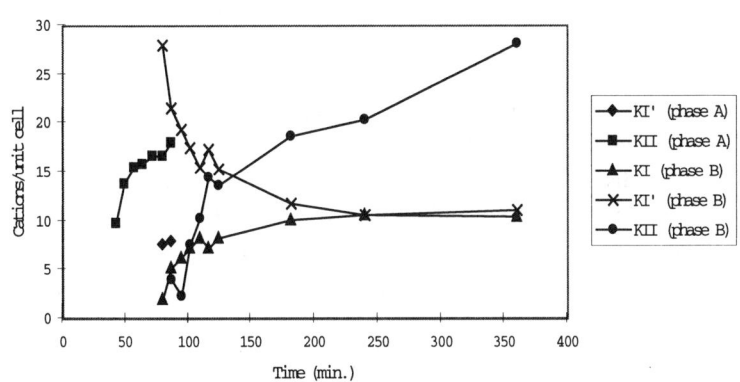

Figure 5. Changes in potassium occupancy at sites I, I′ and II as determined from the Rietveld refinement (data from phase **A** are shown only up to 87 minutes for clarity).

Kinetic information and which initial patterns to use for Rietveld refinement were obtained using Iterative Target Transform Factor Analysis (ITTFA) [12]. Component phases identified from the ITTFA were categorized as four different 'events' (Figure 6a). Event 1 was attributed to the errors associated with imaging plate geometry such as tilt and zero point errors, which were corrected [13]. Events 2 and 4 represented the disappearance of phase **A** and the growth of phase **B** respectively. The remaining event 3 reflected the two-phase region where phases **A** and **B** were present. Events 2, 3, and 4 were combined, and normalized fractions of the combined events were plotted against reaction time (Figure 6b). These data were fit with the Avrami [14] equation [$\alpha=1-\exp(-kt)^n$] where α is a normalized fraction of each component phase. From fits of both curves, a rate constant (k) was determined to be 4×10^{-6} and n = 2.3.

Some information regarding the exchange mechanism may be obtained from these curves. For example, previous *in situ* X-ray diffraction studies have used the exponential factor (n) to propose reaction mechanisms [13, 15]. A value of 2.3, as determined in the present study, is suggestive of a diffusion controlled process with a decreasing rate of nucleation [16]. A more thorough

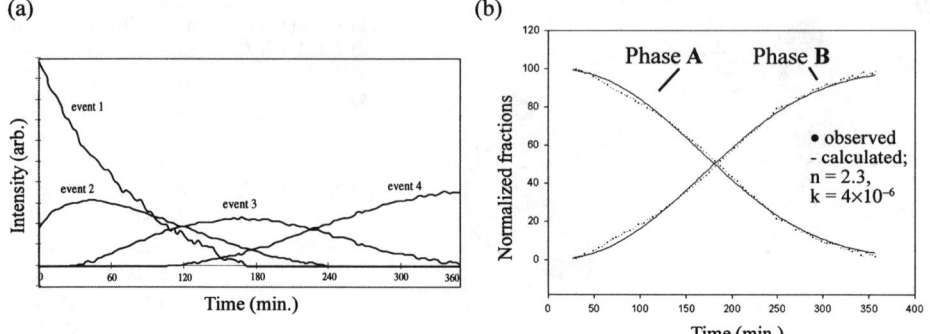

Figure 6. (a) ITTFA results showing the four 'events' of the ion exchange reaction. (b) Normalized fractions of the combined events and fits to the Avrami equation ($\alpha=1-\exp(-kt)^n$, see text for details)

analysis of how ITTFA results can be correlated to reaction mechanism and used to determine critical data for Rietveld analysis is underway.

One disadvantage of the TIP system is the requirement that the plate needs a full exposure before readout. This makes it difficult to observe transient phenomena which can occur in the early stages of reaction. An electronic detector, such as a Charge Coupled Device (CCD) detector, can provide a higher time resolution and real time observation. For this investigation, the CCD detector

Figure 7. Calculated and observed powder diffraction patterns suggesting the initial transient stage where the wall sites (sites II and III′) in the supercage are first replaced by K^+ ions (also note the difference in backgrounds).

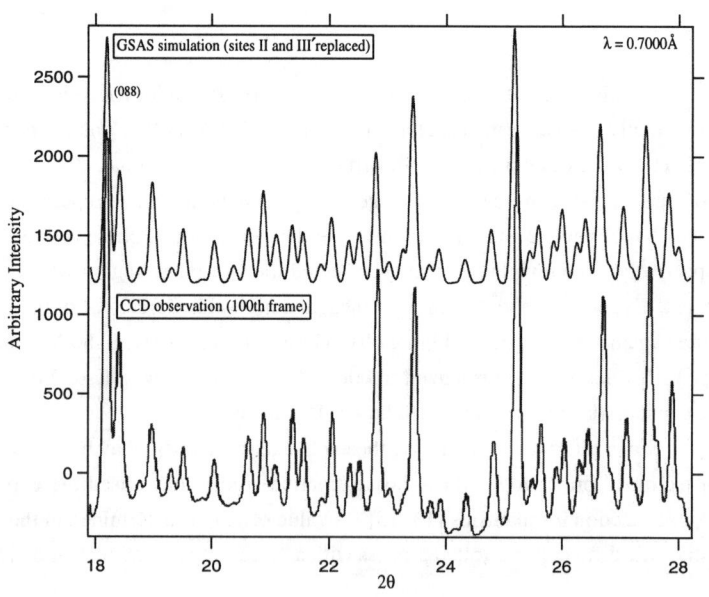

was set for 1 minute exposures during the ion exchange reaction. Two phases were observed during the quasi-real time scan, and a model for the initial stage of the ion exchange was proposed by comparing a calculated powder diffraction pattern with a CCD detector observation (Figure 7).

Considering this simulation and the Rietveld refinement results derived from the TIP data, the full pathway of K^+ ion exchange into Na-LSX can be drawn as follows. Potassium ions first replace sodium ions at sites II and III' in the supercage in the early stage of the ion exchange. This transient step is followed by the diffusion of K^+ ions into the sodalite cage where site I' is then exchanged (Figure 5). With those sites loaded with potassium cations, further K^+ ion exchange into Na-LSX causes a phase transition, expanding the unit cell by positioning K^+ ions into site I, the double 6-ring site (Figure 4).

This work, combined with the previous *ex situ* study [5], illustrates how cation location can influence zeolite structures to a significant degree. The key here is the location of potassium at site I, the center of the double 6-ring. Careful examination of the zeolite X structure indicates that expansion or contraction of the double 6-ring potentially has the most impact on structural parameters.

It is notable to consider, however that the *ex situ* study [5] was performed on several equilibrated samples at each desired K^+-exchange level while the initial transient stage was detected only from the *in situ* work. More detailed kinetic information can be derived by performing *in situ* experiments at different temperatures. For example, information on the activation energy at each site can form a basis to tailor the properties such as the selective adsorption [17] as a function of different cation distribution and the structural phase transitions they induce.

CONCLUSIONS

Time-resolved synchrotron X-ray powder diffraction was utilized to investigate the phase transition occurring upon K^+ ion exchange into Na-LSX. The results from the previous NMR and *ex situ* X-ray powder diffraction study performed on several equilibrated samples in the Na-K solid solution series of LSX formed the basis for this *in situ* work and were consistent with the present findings. Rietveld structural refinement on the two phases formed during the *in situ* ion exchange focused on different cation distribution in each phase. Potassium siting at site I characterized the expanded phase which was last exchanged by K^+ ions.

The extraction of kinetic information from ITTFA treatment of the time-resolved data, although preliminary, demonstrated the potential for this combination of techniques in *in situ* X-ray diffraction experiments. Exploratory use of a CCD detector enabled us to observe the initial stage of the exchange of Na^+ by K^+ where the supercage wall sites, II and III', were first replaced by potassium ions.

ACKNOWLEDGEMENT

Research carried out in part at the National Synchrotron Light Source at Brookhaven National Laboratory is supported by the U. S. Department of Energy, Division of Materials Sciences and Division of Chemical Sciences. The authors gratefully acknowledge the assistance of David E. Cox for his help with data collection at beamline X7A. This work was supported by a grant from the NSF DMR-9713375.

REFERENCES

1. D. W. Breck, Zeolite Molecular Sieves, (Robert E. Krieger, Malabar, FL, 1984).
2. G. H. Kuhl, Zeolites, **7**, 451 (1987).
3. S. W. Carr; F. Hollway and J. D. Hopwood, Unpublished data.
4. C. G. Coe, S. M. Kuznicki, R. Srinivasan and R. J. Jenkins, ACS Symposium Series, **368**, 478 (1988).
5. Y. Lee, S. W. Carr and J. B. Parise, Chem. Mater. (In press) (1998).
6. H. S. Sherry, J. Phys. Chem. **70**, 1158 (1966).
7. R. A. Young, The Rietveld Method (Young, R. A., Ed., Oxford University Press Inc., New York, 1995).
8. A. C. Larson and R. B. VonDreele, "GSAS: General Structure Analysis System", (Report LAUR 86-748, Los Alamos National Laboratory, 1986).
9. P. Norby, C. L. Cahill, C. Koleda and J. B. Parise, J. Appl. Cryst. **31**, 481 (1998).
10. P. Norby, J. Appl. Cryst. **30**, 21 (1997).
11. D. H. Olson, J. Phys. Chem. **74**, 2758 (1970).
12. X. Liang, J. E. Andrews and J. A. Haseth, Anal. Chem. **68**, 378 (1996).
13. Norby, P. J. Am. Chem. Soc. **119**, 5215 (1997).
14. Avrami, M. L. J. Chem. Phys. **9**, 177 (1941).
15. Clark, S. M.; Evans, J. S. O.; O'Hare, D.; Nuttall, C. J.; Wong, H.-V. *J. Chem. Soc., Chem Commun.* 809 (1994).
16. Hulbert, S. F. J. Br. Ceram. Soc. **6**, 11 (1969).
17. C. G. Coe, T. R. Gaffney, J. F. Kirner, H. C. Klotz, J. E. MacDougall and B. H. Toby, 21th National Meeting of the American Chemical Society, New Orleans, LA, 1996, AN 1996:221366.

JOINT X-RAY DIFFRACTION / NMR STRUCTURE ELUCIDATION OF MICROPOROUS FLUORINATED ALUMINO-PHOSPHATES: ULM-3 Al AND ULM-4 Al

F. TAULELLE[a], V. MUNCH[a], C. HUGUENARD[a], A. SAMOSON[b], T. LOISEAU[c], N. SIMON[c], J. RENAUDIN[d] and G. FÉREY[c]

[a]RMN et Chimie du Solide, UMR 7510 ULP-Bruker-CNRS, Université Louis Pasteur, 4 rue Blaise Pascal, 67070 Strasbourg Cedex, France.
[b]Intitute of Chemical-Physics and Biophysics, Akad TEE 23, Tallinn 0026, Estonia
[c]Institut Lavoisier UMR CNRS 173 IREM, Université de Versailles-St Quentin, 45 av. Etats-Unis, 78035 Versailles Cedex, France.
[d]Laboratoire des Fluorures, UPRES-A CNRS 6010, Av. 0. Messiaen, 72000 Le Mans, France
email : taulelle@chimie.u-strasbg.fr

ABSTRACT

Two new alumino-phosphate phases, ULM-3 Al and ULM-4 Al, are reported that have been synthesized hydrothermally. Single crystals as well as powders have been analyzed by diffraction and high resolution solid state 1D and 2D NMR. Several 2D NMR experiments - RFDR, Double Quanta, MQMAS, and high resolution 1D experiments, using very high speed MAS and DOR - have been combined with X-ray diffraction to solve properly the structures of both these compounds and to establish firmly that their symmetries are lower than those of their gallium parents.
The hexameric unit, which is found in both structures with a different topological pattern, is believed to be the structural building unit of a larger class of structures where gallium can be replaced by aluminum or iron.

INTRODUCTION

Since the first series of microporous aluminophosphates, denoted as $AlPO_4$-n, was prepared in 1982[1], these materials have been extensively studied. The introduction of fluoride ions to the synthesis[2] strongly modifies the pH of the solutions and has lead to new open framework topologies. In many cases, the fluorine is incorporated into the structure. It can be found in the centers of the double 4-ring cages as in cloverite[3] or in the coordination sphere of the metal atom (M = Al or Ga) as in the ULM-n[4] or T-GaPO[5],[6] series. Besides the existence of the D4R units stabilized by the fluoride anion, the hexameric cluster $Ga_3(PO_4)_3F_2$ is probably one of the most common building units encountered in the fluorinated gallophosphate[7]. The systematic study has now been extended to the aluminum and iron systems in which the hexameric cluster is also observed.[8],[9]

Solid state NMR has been used extensively to characterize microporous compounds, and especially silicate containing materials as well as aluminophosphates. Up to now the situations in which NMR characterization is involved can be sorted out into two classes. In the first, a single crystal is obtained, in order to establish the structure by X-ray diffraction, and the NMR is then run to check the NMR methods. The other situation is to use NMR when a single crystal can't be obtained and, by combining it with other means of characterization, trying to describe as much as one can of the structure of the material.

We present results on two cases - ULM-3 Al and ULM-4 Al - for which the combined use of both diffraction and NMR has led to a proper structure refinement. The choice of the observable nuclei (^{31}P, ^{27}Al, ^{19}F) as well as the NMR methods, MAS, Radio-Frequency Dipolar Recoupling (RFDR), Double Quanta (DQ), Multiple Quanta MAS (MQMAS) and DOR, have allowed us to reach our goal.

We report here the structural characterization of the fluorinated aluminophosphates ULM-3 Al and ULM-4 Al, analogous to the known gallium compounds. Both phases possess a 3D open framework built up from the connection of the hexameric units, but exhibit a lower symmetry than the Ga analogs.

EXPERIMENTAL

The synthesis of these aluminophosphates was carried out hydrothermally in a 23 ml Teflon-lined Parr bomb under autogeneous pressure. ULM-3 Al and ULM-4 Al were obtained by using 1,4-diaminobutane (DAB) and 1,3-diaminopropane (DAP) as structure directing agent, respectively. ULM-3 Al has been prepared from molar composition: 1 Al_2O_3 ; 2 H_3PO_4; 2 HF ; 1 DAB ; 80 H_2O (heating at 180°C for 3 days). ULM-4 Al is obtained from the molar ratio: 1 $(iPrO)_3Al$; 1 H_3PO_4; 1 HF ; 0.5 DAP ; 80 H_2O (heating at 180°C for 4 days).

The structure of the both compounds was determined by single crystal X-ray diffraction analysis. For ULM-3 Al, the crystal system is orthorhombic (Pbc2$_1$; a = 10.023(2) Å, b = 18.180(3) Å, c = 15.841(3) Å, V = 2886.5 Å^3) whereas ULM-4 Al crystallizes in the monoclinic system (P2$_1$; a = 8.5552(2) Å, b = 10.1133 (2) Å, c = 16.7025(4) Å, β = 94.94(1) °, V = 1439.7(6) Å^3)

NMR has been run on a DSX 500 spectrometer from Bruker. MAS probeheads are of 4mm. and 2.5 mm rotor diameters. Achievable spinning speeds were respectively 20 and 35 kHz. DOR experiments were performed on a probe built in Tallinn, Institute of Chemical-Physics and Biophysics, with a dedicated computer regulated double pneumatic unit.

RESULTS AND DISCUSSION

Our first attempt to describe the structure of ULM-3 Al used the same space group, Pbca, as the gallium parent, and was considered as satisfactory. However a NMR ^{31}P check, showed six inequivalent phosphorus instead of three for the diffraction refinement solution. A RFDR experiment, figure 1, showed that doubling was true for each site even when below the resolution limit. The center of symmetry has therefore been lost. The space group is hence Pbc2$_1$ instead of Pbca. Once the space group fulfilled both diffraction and NMR requirements, the assignment of inequivalent phosphorus was performed using the linear relationship between mean d(P-O) distance around the phosphorus and the isotropic chemical shift. To confirm the structure proposition a DQ ^{31}P spectrum was acquired and the topology was analyzed and compared to the one proposed by diffraction. Both agree, and the phosphorus lattice is considered as being firmly established.

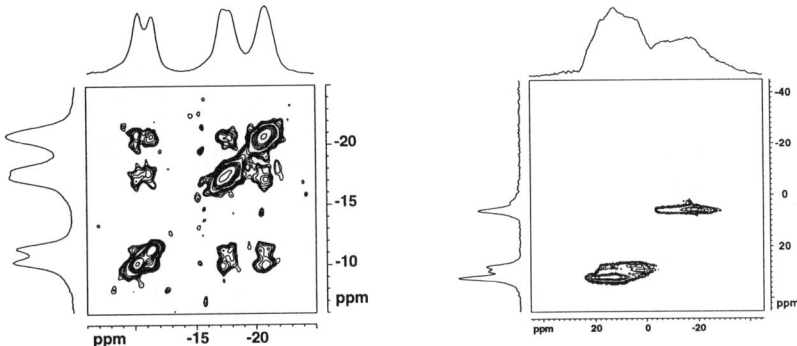

Figure 1: ULM-3 Al ^{31}P NMR RFDR (left) and ^{27}Al MQMAS (3QMAS z-filter) (right) experiments

The ULM-3 Al fluorine sub-network was compared the same way by analysis of ^{19}F MAS multiplicities and then checking by RFDR. A characteristic of fluorine is that it requires very high spinning speeds, larger than 25 kHz, to get a reasonable linewidth for high resolution solid state and good 2D data.

^{27}Al has been a little more difficult to observe in high resolution conditions. MQMAS and DOR experiment have been performed. First, MQMAS spectra didn't show more than one hexacoordinated resonance, separated from two unresolved pentacoordinated peaks. The DOR spectrum exhibits a resolution of the three resonances in the ratio 1: 1: 1. Other MQMAS spectra at 30 kHz and 280 kHz radiofrequency field seem to indicate the same type of resolution as that observed in ^{31}P. The same strategy has been applied to ULM-4 Al and leads to the same conclusion that the aluminum compound is less symmetrical than its gallium parent.

The structures of ULM-3 Al and ULM-4 Al can therefore be described as being built up from the common hexameric entity $Al_3(PO_4)_3F_2$. It is formed from three PO_4 tetrahedra, two AlO_4F trigonal bipyramids and one AlO_4F_2 octahedron. The two types of aluminum polyhedra are linked together through the fluorine atoms. Within this unit, two of the three tetrahedra are linked by comers to one octahedron and one bipyramid, whereas the third connects the three aluminum polyhedra. These comer sharing building units are arranged in two different ways in ULM-3 Al and ULM-4 Al, but generate 10-ring channels in both cases as shown in figure 2. Whereas tunnels are marquise-shaped in ULM-3 Al, they are rectangular in ULM-4 Al .

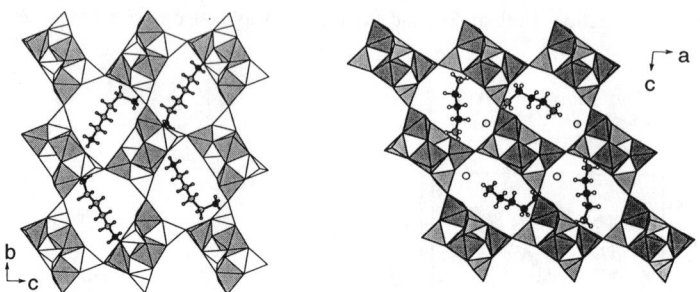

Figure 2 : Polyhedral representation of ULM-3 Al (left) and ULM-4 Al (right) structure types. The aluminum sites are indicated by darker polyhedra.

CONCLUSION

The combined usage of X-ray diffraction and NMR allows finely detailed resolution of inorganic structures. It requires performing NMR for most of the nuclei available with NMR, and advanced high resolution experiments (1 and 2D) included into the refinement process of the diffraction experiment.

By elucidating the motifs building these two structures, this study reinforces the idea that the hexameric units identified in the structure can be a legitimate structural building unit of such systems.

REFERENCES

1. S. T. Wilson, B. M. Lok, C. A. Messina, C. T. R., and E. M. Flaningen, J. Am. Chem. Soc., **104**,1146 (1982).
2. J. L. Guth, H. Kessler, and R. Wey, Stud. Surf. Sci. Catal., **60**, 63 (199 1).
3. M. Estermann, L. B. McCusker, C. Barelocher, A. Merrouche, and H. Kessler, Nature, **352**, 320(1991).
4. G. Ferey, Journal of Fluorine Chemistty, **72**, 187 (1995).
5. S. J. Weigel, University of Santa Barbara, Santa Barbara, California, 1997.
6. S. J. Weigel, S. C. Weston, A. K. Cheetharn, and G. D. Stucky, Chemistry of Materials, **9**, 1293 (1997).
7. T. Loiseau and G. Férey, MRS Spring Meeting Vol **431**, 1996, pp 27.
8. J. Renaudin, T. Loiseau, F. Taulelle, and G. Ferey, C. R. Acad Sci., Ser. IIb, 323, 545 (1996).
9. M. Cavellec, D. Riou, J. M. Greneche, and G. Ferey, Journal of Magnetism and Magnetic Materials, **163**, 173 (1996).

AN ATTEMPT TO LOCATE PROTONS IN THE ZSM-5 STRUCTURE BY COMBINED SYNCHROTRON AND NEUTRON DIFFRACTION

B. TOBY[*a], S. PURNELL[*b], R. HU[b], A. PETERS[b] and D. H. OLSON[c]

[a] NIST Center for Neutron Research, National Institute for Standards and Technology, Gaithersburg, MD 20899; Brian.Toby@NIST.gov

[b] Washington Research Center, W.R. Grace and Company, 7500 Grace Drive, Columbia, MD 21044;

[c] Department of Chemical Engineering, University of Pennsylvania, 311A Towne Building, 220 South 33rd Street, Philadelphia, PA 19104.

*Authors to whom correspondence should be addressed.

ABSTRACT

We have prepared a sample of ZSM-5, MFI framework at a Si/Al ratio of ~12, exchanged with protons and also exchanged with deuterium. A single structural model was used to simultaneously fit neutron diffraction data for both the hydrogen and deuterium exchanged samples, as well as synchrotron diffraction data. This model did not conclusively demonstrate any hydrogen atom positions. A sensitivity analysis for the measurement proves that the hydrogen atoms must be distributed over at least three sites in the structure, and probably more.

INTRODUCTION

While the framework structure of ZSM-5 has been known for almost 20 years [1], the location of the protons hydrogen-exchanged material is not known. In the case of faujasite, the proton location was proposed on the basis of IR and x-ray diffraction data [2,3] and confirmed by several neutron diffraction studies of deuterated and protonated samples [4,5]. These studies take advantage of the fact that both ^1H and ^2H have large scattering cross sections for neutrons, but scatter with opposite phases. The case of MFI is more difficult than the case of FAU. There are fewer acidic sites in ZSM-5, and there are more possible cationic locations. Consequently, there have been several attempts to locate the energetically most probable proton sites. These studies assume that during the crystallization of the structure, aluminum incorporation occurs under conditions of energetic equilibrium. Also assumed is that the equilibrium configurations and distributions of the calcined (dried) and exchanged sample are the same as in the crystallizing slurry containing water and template or sodium. Calculations of the structure do not include contributions from the slurry.

Calculations have given a variety of results. Derouane and Fripiat [6,] have suggested T2 and T12 as the most likely candidates with protons located on O2, O24, and O13. Hay and Redondo [6] have suggested T9(O18), T1(O21), T5(O5), T12(O24), and T6(O18, O5) in that order as being the thermodynamically most likely. However, these authors point out that their calculations suggest that the energy differences from site to site are small, on the order of 20 KJ or less, and a distribution of protons over a wide range of sites seems likely. Schroder, Sauer, Leslie, and Catlow [7] also suggest "that bridging hydroxyl groups in ZSM-5 will be distributed over a broad range of framework sites even at low temperature, provided that this is controlled by normal statistical factors."

In this study we have used neutron and synchrotron powder diffraction results test these suggested sitings, and the results rule out occupancy greater than 0.5 at any single site.

EXPERIMENTAL

A sample of ZSM-5 was prepared at a Si/Al ratio of about approximately 12. This ratio was confirmed by several analyses. By elemental analysis, the sample contained 6.5% Al_2O_3, 93.5% SiO_2, and 0.04% Na_2O giving Si/Al = 12.2. Use of ^{29}Si MAS NMR gave a Si/Al ratio of 10.9. Isopropyl amine TPD gave 1100 μmoles of acidity per gram of zeolite, Si/Al = 14.1.

One aliquot of material was dehydrated at 200°C under vacuum and used for synchrotron and neutron diffraction measurements. A deuterated sample was prepared from the remainder of the material by treating with D_2O in a closed system under helium, drying at 200°C under vacuum, and repeating the treatment a total of four times. Analysis by FTIR showed that at least 80% of the sample was deuterated. An additional set of neutron measurements was made where the zeolite sample was dehydrated and then exchanged *in-situ* to avoid possible exposure to ambient conditions. In this case eight D_2O exposures were used, but dehydration was performed at 100°C. The results from the *in-situ* measurements did not differ significantly from the previous.

Neutron powder diffraction measurements were made at 25 K and at ambient conditions using the BT-1 diffractometer at the NIST NBSR reactor with both materials. Synchrotron powder diffraction measurements were made at the same nominal temperatures using the X7A diffractometer at the Brookhaven National Synchrotron Light Source with only the protonated material. Rietveld fits to the data were performed using the GSAS package [9] where a single structural model was used to simultaneously fit all three data sets collected at each temperature.

RESULTS AND DISCUSSION

Good fits were obtained using an orthorhombic *Pnma* symmetry SiO_2 framework and no H/D atoms (overall $\chi^2=2.3$ at 25 K and $\chi^2=1.6$ at 295 K). Weak soft constraints on T-O bond distances (total soft constraints contributed ~3% of the χ^2) were required. No evidence was seen for a monoclinic distortion. Figures 1(a) and 1(b) show the quality of the Rietveld fits without adding hydrogen atoms to the model.

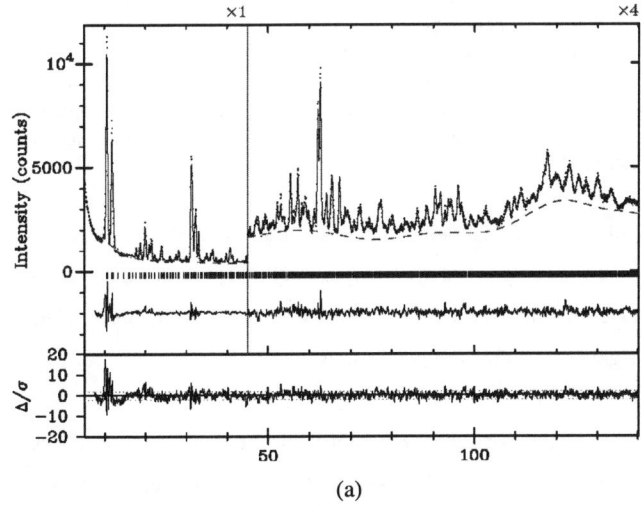

(a)

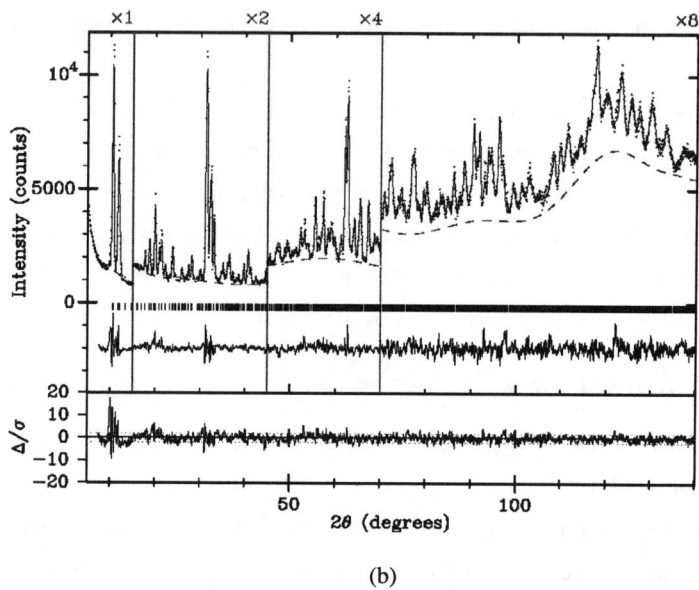

(b)

Figure 1. Rietveld fits for (a) neutron powder diffraction data for D-ZSM-5 and (b) synchrotron powder diffraction data. The upper plot shows the experimental diffraction pattern (black dots) superimposed with the pattern calculated from the derived crystal structure (line) in the top curve. The dashed line shows the fitted background. Below this, vertical lines show the positions of allowed reflections. The middle curve shows the difference between the observed and computed patterns and the lower curve shows the differences relative to the standard uncertainty (esd) for each data point.

A Si:Al ratio of 12:1 gives a nominal unit cell composition $H_{7.4}Al_{7.4}Si_{88.6}O_{192}$. If the H/D atoms are located at a single general symmetry crystallographic site, the expected occupancy would be 0.925. Neutron difference Fourier maps did not show any potential deuterium sites with Fourier density significantly greater than the noise level.

Further, a hydrogen atom for every O atom was inserted into the model at positions calculated in the plane of each T-O-T angle. Refining H/D atom occupancies resulted in values ranging from -0.11(3) to 0.21(5), for the room temperature refinement and -0.08(2) to 0.11(2) for the 25 K refinement. There was poor agreement between the results for each temperature. Attempts were also made assuming an approximate tetrahedral geometry for selected O atoms. Again, no significant occupancies were noted.

The sensitivity of the experiment to hydrogen atom siting was then demonstrated by adding H and D atoms to the model with full occupancy at positions corresponding to a single hydrogen bonded to atoms O6, O10, O20, O24, O18, O21. These models gave significantly worse fits than a model with no protons (for 25 K, χ^2=3.0-3.4 from 2.3 without H/D atoms). Adding pairs of H/D atoms at occupancy 0.46 also gave significantly worse agreement than the model with no H/D atoms. From these refinements, we conclude that the experiment would detect a proton located at a single site and would likely detect a proton if shared between two sites.

Figure 2 shows the impact that adding a single D atom at occupancy 0.95 makes to a section of the simulated neutron diffraction pattern. Effects of smaller magnitude are easily observed in Rietveld refinement.

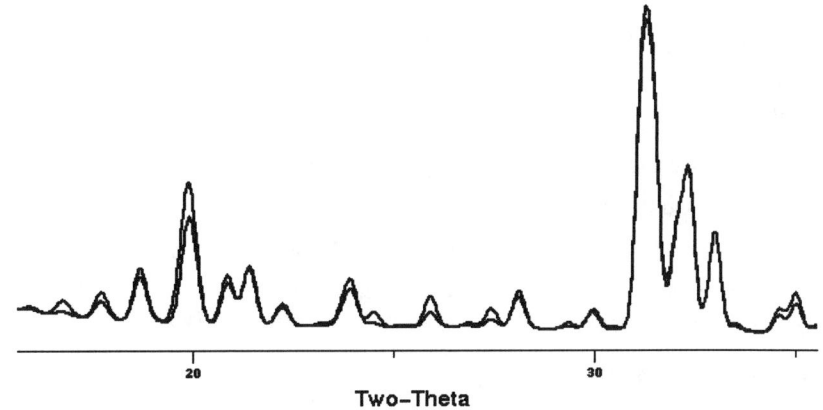

Figure 2. A simulated neutron diffraction pattern for a model with no D atoms and a model with a single D atom added at a single site.

CONCLUSIONS

Based on these results we can make the conservative conclusion that this experiment is sensitive to sites with hydrogen occupancy >0.45 (~4 atoms/unit cell) and is probably significantly more sensitive. This can only be consistent with our results if there are no acidic sites with occupancy greater than 0.45, which requires more than two acidic sites.

We thus have demonstrated that *hydrogen atoms must be present in at least three sites* in high-alumina ZSM-5, and likely more. Since the occupancy is sufficiently low that all hydrogen atoms could be accommodated in a single crystallographic site, this result is likely transferable to lower-silica forms of ZSM-5 as well. This conclusion is consistent with the more recent calculations [7,8].

REFERENCES

1. D. H. Olson, G. T. Kokotailo, S. L. Lawton, *J. Phys. Chem.*, 1981, **85**, 2238.
2. J.L. White, A.N. Jelli, J.M André and J.J. Fripiat, Trans. Faraday Soc. 1967, **63**, 461.
3. D.H. Olson and E. Dempsey, J. Catal. 1969, **13**, 221.
4. M. Czjzek, H. Jobic, A. N. Fitch, and T. Vogt, *J. Phys. Chem.*, 1992, **96**, 1535.
5. V. Bosacek, S. Beran, and Z. Jirak, *J. Phys. Chem.*, 1981, **85**, 3856.
6. E. G. Derouane and J. G. Fripiat, *Zeolites*, 1985, **5**, 165.
7. A. Redondo and P. J. Hay, J. Phys. Chem. 1993, **97**, 11754.
8. K.-P. Schroder, J. Sauer, M. Leslie, and C. R. A. Catlow, *Zeolites*, 1992, **12**, 20.
9. A. C. Larson and R. B. Von Dreele, GSAS, Generalized Structure Analysis System. Report LAUR 86-748; Los Alamos National Laboratory: Los Alamos, NM, 1988.

ACKNOWLEDGEMENT

Work at the National Synchrotron Light Source is supported by the U. S. Department of Energy, Departments of Materials Sciences and Chemistry.

SINGLE CRYSTAL STRUCTURE ANALYSIS OF A MICROCRYSTAL OF ZSM-11 USING SYNCHROTRON X-RAY DATA

H. VAN KONINGSVELD*, M. J. DEN EXTER, J. H. KOEGLER, C. D. LAMAN, S.L. NJO and H. GRAAFSMA[$]

*Laboratory of Organic Chemistry and Catalysis, Delft University of Technology, Julianalaan 136, 2628 BL Delft, The Netherlands; havank@cad4sun.tn.tudelft.nl
[$]European Synchrotron Radiation Facility, B.P. 220, 38043 Grenoble Cedex, France

ABSTRACT

For the first time a single crystal structure analysis of ZSM-11 has been performed. Pure single crystals of ZSM-11, with typical dimensions of 20 x 20 x 20 µm, were grown using N,N-diethyl-3,5-dimethylpiperidinium ions as template ions. The structure of a single microcrystal of as-synthesized ZSM-11 was successfully refined without any constraints on the framework atoms from room temperature synchrotron X-ray data in space group $I\bar{4}m2$ to R(Rw)=0.064(0.075). The strongly disordered template molecules were included at positions obtained from molecular mechanics calculations. The framework accuracy achieved in the analysis of the microcrystal of ZSM-11 is comparable to that obtained in the analysis of large ZSM-5 crystals. An accurate framework structure of ZSM-11 on atomic scale is now available for modelling studies (f.e. site acidity calculations) and further parameterization of force fields.

INTRODUCTION

The zeolites ZSM-5 and ZSM-11, with structure type codes MFI and MEL respectively, can form an infinite number of intermediate structures. These structures all consist of an array of parallel pentasil layers (See Fig.2) which can be connected in two different ways: neighbouring pentasil layers can be connected through inversion or reflection (mirroring). In general, without specific stirring (organic) template molecules, the two symmetry operations between the pentasil layers may be randomly distributed. The pure end-members, MFI and MEL, exlusively have inversion centres (in MFI) or mirror planes (in MEL) between successive pentasil layers. Pure MFI crystals can be grown using tetrapropylammonium-ions (TPA) as template ions. Recently, Nakagawa et al. [1,2] reported on a method to synthesize pure MEL phase using the novel organic template ion N,N-diethyl-3,5-dimethylpiperidinium (DDP; Fig. 1). Molecular mechanics calculations showed [3] that a mixture of cis- and trans-DDP stabilizes MEL best. Apparently,

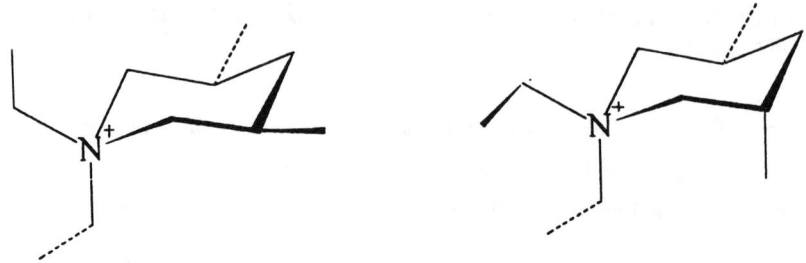

Figure 1. The organic template ion *N,N*-diethyl-3,5-dimethylpiperidinium (DDP): *cis*-DDP (left) and *trans*-DDP (right).

TPA and DDP are able to induce either inversion centres (TPA in MFI) or mirror planes (DDP in MEL) between the pentasil layers. This paper reports on the synthesis of single crystals of ZSM-11 using cis/trans-DDP as template and presents the subsequent structure analysis using synchrotron X-ray data.

EXPERIMENTAL

The procedure to synthesize ZSM-11 used was adopted from the one proposed by Nakagawa [1] using Aerosil 200 as the Si source and a 70/30% mixture of cis/trans DDP-iodide as template. The molar ratios were: 100 SiO_2 : 50 NaOH : 33.7 DDP : 8325 H_2O. The synthesis temperature was 180 °C and the synthesis time was 5 days, resulting in a batch composition of approximately 70% ZSM-11 crystals and 30% quartz [4]. The crystal used in the X-ray analysis measured 18 x 20 x 23 µm. The measurements were performed at the Materials Science beamline 9.8 of the European Synchrotron Radiation Facility. A horizontally and vertically focussed beam of 300 x 300 µm was used, with a wavelength of 0.6199 Å. Data were collected in 0.05° ω slices, with exposure times of 5 seconds per frame, using a SIEMENS 1k x 1k SMART detector. The crystal to detector distance was 3.8 cm. Data were integrated with the SIEMENS SAINT program. Initial positions of the framework atoms were taken from the literature [5]. All T atoms were treated as Si atoms. The structure was refined in $I\bar{4}m2$ by full matrix anisotropic least squares refinement on F without any constraint on the atoms to the conventional R(Rw) = 0.064(0.072) for all reflections with I > 2.5 σ(I) and θ_{max} = 30°. Lowering of the symmetry to I$\bar{4}$ did not improve the description of the model. Attempts to locate the template ions from difference Fourier maps failed. The template ions were therefore placed at positions obtained from molecular mechanics calculations [3] and refined as (disordered) rigid groups. Further details on data collection, refinement and final positional and equivalent isotropic displacement parameters of the template ions are available from the authors on request. Atomic coordinates of the ZSM-11 framework atoms are given in Table 1.

	x/a	y/b	z/c	U	PP
Si(1)	0.07697(3)	X	0	0.0165(2)	0.975(5)
Si(2)	0.12066(3)	0.18805(3)	0.14403(5)	0.0161(2)	0.919(3)
Si(3)	0.07657(3)	0.22544(3)	0.35715(5)	0.0177(2)	0.935(3)
Si(4)	0.27751(3)	0.18986(3)	0.14020(6)	0.0198(2)	0.945(3)
Si(5)	0.30711(3)	0.07607(3)	-0.00710(5)	0.0196(2)	0.951(4)
Si(6)	0.19003(3)	X	1/2	0.0202(2)	0.965(5)
Si(7)	0.07623(3)	0.37981(3)	0.35642(5)	0.0179(2)	0.831(3)
O(1)	0.0895(1)	0	0.0246(2)	0.0325(8)	
O(2)	0.0993(1)	0.11881(9)	0.0956(2)	0.0403(7)	
O(3)	0.0934(1)	0.1882(1)	0.2553(1)	0.0468(8)	
O(4)	0.1991(1)	0.1953(1)	0.1428(3)	0.067(1)	
O(5)	0.3000(1)	0.1214(1)	0.0900(2)	0.0514(8)	
O(6)	0.3038(1)	0	0.0289(2)	0.0321(7)	
O(7)	0.3088(1)	-X+1/2	1/4	0.0425(7)	
O(8)	0.11974(9)	0.1927(1)	0.4431(2)	0.0397(6)	
O(9)	0.2495(1)	0.1942(1)	0.4217(2)	0.0527(8)	
O(10)	0.0933(1)	0.30264(8)	0.3479(2)	0.0440(7)	
O(11)	0	0.2129(1)	0.3850(2)	0.0347(8)	
O(12)	0	0.3909(2)	0.3893(2)	0.0374(9)	
O(13)	0.0903(1)	0.2486(1)	0.0830(2)	0.0542(8)	
O(14)	0.1228(1)	0.4117(1)	0.4399(2)	0.0547(8)	
O(15)	0.0870(1)	-X+1/2	1/4	0.066(1)	

Table 1. Atomic positional, isotropic displacement and site occupation parameters.

RESULTS AND DISCUSSION

Figure 2 shows the pentasil layer in ZSM-11 and the orientation of the template ions. The three-dimensional structure is obtained by connecting mirror related pentasil layers along **b**. In MEL straight intersecting channels run parallel to the **a** and **b** axes. There are two types of intersections. In type 1 (MEL1) the overlap of the intersecting channels is large and in type two (MEL2) this overlap is small. Distances and angles in as-synthesized ZSM-11 are compared in Table 2 with the values in as-synthesized ZSM-5 containing the TPA ion as template. The overall Si-O distance (1.591 Å) and Si-O-Si angle (153.8°) are in good agreement with those from known porous silica phases. Variations of Si-O and Si-O-Si are essentially equal. The geometry of the ZSM-11 framework is slightly more accurate determined than that of the ZSM-5 framework. Subtle differences in the local framework geometry lead to differences in average $Si(OSi)_4$ angles (α). There exists a quantitative relationship between α and the chemical shift (δ) for the $Si(OSi)_4$ signal. In Table 3 α and δ are given. The calculated chemical shifts agree with the experimental values presented in Fig.3b of Ref.[6]. Besides some small differences in the line positions, the only remarkable deviation is the resonance of Si(7) which is calclated at -113.4 ppm but appears at -112.4 ppm in the experimental spectrum. Nevertheless, the good agreement may be regarded as a further independent test for the reliability of the structure refinement. The 10-ring pores at both intersections are elliptical with limiting apertures (measured as diagonal O--O distances) of 7.670 x 8.244 Å in MEL1 and 7.941 x 8.212 Å in MEL2. The same devation from circular pores was observed in as-synthesized ZSM-5.

	ZSM-11	ZSM-5
O-Si-O range (°)	106.6 - 112.4(2)	106.0 - 112.0(3)
Average O-Si-O/ SiO_4 (°)	109.5	109.5
Si-O range (Å)	1.576 - 1.606(3)	1.567 - 1.605(4)
Range of av. $Si(O)_4$ (Å)	1.586 - 1 596	1.580 - 1.591
av. Si-O (Å)	1.591	1.586
Si-O-Si range (°)	144.2 - 170.7(3)	144.9 - 175.9(4)
Range of av. $Si(OSi)_4$ (°)	149.8 - 159.5	150.5 - 162.8
av. Si-O-Si (°)	153.8	155.4

Table 2. Comparison of the framework geometry in as-synthesized ZSM-11 and ZSM-5.

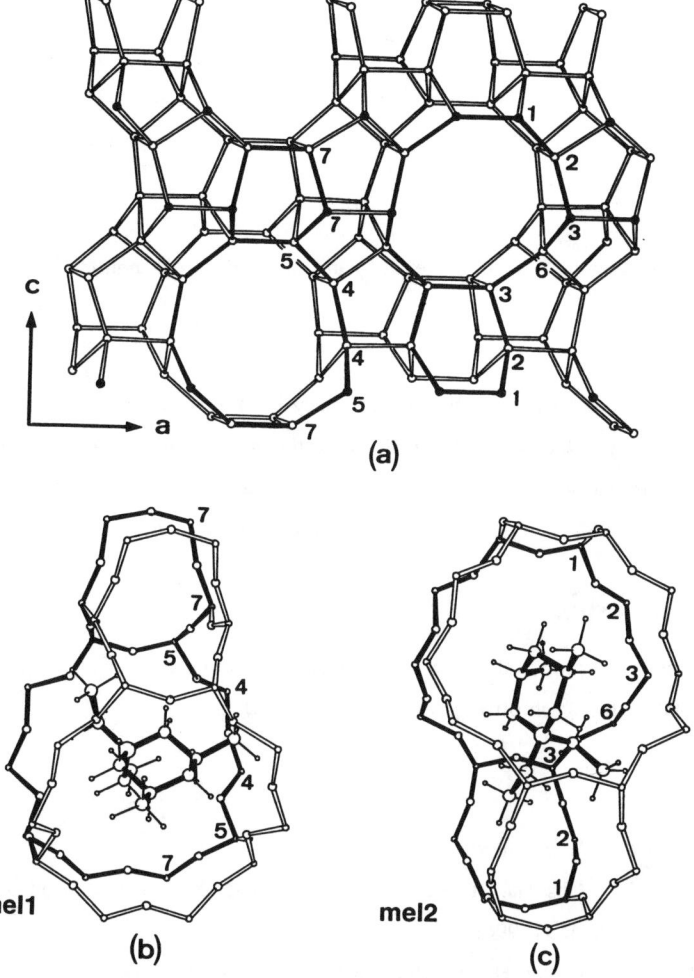

Figure 2. Pentasil layer and intersections in ZSM-11. Parts of the intersections MEL1 and MEL2 are heavily outlined. A neighbouring layer can be connected through T atoms indicated by full circles.

(a): Connectivity and T atom numbering used.

(b): Template ion cis-DDP at the MEL1 intersection.

(c): Template ion trans-DDP at the MEL2 intersection.

Si(n) n =	α	δ	Si(n) n =	α	δ
1	149.9	-112.0	5	153.4	-114.1
2	159.5	-117.0	6	154.0	-114.4
3	150.0	-112.2	7	152.1	-113.4
4	157.5	-116.1			

Table 3. Average Si(OSi)$_4$ angle (α, °) and chemical shift (δ, ppm) in ZSM-11.

CONCLUSIONS

For the first time a single crystal structure analysis of ZSM-11 has been performed. The framework accuracy achieved in the analysis of the microcrystal of ZSM-11 using X-ray diffraction data from a synchrotron source is comparable to the framework accuracy obtained in the analysis of large ZSM-5 crystals using diffraction data from a conventional X-ray tube.

ACKNOWLEDGEMENT

We are grateful to prof. Bill Clegg for his assistance during preliminary measurements on ZSM-11 crystals at the synchrotron source in Daresbury, U.K.

REFERENCES

1. Y. Nakagawa, WO Patent 95/09812 (1995).
2. O. Terasaki, T Ohsuma, H. Sakuma, D. Watanabe, Y. Nakagawa and R.C. Medrud, Chem. Mater. **8**, 463 (1996).
3. S.L. Njo, J.H. Koegler, H. van Koningsveld and B. van de Graaf, Microporous Mater. **8** 223 (1997).
4. M.J. den Exter. Thesis. Delft University of Technology, 1996.
5. C.A. Fyfe, H. Gies, G.T. Kokotailo, C. Pasztor, H. Stroble and D.E. Cox, J. Am. Chem. Soc. **111**, 2470 (1989).
6. C.A. Fyfe, Y. Feng, H. Grondey, G.T. Kokotailo and A. Mar, J. Phys. Chem. **95** 3747 (1991).

LASER-INDUCED LUMINESCENCE INVESTIGATION OF Y ZEOLITES CATALYSTS SURFACES

C. LALO, J. DESON, A. GEDEON and J. FRAISSARD

Laboratoire de Chimie des Surfaces UPMC-CNRS, ESA 7069, 4 place Jussieu 75252 Paris cedex 05; jfr@ccr.jussieu.fr

ABSTRACT

We report two attempts for characterization of species at low concentration in Y zeolites by Laser-induced Luminescence spectroscopy. In reduced copper-exchanged NaY zeolites, laser-induced phosphorescence of Cu^+ ions allows the determination of the density of Cu^+ sites, while in HY zeolites acidic adsorption sites can be identified by laser-induced fluorescence spectroscopy of an aromatic base, quinoline, adsorbed on the surface.

INTRODUCTION

Analysis of Y zeolites surfaces using laser-induced luminescence diagnosis allows the characterization of species at low density level [1]. From the linear dependence of the intensity of the luminescence upon the concentration of the species in this homogeneous medium, relative densities in the sample can be measured.

In reduced copper-exchanged NaY zeolites time-resolved laser-induced phosphorescence of Cu^+ ions allows the determination of the density of different Cu^+ sites [2]. In HY zeolites, laser-induced fluorescence spectroscopy of an aromatic base, quinoline which can be introduced into Y zeolite supercage allows the recognition of acidic surface sites.

EXPERIMENTAL

Samples preparation and characteristics

Cu-exchanged Y zeolites

The initial NaY zeolite used is LZY-52 from (UOP) (Si/Al = 2.4). The chemical composition of the copper-exchanged zeolites corresponds to a Cu^{2+} for Na^+ exchange level (λ) 35 or 58%. The prepared samples were introduced directly in a quartz optical cell and evacuated at ambient temperature for 5 h (10^{-4} Pa). The zeolites were then heated slowly to the desired temperature T_d (623 K) and held at this temperature for time t (12 h or 34 h) and then brought slowly back to 298 K. The dehydrated samples (D) are denoted $Cu\lambda Y-T_d(D/t)$. Some of them were reduced in hydrogen (10^3 Pa) for 30 mn or 1 h at constant temperature (T_r = 623 K or 723 K). The reduced samples (R) are denoted $Cu\lambda Y-T_r(R/t)$.

Acidic Y zeolites

Three types of Y zeolites have been studied, NaY, HY and dealuminated HY. NaY zeolite (LZY52) was supplied by Union Carbide Co. HY zeolite was UOP NH_4Y zeolite, denoted Y64. D-HY zeolite was prepared from NH_4Y. It was partially dealuminated by steaming and subsequently washed twice with Na_2H_2EDTA; its crystallinity is near 90%.

In these samples, adsorption sites are silanol groups SiOH (NaY), Brønsted acidic bridging SiO(H)Al groups (HY) and in the last, both Brønsted sites and electron-acceptor Lewis centres which may be coordinatively unsaturated aluminium ions.

The three samples were activated under « shallow bed conditions » : powder layer less than 5 mm thick; dehydration under dynamic vacuum before and during heating for 14 h at 12 $K.h^{-1}$ to 725 K at less than 10^{-2} Pa. Quinoline vapour was adsorbed at 300 K. The amount of quinoline in each sample was determined by volumetry and gravimetry. The samples prepared contain 0.1 and 1 mg of quinoline per gram of zeolite.

Experimental technique

In a quartz optical cell under vacuum (10^{-4} Pa) zeolites samples were photoexcited at 240 nm using a frequency-doubled dye laser, and the fluorescence was recorded from 300 to 650 nm. The laser emitted 0.3 mJ at 240 nm, the pulse width being about 30 ns. The size of the beam was adjusted by a dispersing lens and a diaphragm so as to obtain maximum coverage of the sample.

Luminescence was detected perpendicularly to the laser beam through a quartz collecting lens by a photomultiplier (Hamamatsu R212UH) at the exit slit of a monochromator and stored in a gated integrator. The signal decay at selected wavelengths was recorded with a digital storage oscilloscope (Tektronics 2432) and transferred to a microcomputer (PC compatible), the limit of the detection system being 1 µsec.

RESULTS

Laser-induced phosphorescence of Cu^+ ions in reduced copper-exchanged Y zeolites

The absorption of UV light by Cu^+ ions occurs as a singlet-singlet transition $3d^{10} \rightarrow 3d^9 4s$; energy is then transferred non-radiatively to the lowest lying triplet state and phosphorescence is observed. The laser-induced photoluminescence spectrum of vacuum dehydrated samples or samples reduced in hydrogen shows a broad-band emission from 450 nm to 600 nm. (Figure1).

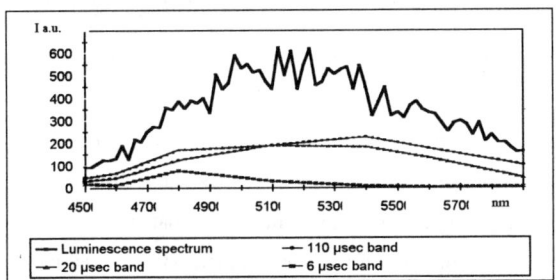

Figure 1. Decomposition of the laser-induced luminescence spectrum for samples reduced in hydrogen Cu35Y-723(R/30min).

By following the luminescence at selected wavelengths, a tri-exponential decay was observed showing the contributions of three components with lifetimes 6 ± 2, 20 ± 5, and 110 ± 10 µs and the luminescence spectra can be decomposed into several bands specific for each component.

From a linear dependence of the intensity of each luminescence band upon the concentration of the species in the sample, different trends for the variation of the concentration with the sample pretreatment are found: the influence of the sample Cu-content, of the reduction time and of the temperature have been studied (Figure2).

Two of the emitting centres with lifetimes of 20 and 110 µsec correspond to Cu^+ ions with different locations in the supercage as has been shown by ^{129}Xe-NMR.

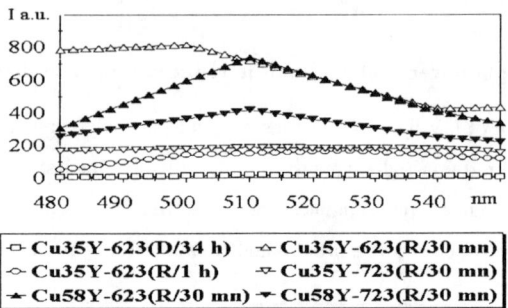

Figure 2. Effect of pretreatment on the laser-induced luminescence spectra of the 20 μs component

It appears that significant concentrations of Cu^+ sites can be detected in samples heated at high temperature under vacuum for a long time, showing that Cu^{2+} ion is to some extend autoreduced to cuprous ion during dehydration and that prolonged reduction in hydrogen diminishes the concentration of Cu^+ ion which can undergo further reduction to Cu^0.

Laser-induced fluorescence of quinoline adsorbed on acidic Y zeolites

Quinoline is well known to exhibit only phosphorescence in non-polar media. In polar media, the fluorescence is attributed to an inversion of the relative disposition of energy levels of n,π^* and π,π^* character. Moreover, this N-heterocyclic base offers interesting luminescence spectra when it interacts with an acid : fluorescence spectra of quinolinium ion and of charge-transfer complexes have been observed.

Under photoexcitation at 240 nm, laser-induced luminescence spectra of quinoline adsorbed on the three samples reveal interactions between the adsorbate and surface sites (Figure3).Unstructured broad bands with lifetimes less than 1 μsec are observed; this short lifetime indicates that it is a fluorescence.

Emitting centres were identified from the emission origin and maxima of the bands :

On NaY, a broad band peaking at 380 nm with an emission origin near 320 nm is characteristic of the formation of the a hydrogen-bonded quinoline complex between quinoline and non-acid silanol groups.

On HY samples, the luminescence is the most intense. The less loaded specimens exhibit two bands peaking at 400 nm and 450 nm. The band peaking at 400 nm with an emission origin at 360 nm which becomes the more intense for more loaded samples, corresponds to the spectrum of the protonated form of quinoline: the protonation of quinoline on BrØnsted sites is characterized by the fluorescence of quinolinium ion .

On dealuminated-HY zeolite, the luminescence is significantly more extended towards longer wavelengths: the complexation with electron-acceptor Lewis sites is characterized by the fluorescence of a charge-transfer complex which is strongly red-shifted compared to the cation.

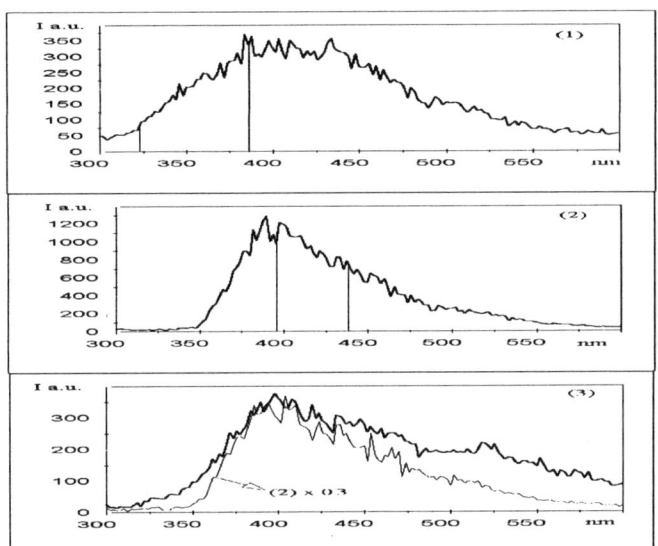

Figure 3. laser-induced fluorescence spectrum of quinoline adsorbed
(1) on NaY (0.1 mg / g);(2) on HY (1 mg / g);(3) on dealuminated D-HY (0.1 mg /g)

An attempt was made to observe a saturation of these sites by increasing the concentration of adsorbate. However, it appears that a quantitative determination of the concentration of acidic sites in the cavities should be only possible for highly cristalline HY zeolites, assuming that the fluorescent probe is uniformely distributed within zeolites.

REFERENCES

1. J. Dedecek and B. Wichterlova J. Phys. Chem. **98**, 5721 (1994)
2 . J. Deson, C. Lalo, A. Gédéon, F. Vasseur and J. Fraissard Chem. Phys. Letters **258,** 381 (1996)

STRUCTURAL INVESTIGATION OF SORBATE-INDUCED PHASE TRANSITIONS IN ZSM-5 BY FT-RAMAN SPECTROSCOPY

Y. HUANG and P. QIU

Department of Chemistry, The University of Western Ontario, London, Ontario, Canada N6A 5B7; yhuang@julian.uwo.ca

ABSTRACT

FT-Raman spectra of several p-dihalogenobenzenes adsorbed on ZSM-5 were measured as a function of loading. A sorbate-induced phase transition in ZSM-5 starting at a loading of 4 molecules/u.c. was detected for both p-dichloro- and p-dibromobenzene/ZSM-5 complexes. Adsorption of p-difluorobenzene in ZSM-5 also caused a sudden structural change in the zeolitic framework at a loading of 5 molecules/u.c. The present work demonstrates that Raman spectroscopy is a very useful new tool for the investigation of sorbate-induced phase transitions in zeolites.

INTRODUCTION

Many zeolites undergo structural changes induced by sorbed organic molecules. Phase transitions of this type are very important in the applications of these materials as sorbents and catalysts. In the past, XRD and ^{29}Si MAS NMR have been extensively utilized to study the sorbate-induced structural changes [1]. Recently, we have demonstrated that FT-Raman spectroscopy is a useful tool for the investigation of sorbate-induced phase changes in zeolites [2]. For the p-xylene/ZSM-5 system, we have shown that the spectral parameters of guest molecules such as band frequency, splitting and linewidth are very sensitive to structural transitions in the host framework. In the present study, FT-Raman investigations of the p-difluorobenzene (pdfb)/ZSM-5, p-dichlorobenzene (pdcb)/ZSM-5 and p-dibromobenzene (pdbb)/ZSM-5 systems have been undertaken to examine the host-guest interactions in these systems. For pdcb/ZSM-5, there are excellent NMR and XRD data available in literature [3-5] against which to check the viability and reliability of the new Raman approach. *p*-Dibromobenzene/ZSM-5 is an incompletely characterized system [6] and pdfb/ZSM-5 complexe has not yet been examined.

EXPERIMENTAL

Samples loaded with *p*-dihalogenobenzenes were prepared by adding precisely measured amounts of *p*-dihalogenobenzenes to weighed aliquots of ZSM-5 (Si/Al=336) in glass vials. The vials were then sealed and placed in an oven at 80 °C (3 hrs), 150 °C (3 hrs) and 190 °C (15 hrs) for pdfb/ZSM-5, pdcb/ZSM-5 and pdbb/ZSM-5 complexes, respectively, to uniformly disperse the sorbate molecules throughout the sample. All Raman spectra were recorded at room temperature on a Bruker RFS-100 FT-Raman spectrometer equipped with a Nd^{3+}:YAG laser operating at 1064.1 nm. The laser power was typically 80 mW at the sample. The resolution was 2 cm^{-1}.

RESULTS AND DISCUSSION

p-Dichlorobenzene/ZSM-5: Previous single crystal X-ray diffraction studies have suggested that the structure of ZSM-5 with a loading of 2.6 pdcb molecules/u.c. (low-loaded form) is orthorhombic, Pnma [3], while the high-loaded form (8 molecules/u.c.) is also orthorhombic, but has a different space group ($P2_12_12_1$) [4]. FT-Raman spectra of pdcb adsorbed on ZSM-5 were measured as a function of loading from 1.5 to 8 molecules/u.c. Within the loading range of 1.5-3 molecules/u.c., the spectra of pdcb were independent of the loading, indicating that the additional pdcb molecules access identical sites inside the framework. This also suggests that the structure of ZSM-5 remains unchanged throughout the loading range mentioned above. When the loading was increased to 4 molecules/u.c. the spectrum of adsorbed pdcb started showing distinct changes. The changes were complete at a coverage of 7 molecules/u.c. These results indicate that the zeolitic host framework undergoes a phase transformation in the 4-6 molecules/u.c. loading range and the transition is complete at a loading of 7 molecules/u.c. At the transition, several vibrational modes exhibited large shifts to higher energies. Fig. 1 illustrates a sudden increase in the frequency of the C-Cl out-of-plane bending mode at a loading of 5 molecules/u.c. The transition from the low- to high-loaded phase was also accompanied by changes in the linewidth. Fig. 2 shows the effect of loading on the linewidth of a ring deformation mode. Upon transition, some bands split into several components. Fig. 3 shows the splitting in the C-H and C-C stretching regions.

p-Dibromobenzene/ZSM-5: The Raman spectra of pdbb/ZSM-5 complex were measured with respect to the loading of sorbate. Similar to the situation for pdcb/ZSM-5, the spectra of pdbb adsorbed in ZSM-5 looked identical in the low coverage range (1.5-3 molecules/u.c.). Marked changes in spectral parameters started taking place at a loading of 4 molecules/u.c., implying the onset of a phase transition. The changes were complete at a loading of 7 molecules/u.c., indicating the completion of the phase transition. The spectrum of the high-loaded phase was quite different from that of the low-loaded phase in that most of the bands were broader and many of them were split into doublets. Also, there was considerable splitting of the C-H stretching modes (Fig. 4A). Within a loading range of 3-7 molecules/u.c., the sorbate spectra changed continuously with increasing loading. For example, the C-Br out-of-plane bending mode appeared as a single band at 266 and 273 cm^{-1} in the low- and high-loaded phases, respectively. But, these two peaks co-existed in the spectrum of each sample with intermediate loading (4, 5, 6 molecules/u.c.) (Fig. 4B). The 273 cm^{-1} band gained intensity with increasing loading at the expense of the 266 cm^{-1} band, indicating that both low- and high-loaded phases of ZSM-5 are simultaneously present in different proportions. From the intensities of deconvoluted spectra, the effect of sorbate concentration on the extent of the high-loaded phase can be

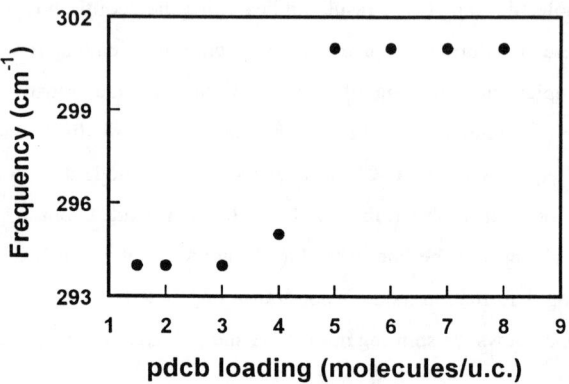

Figure 1. Plot of the frequency of C-Cl bending mode of pdcb adsorbed in ZSM-5 as a function of the loading.

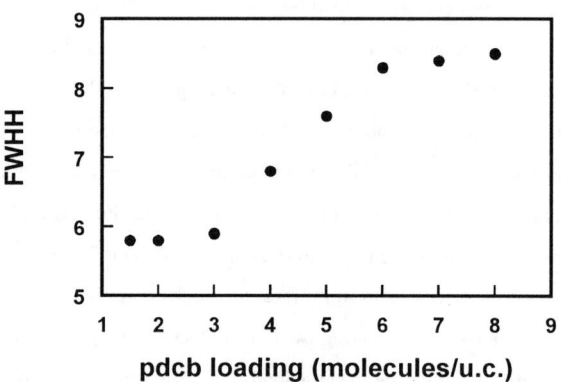

Figure 2. Plot of the full width at half-height in cm^{-1} of a Raman mode at 747 cm^{-1} of pdcb as a function of the loading.

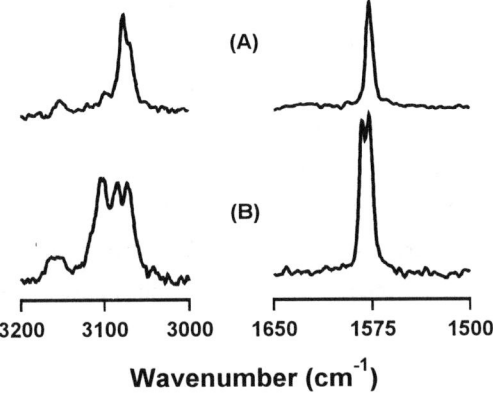

Figure 3. FT-Raman spectra of pdcb adsorbed in ZSM-5 in the regions of 3200-3000 cm^{-1} and 1700-1500 cm^{-1} at a loading of (A) three and (B) seven molecules/unit cell.

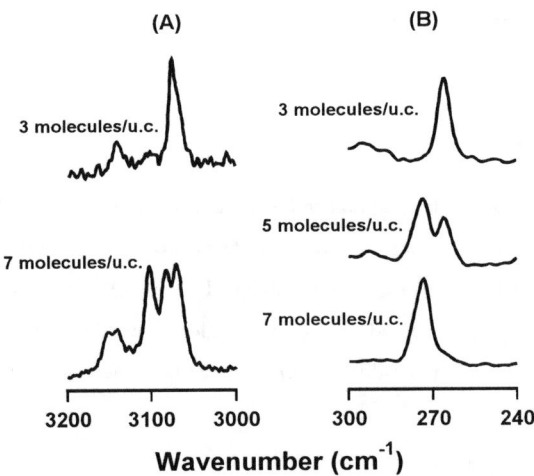

Figure 4. FT-Raman spectra of pdbb adsorbed in ZSM-5 in the regions of (A) 3200-3000 cm^{-1} and (B) 300-240 cm^{-1}.

evaluated and is shown in Fig. 5. The vibrational modes of the zeolitic framework also exhibited some interesting changes. For instance, the most intense Raman band of the zeolitic framework appeared as a broad envelop centered at about 370 cm^{-1} in the low-loaded phase. Upon transition, the width of this profile decreased dramatically from 57 cm^{-1} for low-loaded phase to 30 cm^{-1} for high loaded phase (Fig. 6). The spectral pattern of high-loaded pdbb/ZSM-5 complex looked very similar to that of high-loaded pdcb/ZSM-5. The similarity between the spectra of the high-loaded phases for the two complexes suggests that their high-loaded phases probably have the same structure.

p-Difluorobenzene: Absorption of pdfb in ZSM-5 also induced a phase transition. The Raman spectra of pdfb absorbed in ZSM-5 showed many distinct changes at a loading of 5 molecules/u.c. For instance, Fig. 7 shows the splitting in the C-H stretching, C-H in plane deformation and C-C-C bending regions. In contrast to the situation for the two sorbate/zeolite complexes discussed above where the transitions detected were sluggish, the structural change in pfdb/ZSM-5 appeared to be rather sudden.

CONCLUSIONS

The present work further demonstrates the usefulness of Raman spectroscopy in study of host-guest interaction in zeolitic systems. For the *p*-chlorobenzene/ZSM-5 complex, our results confirmed those obtained from XRD and NMR. A phase transition induced by sorbed guest molecules was identified in both *p*-dibromobenzene/ZSM-5 and *p*-difluorobenzene/ZSM-5 systems.

ACKNOWLEDGEMENT

Y.H. thanks the support of the NSERC of Canada in the form of operating grant.

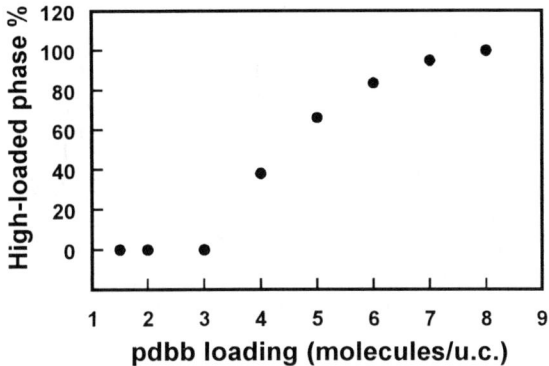

Figure 5. The effect of pdbb concentration on the extent of the high-loaded phase of ZSM-5.

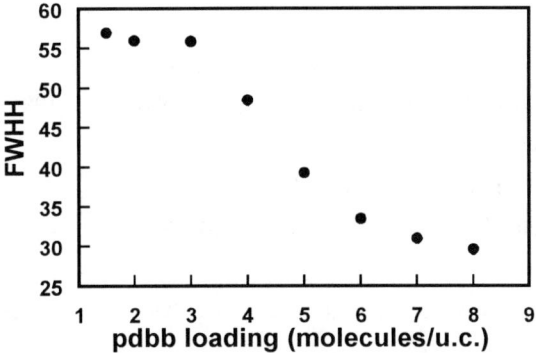

Figure 6. Plot of the full width at half-height in cm^{-1} of the most intense Raman band of the zeolite framework as a function of the loading.

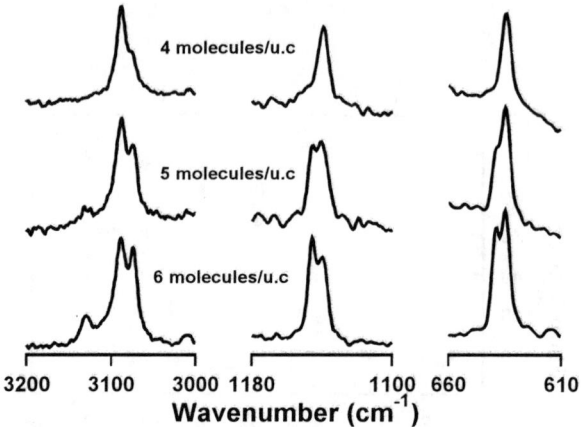

Figure 7. FT-Raman spectra of pdfb adsorbed in ZSM-5 in the regions of (A) 3200-3000 cm^{-1}, (B) 1180-1100 cm^{-1} and (C) 660-610 cm^{-1}.

REFERENCES

1. C.A. Fyfe; K.T. Mueller, G.T. Kokotailo, in <u>NMR Techniques in Catalysis</u>, Edited by A.T. Bell and A. Pines (Marcel Dekker Inc., New York, 1994) pp. 11-67 and the references therein.
2. Y. Huang, J. Am. Chem. Soc., **118**, 7233 (1996).
3. H. van Koningsveld, J.C. Jansen and A.J.M. de Man, Acta Cryst., **B52**, 131 (1996).
4. H. van Koningsveld, J.C. Jansen and H. van Bekkum, Acta Cryst., **B52**, 140 (1996).
5. H. Gies, B. Marler, C.A. Fyfe, G. T. Kokotailo, Y. Feng and D.E. Cox, J. Phys. Chem. Solids, **52**, 1235 (1991).
6. B.F. Mentzen, Mat. Res. Bull., **27**, 831 (1992)

ESCA STUDY OF ZEOLITES

I. JIRKA

J. Heyrovský Institute of Physical Chemistry, Academy of Sciences of the Czech Republic, Czech Republic; JIRKA@JH-INST.CAS.CZ

ABSTRACT

Initial and final state effects influencing core level binding energies, kinetic energies of the KLL Auger transitions of Al, and the line shapes of the Al KLL Auger lines of the series of zeolites with decreasing concentration of Al are discussed. Dependence of the screening of the photoelectron and Auger hole states of Al on the structure of zeolite is demonstrated.

INTRODUCTION

Electron Spectroscopy for Chemical Analysis (ESCA) enables simultaneous investigation of the electron structure of the surface layers of given sample and its (semi)quantitative analysis. The information on electron structure can be obtained by interpretation of the influence of the initial and final state effects affecting the values of the core level binding energies E_b of pertinent elements. However, this interpretation is still rather controversial. Possible initial state effects are related with structural changes in the framework of zeolite (i.e. with the changes of the T-O bond lengths and T-O-T (O-T-O) bond angles). These effects may be influenced by chemical modification of the zeolite framework (concentration of skeletal Al, isomorphous substitution) as well as by ion exchange. Final state effects are related with screening mechanisms of the hole states produced by photoemission and Auger deexcitation. Influence of rehybridisation of valence orbitals appearing in the series of zeolites with decreasing concentration of Al on E_b values of Al 2p photoelectron line, and on the line shapes of Al KLL Auger spectrum is demonstrated.

EXPERIMENTAL

The series of well characterised high crystalline zeolites (sodium forms of zeolite A (Si/Al = 1), dealuminated zeolites Y (Si/Al =2.5 - 16.9) ferrierite (Si/Al = 8.0), mordenite (Si/Al = 5.2), and ZSM-5 (Si/Al = 10.2)) containing no extraframework AlO_x species were used. Details of preparation of the samples may be found elsewhere (see 1 and references therein). The photoelectron and Auger spectra were measured on an ESCA III Mk 2 (VG Scientific) spectrometer with twinning Al/Mg anode. The Al 2p, Si 2p, O 1s and C 1s lines were measured using constant pass energy of 50 eV and nonmonochromatised Al $K\alpha_{1,2}$ line ($h\nu$ = 1486.6 eV). The Al KLL Auger spectrum was detected at a pass energy of 100 eV using the bremsstrahlung component of the Mg radiation. Lines used for calibration of the Al KLL spectra were simultaneously measured applying Mg $K\alpha_{1,2}$ line at the same conditions. Our E_b values (1) are compared with data published by others (2 - 5), which were recalibrated using the E_b of C 1s photoelectron line (284.4 eV) to achieve identical calibration with our data. The base pressure during experiment was typically less than $\sim 10^{-7}$ Pa. Further details on conditions of measurement can be found in (1).

RESULTS AND DISCUSSION

To discuss core level E_b shifts, the modified charge potential model is frequently used:

$$\Delta E_b = E_b^{(x)} - E_b^{(stand)} = -k\Delta q + dV - \Delta R + \Delta \chi \qquad (1)$$

where the first two terms account for initial state effects (the change of charge transfer and the change of Madelung potential), and the second two terms account for final state effects (the change of relaxation energy and the shift of zeolite - spectrometer contact potential). The trends summarised below have been found in the series of zeolites with decreasing concentration of Al (4, 5, 7):

1. Ionicity of T-O bond is increasing.
2. The T-O bond length is decreasing and the O-T-O bond angle is increasing.
3. Madelung potentials are increasing.

4. Relaxation of the final hole states is decreasing.

The trends 1, 3, 4 induce an increase of the E_b values with decreasing concentration of Al, the effect of structural changes of the zeolite framework is qualitatively discussed below. To estimate which trend dominantly influences the ΔE_b values the effect of the change of zeolite-spectrometer contact potential should be known. The shift of E_b can be induced by $\Delta\chi$ only, in principle (i.e. $-k\Delta q + dV - \Delta R = 0$). This limiting case may be simply tested by plotting E_b values of cation species (Al, Si) against the E_b of anion (O) (6). If the E_b values are shifted solely by the $\Delta\chi$, then

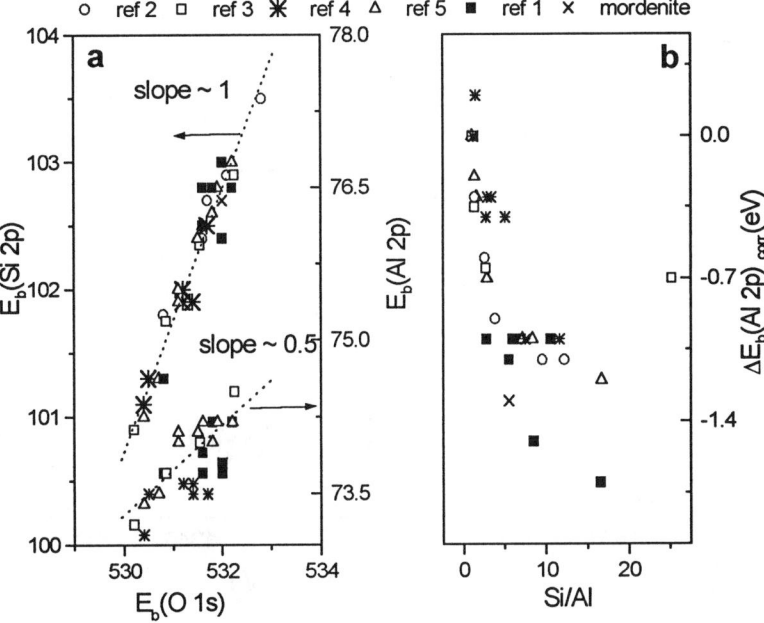

Figure 1

Correlation of the E_b of Si 2p and Al 2p photoelectron lines with the E_b of O 1s photoelectron line (a) and the dependence of ΔE_b(Al 2p) values (ΔE_b(Al 2p) = ΔE_b(Al 2p)$^{(x)}$ - ΔE_b(Al 2p)$^{(A)}$) corrected on $\Delta\chi$ on Si/Al (b).

this plot should be linear with unit slope. Such dependence with high correlation coefficient was observed for the E_b of the Si 2p photoelectron line. Thus, the ΔE_b(Si 2p) values can be used for evaluation of the shift of zeolite/spectrometer contact potential $\Delta\chi$. Only poor correleation exist between the E_b of Al 2p and O 1s lines. The

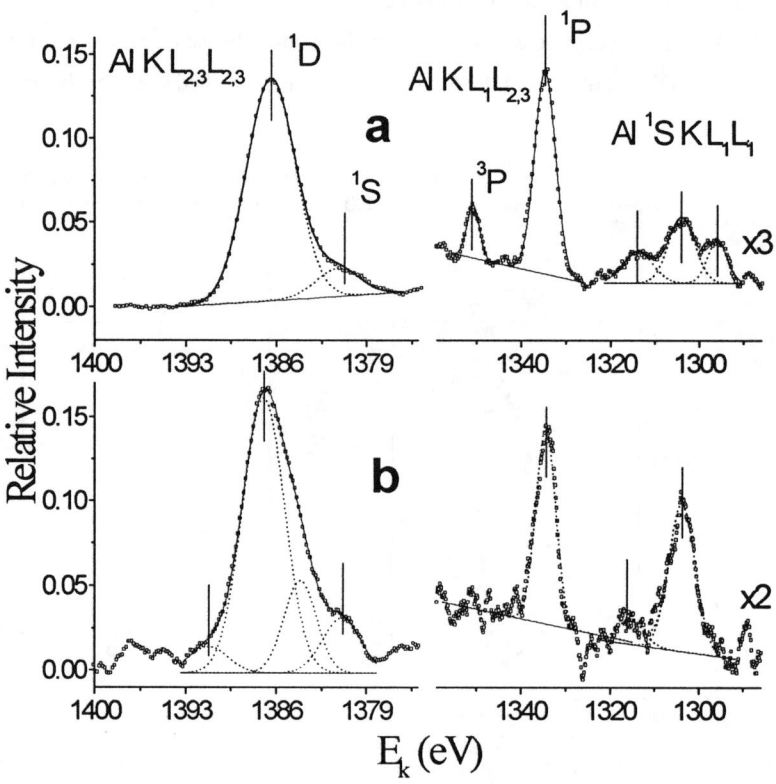

Figure 2
The Al KLL Auger spectra of faujasite (a, Si/Al = 2.5) and mordenite (b, Si/Al = 5.2) slope of this dependence is far from unity (see Figure 1a). Though, together with the shift of reference level some other effects are influencing the value of E_b(Al 2p).

Dependence of ΔE_b(Al 2p) values corrected on $\Delta \chi$ (estimated from ΔE_b of Si 2p line) on Si/Al is depicted in Figure 1b. Some discrepancies among our and published data which typically do not exceed ±0.3 eV were observed (experimental error of estimation of our ΔE_b(Al 2p) values is ±0.2 eV). However, negative shift of the E_b(Al 2p) is unambigously evident with decreasing concentration of Al in the zeolites. This effect is tentatively explainable as a consequence of increase of s character of Al - O bond, which is appearing in the series of zeolites with decreasing concentration of Al in the skeleton.

Analogously may be explained the line shape and intensity changes of the Al K L_1L_1, Al K $L_1L_{2,3}$ and Al K $L_{2,3}L_{2,3}$ Auger spectra, displayed in Figure 2. The Al KLL Auger spectra of faujasite and mordenite are depicted there. Similar Auger line shapes like for mordenite were also observed for ZSM-5 and ferrierite (not shown in the Figure). Dependence of probabilities of Al KLL Auger transitions on Al - O bond length has been demonstrated (1). The shapes of Al KLL Auger lines are affected by decrease of concentration of Al nonequivalently. Two lines are observable in the Al K $L_{2,3}L_{2,3}$ and Al K $L_1L_{2,3}$ Auger spectrum of faujasite. These lines belong to ^{1}D and ^{1}S final states (Al K $L_{2,3}L_{2,3}$ spectral region), and ^{1}P and ^{3}P final states (Al K $L_1L_{2,3}$ spectral region), respectively (8). Three lines are detected in the region of Al ^{1}S K L_1L_1 Auger spectrum of faujasite. Identification of Al K L_1L_1 Auger lines is at present not straightforward. Different line shapes of Auger Al KLL spectrum were observed for mordenite. Two lines were detected in the region of Al ^{1}D K $L_{2,3}L_{2,3}$ and Al ^{1}S K L_1L_1 Auger spectra. The Al ^{3}P K $L_1L_{2,3}$ Auger line was not observable for this zeolite. Only one line at E_b = 73.9 eV was observed in the Al 2p photoelectron spectrum of both mordenite and faujasite. Detected line shape changes in the Al KLL Auger spectra may be explained as a consequence of the change of screening of the Auger final hole states due to increasing s character of Al - O bond. It follows from above interpretation that screening of the hole states after Auger deexcitation is rather local. In the case of delocalised (i.e. purely electrostatic) screening the energy of all the Auger hole states would be influenced identically.

series of zeolites with decreasing concentration of Al is dominantly influenced by increasing participation of the s-type electrons in the Al - O bonds. This rehybridisation causes more effective screening of the hole states produced by photoemission. Screening of the Auger final states localised on Al is dependent on their symmetry.

ACKNOWLEDGEMENT

This work was supported by Grant COST D5/0002/94 and by Ministry of Education of the Czech Republic (0CD5.10/1998). Access to the spectrometer ESCA III Mk 2 provided by dr. Z. Bastl is highly appreciated.

REFERENCES

1. I. Jirka, J. Phys. Chem., **101**, 8133 (1997).
2. J.Z. Shyu, E.T. Skopinski, J.G. Goodwin, and A. Sayari, Appl. Surf. Sci,. **21**, 297 (1985).
3. T.L. Barr and M.A. Lishka, J. Am. Chem. Soc., **108**, 3178 (1986).
4. Y. Okamoto, M. Ogawa, A. Maezawa and T. Imanaka, J. Catal, **112**, 427 (1988).
5. W. Grunnert, M. Muhler, K.-P. Schroder, J. Sauer and R. Schlogl, J. Phys. Chem. **98**, 10920 (1994).
6. M. Casamassima, E. Darque-Ceretti, A. Etcheberry and M. Aucourturier, Appl. Surf. Sci., **52**, 205 (1991).
7. W.J. Mortier and R.A. Schoonheydt, Progress in Solid State Chem., **16**, 4 (1985).
8. K.D. Sevier, <u>Low Energy Electron Spectrometry</u> (Wiley - Interscience, New York, 1972).

FAULTING EFFECTS IN THE CHA-GME GROUP OF ABC-6 MATERIALS

JACQUES PLÉVERT,[a,b] RICHARD M. KIRCHNER[b] and ROBERT W. BROACH[c]

[a] Work done at Manhattan College. Present address: Engineering Research Institute, School of Engineering, The University of Tokyo,Tokyo 113, Japan; plevert@catal.t.u-tokyo.ac.jp

[b] Manhattan College, Chemistry Department, Bronx, NY 10471 USA; rkirchne@manhattan.edu

[c] UOP LLC, 50 E. Algonquin Road, Des Plaines, IL 60017 USA; rwbroach@uop.com

ABSTRACT

This paper describes the effect that faulting has on members of the chabazite-gmelinite (CHA-GME) group of the ABC-6 family. We address variations in shape and position of peaks in powder patterns, generation and distribution of new and different sized cages formed by faulting, and effects of faulting on physical properties such as sorption. Using the methods presented, the degree of faulting in a given material can be estimated by comparing the experimental x-ray powder pattern to the DIFFaX simulated pattern. The types of cages present and their relative numbers can be estimated from probability calculations. The computational methods and results of this study can be applied to identify and characterize faulted materials in other groups of the ABC-6 family.

INTRODUCTION

The ABC-6 family represents the richest, most-diverse family of zeolite polytypes presently known. Numerous isotypic materials are listed for the idealized ABC-6 structure types in "The Atlas of Zeolite Structure Types" [1]. A common disorder in the ABC-6 family of materials is faulting in the stacking sequence. Evidence of faulting is given by streaking in TEM selected area diffraction patterns (SADPs) and by broadening and shifting of various peaks in x-ray powder diffraction patterns. The difficulty of obtaining a fault-free chabazite is well known, and is due to the fact that energy differences between similar stacking sequences are small [2]. Thus, different stacking sequences may result from modifications in synthesis conditions or by adding cations and organic directing agents [3]. Materials that fault are difficult to reproduce, and various preparations can have significantly different properties [4,5,6].

This paper describes the effect that faulting has on members of the chabazite-gmelinite (CHA-GME) group of the ABC-6 family. Specifically, the variations in shape and position of peaks in powder patterns, the generation and distribution of new and different-sized cages formed by faulting, and the effects of faulting on physical properties such as sorption, are described. The computational methods and results of this study can be applied to identify and characterize faulted materials in other groups of the ABC-6 family.

FRAMEWORK TOPOLOGIES AND CAGES IN THE ABC-6 FAMILY

The framework topologies in the ABC-6 family consist of layers of six-rings, arranged in a hexagonal array, that are interconnected by tilted four-rings [7,8,9]. The hexagonal layers are related by translation along c and rotation by ±120°. The three possible six-ring layers are designated A, B, and C. The stacking sequence of layers is used to characterize the topology. An A layer can be followed by an A, B, or C layer. A sequence with three or more identical layers (for example, AAA) is not likely in a four-connected framework because of severe strain in tetrahedral bond angles in layers sandwiched between like layers. The observed polytypes have stacking sequences consisting of single six-ring layers (S6R), double six-ring layers (D6R), and mixtures of single and double six-ring layers (S/D6R).

Different stacking sequences generate different types of cages in the structures. For stacking sequences up to and including eight layers, 26 unique types of cages are theoretically possible. Only 12 (Table 1) of these 26 cage types are found in the observed ABC-6 polytypes included in "The Atlas." However, other cages can be produced when faults are present in these polytypes.

Table 1. The 12 Different Types of Cages Found in Observed ABC-6 Polytypes

Number of Layers	Stacking Sequence	Name of Cage*	Cage Face Symbol	Observed in the Following ABC-6 Polytypes
2	AA	*hpr* (D6R)	$4^6 6^2$	AFT, CHA, EAB, ERI, GME, LEV, OFF, SAT
3	ABA	*can*	$4^6 6^3 6^2$	AFG, CAN, ERI, LIO, LOS, OFF, SAT
4	ABCA	*sod*	$4^6 6^8$	SOD
4	ABBA	*gme*	$4^6 4^3 6^2 8^3$	AFT, EAB, GME, OFF
5	ABCBA	*los*	$4^6 6^6 6^3 6^2$	LIO, LOS
5	ABCCA	*lev*	$4^3 4^3 4^3 6^3 6^2 8^3$	LEV
6	ABCCBA	*eab*	$4^6 4^3 6^6 6^2 8^3$	EAB
6	ABBCCA	*cha*	$4^6 4^6 6^2 8^6$	AFT, CHA
7	ABCBCBA	*lio*	$4^6 6^6 6^6 6^3 6^2$	AFG, LIO
7	ABBCBBA	*eri*	$4^6 4^6 6^3 6^2 8^6$	ERI
8	ABBCBCCA	*sat*	$4^{12} 6^8 8^6$	SAT
8	ABBCCBBA	*aft*	$4^6 4^4 4^3 6^2 8^6 8^3$	AFT

*Cages are named after the shortest sequence idealized topology in which the cage is found.

MODELING XRD PATTERNS OF FAULTED CHA-GME MATERIALS

The amount of faulting present in a member of the CHA-GME group can be estimated by modeling the x-ray diffraction patterns using the program DIFFaX [13]. Bateman and Kirchner [5] successfully modeled the faulting in LZ-276 by defining two layer types: one is "stack up" (i.e., AB) and the other is "stack down" (i.e., BA). Both types consist of two six-ring layers joined by appropriate oxygen bridges. An advantage of this model is that calculations using DIFFaX are simple, since only the probability of changing stacking sequence from stack up to stack down needs to be varied. The model was used to simulate x-ray powder diffraction patterns for variously faulted CHA-GME materials (Figure 1).

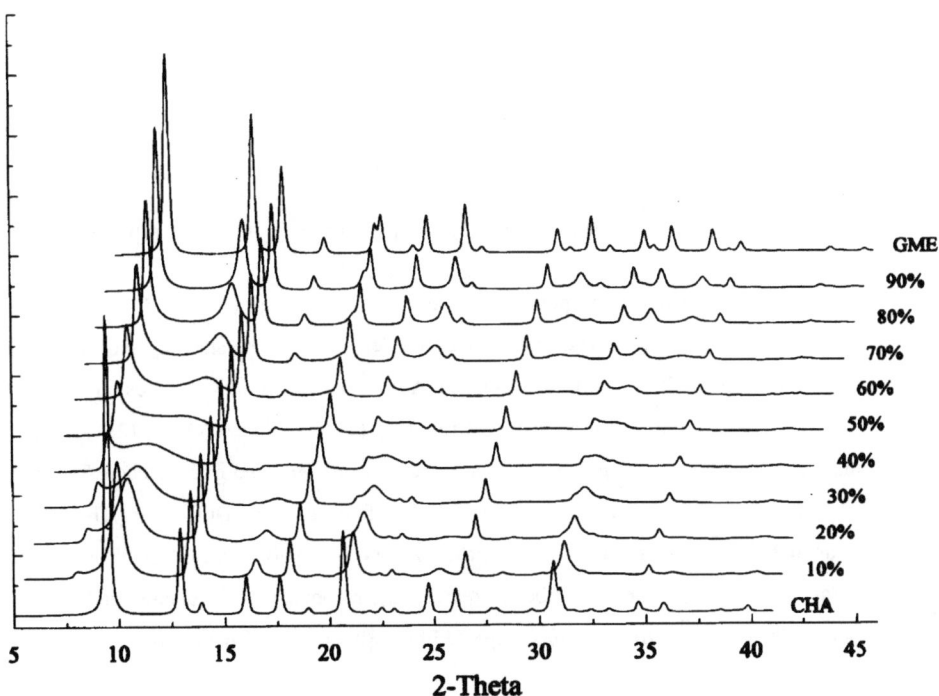

Figure 1. Simulated XRD patterns for various degrees of faulting in CHA-GME materials.

FAULTING IN THE ABC-6 FAMILY AND EFFECT ON POSSIBLE CAGES

A fault in the ABC-6 family can be described as a local mirror plane normal to the stacking sequence direction (the *c* axis of the unit cell, by convention). Faulting in zeolite structures results from (almost isolated) events occurring during the growth of the crystal. What results from a fault depends on the number and types of layers following the fault. A single fault might produce an almost undetectable defect in the crystal or a "twinned" crystal. A modest number of faults would lead to domain intergrowths, or regions in the crystal of different periodicity (regions with different stacking sequences). Because crystallization of the metastable zeolite phase is dictated by kinetics rather than thermodynamics, extensive faulting should be observed in zeolite systems that are not strongly structure directed during crystal growth. In ABC-D6R materials a complete range of behavior from no faulting (for example, in $AlPO_4$-52 [10]) to moderate faulting (for example, in LZ-276 [5]) to faulting at a 50% probability level that gives maximum randomness in the stacking sequence (for example, in babelite [11]) has been observed.

In the following example of a hypothetical stacking sequence in a faulted CHA-GME material, mirror planes associated with each fault are represented by |.

A A B B C C A A B B C |C B B A A C C B B A A C C B B A |A B |B A |A B |B ...

The first part of the sequence corresponds to an ideal CHA topology (AABBCC). This region contains only *hpr* and *cha* cages. The first fault on the left reverses the stacking sequence. The sequence following the fault still corresponds to CHA, but the fault produces cages not found in pure CHA. The region of the fault has local AFX topology (as found in SAPO-56 [12]). Centered on the fault, in adjacent columns parallel to *c*, are one *hpr*, one *gme* and one *aft* cage. The right side of the stacking sequence contains a region with a fault at every D6R layer. This region corresponds to 100% probability of faulting in the CHA-GME family (pure GME). Thus, as the faulting probability in the first end member increases, larger domains of the second end member are obtained. A similar range of behavior is expected for S6R and S/D6R materials.

PES OF CAGES FOUND IN FAULTED CHA-GME MATERIALS

In addition to the cages known in the well-ordered polytypes, faulting can produce new larger cages. As described previously, the occurrence of one isolated fault in the CHA sequence converts two *cha* cages into one large *aft* cage and one *gme* cage. This result of faulting in CHA is noted as follows.

$$2\ cha \rightarrow 1\ aft + 1\ gme \tag{1}$$

Two successive faults generate a larger new cage, designated *aft*(2). Three *cha* cages are converted into 1 *aft*(2) and 2 *gme* cages noted as follows.

$$3\ cha \rightarrow 1\ aft(2) + 2\ gme \tag{2}$$

The case for n successive faults leads to the following general case.

$$(n+1)\ cha \rightarrow 1\ aft(n) + n\ gme \tag{3}$$

Thus faulted members of the CHA-GME group contain *gme*, *cha*, *aft*, and various *aft*(n) cages. The number of six-ring layers that define *gme*, *cha*, *aft*, and *aft*(2) cages are 4, 6, 8, and 10, respectively. An aft(n) cage adds two additional six-ring layers for each additional successive fault (n), and is defined by (2n+6) six-ring layers. The length of a cage is the distance between the six-ring layers that cap each end of the cage. Since the length of the *c*-axis in D6R materials is approximately $n*(2.46\ Å)$ in an idealized SiO_2 topology [1], the interior lengths of *gme*, *cha*, *aft*, and *aft*(n) cages are about $3*2.46$, $5*2.46$, $7*2.46$, and $(2n+5)*2.46$ Å, respectively. When the *aft*(n) cage length equals the crystallite length in the *c*-axis direction, a one-dimensional 12-ring column is produced (as found in GME).

DISTRIBUTIONS OF CAGES IN THE FAULTED CHA-GME GROUP

The distribution of types of cages in a faulted member of the CHA-GME group can be estimated by considering faulting probabilities. An isolated fault occurs when three successive layers (no fault)(fault)(no fault) are present. If the probability for a fault to occur in the CHA-GME group is x, then the probability of having an isolated fault is $(1-x)(x)(1-x)$. Similarly, the probability of formation of an *aft*(n) cage is the probability x^n of finding n successive faults multiplied by the probability $(1-x)$ of finding a nonfaulted layer above and below the layer to close the cage, and is

$$P_{aft(n)} = (1-x)^2(x^n) \tag{4}$$

Figures 2 and 3 show the numbers and types of cages for random faulting in the CHA-GME group. The probabilities for obtaining $aft(n)$ cages are calculated using equation (4). The distribution of *gme* cages is apparent from equations (1) - (3) which show that 1 *gme* cage is produced for each fault. Also, from equations (1) - (3), the probability for obtaining *cha* cages, P_{cha} is

$$P_{cha} = 1 - \sum[(n+1)P_{aft(n)}] \tag{5}$$

Some generalizations can be drawn from the figures. For example, at a given degree of faulting, the probability of finding larger cages decreases with cage size. Also, as the degree of faulting increases, the number of types of cages generated increases rapidly.

Equation (4) presumes crystallites with an infinite number of layers. For real crystallites with a finite number of layers, N, the calculated probabilities are valid when the cage length is small compared to the crystallite lengths $[(2n+6) \ll N]$. However, for cage lengths approaching the length of the crystallites $[(2n+6) \cong N]$, the probabilities given by equation (4) need to be slightly modified. Thus, crystallite size also plays a subtle role in cage distribution.

CONCLUSIONS

We have determined the types and distributions of all cages present in faulted materials in the ABC-6 family. In this paper, examples were drawn from the CHA-GME group. Using the methods presented here, the degree of faulting in a given material can be estimated by comparing the experimental x-ray powder pattern to the DIFFaX simulated pattern. The types of cages present and their relative numbers can be estimated from probability calculations (Figures 2 and 3). These methods can be used to help explain varying physical properties in LZ-276 and other phi-like zeolites, all faulted members of the CHA-GME group. In progress is the modification of equation (4) for very small crystallites, and the calculation of the accessible interior volume in faulted crystallites.

ACKNOWLEDGMENTS

Thanks are expressed to M.M.J. Treacy and J.V. Smith for discussion. RMK thanks Manhattan College for sabbatical leave, and UOP LLC for partial support during this leave.

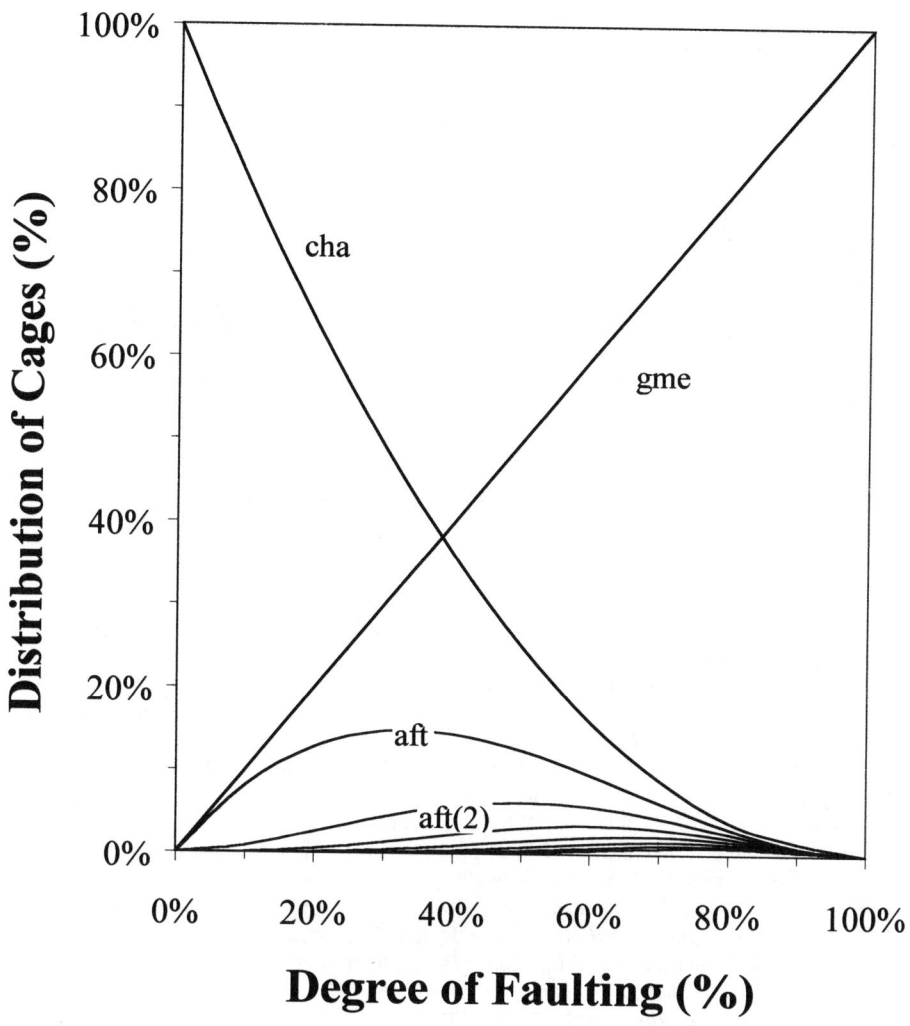

Figure 2. Distribution of number and types of cages found for random faulting in the CHA-GME group. Unlabeled curves correspond to aft(n) distributions with increasing n. As the number of *gme* cages increases, larger and larger 12-ring cavities parallel to c are formed until at 100% faulting in CHA-GME, the open 12-ring channels in GME are obtained.

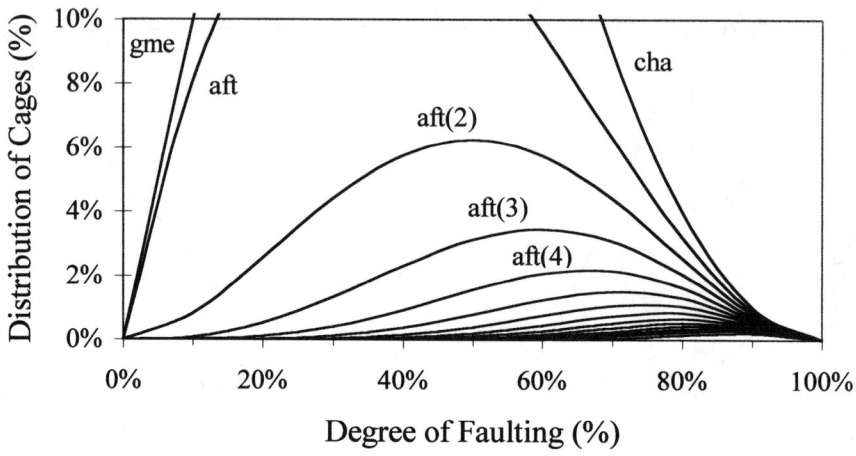

Figure 3. Expansion of Figure 2 to emphasize the distribution of larger sized cages.

REFERENCES

1. Meir, W.M., Olson, D.H. and Baerlocher, Ch., "The Atlas of Zeolite StructureTypes," 4th ed., in *Zeolites*, 1996, **17**.
2. Lillerud, K.P. and Akporiaye, D. In *Zeolites and Related Microporous Materials: State of the Art 1994*, J. Weitkamp, H.G. Karge, H. Pfeifer and W. Holderich, Eds., Studies in Surface Science and Catalysis, **84**, Elsevier Science, Amsterdam, 1994; pp.543-550.
3. Noble, G.W., Wright, P.A. and Kvick, Å., *J.C.S.-Dalton Trans.*, 1997, **23**, 4485.
4. Lillerud, K.P., Szostak, R. and Long, A., *J. Chem. Soc. Faraday Trans.*, 1994, **90**, 1547.
5. Sears, M., Skeels, G.W., Flanigen, E.M., Bateman, C.A., McGuire, N., and Kirchner, R.M., Recent Research Results Poster #RP001 presented at the International Zeolite Conference, Garmisch-Partenkirchen, Germany, July 1994.
6. Skeels, G.W., Sears, M., Bateman, C.A., McGuire, N., Flanigen, E.M., Kirchner, R.M., and Kumar, K., *Microporous and Mesoporous Materials*, 1998, submitted.
7. Kokotailo, G.T. and Lawton, S.L., *Nature* 1964, **203**, 621-623.
8. Smith, J.V. and Bennett, J.M., *Am. Miner.* 1981, **66**, 777.
9. Akporiaye, D.E., *Zeolites*, 1992, **12**, 197-201.
10. McGuire, N.K., Bateman, C.A., Blackwell, C.S., Wilson, S.T., and Kirchner, R.M., *Zeolites* 1995, **15**, 460.
11. Szostak, R. and Lillerud, K.P., *J. Chem. Soc., Chem. Commun.*, 1994, 2357.
12. McGuire, N.K., Blackwell, C.S., Bateman, C.A., Wilson, S.T. and Kirchner, R.M., Recent Progress Reports 10th IZC, Garmish-Partenkirchen, Germany, 1994.
13. MacDIFFaX, version 1.801 (see M.M.J. Treacy, J.M. Newsam, and J.W. Deem, *Proc. R. Soc. Lond.*, 1991, **433**, 499-520).

HYDROTHERMAL SYNTHESIS AND STRUCTURAL STUDY OF A NEW FLUORINATED GALLOPHOSPHATE Ga4(PO4)3(HPO4)F3. T (T = amine).

S. J. WEIGEL[a], T. LOISEAU[b] G. FÉREY[b], V. MUNCH[c], F. TAULELLE[c], R. E. MORRIS[d], G. D. STUCKY[a] and A. K. CHEETHAM[a,b]

[a]Materials Research Laboratory, University of California, Santa Barbara, California 93106. U.S.A.

[b]Institut Lavoisier UMR CNRS 173 IREM, Université de Versailles-St Quentin, 45 av. Etats-Unis, 78035 Versailles Cedex, France.

[c]RMN et Chimie du Solide, UMR 7510 ULP-Bruker-CNRS, Université Louis Pasteur, 4 rue Blaise Pascal, 67070 Strasbourg Cedex, France.

[d]School of Chemistry, University of St Andrews, Purdie Building, St Andrews, KY16 9ST, U.K

email : loiseaugchimie.uvsq.fr

ABSTRACT

A new fluorinated gallophosphate type was synthesized by using several different structure-directing agents : tetramethylammonium hydroxide, pyridine and N,N,N',N'-tetramethylene-1,3-propanediamine. This compound was characterized by single-crystal X-ray diffraction analysis and ^{19}F NMR spectroscopy using RFDR techniques. The tridimensional structure is built up from the connection of octameric units $Ga_4(PO_4)_3(HPO_4)F_3$, giving rise to 8-ring channels.

INTRODUCTION

Since the discovery of the aluminophosphates, $AlPO_4$-n, by Flanigen and co-workers[1], using organic molecules (amines, alkylammoniums, ...) as structure directing agents, a lot of microporous phosphates-based materials have been reported. In these systems, the fluoride method, developed by Guth and Kessler[2] in Mulhouse, facilitates the synthesis of new fluorinated open framework phases, as for example Cloverite[3]. In this case, fluorine is trapped within the double-4-ring cage (D4R) and seems to stabilize this building unit. More recently, a series of fluorinated alumino and gallophosphates, ULM-n[4], T-GaPO (T = amine)[5] and TREN-GaPO[6] was found. For these phases, fluorine belongs to the coordination sphere of the metal (Al or Ga) and therefore increases the tetrahedral coordination to five-fold (trigonal bipyramidal) and six-fold (octahedral) coordination. The incorporation of fluorine in the framework induces strong hydrogen bond interactions between this and the ammonium groups of the amines.

We report here the synthesis and structural characterization of a new oxyfluorinated gallophosphate type $Ga_4(PO_4)_3(HPO_4)F_3$. T (T = amine). This compound can be obtain with tetramethylammonium hydroxide (TMA), pyridine (PYR) or N,N,N',N'-tetramethylene-1,3-propanediamine (TMP).

EXPERIMENTAL

Synthesis of these phases was done hydrothermally in 23 ml Teflon lined Parr digestion bombs. TMA-GaPO has been prepared from the TMA(OH)/pyridine amine mixture with the molar composition: 1 Ga_2O_3 ; 1.25 P_2O_5 ; 5.2 HF; 0.44 TMA(OH); 32.1 PYR; 100 H_2O (heating at 170°C for 3 days). PYR-GaPO has been synthesized in the presence of pyridine and benzyl viologen dichloride with molar composition: 1 Ga_2O_3 ; 1.31 H_3PO_4; 5.21 HF; 0.87 benzyl viologen dichloride; 32.1 PYR; 86.25 H_2O (heating at 170°C for 6 days). TMP-GaPO is obtained from the molar ratio 1 Ga_2O_3 ; 1 P_2O_5 ; 2 HF ; 1.65 TMP ; 80 H_2O (heating at 180°C for 2 days).

The structure of the three compounds was determined by single crystal X-ray diffraction analysis. For TMA-GaPO and PYR-GaPO, the crystal system is monoclinic ($P2_1/n$; a ≈ 12.22 Å, b ≈ 14.28 Å, c ≈ 13.05 Å, β ≈ 91.75°, V ≈ 2276 $Å_3$) whereas TMP-GaPO crystallizes in the triclinic system (P-1 ; a = 8.908(3) Å, b = 8.985(5) Å, c = 14.442(3) Å, α = 91.84(3)°, β = 90.86(3) °, γ = 95.65(3)°, V = 1149.6(2) $Å^3$).

The NMR experiments were carried out with a Bruker spectrometer DSX500 by using a 2.5 mm rotor at spinning rate of 30 kHz.

STRUCTURE DESCRIPTION

The three compounds have a similar structure except that the organic molecules trapped in the channels are different. The three-dimensional framework is built up from an asymmetric unit composed of four tetrahedral phosphate units (PO_4) connected to four trigonal bipyramidal galliums (GaO_4F). Pairs of gallium atoms are linked together through fluorine atoms (F(l) and F(2)). Three of the four phosphate groups link these gallium dimers. The fourth phosphate group is only connected to one of the dimers (figure 1). In the resulting octameric unit, one of the phosphate groups has a P-OH terminal bond.; a Ga-F terminal bond (Ga(2)-F(3)) is also observed for one of the gallium polyhedra. The inorganic network can be generated by connecting the octameric building unit $[Ga_4(PO_4)_3(HPO_4)F_3]$ through single 4-rings. The presence of terminal bonds leads to an interrupted framework structure, reminiscent of cloverite. This results in elliptical 8-ring channels down the [100] direction and a distorted 8-ring channels along [101] direction (figure 2).

The most striking feature of this structure is the interplay between the organics and their segregation into different parts of the pore structure. In TMA-GaPO, the TMA(OH) cations are found in one of the channels and pyridine is found in the other channel. In the structure of PYR-GaPO and TMP-GaPO, both channels contain the pyridine or N,N,N',N'-tetramethylene-1,3-propanediamine, respectively.

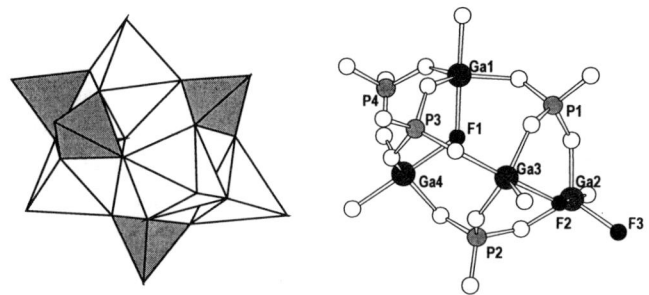

Figure 1 : The $Ga_4(PO_4)_3(HPO_4)F_3$ building unit.
It is composed exclusively of five-fold coordinated gallium atoms.
Fluorine atoms are indicated by black circles.

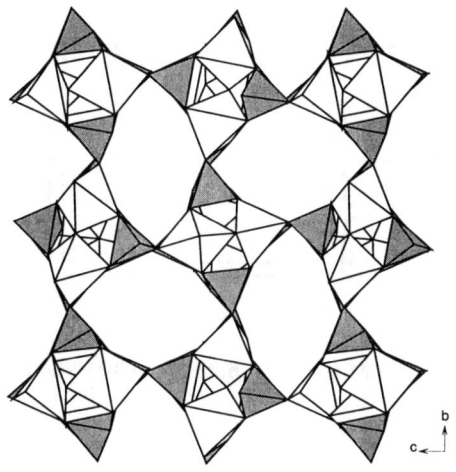

Figure 2 : View down the [100] direction of $Ga_4(PO_4)_3(HPO_4)F_3 \cdot T$ (T = amine) showing the 8-ring channels.

NMR STUDY

Characterization of TMP-GaPO by MAS at very high spinning speed (30 kHz) shows three inequivalent fluorine in a 1:1:1 ratio at c.a. -70, -120 and -180 ppm. A Radio-Frequency Dipolar Recoupling (RFDR) experiment has been run on this compound. The lack of off diagonal peak between the -70 ppm and the -120/180 peaks shows clearly that the distances between the fluorine located at -70 ppm in NMR and the ones located at -120/-180 are noticeably larger than the distance between the -120 and -180 ppm atoms. Furthermore however short we make the mixing time (16 µs), the off diagonal peaks are present between the -120 and -180 ppm components. This indicates quite clearly that both fluorines, are bonded to the same gallium atom. This seems to agree quite well with the proposed structure obtained by X-ray diffraction, we would assign the -70 ppm resonance as being associated with the bridging F(1) atom, and the other two resonances to F(2) and F(3) which belong to the same gallium dimer. Our results would also imply that the ten-ninal OH reported in TMA-GaPO and PYR-GaPO[5] is actually a fluorine.

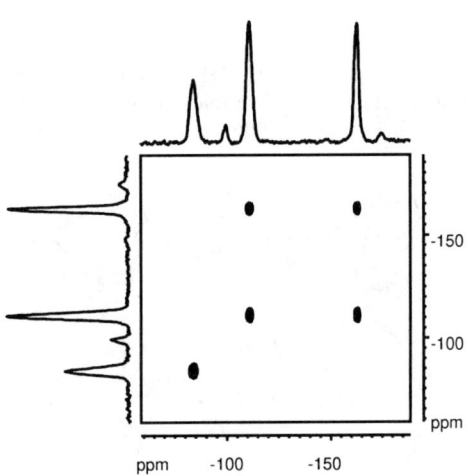

Figure 3 : RFDR ^{19}F MAS NMR spectrum of TMP-GaPO (500 MHz).

REFERENCES

1. S. T. Wilson, B. M. Lok, C. A. Messina, T. R. Cannan and E. M. Flanigen, J. Am. Chem. Soc. **104** (1982) 1146.

2. J. L. Guth, H. Kessler and R. Wey, Stud. Surf. Sci. Catal. **60** (1991). 63

3. M. Estermann, L. B. McCusker, C. Baerlocher, A. Merrouche and H. Kessler, Nature **352** (1991) 320.

4. G. Férey, J. Fluorine Chem. **72** (1995) 187.

5. S. J. Weigel, thesis dissertation, Univ. of Santa Barbara, California (1997).

6. S. J. Weigel, S. C. Weston, A. K. Cheetham and G. D. Stucky, Chem. Mater. **9**(6) (1997) 1293.

CATION SITING IN MICROPOROUS MATERIALS: A 2D TRIPLE-QUANTUM MAS NMR STUDY

J. R. AGGER*, M. W. ANDERSON*, J. ROCHA†, D. P. LUIGI*, M. NADERI* and A. K. BAGGALEY†

*UMIST Centre for Microporous Materials, Manchester, UK; m.anderson@umist.ac.uk
†Department of Chemistry; University of Aveiro, Aveiro, Portugal; j.rocha@dq.ua.pt

ABSTRACT

^{23}Na 2D 3Q MAS NMR spectroscopy and *ab initio* calculation of electric field gradient tensors have been applied to the study of cation locations in a variety of titano-silicates. The phase transformation of ETS-4 to narsarsukite is studied. Evidence is presented for strong preferential siting of K$^+$ in ETS-10, in contrast to ETS-4. Five cation sites are proposed for ETS-10, three of which seem to be largely unaffected by dehydration.

INTRODUCTION

Many of the principal properties of microporous materials are dependent on cation siting: ion exchange, sorption and catalysis. Ostensibly, one of the best techniques to study the local environment around sodium cations is nuclear magnetic resonance (NMR) spectroscopy. However, due to the quadrupolar nature of the ^{23}Na nucleus (I=3/2), simple magic-angle-spinning (MAS) spectra tend to yield broad signals, often comprising superposition of complex quadrupolar line shapes. This renders signal deconvolution virtually impossible in most cases. Dynamic angle spinning [1] and double rotation [2] NMR offer possible solutions for the removal of second order quadrupolar broadening, but both have their own associated difficulties.

In this study we bring to bear the novel technique of two-dimensional triple-quantum [3-5] (2D 3Q) MAS NMR to study sodium cation locations in a variety of microporous materials. Manipulation of both spin-dependent and angular-dependent terms enables averaging of the anisotropic second-order quadrupolar broadening. The isotropic spectra thus obtained offer an improved insight into cationic environments.

EXPERIMENTAL

Na,K-ETS-10 (mole ratio Na$^+$:K$^+$ equal to 3:1) was prepared using titanium (III) chloride as the Ti source following procedures reported by Kuznicki [6], modified by Anderson *et al.* [7]. A potassium free gel was also used to prepare a sample of Na-ETS-10. Na,K-ETS-4 and Na-ETS-4 were prepared following the procedures reported by Kuznicki *et al.* [6,8] The latter sample was shallow-bed calcined under flowing air at 800 °C to yield narsarsukite. In order to match the stoichiometry between ETS-4 and narsarsukite, ETS-4 was also calcined in the presence of fumed silica.

Single- and triple-quantum ^{23}Na MAS NMR spectra were recorded at two fields operating at 105.845 and 105.842 MHz on Bruker MSL-400 spectrometers. 4 mm zirconia rotors were spun at between 10 and 15 kHz. The recycle delay was 0.5 s. Radio-frequency magnetic field amplitudes in the range 73 to 166 kHz were employed. To produce pure-adsorption line shapes in the 3Q MAS spectra, the optimum conditions for excitation and transfer of the (±3Q) coherences using a simple two-pulse sequence were used. The ppm scales are referenced to v_0 in the F2 dimension and to $3v_0$ in the F1 dimension. Spectra are sheared such that the anisotropic axis lies parallel to F2. Aqueous sodium chloride was used as an external chemical-shift standard.

RESULTS AND DISCUSSION

ETS-4

The single-quantum ^{23}Na MAS NMR spectra of Na,K-ETS-4 and Na-ETS-4 samples both show a broad signal at *ca.* -11 ppm with an ill-resolved high-field shoulder. Similarity is also present in the 2D 3Q MAS NMR spectra, shown in figure 1, both of which show three distinct signals.

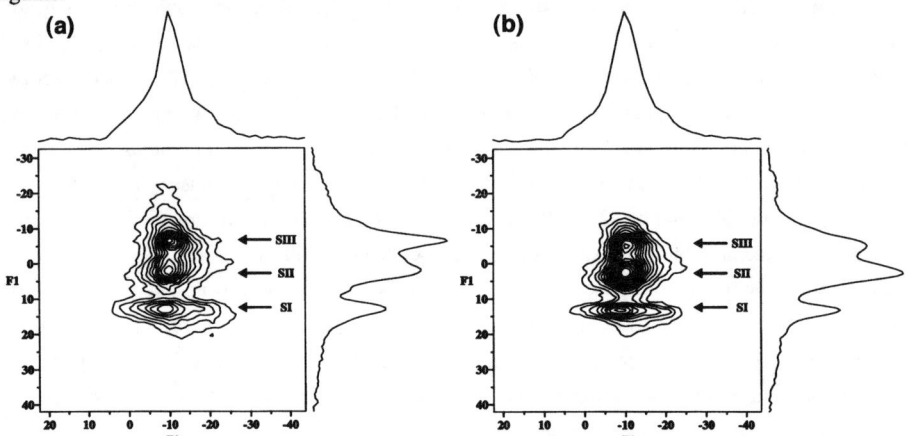

Figure 1. The ^{23}Na 2D 3Q MAS NMR spectra of (a) Na,K-ETS-4 and (b) Na-ETS-4

The structure of ETS-4 is believed to be related to the natural mineral zorite, although certain discrepancies exist between the ratios of silicon sites determined by ^{29}Si MAS NMR, previously reported. The average asymmetric unit of the mineral zorite is reported to contain only two sodium sites [9]. The 2D 3Q MAS NMR spectra, however, clearly reveal a minimum of three sites. The reason for this discrepancy could have at least two origins. First the discrepancies between zorite and ETS-4 structure already mentioned for the silicon sites may well also be reflected at the sodium sites. Second, ETS-4 is inherently a disordered structure with intergrowths in two dimensions. This intergrowth structure may well be responsible for the appearance of extra cation sites. Upon replacement of Na^+ by K^+ there is little change in the relative ratio of the intensity of the three signals which indicates that K^+ has no strong preference for any individual site.

Narsarsukite

Upon calcination of ETS-4 at high temperature a phase change occurs to a synthetic analogue of narsarsukite. Owing to stoichiometric differences between the two phases a large amorphous component is also present. In order to overcome this problem, fumed silica was added to the ETS-4 sample and mixed homogeneously to give a stoichiometric match. Upon heating this mixture gave a pure sample of synthetic narsarsukite. The effect on the sodium cation can be monitored effectively through the ^{23}Na 2D 3Q MAS NMR spectra.

The ideal tetragonal structure of narsarsukite contains one sodium site, the cation being octahedrally co-ordinated to six oxygen atoms in a highly asymmetric fashion [10]. In accordance with this, the narsarsukite sample prepared by calcining ETS-4 with amorphous silica shows one strong signal with a C_Q of 3.1 MHz, and η of 0.15 as shown in figure 2(a). This indicates that the addition of silica prior to calcination of ETS-4 improves the phase purity of the narsarsukite obtained to above 95%. The ETS-4 sample calcined without addition of silica exhibits a more complicated ^{23}Na 2D 3Q MAS NMR spectrum as shown in figure 2(b). In addition to the narsarsukite signal, two other strong signals are present. Both are from the amorphous phase: the narrow signal with a C_Q of *ca.* 0.8 MHz is from a symmetrical site whilst the other signal with a C_Q of *ca.* 2.9 - 3.4 MHz is from an asymmetric site.

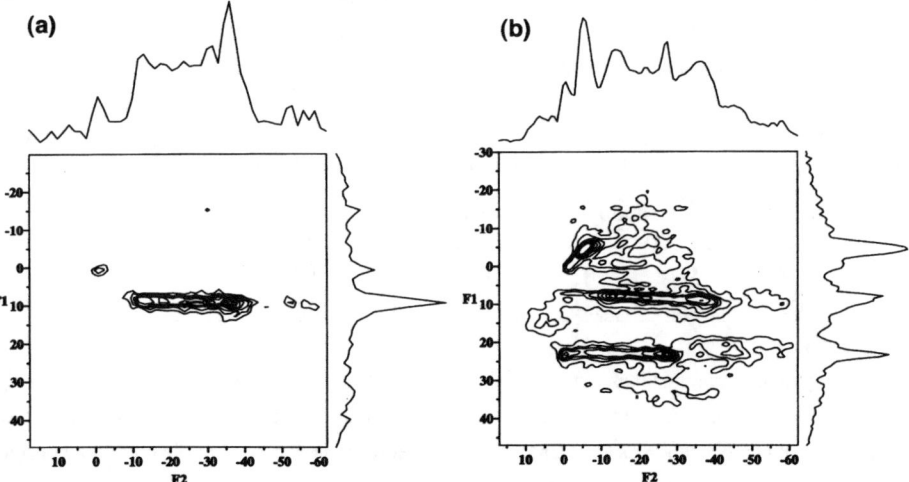

Figure 2. ^{23}Na 2D 3Q MAS NMR spectra of Na,K-ETS-4 after calcination at 800 °C (a) with the addition of fumed silica and (b) without fumed silica.

Calculations of electric field gradients for narsarsukite

In order to begin to assign the multi-quantum ^{23}Na spectra of cations in titanosilicates it is important to be able to calculate the electric field gradient tensors at the sodium site. This has been done previously for a number of zeolitic systems [11], however, the procedure for such calculations on periodic structures is still in its infancy. Our ultimate goal would be to perform such calculations on important structures such as ETS-10 and ETS-4, however, in this communication we have investigated the relatively simple structure of narsarsukite.

The interaction between a quadrupolar nucleus and its electric field gradient (EFG), is given by the quadrupole coupling constant, $Q_c = V_{zz} eQ/h$ (where V_{zz} is the largest principle component of the EFG tensor). The EFG depends upon the charge distribution around the nucleus and its deviation from axial symmetry is defined by the asymmetry parameter $\eta = |(V_{xx} - V_{yy})/V_{zz}|$, which varies between 0 and 1. EFG tensors are normally determined using a point-charge model (PCM). For such a model it is crucial that the charges are accurately determined. This can either be done using an electronegativity approach, a bond valence model or through *ab initio* methods. EFG tensors must also be adjusted for contributions arising from polarisation of the upper core-state *via* Sternheimer antishielding factors. Alternatively the EFG tensors may be calculated fully using an *ab initio* approach.

Table 1 shows the results for quadrupole coupling constants and asymmetry parameters obtained for narsarsukite using two different approaches. First, a full *ab initio* treatment using two different basis sets calculated using the CRYSTAL 95 code. Second, a point charge model where the charges are extracted from the Mulliken charge population derived from a periodic *ab initio* calculation using a 6-21G basis set. The PCM approach was performed with the GULP code on both a full unit cell and a cluster cut at *ca.* 4.0 Å from the sodium nucleus

Table 1. Experimental and theoretical values of the quadrupole coupling constant (C_Q / MHz) and asymmetry parameter (η) for the single sodium cation site in narsarsukite.

Experimental		ab initio STO-3G		ab initio DVAE		PCM full unit cell		PCM 4 Å cluster	
C_Q	η	C_Q	η	C_Q	η	C_Q	η	C_Q	η
3.1	0.15	3.5	0.36	2.8	0.44	0.5	0.87	0.5	0.87

The asymmetry parameter is extremely difficult to calculate accurately. Therefore, comparing the quadrupole coupling constants, the full *ab initio* calculations give a good correlation between experiment and theory whereas the PCM tends to under estimate the quadrupole coupling constant.

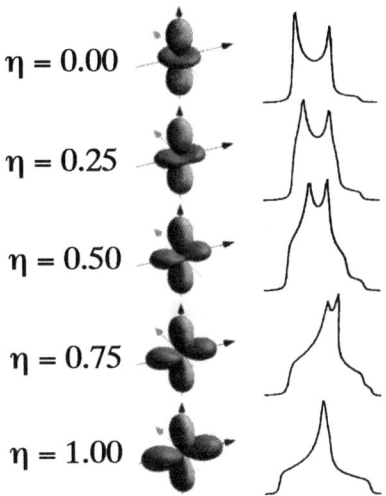

Figure 3. Variation of EFG tensor ovaloid topology and quadrupole spectral lineshape with η.

In order to represent the EFG tensor for such periodic structures it is very convenient to plot the tensor as an ovaloid. The distance from the centre of the ovaloid to the surface in a given direction is proportional to the EFG in that direction. Figure 3 shows how the ovaloid varies in shape for different asymmetry parameters and how this is related to the second-order quadrupolar broadening in the MAS NMR spectrum. The size of the ovaloid is a representation of the quadrupole coupling constant. Figure 4 shows the comparison between experiment and theory for narsarsukite and illustrates how the differences in asymmetry parameter are reflected in only small deviations in the shape of the related ovaloid. The calculations also give the orientation of the EFG tensor and this is superimposed on the structure of narsarsukite in figure 5.

ETS-10

The 2D 3Q ^{23}Na MAS NMR spectrum of hydrated Na,K-ETS-10, shown in figure 6, exhibits two signals in agreement with the work of Ganapathy *et al.* [12]. The stronger signal comprises two sites as evidenced by the shoulder in the isotropic F1 projection. The sheared spectrum, yields isotropic chemical shifts (δ_{iso}) of -7.5 and -7.6 ppm and second order quadrupole effects (SOQEs) of 1.3 and 1.1 MHz respectively. The nature of these quadrupole parameters suggests the sites are very similar. The weaker signal yields δ_{iso} of -1.6 ppm and SOQE of 1.7 MHz. The stronger signal is largely unaffected by dehydration however the weaker signal shifts significantly due to a change in δ_{iso} and SOQE to -3.7 ppm and 2.4 MHz.

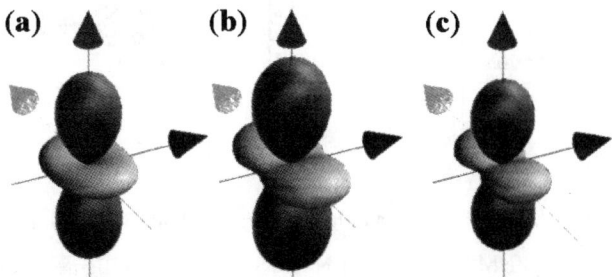

Figure 4. Comparison of (a) experimental, (b) theoretical (full ab initio calculation using the STO-3G basis set) and (c) theoretical (full ab initio calculation using the DVAE basis set) EFG tensor ovaloids for the single sodium site of narsarsukite.

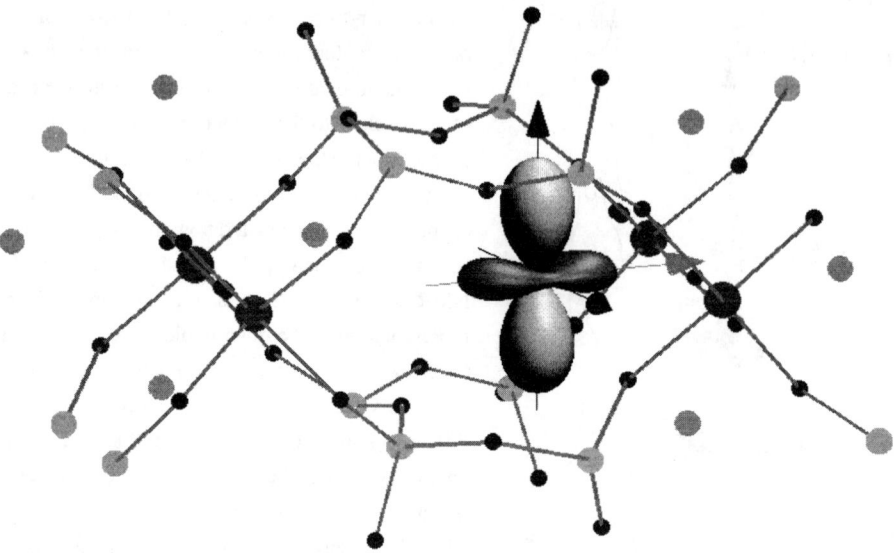

Figure 5. Superposition of the single sodium site EFG tensor ovaloid (full ab initio calculation using the STO-3G basis set) on the structure of narsarsukite.

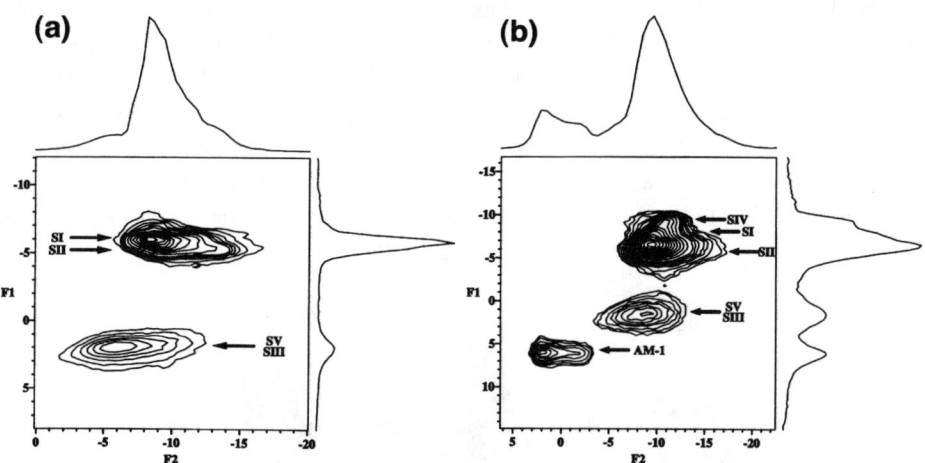

Figure 6. ^{23}Na 2D 3Q MAS NMR spectra of hydrated (a) Na,K-ETS-10 and (b) Na-ETS-10.

The 2D 3Q ^{23}Na MAS NMR spectrum of Na-ETS-10, shown in figure 7, exhibits a second order quadrupole doublet assigned to an impurity phase known as AM-1 [13]. This signal has a δ_{iso} of 3.4 ppm and a SOQE of 1.3 MHz in good agreement with previously published work. Of the two main signals, the weaker one has δ_{iso} and SOQE of -2.6 ppm and 1.7 MHz. The stronger signal now comprises three sites with δ_{iso} of -7.8, -9.0 and -9.8 ppm and SOQEs of 1.2, 1.0 and 0.8 MHz respectively. The main signal again seems largely unaffected by dehydration and the weaker signal shifts corresponding to a change of δ_{iso} and SOQE to -4.2 ppm and 2.1 MHz respectively. Complete dehydration causes the appearance of a further signal at -20 ppm F2 which has a δ_{iso} of -19.0 ppm and a SOQE of 0.9 MHz.

We propose five cation locations for ETS-10 based on a slight modification of the work of Grillo and Carrazza [14] due to over-crowding of site IV cations in the seven ring cage as shown in figure 7. Sites I, II and III remain unchanged. Instead of eight site IV cations per unit cell, we propose only four, with the remaining four cations located in a new site V which is loosely located within the 12-ring pore. The stronger signal which is largely unaffected by dehydration comprises two very similar sites which we believe to be sites I and II, tucked away on either side of the 12-ring pore. In Na-ETS-10, a third site forms part of this signal. This is excellent evidence that K^+ shows a strong preference for this site. As it is not affected by dehydration we believe this to be site IV which is also buried in the seven ring cage. The two remaining sites III and V, which are most accessible to water, are likely to be responsible for the weaker signal and the signal which appears at -20 ppm upon dehydration in Na-ETS-10.

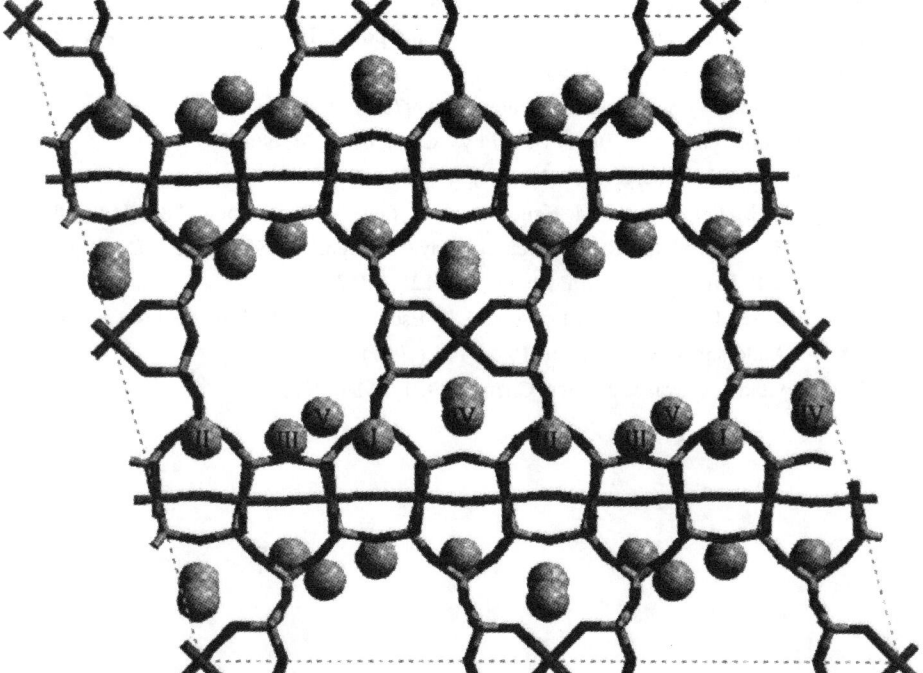

Figure 7. The five proposed cation sites of ETS-10.

CONCLUSIONS

This study highlights the power of 2D multiple-quantum MAS NMR to produce isotropic spectra of the quadrupolar ^{23}Na nucleus. Such enhanced resolution may usefully be employed in the determination of cation locations in microporous materials. Furthermore, the rapid development of *ab initio* modelling techniques now makes calculation of quadrupole parameters a possibility, at least for relatively simple systems. In this study we have applied both techniques to the determination of cation locations in a variety of titanosilicate materials. The results are very promising and further work is currently underway to refine the modelling techniques for more complex systems such as ETS-10.

REFERENCES

1. A. Llor, A and J. Virlet Chem. Phys. Lett. **152**, 248 (1988).
2. A. Samoson, E. Lippmaa and A. Pines Mol. Phys. **65**, 1013 (1988).
3. L. Frydman and J.S. Harwood J. Am. Chem. Soc. **117**, 5367 (1995).
4. C. Fernandez and J-P. Amoureux Chem. Phys. Lett. **242**, 449 (1995).
5. A. Medek, J.S. Harwood and L. Frydman J. Am. Chem. Soc. **117**, 12779 (1995).
6. S.M. Kuznicki, U.S. Patent No. 4 853 202 (1989).
7. M.W. Anderson, O. Terasaki, T. Oshuna, A. Philippou, S.P. MacKay, A. Ferreira, J. Rocha and S. Lidin Nature **367**, 347 (1994).
8. S.M. Kuznicki and A.K. Thrush Eur. Patent 0405978A1 (1990).
9. P.A. Sandomirskii and N.V. Belov Sov. Phys. Crystallogr. **24**, 686 (1979)
10. D.R. Peacor and N.J. Buerger Am. Mineral. **47**, 539 (1962).
11. H. Koller, G. Engelhardt, A.P.M. Kentgens and J. Sauer J. Phys. Chem. **98**, 1544 (1994).
12. S. Ganapathy, T.K. Das, R. Vetrivel, S. Ray, T. Sen, S. Sivasanker, L. Delevoye, C. Fernandez, J.P. Amoureux J. Am. Chem. Soc. **120**, 4752 (1998).
13. Z. Lin, J. Rocha, P. Brandão, A. Ferreira, A.P. Esculcas, J.D. Pedrosa de Jesus, A. Philippou and M.W. Anderson J. Phys. Chem. **101**, 7114 (1997).
14. M.E. Grillo and J. Carrazza J. Phys. Chem. **100**, 12261 (1996).

MAGIC-ANGLE-TURNING NMR AND THEORETICAL STUDIES OF CHEMICAL SHIFT TENSORS ON MICROPOROUS CATALYSTS

A. PHILIPPOU[*], F. SALEHIRAD[*], D. P. LUIGI[*] and M. W. ANDERSON[*]

[*]UMIST Centre for Microporous Materials, Department of Chemistry, UMIST, Manchester, UK;
m.anderson@umist.ac.uk

ABSTRACT

The magic-angle-turning nuclear magnetic resonance (MAT NMR) technique is a new powerful tool in solid-state NMR applications producing 2D high-resolution, isotropic chemical shift and chemical shift anisotropy (CSA) correlation spectra. It is particularly applicable to systems where CSA is large compared to the spinning rate and as a result, the corresponding MAS spectra are overcrowded with spinning sidebands. Two systems were chosen to illustrate the potential of this novel technique in the area of microporous materials. The first system entails stable surface methoxy groups, intermediates observed in several zeolite-catalysed reaction processes. The second system relates to the various phosphorous chemical environments in MgAPO-20, a magnesium aluminophosphate with a sodalite structure. These experimental findings are supported by theoretical studies.

INTRODUCTION

Magic-angle-spinning nuclear magnetic resonance (MAS NMR) experiments on solids yield high-resolution spectra depicting isotropic chemical shifts, which probe, quantitatively, chemical structure. However, the process of MAS, while giving high-resolution spectra, removes all angle-dependent information which can potentially give precise conformational details. There are a number of spectroscopic techniques which can be used to reintroduce this angle-dependent information whilst retaining high resolution and one of the most successful is a two-dimensional magic-angle turning (MAT) experiment [1-3]. Recently a complete analysis of chemical shift information from single crystal work has been used successfully to determine small distortions in naphthalene which cannot be resolved by x-ray diffraction [4]. MAT can provide the same information from a powder and is ideally suited to monitor the fine details of adsorbate structure at catalytically active sites.

Accurate theoretical calculations of chemical shift tensors have recently been achieved using *ab initio* techniques [5-8]. This provides a full representation of not only the chemical shielding tensor but also electric field gradient tensors for quadrupolar nuclei. This information can then be used to simulate solid-state NMR spectra recorded either with MAT technology or with very low temperature experiments. From a catalytic point of view this provides a complete description of the angular distribution of electron density, bond angles and bond distances.

EXPERIMENTAL

Sample Preparation

For the investigation of methoxy groups, the catalysts used were H-SAPO-34 and H-ZSM-5. These catalysts were methylated following the standard experimental procedure described by Salehirad and Anderson [9]. The MgAPO-20 sample was synthesised following the method reported in the literature [10].

Magic-angle-turning NMR

The MAT NMR experiments were recorded using the PHORMAT pulse sequence [2] shown in Figure 1. A Chemagnetics pencil rotor containing the solid material was spun at 30±0.2 Hz and the spinning rate was constantly controlled by a modified Bruker MAS control unit to ensure stable spinning over long time periods. For the ^{13}C MAT experiments, a total of 36 complex blocks (256 transients for each block) of data with a 3.02 ms acquisition time in the F1 dimension and a repetition delay of 5s have produced the 2D spectra shown in this report. For the ^{31}P MAT experiment, a total of 45 complex blocks of data (32 transients for each block) with a 2.7 ms acquisition time in the F1 dimension and a repetition delay of 80 s were used. Because of the natural abundance of ^{31}P, the cross polarisation section in the pulse program was no longer needed. The raw data were manipulated following the routine [2] reported by Hu *et al.* and using NMR software written by P. J. Grandinetti (RMN FAT 1996).

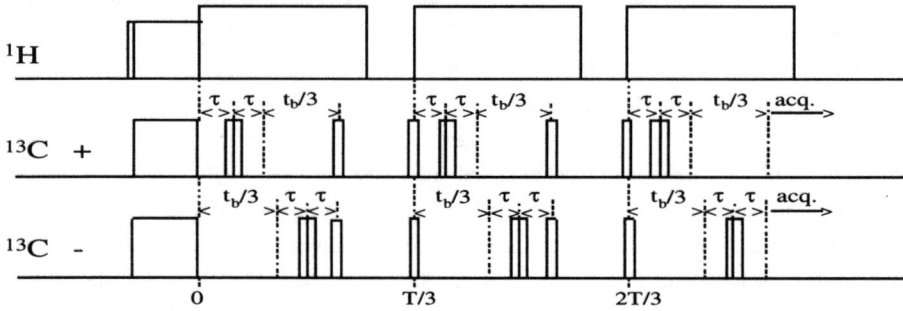

Figure 1. The PHORMAT pulse sequence [2] where 90° pulses are represented by single rectangles and two adjacent rectangles denote 180° pulses; τ is an echo delay time determined by the probe ring down and the receiver recovery time; T is the rotor period and t_b is the time incremented at the end of each complex block.

Theoretical Studies

An *ab initio* geometry optimisation followed by theoretical calculations of ^{13}C shielding tensors were both carried out with Gaussian 94 Inc. Package on zeolitic clusters which mimic methoxy groups. The geometry optimisation was carried out at Hartree Fock (HF) theory level as well as B3LYP Density Functional Theory (DFT) with a standard 6-31G** basis set. The shielding tensors were calculated with the Gauge-Independent Atomic Orbital (GIAO) [11,12] method at both HF and DFT theory level with two combinations of basis sets; i.e. DZP on all atoms except oxygens where TZP was used, or TZP on all atoms.

RESULTS AND DISCUSSION

^{13}C MAT NMR of methylated ZSM-5 and SAPO-34

These studies relate to stable surface methoxy groups, intermediates observed in several zeolite-catalysed reaction processes [9,13,14].

Figure 2 illustrates both ^{13}C MAS and MAT NMR spectra of the methylated ZSM-5. The 2D MAT spectrum depicts the isotropic spectrum along F1 with CSA slices along F2 at the various isotropic positions. As shown in this figure, five slices were obtained at 50.1, 52.5, 58.4, 61.2 and 65.0 ppm along F1. All these spectra were simulated and the principal tensor values are reported in Table 1. Though the assignment of these spectra requires further debate, it is suggested that the line at 50.1 ppm relates to methoxy species formed upon methanol reaction with terminal silanol groups whereas the line at 58.4 ppm is attributed to methoxy attached to framework Al sites. Lewis acidity may also contribute to the line at 50.1 ppm.

Table 1. Principal CSA tensor components of ^{13}C in methylated ZSM-5 (ppm)

δ_{iso}	δ_{11}	δ_{22}	δ_{33}	δ_{cs}	η_{cs}
50.1	81.5	58.7	10.1	-40.0	0.6
52.5	80.2	71.1	6.1	-46.4	0.2
58.4	87.1	78.8	9.3	-49.1	0.2
61.2	94.3	79.0	10.2	-51.0	0.3
65.0	--	--	--	--	--

Figure 3 illustrates the ^{13}C MAS and MAT NMR spectra of the methylated SAPO-34. As shown in this figure, four slices were obtained at 50.9, 56.6, 60.2 and 64.8 ppm along F1. The lines at 56.6 and 50.9 ppm are assigned to methoxy groups formed on framework bridging (Type I) and terminal (Type II) aluminium respectively. Theoretical calculations of ^{13}C shielding tensors on clusters (see Figure 4) which mimic these methoxy types were carried out and the agreement between experiment and theory is best depicted in Table 2 and Figure 4. The small inconsistency rests with (i) conversion of the theoretical absolute shift to the experimental chemical shift scale and (ii) the cluster size limitation.

^{31}P MAT NMR of MgAPO-20

The second system entails the various phosphorous chemical environments in MgAPO-20, a magnesium aluminophosphate with a sodalite structure. The Chemical shift tensors of these chemical environments (see Table 3) are extracted from the ^{31}P MAT NMR spectra of MgAPO-20, shown in Figure 5. A correlation between the symmetry of P(3Al, 1Mg) and P(2Al, 2Mg) is

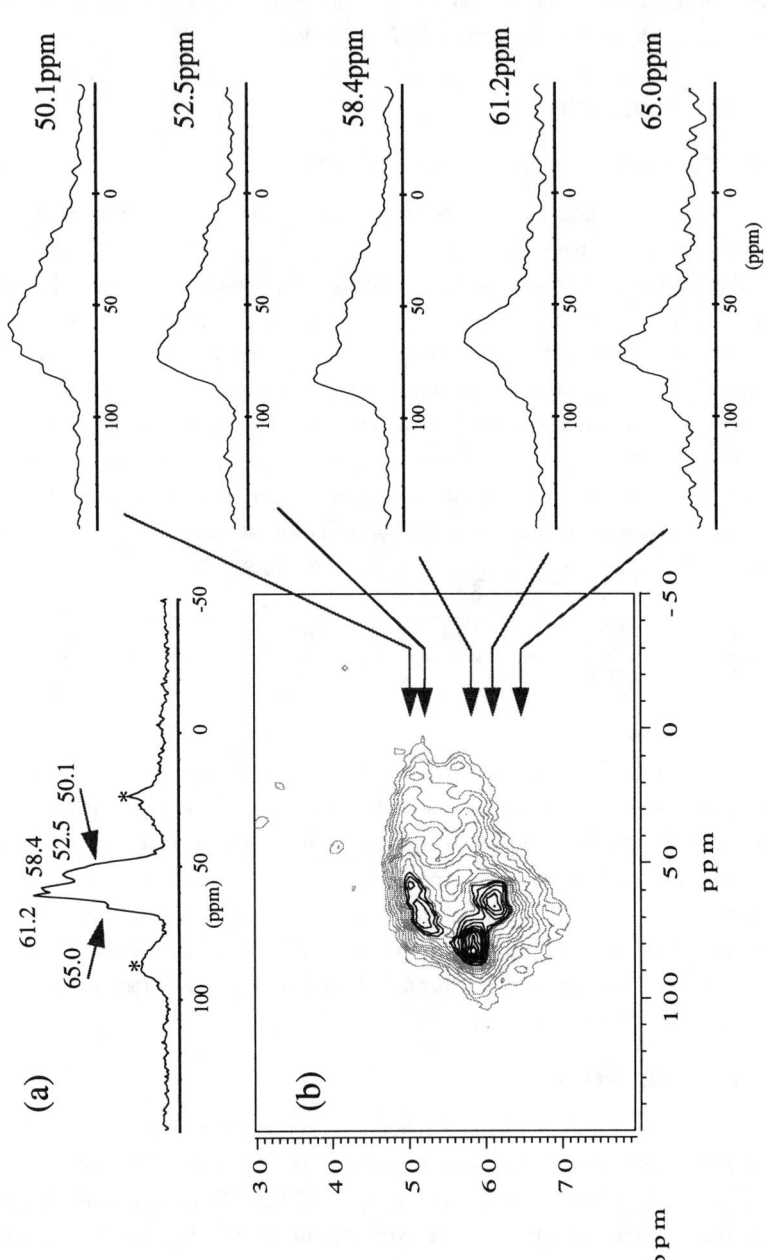

Figure 2. (a) ^{13}C MAS NMR and (b) ^{13}C MAT NMR spectra of methylated ZSM-5.

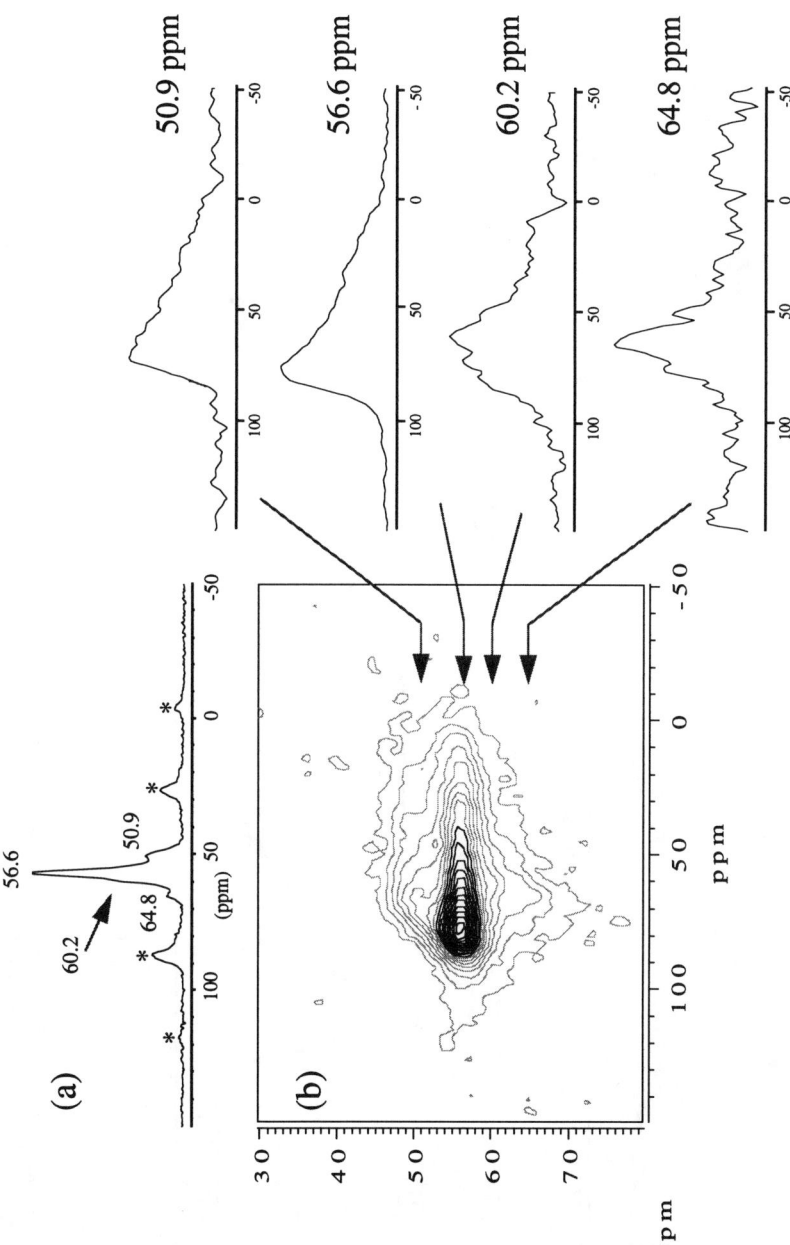

Figure 3. (a) ^{13}C MAS NMR and (b) ^{13}C MAT NMR spectra of methylated SAPO-34.

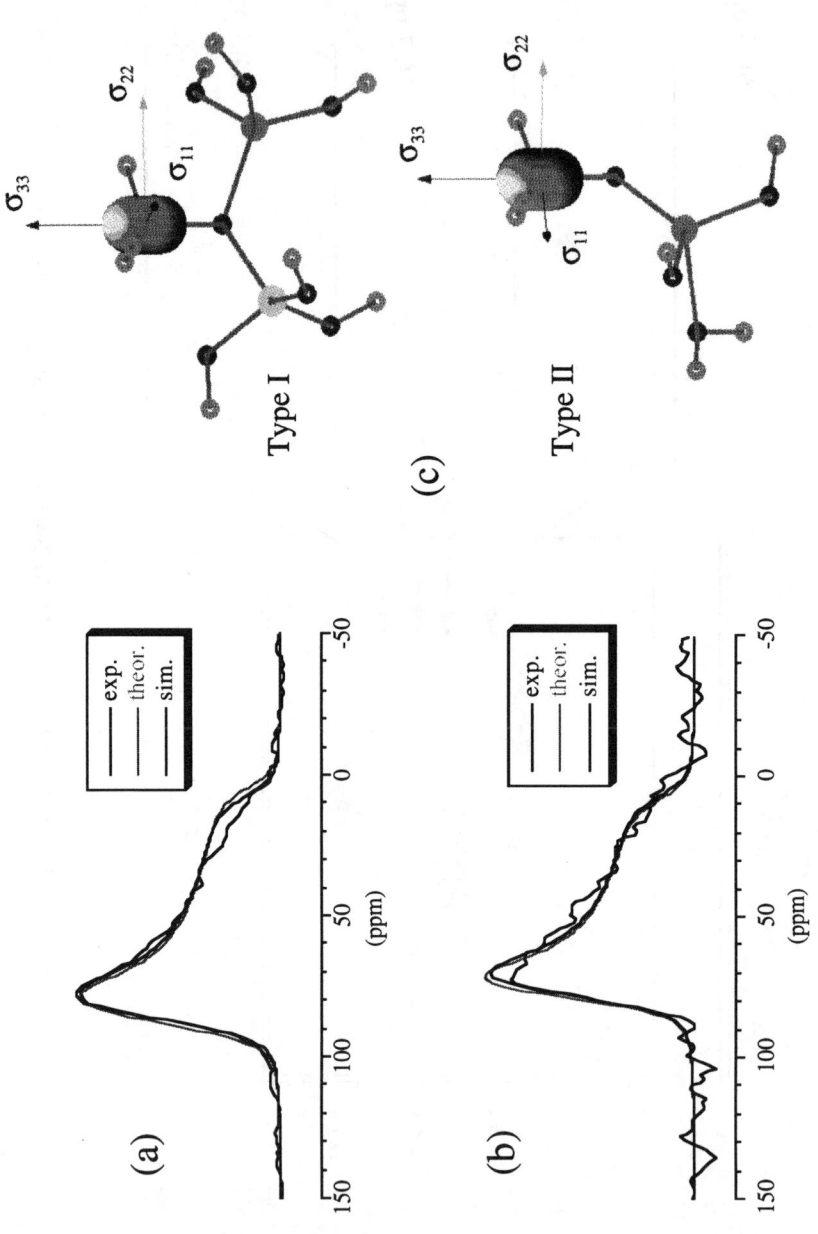

Figure 4. Experimental (exp.), Theoretical (theor.) and Simulated (sim.) lines of the signals at (a) 56.6 and (b) 50.9 ppm; (c) ball and stick representations for the clusters Type I and II where the principal axis system and the ^{13}C shielding tensor ovaloid are superimposed.

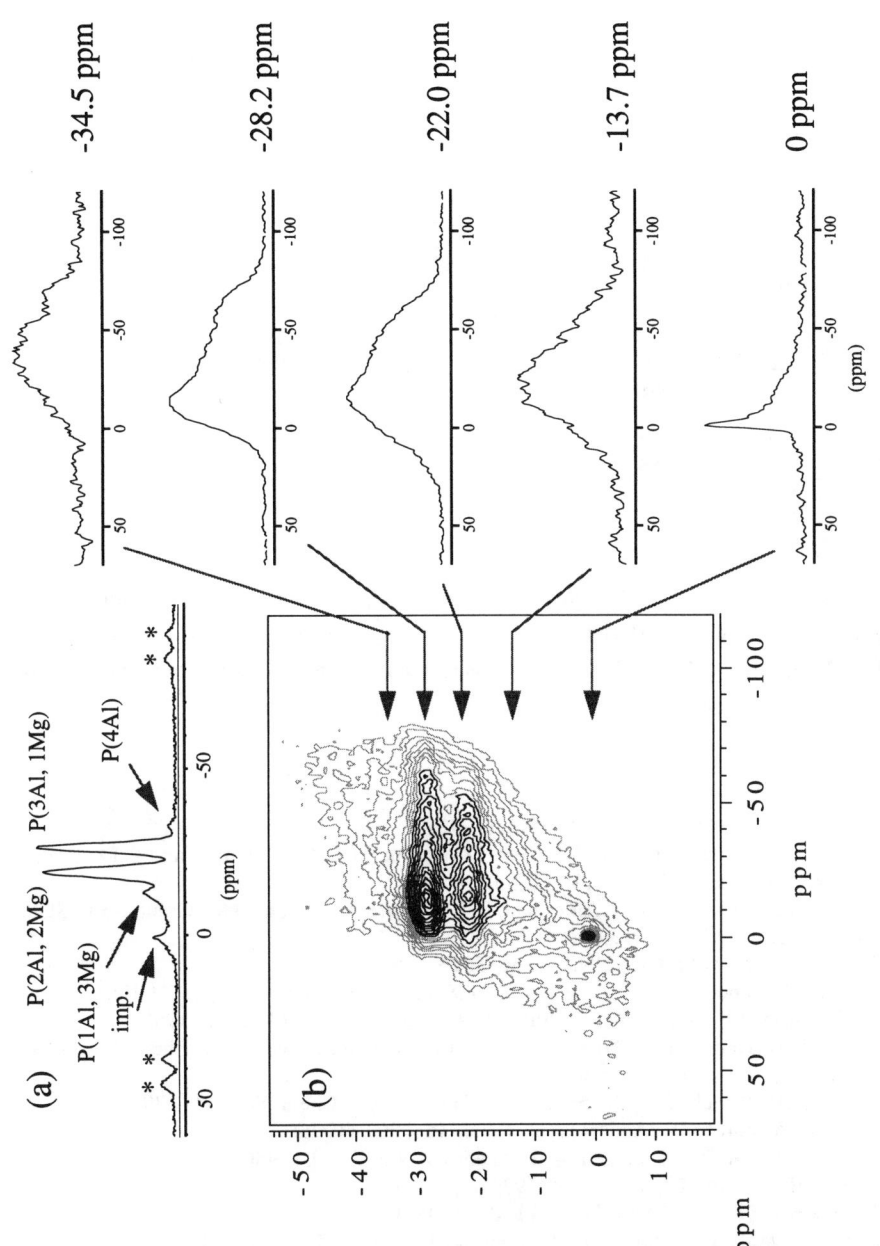

Figure 5 (a) ^{31}P MAS NMR and (b) ^{31}P MAT NMR spectra of MgAPO-20

clearly observed. As expected, phosphorous in P(3Al, 1Mg) is much closer to axial symmetry than phosphorous in P(2Al, 2Mg).

Table 2. Principal CSA tensor components of ^{13}C in methylated SAPO-34 (ppm).

δ_{iso}	δ_{11}	δ_{22}	δ_{33}	δ_{cs}	η_{cs}
Experiment					
50.9	77.0	70.0	5.7	-45.2	0.15
56.6	86.5	77.0	6.4	-50.1	0.19
60.2	--	--	--	--	--
64.8	--	--	--	--	--
Theory					
50.9	75.0	71.9	5.8	-45.1	0.07
56.6	88.2	76.7	4.9	-51.7	0.22

Table 3. Principal CSA tensor components of ^{31}P in MgAPO-20 (ppm).

δ_{iso}	δ_{11}	δ_{22}	δ_{33}	δ_{cs}	η_{cs}
-34.5	-	-	-	-	-
-28.2	1.1	-12.5	-73.2	-45.0	0.3
-22.0	13.8	-15.7	-64.2	-42.2	0.7

CONCLUSIONS

The potential of MAT NMR in the area of microporous materials is illustrated in this presentation. Both, the ^{13}C MAT NMR spectra of methylated ZSM-5 and SAPO-34 and the ^{31}P MAT NMR spectra of MgAPO-20 provide vital information as to the chemical and electronic nature of these systems.

REFERENCES

1. Z. Gan, J. Am. Chem. Soc. **114,** 8307 (1992).
2. J.Z. Hu, W. Wang, F. Liu, M.S. Solum, D.W. Alderman, R.J. Pugmire and D.M. Grant, J. Magn. Res. series A **113,** 210 (1995).
3. J.Z. Hu, A.M. Orendt, D.W. Alderman, R.J. Pugmire, C. Ye and D.M. Grant, Solid State Nucl. Mag. Res. **3,** 181 (1994).
4. J.C. Facelli and D.M. Grant, Nature **365**, 325 (1993).
5. P.E. Sinclair and C.R.A. Catlow, J. Chem. Soc. Faraday Trans. **93**, 333 (1997).
6. K. Wolinski, J.F Hinton and P. Pulay, J. Am. Chem. Soc. **112,** 8251 (1990).
7. T. Xu, D.H. Barich, P.D. Torres, J.B. Nicholas and J.F. Haw, J. Am. Chem. Soc. **119,** 406 (1997).
8. T. Xu, D.H. Barich, P.D. Torres and J.F. Haw, J. Am. Chem. Soc. **119,** 396 (1997).
9. F. Salehirad and M.W. Anderson, J . Catal. **164,** 301 (1996).
10. S.T. Wilson and E.M Flanigen, U.S. Patent No. 4,567,029, (1986).
11. R. Ditchfield, Mol. Phys. **27**, 789 (1974).
12. K. Wolinski, J. Am. Chem. Soc. **112,** 8251 (1990).
13. J. Nováková, L. Kubelková and Z. Dolejsek, J. Catal. **108**, 208 (1981).
14. V. Bosácek, J. Phys. Chem. **97,** 10732 (1993).

MAGNETIC RESONANCE STUDIES ON VAPO-5 AND MgVAPO-5 MICROPOROUS MATERIALS

T. BLASCO*, P. CONCEPCION, L. FERNANDEZ, J. M. LOPEZ NIETO and A. MARTINEZ-ARIAS[§]

I.T.Q. (CSIC-UPV), Valencia, Spain; tblasco@itq.upv.es
[§] I.C.P. (CSIC), Cantoblanco, Madrid, Spain; amartinez@icp.csic.es

ABSTRACT

As synthesised VAPO-5 and MgVAPO-5 molecular sieves were characterised with the use of NMR and ESR spectroscopic techniques. ^{31}P MAS NMR of MgVAPO-5 has allowed to distinguish different P(n Al, 4-n Mg) structural units and calculate the framework composition. From the relative site population it has been deduced that the presence of V does not change the random distribution of Mg^{2+} atoms in the AFI structure.

For VAPO-5 and MgVAPO-5 samples, the spin density calculated from the ESR spectra accounts for less than 30-35 % of the vanadium content. The spectra are consistent with the presence of isolated $(VO)^{2+}$ ions in square pyramidal or distorted octahedral symmetry. The contribution of a broad resonance to the spectra of MgVAPO-5 samples has been assigned to associated V^{4+} cations. The location of $(VO)^{2+}$ ions into the AFI structure in VAPO-5, and the preferential formation of extra framework V^{4+} species in MgVAPO-5 are confirmed by the ^{31}P MAS NMR longitudinal relaxation time T_1. The ^{51}V NMR spectra indicate the presence of both tetrahedral and octahedral V^{5+} in VAPO-5 and octahedral V^{5+} in MgVAPO-5.

INTRODUCTION

While it is generally accepted [1-3] that Mg^{2+} substitutes for Al^{3+} in aluminum phosphate molecular sieves, the incorporation mechanism of vanadium in crystalline aluminophosphates is a matter of debate. Although it is commonly accepted that vanadium is incorporated as V^{4+} (VO^{2+}) and then oxidized to V^{5+} upon calcination [4-6], different models based either on the substitution for P^{5+} [4-5] or on the replacement of framework Al^{3+} [6] have been proposed. As for Mg^{2+} substituted aluminophosphates [1-3], this later mechanism should modify the chemical shift position of neighboring ^{31}P atoms in the MAS NMR spectra, but the strong dipolar coupling with paramagnetic V^{4+} broadens the ^{31}P peaks preventing the resolution of different environments, and no unambiguous evidence for isomorphous substitution of vanadium in $AlPO_4$-5 has been found with the use of this and other spectroscopic techniques.

In this work we have studied the incorporation of Mg and/or V in the AFI structure, with ESR and MAS NMR spectroscopy. We will show that the longitudinal relaxation time T_1 of ^{31}P MAS NMR in combination with ESR and ^{51}V NMR gives valuable information about how the V atoms are introduced into the framework in the as prepared materials.

EXPERIMENTAL

VAPO-5 and MgVAPO-5 materials were obtained from hydrothermal synthesis using triethylamine (TEA) as template following the synthesis procedure described previously [7-8]. The chemical composition of the resulting materials, as well as $AlPO_4$-5, MgAPO-5 and a vanadia supported on $AlPO_4$-5 (V/AlPO) used as references, are summarised in table 1. Chemical analysis of Al, Mg and V was done by atomic absorption, and P was determined by a colorimetric method using the complex formed between phosphorous and molybdovanadic acid.
. The XRD patterns of all samples are characteristic of crystalline AFI structure. The V and/or Mg containing $AlPO_4$-5 samples will be referred to as **xMyV** where x and y correspond to the atomic Mg and V contents, respectively, as wt % in the calcined samples.

^{31}P and ^{51}V solid state NMR spectra were recorded with a Varian VXR-400 S WB spectrometer. The standard inversion recovery pulse sequence was applied to the samples spinning at 7 KHz to determine the longitudinal relaxation time T_1 of ^{31}P. The ESR spectra were recorded at 77 K with a Bruker ER-200 spectrometer at X-Band. DPPH (g = 2.0036) was used as standard to calibrate the g value scale, and $CuSO_4$ to quantify the number of paramagnetic V^{4+} species.

RESULTS AND DISCUSSION

As it has been reported in a previous publication [7], the unit cell volume of the crystalline VAPO-5 materials studied here increases with the vanadium content, indicating the incorporation of V atoms into the crystals. As it was also previously established [8], a further increase of the unit cell volume is observed for MgVAPO-5, strongly suggesting the presence of both Mg and V in the framework structure.

Figure 1 shows representative ^{31}P MAS NMR spectra obtained in this work. . For MgAPO-5 (sample **2.0M**) and MgVAPO-5 the spectra are constituted by a resonance at –30 ppm typical of P(4Al) environments in microporous aluminium phosphates and the contribution of additional resonance signals at c.a. –23, -17 and –11 ppm ascribed to P (3Al, 1Mg), P (2Al, 2Mg) and P (1Al, 3Mg) (see figure 1) , respectively, in agreement with previous results on Mg substituted aluminophosphates [1,2] and silicoaluminophosphates [3]. The decomposition of the ^{31}P spectra into individual Gaussian lines has allowed us to estimate the relative intensity of each P (4-n Al, n Mg) structural unit and, neglecting the vanadium content, to calculate the P/Mg framework ratios shown in table 1. The good agreement with the values found by chemical analysis (table 1) indicates that most Mg occupies framework sites in all samples except **4.4M2.4V**, in which there must be a significant amount of non-framework species. The relative intensity of the P (4-n Al, n Mg) peaks indicates that the presence of vanadium atoms does not modify the random distribution of the Mg atoms in the AFI structure [1]. The spectra of VAPO-5 samples give only a peak typical of P(4Al) unit. However, the absence of other resonance signals which could be assigned to P (4-n Al, n V) environments does not allow to rule out the substitution of vanadium for aluminum due to presence of paramagnetic V^{4+}, as already mentioned in the introduction.

Sample	Metal content (wt %)[b]		(P/Mg) molar ratio		T_1 (s)[d]	V^{4+}/V_t[e]
	Mg	V	CA[b]	NMR[c]		
AlPO$_4$-5	-	-	-	-	45	-
2.0M	2.0	-	10	8.6	47,52	-
0.5V[a]	-	0.5	-	-	14	0.67
1.2V	-	1.2	-	-	7	0.29
2.0V	-	2.0	-	-	6	0.34
2.7M0.6V	2.7	0.6	7.3	9.6	27, 29	0.35
2.9M1.2V	2.9	1.2	7.7	8.0	26, 27	0.27
4.4M2.4V	4.4	2.4	2.8	6.1	18, 20	0.31
4.7M1.5V	4.7	1.5	nd[f]	7.8	34, 36	0.30
V/AlPO[a]	-	2.0	-	-	27	0.77

[a] Synthesised from V^{4+}. [b] chemical analysis. [c] from ^{31}P MAS NMR. [d] T_1 longitudinal relaxation time of ^{31}P. The two values given for Mg containing materials correspond to the P(3Al 1Mg) and P(4Al) resonances, respectively. [e] derived from ESR.. [f] nd= not determined.

Table 1. Chemical composition and main properties of V and/or Mg containing AlPO$_4$-5.

To get information about the co-ordination environment of paramagnetic V^{4+} cations in our samples, we have used the ESR spectroscopy. As indicated in table 1, the calculated spin densities correspond to less than 30-35% of the V content, with the exception of samples **V/AlPO** and **0.5V**. Typical ESR spectra of VAPO-5 and VMgAPO-5 are illustrated in figure 2 for samples **1.2V** and **2.9M1.2V**, respectively. They are mainly constituted by an axially symmetric signal with hyperfine structure characteristic of isolated VO^{2+} species in a distorted octahedral or square pyramidal symmetry [4-5,7,9]. The spectrum of sample **V/AlPO**, shown in figure 2, is the result of the superimposition of a very weak signal of isolated V^{4+} ions and an intense broad band centred at $g \approx 2$ attributed to the presence of an extra framework oxide phase [6,9]. The inspection of the spectra of MgVAPO-5 samples shows a distortion of the spectral baseline (see figure 2b) due to the contribution of a broad resonance of associated V^{4+} cations, which could tentatively be assigned to non framework species. However, from ESR itself it is not possible to get any definite conclusion on the incorporation of V^{4+} into the framework.

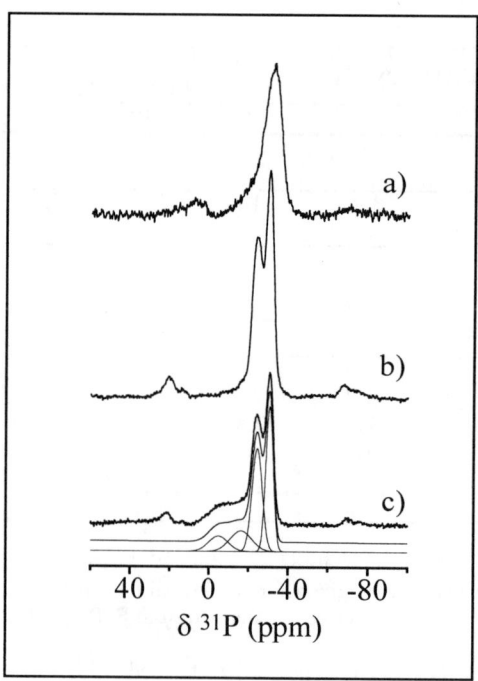

Figure 1. ^{31}P MAS NMR spectra of samples a) 2.0V, b) 2.0M and c) sample 4.4M2.4V compared with the simulation obtained from individual Gaussian peaks also shown.

The presence of extra framework V^{4+} ions in MgVAPO-5 materials is further confirmed by ^{31}P MAS NMR. A decrease on the longitudinal relaxation time T_1 of ^{31}P has been taken as an evidence of the substitution of Al^{3+} by paramagnetic Co^{2+} in CoAPO-5 [10]. The ^{31}P T_1 values measured here (see table 1) indicate that the presence of vanadium induces a shortening on the ^{31}P longitudinal relaxation time, which is specially short for VAPO-5. For MgVAPO-5 samples, the T_1 values are close to that of **V/AlPO** in which V is not incorporated into the framework. The discrepancies on the relaxation behaviour of ^{31}P in VAPO-5 and MgVAPO-5 cannot be ascribed to V^{4+} concentration effects, and then, they must come from the different location of the V^{4+} cations, which will probably occupy framework sites in VAPO-5. The preferential placement of vanadyl ions in non-framework positions in MgVAPO-5 can tentatively be explained by the favoured incorporation of divalent Mg^{2+}.

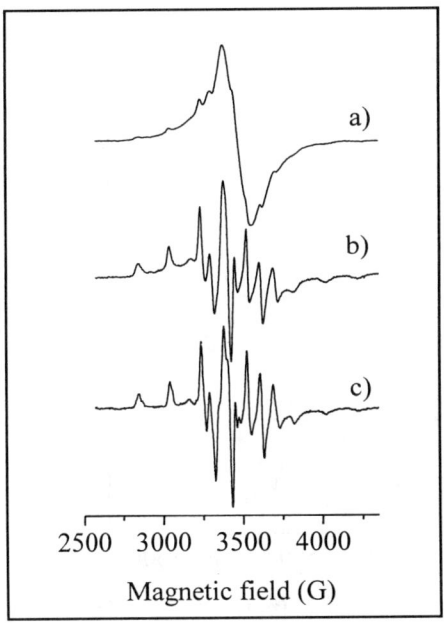

Figure 2. ESR spectra recorded at 77 K of as synthesised a) **V/AlPO**, b) **2.9M1.2V** and c) **1.2V** samples.

Solid state ^{51}V NMR spectroscopy has been shown to be a powerful technique to determine the symmetry environment of vanadium atoms in vanadium containing catalysts [4,

11-12]. Figure 3 displays typical wide line ^{51}V spectra obtained in this work, corresponding to samples **1.2V** and **2.9M1.2V**. The spectra show that most V^{5+} is in octahedral coordination in MgVAPO-5, while both tetrahedral and octahedral V^{5+} are found in VAPO-5. Previous XRD data and catalytic results [7,8] support the successful incorporation of V in MgVAPO-5. Since the results presented here suggest that V^{4+} cations are forming an oxidic phase or are dispersed in non framework positions, probably into the pores of the AFI structure, we can speculate that the octahedral V^{5+} may arise from the coordination of tetrahedral framework V^{5+} to extra molecules [4-6]. The presence of tetrahedrally coordinated metal cations is usually taken as evidence for isomorphous substitution, suggesting the incorporation of V^{5+} into the AFI structure in VAPO-5. Therefore, although both type of environments could correspond to framework V^{5+} cations, we must point out that we have no spectroscopic evidence for the incorporation of V^{5+} into the aluminophosphate structure.

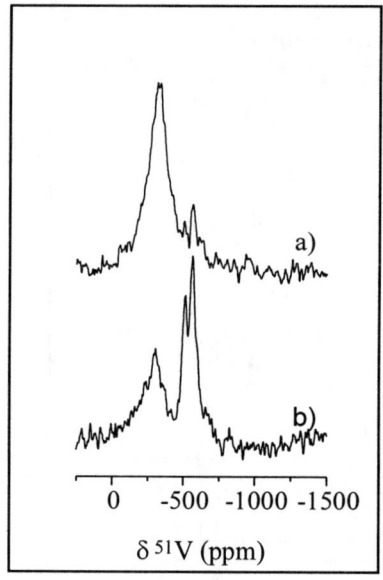

Figure 3. ^{51}V wide-line NMR spectra of as synthesised a) **2.9M1.2V** and b) **1.2V** samples

CONCLUSIONS

The combined use of ESR and NMR spectroscopies gives valuable information on the coordination state and location of Mg and/or V in aiuminium phosphate molecular sieves. This approach has allowed us to establish that in VAPO-5, vanadium atoms are incorporated both as isolated vanadyl $(VO)^{2+}$ ions in square pyramidal or distorted octahedral environments, and probably also as V^{5+} in fourfold or sixfold coordination. In MgVAPO-5, Mg^{2+} are randomly distributed in the AFI structure, and the vanadium atoms must be incorporated to the structure as octahedral V^{5+}, while V^{4+} ions are in non framework positions.

ACKNOWLEDGEMENTS

The authors acknowledge the financial support of this work by the CYCIT (project number MAT 97-0561). Prof. J. Soria is acknowledged for the use of the ESR spectrometer.

REFERENCES

1. F. Deng, Y. Yue, T.C Xiao, Y. Du, C. Ye, L. An and H. Wang, J. Phys. Chem. **99**, 6029 (1995).
2. P. J. Barrie and J. Klinowski, J. Phys. Chem. **93**, 5972 (1989).
3. D. E. Akporiaye, A. Andersen, I. M. Dahl, H. B. Mostad and R. Wendelbo, J. Phys. Chem. **99**, 14142 (1995).
4. C. Montes, M. E. Davis, B. Murray and M. Naryana, J. Phys. Chem. **94**, 6425 (1990).
5. S.H. Jhung, Y.S. Uh and H. Chon, Appl. Catal. **62**, 61 (1990).
6. M.S. Rigutto and H. Van Bekkum, J. Mol. Catal. **81,** 77 (1993).
7. T. Blasco, P. Concepción, J.M. López Nieto, and J. Pérez-Pariente, J. Catal. **152**, 1 (1995).
8. P. Concepción, J. M. López Nieto, A. Mifsud and J. Pérez-Pariente, Appl. Catal. A **151**, 373(1997).
9. B. M. Weckhuysen, I. P. Vannijvel, and R. A. Schoonheydt, Zeolites **15**, 482 (1995).
10. S-H Chen, S-P Sheu and K-J Chao, J. Chem. Soc. Chem. Commun., 1504 (1992).
11. H. Eckert, and I. E. Wachs, J. Phys. Chem. **93**, 6796 (1989).
12. O.B. Lapina,V. M. Mastikhin, A. A. Shubin, V. N. Krasilnikov, and K. I. Zamarev, Progr. NMR Spectrosc. **24**, 457 (1992).

^{31}P AND ^{27}Al MAS NMR OF MAPO-36 AND MAPO-5 WITH HIGH Mg CONTENT

M. V. GIOTTO*, M. da S. MACHADO*, S. P. O. RIOS*, J. PÉREZ-PARIENTE§, D. CARDOSO*

*Chemical Engineering Department, Federal University of S. Carlos, P.O. Box 676
13565-905 S. Carlos, SP, Brazil, FAX: 55-16-2608266 - E-mail: dilson@power.ufscar.br
§Instituto de Catálisis y Petroleoquímica - CSIC - Madrid, Spain

ABSTRACT

MAPO-36 and MAPO-5 with high Mg content were prepared from reaction mixtures with composition xMgO:(1-x/2)Al$_2$O$_3$:P$_2$O$_5$:R:wH$_2$O:$2x$AcOH. Characterization of these molecular sieves was accomplished by means of high-resolution ^{31}P and ^{27}Al MAS NMR, X-ray diffraction and elemental analysis. In the case of MAPO-5 framework phosphor atoms were only observed at coordination P(4Al) and P(3Al,1Mg). Samples of both structures prepared from reaction mixtures with high Mg content (x>0.45) revealed non-zeolitic ^{31}P signals associated to dense phosphates. The ^{27}Al NMR spectra of both structures showed ^{27}Al signals with coordination greater than four, which intensity decreases as Mg content in the solid increases. ^{1}H-^{31}P CPMAS measurements of AlPO-5 enhances strongly two ^{31}P signals indicating the existence of POH groups associated to structural defects.

Keywords: AlPO-5, MAPO-5, MAPO-36, synthesis, NMR, characterization

INTRODUCTION

MAPO-36 (ATS) and MAPO-5 (AFI) are large pore size molecular sieves with one-dimensional channel systems. Due to the high strength of their acid sites [1], especially in the case of MAPO-36 [2], these materials experience increasing catalytic applications. Some ^{31}P and ^{27}Al NMR spectroscopy investigations on these structures have been published in recent years [3-6], although in the case of MAPO-36 these investigations were not systematic.

In this present work, samples of MAPO-36 and MAPO-5 with different Mg content as well as of AlPO-5 were prepared with the aim to determine the effect on the properties of these materials.

EXPERIMENTAL

Synthesis

Samples of both MAPOs and AlPO-5 were prepared using pseudoboehmite (Pural), magnesium acetate and H$_3$PO$_4$ (Aldrich) as source of Al, Mg and P, respectively, from reaction mixtures with composition: xMgO:(1-x/2)Al$_2$O$_3$:P$_2$O$_5$:R:wH$_2$O:$2x$AcOH.   (1).

MAPO-36 was synthesized according to the procedure of Ono [7,8] at 150°C using w=40, R= tripropylamine (TPA) as template and the Mg values, x, listed in Table 1. MAPO-5 and AlPO-5 were synthesized according to the method of Concepción et al. [9], at 170°C using w=100, R= triethylamine (TEA) as template and the Mg values listed in the same Table. In order to explore the presence of slightly soluble solids in the synthesized material, some samples were leached three times in aqueous medium at 50 °C. These samples were identified with their respective code, plus the letter L (Table 1).

Characterization

All MAS NMR spectra were acquired by means of a Varian-Unit 400 MHz spectrometer operating at 104.21 and 161.90 MHz for ^{27}Al and ^{31}P, respectively. ^{27}Al and ^{31}P chemical shifts were recorded in relation to 1M aqueous solution of $Al(H_2O)_6^{3+}$ and H_3PO_4 85% aqueous solution. Quantitative MAS ^{27}Al spectra were obtained with selective excitation using a short pulse of 1 μs (π/20), repetition time of 2 s, 200 transients. Those of ^{31}P were obtained with single-pulse excitation with proton decoupling using a pulse length of 5.3 μs (π/2), 60 transients and repetition time of 30 s. ^{1}H-^{31}P CP MAS were performed using ^{1}H pulse length of 5.0 μs (π/2) with contact time of 2 ms, 300 transients and repetition time of 5 s. The rotation frequency was 8 kHz using a silicon nitride (Si_3N_4) rotor. XRD spectra were obtained using a Rigaku diffractometer, scanning 2θ between 5° and 40° with CuKα radiation at a scan rate of 2 degrees/min. The overall phosphor, magnesium and aluminum content in the solid were determined by induced coupled plasma (Thermo Jarrel Ash).

RESULTS AND DISCUSSION

Figure 1a presents the diffractograms of MAPO-36 and Figure 1b those of MAPO-5 and AlPO-5, with different Mg content in the synthesis gel as indicated in Table 1 (x). The samples with the lowest magnesium content show diffractograms of pure ATS (x<0.45) and AFI (x<0.7) phases. The MAPO-36 samples having higher Mg content present some peaks of AFI phase. The MAPO-36 sample with the highest Mg (sample D in Figure 1a) contains about 30% AFI phase and that of MAPO-5 (sample D Figure 1b) presents traces of MAPO-34 (CHA). These results agree with those obtained by other authors [1,4,9] who also observed the formation of CHA in the MAPO-5 synthesis, however at Mg contents as low as x=0.2.

		^{31}P Chemical shifts (ppm)			Framework * Population (%)	
Sample	x	P(4Al)	P(3Al)	NZ	P(4Al)	P(3Al)
ATS [A]	0.13	-27.3	-20.7	-7.1	77.3	22.7
ATS [B]	0.29	-27.6	-21.7	-7.6 / -13.2	58.0	42.0
ATS [C]	0.45	-28.7	-22.7	-7.6 / -13.2	56.6	43.4
ATS [D]	0.69	-28.7	-22.7	-7.6 / -13.2	56.2	43.8
ATS [D-L]	0.69	-28.7	-22.7	-7.6 / -13.2	56.3	43.7
AFI [A]	0.1	-29.5	-24.4	-----	67.5	32.5
AFI [B]	0.5	-29.7	-23.7	0.29	63.1	36.9
AFI [C]	0.7	-29.7	-23.7	0.27	60.4	39.6
AFI [D]	1.0	-29.5	-23.7	0.16	59.5	40.5
AFI [C-L]	0.7	-30.1	-23.6	0.15	63.7	36.3

NZ = phosphor from non zeolitic material,   * accuracy of 5% of each peak area.

Table 1: ^{31}P MAS NMR data for MAPO-36 and MAPO-5

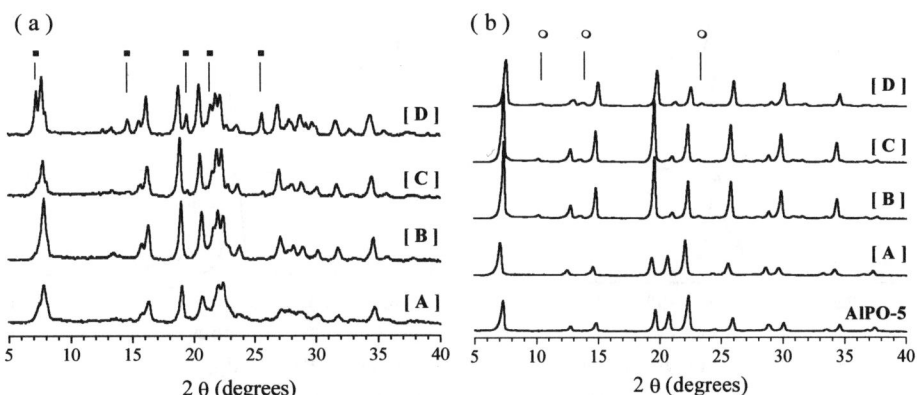

Figure 1: X-Ray Diffractograms of (a) MAPO-36 and (b) MAPO-5 and AlPO-5. Contamination: (■) AFI and (○) CHA

Figure 2 shows the ^{31}P and ^{27}Al NMR spectra of MAPO-36 with different Mg content. ^{31}P MAS NMR spectrum of sample A (Figure 2a) exhibits peaks around -27.3 ppm and -20.7 ppm, which are attributed to P(4Al) and P(3Al,1Mg) sites, respectively. As the Mg content in the reaction mixture was increased, both signals are shifted upfield (Table 1). The relative populations of each ^{31}P environment, calculated by deconvolution of the spectra, indicate that increasing the Mg content in the gel up to $x=0.45$, the intensity of P(4Al) signal is reduced and of P(3Al,1Mg) increased. This indicates an increasing Mg incorporation into the molecular sieve structure in this range, and that higher Mg content in the gel doesn't favor its incorporation into the ATS framework. No signal that could be attributed to P(2Al,2Mg) was found. The samples also exhibit a clear signal at around -7.5 ppm, not observed by other authors for this structure [5,6,7]. Measurements of ^{31}P MAS NMR of MgHPO$_4$.3H$_2$O yields a signal at -7.8 ppm and the double phosphates TPAMgPO$_4$, TEAMgPO$_4$ and NH$_4$MgPO$_4$ also present a strong signal around –7.5 ppm. So, the signal found together with MAPO-36 may be attributed to non-zeolitic Mg-phosphate compounds. Moreover, some spectra exhibit a poor defined signal at –13.2 ppm. Nakashiro et al. [7] found a ^{31}P signal at -11 ppm attributed, without any evidence, to the interaction effects of occluded organic molecules with framework phosphorous.

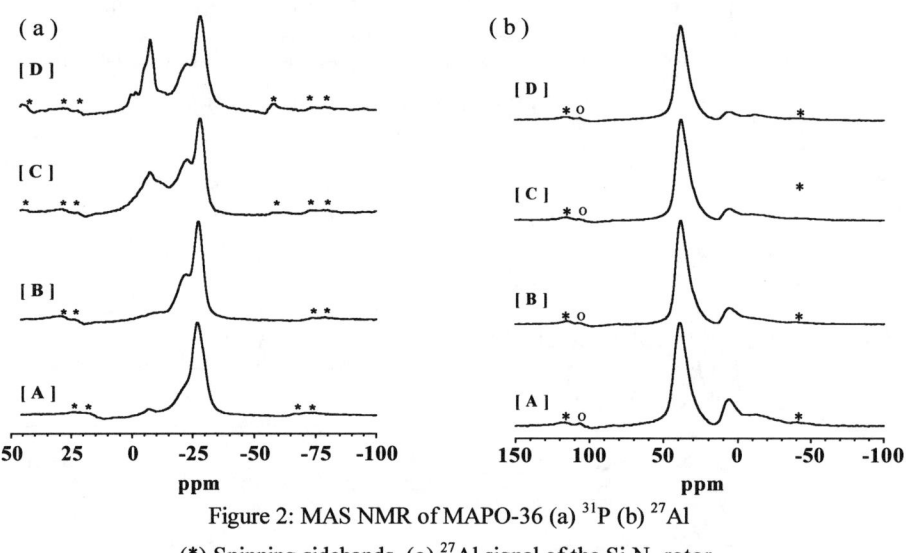

Figure 2: MAS NMR of MAPO-36 (a) ^{31}P (b) ^{27}Al
(*) Spinning sidebands, (o) ^{27}Al signal of the Si$_3$N$_4$ rotor.

Figure 2b shows ^{27}Al MAS NMR spectra of MAPO-36, which exhibit a signal at 38.5 ppm in a region that is assigned to tetrahedrally coordinated aluminum in AlPOs [4,6]. The spectra also present two other signals of lower intensity at 6.5 ppm and -12 ppm, which according to some authors might be a result of water or amine molecules interaction with Al [3-5]. The signal at -12 ppm in MAPO-36 was attributed by Akolekar et al. [6] to the pentacoordinated Al.

However, detailed studies with AFI, containing [4] or not Mg [10], attribute this signal to hexacoordinated Al in Al(4P) environment and the signal at 6.5 ppm to pentacoordinated Al in Al(4P) environment. In addition, this latter signal may also be attributed to hexacoordinated Al in Al(4Al) environment present in pseudoboehmite, employed in the synthesis of this molecular sieve.

Figure 3 shows the ^{31}P and ^{27}Al MAS NMR spectra of MAPO-5 and AlPO-5. ^{31}P MAS NMR spectrum of sample A (Figure 3a) presents peaks around -29.7 ppm and at -23.7 ppm, which can be attributed to P(4Al) and P(3Al,1Mg) sites, respectively. On increasing the Mg content in the gel there is a proportional increase in the P(3Al,1Mg) relative population (Table 1) but contrary to MAPO-36, the chemical shift of these signals did not show a dependence on Mg content.

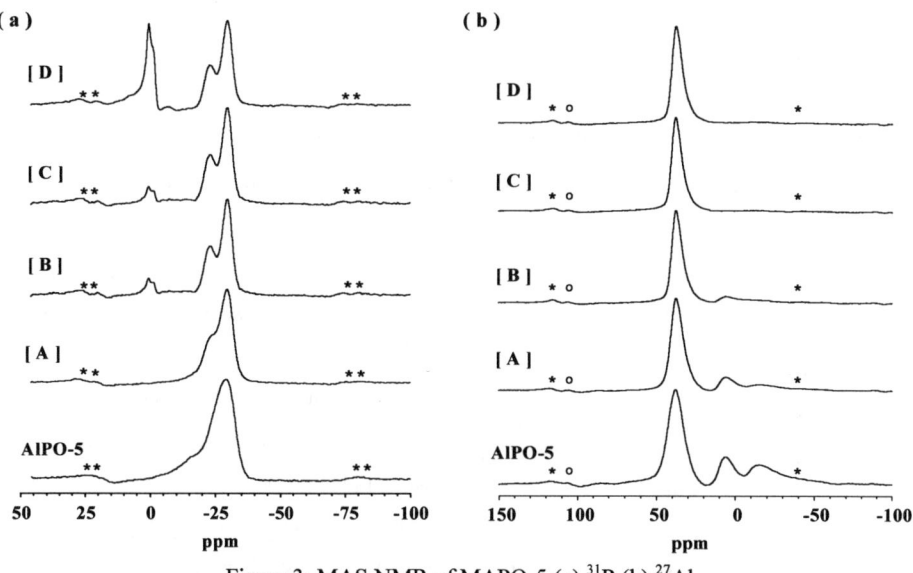

Figure 3: MAS NMR of MAPO-5 (a) ^{31}P (b) ^{27}Al
(*) Spinning sidebands, (o) ^{27}Al signal of the Si$_3$N$_4$ rotor

The AFI samples B, C and D, prepared using $x \geq 0.5$ present in addition a ^{31}P signal around 0 ppm, whose intensity increases with Mg content. As this signal does not correspond to the previously described Mg double phosphates, ^{31}P MAS NMR of (NH$_4$)$_2$HPO$_4$ and of (NH$_4$)H$_2$PO$_4$ were performed, which presented signals around 1 ppm. Hence, the signal encountered at 0 ppm in MAPO-5 possibly belongs to TEA phosphate, contrarily to statements of Shea [3] who assigned they to dense Mg phosphates.

In Figure 3b, the ^{27}Al spectra of AFI samples present a resonance at 37.5 ppm typical of the tetrahedral environment of Al(4P). The signals at 7.2 ppm and -12 ppm occur at similar chemical shifts to those of MAPO-36. They may therefore be assigned to penta and hexacoordinated Al species as already discussed for Figure 2b. The reduction in intensity of the signal at -12 ppm with the increase in Mg content could be a consequence

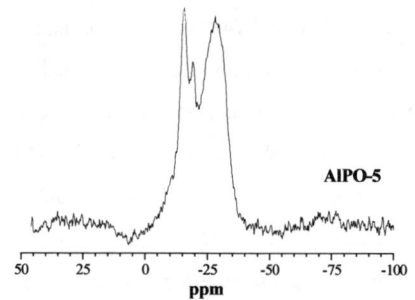

Figure 4: ^{1}H-^{31}P CP NMR MAS of AlPO-5 using contact time of 2 ms.

of less not reacted pseudoboehmite due to the decrease of aluminum in the reaction mixture (Equation 1).

Table 2 shows the global and framework composition of the samples obtained by elemental analysis and deconvolution of ^{31}P MAS NMR spectra, respectively. The sample A of both structures shows [Mg]$_{NMR}$>[Mg]$_{solid}$. This result could be explained considering that some additional resonances are present on these MAPOs, which are not due to the presence of framework magnesium. Indeed, the ^{31}P spectrum of AlPO-5 (x=0) show a broad shoulder at around -14 ppm (Figure 3a) which occurs in a chemical shift range that does not correspond to the expected P(4Al) signal. In order to obtain more information about the origin of these resonances, ^{1}H-^{31}P CPMAS experiments of AlPO-5 were performed. Besides the framework P(4Al) signal at -29.9 ppm, the spectrum (Figure 4) shows two sharp signals at -20.2 and -14.1 ppm, which could be assigned to POH groups [7]. As consequence, to improve the precision of P(nAl, mMg) (n=4-m, m=0-4) calculus, the -20.1 and -14.1 ppm signals were discounted in the case of MAPO-5 with lower Mg content.

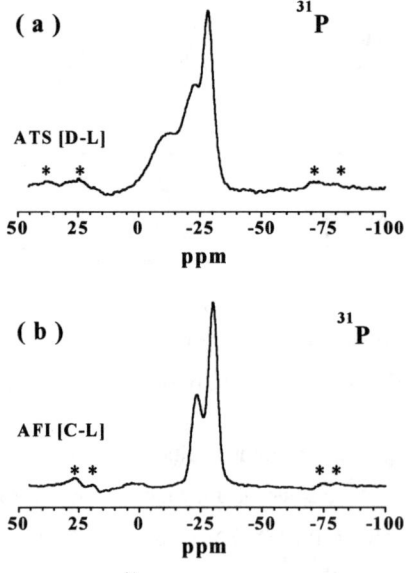

Figure 5: ^{31}P MAS NMR of leached (a) ATS [D-L], (b) AFI [C-L]

Despite this consideration, the sample AFI [A] maintains a clear incoherent result of magnesium molar fraction $[Mg]_{NMR}>[Mg]_{bulk}$. Similar results for samples with low Mg content were found by Prasad et al. [11] and were justified by the presence of extra-framework Al from not reacted pseudoboehmite, which reduces the magnesium content in the bulk composition. The molar fraction of Mg in samples B, C and D calculated by NMR is lower in both structures than that obtained by elemental analysis, indicating the presence of extra-framework Mg. This excess Mg can be present at cationic positions compensating the framework charge or, as already established in the case of MAPO-36, in the form of dense Mg-phosphates. For both structures the maximum molar fraction of Mg found in the framework was around 0.05.

Figure 5 presents ^{31}P and ^{27}Al MAS NMR spectra of leached MAPO-36 and MAPO-5 samples. Comparing sample ATS D-L with D (Figure 5a and Figure 2a) a drastic reduction in the ^{31}P signal at -8ppm is observed after leaching. A similar behavior was observed on comparing sample AFI C-L and C in Figure 5b and Figure 3a, respectively. The ^{27}Al MAS NMR spectra are not affected by the leaching of the samples. It may thus be concluded that the extra-framework phosphate species formed together with ATS and AFI may be dissolved in aqueous medium at 50 °C. This behavior agrees with the former hypothesis that besides the ATS and AFI there is the formation of dense phosphates.

Sample	Bulk composition [a]	Framework [b]	[Mg] [b]	Mg/Al [b]
ATS [A]	$Mg_{0.025} Al_{0.464} P_{0.501}$	$Mg_{0.028} Al_{0.472} P_{0.5}$	0.028	0.060
ATS [B]	$Mg_{0.050} Al_{0.469} P_{0.481}$	$Mg_{0.051} Al_{0.449} P_{0.5}$	0.051	0.114
ATS [C]	$Mg_{0.074} Al_{0.433} P_{0.483}$	$Mg_{0.054} Al_{0.446} P_{0.5}$	0.054	0.121
ATS [D]	$Mg_{0.160} Al_{0.357} P_{0.483}$	$Mg_{0.055} Al_{0.445} P_{0.5}$	0.055	0.124
ATS [D-L]	$Mg_{0.079} Al_{0.439} P_{0.482}$	$Mg_{0.055} Al_{0.445} P_{0.5}$	0.055	0.124
AFI [A]	$Mg_{0.027} Al_{0.464} P_{0.509}$	$Mg_{0.041} Al_{0.459} P_{0.5}$	0.041	0.090
AFI [B]	$Mg_{0.100} Al_{0.391} P_{0.508}$	$Mg_{0.046} Al_{0.454} P_{0.5}$	0.046	0.101
AFI [C]	$Mg_{0.128} Al_{0.362} P_{0.510}$	$Mg_{0.049} Al_{0.451} P_{0.5}$	0.049	0.109
AFI [D]	$Mg_{0.222} Al_{0.284} P_{0.494}$	$Mg_{0.051} Al_{0.449} P_{0.5}$	0.051	0.114
AFI [C-L]	$Mg_{0.086} Al_{0.414} P_{0.499}$	$Mg_{0.045} Al_{0.455} P_{0.5}$	0.045	0.100

[a] Calculated from chemical analysis
[b] Calculated from ^{31}P MAS NMR spectra (assuming [P] =0.5)

Table 2: Framework and Chemical composition of MAPO-36 and MAPO-5

CONCLUSIONS

MAPO-5 with high Mg content was prepared and, contrary to observations of other authors, framework phosphor atoms were only observed at coordination P(4Al) and P(3Al,1Mg). Samples of both MAPO-5 and MAPO-36 structures prepared from reaction mixtures with higher Mg content ($x>0.45$) showed non-zeolitic ^{31}P signals associated with dense phosphates. The intensity of these signals was reduced after leaching, indicating that these dense phosphates are partially soluble in aqueous medium. In MAPO-36 the non-zeolitic phosphor is related to Mg phosphates with ^{31}P resonance around –8 ppm. In MAPO-5 the non-zeolitic phosphor apparently belongs to alkylammonium phosphates whose ^{31}P resonance occurs around 0 ppm. The ^{27}Al NMR spectra of both structures indicate that increase in Mg content reduces the ^{27}Al signals with coordination greater than four, disappearing completely in the case of MAPO-5. ^{1}H-^{31}P CPMAS measurements of AlPO-5 enhances strongly two ^{31}P signals indicating the existence of POH groups associated with structural defects.

ACKNOWLEDGEMENTS
FAPESP and CNPq are acknowledged for financial support of this project.

REFERENCES
1. D. J. Parrillo, C. Pereira, G. T. Kokotailo, R. J. Gorte, J. Catal., 138, 377-385, 1992.
2. D. B. Akolekar, Zeolites, 14, 53-57, 1994.
3. W. Shea, R. B. Borade, A. Clearfield, J. Chem. Soc. Faraday Trans., 89, 3143-3149, 1993.
4. F. Deng, Y. Yue, T. Xiao, Y. Du, C. Ye, L. An, H. Wang, J. Phys. Chem., 99, 6029-6035, 1995.
5. S. Prasad, D. H. Barich, J. F. Haw, Catal. Letters, 39, 141-146, 1996.
6. D. B. Akolekar, S. Bhargava, J. Mol. Catal. A: Chemical, 122, 81-90, 1997.
7. K. Nakashiro, Y. Ono, S. Nakata, Y. Morimura, Zeolites, 13, 561-564, 1993.
8. M. da S. Machado, D. Cardoso, J. Pérez-Pariente, to be published
9. P. Concepción, J. M. López Nieto, A. Mifsud, J. Pérez-Pariente, Zeolites, 16, 56-64, 1996.
10. C. A. Fyfe, K. C. Wong-Moon, Y. Huang, Zeolites, 16, 50-55, 1996.
11. S. Prasad, J. F. Haw, Chem. Mater., 8, 4, 861-864, 1996.

PROBING THE STRUCTURE OF METAL-SUBSTITUTED MOLECULAR SIEVES BY SOLID-STATE NMR

ANDREA LABOURIAU, SUSAN NEUGEBAUER CRAWFORD, KEVIN C. OTT, and WILLIAM L. EARL

Chemical Science and Technology Division, Los Alamos National Laboratory, Los Alamos, New Mexico 87545

ABSTRACT

We have used the paramagnetic influence of metals in $AlPO_4$ and TS-1 to obtain information about the state of the metal in substituted molecular sieves. The hyperfine interaction produces chemical shifts and broadening that are different from the dipole interaction, which obtains when the metal is simply exchanged in the pore. Our data strongly indicate that the titanium in TS-1 is substituted in the framework.

INTRODUCTION

We are interested in the catalytic activity of zeolites with heteroatoms in the framework, especially those that are oxidation-reduction catalysts. The titanium substituted zeolites are of particular interest because of their proven activity for selective oxidation. In spite of numerous spectroscopic and structural studies, there are virtually none that conclusively place titanium in the framework. Diffuse reflectance UV demonstrates that the titanium is not in the form of anatase but not that it is in the framework. Infra-red spectroscopy of these samples has a band at about 960 cm^{-1}, the source of which is debated. X-ray spectroscopies (EXAFS and XANES) are equivocal, with different authors offering different interpretations [1-3]. The ^{29}Si NMR spectrum has no peak which can be assigned to silicon that is next nearest neighbor to titanium [4]. In summary, these interesting catalysts elude all attempts at spectroscopic structural characterization. Additionally if a structural tool were to present itself, it would be most useful to follow the structure of titanium under reaction conditions (structure function relationships). To those ends we have attempted several different types of NMR experiments (^{47,49}Ti and ^{29}Si) all with negative results. However, an EPR experiment by Tuel, et al [5] combined with recent ^{31}P NMR experiments on cobalt-containing aluminum phosphate molecular sieves by Peeters, et al, [6,7] as well as work by Sananes, et al [8,9] and Canesson, et al [10,11] led us to think about the potential of using paramagnetism in the TS-1 structure to get information about the nature of titanium in the zeolite.

In the cobalt-AlPO$_4$ work [6-11] much of the ^{31}P signal is rendered "NMR invisible" by the Co paramagnetism. It was realized that if the cobalt is in the framework, the reason for loss of signal is that the "contact" or hyperfine interaction between the ^{31}P and the paramagnetic electrons both shifts and broadens the ^{31}P resonance. It is possible to recover this signal through a spin echo NMR technique which has been, unfortunately, dubbed SEM NMR for spin-echo mapping NMR [12]. In Co-AlPO$_4$ materials the ^{31}P resonance of the "invisible" phosphorus can be quantitatively recovered. The French group has pursued this methodology and refined the spin-echo technique to the point where they can reconstruct a quantitatively accurate and interpretable NMR spectrum.

Analogous NMR experiments should be possible using ^{29}Si NMR rather than ^{31}P. The difficulty is that the intrinsic sensitivity of ^{31}P is about 15 times that of ^{29}Si which translates to a factor of 225 more time required for the ^{29}Si experiment. To get around this time requirement we resorted to quantitative ^{29}Si NMR and quantified the loss of signal between paramagnetic and diamagnetic samples. This is where Tuel's EPR experiment becomes interesting. We can measure the ^{29}Si NMR spectrum of TS-1 before and after reduction of the Ti^{4+} (diamagnetic) to Ti^{3+} (paramagnetic). The reduction in ^{29}Si NMR signal is a direct measure of the fact that the Ti is actually in the TS-1 framework.

EXPERIMENTAL

The samples used were AlPO$_4$-5 and [Co]-AlPO$_4$-5 synthesized using literature procedures and TS-1 synthesized with two different levels of incorporation of titanium. All samples were characterized by powder x-ray diffraction (XRD), diffuse reflectance UV, and elemental analysis. The analyses of the TS-1 samples give Si:Ti ratios of 30:1 and 120:1. The [Co]-AlPO$_4$-5 sample has a P:Al:Co ratio of 14:13:1. The [Co]-AlPO$_4$-5 was also steamed for 72 hours at 720 K, which removes most of the cobalt from the structure, leaving it in the pores of the sieve. The XRD diffractograms were obtained for AlPO$_4$-5, [Co]-AlPO$_4$-5 prior to steaming, [Co]-AlPO$_4$-5 post steaming, and both samples of TS-1 prior to any treatment. No attempt was made to obtain XRD data on the reduced TS-1 samples nor did we obtain diffractograms of the TS-1 after reduction and reoxidation.

NMR spectra were obtained on a Varian Unity-400 with a 9.4 T magnet and a Varian (7 mm) MAS probe. The nominal resonance frequencies are 161.9 and 79.5 MHz for phosphorus and silicon. Many spectra were obtained without MAS (static). Spectra were obtained using standard Bloch decays (single pulse) and modified Hahn echoes. We used small chips of the titanosilicate mineral, titanite, (CaTiOSiO$_4$) and AgPF$_6$ as internal intensity reference standards for ^{29}Si and ^{31}P, respectively. The titanium in TS-1 was reduced by soaking in CO at 673 K for several hours in a glass tube on a vacuum line [5]. The samples were packed in NMR rotors in a dry, oxygen free atmosphere. The Varian NMR rotors have a double o-ring cap which effectively excludes air during transfer of the sample from the glove box to the spectrometer. We used dry nitrogen as a drive gas for the NMR turbine to minimize the potential of reoxidation of the Ti^{3+}. After obtaining ^{29}Si spectra of the reduced samples they were exposed to room air to reoxidize and another ^{29}Si NMR spectrum obtained.

RESULTS AND DISCUSSION

^{31}P NMR experiments

The ^{31}P NMR experiments on AlPO$_4$-5 were done to establish the technique and demonstrate that a decrease in the ^{31}P resonance intensity is a useful means of establishing that Co is in the AlPO$_4$ framework. Figure 1 contains several NMR spectra for pure AlPO$_4$-5 and [Co]-AlPO$_4$-5. The spectrum of AlPO$_4$-5 (figure 1A) has a relatively sharp peak at about - 30 ppm and only a few spinning sidebands. The signal corresponding to the internal reference is at -145.6 ppm. Knowing the weights of the sample and reference, it is possible to quantify the observed ^{31}P MAS NMR signal by integrating the peaks and all corresponding spinning sidebands. In the AlPO$_4$-5 sample 100% of the phosphorous NMR signal is "visible". Figure 1B shows the ^{31}P MAS spectrum obtained for the [Co]-AlPO$_4$-5 sample. This spectrum contains a large number of spinning sidebands. From the integrals, we conclude that ca. 50% of the ^{31}P NMR signal is "invisible" relative to the pure AlPO$_4$-5 sample. This agrees with Peeters et al's results. Figure 1C is the ^{31}P spectrum of [Co]-AlPO$_4$-5 after steaming. The normalized intensity of the ^{31}P signal is the same as that of the pure AlPO$_4$-5. Note that the peak height of the central resonance is less than in 1A but there is significant intensity in the sidebands, which must be accounted for in the spin counting. The increased sideband intensity is expected in samples with paramagnetic "impurities". This data demonstrates that steaming removes the cobalt from the framework.

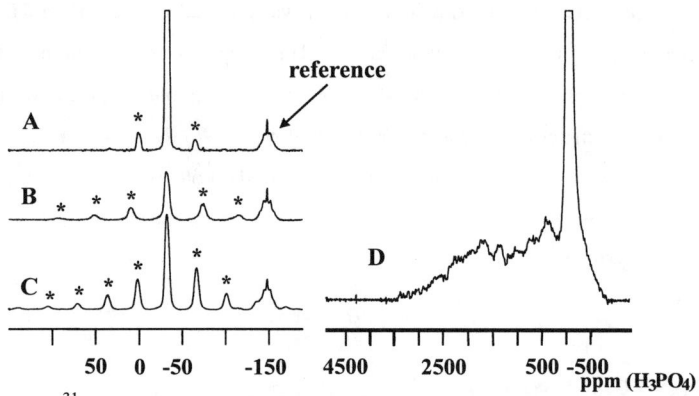

FIGURE 1: The ^{31}P NMR spectra of AlPO$_4$-5. Spectra are: 1A pure AlPO$_4$-5, 1B [Co]-AlPO$_4$-5. 1C [Co]-AlPO$_4$-5 after steaming and removal of Co from the structure. Figure 1D is the spin echo mapping (SEM) spectrum of [Co]-AlPO$_4$-5 (static).

Additionally, we applied the spin echo-mapping technique (SEM) as proposed by Sananes and co-workers [8,9] and explained by Tong [12] to the [Co]-AlPO$_4$-5 sample. This technique is applied to static samples. It is essentially a series of Hahn echo spectra, each shifted to a different carrier frequency and the resulting spectra are co-added to give a resulting spectrum.. We used a 90 pulse length of 6 μs, a 180 pulse length of 12 μs and an echo time, tau, of 20 μs.

The quantitative ^{31}P NMR spectrum is the sum of all the spectra obtained at different irradiation frequencies. Figure 1D shows the total ^{31}P NMR spectrum obtained for the [Co]-AlPO$_4$-5 sample using the SEM technique. As can be seen, the total ^{31}P NMR signal spreads over 2000 ppm. The sharp (truncated) peak at the right of the spectrum corresponds to the center peak in spectrum 1B. We did not attempt to rigorously follow the methods outlined by Sananes so the intensities in spectrum, 1D should not be taken to be quantitative.

^{29}Si MAS NMR experiments

Figure 2 contains ^{29}Si MAS NMR spectra of two TS-1 samples before reduction, after reduction, and reoxidized. Tuel and co-workers noted [5] that a TS-1 sample exposed to carbon monoxide in the absence of oxygen and water can be reduced. Virtually all of the diamagnetic Ti^{4+} atoms are reduced to paramagnetic Ti^{3+}.

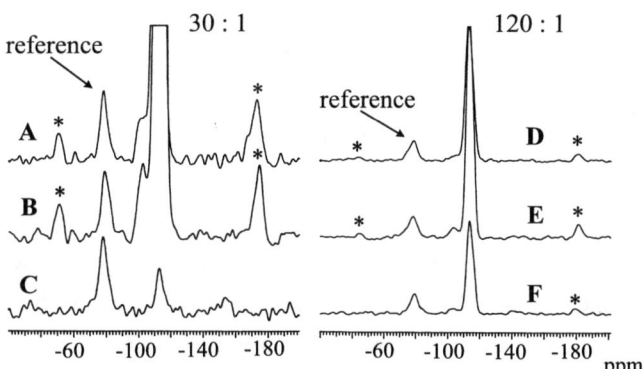

FIGURE 2: Quantitative ^{29}Si NMR spectra of the two TS-1 samples. Asterisks indicate spinning sidebands. The spectra are: 2A prior to reduction, 2B after reduction and reoxidation, 2C after reduction, 2D prior to reduction, 2E after reduction and reoxidation, and 2F after reduction.

In order to make good comparisons, we obtained spectra of the TS-1 samples prior to reduction to establish the intensity of 100% of the ^{29}Si NMR signal. Then we reduced the titanium after which some of the ^{29}Si NMR signal becomes "NMR invisible". In the TS-1 sample with a Si:Ti ratio of 30:1 ca. 95% of the silicon NMR signal becomes "invisible" following CO reduction.

The spectra are normalized to give equal intensity for the reference peak and plotted at high gain to show the small ^{29}Si peak for the post reduction signal in figure 2C. Reoxidation with O_2 or air changes Ti^{3+} back to Ti^{4+} and the entire ^{29}Si NMR signal becomes observable again. Spectra 2A and 2B essentially overlay each other.

Because it is difficult to interpret the data when so little signal remains, we repeated these experiments with a TS-1 sample with a Si:Ti ratio of 120:1. In the reduced sample 50% of the ^{29}Si signal vanishes but it is quantitatively recovered upon reoxidation. This is not entirely clear from the figure because the major peaks in spectra 2D and 2E are both truncated in the plot. However, the integrals of these peaks and their associated spinning sidebands are the same, within experimental error. It would certainly be satisfying to obtain ^{29}Si SEM spectra of these TS-1 samples but as noted above, the sensitivity is so low as to make this extremely costly of spectrometer time. Spin-echo mapping is normally applied to static samples however, we believe that it can also be combined with magic angle spinning, if the pulses are synchronized to the sample rotation. We are starting to try that experiment in order to improve the effective sensitivity for reduced TS-1 samples.

We interpret the disappearance of ^{29}Si NMR intensity to indicate that the titanium in TS-1 is in the lattice. It is worth making a few qualitative comments about the paramagnetic effects on NMR spectra of solids. In virtually all of the other experiments we have done on samples with dilute paramagnets, e.g. [13], we find that the NMR relaxation times are often shortened and the signal may be broadened but it does not disappear as is seen in both the ^{31}P and ^{29}Si experiments reported here. We believe that the signal loss is due to hyperfine interactions between the ^{29}Si nuclei and the unpaired electrons. In other words, those silicon nuclei that are bonded, through a small number of bonds, to the reduced titanium are chemically shifted and broadened by the interaction. This is analogous to spectrum 1D. However, we cannot see the broad and shifted peak without an extremely extended and difficult spin echo experiment. Nuclei that are distant from the paramagnetic titanium are nearly unaffected, so they are not chemically shifted. The details of the electron-nuclear hyperfine interaction determine how far the hyperfine reaches. Those details include the strength of the Ti-O and Si-O bonds, the number of unpaired electrons, which orbitals they occupy, and the electronic relaxation times, T_1 and T_2. The types of information that would aid in the interpretation of all of this data are the electron relaxation times and some knowledge of the hyperfine coupling constant. There are obvious extensions of this method to other metal containing zeolites such as silicalite with iron in the framework.

CONCLUSIONS

The relevant conclusion to TS-1 chemistry is that we have determined that the titanium is in the zeolite framework. With further quantitative NMR we expect to be able to quantify this interaction to correlate the NMR data with elemental analysis to tell whether all or part of the titanium is in the zeolite framework. The spin echo mapping technique potentially contains more information than our ^{29}Si experiments on TS-1. However, careful, quantitative NMR experiments are also useful in characterizing framework substitution with paramagnetic or potentially paramagnetic metals.

REFERENCES.

1. R. J. Davis, Z. Liu, J. E. Tabora, and W. S. Wieland, Catal. Lett. **34**, 101 (1995).
2. O. D. Trong, A. Bittar, A. Sayari, S. Kaliaguine, and L. Bonneviot, Catal. Lett. **16**, 85 (1992).
3. P. Behrens, J. Felsche, S. Vetter, G. Schulz-Ekloff, N. I. Jaeger, and W. Niemann, J. Chem. Soc. ,Chem. Comm. 678 (1991).
4. A. Tuel and Y. Ben Taarit, J. Chem. Soc. , Chem. Commun. 1578 (1992).
5. A. Tuel, J. Diab, P. Gelin, M. Dufaux, J.-F. Dutel, and Y. Ben Taarit et al, J. Mol. Catal. **63**, 95 (1990).
6. M. P. J. Peeters, L. J. M. van den Ven, J. W. de Haan, and J. H. C. van Hooff, Colloid Surf. A: Physicochem. Eng. Aspects **72**, 87 (1993).
7. M. P. J. Peeters, J. H. C. van Hooff, R. A. Sheldon, V. L. Zholobenko, L. M. Kustov, and V. B. Kazansky, in Proceedings from the Ninth International Zeolite Conference. R. von Ballmoos, J. B. Higgins and M. M. J. Treacy, Eds. (Butterworth-Heinemann, Boston, 1993), p. 651.
8. M. T. Sananes, A. Tuel, G. J. Hutchings, and J. C. Volta, J. Catal. **148**, 395 (1994).
9. M. T. Sananes, and A. Tuel, Solid State Nucl. Magn. Reson. **6**, 157 (1996).
10. L. Canesson and A. Tuel, J. Chem. Soc. , Chem. Commun. 241 (1997).
11. L. Canesson, Y. Boudeville, and A. Tuel, J. Am. Chem. Soc. **119**, 10754 (1997).
12. Y. Y. Tong, J. Magn. Reson. ,A **119**, 22 (1996).
13. A. Labouriau, Y.-W. Kim, and W. L. Earl, Phys. Rev. B **54**, 9952 (1996).

THE REVERSIBLE COORDINATION OF FRAMEWORK ALUMINUM IN ZEOLITES

B. H. WOUTERS[*], T.-H. CHEN and P. J. GROBET

Center for Surface Chemistry and Catalysis, Catholic University of Leuven, Belgium; bart.wouters@agr.kuleuven.ac.be

ABSTRACT

The coordination state of the aluminum in zeolite Y has been investigated for samples calcined at different temperatures. Part of the octahedral peak observed in the ^{27}Al MAS NMR spectrum of the calcined zeolite Y can be transformed to the tetrahedral coordination by means of gaseous NH_3.

For the sample calcined at 673 K a good agreement between the Si/Al ratio calculated from the ^{27}Al MAS NMR spectrum and the ^{29}Si MAS NMR spectrum is obtained after ammonia adsorption. Some framework Al species that interact with the hydration water are proposed as octahedral reversible Al. The adsorption of ammonia prevents these sites to complete their coordination.

At higher calcination temperatures, besides the reversible octahedral Al, NMR invisible aluminum species are formed. This kind of aluminum is inaccessible to water and ammonia adsorption and is therefore in a shielded environment. Part of this invisible Al should be considered as framework species.

INTRODUCTION

The aluminum content of zeolites as well as the coordination state of this aluminum, are related to the catalytic properties of these aluminosilicates. Commonly, the octahedral aluminum, observed at 0 ppm in the ^{27}Al MAS NMR spectrum, is assigned to extra-framework aluminum species generated during the calcination process. Recent work on zeolite beta and ZSM-5 demonstrated the existence of octahedral aluminum, which is part of the framework. [1-4].

EXPERIMENTAL

A zeolite Y, obtained from Zeocat, was three times ion exchanged in a 1 M solution of NH_4Cl under reflux. Next this sample was washed until it was chloride-free. This sample was calcined for 4 hour (under deep bed conditions) in an oxygen flow of 80 cm^3 min^{-1} at 673 K,

773K and 823 K, respectively. The temperature was ramped at a rate of 5 K min^{-1} until the final calcination temperature was reached. After calcination, the samples were hydrated in a desiccator with a relative humidity of 79 %. Finally, the samples were heated at 388 K and exposed to an ammonia vapor for 1 hour and subsequently hydrated.

The ^{27}Al and ^{29}Si MAS NMR measurements were performed on a BRUKER MSL400 and a BRUKER AMX300 respectively. For the quantitative aluminum measurements an excitation pulse of 0.6 µs was used which corresponds to a flip angle of π/18. The recycle delay was set to 0.1 s. The Si spectra were recorded with 45° pulses and a repetition time of 20 s.

In the following, the samples will be labelled according to the temperature of calcination, the label A is added for the ammonia-adsorbed samples.

RESULTS

In the Figure 1 the ^{27}Al MAS NMR spectra of the sample calcined at 673 K before (C673) and after ammonia (C673 A) are represented. In the spectrum of sample C673, we observe a NMR line at 59 ppm attributed to the framework Al and a signal at 0 ppm, normally assigned as extra-framework Al. When a comparison between the total aluminum intensity of C673 and the parent ammonium Y sample is made, no intensity loss in observed. The samples calcined at 773 K and 823 K show a similar spectrum (not displayed), of which only the relative intensities to the parent sample differ as is indicated in Table 1. A linear decrease of the intensity of the tetrahedral line in the ^{27}Al MAS NMR spectrum with respect to the

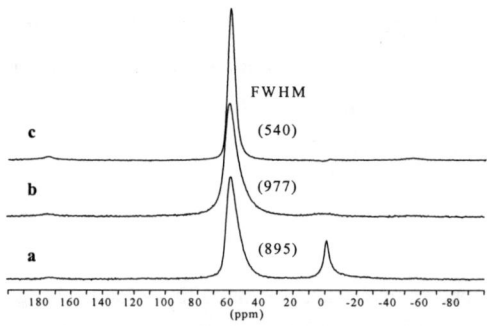

Figure 1. ^{27}Al MAS NMR spectra of a) C673, b) C673A and c) the parent ammonium Y

calcination temperature is observed. When the calcination temperature is raised to 773 and 823 K, the build up of the octahedral site does not compensate the loss of intensity of the tetrahedral site; NMR invisible aluminum is formed.

The ^{29}Si MAS NMR spectrum of C673 is displayed in figure 2a. Some dealumination, which is evidenced by the decreased intensity of the Si(1Al) line, is observed. Calcination at 773 and 823 K further reduces the intensity of this line.

For all the samples, the Si/Al ratio is calculated based on the ^{29}Si MAS NMR spectrum [5] and by means of the relative intensity of the 59 ppm band in the ^{27}Al MAS NMR spectrum, assuming a 100 % crystalline zeolite Y structure. For the calcined samples, we report Si/Al ratios which are higher when calculated from the ^{27}Al MAS NMR spectrum compared to the values obtained from the ^{29}Si MAS NMR spectrum (Table 1).

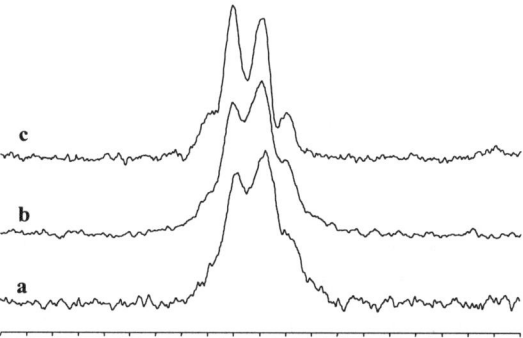

Figure 2. ^{29}Si MAS NMR spectra of a) C673, b) C673 A and c) the parent ammonium Y.

Table 1. Si/Al ratio calculated from the ^{27}Al and ^{29}Si MAS NMR spectrum

Sample	^{27}Al MAS NMR				^{29}Si MAS NMR
	T int (%)^a	O int (%)^b	Inv Al (%)^c	Si/Al	Si/Al
NH$_4$Y	100	0	0		2.65
C673	76	24	0	3.81	2.94
C773	71	16	13	4.1	3.09
C823	69	17	14	4.2	3.15
C673 A	93	7	0	2.92	2.95
C773 A	86	3	11	3.2	3.04
C823 A	80	8	12	3.7	3.12

After ammonia adsorption, the intensity of the tetrahedral site in the ^{27}Al MAS NMR spectrum has increased for each sample, while a reduction of the signal intensity of the 0 ppm peak is observed (Figure 1b). The intensity loss of the octahedral site is in all cases comparable to the gain at the tetrahedral aluminum sites. The line width of the tetrahedral site has increased relative to the parent NH_4Y sample. For the samples calcined at 773 and 823 K, the amount of invisible aluminum remains unchanged when adsorbing ammonia on the sample. The Si/Al ratio calculated from the intensity of the 59 ppm band in the ^{27}Al MAS NMR spectrum decreases relative to the calcined material, where as the ^{29}Si MAS NMR spectrum, and therefore the Si/Al ratio determined from this spectrum, remain unchanged.

From the $\{^1H\}$-^{29}Si cross polarization (CP) MAS NMR spectra of the calcined samples and from the quantitative ^1H MAS NMR measurements we concluded that only a small amount of Si-OH species was present in the samples.

DISCUSSION

Calcination at 673 K

From Table 1 it can be seen that, compared to the results obtained from the ^{29}Si MAS NMR spectrum, a higher degree of dealumination is expected for the sample C673 when the ^{27}Al MAS NMR spectrum is considered. This difference could easily be explained when a lot of Si-OH species were present in the sample, since these species mainly give rise to a band at –100 ppm in the ^{29}Si MAS NMR spectrum which coincides with Si(1Al) line. And so the calculation of the framework Si/Al ratio, which is based on the different intensities of the Si(nAl) species will be erroneous. By both the ^1H MAS NMR data and the $\{^1H\}$-^{29}Si MAS NMR cross-polarization results it was shown that this silanol content is small. The contribution of the Si-OH line to the peak at –100 ppm is therefore too small to account for the different Si/Al ratio we got from our ^{27}Al MAS NMR measurement. The Si/Al ratio determined from the ^{29}Si MAS NMR spectrum is sufficiently accurate to state that the degree of dealumination of this C673 sample is overestimated when calculating the framework Al content from the ^{27}Al MAS NMR spectrum.

When ammonia is adsorbed on this calcined and hydrated sample C673, part of the octahedral Al is converted into a tetrahedral coordination (Figure 1). From the 60 ppm line of this sample in the ^{27}Al MAS NMR spectrum we estimate a Si/Al ratio of 2.92. This way a good correlation is achieved with the aluminum content determined from the ^{29}Si MAS NMR spectrum. Some dealumination can be seen from the decreased Si(2Al) peak intensity in the ^{29}Si MAS NMR spectrum. In the ^{27}Al MAS NMR spectrum this dealumination is observed by some residual octahedral Al after ammonia adsorption.

As a consequence, since the data from the ^{29}Si MAS NMR measurement give us the framework Al content or the framework Si/Al ratio, we have to assume that some framework Al species exist in the Y CH sample which coordinate octahedrally giving the line at a line at 0 ppm in the ^{27}Al MAS NMR spectrum but can be transformed to a tetrahedral state by the adsorption of ammonia.

These results are in agreement with the observations of Bourgeat-Lami et al. [1] Besides the ammonia adsorption, the authors also ion exchanged their calcined zeolite beta samples with different monovalent cations and observed the reversibility of the octahedral site to a tetrahedral coordination. We can thereby conclude that the big cations, compared to the protons, prevent some framework Al sites to complete their coordination shell with hydration water. The reversible part of the octahedral sites in the ^{27}Al MAS NMR spectrum can thus be seen as framework tetrahedral sites which coordinate with the hydration water to form six-fold coordinations.

A distorted framework Al has been proposed for zeolite beta [1] as the origin of the octahedral framework Al site. Woolery et al. [3] ascribe this to partial hydrolysis of the Al-O bonds in the framework. For our zeolite Y sample, we report an increased line width of the 59 ppm line in the ^{27}Al MAS NMR of the calcined and ammonia adsorbed sample (C673 A) compared to the parent ammonium Y sample. When the reversible part of the octahedral coordination would result from a distorted site, we can expect the same line width after ammonia adsorption on the calcined sample as is measured for the ammonium Y sample. This indicates that distinct new species are formed. Therefore we are in favor of the partial hydrolysis of framework Al-O bonds in case of zeolite Y.

Calcination at 773 and 823 K

The sample calcined at 773 and 823 K show a decrease in the tetrahedral site intensity in the ^{27}Al MAS NMR spectrum. The amount of invisible Al reaches up to 14 % of the initial Al intensity. After ammonia adsorption on these samples, only the octahedral Al is converted into tetrahedral Al, the NMR invisible species are not influenced since the amount of invisible aluminum is unchanged (Table 1). The Si/Al ratios calculated from the Al spectrum of the calcined and ammonia adsorbed samples are for both samples higher than those obtained from the Si measurement. Our ^{29}Si MAS NMR spectra do not indicate a large increase in the amount of an amorphous siliceous phase and the silanol concentration is low, as was verified by both ^{1}H MAS NMR and {^{1}H}-^{29}Si CP MAS NMR. Therefore we consider the Si/Al ratio determined from the ^{29}Si MAS NMR spectrum as a reliable measure for the framework Al content. This would then mean that again the degree of dealumination is overestimated by the ^{27}Al MAS NMR measurement, but secondly since the ^{27}Al MAS NMR spectrum of the ammonia adsorbed samples still reflect a higher dealumination degree relative to the results from the Si NMR data, we have to consider the possibility of framework Al sites which are ^{27}Al NMR invisible. These species are probably in a shielded environment since neither the hydration water nor ammonia adsorption can interact and coordinate with the to lower the distortion around this Al.

We obtained similar results with a mordinite sample. Together with the results for beta [1,3] and ZSM-5 [2], we can state the presence of this reversible Al is not restricted to a certain structure type with a well defined Si/Al ratio.

CONCLUSION

From the sample calcined at 673 K it is clear that the Si/Al ratio calculated from the Al spectrum can only be explained if we consider the reversible part of the octahedral sites as framework aluminum.

Calcination at higher temperature generates, besides this reversible octahedral framework Al, NMR invisible Al species that belong to the framework.

The formation of this tetrahedral framework Al that coordinates with hydration water probably initiates the dealumination process generating non-framework Al.

REFERENCES

1. E. Bourgeat-Lami, P. Massiani, F. Di Renzo, P. Espiau and F. Fajula, Appl. Catal **72**, 139 (1992).
2. G. L. Woolery, G. H. Kuehl, A. W. Chester and J. C. Vartuli, Zeolites **19**, 288 (1997).
3. K. H. C. Timken, G. Kuehl, in <u>Book of Abstracts, 11th International Zeolite Conference</u>, Korea (1996), RP62.
4. L. W. Beck and J. F. Haw, J. Phys. Chem. **99**, 1076 (1995).
5. G. Engelhardt and .D. Michel, <u>High-Resolution Solid State NMR of Silicates and Zeolites</u>, (John Wiley & Sons, Chichester, 1987) pp. 212.

INFLUENCE OF GUEST COMPOUNDS ON THE BASE STRENGTH OF ZEOLITES Y AND X INVESTIGATED BY NMR SPECTROSCOPY

M. HUNGER*, U. SCHENK, B. BURGER and J. WEITKAMP

Institute of Chemical Technology I, University of Stuttgart, D-70550 Stuttgart, Germany; fax: +49/711/685-4065, e-mail: michael.hunger@po.uni-stuttgart.de

ABSTRACT

Impregnation of cesium exchanged zeolites Y and X with cesium hydroxide followed by calcination not only creates a basic guest compound, but these treatments also affect the chemical properties of the zeolite framework. By using methoxy groups, formed at framework oxygen atoms, as ^{13}C NMR spectroscopic probes, a decrease in the mean electronegativity of the zeolite framework was observed upon cesium-exchange and impregnation with cesium hydroxide, which corresponds to an increase in the base strength of the zeolite framework. However, the strongest base sites are created by the oxidic guest compound. On zeolite Y, this guest compound is highly dispersed. ^{27}Al DOR NMR spectroscopy of dehydrated zeolites Y and X in different magnetic fields revealed a cation- and impregnation-induced change of the ^{27}Al quadrupolar interaction and, hence, of the local structure of the zeolite framework.

INTRODUCTION

Until recently, only scarce attention has been paid to basic zeolite catalysts, even though such materials show great potential for a number of industrially important reactions - such as dehydrogenation of alcohols, isomerization of olefins, side-chain alkylation of toluene or the synthesis of 4-methylthiazol [1]. The chemical properties of basic zeolites are determined by the alkali metal cations, which act as weak Lewis acid sites, and the basic framework oxygen atoms [2]. Exchanging sodium ions by rubidium or cesium cations leads to a decrease in the mean electronegativity of the framework, and the basicity of the oxygen atoms increases. Strong base sites can be created by incorporating alkali metal oxides into the zeolite pores [3, 4]. Multinuclear NMR spectroscopy, successfully used to characterize basic zeolites [5, 6], was employed to investigate the host-guest interactions of oxidic nanoparticles in zeolites Y and X prepared by exchange with cesium cations followed by impregnation with cesium hydroxide and calcination.

EXPERIMENTAL

Zeolites containing base sites were prepared according to reference [3]. The parent zeolites NaY and NaX (Union Carbide Corp., Tarrytown, N.Y., USA, n_{Si}/n_{Al} = 2.6 and 1.3, respectively) were fivefold ion-exchanged at 353 K in a surplus of a 0.4 M aqueous CsCl solution, which led to a degree of sodium exchange of 70 % and 55 % (zeolites CsNaY-70 and CsNaX-55), respectively. Subsequently, samples of these materials were suspended in 0.2 M solutions of CsOH. The suspensions were stirred to dryness at 353 K and, subsequently, calcined at 673 K (zeolite Y) or 773 K (zeolite X). For these impregnated materials, the codes CsNaY-70/nCsOH and CsNaX-55/nCsOH were used, where n (n = 4, 8, 16) stands for the number of CsOH per unit cell (u.c.). The chemical compositions of the samples were determined by AAS and ICP-AES. The prepared materials were activated for 12 h at a temperature of 673 K and a pressure below 10^{-2} Pa. The adsorption of probe molecules on the calcined samples was carried out in a vacuum line. The NMR spectra were recorded at resonance frequencies of 400.1 MHz and 100.5 MHz for ^{1}H MAS NMR and ^{13}C CP/MAS NMR (Bruker MSL 400), respectively, and of 104.2 MHz (Bruker MSL 400) and 195.4 MHz (Bruker DMX 750) for ^{27}Al DOR NMR spectroscopy.

RESULTS AND DISCUSSION

Bosacek et al. [7, 8] proposed a new method allowing the characterization of the base strength of the zeolite framework. After formation of methoxy groups bound to the framework of different zeolites by adsorption of methyl iodide, Bosacek found a correlation between the isotropic ^{13}C NMR shifts of these methoxy groups and the mean Sanderson electronegativities, S^m [9], of the zeolite framework. Figure 1 shows the ^{13}C CP/MAS NMR spectra of the dehydrated zeolites NaY (a), CsNaY-70 (b), NaX (c) and CsNaX-55 (d) studied in the present work after adsorption of 16 CH$_3$I molecules per unit cell. The signals marked by asterisks are

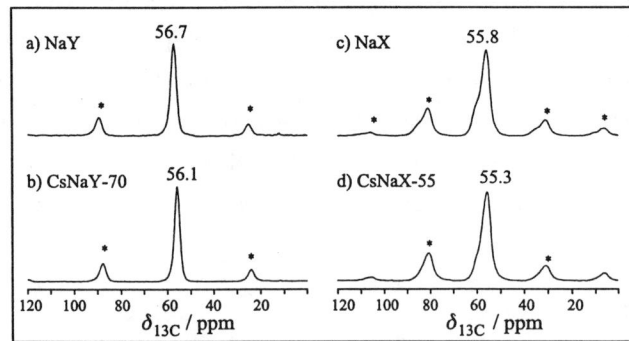

Figure 1. ^{13}C CP/MAS NMR spectra of calcined (673 K) zeolites NaY (a), CsNaY-70 (b), NaX (c) and CsNaX-55 (d), loaded with 16 CH$_3$I/u.c. and recorded with a sample spinning rate of 2.0 kHz.

due to spinning sidebands and indicate that the species causing the sideband pattern are rigidly bound to the zeolite framework which is typical for methoxy groups. The quantitative analysis of the MAS sideband intensities, applying the method introduced by Herzfeld and Berger [10], allows a determination of the anisotropy, $\Delta\sigma$, of the chemical shift, a characteristic parameter of rigid species. Simulation of the MAS sideband patterns, which are due to the ^{13}C MAS NMR signals at 55 to 57 ppm, leads to a chemical shift anisotropy of $\Delta\sigma = -41 \pm 2$ ppm. This value is in good agreement with the chemical shift anisotropy of methoxy groups in zeolite NaX ($\Delta\sigma = -42.7$ ppm) found by Bosacek et al. [7, 8].

In Figure 2, the isotropic chemical shifts, δ_{13C}, of methoxy groups at zeolite frameworks published by Bosacek [7] (open circles) are depicted as a function of the mean Sanderson electronegativity, S^m. The chemical shifts of the methoxy groups in zeolites NaY (56.7 ppm), CsNaY-70 (56.1 ppm) and NaX (55.8 ppm) found in the present work (closed circles) correlate well with their mean electronegativities. Only the chemical shift observed for methoxy groups in zeolite CsNaX-55 (55.3 ppm) significantly deviates from the dotted line. In the ^{13}C CP/MAS NMR spectra of methoxy groups in zeolite NaX, Bosacek [7] found signals at 53.5 ppm and 58.8 ppm which the author explained by methoxy groups bound to different framework oxygen atoms such as O1 and O4. The above-mentioned signal at ca. 58 ppm can also be observed in Figures 1c and 1d as low-field shoulders. The deviation of the chemical shift of methoxy

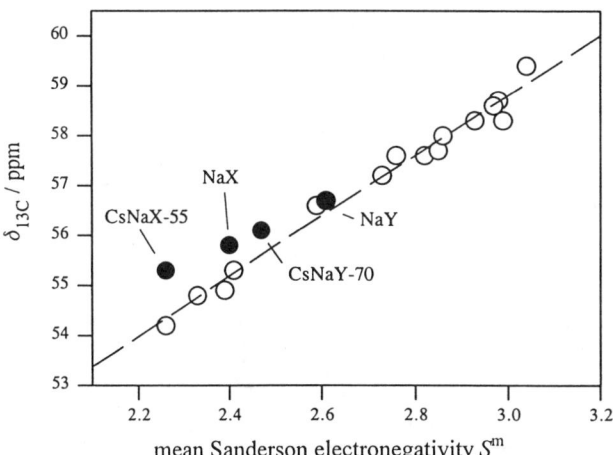

Figure 2. Dependence of the ^{13}C NMR shift of surface bound methoxy groups in zeolites on the mean electronegativity, S^m, of the zeolite framework (○: values taken from Ref. [7], ●: values taken from Figure 1).

groups in zeolite CsNaX-55 from the dotted line in Figure 2 is, therefore, caused by a superposition of the signals due to different types of methoxy groups. In spite of this fact, the ^{13}C NMR high-field shift of methoxy groups in zeolites NaY, CsNaY-70, NaX and CsNaX-55 indicates an increase in the base strength of these materials in the given order. Hence, ^{13}C NMR shifts of methoxy groups are a qualitative measure of the base strength of the cesium-exchanged samples under study. In the following, this method was applied to characterize zeolites Y and X impregnated with cesium hydroxide. To assign the ^{13}C CP/MAS NMR signal of methoxy groups bound to the guest compound, silica gel (Fluka, 500 m^2/g) was impregnated with cesium hydroxide and subsequently calcined at 673 K. The ^{13}C CP/MAS NMR spectrum of this sample loaded with CH$_3$I (Fig. 3a) consists of a single signal at 50.3 ppm. Obviously, this signal is due to methoxy groups bound to the strongly basic guest compound introduced by impregnation. The spectra of zeolite CsNaY-70 impregnated with 8 and 16 CsOH/u.c. (Figs. 3b and 3c) and loaded with CH$_3$I consist of signals at 50.3 ppm, 56.0 to 55.4 ppm and 61.0 ppm due to methoxy groups bound to the guest compound and the zeolite framework and due to dimethyl ether as suggested by Bosacek [7], respectively. An additional weak shoulder at 64 ppm is caused by physisorbed dimethylether. Interestingly, an increasing amount of cesium hydroxide, impregnated onto zeolite CsNaY-70, leads to a high-field shift of the ^{13}C NMR signal of methoxy groups. This finding indicates a change in the base strength of the host framework. However, the strongest base sites in this material are caused by the guest compound (*vide supra*). Since the spectrum depicted in Figure 3c does not show a signal at 56.1 ppm (methoxy groups in zeolite CsNaY-70), the whole framework of zeolite CsNaY-70/

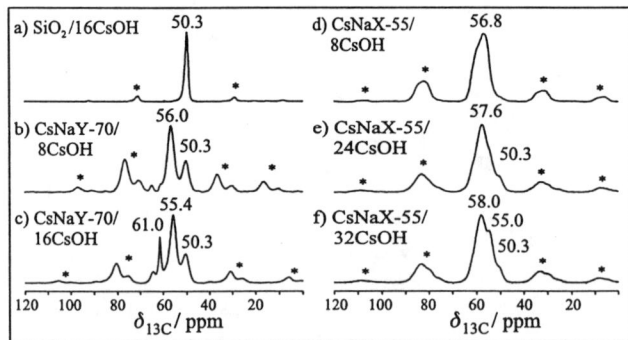

Figure 3. ^{13}C CP/MAS NMR spectra of calcined (673 K) silica gel impregnated with 16 wt.-% CsOH (a), of zeolite CsNaY-70 impregnated with 8 (b) and 16 CsOH/u.c (c) and of zeolite CsNaX-55 impregnated with 8 (d), 24 (e) and 32 (f) CsOH/u.c. All samples were loaded with 16 CH$_3$I/u.c.

16CsOH is affected uniformly by the impregnation with cesium hydroxide. The simulation of the MAS sideband pattern of the signal at 50.3 ppm, observed with increasing intensities in the ^{13}C MAS NMR spectra of zeolites Y after impregnation, yields a chemical shift anisotropy of - 42 ± 2 ppm. The high intensity of this signal, observed for zeolite CsNaY-70/16CsOH (32 ± 2 % of the total intensity), points to a large surface and, therefore, a high dispersion of the guest compound on this sample.

The ^{13}C CP/MAS NMR spectra of zeolite CsNaX-55 impregnated with cesium hydroxide (Figs. 3d, 3e and 3f) consist of broad signals with a line width of up to 10 ppm. With increasing amount of cesium hydroxide introduced by impregnation, the peaks of the broad signal shift from 56.8 ppm (CsNaX-55/8CsOH) to 58.0 ppm (CsNaX-55/32CsOH) while high-field shoulders appear at 55.0 and 50.3 ppm. The chemical shifts of the latter two signals correspond to the resonance positions expected for methoxy groups bound to basic framework sites of zeolite X (see Figs. 1c and 1d) and to the basic guest compound (see Fig. 3a). However, the low intensities of these signals of 25 ± 2 % and 10 ± 2 %, respectively, for zeolite CsNaX-55/32CsOH indicate that the guest compound is not highly dispersed in the zeolite particles. This may be due to sterical reasons originating from the large number of cesium cations located in the supercages. On the other hand, a significant portion of methoxy groups is bound to framework oxygen atoms with a lower base strength responsible for the low-field shift of the peak from 56.8 to 58.0 ppm. These methoxy groups are identical with those causing the shoulder at 58.8 ppm observed in the ^{13}C CP/MAS NMR spectra of zeolites NaX and CsNaX-55 in Figures 1c and 1d. Hence, impregnated zeolites CsNaX-55 possess a variety of base sites with different base strength.

To clarify whether hydroxyl groups are formed on zeolites Y and X as a result of cation exchange and impregnation with cesium hydroxide, the calcined samples were characterized by ^{1}H MAS NMR spectroscopy. The spectra shown in Figure 4 can be decomposed into signals at ca. 0 ppm due to unperturbed metal OH groups, at 1.3 to 2.0 ppm caused by silanol groups and at 2.3 to 3.0 ppm originating from metal OH groups located in small cavities [11]. A weak signal of acidic bridging OH groups can be observed only in the spectrum of zeolite CsNaX-55 at ca. 4 ppm (Fig. 4e). As a result of cesium exchange the amount of silanol groups at framework defects increases from 1.4 ± 0.2 SiOH/u.c. for zeolite NaY to 1.8 ±0 .2 SiOH/u.c for zeolite CsNaY-70 and from 0.8 ± 0.2 SiOH/u.c. for zeolite NaX to 3.6 ± 0.2 SiOH/u.c. for zeolite CsNaX-55. Surprisingly, impregnation with CsOH leads to a decrease of the amount of

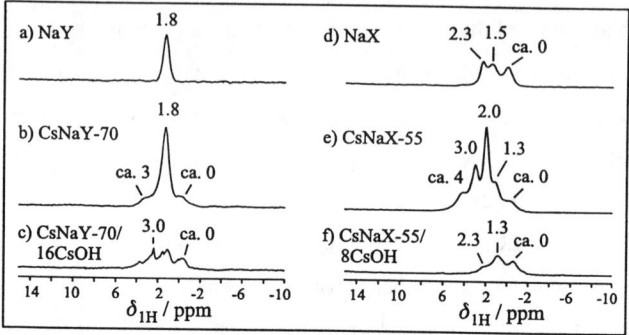

Figure 4. ^{1}H MAS NMR spectra of calcined (673 K) zeolites NaY (a), CsNaY-70 (b), CsNaY-70/16CsOH (c), NaX (d), CsNaX-55 (e) and CsNaX-55/8CsOH (f), recorded with a sample spinning rate of 10 kHz.

silanol groups to 0.4 ± 0.2 SiOH/u.c. in zeolite Y and 0.8 ± 0.2 SiOH/u.c. in zeolite X. No change in the OH concentrations was found after increasing the impregnation of zeolite CsNaX-55 from 8 CsOH/u.c. to 32 CsOH/u.c. (not shown). Generally, after impregnation of cesium-exchanged zeolites Y and X a significant decrease of defect OH groups was found. Probably, this effect is caused by a bonding of the guest compound to initial framework defects.

In a recently published work [5], static ^{27}Al spin-echo NMR spectroscopy was applied to investigate dehydrated basic zeolites Y. In this study, a variation of the local geometry of framework aluminum atoms after cation exchange and impregnation was found. In the present work, ^{27}Al DOR NMR spectroscopy of dehydrated zeolites Y and X was performed in two different magnetic fields. The ^{27}Al DOR NMR spectra of zeolites NaY, CsNaY-70 and CsNaY-70/16CsOH shown in Figure 5 and of zeolites NaX, CsNaX-55 and CsNaX-55/8CsOH shown in Figure 6 were recorded with a spinning rate of the outer DOR rotor of 1400 Hz and at resonance frequencies of 104.2 MHz (left) and 195.4 MHz (right). The dominating solid state interaction of aluminum atoms (spin $I = 5/2$) in dehydrated zeolites is the quadrupolar interaction between the electric quadrupole moment and the electric field gradient at the sites of the resonating nuclei [12]. The strength of the quadrupolar interaction is described by the *SOQE* parameter (*SOQE*: second-order quadrupolar effect). The *SOQE* parameter can be obtained by a evaluation of the field-dependent second-order quadrupolar shift [13]. Applying this method, the following parameters were calculated for zeolites Y: *SOQE* = 5.8 MHz for

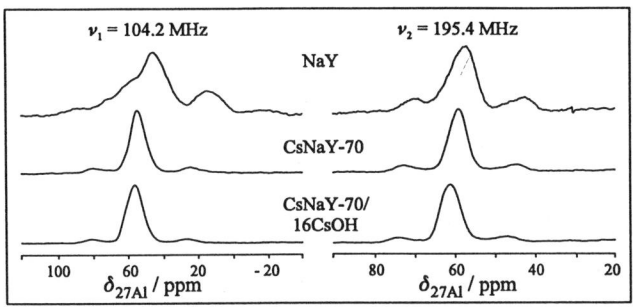

Figure 5. ^{27}Al DOR NMR spectra of calcined (673 K) zeolites NaY (top), CsNaY-70 (middle) and CsNaY-70/16CsOH (bottom), recorded at resonance frequencies of 104.23 MHz (left) and 195.43 MHz (right).

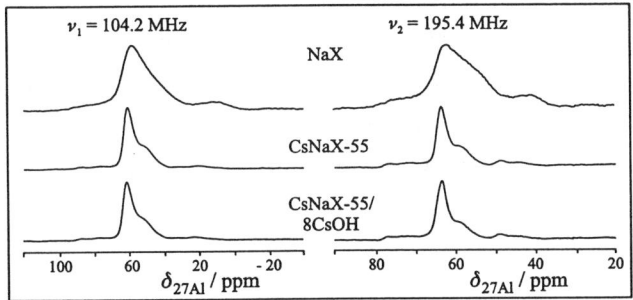

Figure 6. ^{27}Al DOR NMR spectra of calcined (673 K) zeolites NaX (top), CsNaX-55 (middle) and CsNaX-55/8CsOH (bottom), recorded at resonance frequencies of 104.23 MHz (left) and 195.43 MHz (right).

NaY, *SOQE* = 3.7 MHz for CsNaY-70 and *SOQE* = 3.4 MHz for CsNaY-70/16CsOH. Interestingly, the ^{27}Al DOR NMR spectra of zeolite X consist of at least two signals, a strong low-field line and a weak high-field shoulder with an experimental intensity ratio of 60 ± 5 % : 30 ± 5 % (Fig. 6). The parameters determined for the low-field line of zeolite X are: *SOQE* = 3.7 MHz for NaX, *SOQE* = 2.8 MHz for CsNaX-55 and *SOQE* = 2.8 MHz for CsNaX-55/8CsOH. The parameters obtained for the high-field shoulder are: *SOQE* = 5.2 MHz for NaX, *SOQE* = 4.0 MHz for CsNaX-55 and *SOQE* = 3.7 MHz for CsNaX-55/8CsOH. In all cases, the exchange of the sodium form zeolites Y and X with cesium cations and subsequent impregnation with cesium hydroxide leads to a decrease of the quadrupolar interaction of the framework aluminum atoms. According to Ghose and Tsang [14], the above-mentioned decrease of the *SOQE* parameters can be explained by a cation exchange- and guest-induced variation of the local geometry of the framework aluminum atoms.

CONCLUSIONS

By application of multinuclear solid state NMR spectroscopy it is shown that both cesium exchange and impregnation of zeolites Y and X with cesium hydroxide increase the base strength of the host framework and affect the local geometry of the framework aluminum atoms. While impregnation of zeolite CsNaY with cesium hydroxide leads to the formation of a well dispersed and strongly basic guest compound, only a small amount of strong base sites is formed on impregnated zeolite CsNaX.

ACKNOWLEGDEMENTS

Support of this work by Deutsche Forschungsgemeinschaft, Max-Buchner-Forschungsstiftung and Fonds der Chemischen Industrie is gratefully acknowledged. In addition, we are grateful to Prof. Dr. Dieter Michel (University of Leipzig) for providing the solid-state NMR facilities on the spectrometer Bruker DMX 750 (DFG project Mi 390/5) and Ago Samoson (Estonian Academy of Sciences, Tallinn) for support in recording high-field ^{27}Al DOR NMR spectra.

REFERENCES

1. J. Weitkamp, U. Weiß and S. Ernst, in Catalysis by Microporous Materials, Edited by H.K. Beyer, H.G. Karge, I. Kiricsi and J.B. Nagy, Stud. Surf. Sci. Catal., Vol. 94 (Elsevier, Amsterdam, 1995) p. 363.
2. D. Barthomeuf, J. Phys. Chem. **88**, 42 (1984).
3. P.E. Hathaway and M.E. Davis, J. Catal. **116**, 263 (1989).
4. D. Barthomeuf, Catal. Rev. **38**, 521 (1996).
5. M. Hunger, U. Schenk, B. Burger and J. Weitkamp, Angew. Chem. Int. Ed. Engl. **36**, 2504 (1997).
6. M. Hunger, U. Schenk and J. Weitkamp, J. Mol. Catal. A: Chemical, in press.
7. V. Bosacek, J. Phys. Chem. **97**, 10732 (1993).
8. V. Bosacek, H. Ernst, D. Freude and T. Mildner, Zeolites **18**, 197 (1997).
9. R.T. Sanderson, J. Am. Chem. Soc. **105**, 2259 (1983).
10. J. Herzfeld and A. Berger, J. Chem. Phys. **73**, 6021 (1980).
11. M. Hunger, Catal. Rev.-Sci. Eng. **39**, 345 (1997).
12. D. Freude and J. Haase, NMR, Basic Principles and Progress **29**, 1 (1993).
13. M. Hunger, G. Engelhardt, H. Koller and J. Weitkamp, Solid State Nucl. Magn. Reson. **2**, 111 (1993).
14. S. Ghose and T. Tsang, Am. Mineral. **58**, 748 (1973).

THE USE OF TERTIARY AMINES IN THE ELEMENTAL CHARACTERIZATION OF ZEOLITES AND CATALYSTS

MARK E. TATRO

Spectrasol, Inc., PO. Box 1126, Warwick, NY 10990-11226

ABSTRACT

The determination of SiO_2/Al_2O_3 ratios in zeolites requires very accurate and precise measurement of elemental silicon and aluminum in the zeolite solution. To meet this requirement we developed a reagent, Spectrasol Z-B, comprised of tertiary amines to neutralize hydrofluoric acid prior to ICP analysis using an all glass sample introduction system. We extended the use of this reagent to include the ICP analysis of fluid cracking catalysts and clays. By avoiding the use of boric acid to neutralize HF we were able to use the quartz concentric ICP nebulizer which improved the precision of ICP analysis.

INTRODUCTION

Due to its speed, large linear range and detection limits the instrument of choice for the analysis of major and minor elements in silicate materials such as zeolites, catalysts and coal ash is an Inductively-Coupled Plasma - Optical Emission Spectrometer (ICP-OES). For optimal precision analysis an ICP should be equipped with a glass Meinhard concentric nebulizer, a glass spray chamber and a quartz torch. Since the resulting HT solutions from the preparation of silicate materials cannot be aspirated through glass, analysts have resorted to HF resistant ICP nebulizers and torches or the use of boric acid to complex free fluoride in solution (1-3). Unfortunately, the HF resistant transport systems tend to degrade ICP precision and the resulting 3 - 5% boric acid solution clogs glass concentric nebulizers and does not prevent the etching of glassware (4). During studies to determine the SiO_2/Al_2O_3 ratios of zeolites and to characterize the composition of fluid cracking catalysts we developed a reagent comprised of a mixture of tertiary amines to neutralize the pH of the resulting HF solution and to complex any free fluoride without the addition of boric acid to the sample. A study comparing the use of both boric acid and the Spectrasol Z-B tertiary an-tines reagent for ICP analysis of zeolites has been published elsewhere (5).

EXPERIMENTAL

The zeolite SiO_2/Al_2O_3 study used triplicate weighings of 100 mg of a NaY zeolite sample to which was added 1 ml 18 meg-ohm. water, 2 mls conc. HCl and 8 mls conc. (48%) HF. Upon room temperature dissolution, 50 mls of the Spectrasol Z-B neutralizing reagent was added to the samples which were diluted to 100 gms with 18 meg-ohm water. We had previously determined that a ratio of 5: 1 of Spectrasol Z-B reagent to total acid volume (50: 10 in this case) was required to reach a final pH of 7.2 ± 0.2 for the neutralized solution. Analysis of Al, Na and Si was performed on an ARL model 3410 sequential ICP equipped with a quartz Meinhard concentric nebulizer, quartz spray chamber and quartz torch and using the wavelengths of 396.152 nm for Al, 251.920 nm for Si and 589.592 nm for Na analysis.

The fluid cracking catalyst study was a long term correlation study (n = 50) comparing the typical hot plate catalyst acid preparation with the simpler Spectrasol preparation. The hot plate acid preparation consisted of the transfer of 0.5 gm of the catalyst or clay sample into a platinum dish followed by the addition of 20 mls conc. $HClO_4$ and 25 mls conc. HF, heated to $HClO_4$ fumes, the addition of 15 mls 4% H_3BO_3, heated to $HClO_4$ fumes again, the addition of 100 mls 18 meg-ohm water and 10 mls conc. HCl, heated to boiling to dissolve Al_2O_3, the addition of 2.5 mls of 10,000 ug/ml Co to serve as an internal standard and diluted to volume in a 250 mls Nalgene volumetric flask.

The Spectrasol preparation consisted of the transfer of 200 mg of the catalyst or clay sample into 120 mls PFA teflon screw capped pressure vessels to which was added 2 mls 18 meg-ohm water, 0.5 ml conc. HNO_3, 3 mls conc. HCl, 12 mls conc. (48%) HF and 1 ml of 10,000 ug/ml Co. The Teflon vessels were capped and placed in a hot water bath set at 95°C for 30 minutes after which they were cooled in an ice bath before opening to prevent loss of volatile fluorides. The cooled digests were transferred to a tared plastic bottle and 75 mls of the Spectrasol Z-B neutralizing reagent was added. Analysis of all analytes was performed on a J-Y 48 simultaneous ICP equipped with a quartz concentric nebulizer.

RESULTS AND DISCUSSION

From the zeolite SiO_2/Al_2O_3 study we were looking to confirm that the neutralization of HF with the tertiary amines reagent, instead of with boric acid, prevented etching of the ICP glassware, prevented clogging of the concentric nebulizer and resulted in precise Al, Na and Si analysis. Table 1 confirms excellent precision for Si and Al and therefore a very precise measurement of the resulting SiO_2/Al_2O_3 ratio. To investigate the accuracy of analyses we used a reference zeolite that was characterized at PQ Corporation (Philadelphia, PA) for SiO_2 and Al_2O_3 content by gravimetric analysis (6). Table 2, which shows the results of triplicate analysis on a dry weight basis using the Spectrasol Z-B preparation, confirms the accuracy of the ICP measurements.

Our next endeavor was to extend the use of the Spectrasol Z-B neutralization reagent to the more refractory fluid cracking catalysts and clays. The staff at W. R. Grace's Davison Division (Curtis Bay, MD) undertook a long term study to correlate the ICP results obtained from their hot plate acid prep and the Spectrasol preparation (7). Table 3 shows the results of this study with the correlation of linear fit (r) between the ICP results from the two preparations. The hot plate acid prep requires the use of a perchloric acid fume hood, is tedious and requires that the measurement of Si be done by difference since Si is lost as SiF_4 during the preparation. We set our goal at a correlation fit of at least 0.99. The lower correlation for SiO_2 was attributed to errors from its determination by difference using the hot plate method. The lower correlation for Cr and Cu were attributable to the fact that the concentrations of these elements were close to the ICP detection limit. The excellent correlation for the rare earths (cerium, lanthanum, praseodynium, neodymium, samarium and gadolinium) was viewed as further confirmation that all excess HF was neutralized by the tertiary amines.

Since tertiary amines tend to be quite viscous the viscosity of the solutions neutralized with the Spectrasol Z-B reagent typically has a viscosity relative to water of 3.2 cps (5). One of the outcomes of this increased viscosity is that standards must be matrix matched to samples or an internal standard is required. Another outcome is that the precision of ICP analysis appears to be enhanced due to the surface tension of the nebulized aerosol droplets entering the ICP torch. We are following up on this hypothesis.

The use of the Spectrasol neutralizing reagent has been extended recently to include the ICP analysis of low silicon concentrations in food and coral soil (8) with much the same advantages as outlined herein.

Table 1 Results of ICP analysis of a NaY zeolite sample where the HF was neutralized with the Spectrasol Z-B tertiary amines reagent (n = 3).

Analyte	Average (%)
SiO_2	41.8 ± 0.3
Al_2O_3	15.5 ± 0.2
Na_2O	9.8 ± 0.3
SiO_2/Al_2O_3 (m)	4.58 ± 0.03

Table 2 Comparison of ICP results of a NaY zeolite where the HF was neutralized with the Spectrasol Z-B tertiary amines reagent with gravimetric results (n =3).

Analyte	ICP (%)	Gravimetric (%)	Recovery (%)
SiO_2	61.6 ± 0.2	62.3	99
Al_2O_3	22.5 ± 0.1	22.5	100
SiO_2/Al_2O_3 (m)	4.65 ± 0.02	4.70	99

Table 3 Correlation (r) of hot plate acid preparation with Spectrasol preparation of fluid cracking catalyst and clay samples (n=50).

Analyte	Concentration Range	Correlation (r)
SiO_2	41.2 - 81.7%	0.985
Al_2O_3	12.0 - 55.2%	0.998
CaO	0.02 - 1.61%	0.997
Fe_2O_3	0.48 - 3.45%	0.997
MgO	0.01 - 2.98%	0.999
Na_2O	0.01 - 1.07%	0.998
P_2O_5	0.01 - 1.32%	0.998
Re_2O_3	0.01 - 3.88%	0.998
SO_4	0.06 - 2.10%	0.992
TiO_2	0.72 - 3.17%	0.998
Cr	1.0 - 210 ppm	0.900
Cu	5.0 - 140 ppm	0.950
Ni	1.0 - 4100 ppm	0.999
Sb	183 - 1900 ppm	0.999
V	1.0 - 3200 ppm	0.999

REFERENCES

1. R. Bock, in Handbook of Decomposition Methods in Analytical Chemistry, Blackie,. Glasgow, p.55, 58 (1979).
2. R. A. Nadkarni, Anal. Chem., **56**, 2233 (1984).
3. L. A. Fernando, W. D. Heavner and C. C. Gabrielli, Anal. Chem., **58**, 511 (1986).
4. K. E. Jarvis, A. L. Gray and R. S. Houk, in Handbook of ICP-MS, Blackie, Glasgow, p. 194 (1992).
5. D. A. Peru and R. J. Collins, Fres. J of Anal. Chem., **346**, 909 (1993).
6. I. L. Bass and M. E. Tatro, Spectroscopy, **2**(11), 22 (1987).
7. M. E. Tatro, Spectroscopy, **5**(2), 16 (1988).
8. A. P. Krushevska and R. M. Barnes, J. of Anal. At. Spec., **9**, 981 (1994).

A DETAILED NMR STUDY TO THE POLARISATION OF NON-FRAMEWORK LA^{3+} CATIONS WITH THE FRAMEWORK Y ZEOLITE: APPLICATION OF ^{29}SI, ^{27}AL MAS AND ^{27}AL MQ MAS NMR.

J. A. van BOKHOVEN, A. L. ROEST, A. P. M. KENTGENS[#] and D. C. KONINGSBERGER

Debye institute, Utrecht University, Sorbonnelaan 16, Utrecht, The Netherlands,

[#]NSR Center, University of Nijmegen, Toernooiveld 1, Nijmegen, The Netherlands.

ABSTRACT

In order to understand the effect of high-valent cations on the framework in Y zeolite a detailed ^{27}Al NMR and ^{29}Si NMR study to Y zeolite has been performed. ^{27}Al MQ MAS NMR has been applied to La exchanged Y zeolite in order to get a resolution of quadrupolar interactions. ^{29}Si NMR shows no dealumination has taken place after ion exchange of $La(NO_3)_3$ to different extends (up to 16.7 wt% La^{3+}). ^{27}Al MQ MAS NMR shows the 40 ppm peak visible in ^{27}Al NMR is a distorted tetrahedral coordinated Al in the framework, having a distorted electrical field around the Al atom caused by the three-valent cations. Linear correlations between the fraction of charge compensated by La^{3+} versus Na^+ ions and the fraction of distorted Si and Al atoms in the zeolite framework are found. Now, it becomes clear how the multi-valent cations polarise the framework.

INTRODUCTION

Zeolites are a common used catalyst. However, the detailed relationship between local structure and catalytic activity remains in many cases unclear. The effect of steaming of zeolites, which often leads to enhanced activity, is poorly understood. Many models exist, such as increased acidity by polarisation of the acid sites by extra-framework species[1], or to the catalytic effect of Lewis acid sites on extraframework species[2]. In this respect, the co-ordination of aluminium. species in zeolite has been subject of much research. Frequently used techniques are ^{27}Al NMR and ^{29}Si NMR. ^{29}Si NMR allows for the determination of the Si/Al ratio of the framework of the zeolite, while taking respect of Löwensteins Rule. ^{27}Al NMR allows for the determination of the co-ordination of the aluminium species in the zeolite. Four, five and six-co-ordinated Al can be distinguished. However, since Al is a quadrupolar nucleus, peaks may be broadened as well as shifted, which makes interpretation of Al NMR spectra difficult[3].

Here we show a technique called Multiple Quantum (MQ) MAS, which gives resolution of the quadrupolar interactions by choosing a pulse scheme proposed by Frydman et al.[4]. The size of the quadrupolar coupling constant can be made and this value is used to fit the 1D ^{27}Al NMR spectra using real quadrupolar line-shapes. Using this technique the interaction of extra-framework ionic species with the framework can be investigated. Distortions of the framework become visible and it will be shown that the three-valent La ions polarise the framework.

EXPERIMENTAL

NaY was obtained (LZ-Y54) and ion exchanged with La(NO$_3$)$_3$ and calcined at 300°C for 3 hours. Using different La(NO$_3$)$_3$ concentrations and a second exchange for the highest concentration different La^{3+} concentrations are obtained. The notation La(Na)Yx is used, where x stands for weight % of La in the sample.

1D ^{27}Al MAS NMR and ^{29}Si MAS NMR data were taken on a Bruker 500AM with spinning speeds of 12.4 and 5 kHz respectively; MQ MAS was taken on a Bruker 300AM. Samples were overnight saturated over a 1 M NH$_4$Cl solution H$_2$O environment.

RESULTS

Figure 1 shows the ^{29}Si and ^{27}Al MAS NMR spectra of NaY and La(Na)Y16.7. The Na Y shows a characterestic pattern and a Si/Al ratio of 2.6 is calculated. The La(Na)Y 16.7 spectrum is deconvoluted using a NaY part and an additional LaY part, which is shifted about 5 ppm from the NaY part. The ratio of intensity for both parts is equal to the ratio of charge compensated by each of the two cations (calculated on basis of Na content). Thus a Si/Al ratio of 2.6 has been found.

The ^{27}Al NMR spectra show the typical tetrahedral framework peak at 60 ppm for NaY. In La(Na)Y16.7, in addition to this framework tetrahedral Al peak, a broad component at 40 ppm is visible, which must be assigned to framework aluminium, since all Al is in the framework.

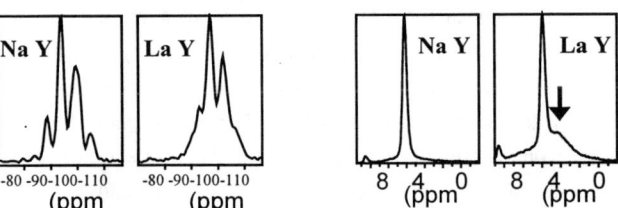

Fig1. ^{29}Si MAS NMR (left) and ^{27}Al MAS NMR (right) spectra of Na Y and La Y.

In Figure 2 the ^{29}Si MAS NMR (bottom) and ^{27}Al MAS NMR (top) of La(Na)Yx spectra are given. Each ^{29}Si MAS NMR can be fitted using a NaY part and a LaY part and in all cases a Si/Al ratio of 2.5/2.6 is found. The relative intensity of the LaY part of the spectra is an almost one to one function of the fraction of charge compensated by La^{3+} ions acting as charge balancing cations (calculated on the basis of each Na$^+$ ion balancing one negative charge, La^{3+} species balancing the rest). The ^{27}Al MAS NMR spectra show a clear increase in the intensity of the 40 ppm component upon increase of amount of La^{3+}.

The multiple quantum MAS ^{27}Al NMR is presented of La(Na)Y16.7 in Figure 3. This plot shows the 40 ppm peak for LaY to be caused by a quadrupolar broadened and shiftied tetrahedrally co-ordinated Al. This plot makes an estimation of the quadrupolar coupling constants possible and these values are used to fit the quantitative 1D ^{27}Al MAS NMR spectra, being able to use quadrupolar line-shaped peaks. In this manner a linear relationship between the relative intensity of the 40 ppm peak and the percentage of charge compensated by the La^{3+} ions is found.

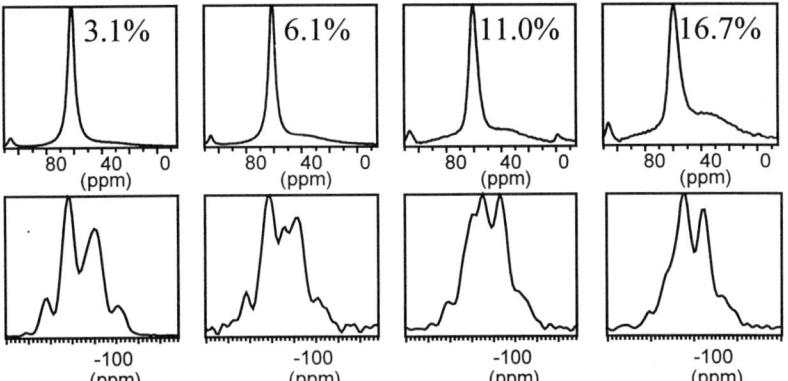

Fig. 2. ^{29}Si MAS NMR (bottom) and ^{27}Al MAS NMR (top) spectra of La(Na)Yx. The weight percentages La are given.

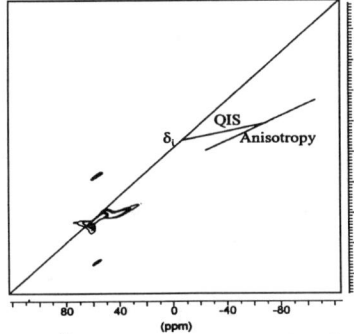

Fig. 3. ^{27}Al MQMAS of LaY showing the origin of the 40 ppm peak. Projection on the x-axis reflect the 1D ^{27}Al NMR spectra.

DISCUSSION

The ^{29}Si NMR spectra indicate that for any La(Na)Yx all Al is in the framework after ion exchange of NaY with La(NO$_3$)$_3$. Thus all peaks observed in the ^{27}Al NMR must be framework Al. This is indicated as well by the absence of a 0 ppm peak which is in general attributed to non-framework octahedrally co-ordinated Al. The ^{29}Si NMR shows the effect of La^{3+} ions on the framework of the zeolite. The fraction of all Si atoms which is distorted by the La^{3+} ions is equal to the fraction of charge compensated by the La^{3+} ions. This shows the overall effect the La^{3+} ions have on the framework.

The effect of La^{3+} ions on the Al atoms in the zeolite is shown by the ^{27}Al MQMAS NMR. The bending off the diagonal in the ^{27}Al MQMAS plot in the direction of the quadrupolar induced shift (QIS) as given by the arrows indicates the 40 ppm peak in LaY is a separate peak from the 60 ppm peak. The ^{27}Al MAS MQ NMR shows it can be attributed to a tetrahedrally co-ordinated Al species, experiencing a large electrical field gradient, which is caused by the La^{3+} ions present in the sodalite cages. The pretreatment of the samples will have caused the La^{3+} to migrate into the sodalite cages.

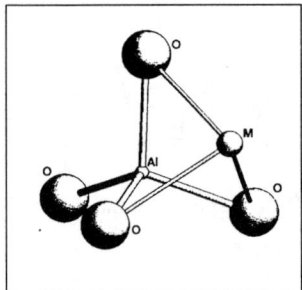

Fig 4. Schematic plot of tetrahedron in interaction with an AlO_4^{5-} tetrahedron of the framework.

Figure 4 shows a model of the interaction of the three-valent La ions with the AlO_4^{5-} tetrahedra in the framework. It has been proposed in the literature that increasing polarisation power (charge divided by radius) of cations cause a linear increase in distortion of the electrical field in AlO_4^{5-} tetrahedra in aluminosilicates'.

As one of the models presented in the introduction suggested that the La^{3+} ions polarise the framework, hence have a positive effect on the catalytic activity, it is now shown to what extend the ions polarise the framework. In the future more definite correlations between the polarisation of the framework and catalytic activity will be investigated.

REFERENCES

1. R. Carvajal, P. J. Chu and J. H. Lunsford, J.Catal. **125** (1990) 123, Topics in Catalysis **4**, (1997) 2742, D.C.Koningsberger, J. T.Miller, Catal. Lett., **29**, (1994) 77
2. eg. Mirodatos and D. Barthomeuf, J. Chem. Soc., Chem. Commun., (1981) 39
3. A. P. M. Kentgens, Geoderma **80** (1997) 271-306
4. Frydman et al., J. Am. Chem. Soc. **117**, 5367.
5. P. J. Dirken, G. H. Nachtegaal, A. P. M.Kentgens, Solid State NMR, **5** (1995) 189.

AS-SYNTHESIZED ITQ-1, THE ALL-SILICA ANALOG OF MCM-22(P): ORDERED, DISORDERED OR SOMETHING IN BETWEEN?

S.L. NJO[*], H. VAN KONINGSVELD[*], B. VAN DE GRAAF[*], CH. BAERLOCHER[§] and L.B. MCCUSKER[§]

[*]Laboratory of Organic Chemistry and Catalysis, Delft University of Technology, Julianalaan 136, 2628 BL Delft, The Netherlands; l.njo@stm.tudelft.nl
[§]Laboratory of Crystallography, ETH-Zentrum, 8092 Zürich, Switzerland

ABSTRACT

An effort has been made to elucidate the structure of the as-synthesized form of ITQ-1, the all-silica analog of the MCM-22 precursor, using synchrotron powder diffraction data and molecular simulations. In contrast to the calcined materials, the as-synthesized forms of MCM-22 and ITQ-1 are not fully 4-connected. Although a certain amount of disorder is present, additional insight into the structure could be obtained. The structure consists of an ordered array of the double layers present in the calcined form, which have deep pockets in their surfaces and which contain a 10-ring sinusoidal channel system. Not only are the double layers not connected to one another, but some of the Si(1) and O(1) atoms at the surface of these layers are missing. Contrary to expectations, the unusual Si site 'buried' inside the small cages is fully occupied. Each unit cell was found to contain two N,N,N-trimethyl-(1-adamantyl)ammonium (TMADA$^+$) ions in the pockets at the surface (adamantyl group pointing into or out of the pocket) and three hexamethyleneimine (HMI) molecules in the sinusoidal channels within the double layer (located sideways between the 4-rings of adjacent double 6-rings). Furthermore, the interlayer region seems to contain approximately four disordered species, whose identity and comformation are not totally clear. The most likely interpretation is that there are three HMI molecules (located in the cleaved 10-ring aperture) and one anionic 3-ring ($H_4Si_3O_9$) species (located between TMADA$^+$ ions in adjacent layers). The presence of such a 3-ring between the organic cations would explain why the layers stack in such an ordered manner although they are not directly connected. One might speculate that these silicate 3-rings disintegrate upon calcination, and that their components then fill the unoccupied Si(1) and O(1) sites to form the fully 4-connected **MWW** framework.

INTRODUCTION

Since the synthesis of the calcined form of MCM-22 was first reported in 1990 [1], considerable effort has gone into the further characterization of this material, because it exhibits remarkable catalytic properties for a variety of hydrocarbon conversion reactions [2, 3]. Another peculiar characteristic of MCM-22 is the fact that the fully 4-connected framework structure is only formed upon calcination. ^{29}Si MAS NMR experiments on ITQ-1, the all-silica analog of the MCM-22 precursor (MCM-22(P) [4]), show that approximately 1/3 of the Si atoms (24 per unit cell) are only 3- or 2-connected in the as-synthesized form [5].

The fully 4-connected calcined framework, having **MWW** topology (Figure 1), can be viewed as a stacking of double layers that are joined via single oxygen bridges. There are two independent, 2-dimensional, 10-ring channel systems. One lies in the double layer, and has sinusoidal channels, while the other lies between the double layers, and has straight channels with large side pockets at the channel intersections. The side pockets (with 12-ring access from the channel) are present on both sides of the channel, and form a large cage [$5^{12}6^{14}10^6$] that can accommodate bulky intermediate species such as 4-*tert*-butylcyclohexanone [3].

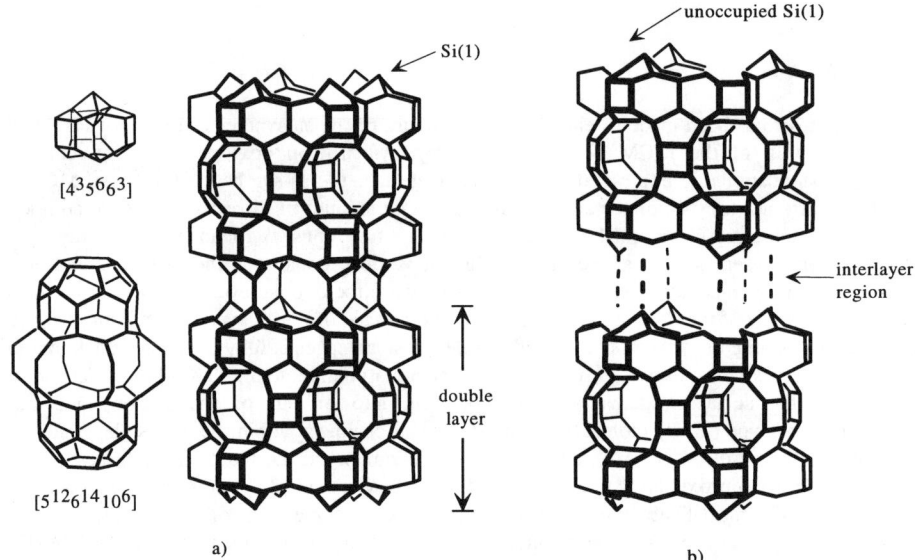

Figure 1. a) The **MWW** topology. The small cage with its 'buried' T-atom and large cage formed at the straight channel intersections are shown separately; b). the proposed as-synthesized structure. Oxygen atoms have been omitted for clarity.

It is generally assumed that the double layers, which contain the sinusoidal channels and have deep pockets at the surface, are already present in the precursor MCM-22(P), and that during the calcination process, the layers are joined by condensation at the surface to form the 3-dimensional structure. The precursor MCM-22(P) can also be used as a starting material in the synthesis of other exciting materials [6]. For example, MCM-22(P) can be treated in such a way that the access to the side pockets, which are assumed to contain excellent catalytic sites, is enhanced by enlarging the space between the double layers.

EXPERIMENTAL

The sample investigated was synthesized [5] in the presence of N,N,N-trimethyl-(1-adamantyl)ammonium (TMADA$^+$) ions and hexamethyleneimine (HMI) molecules (Figure 2). It was kindly provided by Dr. M.A. Camblor (Instituto de Tecnología Química, CSIC-UPV, Universidad Politécnica de Valencia, Spain). Synchrotron powder data were collected on the Swiss Norwegian Beamline at the ESRF in Grenoble. Rietveld refinement was performed with the XRS-82 program package [7]. Molecular mechanics (MM) calculations were carried out using the DMM force field incorporated in DELPHI [8] and molecular dynamics (MD) calculations were conducted employing a simple force field, a combination of DMM and DREIDING [9].

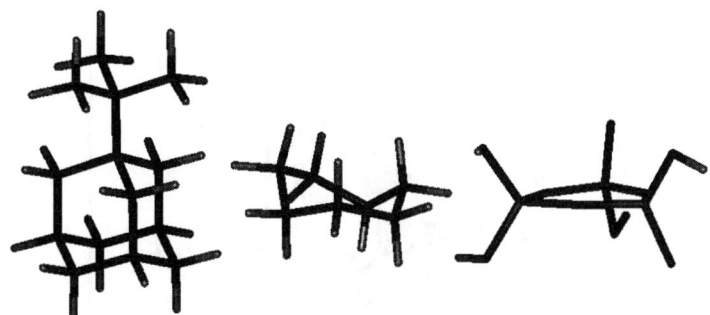

Figure 2. The templates TMADA+ and HMI, and the anionic silicate cluster.

RESULTS AND DISCUSSION

Lattice parameters

Indexing of the powder pattern revealed that the as-synthesized material also exhibits hexagonal symmetry (a = 14.208 Å, c = 27.503 Å), but that the c-axis is 2.5 Å longer than that of the calcined material. This suggests that the structure is indeed broken into layers.

Which bonds are broken?

Since the dimensions of the unit cell along a (and b) are very similar to those of the calcined material, it was assumed that the structure of the single layers was likely to be unchanged. A refinement was carried out in the space group *P6mm* to determine which of the connecting bonds (between the double layers or within the double layers) were broken. One single layer was kept fixed while the other was allowed to move in the c-direction from several different starting points. A model comprising the double layer as shown in Figure 1 appeared to be the most likely one. This model was then used for the initial location of the organic species.

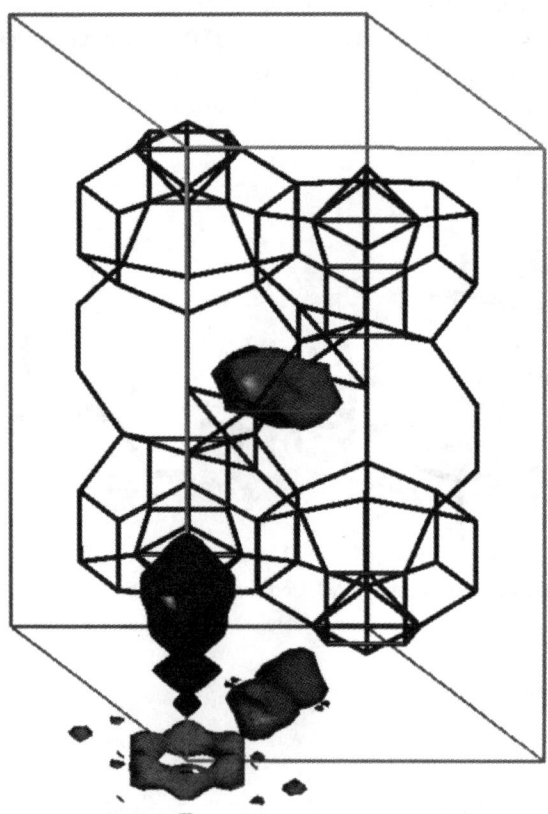

Figure 3. Residual electron densities in the pores of uncalcined ITQ-1. For clarity, symmetry related contours have been omitted.

Location of the Organic Species

A series of difference Fourier maps showed electron density in the pockets on the outer surface of the double layer, in the sinusoidal channels, and in the region between the double layers (Figure 3). TMADA$^+$ ions could be identified in the pockets and HMI molecules in the sinusoidal channels. Both the refinement and the MM calculations proved to be insensitive to the orientation of the TMADA$^+$ ion (i.e. whether the adamantyl group points into or out of the pocket). Both orientations are probably present. The HMI molecules in the sinusoidal channels were found to lie sideways (short axis perpendicular to c) between the 4-rings of adjacent double 6-rings. Both the refinement and the MM calculations show this orientation quite clearly, but again both are insensitive to the exact location of the individual atoms of the HMI molecule. The molecules are quite likely to be freely rotating with several conformations. The electron density in the interlayer region is rather diffuse. Its shape seems to be more consistent with HMI molecules than with TMADA$^+$ ions, but the molecules are disordered. MM and MD calculations indicate that HMI molecules might be present in the cleaved 10-ring aperture, but that they are too large to be located between TMADA$^+$ ions in adjacent layers. However, an anionic 3-ring ($H_4Si_3O_9$) species (Figure 2) does seem to fit between two TMADA$^+$ ions.

Search for Additional Broken Bonds

Using the structure described above as a basis, a series of structure refinements were performed in the space group *P6/mmm* in an attempt to find a framework model consistent with the NMR results, which indicate that 24 Si per unit cell are less than 4-connected [5]. For example, models were tested in which the $\{4^35^66^3\}$ cages with a 'buried' Si atom were replaced by simple $\{4^35^66^3\}$ **DOH** cages, or in which a number of equivalent bonds were broken to yield 24 $Q_{n(n<4)}$ sites. During the course of these refinements, it became apparent that the Si(1) (see Figure 1) and O(1) sites were only partly occupied. However, none of the refinements resulted in a model totally consistent with the NMR data. Models with partly occupied Si(1) and O(1) sites were refined in a number of different space groups (*P6/m*, *Cmmm*, *P6mm*, *Cmm*2 and *P-3m1*) to see if an ordered description of these sites was possible, but no significant improvement over the results obtained in the higher space group *P6/mmm* was found (final R-values in *P6/mmm*: R_F = 0.082 and R_{wp} = 0.241 (R_{exp} = 0.127). With this model, approximately 12 sites per unit cell are less than 4-connected. It appears that the formation of the fully 4-connected framework upon calcination is more complicated than a simple condensation as the organic material is removed and the layers move together. One might speculate that the anionic 3-ring silicate species located in the interlayer region between TMADA$^+$ ions in adjacent layers disintegrate upon calcination, and that their components then fill the unoccupied Si(1) and O(1) sites to form the fully 4-connected **MWW** framework. Furthermore, the presence of such a 3-ring between the organic cations would explain why the layers stack in such an ordered manner although they are not directly connected to one another.

ACKNOWLEDGEMENTS

Financial support from the Netherlands Organisation for Scientific Research (NWO) is gratefully acknowledged (SLN).

REFERENCES

1. M.K. Rubin and P. Chu, U.S. Patent No. 4 954 325 (1990).
2. A. Corma and J. Martinez-Triguero, J. Catal. **165**, 102 (1997).
3. M.J. Verhoef, E.J. Creyghton. J.A. Peters and H. van Bekkum, J. Chem. Soc., Chem. Commun. 1989 (1997).
4. S.L. Lawton, A.S. Fung, G.J. Kennedy, L.B. Alemany, C.D. Chang, G.H. Hatzikos, D.N. Lissy, M.K. Rubin, H.-K. C. Timken, S. Steuernagel and D.E. Woessner, J. Phys. Chem. **100**, 3788 (1996).
5. M.A. Camblor, A. Corma, M-J. Díaz-Cabañas and C. Baerlocher, J. Phys. Chem. B. **102**, 44 (1998).
6. W.J. Roth, C.T. Kresge, J.C. Vartuli, M.E. Leonowicz, A.S. Fung, and S.B. McCullen, Stud. Surf. Sci. Catal. **94** 301 (1995).
7. Ch. Baerlocher, X-Ray Rietveld System XRS-82, Laboratory of Crystallography, ETH Zürich, Switzerland, (1982).
8. S.L. Njo, J.H.Koegler, H. van Koningsveld and B. van de Graaf, Microporous Mater. **8**, 223 (1997).
9. S.L. Mayo, B.D. Olafson and W.A. Goddard III, J. Phys. Chem. **94**, 897 (1990).

STRUCTURE ANALYSIS OF THE ION EXCHANGED FORMS OF THE NEW MICROPOROUS TITANOSILICATE $M_2TiSi_3O_9 \cdot 2.5H_2O$ ($M = NH_4^+$, K^+ and Li^+)

J-L PAILLAUD*, V. VALTCHEV*,†, S. MINTOVA† and H. KESSLER*

* Laboratoire de Matériaux Minéraux, UPRES-A 7016, École Nationale Supérieure de Chimie de Mulhouse, Université de Haute Alsace, 3, rue Alfred Werner, 68093 Mulhouse Cedex, France; JL.Paillaud@univ-mulhouse.fr

† Central Laboratory of Mineralogy and Crystallography, Bulgarian Academy of Sciences, 92 Rakovski Street, 1000 Sofia, Bulgaria

ABSTRACT

Structure features of the NH_4- and Li,K-forms of the new microporous titanosilicate $K_2TiSi_3O_9 \cdot H_2O$ were investigated. The material can be totally exchanged by NH_4^+ and to some extent with the other alkaline cations. Rietveld refinement showed that in the latter case the potassium present in the eight-membered-ring channels is totally exchanged, while the other one located near the centre of the seven-sided windows is partially substituted.

INTRODUCTION

Open structures built from common TiO_6 and SiO_4 units have attracted considerable interest in the past few years. The first members of this new family of microporous materials, ETS-4 and ETS-10, have been discovered by Kuznicki (1). ETS-4 is related to the mineral zorite (2), while ETS-10 has a unique framework topology (3). Other members of the family are the synthetic counterparts of the minerals pharmacosiderite (4), nenadkevichite (5) and pollucite (6). Several new titanosilicates with new framework topologies like GTS-1 (7) and UND-1 (8) have also been synthesised. Another interesting example of a titanosilicate with mixed Ti coordination is the layered structure named JDF-L1, which contains five-coordinated Ti(IV) in the form of TiO_5 square pyramids (9).

We have synthesised a new member of this family of microporous materials, named STS (Sofia Titanium Silicate) (10). A preliminary study of the material showed that the water in the structure is strongly bonded and differs from the typical one in zeolite type materials (11). Recently the structure of this phase was determined by Dadachov and Le Bail (12), it is isostructural with the mineral umbite. The newly synthesised material is also an orthorhombic homologue of the monoclinic titanosilicate UND-1 (8) described by Lui et al.. Lin et al. also published data on the structure of this new titanosilicate STS that they named AM-2 (13).

We have started a study of the ion-exchange properties of the STS phase, indeed this family of microporous materials has shown interesting ion-exchange properties (14). The ion-exchange experiments showed that the new material can be exchanged totally by NH_4^+ and to different extents with the other alkaline cations. Herein we report preliminary results on the structural features of the NH_4- and Li,K-forms of the new microporous titanosilicate.

EXPERIMENTAL

Synthesis

The starting materials for the synthesis were amorphous silica (Aerosil 130, Degussa), titanium tetrachloride (Rhône-Poulenc, 25 mass% in HCl), potassium hydroxide (KOH, Merck), sodium hydroxide (NaOH, Merck) and distilled water. The new titanosilicate was synthesised according to the procedure described in ref. 10 from a starting gel with the molar composition 9 K_2O : TiO_2 : 10 SiO_2 : 675 H_2O. A well crystallised sample was obtained at 200°C for 12 hours. The product was converted in the M-forms by treating it with MCl solutions where M^+ is NH_4^+, Li^+, Na^+, Rb^+ or Cs^+. One gram titanosilicate was treated three times for 24 hours at 100°C with a 30 ml 1 molar MCl (all from Merck) solution. The M-forms of the titanosilicate were saturated with H_2O over a 50 wt. % solution of NH_4Cl. The ion exchanged forms obtained have the following formulae : $(NH_4)_2TiSi_3O_9 \cdot 2.5H_2O$, $K_2TiSi_3O_9 \cdot H_2O$, $Li_{1.1}K_{0.9}TiSi_3O_9 \cdot 2.5H_2O$, $Na_{1.37}K_{0.63}TiSi_3O_9 \cdot 1.3H_2O$, $Rb_{1.35}K_{0.65}TiSi_3O_9 \cdot H_2O$ and $Cs_{0.49}K_{1.51}TiSi_3O_9 \cdot H_2O$.

Characterisation

The XRD powder data of $(NH_4)_2TiSi_3O_9 \cdot 2.5H_2O$ (sample A) and $Li_{1.1}K_{0.9}TiSi_3O_9 \cdot 2.5H_2O$ (sample B) were collected on a STOE STADI-P diffractometer in Debye-Scherrer geometry equipped with a linear position-sensitive detector (6° in 2θ) and employing Ge monochromated CuKα$_1$ radiation (λ = 1.5406 Å, 2θ range 8-80 °, step scan 0.01 °, T = 293 K). The ^{29}Si MAS and ^{29}Si CP-MAS NMR spectra were recorded on a Bruker MSL 300 (ω$_0$ = 59.631 MHz, MAS π/2 pulse of 4.65 µs with a recycling time of 3 min., CP-MAS contact time of 5 ms with a recycling time of 4s). Micrographs of the as-synthesised sample were taken on a Philips XL 30 LaB$_6$ scanning electron microscope (SEM).

Rietveld refinement

The X-ray powder diffraction patterns were indexed using the Werner indexing method (15). An orthorhombic cell was found for both NH$_4$- and Li,K-samples. Systematic absences indicated the same space group P 2$_1$2$_1$2$_1$ as for the K-form. Subsequent analysis and structure refinement were performed using the EXTRA (16), direct method program SIRPOW (17) and GSAS (18) packages. The intensities were extracted using EXTRA with the method of Le Bail (19) and SIRPOW used in the default mode on the extracted data (atom relabelling was necessary). By this way silicon, titanium, framework oxygens and the compensating cations were located. The resulting atomic positions were introduced in GSAS for a Rietveld refinement. From difference-Fourier maps, it was possible to locate two molecules of water out of 2.5 in both cases. During the refinement, the thermal parameters of the three silicon atoms were constrained to be equal as well as the ones of the nine framework oxygen atoms. The Rietveld refinement gave the cell dimensions a = 12.9656(12) Å, b = 10.1229(13) Å, c = 7.2295(8) Å, V = 948.884(18) Å^3 for $(NH_4)_2TiSi_3O_9 \cdot 2.5H_2O$ and a = 12.8652(15) Å, b = 10.1414(15) Å, c = 7.2471(10) Å, V = 945.532(22) Å^3 for $Li_{1.1}K_{0.9}TiSi_3O_9 \cdot 2.5H_2O$. The reliability factors are wRp = 0.0388, Rp = 0.0506, R$_F$ = 0.0845, χ^2 = 2.76, D$_{wd}$ = 0.462, and wRp = 0.0526, Rp = 0.0416, R$_F$ = 0.0751, χ^2 = 1.696, D$_{wd}$ = 0.597 for samples A and B, respectively, and the atomic parameters are reported below in Tables 1 and 2.

TABLE 1: Atomic parameters for sample A. Standard deviations are given in parenthesis.

Atoms	X	Y	Z	Occ. Factor	$U_i \times 100$ (Å^2)
Ti(1)	.5387(5)	.7060(3)	1.0015(13)	1	.73(10)
Si(1)	.7991(3)	.6988(5)	1.0061(19)	1	.85(10)
Si(2)	.5698(8)	.4119(9)	1.1956(15)	1	.85(10)
Si(3)	.5779(8)	.4186(8)	.7856(13)	1	.85(10)
O(1)	.5571(15)	.5598(14)	1.1862(17)	1	0.45(11)
O(2)	.5143(12)	.8208(14)	1.2232(20)	1	0.45(11)
O(3)	.6815(7)	.7564(7)	.9947(34)	1	0.45(11)
O(4)	.6843(12)	.4209(12)	.6540(16)	1	0.45(11)
O(5)	.8167(13)	.6299(11)	.7937(18)	1	0.45(11)
O(6)	.5505(14)	.5838(15)	.7909(20)	1	0.45(11)
O(7)	.5967(6)	.3358(9)	.9829(30)	1	0.45(11)
O(8)	.4739(14)	.3498(13)	.6828(20)	1	0.45(11)
O(9)	.6211(6)	1.1760(8)	.5015(33)	1	0.45(11)
N1	.6598(6)	.6929(10)	.4868(32)	1	2.94(37)
N2	.3240(13)	.5308(17)	.2861(26)	1	15.94(57)
Ow1	.3942(9)	.5687(11)	.5119(42)	1	8.89(59)
Ow2	.1774(10)	.5663(12)	.4103(17)	1	6.13(61)

TABLE 2: Atomic parameters for sample B. Standard deviations are given in parenthesis.

Atoms	X	Y	Z	Occ. Factor	$U_i \times 100$ (Å^2)
Ti(1)	.5288(3)	.6958(5)	.9998(21)	1	1.12(14)
Si(1)	.7928(5)	.7024(7)	1.0044(38)	1	.95(12)
Si(2)	.5747(14)	.4037(19)	1.2105(19)	1	.95(12)
Si(3)	.5828(14)	.4099(18)	.7956(19)	1	.95(12)
O(1)	.5459(26)	.5538(44)	1.1919(38)	1	0.67(14)
O(2)	.5108(20)	.8267(34)	1.1876(32)	1	0.67(14)
O(3)	.6761(10)	.7560(10)	.9948(72)	1	0.67(14)
O(4)	.6901(21)	.3742(26)	.6918(34)	1	0.67(14)
O(5)	.8170(20)	.6042(26)	.8387(35)	1	0.67(14)
O(6)	.5507(28)	.5703(44)	.8064(38)	1	0.67(14)
O(7)	.5984(9)	.3389(11)	1.0174(42)	1	0.67(14)
O(8)	.4781(21)	.3283(35)	.7173(33)	1	0.67(14)
O(9)	.6292(9)	1.1739(13)	.4872(56)	1	0.67(14)
Li1	.2817(56)	.4681(65)	.8116(98)	1.	2.76(305)
K	.3477(4)	.1921(6)	.0018(29)	.9	2.40(19)
Li2	.3477(4)	.1921(6)	.0018(29)	.1	2.40(19)
Ow1	.0964(10)	.4366(13)	1.0464(24)	1	2.33(68)
Ow2	.3225(20)	.4543(23)	.1428(29)	1	5.33(132)
Ow3	.2132(40)	.3839(45)	-.0640(77)	.408(40)	8.82(365)

RESULTS AND DISCUSSION

The SEM picture of the as-synthesised potassium form is shown in Figure 1. STS forms essentially spherical bundles with size between 5 and 15 µm. The aggregates are built from parallel plate-like crystals 0.1-0.2 µm thick and up to 1µm long. Neither SEM nor XRD analysis detected other crystalline material in the sample.

The $(NH_4)_2TiSi_3O_9 \cdot 2.5H_2O$ and $Li_{1.1}K_{0.9}TiSi_3O_9 \cdot 2.5H_2O$ samples are isostructural with the parent potassium analogue (12). The framework is built from TiO_6 octahedra and SiO_4 tetrahedra in the form of a three-dimensional corner-sharing mixed arrangement. Three-, six-, seven- and eight-membered rings are present in the structure. Seven-sided windows interconnect the eight- and six-membered-ring channels running along the c-axis (Figure 2).

There are two cationic sites in $K_2TiSi_3O_9 \cdot H_2O$, M1 and M2, which are located in the eight-membered-ring channels and near the centre of the seven-sided windows, respectively (12). As for the parent K-form, the ammonium or the lithium cations occupy the same crystallographic sites M1 and M2. However, the M1 site is slightly shifted towards the centre of the eight-membered-ring channel in both samples A and B. This shift is more pronounced when lithium occupies this position. For sample A, the very high value of the N2 thermal parameter probably originates from a disorder of this site. In fact, during the refinement difference-Fourier maps revealed electronic density peaks close to N2 which might be attributed to several M2 sites with low occupancy factors.

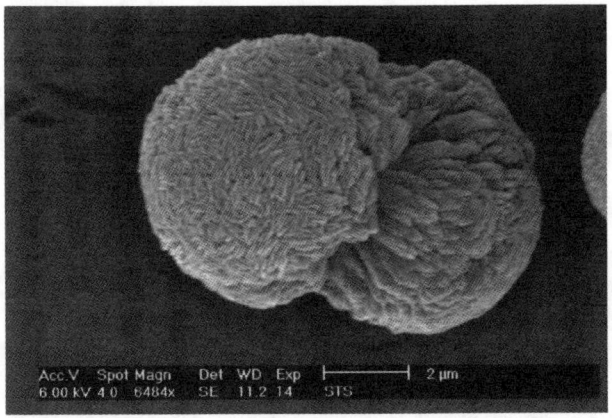

Figure 1. SEM micrograph of the as-synthesised K-form.

For both samples A and B, the M2 site is situated near the centre of the seven-sided windows. The potassium in this position is much more difficult to exchange and only 10 % of the potassium can be replaced by lithium as it is proved by the Rietveld refinement. Keeping in mind the results of the chemical analysis, we can speculate that this position is also partially available for the sodium or rubidium cations and not available for the cesium.

The ammonium and lithium exchanged products have similar unit cell dimensions but the volume is ≈ 40 Å^3 larger than in $K_2TiSi_3O_9 \cdot H_2O$. More likely it is due to the higher water content in $(NH_4)_2TiSi_3O_9 \cdot 2.5H_2O$ and $Li_{1.1}K_{0.9}TiSi_3O_9 \cdot 2.5H_2O$. From their chemical formula ten water molecules are present in the unit cell, which corresponds to at least three water sites with different occupancy factors. Two water oxygen sites (Ow1, Ow2) with an occupancy factor of one have been located inside the large eight-membered-ring channel in the case of samples A and B (Figure 2). The interatomic distances between the water oxygens and framework oxygens O(1-9) and cations are between 2.3 and 3.1 Å. These distances suggest hydrogen bonding in both cases. Moreover a short Ow2-Si (2.5 Å) distance was found by the Rietveld refinement in sample A. The strong cross-polarisation effect observed in the ^{29}Si NMR spectrum of sample A is probably not only due to the close interaction of the water hydrogens but also to the ammonium protons. In fact, the ^{29}Si MAS NMR signal of the ammonium form (Fig. 3b) is broader than those of the K- and Li,K-forms (Fig. 3 a and c), whereas the ^{1}H-^{29}Si cross-polarisation improves only the resolution of the three silicon crystallographic sites of $(NH_4)_2TiSi_3O_9 \cdot 2.5H_2O$ (Fig. 3 d and e).

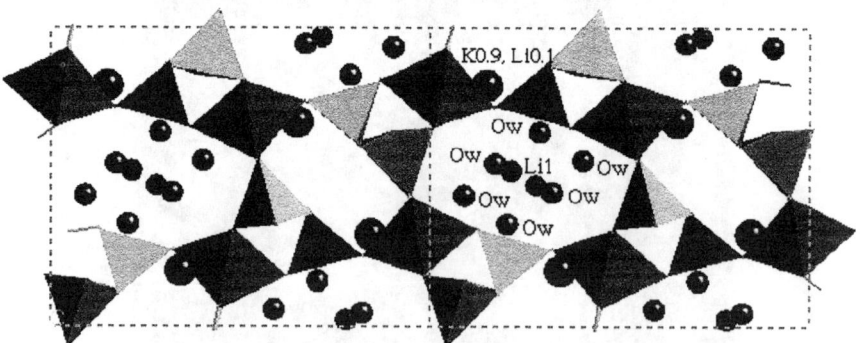

Figure 2. Projection of the structure of $Li_{1.1}K_{0.9}TiSi_3O_9 \cdot 2.5H_2O$ along [001].

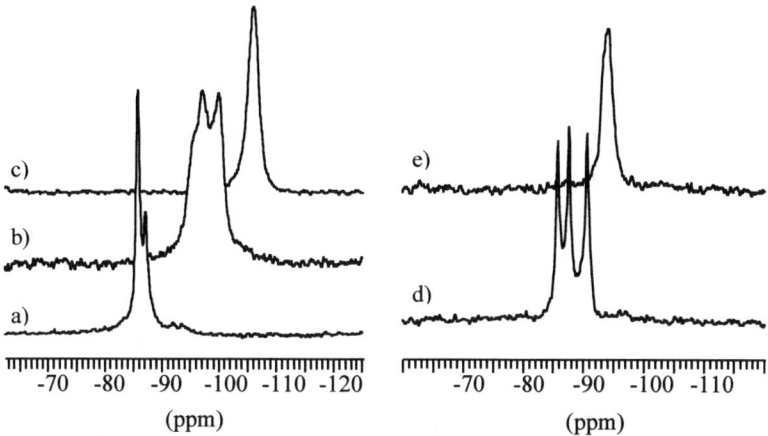

Figure 3. ^{29}Si MAS NMR spectra of a) $K_2TiSi_3O_9 \cdot H_2O$, b) $(NH_4)_2TiSi_3O_9 \cdot 2.5H_2O$ and c) $Li_{1.1}K_{0.9}TiSi_3O_9 \cdot 2.5H_2O$ and the corresponding ^{29}Si CP-MAS spectra of d) $(NH_4)_2TiSi_3O_9 \cdot 2.5H_2O$ and e) $Li_{1.1}K_{0.9}TiSi_3O_9 \cdot 2.5H_2O$.

CONCLUSION

The ion-exchange behaviour of the K-form of the STS material depends on the nature of the alkaline cations and on the position of the cationic sites in the structure. The structure analysis of $(NH_4)_2TiSi_3O_9 \cdot 2.5H_2O$ and $Li_{1.1}K_{0.9}TiSi_3O_9 \cdot 2.5H_2O$ showed that the potassium in the M1 site can be totally exchanged, while the one in the M2 position can be only partially substituted by lithium whereas it is fully occupied by ammonium. The M1 position is shifted towards the centre of the eight-membered-ring channels when potassium is substituted by ammonium or lithium. The STS molecular sieve clearly shows a high selectivity with respect to ammonium (100 % exchange).

ACKNOWLEDGMENTS

We thank Drs C. Marichal-Westrich and L. Delmotte for the NMR experiments.

REFERENCES

1. S. M. Kuznicki, U.S. Patent No. 4 853 202 (1989).
2. A. Phillippou and M. Anderson, Zeolites **16**, 9 (1996).
3. M. W. Anderson, O. Terasaki, T. Ohsuna, A. Phillipou, S. P. Mckay, A. Ferreira, J. Rocha and S. Lidin, Nature **367**, 347 (1994).
4. W. T. A. Harrison, T. E. Gier and G. D. Stucky, Zeolites **15**, 408 (1995).
5. J. Rocha, P. Brandao, Z. Lin, A. Kharlamov and M. W. Anderson, Chem. Commun., 669 (1996).
6. D. E. Cready, M. Lou Balmer and K. D. Keefer, Powder Diffraction **12**, 40 (1997).
7. D. M. Chapman and A. L. Roe, Zeolites **10**, 730 (1990).
8. X. Liu, M. Dhang and J. K. Thomas, Microporous Materials **10**, 273 (1997).
9. M. A. Roberts, G. Sankar, J. M. Thomas, R. H. Jones, H. Du, J. Cheng, W. Pang and R. Xu, Nature **381**, 401 (1996).
10. S. Mintova and V. Valtchev, Bulg. Pat. Appl. No. 100 465 (29 March 1996).
11. B. Mihailova, V. Valtchev, S. Mintova and L. Konstantinov, J. Mater. Sci. Lett. **16,** 1303 (1997).
12. M. S. Dadachov and A. Le Bail, Eur. J. Solid State Inorg. Chem. **34**, 381 (1997).
13. Z. Lin, J. Rocha, P. Brandao, A. Ferreira, A. P. Esculcas, J. D. Pedrosa de Jesus, A. Philippou and M. W. Anderson, J. Phys. Chem. B **101**, 7114 (1997).
14. E. A. Behrens and A. Clearfield, Microporous Materials **11**, 65 (1997).
15. P. E. Werner, L. Eriksson and M. Westdhal, J. Appl. Crystallogr. **18**, 367 (1995).
16. A. Altomare, M. C. Burla, G. Cascarano, C. Giacovazzo, A. Guagliardi, A. G. G. Moliterni and G. Polidori, J. Appl. Crystallogr. **28**, 842 (1995).
17. A. Altomare, M., G. Cascarano, C. Giacovazzo, A. Guagliardi, M. C. Burla, G. Polidori and M. Camalli, J. Appl. Crystallogr. **27**, 435 (1994).
18. A. Larson and R. Von Dreele, Los Alamos National Laboratory, NM 87545 (1995).
19. A. Le Bail, J. Solid State Chem. **83**, 267 (1989).

Part II
Catalysis and Characterization

ADSORPTION COMPLEXES OF METAL HALIDES AT THE BRONSTED ACID SITES IN ZEOLITES

A. I. BIAGLOW

Department of Chemistry, United States Military Academy, West Point, NY, USA 10996; ma7196@usma.edu

ABSTRACT

This work discusses the formation of adsorption complexes of metal halides such as BF_3, BCl_3, $AlCl_3$, and SbF_5 at the Bronsted acid sites in zeolites. The acidic properties of the complex are characterized with alkane cracking reactions. We find that metal halide-doped zeolites are capable of decomposing alkanes even below room temperature. Deuterium incorporation suggests the formation of pentacoordinate carbonium ions in this system. The modified zeolites were also examined using carbonylation of benzene and toluene with CO. We have found that the carbonylation reaction occurs at room temperature with CO pressures less that 1.5 atm. Untreated zeolites do not catalyze this reaction, even at elevated temperatures and pressures. The acidic properties of the modified zeolites were further examined using adsorption and ^{13}C NMR of acetone as well as with simple cluster calculations. Finally, we discuss the conditions under which these materials can be stabilized.

INTRODUCTION

Combinations of Bronsted and Lewis acids, such as HF/BF_3, FSO_3H/SbF_5, and HF/SbF_5 are extremely acidic and capable of protonating a wide range of organic molecules. Many studies have focused on the idea of supporting metal halides on oxides such as silica, alumina, and zeolites, to produce a porous solid form of the metal halide.[1-3] Historically, a significant amount of work has focused on $AlCl_3$.[4] The materials prepared from $AlCl_3$ are generally considered to be $AlCl_2$-functionalized, since they release HCl during synthesis.[3,4] These materials are reportedly stable and nearly superacidic, possessing both Bronsted and Lewis acid sites.[3,4] In the present study, we have focused on the initial complex that forms when the metal halide first comes into contact with the zeolite. By carefully exposing the zeolite to metal halide vapor, it is possible to produce complexes with a stoichiometry of one molecule per zeolite

Bronsted site.[5] This complex seems to be very acidic, catalyzing both the carbonylation reactions and the cracking of n-butane and n-hexane, at temperatures as low as 273 K.

EXPERIMENTAL

Two different zeolite samples, both obtained from The PQ Corporation, were used in this study. The first sample was obtained in the ammonium form and had a bulk Si/Al_2 value of 5.3. This sample had been steamed and acid leached until the lattice parameter was 24.52Å, corresponding to ~30 Al atoms per unit cell. The Bronsted site concentration of this material, determined using TGA of isopropylamine, was 850 µmol/g. The second sample was obtained in the sodium form and also had a bulk Si/Al_2 ratio of 5.3. This sample was not steamed and does not contain Bronsted acid sites as measured with the isopropylamine TGA technique.

The batch alkane cracking measurements were conducted over zeolite samples that were sealed in small glass ampoules. The zeolite was first weighed under vacuum at 750 K, transferred to L-shaped glass sample tubes, and then outgassed on a high vacuum manifold to 10^{-5} Torr. The samples were always spread in thin layers, less than 1 mm deep, to avoid bed effects on adsorption. The samples were cooled and then exposed to 1.0 molecule of metal halide vapor, and then to 70 Torr of n-hexane or n-butane vapor. Samples for the carbonylation reaction were prepared in a similar fashion, with 1.0 molecule of benzene or toluene per site and varying amounts of CO gas. The samples were then sealed in the glass tubes and aged for various times and temperatures prior to analysis. The samples, sealed inside the ampoules, could be analyzed directly using NMR, or broken open inside a special container under inert atmosphere for analysis using gas chromatography and mass spectrometry. The special container is a closed steel tube into which a steel rod is pushed through a Cajon fitting to break the ampoule, and from which gases can be withdrawn through a septum.

In order to study the role of the zeolite Bronsted acid site in the proton transfer process, the cracking reactions described above were repeated using deuterated zeolite. The HY sample was first outgassed as described above, cooled to room temperature, and then repeatedly exposed to 15 Torr of D_2O vapor. The deuterated sample was then outgassed at 750 K, treated with metal halide vapor, exposed to n-hexane, and then sealed as described earlier.

RESULTS

In this short paper, we will only discuss the results of the alkane cracking and deuteration experiments. The evidence for the formation of the 1:1 complex at the zeolite Bronsted site can be found in Reference 5. The carbonylation of benzene with CO in 1:1 HY/AlCl$_3$ is discussed in Reference 6. We also have unpublished data showing carbonylation of toluene in 1:1 HY/BCl$_3$, with CO pressures less than 1.5 atm and with yields above 90%.

The results of the batch n-hexane cracking and isomerization reactions are reported in Table 1 for HY treated with 1 molecule per Bronsted acid site of SbF$_5$, BCl$_3$, and BF$_3$, as well as untreated HY. The conversion over 1:1 HY/SbF$_5$ after 1 hour at room temperature is 71.7% and increases to 91.0% if the sample is warmed to 373 K. The conversion over HY treated with BCl$_3$ or BF$_3$ is somewhat lower, 31.0% and 15.7%, respectively. Under identical conditions but without the metal halide, pure undoped HY shows a conversion of 0.6%, and this is probably due to a small amount of thermal cracking which occurs on flame sealing the glass ampoules. The untreated HY shows significant conversion only at elevated temperatures for prolonged periods of time. We observe 9.3% conversion when the sample is heated to 373 K for 15 hours.

Table 1. Results of the batch n-hexane cracking experiments.

Sample	Reaction Conditions	% Conversion[a]
HY/SbF$_5$	295 K, 1 hr	71.7
HY/SbF$_5$	373 K, 4 hrs	91.0
HY/BCl$_3$	295 K, 1 hr	31.0
HY/BF$_3$	295 K, 1 hr	15.7
HY	295 K, 1 hr	0.6
HY	373 K, 15 hrs	9.3
NaY/SbF$_5$	295 K, 1 hr	10.2
NaY/BCl$_3$	295 K, 1 hr	0.5
NaY/BF3	295 K, 1 hr	0.3
NaY	295 K, 4 hrs	0.2

[a] The GC analysis was conducted on the vapor phase after opening the ampoules. The uncertainty in the measurement is ~0.9%.

The products of the n-hexane cracking reaction on the various metal halide-treated zeolites were primarily 2-methylpropane and 2-methylbutane. We also observed much smaller amounts of methane, ethane, propane, n-butane, and n-pentane. Significant quantities of 2-methylpentane and 2,2-dimethylbutane also form during the reaction. The relatively small amounts of methane, ethane, and propane suggest that we are not observing the primary cracking products of n-hexane, but that the reaction is turning over many times within the catalyst pores before the gaseous products diffuse out.

Previous thermogravimetric studies demonstrate that the metal halide molecules are coordinating with the zeolite Bronsted sites on adsorption from the gas phase.[5] An important question is whether this complex is responsible for the observed catalytic activity of the composite material in the cracking and carbonylation[6] reactions. To address this question and to examine the role of the zeolite Bronsted site, we substituted NaY for HY in the cracking reaction. The results are shown in Table 1 for comparison with HY under identical metal halide loading. Of the samples examined, only NaY/SbF$_5$ shows appreciable reactivity, with a conversion of 10.2%, most likely due to the activity of SbF$_5$ as a strong Lewis acid. The BCl$_3$ and BF$_3$ treated NaY samples were indistinguishable from the pure NaY control.

To further explore the role of the zeolite Bronsted site, we compared the GCMS results obtained from 1:1 HY/SbF$_5$ to those obtained from 1:1 DY/SbF$_5$ from the n-hexane cracking and isomerization reactions. Deuterium exchange into hydrocarbons from deuterated HY is evidence for the formation of pentacoordinate carbonium ions from saturated hydrocarbons, as possible intermediates in the cracking and isomerization reactions. Each of the reaction products was analyzed with an online mass spectrometer, and each showed an increase in the characteristic m/e values, consistent with deuterium incorporation. For example, in the case of the undeuterated HY/BCl3 sample, we obtain significant signal in the parent peak region of the isohexane product, with m/e values at 85, 86, and 87. When HY is replaced with DY, the same m/e region has features with m/e values at 85, 86, 87, 88, and 89. As another example, we observe large features at m/e values of 70, 71, and 72 for the undeuterated zeolite, and at 70, 71, 72, 73, and 74 for the deuterated zeolite.

DISCUSSION

The proposed structure of the 1:1 complex is shown in **1** below. This structure is based on similar models presented in the literature,[3] as well as on electronic calculations conducted by our group.

M = B, Al, Sb
X = F, Cl

1   **2**

The energy difference between **1** and the isolated, gas-phase molecules is -26 kcal/mol for BCl_3, determined from a Hartree-Fock calculation using the 6-31g* basis set and MP2 correlation energy (but not accounting for vibrational energies). The cluster is as shown in 1, with H-atoms terminating the structure. Another interesting comparison is found with a site representing silicalite, **2**. The energy difference between **2** and the isolated gas-phase species is +6 kcal/mol, suggesting that this structure does not form. These calculations are completely consistent with and perhaps explain the experimental observation of the 1:1 complex on adsorption of gas-phase BCl_3.[5] Finally, we note that these results are different for other metal halides, suggesting the possibility of identifying a complex which will be stable under reaction conditions. We are currently attempting to use the cluster calculations as a guide to which metal halides are likely to form 1:1 complexes and to estimate the stability of the complex. Microcalorimetric measurements to further characterize the nature of the adsorption complex are currently underway.

CONCLUSIONS

Metal halide molecules form stoichiometric 1:1 adsorption complexes at the Bronsted acid sites in zeolites. In some cases, these complexes appear to be very stable and very acidic. The metal-halide doped zeolites examined by us are capable of catalyzing both carbonylation and cracking reactions at room temperature. We have also observed that the metal halide-zeolite composite materials are capable of deuterating alkanes at room temperature. These are reactions which are known to require extremely active solid catalysts. Results from the ^{13}C NMR of coadsorbed acetone show that the ability of the complex to transfer a proton to acetone approaches that of magic acid solutions.

REFERENCES

1. K. Tanabe, H. Hattori, and T. Yamaguchi, Crit. Rev. Surf. Chem. **90**, 1 (1990).
2. E.E. Getty and R.S. Drago, Inorg. Chem. **29**, 1186 (1990).
3. M.A. Makarova, S.P. Bates, and J. Dwyer, J. Am. Chem. Soc. **117**, 11309 (1995).
4. T. Xu, N. Kob, R.S. Drago, J.B. Nicholas, and J.F. Haw, J. Am. Chem. Soc. **119**, 12231 (1997).
5. W.P. Fletcher, C.S. Gilbert, and A.I. Biaglow, Catal. Lett. **47**, 135 (1997).
6. T.H. Clingenpeel and A.I. Biaglow, J. Am. Chem. Soc. **119**, 5077 (1997).

ACKNOWLEDGMENTS

We gratefully acknowledge financial support from the National Science Foundation, Grant CTS-95-20920, as well as from the Hoechst Celanese Corporation. We would also like to acknowledge helpful discussions with Drs. Gary Washington and Tania Wessel during the course of this work.

SKELETAL ISOMERIZATION OF C_4-C_8 PARAFFINS: COMPARISON OF ZEOLITES AND SULFATED OXIDES

L. M. KUSTOV*, T. V. VASINA*, O. V. MASLOBOISHCHIKOVA*, A. V. IVANOV*, E. G. KHELKOVSKAYA-SERGEEVA* and P. ZEUTHEN[#]

*N.D. Zelinsky Institute of Organic Chemistry, Russian Academy of Sciences, Moscow, Leninsky prosp. 47, 117334 Russia, FAX: (7095)135-5328, E-mail: lmk@ioc.ac.ru
#Haldor Topsoe A/S, Nymollevej 55, P.O. Box 213, DK-2800, Lyngby, Denmark

ABSTRACT

Various zeolite- and oxide-based acid-type catalysts were studied in the reaction of skeletal isomerization of C_4-C_8 n-paraffins. The mechanism of n-butane isomerization on both zeolites and sulfated oxides is shown to be identical and involves the formation of a C_8 intermediate via the bimolecular reaction. Unlike C_4 isomerization, transformation of higher paraffins occurs mostly via the monomolecular mechanism, and Pt-containing systems are the most active catalysts for the skeletal isomerization of C_5-C_8 paraffins. The zeolite catalysts were found to be more stable and selective, especially in the case of C_7-C_8 paraffins. Additives of various compounds to the feed were shown to influence the performance of the isomerization catalysts. Small additives of water, isooctane, butenes and other agents enhance either the activity or selectivity of the catalysts, thus confirming the acid mechanism of n-butane isomerization. Both zeolites and sulfated oxides demonstrate high sulfur tolerance in the isomerization process.

INTRODUCTION

Zeolites and sulfated oxides are considered as perspective catalysts for n-paraffin isomerization. Recently, a great number of papers related to the study of the nature of the acidity of sulfated zirconia and other modified sulfated oxides and their activity in skeletal isomerization of normal alkanes were published [1—4]. Earlier [4, 5], we studied in detail the activity of sulfated zirconia and other metal-modified sulfated oxides in n-butane and n-hexane isomerization. In the case of n-butane isomerization, a bimolecular mechanism with a predominating acid function has been confirmed, whereas in the case of higher paraffins, a monomolecular mechanism with a predominating metal (dehydrogenating) function has been established. However, almost no comparison of the sulfated oxides with the known zeolite catalysts for paraffin isomerization was made. There are two more important problems that deserve a comprehensive study: (1) the performance of zeolites and sulfated oxides in the skeletal isomerization of higher paraffins (C_7—C_8) and (2) the influence of additives to the feed (olefins, branched paraffins, alcohols, water, sulfur-containing compounds, etc.) that may improve the catalyst performance in a certain manner.

The goal of this study was to compare the performance of platinated zeolites in skeletal isomerization of C_4-C_8 alkanes with that of sulfated oxides. The additional goal was to study the influence of various additives to the feed comprising an n-paraffinic hydrocarbon and hydrogen or helium with the purpose to gain more information about the reaction mechanism, as well as to search for the approaches to enhancing the isoparaffin yield.

EXPERIMENTAL

The samples of 0.5%Pt/HZSM-5 (Si/Al = 20), 0.5%Pt/RE-FAU (Si/Al = 1.3—3.0), and 0.5%Pt/H-MOR (Si/Al = 5.0) zeolites were used as catalysts for n-paraffin isomerization. For comparison, 5%SO_4/ZrO_4, 1%Fe + 5%SO_4/ZrO_2, and 0.5%Pt + 5%SO_4/ZrO_2 prepared as described in [5—7] were also studied. Platinum was supported on zeolites and sulfated zirconia from an aqueous solution of H_2PtCl_6 followed by drying at 120°C for 6 h and calcination in an air flow at 450°C for 2 h.

The reaction of isomerization of alkanes was performed in a flow catalytic setup at atmospheric or high pressures (10—30 atm). The catalysts were preliminarily activated at 350°C in an air or hydrogen flow for 2 h. The gas hourly space velocity was 300 h^{-1} in the case of n-butane and the liquid space velocity was 1.0—2.0 h^{-1} in the case of n-C_5—C_8 paraffins. The alkane-to-hydrogen (or inert gas) ratio was 1 : (3—4). The reaction was carried out at 100—350°C with GC analysis of the products. Regenerations were carried out in an air flow at 450—550°C for 2 h. Additives (0.05—3.0 wt %) of isooctane, water, CO_2, olefins, methanol, air, and thiophene were introduced into the gas or liquid feed.

RESULTS AND DISCUSSION

Table 1 shows the data on the activity and selectivity of the Pt/RE-FAU zeolite catalyst and of the modified sulfated zirconia samples in isomerization of C_4—C_8 paraffins. The data were chosen so as to show the optimal yields and selectivities for the isomerization products. At the same time, the yield of the gas products (C_1—C_3 in the case of butane and C_1—C_4 in the case of the higher paraffins) was minimized. It is seen from these data that the activity of the sulfated

Catalysts	Feed composition	T, °C	Conversion wt. %	Product yields, wt. %		Selectivity to
				gas	iso-C_4—C_8	iso-C_4—C_8
Pt/RE-FAU	C_8H_{18}+H_2	190	17.2	-	17.2	100
		200	37.3	2.1	35.2	94.4
		210	65.7	9.5	56.2	85.5
		220	87.4	39.0	48.4	55.4
-,-	C_7H_{16}+H_2	240	59.8	4.4	55.4	92.6
-,-	C_6H_{14}+H_2	220	55.7	-	55.7	100
		240	78.4	1.3	77.1	98.3
		260	80.6	5.2	75.4	93.5
-,-	C_5H_{12}+H_2	260	51.9	0.3	51.6	99.4
		280	66.1	3.4	62.7	94.9
-,-	C_4H_{10}+H_2	380	42.1	9.3	32.8	77.9
Pt/SO$_4$/ZrO$_2$	C_8H_{18}+H_2	150	39.5	9.6	28.4	71.9
-,-	C_7H_{16}+H_2	160	65.4	28.5	36.9	56.4
-,-	C_6H_{14}+H_2	160	81.8	1.8	80.0	97.8
		190	83.0	5.9	77.1	92.9
Pt/Ga/SO$_4$/ZrO$_2$	C_5H_{12}+H_2	240	57.5	9.5	48.0	83.5
SO$_4$/ZrO$_2$	C_4H_{10}+N_2	230	34.0	6.3	27.7	81.5
Fe/SO$_4$/ZrO$_2$	C_4H_{10}+He	230	40.2	7.1	33.1	82.3
Ga/SO$_4$/ZrO$_2$	C_4H_{10}+He	190	40.5	4.9	35.6	87.9

Table 1. Performance of zeolite and sulfated oxide catalysts in isomerization of C_4—C_8 n-paraffins (LHSV = 1.0—1.1 h^{-1}, alkane-to-hydrogen ratio is 1 : (3—4).

zirconia catalysts is essentially shifted towards lower reaction temperatures as compared to zeolites. The most pronounced difference is found for n-butane isomerization: the zeolite catalyst operates at ~350°C, whereas the sulfated catalyst is active at as low temperatures as 150—230°C. This fact is consistent with stronger acidic properties of Brönsted acid sites in SO_4/ZrO_2 as compared with those in zeolites and is in agreement with our earlier IR-spectroscopic data [4].

The temperatures required for skeletal isomerization of n-paraffins decrease with increasing number of carbon atoms in the linear paraffin chain from 4 to 8. The products distribution in the case of n-butane and higher paraffins is quite different in terms of the contribution of the cracking and disproportionation routes (Table 2). For n-butane isomerization, the main process is accompanied by the reaction of n-butane disproportionation into C_3 and C_5 products, which occurs via the same C_8 intermediate. Unlike C_4 isomerization, the conversion of the higher paraffins involves no disproportionation reactions leading to the products with the number of carbon atoms exceeding the number of carbon atoms in the parent paraffin molecule.

Catalyst	T, °C	Conversion, %	iso-Selectivity, %	Product yields, wt. %				
				C_1	C_2	C_3	i-C_4	C_5
Pt/HZSM-5	350	40.3	52	1.9	5.3	10.0	20.9	2.2
Pt/LaCa-FAU	380	34.7	74	0.4	0.8	3.8	25.7	4.0
SO_4/ZrO_2	230	33.4	82	<0.1	<0.1	2.8	27.7	2.9
Pt/SO_4/ZrO_2	240	11.3	~100	-	-	-	11.3	-

Table 2. Product distribution in n-butane isomerization on Pt/zeolites and sulfated oxides

Thus, analysis of the data on the product distribution for different paraffins and the comparison of the performance of platinum-containing catalysts (both zeolite- and sulfated zirconia-based systems) and non-platinated catalysts allow us to propose different mechanisms for isomerization of n-butane on zeolites and sulfated oxides and transformation of the higher paraffins on the same catalysts, which have been already discussed earlier [4, 5, 7—9]: the bimolecular mechanism in the case of n-butane involving a C_8 intermediate and the monomolecular mechanism in the case of C_5—C_8 n-paraffin conversion involving an olefin or cyclopropane intermediates. Both mechanisms seem to require bifunctional catalysts. The role of acid sites is most important for the former mechanism, whereas the contribution of the metal via the dehydrogenating function is likely more significant for the latter mechanism.

Further, we attempted to use various additives to the feed in order to improve either the activity or selectivity of both zeolite and sulfated oxide catalysts, as well as to shed some more light on the reaction mechanism (Table 3). Such an additive should not be necessarily involved in the reaction mechanism, but it may (1) either change the probability of different pathways and, as a result, change the process selectivity, or (2) act as a «scavanger» or «terminator» of intermediate species leading to side products, such as coke precursors, or (3) change the distribution of the active sites

Catalyst	Feed composition	T, °C	Conversion wt %	Product yields, wt. %			Selectivity to
				gas	iso-C_4—C_8		iso-C_4—C_8
Pt/RE-FAU	C_6H_{14}	260	80.5	5.0	75.5		93.8
	C_6H_{14}+1-C_6H_{12} (1%)	260	80.6	5.2	75.4		93.5
-„-	C_4H_{10}	350	22.0	6.3	15.7		71
	C_4H_{10}+i-C_8H_{18} (1%)	-„-	23.9	5.8	17.4		73
-„-	C_4H_{10}	350	20.7	4.4	15.7		76
	C_4H_{10}+CH_3OH (1%)	-„-	17.2	1.0	15.2		88
-„-	C_4H_{10}	350	20.5	2.6	17.1		83
	C_4H_{10}+H_2O (1%)	-„-	19.4	2.0	17.4		90
-„-	C_4H_{10}	350	20.6	2.7	17.2		83
	C_4H_{10}+thiophene (0.05%)	-"-	18.4	1.3	16.2		88
Pt/HZSM-5	C_4H_{10}	350	16.4	1.6	14.8		90
	C_4H_{10}+i-C_8H_{18} (0.7%)	-„-	40.4	15.7	24.7		61
Fe/SO_4/ZrO_2	C_4H_{10}	215	27.7	4.0	23.7		85.6
	C_4H_{10}+n-C_4H_8 (1.4%)	-„-	29.1	4.8	24.3		83.5
Ga/SO_4/ZrO_2	C_4H_{10}+He	190	28.8	3.7	25.1		87.2
	C_4H_{10}+He+air	-„-	34.1	4.8	29.3		85.9
Pt/SO_4/ZrO_2	C_6H_{14}	160	82.0	3.0	79.0		96.3
	C_6H_{14}+thiophene (0.05%)	-„-	72.6	1.5	71.1		98.0

Table 3. Influence of additives on the performance of zeolite and sulfated oxide catalysts in isomerization of n-paraffins

in a certain way that may improve the selectivity, for instance via poisoning of the strongest acid sites or other unnecessary active centers. First, it should be noted that only small amounts of the studied additives (~1 wt. %) may exert a positive effect on the performance of zeolites and sulfated zirconia, whereas higher concentrations of the same additives to the feed (2—3 wt. % and higher) show, as a rule, a negative effect on the activity and the yield of the isomerization products. Small additives of isooctane improve the activity of zeolite catalysts, whereas methanol additives suppress the activity but improve the selectivity (from 70—75 to 93—98% for the zeolite catalyst).

Other additives demonstrate different influence on the process of n-paraffin isomerization: most of the additives increase the selectivity of zeolites, but only a few examples were found when the activity increases in parallel to the selectivity increase. The data obtained shed some light on the mechanism of paraffin isomerization on heterogeneous acid-type catalysts. The catalysts tested exhibit high sulfur tolerance and resistance to other poisoning additives, like water, air, olefins etc.

CONCLUSIONS

The performance of Pt/zeolite and sulfated zirconia based catalysts in the reaction of normal paraffins isomerization was studied.
1. The mechanism of n-butane isomerization on both zeolites and sulfated oxides is identical and involves the formation of a C_8 intermediate via the bimolecular reaction.
2. Transformation of higher paraffins occurs mostly via the monomolecular mechanism
3. Pt-containing systems are the most active catalysts for the skeletal isomerization of C_5-C_8 paraffins.
4. The zeolite catalysts were found to be more stable and selective, especially in the case of C_7-C_8 paraffins.
5. Small additives of water, isooctane, butenes and other agents enhance either the activity or selectivity, thus confirming the acid mechanism of n-butane isomerization.
6. Both zeolites and sulfated oxides demonstrate high sulfur tolerance in the isomerization process.

REFERENCES

1. A. Corma, Chem. Rev., **95**, 559 (1995).
2. X. Song and A. Sayari, Catal. Rev. - Sci. Eng., **38**, 329 (1996).
3. T. Okuhara, N. Mizuno, and M. Misono, Adv. Catal., **41**, 113 (1996).
4. L.M. Kustov, T.V. Vasina, A.V. Ivanov, et al., Stud. Surf. Sci. Catal., **101**, 821 (1996).
5. A.V. Ivanov, T.V. Vasina, O.V. Masloboishchikova, et al., Kinet. Catal., **39,** 176 (1998).
6. A.V. Ivanov, L.M. Kustov, T.V. Vasina, et al., Kinet. Catal., **38**, 416 (1997).
7. V. Adeeva, G.D. Lei, and W.M.H. Sachtler, Catal. Lett., **33**, 135 (1995).
8. K.T. Wan, C.B. Khouw, and M.E. Davis, J. Catal., **158**, 311 (1996).
9. A.S. Zarkalis, C.-Y. Hsu, and B.C. Gates, Catal. Lett., **29**, 235 (1994).

STRONG ACID SITES FORMED BY A COMBINATION OF FRAMEWORK ACID SITE AND EXTRA-FRAMEWORK CATION IN POROUS MATERIALS AS MEASURED BY TEMPERATURE PROGRAMMED DESORPTION OF AMMONIA

TAKEHISA KUNIEDA, NAONOBU KATADA, and MIKI NIWA

Department of Materials Science, Faculty of Engineering, Tottori University, Koyama, Tottori 680-8552 Japan; mikiniwa@chem.tottori-u.ac.jp

ABSTRACT

Very strong acid site was observed on the zeolite ZSM-5 in measurements of temperature programmed desorption of ammonia. This site was ascribable to the non-framework aluminum which interacted with the usual framework acid site. By impregnation of aluminum cation, the strong acid site was created at the expense of the acid site in a zeolite framework. Lewis-type acid character was enhanced by the formation of the strong acid site. When it was on the external surface, the shape-selectivity was deteriorated as an isomerization of selective product; when inside the pore, the catalyst life was decreased. This kind of acid site could be observed on broad kinds of zeolite and mesoporous material.

INTRODUCTION

Solid acidity is an important character of zeolite catalyst. Temperature programmed desorption (TPD) of ammonia is frequently used as a characterization technique of the zeolite acidity. We have studied this technique in order to establish it as the method on which the qualitative and quantitative measurement is possible [1]. We have already proposed a new method to determine the strength of acidity and its distribution based upon the derived theoretical equation for the conditions of freely readsorption of ammonia [2]. Furthermore, the water vapor treatment is recently applied to remove the unnecessary low temperature peak [3,4,5]. As an extension of the study, in this presentation, the finding of a very strong acid site on ZSM-5 will be reported. Measurements of the acid site and its characterization will be described together with the influence upon the catalytic activity and selectivity.

EXPERIMENTAL

Zeolite ZSM-5 species were synthesized in our previous study [6], and three of them, named No. 5, 6, and 15 species, were selected, since these were different each other in character (*vide infra*). Silica to alumina molar ratios of No. 5, 6, and 15 were 63, 67, and 91, respectively. TPD of ammonia was measured by a method previously described [1]. The measurement was however improved by important alterations; mass spectrometer with 16 of m/e was used to detect ammonia, and the treatment by water vapor was made after the ammonia adsorption, where both procedures, ammonia adsorption and water vapor treatment, were done at 373 K. The water vapor treatment was made to remove the low temperature peak, and the mass spectroscopy was required for the measurement free from overlapped water vapor.

The post-synthesis modification of zeolite was made by the impregnation of $Al(NO_3)_3$ or $Al_2(SO_4)_3$. After the impregnation, the sample was calcined in a stream of oxygen at 773 K for 2 h.

Infrared spectroscopy was measured using the wafer of zeolite on a Jasco FTIR 5300, after it was evacuated at 773 K. The intensity of adsorbed pyridine band was measured, the Si-O band at 1880 cm^{-1} being used as a standard in order to cancel the difference in the thickness of sample.

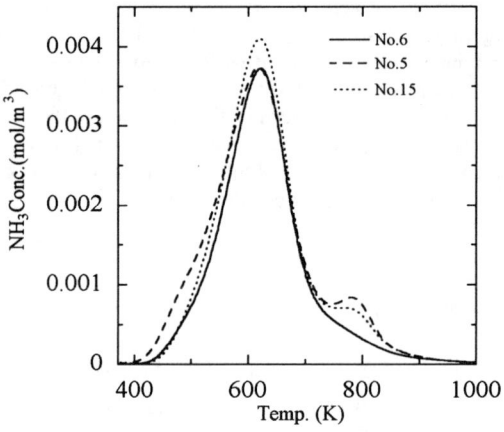

Fig. 1 TPD of ammonia from synthesized HZSM-5 species.

Alkylation of toluene with methanol was measured by a continuous-flow method. Rate of adsorption of *o*-xylene was measured at 373 K by a quartz micro-balance. External surface acidity was evaluated from the cracking of 1,3,5-trimethylbenzene using usual pulse technique. Details of these experiments were reported elsewhere [6].

RESULTS AND DISCUSSION

Finding of additional high temperature peak in TPD measurement

In Fig. 1 shown are TPD spectra of synthesized zeolite ZSM-5 species, Nos 5, 6 and 15. Two peaks have been usually observed at low and high temperatures, named l- and h-peaks, respectively; but in this Figure, the l-peak was rubbed out by the water-vapor treatment, since it was not the peak of ammonia desorbed from the acid sites. Though only the h-peak was observed on No.6 species, another small shoulder peak was detected at higher temperature 790 K on Nos. 5 and 15. Water was desorbed at the same temperature region to strongly overlap the desorbed ammonia; it was therefore impossible to detect the desorbed ammonia with a thermal conductivity detector (TCD). The appearance of the ultra high temperature peak was closely related to the characterization data which was reported in our previous paper [6]. The species of Nos. 5 and 15 deviated strongly from the relationship between the aluminum concentration and the d-spacing of (1000) plane of ZSM-5; thereby, these samples Nos. 5 and 15 were characterized to have extra-framework aluminum atom. Thus, it seemed that the additional higher temperature peak ($h+$-peak, hereafter) was ascribed to the strong acid site due to non-framework aluminum.

In order to further qualify the strong acid site, an addition of aluminum cations to the zeolite was studied. The clean- and well-crystallized species No.6 was immersed into the solution containing aluminum cations, and water was evaporated to dryness on a hot plate. Fig. 2 shows

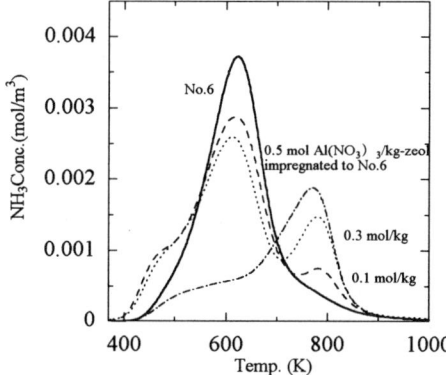

Fig. 2 Change of TPD spectra by impregnation of aluminum nitrate.

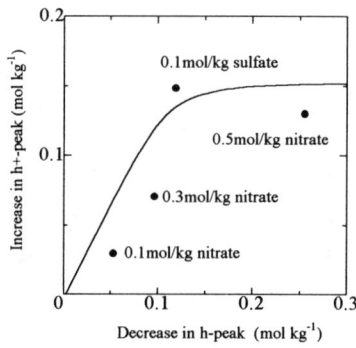

Fig. 3 Complementary change in the h- and $h+$-peaks intensities by impregnation of aluminum nitrate or sulfate to the No. 6 species.

the change of TPD of ammonia by the impregnation of aluminum nitrate. Thus, the intensity of $h+$-peak increased with increasing the concentration of impregnated aluminum nitrate; and simultaneously the intensity of h-peak decreased. With the addition of 0.5 mol/kg-zeolite aluminum nitrate, the intensity of $h+$-peak was prevailing in compared with that of h-peak; this will be utilized by an ir observation which will be discussed later. Quantitative comparison between the changes in intensity, as shown in Fig. 3, suggests approximately the generation of one site of $h+$-peak at the expense of one site of h-peak particularly in the low concentration of impregnated cation. The addition of aluminum cation on a silicalite did not create the $h+$-peak (not shown). It was thus concluded that the $h+$-peak was created by a combination of extra-framework aluminum with framework acid site. The $h+$-peak appeared even by the treatment with HCl, HNO_3, or H_2SO_4 without the aluminum cations added. Some aluminum atoms were dislocated from framework and dissolved in the acidic solution or loaded on another site of zeolite, thus creating the $h+$-peak. However, the $h+$-peak with only a small intensity was obtained by the self-modification.

Yashima et al reported the creation of strong acid site by the alumination with $AlCl_3$ vapor; they observed another high temperature peak by TPD of ammonia, like in this paper [7]. Sendoda and Ono reported that very strong acid sites were created on the ZSM-5 upon calcination at the temperature above 773 K, and these were active for the cracking of paraffin; they concluded that the site was caused by the interaction of the dislodged aluminum with acid OH [8]. Their observed strong acid site is, most probably, the same one that was found in the present study. Corma et al also proposed the generation model of very strong acid sites based on the same idea, as shown in Fig. 4 [9]. The formation of strong acid site due to the interaction of extra-framework Al with the framework acid was first indicated by Mirodatos and Barthemeuf [10]. The ultra-stable Y (USY) is activated by the high temperature steaming, and the activity is believed to be due to the presence of non-framework aluminum, though it was not directly detected in previous studies [11]. These reported active sites in the zeolites may be possibly identified as the strong acid sites due to the

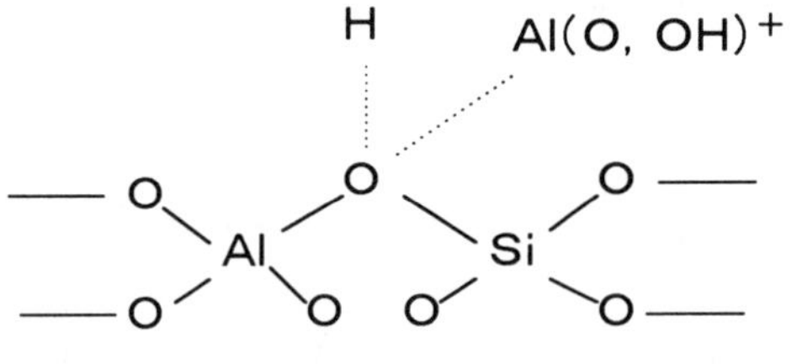

Fig. 4. A model of strong acid site shown in ref. [9].

non-framework aluminum. However, Miller et al. reported the weakening of acid site by the steaming of mordenite based on the TPD measurements [12]. Inconsistent observations might be due to the difference in the location of non-framework aluminum cation.

The strength of $h+$-peak was calculated on the basis of the curve-fitting method previously proposed by us [1]. These were 189, 182, and 176 kJmol^{-1} for the $h+$-peaks shown in Fig. 2 that were created on the No. 6 sample by the addition of 0.1, 0.3 and 0.5 mol/kg of $Al(NO_3)_3$, respectively. The high value of heat of ammonia adsorption on H-ZSM-5 is approximately in agreement with that detected by the micro-calorimetry (185-160 kJmol^{-1}) for a small amount of Lewis acid site [13], thus showing the validity of the measurements.

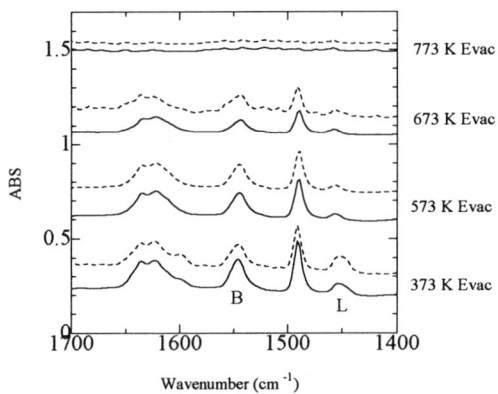

Fig. 5. IR of pyridine adsorbed on ZSM5 No. 6 unmodified (solid line) and modified by 0.5 mol/kg $Al_2(NO_3)_3$ added (dotted line).

Infrared study

The acid sites thus created were characterized by an infrared study of adsorbed pyridine, as shown in Fig. 5. By the modification, the intensity of Brønsted acid site at 1545 cm^{-1} decreased, whereas that of Lewis acid site at 1450 cm^{-1} increased. Table 1 shows the relative intensity of Brønsted and Lewis acid sites on the No. 6 species unmodified and modified by 0.5 mol/kg-zeol. aluminum nitrate solution. However, the Brønsted acid site

Table 1. Relative intensity of B and L acid intensity* of ZSM-5 No.6 and modified by the impregnation.

	Brønsted acid at 1545 cm^{-1}		Lewis acid at 1450 cm^{-1}	
Evac. Temp / K	No. 6	0.5 mol/kg Al added	No. 6	0.5 mol/kg Al added
373	0.27	0.20	0.08	0.16
573	0.21	0.18	0.03	0.05
673	0.11	0.07	0.02	0.04
773	0.01	0.00	0.00	0.01

*arbitrary unit

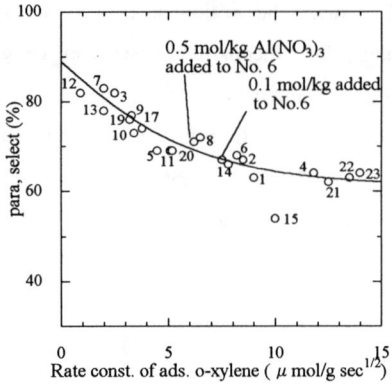
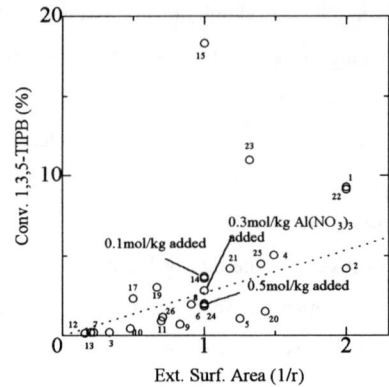

Fig. 6. Plot of para-selectivity against the rate constant of o-xylene adsorption on various synthesized ZSM-5 with those on modified ones with aluminum nitrate

Fig. 7. Plot of external surface acidity against the external surface area with those of modified ones with aluminum nitrate

remained, even when 0.5 mol/kg Al which was larger than the concentration of framework Al was impregnated, and the intensity of $h+$-peak was prevailing, as shown in Fig. 3. Thereby, the acid site seemed to be characterized as a combination of Lewis and Brønsted acid, and the Lewis type character was enhanced.

Catalysis caused by the strong acid sites of $h+$-peak

Finally, the influence of the acid sites showing the $h+$-peak on the catalytic activity was studied in the toluene alkylation with methanol. The selectivity to form p-xylene was studied in our previous paper, and the species No.15 showed the exceptionally low selectivity [6]. The selectivity at a constant conversion of toluene, 20 %, was closely related with the rate constant of o-xylene adsorption except for No. 15, as shown in Fig. 6. The external surface acidity was measured using the cracking of 1,3,5-triisopropylbenzene which was too large to enter the pore of zeolite, and it was clearly found in Fig. 7 that the external surface acid sites were exceptionally abundant on the No.15. Thus, we arrived at a conclusion: it was the strong acid site showing the $h+$-peak that concentrated on the external surface and deteriorated the selectivity of p-xylene formation.

On the other hand, the No. 5 species did not show the low selectivity, though it had the strong acid site of $h+$-peak, because the strong acid site of $h+$-peak on the No.5 species resided inside the pore. On the other hand, the zeolite species modified with the aluminum nitrate, shown above, did not show the decrease in the selectivity (Fig. 6). The impregnated aluminum is inside the pore, because the external surface acidity was not enhanced as was found from the measurement of external surface acidity (Fig. 7); thereby, the selectivity is not decreased. However, the life of catalyst was influenced significantly, and the activity decreased rapidly, as shown in Fig. 8. The rapid deactivation could be elucidated by the rapid formation of coke inside the pore due to the strong acid

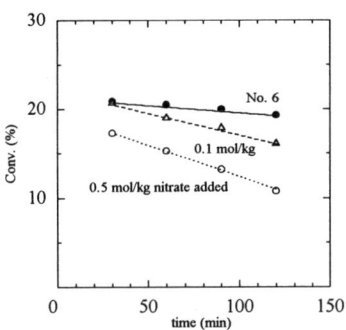

Fig. 8. Change of conversion with the time-on stream on the No.6 species unmodified and modified with aluminum nitrate.

site of $h+$-peak. Thus, we found that the acid sites to show the $h+$-peak, when on the external surface, decreased the selectivity, and, when inside the pore, shortened the catalyst life. From both points of view, the acid sites of $h+$-peak are useless, and should be removed from the ZSM-5 zeolite.

In our recent study, the $h+$-peak is observed on broad kinds of zeolite and mesoporous material [14] as a sharp additional peak or a broad shoulder.

CONCLUSIONS

By the measurement of TPD of ammonia, very strong acid sites were detected on HZSM-5. It was created from a combination of extra-framework Al cation and usual solid acid site, and the strength of acidity was estimated to be 189 - 176 kJmol^{-1}. Lewis acid character of the site was remarkable. When it was on the external surface, the shape selectivity was seriously dropped. Inside the pore, it shortened the life of catalyst.

ACKNOWLEDGMENT

This study was supported by a Grant-in-Aid for Scientific Research (B) from the Ministry of Education, Science and Culture, Japan (No. 10555278).

REFERENCES

1. M. Niwa and N. Katada, Catalysis Surveys from Japan **1**, 215 (1997).
2. N. Katada, H. Igi, J.-H. Kim, and M. Niwa, J. Phys. Chem. B, **101**, 5969 (1997).
3. G. Bagnasco, J. Catal. **159**, 249 (1996).
4. G. L. Woolery, G. H. Kuehl, H. C. Timken, A. W. Chester and J. C. Vartul, Zeolites, **19**, 288 (1997)
5. H. Igi, N. Katada, and M. Niwa, in Proceedings of the 12th International Zeolite Conference, Edited by M. M. J. Treacy, B. Marcus, J. B. Higgins and M. E. Bisher (Materials Research Society, Warrendale, PA, 1998).
6. T. Kunieda, J.-H. Kim, and M. Niwa, J. Catal. **173**, 433 (1998).
7. T. Yashima, K. Yamagishi, S. Namba, S. Nakata and S. Asaoka, Stud. Surf. Sci. Catal., **37**, 175 (1987).
8. Y. Sendoda and Y. Ono, Zeolites **8**, 101 (1988).
9. A. Corma, A. Martinez, and C. Martinez, Appl. Catal. A: General **134**, 169 (1996).
10. C. Mirodatos and D. Barthomeuf, J. Chem. Soc., Chem. Commun 39 (1981)..
11. F. Lonyi and J. H. Lunsford, J. Catal. **136**, 566 (1992).
12. J. T. Miller, P. D. Hopkins, B. L. Meyers, G. J. Ray, R. T. Roginski, G. W. Zajac, and N. H. Rosenbaum, J. Catal. **138**, 115 (1992).
13. L. C. Jozefowicz, H. G. Karge and E. N. Coker, J. Phys. Chem. **98**, 8053 (1994).
14. K. R. Reddy, N. Araki, and M. Niwa, Chem. Lett. 637 (1997).

THE EFFECT OF OXYGENATES ON THE n-HEXANE AROMATIZATION ACTIVITY OF Pt/KL

M. E. DRY, R. J. NASH and C. T. O'CONNOR

Catalysis Research Unit, Department of Chemical Engineering, University of Cape Town, Private Bag, Rondebosch 7701, South Africa. email: med@chemeng.uct.ac.za

ABSTRACT

The object of this work was to investigate whether Fischer-Tropsch (FT) hexane, being sulphur free but containing oxygenated compounds, could be a suitable feedstock for benzene production. The oxygenates, ethanol, MEK and butanal, were individually co-fed with n-hexane over Pt/KL at various hydrogen pressures. At the same concentration levels all these oxygenates had similar quantitative effects, namely, they depressed hexane conversion as well as the selectivities of benzene and the C_1-C_5 alkanes. The higher the level of the oxygenates, the more marked the effect. When the oxygenates were removed from the feed the performance of the catalyst apparently recovered. In all cases CO and H_2O were present in the product stream during oxygenate addition. Co-feeding CO had similar effects to co-feeding the oxygenates but water had no effect. The extent of the negative effects of the oxygenates on conversion was reduced by increasing the H_2 pressure but as expected the benzene selectivity was lowered. The yield of benzene peaked between 2 and 6 bar. It was concluded that FT hexanes with oxygenate levels in the region of 1% could be a feasible feedstock for the production of benzene over Pt/KL.

INTRODUCTION

In 1980 Bernard [1] reported that platinum supported on zeolite KL selectively converted n-hexane to benzene. A benzene selectivity of 95% at 99% n-hexane conversion was claimed [2]. The Pt/KL catalyst was reported to be mono-functional and all the activity occurred on Pt metal clusters inside the zeolite pores. Iglesia and Baumgarten [3,4] ascribed the stability of the Pt/KL catalyst to a low rate of coke formation in the narrow zeolite pores. The Aromax process developed by Chevron, using a Pt/BaKL catalyst was reported to operate continually for the equivalent of one year in an accelerated deactivation test run [5]. The activity of the catalyst is, however, very sensitive to the pressure of sulphur compounds in the feed. This results in irreversible deactivation due to sintering of the platinum clusters and therefore it was recommended that the sulphur in the feed be kept below 50 ppb [6]. This places a very stringent requirement when the feed is based on petroleum naphtha. Since the hydrocarbons produced in the Fischer-Tropsch (FT) process is sulphur free it appears that the FT naphtha could be suitable feedstock for the production of benzene. However, the FT naphtha cut contains significant amounts of oxygenated products such as alcohols, aldehydes, ketones and fatty acids [7], the influence of which on the Pt/KL catalyst was not known. The objective of the work reported here was to investigate the influence of the individual oxygenate types on the conversion of n-hexane to benzene under various process conditions. The oxygenates used were those having boiling points similar to that of n-hexane.

EXPERIMENTAL

The catalyst was a commercially sourced sample of zeolite KL which was pre-dried at 150°C. The platinum was then added by incipient wetness impregnation with a solution of Pt $(NH_3)_4 Cl_2 H_2O$. The Pt/zeolite was first dried in a dessicator at ambient temperature, then at 100°C followed by calcination in air at 350°C. Prior to each aromatization reaction the catalyst was pre-reduced with H_2 at 450°C. Carbon monoxide chemisorption indicated the platinum dispersion to be 90%. The absence of visible Pt clusters in transmission electron microscopy supported the conclusion that the Pt was well dispersed.

Aromatization reactions were carried out in a stainless steel down-flow reactor (4.5mm ID). All gases were fed via mass flow controllers. The n-hexane and oxygenates were introduced via saturators (maintained at the appropriate temperatures) with hydrogen as carrier gas. Air and/or nitrogen was fed for calcination/regeneration and flushing purposes. The temperature of the catalyst bed was measured by a thermocouple located in an axial thermowell. The reaction products were analysed by online gas chromatography. Dimethylether was used as the internal standard and was fed downstream of the reactor. Carbon balances were within ± 5% accuracy. Carbon monoxide was analysed by on-line infrared spectroscopy. Tests carried out at different gas linear velocities but at a fixed residence time indicated the absence of film diffusion constraints.

When investigating the effect of co-feeding oxygenates, hexane and H_2 were fed for two to three days, after which the oxygenate was added while maintaining the same residence time and H_2 partial pressure. At the end of this period the co-feeding was stopped and the extent of recovery monitored. A typical run is illustrated in Figure 1. The oxygenates were also fed individually, at similar process conditions, in the absence of n-hexane in order to determine the nature of their hydrocracked products and these results were used to correct the observed C_1 to C_4 alkanes when co-feeding the oxygenates with the n-hexane.

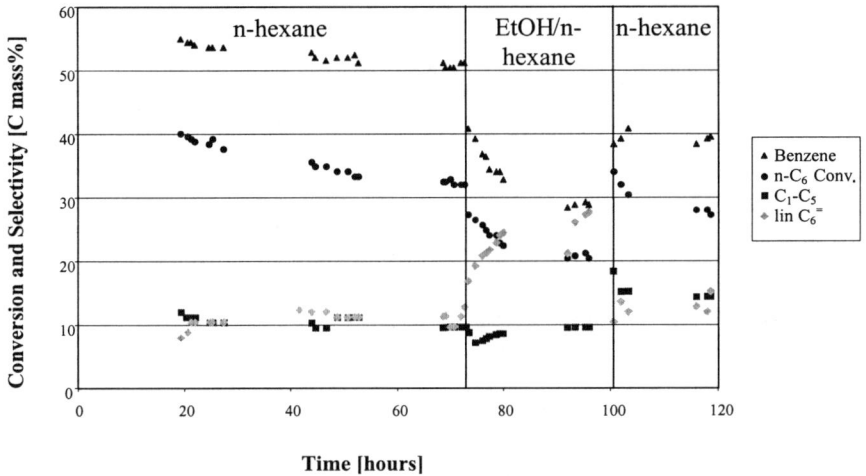

Figure 1. Co-feeding of ethanol with n-hexane.

RESULTS AND DISCUSSION

When feeding n-hexane and H_2 only, the products included C_1 to C_5 alkanes (the hydrocracked products), various linear hexene isomers, iso-hexanes, methyl cyclopentane (MCP), 2- and 3-methylpentane and benzene. A typical distribution obtained at 1 bar H_2 and 450°C is shown in Table 1. The hexenes and isohexanes can be considered as reversible intermediate products while benzene and the $C_1 - C_5$ alkanes can be taken as final products.

Compound	Selectivity	Compound	Selectivity
Methane	2	Hexenes	10.9
Ethane	1.8	MCP	12
Propane	2.4	2-MP	5.3
Butanes	2.3	3-MP	4
Pentanes	3.7	Benzene	55

Table 1: Typical product selectivity (C atom basis) at 450°C and 1 bar H_2.

From Figure 1 it can be seen that on the introduction of the oxygenate the conversion of n-hexane decreased, the selectivity to linear hexenes increased while the selectivity to benzene and the cracked products (C_1 to C_5 alkanes) decreased. The isohexanes also increased. Thus, as expected, depression of the catalyst's activity by the oxygenate resulted in an increase in the amount of intermediate products and a decrease in the final products. On removal of the oxygenate from the hexane feed the conversion of n-hexane increased and returned to the value it probably would have been at that time on stream. The selectivities also appeared to recover. The extent of the performance recovery is discussed further below.

Table 2 compares the effects of four oxygenates, viz. ethanol (EtOH), methylethylketone (MEK), n-butanal (n BuHO) and isobutanal (i BuHO), when co-fed with n-hexane.

Oxygenate	EtOH				MEK	nBuHO	iBuHO	CO
Mole% (of Hexane) co-fed	1.1	2.8	14	23	20	14	13	18
n-Hexane Conversion[a]	-11	-28	-38	-49	-49	-42	-38	-42
Benzene Selectivity[a]	-9	-18	-44	-51	-45	-38	-40	-46

[a]: The percentage change from the performance just before the addition of the oxygenate to the performance after 24h with oxygenate co-feeding.

Table 2: The effect of oxygenates at 450°C and 1 bar H_2.

The n-hexane conversion and the benzene selectivity decreased as the amount of ethanol co-fed was increased. This is in keeping with observations made in this work that in general there was a near-linear relationship between % hexane conversion and benzene selectivity irrespective of whether the changes were brought about by deactivation, oxygenate co-feeding or changes of space velocity or temperature (in the range 375 to 500°C). This is as expected if one assumes that benzene is one of the final products and that the transformation of hexanes to benzene is a relatively slow process step.

Comparing the effects that the individual oxygenated compounds have on conversion and selectivity (see Table 2) it appears that at the same relative molecular concentration they all yield very similar results. This suggests that some common factor is involved. When feeding the oxygenates alone or co-feeding with n-hexane at 450°C and 1 bar H_2, at the residence times used in this work, it was always observed that in addition to the expected alkanes and water, carbon monoxide was always present in the effluent stream. The influence of co-feeding different levels of CO and H_2O individually was subsequently investigated and it was found that while different levels of H_2O addition had no effect whatsoever the addition of CO had an effect very similar to that of the oxygenates (see Table 2). It appears feasible that the "common factor" could well be the CO released as a result of the decomposition of the oxygenates. The temporary occupation of platinum sites by chemisorbed CO should decrease the number of Pt sites available and hence negatively affect the catalyst's performance. Alternatively, or in addition, the temporary occupation of sites by chemisorbed oxygenates would have the same effect if the different oxygenates all chemisorbed to similar degrees.

The influence of increasing the hydrogen pressure on the catalyst's performance when co-feeding oxygenates was investigated and the results are shown in Table 3. The residence times were maintained constant. The different oxygenates again behaved similarly and hence Table 3 represents typical results for any of the oxygenates co-fed.

H2 Pressure	1 bar	2 bar	6 bar
n-Hexane conversion change[a]	-42%	-20%	-11%
Benzene selectivity change[a]	-43%	-22%	-13%
Benzene Yield	6	28	20
C1-C5 Yield	2	6	13
CO produced (mole %)[b]	51%	30%	0

[a] as defined in Table 2

[b] moles CO produced as % of moles oxygenate co-fed

Table 3: Influence of H_2 pressure on the relative change in performance due to oxygenate co-feeding and on the product yields.

As can be seen from Table 3 the negative effect of oxygenates on n-hexane conversion and benzene selectivity was decreased by increasing the H_2 pressure. The actual C_1-C_5 alkane selectivities as well as their yields increased with increasing H_2 pressure as would be expected since the alkanes are hydrocracking products. The benzene selectivity and yield at 2 bar was higher than at 1 or at 6 bar with the maximum probably between 2 and 6 bar. At 2 bar H_2 and higher no hexenes were present in the products. The amount of CO that emerged from the reactor decreased as the pressure increased. This was due to it being hydrogenated to methane. This removal of the CO in the catalyst bed was probably the reason for the lower depression of activity with increasing H_2 pressure.

The gradients of the logarithim plots of hexane conversion rate against time was taken as a measure of the deactivation of the catalyst and the influence of co-feeding oxygenate was thus evaluated. It was found that all the oxygenates co-feeding runs produced very similar results and therefore in Table 4 these results are lumped together.

Feed	Hydrogen Pressure (bar)			
	1	2	3	4
n-Hexane only	-0.12		-0.08	-0.07
Oxygenates co-fed	-0.15	0.1		-0.04

Table 4: Deactivation exponents, n, $r=At^n$, where A is a constant.

Bearing in mind the reliability of the experimental data it appears that the presence of oxygenates in the feed may not have a marked effect on the *deactivation rate* of hexane conversion. The decline in conversion with time is probably due to the slow deposition of coke and not due to sintering of the Pt clusters as it was repeatedly observed that treatment with air at 450°C followed by H_2 reduction restored the catalyst's activity. Permanent loss of activity was observed only when reaction and regeneration cycles were performed at temperatures above 475°C. Table 4 does indicate that the deactivation rate is lower at higher H_2 pressures and this is consistent with the assumption that the deactivation is coke related.

Since hexenes are intermediate products and also since the Fischer-Tropsch process produces large amounts of alkenes it was decided to investigate feeding 1-hexene only at 1 bar H_2 over a wide range of temperatures. Below 70°C only double bond isomerization was observed. Between 70 and 350°C only n-hexane was present. Only above 350°C did the other hexane isomers, the C_1-C_5 alkanes and benzene make their appearances. In general the behaviour at 450°C was very similar to that observed when feeding n-hexane.

CONCLUSIONS

If it is assumed that all hexenes and iso-hexanes are intermediates and can be recycled to extinction then benzene and the C_1-C_5 alkanes will be the only final products. On this basis the benzene selectivity at 1 to 2 bar H_2 pressure was in the vicinity of 80% whether or not oxygenates were being co-fed. At 6 bar, however, hydrocracking is more extensive and the benzene selectivity is only about 60%. The overall impression is that if the oxygenate levels were in the vicinity of 1 mole % of the hexane feed, and also if hexenes were present, the catalyst's performance would not be markedly inferior to the situation when feeding highly purified hexane. This would mean that the purification required for a Fischer-Tropsch hexane feedstock need not be very stringent. Further life-time testing on a bench scale is, however, recommended.

REFERENCES

1. J.R. Bernard in Proceedings of the 5th International Zeolite Conference, Edited by L.V.C. Rees (Heyden, London 1980), p686.
2. C. Besoukhanova, M. Breyesse, J.R. Bernard and D. Barthomeuf in Proceedings of the 7th International Conference on Catalysis, Edited by T. Seiyama and K. Tanabe (Kadanska/Elsevier Scientific Ltd, Tokyo/Amsterdam 1981), p1410.
3. E. Iglesia and J.E. Baumgartner in Proceedings of the 9th International Zeolite Conference, Edited by R. von Ballmoos, J.B. Higgins and M.M.J. Treacy (Butterworth-Heinemann 1993), p421.
4. E. Iglesia and J.E. Baumgartner in Proceedings of the 10th International Congress on Catalysis, Edited by L. Guczi, (Elsevier Science Publishers 1993), p993.
5. T.R. Hughes, W.C. Buss, P.W. Tamm and R.L. Jacobson in Proceedings of the 7th International Zeolite Conference, (Elsevier Scientific Ltd., Amsterdam 1987), p725.
6. G.B. Vicker, J.K. Lao. J.J. Ziemiak, W.E. Gates, J.L. Robbins, M.M.J. Treacy, S.B. Rice, T.H. Vanderspurt, V.R. Cross and A.K. Ghosh, J. Catalysis, 143, 48 (1993).
7. M.E. Dry in Catalysis - Science and Technology, Edited by J.R. Anderson and M. Boudart (Springer-Verlag, Berlin 1981), vol 1, p159.

ZEOLITE BETA: THE RELATIONSHIP BETWEEN CALCINATION PROCEDURE, ALUMINUM CONFIGURATION AND CATALYTIC ACTIVITY IN THE MPV REDUCTION OF KETONES

P.J. KUNKELER, B.J. ZUURDEEG AND H. VAN BEKKUM

Department of Organic Chemistry and Catalysis, Delft University of Technology, Julianalaan 136, 2628 BL, Delft, The Netherlands (e-mail: P.J.Kunkeler@stm.tudelft.nl)

ABSTRACT

Zeolite Beta was calcined under a variety of carefully controlled conditions to study the influence of (hydro-) thermal treatments on the catalytic activity of zeolite Beta in the Lewis acid catalyzed Meerwein-Ponndorf-Verley reduction of ketones. A correlation between the pretreatment conditions and the catalytic activity was found which is explained by different coordination environments of the framework aluminum atoms. The activity of (H) Beta can be increased by several orders of magnitude by controlled mild steaming. The changes induced by the (hydro-) thermal procedures, were investigated by FT-IR, ^{27}Al and ^{29}Si MAS NMR.

INTRODUCTION

Several studies point to the flexibility of the coordination sphere of the aluminum atoms in zeolite Beta [1-4]. Besides tetrahedrally coordinated framework aluminum, octahedrally or otherwise coordinated aluminum has been proposed, being present as framework-connected or as extra-framework species [1]. Based on results reported in the literature it can be concluded that framework aluminum adopts a tetrahedral symmetry when protons are *not* the charge-compensating cations. Aluminum atoms which have adopted another coordination symmetry, but are still connected to the framework, can revert to their tetrahedral coordination sphere by ion-exchange with cations like Na^+ and K^+ or by the action of a base, *e.g.* ammonia [1]. As was recently shown, the calcination of zeolite Beta has an impact on its catalytic activity in the Meerwein-Ponndorf-Verley (MPV) reduction of ketones [5]. Calcination parameters such as the use of shallow- or deep-bed [3], the temperature and the number of calcination steps were found to influence the activity. The highest activity resulted from a deep-bed, high-temperature calcination [5]. The nature of the active site is still a matter of debate. Creyghton *et al.* proposed a five-coordinated Al in which both the alcohol and ketone are simultaneously coordinated to the Al atom [5].

In the present work the influence of the calcination procedure on the catalytic activity of zeolite beta in the MPV reduction was investigated in more detail. Instead of the poorly defined, inhomogeneous deep-bed calcination procedure all calcination procedures were performed using a shallow calcination bed while varying the atmosphere above the zeolite sample. Hot-spots and auto-steaming [3], intrinsically related to deep-bed calcination procedures under an oxidative atmosphere, will be prevented in this way. Based on the combined results of FT-IR, NMR and the catalytic experiments, a mechanism is proposed which accounts for the effects observed as a function of the pretreatment.

EXPERIMENTAL

Zeolite synthesis

Macrocrystalline zeolite Beta (truncated, octahedrally shaped crystals) was prepared according to Kunkeler *et al.* [6] and microcrystalline zeolite Beta *via* the Wadlinger synthesis [7].

Calcination procedure

In order to preclude dealumination of the zeolites, the following procedure was adopted: the as-synthesized (a.s.) materials were first heated under a pure ammonia atmosphere from ambient temperature to 400 °C. Subsequently, the samples were sodium exchanged using 1 M aqueous NaCl under reflux conditions, further calcined at 120 °C in an oxygen / ozone atmosphere and finally subjected to a third calcination at 400 °C. The calcined sodium Beta's were ammonium-exchanged three times using 0.1 M aqueous NH_4NO_3. The ammonium-exchanged Beta's had a Si/Al ratio of 11.6 and a Na/Al ratio of 0.01. ^{27}Al MAS NMR proved the absence of any octahedrally coordinated aluminum [1] and the ^{29}Si MAS NMR showed the framework Si/Al ratio to be 11.7, in good correlation with the wet-analysis data.

Small portions (200 mg) of the (NH$_4$)Beta were activated in a glass tube. Nitrogen, dried or saturated with a 4.2 kPa water pressure, was led continuously over the sample. The samples were spread uniformly to obtain a shallow calcination bed.

Catalytic testing

Meerwein-Ponndorf-Verley (MPV) reductions of 4-*tert*-butylcyclohexanone (2.5 mmol) with 2-propanol (25 ml) as the hydrogen donor were performed under reflux conditions as described by Creyghton *et al.* [5]. *Cis*- and *trans*-4-*tert*-butylcyclohexanol were formed as the only products see Figure 1. The selectivity to the *cis*-alcohol - which isomer is of industrial relevance - was >95%. The catalyst samples were subjected to activation at 450 or 550 °C for different periods of time.

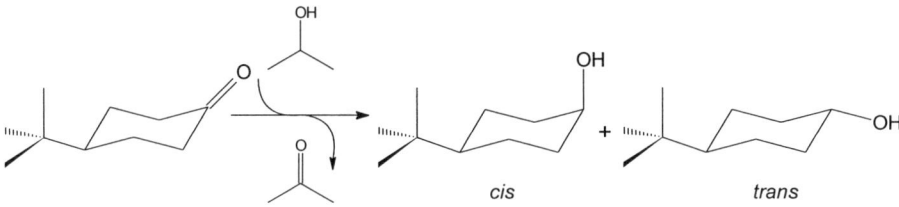

Figure 1. MPV reduction of 4-*tert*-butylcyclohexanone using 2-propanol resulting in *cis*- and *trans*- 4-*tert*-butylcyclohexanol.

RESULTS AND DISCUSSION

Catalytic activity

For both macro- and microcrystalline Beta samples the activation procedure was found to have a significant impact on the activity. As was to be expected, a higher activation temperature causes an increase in the reaction rate (Table 1). Also the duration of the treatment is important for the catalytic performance; an extended activation time boosts the catalytic performance. Steaming, however, causes the most substantial increase in the reaction rate. An extended steaming time results in very active catalysts.

Table 1. Results for the MPV reduction of 4-*tert*-butylcyclohexanone using microcrystalline (H)Beta samples (conversion (%) after 60 min)

sample treatment	conversion (%)	sample treatment	conversion (%)
450 °C (1h) dry	1.4	450 °C (6h) steam	37.2
550 °C (1h) dry	3.6	550 °C (1h) steam	90.6
550 °C (10h) dry	3.8	550 °C (10h) steam	100
450 °C (1h) steam	6.7	550 °C (3d) steam	100 (within 5 min)

Similar results were obtained with the macrocrystalline Beta samples.

Spectroscopic characterisation

According to Kiricsi *et al.*, five types of hydroxyl groups in zeolites can be distinguished by IR spectroscopy: Brønsted acidic OH (3605 cm^{-1}), OH groups attached to extra-framework Al (3660-3680 cm^{-1}), internal silanol groups (3730 cm^{-1}), terminal silanol groups (3745 cm^{-1}) and OH attached to an aluminum species which is believed to be a transient species (3782 cm^{-1}, so called VHF band) [2].

The parent (H)Beta (both micro- and macrocrystalline samples) showed absorptions at 3605 cm^{-1} and 3740 cm^{-1}, super-imposed on a broad band spanning 3700-3000 cm^{-1} due to hydroxyl nests [1]. Upon steaming, the 3605 cm^{-1} absorption decreased while the 3745 cm^{-1} absorption increased. Between these absorption maxima a band appeared at 3666 cm^{-1} which may be attributed to extra-framework Al species. More importantly, the VHF absorption appeared.

Upon dry activation, the ^{27}Al MAS NMR spectra showed the development of an octahedrally coordinated aluminum species (0 ppm). Steaming resulted in a decrease of the octahedral peak intensity and a broadening of the tetrahedral peak (60 ppm). From the ^{29}Si MAS NMR data it was concluded that partial dealumination occurred upon steaming.

We propose the following mechanism for the formation of the catalytic site, in which the aluminum atom bearing the "VHF-hydroxyl" group is believed to be a Lewis acidic framework aluminum as depicted in Figure 2. Upon hydrolysis of an Al-O-Si bond the Al is thought to invert its geometry and to become sufficiently exposed into the zeolite pore to coordinate water and 2-propanol.

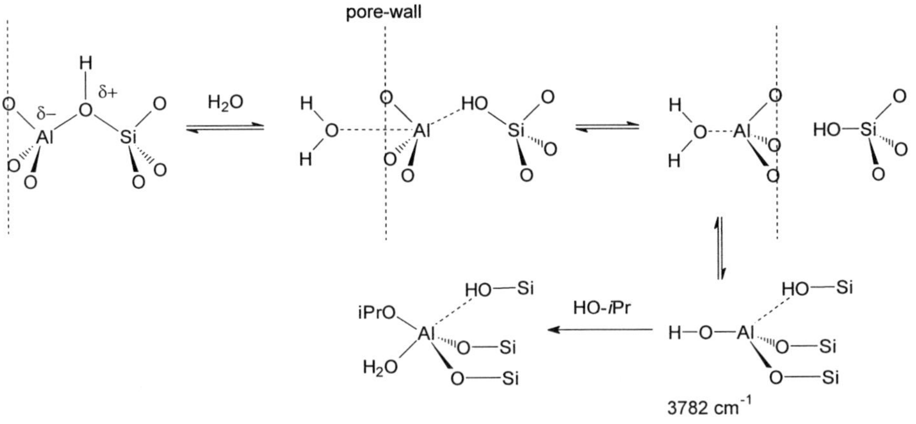

Figure 2. Proposed mechanism for the formation of the catalytic active site (Al-isopropoxide species) in zeolite Beta for the MPV reduction.

Coordination of the ketone to the Al-isopropoxide species initiates the MPV reaction-cycle. Subsequent hydride transfer, acetone elimination and finally alcoholysis of the product and concomitantly regeneration of the active site complete the reaction-cycle, as was proposed recently by van der Waal et al. [8].

ACKNOWLEDGEMENT

The Dutch Institute for Catalysis Research (NIOK) is thanked for financial support (TUD 98-1-01) and Dr. J.C. van der Waal is thanked for valuable discussions. Mr. V. Veefkind (prof. J.A. Lercher's group, University of Twente, The Netherlands) is gratefully acknowledged for conducting the FT-IR measurements. Mr. J.A. van Bokhoven is thanked for the performance of the ^{27}Al-NMR measurements.

REFERENCES

1. E. Bourgeat-Lami, P. Massiani, F. Di Renzo, P. Espiau and F. Fajula and T. Des Courières, Appl. Catal. **72,** 139 (1991).
2. I. Kiricsi, C. Flego, G. Pazzuconi, W.O. Parker, Jr., R. Millini, C. Perego and G. Bellussi, *J.* Phys. Chem. **98**, 4627 (1994).
3. C. Jia, P. Massiani, and D. Barthomeuf, J. Chem. Soc., Faraday Trans. **89**, 3659 (1993).
4. B.Su and V. Norberg, Zeolites **19,** 65 (1997).
5. E.J. Creyghton, S.D. Ganeshie, R.S. Downing and H. van Bekkum, J. Mol. Catal. *A* **115**, 457 (1997).
6. P.J. Kunkeler, D. Moeskops and H. van Bekkum, Microporous Mater. **11**, 313 (1997).
7. R.L Wadlinger, G.T. Kerr and E.J. Rosinski, US Patent 3,308,069 (1967).
8. J.C. van der Waal, P.J. Kunkeler, K. Tan and H. van Bekkum, J. Catal. **173**, 74 (1998).

THE VIBRATIONAL SPECTROSCOPY OF ACID-BASE INTERACTIONS IN ZEOLITE CAVITIES AS A TOOL FOR ACID-STRENGTH INVESTIGATION

A. ZECCHINA, D. SCARANO, C. LAMBERTI, G. SPOTO, C. PAZE' and S. BORDIGA.

Dipartimento di Chimica Inorganica Chimica Fisica e Chimica dei Materiali, Torino, Italy; zecchina(scarano/lamberti/spoto/pazé/bordiga)@ch.unito.it.

ABSTRACT

In order to clarify the role of zeolitic environment in determining the acidic properties of Brønsted groups of protonic zeolites, a spectroscopic study of the properties of HCl molecules entrapped in a zeolitic structure has been performed. In particular the IR spectra of HCl-B (B= benzene, CH_3CN, $(CH_3)_2O$, $(CH_3CH_2)_2O$, THF) adducts adsorbed in silicalite channels at $\cong 210$ K has been studied. The results obtained for HCl-B adducts in silicalite are compared with those obtained in criogenic matrix and in solution and with those obtained upon interaction of the same bases with acidic zeolites (H-ZSM-5, H-MORD, H-Y, H-β). It is inferred that the acid strength of HCl entrapped (and then the acid strength of the structural Brønsted sites) is influenced by the presence of the siliceous framework. This result allows to discuss the effect of long-distance interaction (framework effect) on the acid strength of structural Brønsted sites of zeolites.

INTRODUCTION

In presence of a base, the polarity of the acidic hydroxyl groups (Si(OH)Al) present in a zeolite, is increased because of the formation of hydrogen bonds with the negative end of the basic molecule. The formation of 1:1 hydrogen bonded adducts is accompanied by a perturbation of the OH stretching frequency of the Brønsted sites which is roughly proportional to the interaction enthalpies. The evaluation of the individual red-shifts can be consequently considered as a method for the evaluation of the acid strength. Different probe molecules, with proton affinities ranging in a wide interval, have been used to test the acidity of single zeolites [1].

In Figure 1 the frequency shifts of the ν(ZH···B) modes of various 1:1 adducts formed inside the channels and cavities of a series of zeolites (H-ZSM-5, H-β, H-MORD, H-Y) with a series of bases B of increasing proton affinity (from 118.2 Kcal mol^{-1} (N$_2$) to 202 Kcal mol^{-1} (THF)) are plotted against the frequency shifts of the corresponding SiOH···B adducts formed on silica or on the silanols located on the external surfaces of zeolites. Due to the small size of the probe molecules, the strenght of the hydrogen bonds is not specially affected by steric constrains. The data concerning the $\Delta\nu_{HCl}$ (as measured in cryogenic matrices) induced by the same bases are also reported for sake of comparison.

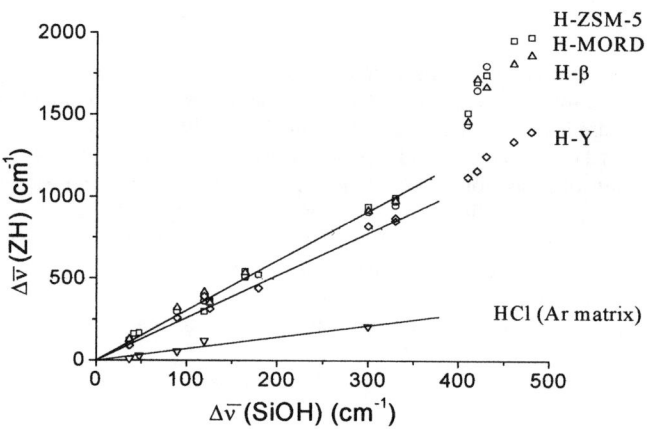

Figure 1. Relation between the frequency shift of the ν (ZH···B) modes (where Z = H-ZSM5, H-β, H-Mord., H-Y and B = benzene, CH$_3$CN, (CH$_3$)$_2$O, (CH$_3$CH$_2$)$_2$O, THF) and the frequency shifts of the corresponding SiOH···B adducts on silica or on the silanols located on the external surfaces of zeolites.

The following can be commented: i) For $\Delta\bar{\nu}$(AH) < 1100 cm^{-1} a straight correlation exists between the $\Delta\bar{\nu}$(AH···B) and the $\Delta\bar{\nu}$(SiOH···B) shifts for all the investigated zeolites. ii) The angular coefficient (α) of the straight lines follows the series: $\alpha_{\text{H-ZSM-5}} \approx \alpha_{\text{H-Mord}} \approx \alpha_{\text{H-}\beta} > \alpha_{\text{H-Y}} > \alpha_{\text{HCl}}$. From these α values a quantitative acidity scale can be derived, which is independent of the molecule used to probe the acidity. iii) For shifts > 1100 cm^{-1} the straight correlation no longer occurs. It is demonstrated that a straight correlation between $\Delta v(\text{AH}\cdots\text{B})$ and $\Delta v(\text{SiOH}\cdots\text{B})$ is verified only for proton potential shapes with single minimum, corresponding to neutral 1:1 adducts. When stronger bases are used, the potential in the AH$\cdots$B adducts shows two minima separated by a low barrier and the complex become progressively more ionic. In such case the linear correlation is no more valid. *Vice versa* a deviation from the linear correlation is indicative of substantial ionicity of the AH$\cdots$B bond [2].

The comparison of the data relative to the strong acid sites of the different zeolites, all having the common bridged Si(OH)Al structure, has therefore shown the following trend of acidity: H-Y (Si/Al=3) < H-MORD (Si/Al= 5) $\cong$ H-β (Si/Al=12.5) $\cong$ H-ZSM-5 (Si/Al=14). Since the Si/Al ratio of the considered zeolites is spanning from 14 (H-ZSM-5), to 12 (H-β), to 5 (H-MORD) and to 3 (H-Y), it is possible to conclude that the Brønsted acidity is not related in a simple way to the Al concentration. The comparison of the various zeolites is also made difficult by the fact that the strong acidic Brønsted groups are present in environments different not only for the chemical composition (Si/Al ratio and possible presence of extraframework aluminium), but also for the local geometry of the sites (dimensions and arrangements of the channels and supercages). In order to estimate the weight of each of these two effects, special data concerning zeolitic systems with one of the two variables kept fixed, should be compared. While studies comparing samples of the same zeolite (*i.e.* with the same structure) but with different Si/Al ratio have been already made, more complex is the study of the role that the zeolitic solvent (framework) itself plays in determining the acidity of a site.

A way to deal with this kind of problem is represented by an *ad hoc* experiment, where the properties of an acid molecule of well known properties inserted in neutral zeolitic frameworks, *i.e.* a system that mimics the properties of an acidic zeolite, is studied. In fact the comparison of the properties of this acid molecule in a zeolitic environment and in the gas phase or in cryogenic matrix, could give new information about the influence of the specific framework (considered as a solvent) on the properties of the acid molecule and hence, indirectly, on the effect of the environment on the acidic properties of constitutional Brønsted sites. This can be done (as far as MFI structure is concerned) [3] by studying the HCl-B interaction in the channels and cavities of silicalite, a purely siliceous material with the same structure of H-ZSM-5. In particular the spectroscopic properties of these 1:1 complexes are compared with those of HCl-B complexes in an Argon matrix, with the hope to clarify the role of zeolitic environment in determining the acidic properties of HCl entrapped in the internal voids and more generally to estimate, at least on a qualitative basis, the effect of the environment on the properties of constitutional Brønsted groups present in the framework.

EXPERIMENTAL

The protonic forms of the zeolites were obtained by thermal decomposition (673 K) *in vacuo* of NH_4^+ precursors, while the silicalite sample was activated at 973 K. All the zeolites were supplied by ENICHEM laboratories (Novara). The IR spectra have been obtained with a Bruker 66 instrument working at 2 cm^{-1} resolution by using an in situ cell allowing thermal treatments (T ranging from 300 to 100 K) under controlled atmosphere and gas dosages.

RESULTS AND DISCUSSION

In Figure 2 the $\Delta\bar{v}$(HCl) values of the intrazeolitic HCl-B complexes (B = benzene, CH_3CN, $(CH_3)_2O$, $(CH_3CH_2)_2O$, THF) are plotted against the frequency shifts of the corresponding SiOH···B adducts. It is usefull to recall that the HCl molecule has been extensively studied in the past, both in the gaseous phase and in criogenic matrices [4] or in solution [5], and that a lot of data about the vHCl frequencies in various HCl-B complexes exist. These data are also reported in Figure 2 for comparison.

From this Figure it is clearly emerging that the data of the HCl complexes in the three different environments (in silicalite, in matrix, in apolar solvents) leads to three different straight lines and that the slope of the HCl-B complexes encapsulated in silicalite is greater than that of HCl-B complexes isolated in a cryogenic matrix. It is so concluded that the effect of the solid environment represented by the silicalite framework on the spectroscopic properties of the HCl-

B adducts entrapped in the channels is relevant, and consequently play an important role in determining the strength of the Bronsted groups. This conclusion has been verified for the neutral framework of silicalite, i.e. when the interaction between the acid molecule and the zeolite is prevalently of the Van der Waals type. The same conclusion (i.e. an increase of acidity induced by the zeolitic solvent) is presumably present also when the involved acid group is the structural Al(OH)Si one.

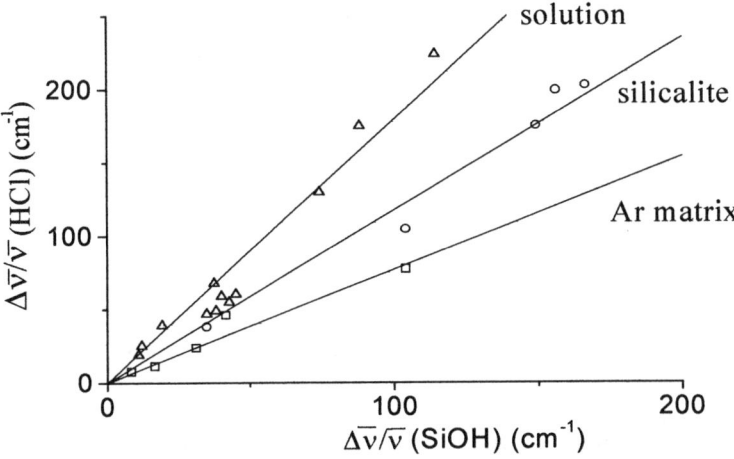

Figure 2. Relation between the frequency shifts of the ν (HCl···B) modes (where B = benzene, CH_3CN, $(CH_3)_2O$, $(CH_3CH_2)_2O$, THF) and the frequency shifts of the corresponding SiOH···B adducts.

The general trend of the slopes of the three straight lines (that is: m(HCl in solvent)>m(HCl in silicalite)>m(HCl in Ar matrix)) can be explained in terms of effects of the electric field associated with the presence of dipoles on the atoms of the "solvent" (cryogenic matrix, zeolite and solution): in fact the electric field induces an increase of the ionic character of the hydrogen bond. This effect is well known in solution as a "solvent effect" and in cryogenic matrices as a "matrix effect". In the macroscopic scale, it is connected with the dielectric constants of the media where the complexes are formed.

These results for entrapped HCl, when extended to the structural acid groups of the acidic zeolites, indicate that the "effective" acid strength of a zeolite does not only depend upon the local structure of the Bronsted groups but also upon the tridimensional structure of the zeolite.

CONCLUSIONS

The effect of the solid environment of silicalite on the spectroscopic properties of B···HCl adducts entrapped in the channels is significant; the framework noticeably enhances the strength of the hydrogen bonding interaction with weak bases, with respect to that observed for the same complexes in a cryogenic matrix.

REFERENCES

1. A. Corma, *Chem. Rev.*, 1995, **95**, 559.
2. A. Zecchina, S. Bordiga, G. Spoto, D. Scarano, G. Spanò and F. Geobaldo, *J. Chem. Soc. Faraday Trans.*, 1996, **92**, 4863.
3. C. Pazè, S. Bordiga, G. Spoto, C. Lamberti and A. Zecchina, *J. Chem. Soc. Faraday Trans*, 1998, **94**, 309.
4. A.J. Barnes, H. E. Hallam and G. F. Scrimshaw., *Trans. Faraday Soc.*, 1969, **65**, 3172.
5. A. Josien and B Sourisseau, *et al., Bull. Soc. Chim.*, 1955, 178.

FT-IR STUDY OF THE REACTIVITY OF THE SURFACE METHOXY SPECIES FORMED BY THE REACTION OF METHANOL ON H-ZSM-5

F. WAKABAYASHI*, J. N. KONDO[†], C. HIROSE[†] and K. DOMEN[†]

* Department of Science and Engineering, National Science Museum,
3-23-1 Hyakunin-cho, Shinjuku-ku, Tokyo 169-0073, Japan; f-waka@kahaku.go.jp
[†] Research Laboratory of Resources Utilization, Tokyo Institute of Technology,
4259 Nagatsuta, Midori-ku, Yokohama 226-8503, Japan

ABSTRACT

The thermal stability and the reactivity of the surface methoxy species formed in the course of the dehydration of methanol (MeOH) to dimethyl ether (DME) on H-ZSM-5 were investigated by means of transmission FT-IR spectroscopy. It was revealed that three sorts of methoxy species are formed on H-ZSM-5 and their thermal stability and reactivity with MeOH are different from each other. These species were assigned to ≡Si(OMe)Al≡ species formed on the acidic OH groups, ≡Si(OMe) species on the internal defect SiOH sites, and ≡Si(OMe) species on the external SiOH sites in the order of decreasing reactivity.

INTRODUCTION

Since H-ZSM-5 is an effective catalyst for the methanol-to-gasoline (MTG) process, adsorption and reaction of MeOH on the zeolite have attracted much attention from a catalytic point of view [1]. It is well known that the dehydration of MeOH takes place on H-ZSM-5 to form DME near 473 K. Since the formation of DME has been regarded as a first step to form gasoline [2], it is quite interesting to study the reaction mechanism of the DME formation from MeOH on H-ZSM-5 as a first step to clarify the mechanism of the MTG process. Although extensive experimental and theoretical studies have been carried out on the subject, many questions remain to be solved.

In the present study, the thermal stability and the reactivity of methoxy species formed in the course of the MeOH conversion to DME on H-ZSM-5 were investigated by means of transmission FT-IR spectroscopy.

EXPERIMENTAL

H-ZSM-5 sample (Si/Al = 27) was pressed into a self-supporting disk ($\rho \cong 7$ mg/cm^2) and was placed in a quartz-made IR cell attached to a conventional closed-circulation system, which allows both sample treatment and IR measurement in situ. The disk was treated with O_2 of 6.7 kPa at 773 K for 30 min and was then evacuated at the same temperature for 30 min. IR spectra were recorded on a Mattson RS-2 FTIR spectrometer using a narrow-band MCT detector at a resolution of 2 cm^{-1} and averaging of 32 scans (in continuous-measurement mode) or 64 scans (in conventional-measurement mode) [3].

RESULTS AND DISCUSSION

Formation of methoxy species on H-ZSM-5 and their thermal stability

MeOH was dosed to H-ZSM-5 in the temperature range between 433 K and 513 K. It was observed that the adsorbed MeOH species were progressively converted to the adsorbed DME species on Brønsted acid sites (BAS), i.e., $\equiv$Si(OH)Al$\equiv$ sites, as reported previously [3]. An example of the spectral change with time is shown in Figure 1: the bands at 3006, 2958, 2925, and 2857 cm^{-1} have been attributed to the MeOH species hydrogen-bonded to BAS [3] and those at 2968, 2946 and 2845 cm^{-1} to the DME species hydrogen-bonded to BAS [3, 4]. The Arrhenius plot of the rate of the surface reaction is shown in Figure 2. From the slope of the fitted line, the activation energy of the surface reaction was calculated to be 97 KJ/mol, which is slightly greater than the apparent activation energy of the MeOH conversion to DME on H-mordenite (80 kJ/mol) obtained from kinetic data [5].

When the gaseous components were evacuated after the surface reaction had reached its steady state, it was observed that the bands due to the adsorbed DME decreased progressively their intensity accompanied by the appearance of new bands at 2969 and 2856 cm^{-1} in the case of 473 K and additional bands at 2978 and 2868 cm^{-1} appeared at 493 K, as shown in Figure 3. The former two bands were observed above 433 K and were found to decrease rapidly with increase of the evacuation temperature. However, the latter two bands became apparent at 493 K because the former bands became relatively small at that temperature. Hence, these two sets of IR bands would have their origin in different species. We call here the species which gives the former bands as Species A and that giving the latter bands as Species B. The effect of evacuation at 453 K on the IR spectra in the $\nu(CH_3)$ region of the CH_3O-formed H-ZSM-5 are shown in Figure 4. This figure clearly shows that

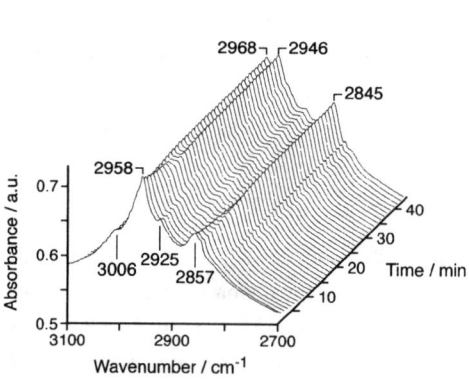

Figure 1. Time course of the spectral change of methanol-adsorbed H-ZSM-5 at 473 K.

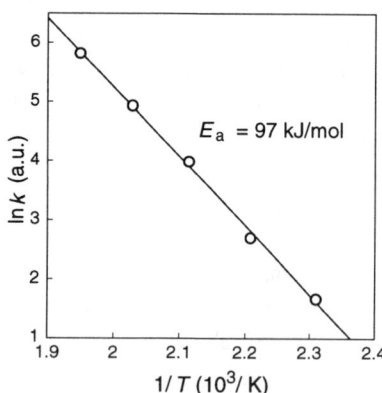

Figure 2. Arrhenius plot of the surface reaction of methanol to dimethyl ether on H-ZSM-5 between 433 K and 513 K.

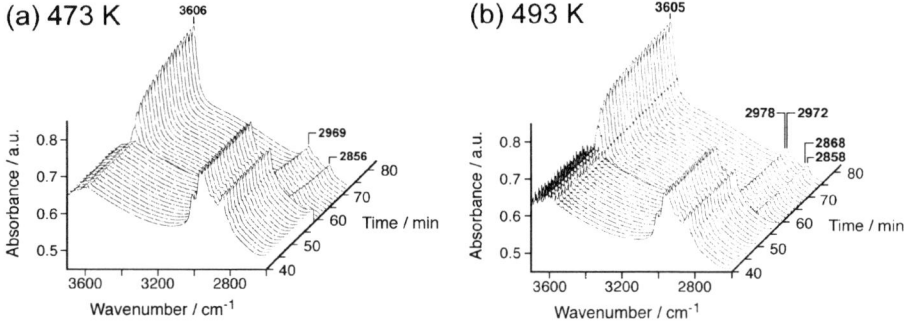

Figure 3. Time course of the spectra change of methanol-adsorbed H-ZSM-5 at 473 K (a) and 493 K (b): Effect of evacuation..

the bands due to Species A (2969 and 2855 cm^{-1}) decrease in intensity with evacuation time. However, those due to Species B (2978 and 2868 cm^{-1}) became apparent after evacuation for 60 min. These facts indicate that Species A is thermally less stable than Species B. Figure 5 shows the change of the difference spectra in the ν(OH) region of the CH$_3$O-formed H-ZSM-5. From this figure, it is clearly seen that the 3606 cm^{-1} band due to the free acidic OH groups regains its intensity, which is indicated by the decrease in the negative band, by evacuation at 453 K. This behavior of the acidic OH groups coincides well with that of Species A. However, it is also seen that the bands at 3744 and 3720 cm^{-1}, which are due to the silanol groups (see below), are almost unaffected by evacuation at 453 K. This fact suggests that Species A can be associated with the acidic OH sites and Species B with the silanol sites.

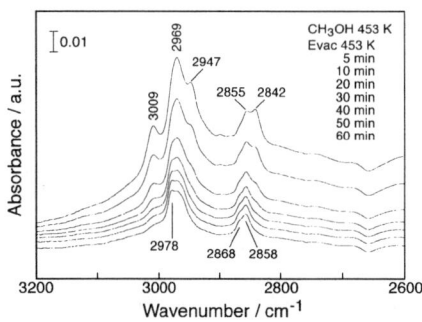

Figure 4. Effect of evacuation at 453 K on the IR spectra of the CH$_3$O-formed H-ZSM-5: ν(CH$_3$) region.

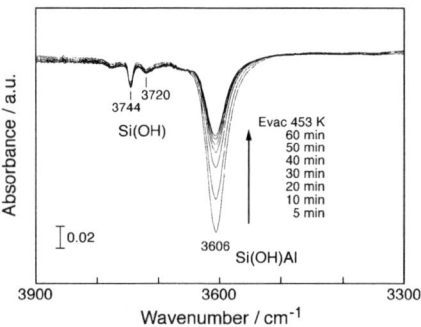

Figure 5. Effect of evacuation at 453 K on the IR spectra of the CH$_3$O-formed H-ZSM-5: ν(OH) region of the difference spectra.

The two sets of IR bands due to Species A and B have previously been observed for the adsorption and the reaction of MeOH on H-ZSM-5 and both sets have been attributed to methoxy species formed on the acidic OH sites, i.e., $\equiv$Si(OMe)Al$\equiv$ [6, 7] : the bands at 2970 and 2857 cm^{-1} were observed by Mirth and Lercher [6] and the bands at 2978 and 2867 cm^{-1} by Forester and Howe [7]. However, our result indicates that these two sets of IR bands have their origin in different species, i.e., Species A and B, respectively. Considering their thermal stability as well as their reactivity as shown below, we attribute Species A (2969 and 2856 cm^{-1}) to the methoxy species formed on the acidic OH groups, i.e., $\equiv$Si(OMe)Al$\equiv$, and Species B (2978 and 2868 cm^{-1}) to the methoxy species formed on silanol sites, i.e., $\equiv$Si(OMe) (see below). This assignment is also supported by the behavior of the ν(OH) as shown in Figure 5.

Reactivity of methoxy species

The methoxy species (CH$_3$O-) were produced on H-ZSM-5 by the reaction of CH$_3$OH at 453 K (Figure 6a) and their reactivity was examined by switching the reactant from CH$_3$OH to CD$_3$OD. A spectrum shown in Figure 6b was measured in the presence of gaseous CD$_3$OD: the bands of Species A decreased in intensity and the bands at 2946 and 2844 cm^{-1} due to the ν(CH$_3$) bands of the adsorbed DME species [3, 4] were observed, which means that CH$_3$OCD$_3$ was produced from Species A. CD$_3$OD was, then, evacuated at 453 K for 30 min and a spectrum shown in Figure 6c was obtained: the band at 2968 cm^{-1} had disappeared and two new bands at 2978 and 2961 cm^{-1} were observed. The 2978 cm^{-1} band can be attributed to Species B although the 2961 cm^{-1} band had not been observed by simple evacuation at 453 K. We attribute the 2961 cm^{-1} band as well as the remaining 2857 cm^{-1} band to another species (Species C) that was not observed by simple evacuation. This assignment is supported by the fact that these two bands remained almost unaffected by the reaction with CD$_3$OD at 493 K (see below: Figure 7). From these results, we conclude that

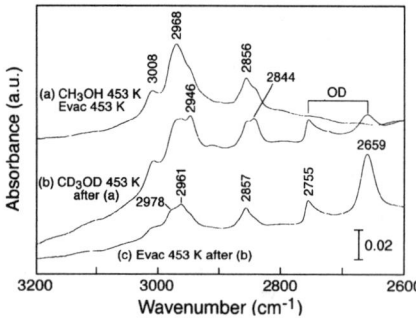
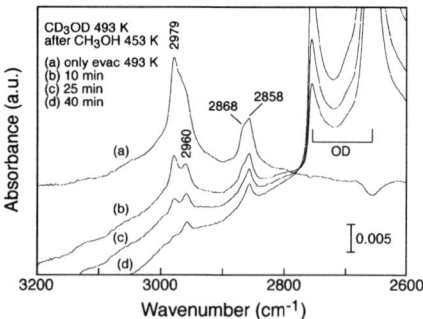

Figure 6. Effect of CD$_3$OD treatment and evacuation at 453 K on CH$_3$O- species formed on H-ZSM-5: (b) with gaseous CD$_3$OD.

Figure 7. Effect of CD$_3$OD treatment and evacuation at 493 K on CH$_3$O- species formed on H-ZSM-5.

Species A is reactive to form DME by reaction with MeOH although Species B and C are less reactive.

Figure 7a shows a spectrum observed after the CH_3OH adsorption on H-ZSM-5 at 453 K and the subsequent evacuation at 493 K. The bands of Species B (2979 and 2868 cm^{-1}) and those of Species C (2960 and 2858 cm^{-1}) were observed. CD_3OD was, then, dosed to the sample at 493 K for the desired period and evacuated at the same temperature. As shown in Figure 7b-d, the bands of Species B decreased gradually in intensity with time although the bands of Species C remained almost unaffected. The $v(CH_3)$ bands of the adsorbed DME were observed in the presence of gaseous CD_3OD (its spectrum is not shown here). Thus, it was concluded that Species B are reactive at 493 K to form DME by reaction with MeOH and that Species C are unreactive even at 493 K. A similar examination at 513 K revealed that Species C is not reactive even at that temperature.

Assignment of the methoxy species

The IR bands and the reactivity of the methoxy species observed in the present study as well as literature data [6-9] are summarized in Table 1. As seen in Table 1, the pair of bands at ca. 2960 and 2858 cm^{-1}, which corresponds to the bands of Species C, was observed for the MeOH adsorption on SiO_2 and was attributed to the methoxy species formed on the silanol sites i.e., ≡Si(OMe) [8]. These bands were also observed for the MeOH adsorption on H-ZSM-5 and were also attributed to the ≡Si(OMe) species [6, 7, 9]. It is known that there are two sorts of silanol sites on H ZSM-5, namely, the external SiOH sites and the internal defect SiOH sites [10]. Considering the fact that the pair of band at ca. 2960 and 2858 cm^{-1} was also observed on pure SiO_2, we assign the Species C to the methoxy species formed on the external SiOH sites. This assignment is consistent with the fact that Species C are the least reactive.

The bands at ca. 2978 and 2968 cm^{-1}, which correspond to Species B, were also observed on H-ZSM-5 and were attributed to the ≡Si(OMe)Al≡ species [7]. However, Species B are less reactive at 453 K, although reactive at 493 K, as revealed in the present study. Furthermore, the behavior of Species B does not coincide with that of the acidic OH groups (Figure 5). Therefore, Species B should not be the reactive ≡Si(OMe)Al≡ species but should be less reactive ≡Si(OMe) species. Therefore, we assign Species B to the ≡Si(OMe) species formed on the internal defect SiOH sites; there are two sorts of ≡Si(OMe) species formed on the external sites (Species C) and on the internal defect sites (Species B). The proposed existence of the two sorts of ≡Si(OMe) species is supported by the fact that two silanol bands at 3744 cm^{-1} (external SiOH sites) [10, 11] and 3720 cm^{-1} (internal defect SiOH sites) [11] were observed as negative bands in the difference spectra between the MeOH-reacted H-ZSM-5 and the background H-ZSM-5 (see Figure 5).

The pair of bands at 2969 and 2856 cm^{-1}, which corresponds to Species A, was also observed on H-ZSM-5 and were attributed to either the ≡Si(OMe)Al≡ species [6] or the methoxy species formed on the non-framework AlOH sites, i.e., Al_{nf}(OMe) species (Al_{nf}: non-framework Al) [9]. The OH band due to non-framework AlOH sites is usually observed at ca. 3660 cm^{-1} [10]. However,

Table 1: IR bands of methoxy species formed by the adsorption or the reaction of methanol on H-ZSM-5 or SiO_2 above 433 K.

Assignment	This work H-ZSM-5 $\nu(CH_3)$	Reactivity	Lercher (1991)[a] H-ZSM-5 $\nu(CH_3)$	Kubelková (1990)[b] H-ZSM-5 $\nu(CH_3)$	Howe (1987)[c] H-ZSM-5 $\nu(CH_3)$	Zecchina (1967)[d] SiO_2 $\nu(CH_3)$
≡Si(OCH₃)Al≡	2969 2856	high (< 473 K)	2970 2860	2977 2871	2980 2868	
≡Si(OCH₃) (I)[e]	2978 2868	medium (>= 493 K)				
≡Si(OCH₃) (II)[f]	2960 2858	low (>> 513 K)	2958 2856	2960 2858	2959 2855	2959 2859
$Al_{nf}(OCH_3)$[g]				2968 2868		

[a] ref. 6   [b] ref. 9   [c] ref. 7   [d] ref. 8
[e] methoxy species formed on the internal defect SiOH sites
[f] methoxy species formed on the external SiOH sites
[g] methoxy species formed on non-framework Al sites

the sample used in the present study has only a very small band (not an apparent band) at ca. 3660 cm^{-1} as reported previously [12], and this band region was almost unaffected by the MeOH reaction (see Figure 5). Thus, Species A observed in the present study cannot be attributed to the Al_{nf}(OMe) species. Since the behavior of the bands of Species A seems to correspond to that of the OH band at 3606 cm^{-1} (due to the acidic OH groups), it is reasonable to relate this pair of bands to the species formed on the acidic OH sites. We, therefore, assign Species A to the ≡Si(OMe)Al≡ species, in accordance with Mirth and Lercher [6].

CONCLUSION

It is revealed out that three sorts of methoxy species are formed on H-ZSM-5 zeolite by the reaction of MeOH. They are attributed to ≡Si(OMe)Al≡ formed on the acidic OH sites (2968, 2856 cm^{-1}), ≡Si(OMe) on the internal defect SiOH sites (2978, 2868 cm^{-1}), and ≡Si(OMe) on the external SiOH sites (2960, 2858 cm^{-1}) in order of decreasing reactivity. It is also revealed that DME can be readily formed from Species A at 453 K, and from Species B at 493 K although Species C are unreactive even at 513 K. These temperatures are well in accordance with the temperature at which the dehydration of MeOH to DME proceeds successfully on H-ZSM-5. This means that methoxy species formed on BAS or internal defect silanol sites might be involved in the reaction. This result is not in accordance with the recent theoretical studies which suggest that the path through the methoxy species is unfavorable compared to the one step formation of DME from two MeOH molecules [13, 14]. Another theoretical study based on first principles calculations suggests that the pathway via methoxy intermediates is also energetically feasible although the product formed by the direct condensation of two MeOH molecules is more stable [15]. The reason for this discrepancy between our experimental result and the theoretical calculation result must be investigated further.

ACKNOWLEDGMENTS

The authors express their grateful acknowledgment to Asahi Chemical Industry Co., Ltd. for providing the H-ZSM-5 sample. This work was partly supported by the Grant-in-Aid for Scientific Research (No. 09440248) from the Ministry of Education, Science, Sports and Culture, Japan.

REFERENCES

1. J. Sauer, P. Ugliengo, E. Garrone and V. R. Saunders, Chem. Rev. **94**, 2095 (1994).
2. M. Jayamurthy and S. Vasudevan, Catal. Lett. **36**, 111 (1996).
3. F. Wakabayashi, M. Kashitani, T. Fujino, J. N. Kondo, K. Domen and C. Hirose, Stud. Surf. Sci. Catal. **105**, 1739 (1997).
4. T. Fujino, M. Kashitani, J. N. Kondo, K. Domen, C. Hirose, M. Ishida, F. Goto and F.

Wakabayashi, J. Phys. Chem. **100**, 11649 (1996).
5. J. Bandiera and C. Naccache, Appl. Catal. **69**, 139 (1991).
6. G. Mirth and J. A. Lercher, Stud. Surf. Sci. Catal. **61**, 437 (1991).
7. T. R. Forester and R. F. Howe, J. Am. Chem. Soc. **109**, 5076 (1987).
8. E. Borello, A. Zecchina and C. Morterra, J. Phys. Chem. **71**, 2938 (1967).
9. L. Kubelková, J. Novakova and K. Nedomova, J. Catal. **124**, 441 (1990).
10. A. Zecchina, S. Bordiga, G. Spoto, D. Scarano, G. Petrini, G. Leofanti, M. Padovan and C. O. Areán, J. Chem. Soc., Faraday Trans. **88**, 2959 (1992).
11. I. Kiricsi, C. Flego, G. Pazzuconi, W. O. Parker Jr, R. Millini, C. Prego and G. Bellussi, J. Phys. Chem. **98**, 4627 (1994).
12. F. Wakabayashi, J. N. Kondo, K. Domen and C. Hirose, J. Phys. Chem. **99**, 10573 (1995).
13. S. R. Blaszkowski and R. A. van Santen, J. Phys. Chem. B **101**, 2292 (1997).
14. P. E. Sinclair and C. R. A. Catlow, J. Phys. Chem. B **101**, 295 (1997).
15. R. Shah, J. D. Gale and M. C. Payne, J. Phys. Chem. B **101**, 4787 (1997).

COMPARATIVE STUDIES OF CATALYTIC EFFECT FOR THE OXIDATIVE METHYLATION OF TOLUENE WITH METHANE OVER BASIC ZEOLITE CATALYSTS *

L. ZHOU, W. LI*, Q. FU, N. GUAN, S. ZHENG and K. TAO

*Department of Chemistry, Nankai university, Tianjin 300071, P.R. China; weili@public1.tpt.tj.cn

ABSTRACT

The catalytic effects for the oxidative methylation of toluene with methane over the basic zeolite catalysts prepared by promoting KY, KX, KM, KZSM$_5$ or Kβ with alkali-metal bromides and alkali-metal oxides have been investigated comparatively. The alkali-metal bromides promoted substrates are more effect than the alkali-metal oxides promoted systems. Moreover, the NaBr promoted systems are better than KBr promoted systems. The most effective catalytic system(8wt%NaBr)/KZSM$_5$ gave a toluene conversion as high as 76.3% and a total C$_8$ selectivity of 76.4%(styrene/ethylbenzene=9.76) resulting in a total C$_8$ yield of 58.3%.

INTRODUCTION

Styrene is one of the most important industrial monomers. At present, styrene is mainly produced by alkylation of benzene with ethylene followed by oxidative dehydrogenation of the resulting ethylbenzene. However, this work process is very expensive and complex. A new method of synthesizing ethylbenzene and styrene directly from toluene and methane was discovered by Khcheyan et al[1] and Yakovich and Bakareva[2]. The process would potentially reduce the cost because of the production of styrene directly from toluene and methane.

Recently, several research groups have employed some metal oxides for the oxidative methylation of toluene with methane[3-6].

It is known that zeolite is the focus of one part of the catalytic chemistry. In contrast to extensive studies of acidity of zeolite, less attention has been given to the studies of base catalysis of zeolite. Applying basic zeolite to the oxidative methylation of toluene with methane is a new field which has potential economical effectiveness and great theory value. In this paper, alkali metal exchanged zeolites have been used in the oxidative methylation of toluene with methane. It was found that these basic zeolites were effect catalyst for the oxidative methylation of toluene with methane.

EXPERIMENTAL

Zeolite NaX, NaY, NaM, Naβ and NaZSM-5 raw poder were supplied from Wenzhou Chemical Plant, Fushun Chemical Plant, Nankai University Chemical Plant, respectively. Samples KX, KY, KM, KZSM5 and Kβ were prepared from NaX, NaY, NaM, NaZSM$_5$ and Naβ raw powder respectively by exchanging with 0.5M KNO_3 aqueous solution at 358K. The exchange procedures were repeated four times. The time of each exchange was 2h. On completion of four times exchange, the sample was filtered and washed with distilled water. The sample was then dried at 398K followed by calaination in air at 773K for 5h. Loaded catalysts were prepared by impregnating aqueous solution of MNO_3(M=Li+, Na+, K+ or Cs+), NaBr and KBr onto KY zeolite described above respectively. The promoters loading was 8wt% for each catalyst. The catalyst was calcined in air at 773K for 5h. The powdered catalyst was made into pellets, crushed, and then sieved(40-60mesh).

The experiments in this work were carried out using a conventional fixed bed flow type quartz glass reactor(6-mm i.d.). The typical experimental conditions were as follows: reaction temperature=973K, CH_4 flow rate =24ml/min, toluene flow rate =2.33ml/min(g), weight of catalyst/total feed rate(W/F) = 2.14g.h./mol. $CH_4/O_2/C_6H_5CH_3/N_2$(diluent)=24:6:2.33:20 ml.min^{-1}. Toluene was introduced by passing the reactant gas mixture into a toluene vapor saturator equipped with just before the inlet of reactor. Liquid products were trapped in an ice-salt bath, and were analyzed by gas chromatography equipped a SE-30 Capillary column and flame ionization detection (FID). The main products were styrene, ethylbenzene, benzene and small amounts of xylene, water.

RESULTS AND DISCUSSION

Table 1 shows the results of the reaction at 973K. Effects of additive alkali metal oxides and alkali metal bromides over KY zeolite on the toluene conversion, C_8 yield and C_8 selectivity were first examined. Among these alkali metal oxides and alkali metal bromides tested, alkali metal bromides promoted KY are more effective catalytic systems than alkali metal oxides promoted KY. Moreover, the ratios of styrene to ethylbenzene of alkali metal bromides promoted KY catalyst are larger than that of alkali Metal oxides promoted KY catalysts. The results which are presented in Table 1 show that the yield of C_8 compounds(ethylbenzene + styrene) decreased in the following order: NaBr/KY > KBr/KY > Cs_2O/KY > K_2O/KY > Na_2O/KY > Li_2O/KY and the selectivity of C_8 compounds (ethylbenzene + styrene) decreased in the following order: NaBr/KY > KBr/KY > Li$_2$O/KY > Cs$_2$O/KY > Na$_2$O/KY > K$_2$O/KY.

Table 1. Oxidative Methylation of Toluene with Methane over Modified Zeolite Catalysts [a]

Catalyst	Conv. (mol%) C$_7$H$_8^b$	Selectivity(%)				ST/Eb[g]	Yield(%)			
		C$_6$H$_6^c$ (BZ)	C$_8$H$_{10}^d$ (EB)	C$_8$H$_8^e$ (ST)	TotalC$_8^f$ (ST+EB)		C$_6$H$_6$ (BZ)	C$_8$H$_{10}$ (EB)	C$_8$H$_8$ (ST)	TotalC$_8$ (ST+EB)
Li$_2$O/KY	25.5	28.1	32.3	25.8	58.1	0.80	7.2	8.2	6.6	14.8
Na$_2$O/KY	37.5	31.5	28	21.6	49.6	0.77	11.8	10.5	8.1	18.6
K$_2$O/KY	38.4	40.1	26.1	16.9	43	0.65	15.4	10	6.5	16.5
Cs$_2$O/KY	57.1	40.0	15.9	36.4	52.3	2.29	9.1	9.1	20.8	29.9
NaBr/KX	65.0	11.8	11.3	64.4	75.7	5.70	7.7	7.4	41.9	49.3
NaBr/KY	65.0	7.5	8.3	73.4	81.7	8.89	4.9	5.4	47.7	53.1
NaBr/KM	62.0	11.2	10.8	68.0	78.8	6.30	6.9	6.7	42.2	48.9
NaBr/KZSM5	76.3	9.9	7.1	69.3	76.4	9.76	7.6	5.4	52.9	58.3
NaBr/Kβ	61.7	16.7	7.1	68.7	75.8	9.68	10.3	4.4	42.4	46.8
KBr/KX	52.2	21.5	18.1	46.8	64.9	2.59	11.2	9.5	24.4	33.9
KBr/KY	58.9	14.8	17.9	51.0	68.9	2.85	8.7	10.5	30.0	40.5
KBr/KM	57.4	16.3	16.1	56.4	72.5	3.5	9.4	9.2	32.4	41.6
KBr/KZSM5	58.2	15.6	13.6	58.6	72.2	4.31	9.1	7.9	34.1	42.0
KBr/Kβ	60.1	14.0	13.3	58.0	71.3	4.36	8.4	8.0	34.9	42.9

Note: a. T$_R$= 973K,W=0.3g,CH$_4$/O$_2$/C$_6$H$_5$CH$_3$/N$_2$ =24:6:2.33:20(ml.min^{-1}). The results presented were obtained after 2h of reaction. The yield is on the basis of moles of toluene converted. b. toluene. c. benzene. d. ethylbenzene. e.styrene.
f. ethylbenzene +styrene. g. ST/EB=selectivity ratio of styrene to ethylbenzene.

The effects of different zeolites promoted with alkali metal bromides were compared. The results showed that NaBr promoted systems were more effective than KBr promoted systems. Among NaBr promoted systems, NaBr promoted KZSM$_5$ gave the highest C$_8$ yield of 58.3% with the C$_8$ selectivity of 76.4%[styrene/ethylbenzene=9.76(mol)]. The rest four type zeolites promoted with NaBr showed the similar toluene conversion, C$_8$ yield and C$_8$ selectivity, but the ratio of styrene to ethylbenzene over the NaBr/KY catalyst was largest(13.3) . Comparing the effects of different zeolites promoted with KBr, the results presented in Table1 showed that the KBr/KZSM-5 and KBr/Kβ catalyst showed similar catalytic performance for the oxidative methylation of toluene. The toluene conversion and C_8 Yield and C_8 selectivity of KBr/KY catalyst was also similar to that of KBr/KZSM-5 and KBr/Kβ catalysts. The lowest performance was exhibited by KBr/KX with a total C_8 selectivity of 64.9% and C_8 yield of 33.9%.

The stability with time-on-stream of NaBr(8wt% NaBr) KY zeolite catalyst was also investigated. The results showed that the toluene conversion suffered 40~45% decreased compared to the initial value after 50h of reaction. In contrast, the total C_8-selectivity remained almost unchanged.

CONCLUSIONS

In conclusion, the basic zeolites are effective catalysts for the oxidative methylation of toluene with methane. The alkali-metal bromides promoted substrates are more effect than the alkali-metal oxides promoted systems. Moreover, the NaBr promoted systems are better than KBr promoted systems.

ACKNOWLEDGMENTS

Financial support for this project was partially obtained form National Natural Science Foundation of China(No.29673022) and Tianjin Natural Science Foundation (No.983700911).

REFERENCES

1. Khcheyan, Kh. Ye., Revenko, O.M., Borisoglebskaya, A.V., Neftekhimiya **21**, 83(1981).
2. Yakovich, N.T. and Bakareva, L.P., Neftekhimiya **23**, 317(1983).
3. T. Suzuki, K.Wada and Y. Watanable, Ind. Eng. Chem. Res **30**, 1719(1991).
4. Y. Osada, K. Enomoto, T. Fukushima. J. Chem. Soc., Chem. Commen., 1156(1989).
5. H. Kim, H. Suh and H. Paik, Appl. Catal. A. **87**, 115(1992).
6. A.Z.Khan and E.Ruckenstein, Appl. Catal. A. **102**, 233(1993).

LEWIS ACID SITES IN ZEOLITE Y STUDIED BY ADSORPTION, EPR AND NMR TECHNIQUES

A. SEIDEL*, A. GUTSZE[†] and B. BODDENBERG*

*Lehrstuhl für Physikalische Chemie II, Universität Dortmund, D-44221 Dortmund, Germany
FAX: +49 231-755-5367; E-mail: bod@pcii.chemie.uni-dortmund.de
[†]Department of Biophysics, Ludwig Rydygier's University School of Medical Science, Bydgoszcz, Poland

ABSTRACT

Adsorption isotherms of CO, NO, and Xe, EPR spectra of NO, and NMR spectra of Xe in zeolites HY and ZnY were measured. The applicability of these experimental techniques for the characterization and quantification of different types of Lewis acid sites is critically discussed.

INTRODUCTION

The Lewis acid properties of zeolites have been the matter of scientific investigations for a long time, mainly because they are considered responsible for the catalytic activity of the aluminosilicates with respect to a variety of economically interesting petrochemical conversions [1]. The origins of Lewis acidity in zeolites are diverse and depend on the actual system under investigation. Transition metal exchanged zeolites may exhibit Lewis acidity due to the presence of cationic species located at crystallographic positions which are accessible to basic probe molecules. Lewis acid properties observed for zeolites in H-form generally are considered to be caused by dehydroxylation processes which result in the formation of coordinatively unsaturated and/or cationic extraframework aluminium species acting as so-called 'true' Lewis acid sites (LAS) [1,2].

Much work has been devoted to the development and evaluation of analytical techniques which allow the characterization and quantification of different types of LAS [3]. Most of these techniques rely on either adsorption measurements of probe molecules, on spectroscopic investigations of the sorbed probes or of the bare zeolite, or on catalytic test reactions. The present investigation demonstrates that the combination of several of these techniques is well suited to obtain reliable information about the LAS and their concentration in zeolite Y type catalysts.

EXPERIMENTAL

Zeolite NH$_4$NaY (Y64, UOP) was used as starting material for the preparation of HY by first increasing the NH$_4$-content to NH$_4$/Al>0.95 using conventional ion-exchange techniques, and subsequently dehydrating and deammoniating the material in vacuo at 420 °C. The vacuum treatment of HY at 600 °C (24 h) resulted in the dehydroxylated sample HY(600). Zeolite ZnY was obtained by reaction of HY with zinc metal vapor (420 °C) and subsequent application of a rehydration/dehydration cycle according to a procedure described elsewhere [4].

Adsorption isotherms were measured volumetrically at 25.0 °C using all-steel devices. The ^{129}Xe NMR investigations were performed at ambient temperature with a solid-state NMR spectrometer CXP100 (Bruker, Germany) operating at the resonance frequency 21.4 MHz. Chemical shifts are reported with xenon gas at $p \to 0$ as the reference. EPR measurements were performed at 77 K with aid of an X-band spectrometer operating at 9.6 GHz. The zeolites were loaded with NO in multiple excess of the number of adsorption sites present within the sample [6 NO per unit cell (uc)] in order to guarantee full occupancy of the LAS by the probe molecules. The concentration of unpaired spins of NO, N_{spin}, was determined by double integration of the derivative EPR signals. Calibration was performed with a strong pitch standard sample (Varian).

RESULTS AND DISCUSSION

Fig. 1 a-c shows adsorption isotherms of CO and Xe as well as ^{129}Xe NMR chemical shifts of xenon, respectively, in zeolites HY, HY(600), and ZnY. In contrast to zeolite HY which exhibits linear adsorption isotherms of CO, Xe, and NO (not shown) as well as linear ^{129}Xe shifts as function of equilibrium Xe pressure, the dehydroxylated and zinc-exchanged materials show more or less pronounced non-linear behavior. It has been shown that the adsorption and NMR data for a variety of ion-exchanged zeolites Y prepared under different conditions, can be well reproduced with eqns. 1 and 2 [4-8]

$$N_J = \sum_i \frac{n_i k_i^J p}{1 + k_i^J p} + K_H^J p, \qquad (1)$$

$$\delta = \frac{1}{N_{Xe}} \left[\sum_i \frac{n_i k_i^{Xe} p}{1 + k_i^{Xe} p} \delta_i + K_H^{Xe} p \delta_H \right] + FN_{Xe}, \qquad (2)$$

assuming localized (Langmuir-type) and non-localized (Henry-type) adsorption of the probes J (J=CO, Xe, NO) on various types of LAS and at the residual internal zeolite surface, respectively.

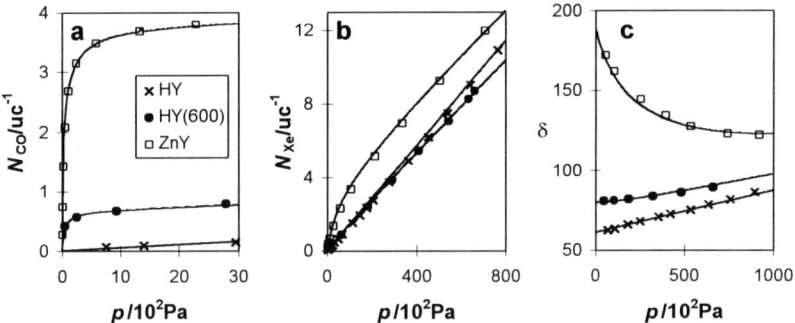

Figure 1. Adsorption isotherms of CO (a) and Xe (b) as well as ^{129}Xe NMR chemical shifts (c) of xenon in zeolites HY, HY(600), and ZnY; the solid curves are fits according to the model discussed in the text (eqns. 1 and 2).

Eqn. 2 is based on the assumption of rapid exchange of Xe between the different adsorption environments. It has been shown that the experimental data of dehydroxylated zeolite HY can be reasonably reproduced assuming a single type of LAS [7,8] which are believed to be some extra-framework cationic aluminate species, e.g. AlO$^+$ [2]. In case of zinc-exchanged zeolites, the assumptions of single site adsorption turned out to be unsuited to reproduce the adsorption and ^{129}Xe NMR chemical shift data. Here, the assumption of two types of LAS, identified to be Zn^{2+} ions located at the crystallographic positions SIII and SII, yielded very good agreement of the simulations with the experimental results for all zinc-exchanged zeolites Y investigated so far [4-6]. The adsorption constants k_i^J and the local isotropic ^{129}Xe shifts δ_i are site-characteristic parameters which depend both on the chemical nature and the location of the LAS. The values of these parameters for the different types of LAS are collected in Table 1.

TABLE 1: Site-characteristic adsorption constants and isotropic ^{129}Xe NMR shifts (25 °C).

probe (J)	site (i)	k_i^J / 10^{-2} Pa^{-1}	δ_i
CO	'true' LAS	2.4	-
CO	Zn^{2+}(SIII)/Zn^{2+}(SII)	13/0.74	-
Xe	'true' LAS	0.004	115
Xe	Zn^{2+}(SIII)/Zn^{2+}(SII)	0.023/0.008	220-235/130-140
NO	'true' LAS	0.20	-
NO	Zn^{2+}(SIII)	0.05	-

Whereas the CO adsorption is highly dominated by the contribution of localized adsorption, the Xe sorption isotherms are mainly influenced by Henry-type adsorption and therefore reflect sensitively any modification or destruction of the alumino-silicate framework [8]. The properties of NO can be classified in some way between CO and Xe [7].

With the hypothesis that different probe molecules detect the same LAS, and that adsorption and spectroscopic techniques yield the same concentrations n_i of LAS, the reliability of such concentration figures is expected to increase with the number of methods applied. However, this hypothesis may not be valid in every case. Inspection of Table 2 reveals that both for the dehydroxylated zeolite HY(600) and the zinc-exchanged zeolite ZnY, the number of NO spins detected by the EPR technique (N_{spin}) underestimates by a factor of 5-15 the concentration of sorption sites (n_i) derived from the adsorption isotherms. These findings can be explained with the presence of distributions of sorption complex geometries and sorption strengths of the LAS which lead to quenching of the orbital motion of the NO - the prerequisite for EPR detectability [2,7] - for only part of the adsorbed molecules. The comparison of the adsorption site concentrations n_i for CO and NO in zeolite ZnY (Table 2) shows that CO at 25 °C detects the Zn^{2+} ions at both SIII and SII, whereas NO sees only part of these sites, namely the Zn^{2+} sites revealing the strongest adsorption strength (SIII). It is only at these sites that the degeneracy of the $^2\Pi_{1/2}$ state of NO is lifted through quenching of the orbital motion. Obviously, NO adsorption on the SII cations at ambient temperature (and even at -25 °C [7]) is such weak ($k_2^{NO} p \ll 1$) that no deviations from Henry type adsorption behavior are observed.

TABLE 2: Site concentrations determined from adsorption isotherms of CO and NO (25 °C), and spin concentrations determined by EPR of sorbed NO (−196 °C).

sample	i	$n_i(CO) / uc^{-1}$	$n_i(NO) / uc^{-1}$	N_{spin} / uc^{-1}
HY(600)	'true' LAS	0.45(5)	0.45(5)	0.03(1)
ZnY	Zn^{2+}(SIII)	2.1(1)	2.1(1)	0.4(1)
	Zn^{2+}(SII)	1.7(1)	0	

Conclusion

Is has been demonstrated that low concentrations of various kinds of Lewis acid sites in zeolite Y can be determined quantitatively from the simultaneous analysis of adsorption isotherms of CO, NO, and Xe and ^{129}Xe NMR chemical shifts of the encaged xenon atoms. It is shown that EPR line intensities of adsorbed NO underestimate the concentration of Lewis acid sites considerably.

Acknowledgement

Financial support of this work by Fonds der Chemischen Industrie is gratefully acknowledged.

References

1. D. W. Breck, *Zeolite Molecular Sieves*, R. E. Krieger Publishing Company, Malabar, 1984.
2. A. Gutsze, M. Plato, H. G. Karge, F. Witzel, *J. Chem. Soc., Faraday Trans.*, 1996, **92**, 2495.
3. H. Karge, *Stud. Surf. Sci. Catal.*,1991, **65**, 133.
4. A. Seidel, B. Boddenberg, *Chem. Phys. Lett.*, 1996, **249**,117.
5. A. Seidel, F. Rittner, B. Boddenberg, *J. Chem. Soc., Faraday Trans.*, 1996, **92**, 493.
6. F. Rittner, A. Seidel, T. Sprang, B. Boddenberg, *Appl. Spectr.*, 1996, **11**, 1389.
7. A. Seidel, A. Gutsze, B. Boddenberg, *Chem. Phys. Lett.*, 1997, **275**, 113.
8. A. Seidel, B. Boddenberg, *J. Chem. Soc., Faraday Trans.*, 1998, **94**, 1363.

THE DISTRIBUTION OF ACID STRENGTH OF OH GROUPS IN STEAMED HY ZEOLITES STUDIED BY IR SPECTROSCOPY

J. DATKA*, B. GIL*, J. FRAISSARD[+], P. MASSIANI[+], P. BATAMACK[+]

* Faculty of Chemistry, Jagiellonian University, Ingardena 3, Cracow, Poland; datka@trurl.ch.uj.edu.pl

[+] Université Pierre et Marie Curie, 4 pl. Jussieu, Paris, France, CNRS URA 1106 and 1428

ABSTRACT

High acidity of OH groups in steamed zeolites is due to two reasons: increase of framework Si/Al and interaction of bridging hydroxyls with extraframework Al. We undertook IR studies of the first of these reasons: the effect of increase of framework Si/Al (by steaming) on the distribution of acid strength of OH groups in HY zeolite. Extraframework Al was removed by EDTA treatment. Experiments of ammonia desorption showed, that OH groups were more homogeneous and more acidic in steamed HYSD zeolite than in non steamed one. It was explained by removal of Si(3Al) and Si(2Al) and therefore of the less acidic $(SiO)(AlO)_2$**Si-OH-Al**$(OSi)_3$ and $(SiO)_2(AlO)$**Si-OH-Al**$(OSi)_3$. The most acidic hydroxyls $(SiO)_3$**Si-OH-Al**$(OSi)_3$ corresponding to Si(1Al) are the most numerous.

INTRODUCTION

Steamed zeolites attract a great attention because of industrial application and interesting physicochemical properties. The high acid strength of OH groups may be explained in two ways. The first reason is the increase of framework Si/Al, which affects the distribution of their acid strength. The interaction of bridging hydroxyls with extraframework Al results in a further increase of their acidity [1-3]. We undertook an IR spectroscopic study of acid properties of steamed zeolites. In the present study, our attention was focussed on the explanation for the high acidity of steamed HY zeolite from the study of the effect of steaming on the distribution of acid strength of bridging hydroxyls. The effect of the extraframework Al was eliminated using zeolite which was steamed and subsequently treated with EDTA. This is a new approach, because in practically all previous studies the zeolites contained extraframework Al so both the presence of extraframework Al and increase of framework Si/Al influenced the properties of bridging hydroxyls. In our study the average acid strength and the distribution of the acid strength was studied by ammonia thermodesorption followed by IR spectroscopy. The information on the status of Al and Si atoms was obtained in MAS NMR experiments.

EXPERIMENTAL

Two Y zeolites were studied:
i) commercial, non dealuminated Linde LZY-62 (denoted as HY). Si/Al was 2.6 and the composition of a unit cell corresponded to the formula: $Na_{10}(NH_4)_{43}[(SiO_2)_{139}(AlO_2)_{53}]$,
ii) zeolite HYSD which was steamed under water vapour at 920 K and subsequently dealuminated by the treatment with an EDTA solution. Si/Al was 11.4 and the unit cell composition corresponded to the formula: $H_{16}[(SiO_2)_{176}(AlO_2)_{16}]$.

^{27}Al MAS NMR experiments were performed on a BRUKER 500 MHz spectrometer at the spin rate of 12 kHz and ^{29}Si MAS NMR on a BRUKER 400 MHz at a spinning rate of 4 kHz. For IR studies, zeolites were activated in situ in a transmission IR cell at 770 K under vacuum for 1 hour. Spectra were recorded with a BRUKER IFS 48 PC spectrometer equipped with an MCT detector.

RESULTS AND DISCUSSION

^{27}Al MAS NMR spectra of HY and HYSD zeolites are presented in Figure 1A. Steaming resulted (spectrum not shown) in an expulsion of some Al from the framework to the extraframework positions. The EDTA treatment removed these extraframework Al species. The spectra of both HY and HYSD zeolites show only one signal of tetrahedral Al.

^{29}Si MAS NMR spectrum of non steamed HY zeolite (Figure 1B) shows typical four signals Si(0Al), Si(1Al), Si(2Al) and Si(3Al). In steamed HYSD zeolite the most important signals are Si(0Al), then Si(1Al), the contribution of Si (2Al) and Si(3Al) being much lower.

IR spectra of non steamed HY, as well as of steamed and subsequently dealuminated HYSD zeolites are presented in Figures 1C and D. Spectra shown in Figure 1C are normalized to the sample mass and these shown in Figure 1D are normalized to the intensity of HF bands. Steaming resulted (spectrum not shown) in a distinct decrease of low frequency (LF) hydroxyls band at 3550 cm^{-1} and high frequency (HF) hydroxyls band at 3640 cm^{-1}, an increase of Si-OH band, due to removal of extraframework Al and also in appearing of new bands: 3525, 3600, 3660 and 3690 cm^{-1}, related to the presence of extraframework Al. Dealumination by EDTA of steamed zeolite (Figures 1C, D, spectra b) resulted in the disappearing of the bands related to extraframework Al. HF and LF hydroxyl bands (besides of Si-OH 3740 cm^{-1}) are the only bands existing in both HYSD and non dealuminated HY. The only difference is the lowering (of 20 cm^{-1}) of the frequency of HF OH band in HYSD compared to HY. This difference suggests higher acid strength of hydroxyls in HYSD zeolite (due to higher polarization of O-H bond - ref. 4)

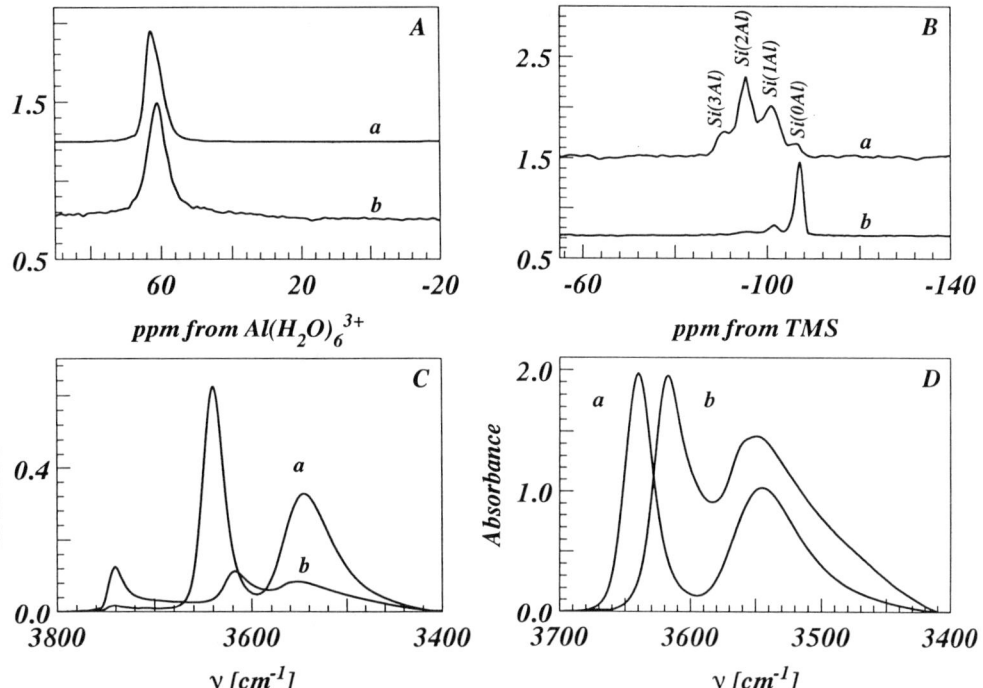

Figure 1. ^{27}Al MAS NMR (A), ^{29}Si MAS NMR (B) and IR spectra (C, D) of HY (a), and HYSD (b) zeolites. Spectra in Fig. C are normalized to the sample mass, and in Fig. D are normalized to the intensity of HF OH band.

Further information on the acid strength of OH groups was obtained in ammonia thermodesorption experiments. An excess of ammonia (sufficient to neutralize all acid sites) was first sorbed at 320 K in both zeolites. Ammonia was next desorbed step-by-step by pumping at increasing temperatures (420 - 620 K). Desorption was realized by ramping the temperature, periodically stopping the ramp for 30 min, cooling to 320 K and recording the spectra at this temperature. The ratio A_{des}/A_0 (where A_0 and A_{des} are the intensities of ammonium ions band at 1450 cm^{-1} upon saturation and desorption, respectively) representing which fraction of ammonium ions survived the desorption, was taken as the measure of acid strength. These values are presented in Figure 2A. The acid strength of OH groups in steamed and dealuminated HYSD zeolite is higher than in non steamed one (HY). The same conclusion was also drawn by comparing the O-H frequencies in Figures 1C, D. The higher acid strength of OH groups in the absence of any interaction with extraframework Al in the steamed zeolite is due to higher framework Si/Al.

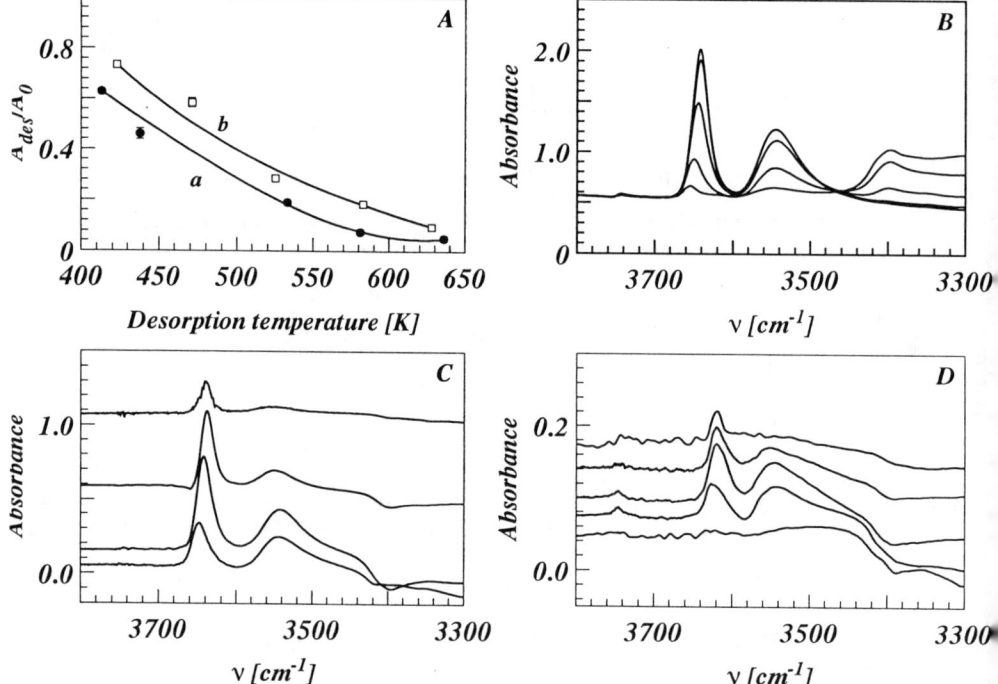

Figure 2. Ammonia desorption from HY and HYSD zeolites

A - A_{des}/A_0 values as a function of desorption temperature: a - HY, b - HYSD

B - IR spectra of OH groups in HY after ammonia desorption at 420, 470, 520, 570, 620 K (from bottom to top)

C, D - difference spectra (difference between spectra recorded upon two consecutive desorption steps, the desorption temperature increases from bottom to top), C - zeolite HY, D - zeolite HYSD

The information on the distribution of acid strength and the heterogeneity of OH groups was obtained by comparing the frequencies of bands of OH groups which are restored by ammonia desorption. If OH groups are heterogeneous, the restoring OH band should shift to the lower frequency with an increase of ammonia desorption temperature (because the less acidic hydroxyls of higher stretching frequency restore at lower temperatures than more acidic ones of lower stretching frequency). On the other hand, no such a shift is expected in the case of homogeneous hydroxyls. Figure 2B presents the spectra of OH groups recorded upon ammonia desorption and Figures 2C, D, the difference spectra (differences between two spectra recorded upon two consecutive desorption steps, differing by 50 K). The difference spectra represent OH groups restored at a given desorption step. In the non steamed HY zeolite, the frequency difference ($\Delta\nu$) between hydroxyls restoring at the lowest temperature (bottom spectrum) and ones restoring at the highest temperature (top spectrum) is 13 cm^{-1}. In steamed and dealuminated HYSD $\Delta\nu$ was 4 cm^{-1}. This indicates a greater heterogeneity of acid strength of OH groups in HY than in HYSD.

Heterogeneity of OH groups in HY zeolites was also evidenced in our earlier studies of benzene and chlorobenzene sorption [5-8]. It can be explained on the basis of ^{29}Si MAS NMR results (Figure 1B): presence of Si(1Al), Si(2Al) and Si(3Al) and therefore (SiO)$_3$Si-OH-Al(OSi)$_3$, (SiO)$_2$(AlO)Si-OH-Al(OSi)$_3$ and (SiO)(AlO)$_2$Si-OH-Al(OSi)$_3$ of different acidity. Steaming decreased distinctly the population of Si(2Al) and Si(3Al) (as evidenced by ^{29}Si MAS NMR study - Figure 1B) and therefore of (SiO)(AlO)$_2$Si-OH-Al(OSi)$_3$ and (SiO)$_2$(AlO)Si-OH-Al(OSi)$_3$ hydroxyls of lower acidity. The most acidic (SiO)$_3$Si-OH-Al(OSi)$_3$ are the most numerous. This explains homogeneity and high acid strength of HF hydroxyls non interacting with extraframework Al in steamed HYSD.

CONCLUSIONS

Steaming decreased the population of (SiO)(AlO)$_2$Si-OH-Al(OSi)$_3$, and (SiO)$_2$(AlO)Si-OH-Al(OSi)$_3$ hydroxyls of lower acidity. The most acidic (SiO)$_3$Si-OH-Al(OSi)$_3$ are the most numerous. These findings explain the increase of homogeneity and high acid strength of HF hydroxyls non-interacting with extraframework Al in steamed zeolite.

ACKNOWLEDGMENTS

This study was sponsored by Komitet Badań Naukowych (grant nr 257/T09/109 09). One of us (B.G.) was a bursar of Fundacja na Rzecz Nauki Polskiej.

REFERENCES

1. R.M. Lago, W.O. Haag, R.J. Mikowsky, D.H. Olson, S.D. Hellring, K.D. Schmitt and G.T. Kerr; in Proceedings of the 7th International Zeolite Conference, Edited by Y. Murakami et al. (Kondansha, Tokyo, 1986), p. 677.
2. M.A. Makarova, A. Garforth, V.L. Zholobenko, J. Dwyer, G.J. Earl and D. Rawlence, Zeolites and Related Microporous Materials: State of the Art 1994 Edited by J. Weitkamp, H.G. Karge, H. Pfeifer and W. Holderich, Studies in Surface Science and Catalysis, Vol. 84 (Elsevier Science B.V. 1994) p. 365
3. M.A. Makarova, S.P. Bates and J. Dwyer, Zeolites: A Refined Tool for Designing Catalytic Sites, Edited by L. Bonneviot and S. Kaliaguine (Elsevier Science B. V. 1995), p. 79.
4. P. A. Jacobs, Catal. rev. Sci. Eng., 1982, **24A**, 415
5. J. Datka, B. Gil, J. Catalysis **145**, 372 (1994).
6. B. Gil, E. Brocławik, J.Datka, J. Klinowski, J. Phys. Chem. **98**, 930 (1994).
7. J. Datka, E. Brocławik, B. Gil, J. Phys. Chem. **98,** 5652 (1994).
8. J. Datka, M. Boczar, B. Gil, Colloids and Surfaces A: Physicochemical and Engineering Acpects **105,**1 (1995).

THE PROPERTIES OF ALKOXYL GROUPS IN ZEOLITES STUDIED BY IR SPECTROSCOPY

J. DATKA*, J. RAKOCZY[+], G. ZADROŻNA[+]

*Faculty of Chemistry, Jagiellonian University, Ingardena 3, 30-060 Cracow, Poland, fax nr (48-12) 634 05 15; datka@trurl.ch.uj.edu.pl

[+]Institute of Organic Chemistry and Technology, Technical University of Cracow, Warszawska 24, 31-155 Cracow, Poland

ABSTRACT

The properties of "isolated" alkoxyl groups (in the absence of unreacted methanol and water) were studied. It has been found that the properties of C-H and O-C bond in zeolitic alkoxyl groups, their reactivity in benzene alkylation and thermal stability depend on composition of zeolites. It is discussed considering electrostatic effects in zeolite framework. The catalytic experiments suggest that formation of zeolitic methoxyl groups is an indispensable step in benzene alkylation by methanol. Y zeolites containing large amounts of alkoxyl groups (formed by preadsorption of methanol) show shape selectivity in toluene alkylation what is explained by narrowing of zeolitic pores by methoxyl groups.

INTRODUCTION

Alkoxyl groups participate in numerous reactions catalyzed by zeolites [1-3]. They are formed by reactions of alcohols with zeolitic hydroxyls [4,5]. We undertook studies of the formation, properties, reactivity and decomposition of methoxyl and ethoxyl groups in faujasite type zeolites. It is known, that the composition of zeolite determines the properties of zeolitic hydroxyls. It was interesting to know whether the zeolite composition influenced also the properties of alkoxyl groups (which had replaced hydroxyls). We studied the properties of C-H and O-C bond, bonding alkoxyl group with zeolite framework in zeolites of various composition (NaHX and NaHY zeolites of various Na/H exchange degrees). The reactivity of zeolitic methoxyl groups with water, methanol, benzene and toluene was followed. It was also interesting to study the role of methoxyl groups in benzene alkylation. We studied also the effect of zeolitic methoxyl groups on the selectivity in toluene alkylation. Most of information on alkoxyl groups was obtained in IR studies, some data was taken from TPD experiments and catalytic tests. A special attention was paid to study the properties of "isolated" alkoxyl groups in the absence of unreacted alcohol and water.

EXPERIMENTAL

NaNH$_4$X (exchange degree 40%) and NaNH$_4$Y (exchange degrees 40% and 77%) zeolites were activated in situ in an IR cell at 620 K (zeolite X) and at 720 K (zeolites Y). They will be denoted as NaHX, NaHY/40 and NaHY/77, resp. Methoxyl and ethoxyl groups were formed by the reaction of methanol or ethanol at 420 K. After 30 min of contact, unreacted alcohols and water were removed by 30 min evacuation at the same temperature. IR studies have shown that the bands of alcohols and water disappeared upon such an evacuation and only the bands of alkoxyl groups remained. The most characteristic of methoxyl groups was the band at 2977 cm^{-1} and of ethoxyl groups the band at 2996 cm^{-1}.

TPD experiments of alkoxyl groups decomposition were done in a flow system with helium as carrier gas. Alkoxyl groups were formed by an injection of alcohols into carrier gas at 420 K, followed by flushing at 450 K.

Catalytic tests were done at 550 - 700 K in a pulse microreactor as well as in a continuous flow reactor.

RESULTS AND DISCUSSION

Acid strength of OH groups in zeolites.

As the acid strength of zeolitic hydroxyls influences their reactivity with alcohols, the properties of alkoxyl groups will be compared with the properties of OH groups present in zeolite before alkylation. The data concerning acid strength of OH groups are presented in Table 1. The values of frequency shifts $\Delta \nu$ of OH interacting with benzene, average electronegativites S_{av} as well as values $|q_O - q_H|$ (differences between charges on oxygen and hydrogen), calculated from the electronegativities according to Sanderson theory are given. The acid strength increased in the order NaHX < NaHY/40 < NaHY/77.

Formation of methoxyl and ethoxyl groups.

The rate of formation of alkoxyl groups was studied by following the increase of intensity of IR bands at 2977 cm^{-1} (methoxyl groups) and 2996 cm^{-1} (ethoxyl groups) as a function of contact time of zeolite with alcohols at 320 K. It has been found that the rate of alkoxyl groups formation increased with the acidity of zeolite and with basicity of alcohol, what agrees with the mechanism assuming that the first step of alkylation is the addition of protons to alcohol molecules and formation of oxonium ions.

Our earlier studies evidenced, that bridging hydroxyls in NaHY zeolites were heterogeneous [6,7] and three kinds of OH groups of various number of Al near the bridge (and therefore of various acidity) were found: $(SiO)_3Si$-OH-$Al(OSi)_3$, $(SiO)_2(AlO)Si$-OH-$Al(OSi)_3$ and $(SiO)_3Si$-OH-$Al(OSi)_3$. According to the results obtained in the present study, methanol molecules reacted with most acidic hydroxyls in the first order. It was evidenced in experiments in which some hydroxyls were consumed in the reaction with methanol. The acid strength of remaining hydroxyls was lower (higher stretching frequency and lower $\Delta\nu$ of hydroxyls hydrogen bonded with benzene) than of hydroxyls in zeolite before alkylation.

Properties of C-H bonds in alkoxyl groups.

The frequencies of C-H stretching vibrations of CH_3 groups in methoxyl and ethoxyl groups in NaHX and NaHY zeolites are presented in Table 2. They increase in the order: NaHX < NaHY/40 < NaHY/77, indicating that the composition of zeolite influences the properties of C-H bond in both methoxyl and ethoxyl groups. The most distinct difference was observed between NaHX and NaHY (differing strongly in acidity), and less distinct difference between NaHY/40 and NaHY/77 (differing less in acidity). Similar effects were observed [8] if methoxyl groups were formed in MHNaY zeolites (M = Li, K, Rb, Cs). C-H stretching frequency increased in the order: Cs < Rb < K < Na < Li. In both cases, C-H frequency increased with the average electronegativity of zeolite.

Stability of O-C bond.

The properties of O-C bond, bonding alkyl groups with zeolite framework are very important, because the stability of this bond influences the stability of alkoxyl groups at higher temperatures (at which catalyzed reactions occur) and also their reactivity (O-C bond must be broken during the reaction). The stability of O-C bond was studied by following the decomposition of alkoxyl groups at high temperatures by IR spectroscopy and by TPD. In IR experiments the alkoxyl groups were formed in zeolites and next decomposed by evacuation at 570 K (methoxyl groups) or 470 K (ethoxyl groups). The values A_d/A_0 (A_0 and A_d are intensities of alkoxyl groups IR bands before and after evacuation) expressing which fraction of alkoxyl groups survived 30 min. evacuation, as well as temperatures of TPD peak are presented in Table 2. The TPD peak temperatures and A_d/A_0 values decrease in the order NaHX > NaHY/40 > NaHY/77, indicating that the stability of O-C bond does also decrease. This effect will be discussed considering the polarization of O-C bond. The values $|q_0 - q_C|$

where q_O and q_C are charges of O and C atoms calculated from Sanderson theory (presented in Table 2). They increase in the order NaHX < NaHY/40 < NaHY/77. The increase of the O-C bond polarization (bonding alkoxyl groups with zeolite framework) results in a decrease of the stability of alkoxyl groups observed in this study. According to data presented in Tables 1 and 2, the polarization of O-C bond increases in the same order as the polarization of O-H bond in hydroxyls which were present in zeolites before alkylation. It indicates that the same factors (increase of average electronegativity of zeolite) influence the polarization of either O-H or O-C bond. The consequence of the variation of O-C bond polarization on the reactivity of methoxyl groups will be discussed in following chapters.

Reactivity of zeolitic methoxyl and ethoxyl groups with water and methanol.

The reaction of methoxyl and ethoxyl groups with water produced methanol or ethanol and restored hydroxyls (zeol-O-CH_3 + H_2O = zeol-OH + CH_3OH).

The reaction of methoxyl groups with methanol formed dimethyl ether and also restored hydroxyl groups (zeol-O-CH_3 + CH_3OH = zeol-OH + CH_3OCH_3).

The reaction of zeolitic methoxyl groups with benzene and their role in alkylation.

Two series of experiments have been done. In one series, the reaction was carried out in an IR cell, methoxyl groups were formed in NaHY/77 zeolite by the reaction of methanol at 420 K, unreacted methanol and water were subsequently removed by evacuation at the same temperature (only the bands of methoxyl groups were present upon such treatment). In another series of experiments, the reaction was carried out in a catalytic microrector by "separated impulses method": the pulse of methanol was injected into reactor (at 550-700 K) the reactor was next flushed with a carrier gas at the same temperature. In both series of experiments the "isolated" methoxyl groups were present (without unreacted methanol and water). The injection of a pulse of benzene (both into the IR cell and into reactor) resulted in the formation of toluene (Figure 1) indicating, that methoxyl groups bonded to zeolite framework can alkylate benzene. In the experiments made by "separated impulses method" (impulse of methanol followed by flushing and subsequently the impulse of benzene) the prolongation of flushing time decreased the yield of toluene (Figure 1B). It can be explained by partial decomposition of methoxyl groups during prolonged flushing at the reaction temperature.

The reaction of methoxyl groups in NaHY/77 zeolite with benzene was studied at

570-670 K. The conversion to toluene reached a maximum at 620 K. The decrease of conversion above the maximum point was due to the decomposition of methoxyl groups.

The role of zeolitic methoxyl groups in benzene alkylation was studied comparing the results obtained in experiments in which methoxyl groups were produced in NaHY/77 zeolite before benzene introduction (unreacted methanol and water were removed) and in experiments in which the mixture of methanol and benzene was introduced simultaneously (Figure 1A; a, c). The toluene yield was distinctly higher if methoxyl groups were present in zeolite before benzene injection. It indicates that the formation of methoxyl groups is an indispensable step in aromatic hydrocarbons alkylation and that the alkylation is realized by the formation of zeolitic methoxyl groups which react subsequently with benzene.

The effect of zeolite composition on the reactivity of methoxyl groups.

The effect of zeolite composition on the reactivity of methoxyl groups was studied comparing NaHY/77 and NaHX zeolites (Figure 1A; a, b). In both zeolites comparable amounts of methoxyl groups were formed by reaction with alcohols (water and nonreacted methanol were removed) and pulse of benzene was introduced. Toluene yield was much lower in the case of NaHX, indicating that methoxyl groups in NaHX were less reactive towards benzene than in NaHY/77. This effect can be due to higher stability of O-C bond as evidenced in IR and (first of all) TPD experiments (Table 2), because of lower polarization of this bond (as shown by calculations - Table 2).

Shape selectivity of NaHY zeolite with methoxyl groups.

Catalytic tests evidenced also that Y zeolites became shape selective in the alkylation of toluene to xylenes if large amounts of methoxyl groups were present. It was proved in experiments in which methanol was preadsorbed (at 623 K) before the mixture of methanol and toluene was injected (at the same temperature) into the reactor (Table 3). The p-xylene/o-xylene ratio was 0.62 - 0.70 what was typical for non-shape selective acid catalysts. This ratio increased twice (to 1.26 - 1.30) if methanol was preadsorbed. The increase of p-xylene/o-xylene ratio (from 0.7 to 2.2) was also observed in a fixed bed continuous flow microreactor (at 553 K) if the methanol content in the reaction mixture increased (from 10 to 80 mol%).

We suppose that shape selectivity of large pores Y zeolite is due to the high concentration of methoxyl groups what results in narrowing zeolite pores.

zeolite	$\Delta\nu$	S_{av}	$|q_O - q_H|$
NaHX	184	3.57	0.350
NaHY/40	296	3.76	0.360
NaHY/77	316	3.92	0.369

Table 1: The data concerning acid strength of OH groups: $\Delta\nu$ of OH interacting with benzene, average electronegativity S_{av} and $|q_O - q_H|$

	zeolite	C-H bond	O-C bond		
		$\nu_{as} CH_3$	A_d/A_0	$T_{TPD\ peak}$	$|q_O - q_C|$
methoxyl	NaHX	2968	0.60	485	0.291
	NaHY/40	2976	0.35	385	0.298
	NaHY/77	2977	0.30	345	0.303
ethoxyl	NaHX	2985	0.60	315	0.291
	NaHY/40	2994	0.35	290	0.298
	NaHY/77	2996	0.10	280	0.302

Table 2: The data concerning the properties of C-H and O-C bonds in alkoxyl groups.

toluene /methanol ratio	methanol preadsorption	p-xylene/o-xylene ratio
2	no	0.62
2	yes	1.26
4	no	0.70
4	yes	1.30

Table 3: The effect of methanol preadsorption on the shape selectivity in toluene alkylation at 623 K

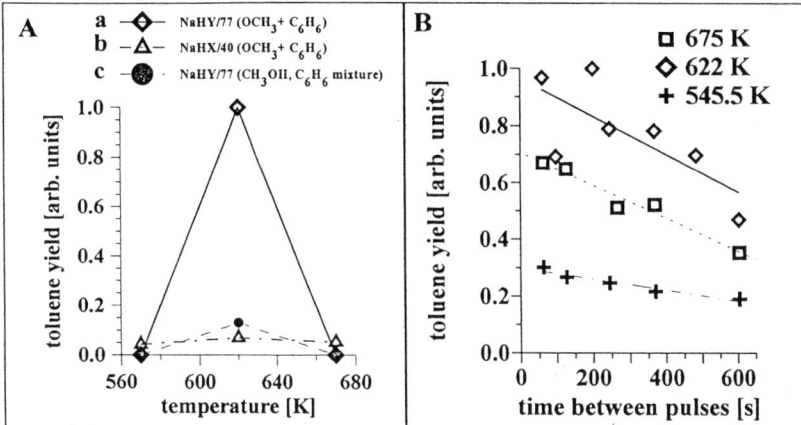

Figure 1. Toluene yield in reaction of methoxyl groups with benzene.

A - impulse of benzene was injected into an IR cell containing NaHY/77 (a) and NaHX (b) in which comparable amounts of methoxyl groups were found, or the mixture of methanol and benzene (1 : 5) was injected into the cell containing NaHY/77 (c).

B - impulse of benzene was injected into microreactor containing NaHY/77 in which methoxyl groups were formed by the impulse of methanol. Toluene yield is presented as a function of time period between impulses of methanol and benzene.

Decomposition of alkoxyl groups.

As mentioned in preceding section, the stability of alkoxyl groups was influenced by the composition of zeolites (it decreased with the average electronegativity). The thermal

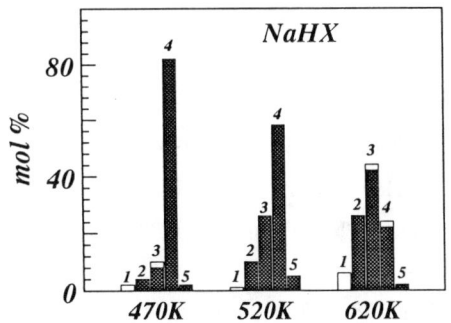
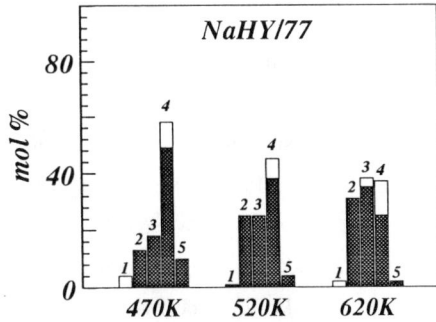

Figure 2. The products of methoxyl groups decomposition as a function of temperature:
1 - 5 = C_1 - C_5 ■ alkenes, □ alkanes

decomposition of ethoxyl groups restored hydroxyl groups and produced ethene. The decomposition of methoxyl groups also restored hydroxyl groups and produced alkenes (C_2 - C_5) as well as a smaller amount of alkanes (Figure 2). It seems that alkenes were primary products of methoxyl groups decomposition (n zeol-O-CH_3 =n zeol-OH + C_nH_{2n}). Some of them underwent disproportionation forming alkanes and coke (IR band at 1590 cm^{-1} appeared). The composition of products of alkoxyl groups decomposition and of further reactions depended on the temperature and on acidity of zeolites (Figure 2). At higher decomposition temperatures smaller hydrocarbons molecules (C_2 - C_3) were formed than at lower temperatures (mostly C_4). The contribution of alkanes increased with the acidity of zeolites, suggesting that disproportionation was catalyzed by acidic hydroxyls [9].

CONCLUSIONS

The properties of zeolitic alkoxyl groups depend on the composition of zeolites. The stretching frequency of CH_3 groups increased with the average electronegativity of zeolite. The increase of average electronegativity caused also higher polarization of C-O bond (bonding alkyl group with zeolite) and therefore lower thermal stability of methoxyl groups (as evidenced in our IR and TPD experiments) and increased their reactivity with benzene (as shown in our catalytic experiments). Catalytic tests evidenced also that large pores HY zeolite became shape selective in toluene alkylation by methanol (more p-xylene was formed) if significant amount of methoxyl groups was formed before the reaction. It seems that this para selectivity is due to narrowing of zeolite pores by methoxyl groups.

REFERENCES

1. C.D. Chang, Catal. Rev. - Sci. Eng. **25**, 1 (1983).
2. J. Rakoczy, T. Romotowski, Zeolites **13**, 256 (1993).
3. P.G. Smirniotis, E. Ruckenstein, Ind. Eng. Chem. Res. **34**, 1517 (1995).
4. P. Salvador, W. Klading, JCS Faraday I **73**, 1153 (1977).
5. C.M. Zicovich-Wilson, P. Viruela, A. Corma, J. Phys. Chem. **99**, 13224 (1995).
6. J. Datka, B. Gil, J. Catalysis **145**, 372 (1994).
7. J. Datka, E. Brocławik, B. Gil, J. Phys. Chem. **98**, 5652 (1994).
8. M. Ziółek, J. Czyżniewska (to be published).
9. J. Datka, Zeolites **1**, 145 (1981).

VIBRATIONAL SPECTROSCOPIC INVESTIGATIONS OF PYRROLE ADSORPTION IN FAUJASITES: STUDIES BY INFRARED, RAMAN AND NEUTRON SPECTROSCOPY

E. GEIDEL[1], H. JOBIC[2] and S. F. PARKER[3]

[1] Institute of Physical Chemistry, University of Hamburg, 20146 Hamburg, Germany;
geidel@chemie.uni-hamburg.de
[2] Institut de Recherches sur la Catalyse-CNRS, 69626 Villeurbanne Cedex, France
[3] ISIS Facility, Rutherford Appleton Laboratory, Chilton, Didcot, Oxon OX11 0QX, UK

ABSTRACT

The nature and strength of interaction of pyrrole adsorbed in alkali-metal cation-exchanged X and Y zeolites was investigated by vibrational spectroscopic techniques. In order to obtain information about the vibrational behaviour of the free probe, measurements by infrared, Raman and inelastic neutron scattering spectroscopy (INS) of pyrrole and some deuterated derivatives were carried out accompanied by normal mode analyses and quantum mechanical calculations. In a second step, vibrational spectra of adsorbed pyrrole were recorded. The results indicate strong host-guest-interactions of pyrrole via the NH-group to the lattice oxygen atoms and simultaneously via the aromatic system to the cations of the zeolites. To verify the experimentally obtained results, and to find out preferred adsorption sites, Monte Carlo simulation techniques were applied. The results were taken as the starting point for quantum mechanical calculations of pyrrole in interaction with cluster models of the zeolites. Finally, the inelastic neutron scattering spectrum of pyrrole adsorbed in NaY was simulated in this way. The calculated profile is in good agreement with the experimental data.

INTRODUCTION

Vibrational spectroscopic techniques (i.e. infrared (IR), Raman spectroscopy and inelastic neutron scattering (INS)) are widely applied for the investigation of adsorbed probe molecules in zeolites. In most cases the aim of such studies is the characterisation of acidic or basic centres at the internal surfaces of zeolites. Pyrrole as an amphoteric molecule is a suitable probe to characterise acidic and basic sites simultaneously. In particular the NH stretching band of adsorbed pyrrole was often used in previous infrared studies as a measure of the basic strength of zeolites, e.g. [1-3]. However, to distinguish between various adsorption sites, one spectroscopic technique alone or monitoring only one characteristic mode often yields insufficient information. Therefore, it is the aim of this study to obtain information about adsorption sites of pyrrole in alkali-metal cation-exchanged faujasites by different vibrational spectroscopic techniques and to verify the experimentally obtained results by classical force field and quantum mechanical calculations.

EXPERIMENTAL AND COMPUTATIONAL PROCEDURE

The parent X and Y samples (NaX: Si/Al=1.18 and NaY: Si/Al=2.6), both from Chemie AG Bitterfeld-Wolfen (Germany), were threefold ion-exchanged in aqueous solutions of KCl, RbCl and CsCl at 353 K for 8 h for each exchange followed by washing and drying.

For IR studies, samples were pressed into self-supporting wafers (for measurements in the ranges 4000-1200 cm^{-1} and 400-50 cm^{-1}) and on silicon wafers (for measurements in the range 900-400 cm^{-1}) and activated at 623 K in vacuo. Probe molecules were adsorbed at room temperature followed by evacuation to desorb the physisorbed amount. The infrared spectra were recorded using a Digilab FTS-20E spectrometer with a resolution of 2 cm^{-1} ratioed against the backround of the activated zeolites. Infrared and Raman spectra of the pure probes (pyrrole, pyrrole-d_1 and pyrrole-d_5) were recorded on a PE System 2000R with a resolution of 1 cm^{-1}.

INS spectra of bulk pyrrole were obtained with the spectrometer IN1BeF at the Institut Laue-Langevin in Grenoble (France) and with the spectrometer TFXA at the Rutherford Appleton Laboratory (UK). The INS spectra of pyrrole adsorbed in NaY and RbY were recorded only on IN1BeF. The spectra were taken under the same experimental conditions as described in [4].

Normal coordinate analyses of the free probe molecules were performed by means of Wilson's GF matrix method with the force constants and the structural parameters described by Boggs et al. [5]. The force constants were fitted by a least-squares refinement to reproduce our own measured experimental frequencies. After force constant adjustment the mean deviations between observed and calculated frequencies were less than 0.85% for all three isotopomers under study.

In order to determine possible adsorption sites of pyrrole in faujasites Monte Carlo (MC) simulations using the CFF91 force field with the extensions for zeolites [6] were performed. One pyrrole molecule per crystallographic unit cell of NaY was taken as the simulation box. The computational details were described elsewhere [7].

The preferred adsorption site calculated by MC was taken as the starting geometry for quantum mechanical calculations based on Hartree-Fock (HF) and second order Møller-Plesset (MP2) methods. As finite models of the sites of interest a pyrrole molecule together with the cation (Na^+, K^+ and Rb^+) and/or together with a framework cutout ($Si_3AlO_4(OH)_8^-$ Na^+ and $Si_3AlO_4(OH)_6O_2Si_2H_6$ M^+) (HF) were taken. To simulate vibrational spectra of adsorbed pyrrole, scaling factors were transferred from ab initio calculations of the free probe.

RESULTS AND DISCUSSION

In the infrared spectra a significant downward shift of the NH/D stretching mode of pyrrole and pyrrole-d_1 from 3530 cm^{-1} and 2608 cm^{-1} (gas phase) was observed during adsorption. The observed shifts are depicted in Fig. 1 demonstrating the dependence on the kind of cation and on the Si/Al-ratio of the zeolite. As can be seen, the observed shifts follow the sequences

$Cs^+ > Rb^+ > K^+ > Na^+$ and $X > Y$. These sequences are well correlated with the basic strength sequences of the lattice oxygens [8] and indicate the direct interaction between the imino hydrogen of pyrrole with the lattice oxygen via hydrogen bonding. The larger deviations from the correlation in the case of CsX can be ascribed to the insufficient degree of ion exchange of this probe (45%).

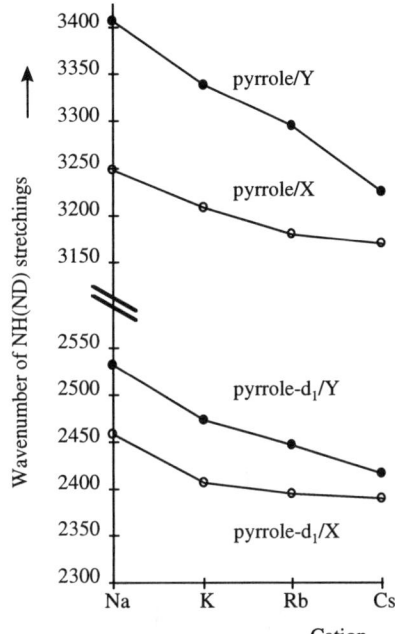

Figure 1. Observed wavenumbers of the NH/D stretching mode of adsorbed pyrrole showing the dependence on the kind of cation and the Si/Al-ratio of the zeolite.

The general trend is confirmed by the reversed trend obtained for the NH wagging mode. Due to the increasing strength of interaction, the NH wagging mode shifts from 479 cm^{-1} (gas phase, IR) to 559 cm^{-1} (liquid, IR) to 594 cm^{-1} (bulk, INS), 613 cm^{-1} (pyrrole in NaY, INS) and 620 cm^{-1} (pyrrole in RbY, INS). The INS spectra are shown in Fig. 2, the peak assigned to the NH wagging mode is indicated by an arrow.

Simultaneously, with adsorption a blue shift enhanced from Cs^+ to Na^+ zeolites was observed for bands in the range 880-740 cm^{-1} in the infrared and especially in the INS spectra (IN1BeF) of pyrrole in RbY and NaY (indicated by lines in Fig. 2). Following the results of our normal mode analyses, these bands have to be assigned to CH wagging modes of pyrrole. This effect should be interpreted as an additional interaction between the aromatic system of pyrrole with the cations. This is in agreement with results observed by temperature-programmed desorption of pyrrole, which have shown, that the effective strength of interaction depends on both the Lewis basic properties of the lattice oxygen atoms and the Lewis acidic properties of the cations [9]. The observed wavenumbers of the characteristic key bands of interest are summarised in Tab.1.

The conclusion of simultaneous interactions can further be confirmed by MC simulations. In the MC calculation of pyrrole in NaY the site of minimal interaction energy is located at a distance of 2.86 Å between the centre of mass of pyrrole and a Na^+-ion at SII site whereas the distance between the nitrogen atom and the next nearest oxygen atom (O1) was predicted to be 3.21 Å. These values are very close to the distances observed by X-ray diffraction for pyrrole in NaY (2.84 and 3.2 Å) [10]. Both - diffraction and simulation techniques - indicate strong host-guest interactions of pyrrole in NaY.

Table 1. Observed wavenumbers (cm^{-1}) for characteristic key bands of pyrrole in different states.

	NH stretching	CH wagging		NH wagging
Gas phase (IR)	3530	868, 826	721, 712	479
Liquid (IR)	3403	840	732	559
Bulk (INS) IN1BeF	-	862	740	594
TFXA	-	873, 848	755, 729	591
Pyrrole in NaY (INS)	3406 (IR)	871	742	613
Pyrrole in RbY (INS)	3294 (IR)	877	742	620

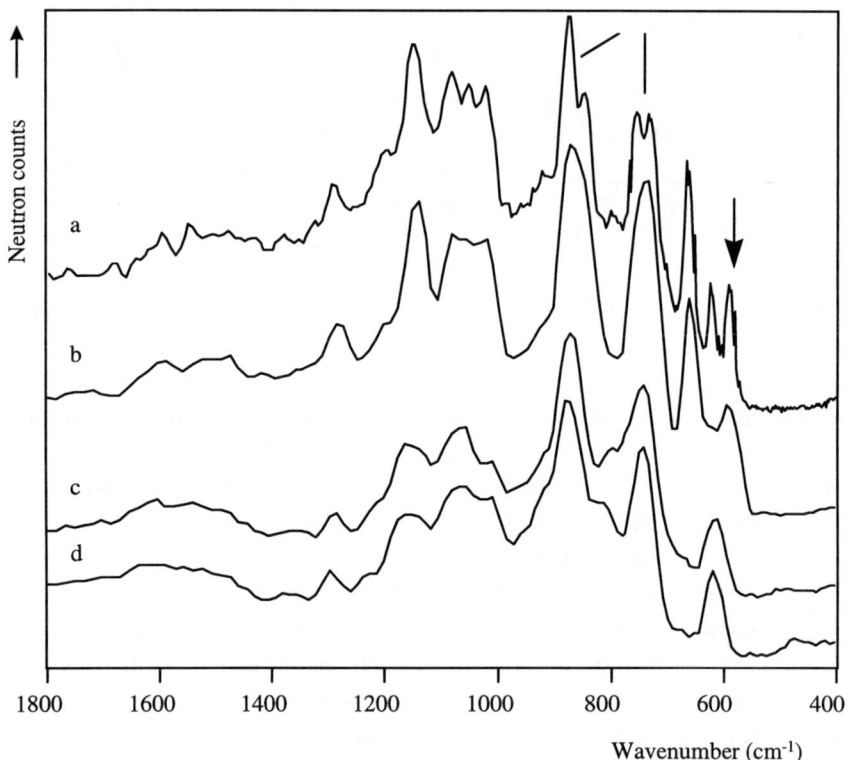

Figure 2. Experimental INS spectra of bulk pyrrole recorded with the spectrometers TFXA (a) and IN1BeF (b) and of pyrrole adsorbed in NaY (c) and in RbY (d).

Finally, the preferred adsorption site obtained by MC was taken as the starting geometry for ab initio calculations. Three kinds of models were chosen as depicted in Fig. 3. With model A the pure interaction between pyrrole and the cations (Na^+, K^+ and Rb^+) was simulated, whereas model B stands for the pure hydrogen bonding. In this case the cation at SI position (Na^+) serves only for charge compensation of the framework cutout. In model C both kinds of interaction were taken into account simultaneously together with a partial suppression of the cation charges (Na^+, K^+, Rb^+) by framework oxygens.

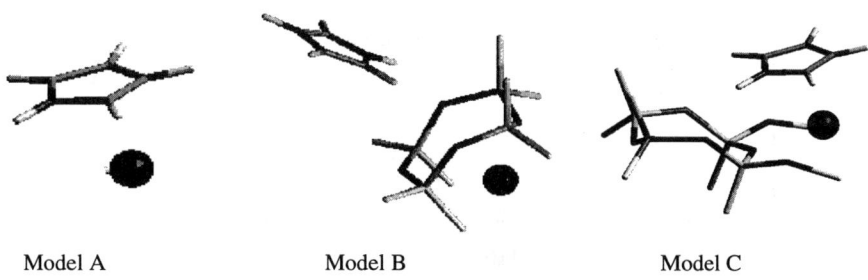

Model A   Model B   Model C

Figure 3. Models under study in quantum mechanical calculations.

For all cluster models at first geometry optimisations were carried out followed by calculations of the vibrational frequencies. A realistic description of the experimentally observed frequency shifts of adsorbed pyrrole could only be achieved for models C. From a comparison of the results for clusters of different size it follows, that obviously both kinds of interactions influence each other. So it was found, that the structure and the vibrational frequencies of the NH-group of pyrrole are not only determined by hydrogen bonding but by the kind of cation at SII site as well. The profile calculated by HF for the pyrrole + $Si_3AlO_4(OH)_6O_2Si_2H_6$ Na^+-cluster is shown in Fig. 4 in comparison with the experiment. As can be seen, the calculated profile is in good agreement with the experimental spectrum of pyrrole in NaY in the fingerprint region. In the simulated INS spectrum frequencies were scaled by a factor of 0.903, intensities were calculated from displacement vectors for the hydrogen atoms. The prediction of NH/D stretching frequencies is not as good because of the known overestimation of hydrogen bonding by the HF-method on the one hand and the transfer of the uniform scaling factor from the spectrum of the gas phase on the other.

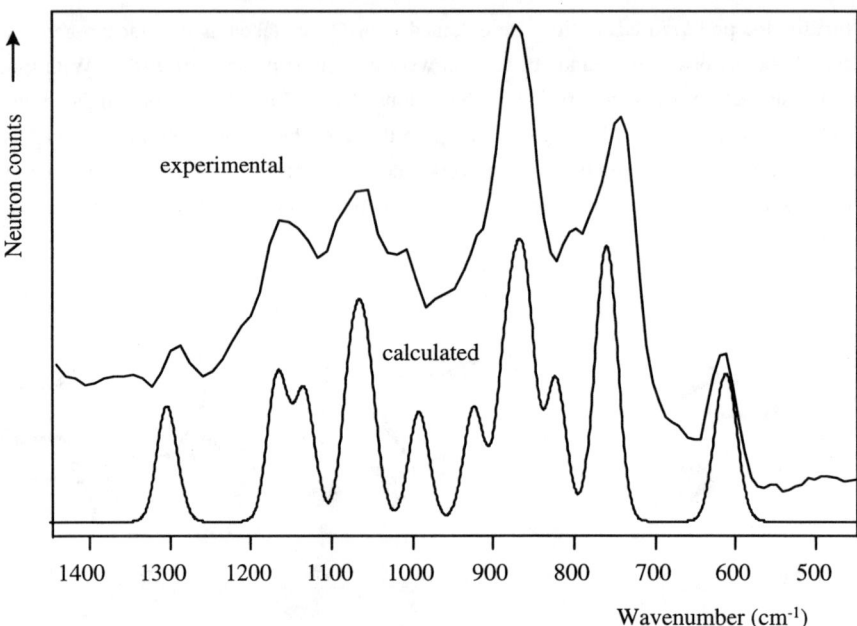

Figure 4. Comparison of experimental (IN1BeF) and calculated inelastic neutron scattering spectra of pyrrole adsorbed in NaY.

REFERENCES

1. P.O. Scokart and P.G. Rouxhet, Bull. Soc. Chim. Belg. **90**, 983 (1981).
2. M. Huang and S. Kaliaguine, J. Chem. Soc., Faraday Trans. **88**, 751 (1992).
3. D. Murphy, P. Massiani, R. Franck and D. Barthomeuf,
   Stud. Surf. Sci. Catal. **105**, 639 (1997) (and references therein).
4. H. Jobic and A.N. Fitch, Stud. Surf. Sci. Catal. **105**, 559 (1997).
5. Y. Xie, K. Fan and J.E. Boggs, Mol. Phys. **58**, 401 (1986).
6. J.-R. Hill and J. Sauer, J. Phys. Chem. **99**, 9536 (1995).
7. E. Geidel, K. Krause, J. Kindler and H. Förster,
   Stud. Surf. Sci. Catal. **105**, 575 (1997).
8. R. Heidler, G.O.A. Janssens, W.J. Mortier and R.A. Schoonheydt,
   Microp. Mater. **12**, 1 (1997).
9. E. Geidel, K. Krause, O. Klepel and B. Hunger,
   9. Deutsche Zeolithtagung, Po52, Halle (Germany) 1997.
10. E. Geidel, C. Kirschhock, R. Heidler, R.A. Schoonheydt, O. Klepel and
    B. Hunger, 10. Deutsche Zeolithtagung, Bremen (Germany) 1998.

NOVEL FREQUENCY RESPONSE TECHNIQUES FOR THE STUDY OF THE KINETICS OF HETEROGENEOUS CATALYSIS

I.R. HARKNESS, M. CAVERS, L.V.C. REES, J.M. DAVIDSON and G.S. McDOUGALL

Chemistry Department, University of Edinburgh, King's Buildings, West Mains Road Edinburgh, EH9 3JJ, Scotland, U.K.,

ABSTRACT

This work demonstrates the design of novel apparatus and its use to apply frequency response techniques to sorption and diffusion processes over zeolites under flow conditions. This approach has the potential to yield rate constants for adsorption, desorption and ultimately *elementary reaction steps* at atmospheric pressure. In order to achieve this aim an integrated DRIFTS/mass spec. frequency response flow reactor has been designed and constructed. The infra red and mass spectrometers are synchronised and a phase reference is provided by the modulations in an inert marker gas in the mass spec. By using the mass spectrometer in conjunction with a tube reactor it has been demonstrated that good quality oscillations can be produced, controlled and monitored over >2 orders of magnitude in frequency. This has been used to study the adsorption and diffusion of propane in silicalite-1. The complete system has been used to study the adsorption-desorption of CO on cation sites in zeolite Na-Y. Clear modulations in the intensity of the DRIFTS absorption bands are detected simultaneously with modulations in the mass spectrum.

The apparatus has thus been demonstrated to work well with simultaneous synchronised measurement of the dynamic responses of the gas and adsorbed phases.

INTRODUCTION

Frequency response (FR) is the study of the response of a system at equilibrium to a periodic perturbation and, in particular, the dependence of this response on the frequency of perturbation. It is an established method for the study of process systems [1], and is applicable to any system which can be described by a set of differential equations linear in the perturbation. In the limit of a small perturbation (in pressure, volume, or temperature) the kinetic equations describing a gas/zeolite system can be linearised and therefore frequency response methods can be applied. FR has been applied successfully to the study of diffusion [2] and sorption [3] in zeolites. The methods of FR can however also be applied to the dynamic response of a catalytic reaction when subjected to periodic perturbation. This potentially rewarding field is under-developed and the aim of this work has been to extend the applicability of FR methods to the study of zeolite catalysis. The end goal of this work is to carry out frequency response studies of catalysed reactions which would enable the extraction of elementary step rate constants under reaction conditions as has been demonstrated theoretically [4]. This would be of academic interest as well as being of great use in the modelling of industrial processes. Transmission infra red of pressed discs has been used previously to monitor adsorption-desorption FR [5] but DRIFTS allows powdered samples to be used thus removing macropore diffusion artefacts.

Propane adsorption and diffusion in silicalite-1 is a system well characterised both by frequency response [6] and by other methods [7] and is thus a suitable trial system to study.

IR spectroscopy of CO adsorbed on cation sites in zeolites is an established method of characterisation [8] and the combination of this with the dynamic information available from frequency response will give a thorough description of the sorption site.

EXPERIMENTAL

The experimental apparatus is depicted in Figure 1.

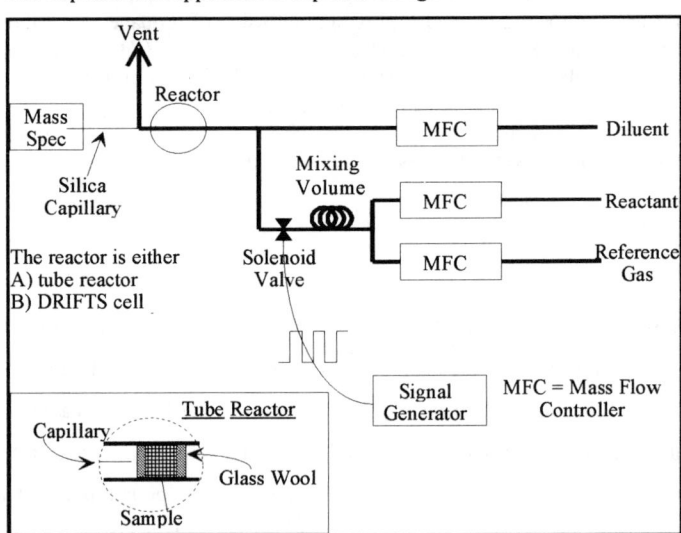

Figure 1. Schematic of the apparatus (Inset shows internal view of tube reactor)

By driving the solenoid valve with a square wave generator, flow modulations can be produced in the reactant and reference gas stream. This is then injected into a much larger constant flow of diluent gas. The resultant gas stream has an approximately constant volumetric flow rate but an oscillating composition. With appropriate settings of the signal generator the small amplitude (<10%) changes necessary to fulfil the linearity condition are produced. The maximum frequency detectable is dependent on the downstream flow characteristics but can be in the order of 3-5 Hz if care is taken to reduce tubing lengths and cross sections. The minimum frequency is determined by the long term stability of the flow rates and is <0.02 Hz. Therefore, more than 2 orders of magnitude of frequency can be investigated. The phase and amplitude of the partial pressure oscillations are calculated by a fast Fourier transform. The reference gas signal is used both as a phase and amplitude reference and results calculated as an amplitude ratio and phase lag relative to this reference. Control experiments have shown that in the absence of any sample there is good agreement between the 2 signals throughout the frequency range studied.

The DRIFTS cell used in this work is an in-house constructed design depicted in Figure 2.

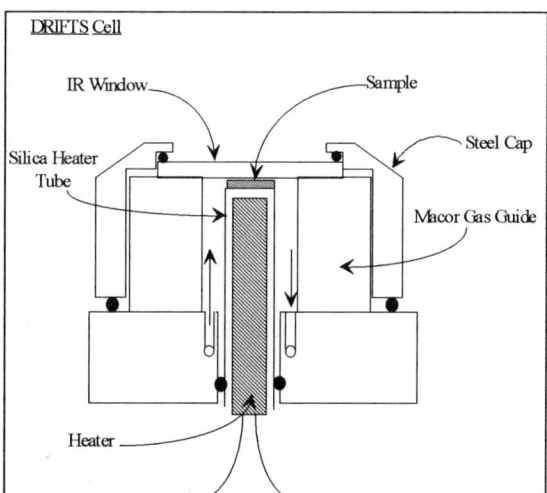

Figure 2. Cross Section of the fast response DRIFTS environmental cell

The sample ($\approx$30mg) is placed in a depression on a silica sample post. The gas enters the cell at the base and passes up a slot in the gas guide next to the sample post. It then flows across the sample and down a similar slot to exit. Due to the small dead volume of the cell the residence time is very low. It has been estimated from transient experiments to be ≈ 150ms, which is far better than commercially available cells. Sample heating is carried out by a heater coil inserted in the silica sample post. This can produce sample temperatures, measured by a thermocouple placed in the sample bed, in excess of 450°C without degradation of the seal. The low path length of the infra red radiation through the gas phase means that even at partial pressures of up to 70 torr, no interfering gas phase signal is observed; only signals from adsorbed species.

THEORY

The sample is treated as a column of particles in which Fickian diffusion for an isotropic sphere is assumed. Therefore in the column (neglecting gas phase diffusion)

$$-v\left(\frac{\partial C}{\partial z}\right) + \frac{3}{R}(1-\varepsilon)\left[-D\left(\frac{\partial q}{\partial r}\right)_{r=R}\right] = \varepsilon\left(\frac{\partial C}{\partial t}\right)$$

where C = propane gas concentration, q=adsorbed propane concentration v=gas velocity, z=column coordinate, r=particle radial coordinate, R=size of particles, ε=fractional void space, t=time

Taking Laplace transforms

$$-v\left(\frac{\partial \overline{C}}{\partial z}\right) + \frac{3}{R}(1-\varepsilon)\left[-D\left(\frac{\partial \overline{q}}{\partial r}\right)_{r=R}\right] = s\varepsilon\overline{C}$$

This is an *ordinary* differential equation and thus can be solved.

$-D\left(\frac{\partial \overline{q}}{\partial r}\right)_{r=R}$ can be obtained from solution of the standard equations for Fickian diffusion

Since the perturbation is harmonic, the Laplace variable can be replaced by Iω

$$\frac{Ar}{C_3H_8} = \left[1 + \frac{I}{\omega\varepsilon R}\left(\left(3D(\varepsilon-1) + Ie^{(-I\omega L/v)}\left(3\frac{D}{R}(1-\varepsilon)-\frac{I}{\omega}\chi\right)\omega\varepsilon R\right)\right)\right]e^{I\omega L(\varepsilon-1)/v}$$

$$\chi = -\frac{K}{R} - IK\coth\left(\sqrt{\frac{-I\omega R}{D}}\right)\sqrt{\frac{-I\omega R}{D}}$$

where ω=angular frequency of perturbation, L= length of sample bed and K = gradient of the isotherm at the equilibrium conditions

The amplitude ratio is equal to the magnitude of the complex transfer function whereas the phase lag is equal to the argument.

RESULTS AND DISCUSSION
Propane on silicalite-1

The influence of adsorption/desorption processes on the propane wave form is immediately evident from Figure 3. Whereas the reference gas (argon) retains it square wave character after passing through the sample bed, the propane wave becomes more rounded. The relative amplitude of the propane modulation is also reduced and there is a clear phase lag relative to the argon wave.

Both of these observations are in agreement with expectations. Recently published theoretical work [9] suggests that the amplitude ratio (reference/propane) should decrease with increasing frequency, as is observed with frequency response under batch conditions. However, in all experiments the opposite was observed, namely an increase in amplitude ratio as the frequency was increased. This is depicted in Figure 4. In addition to this unexpected behaviour, phase lags in excess of $\pi/2$ were observed, these cannot be explained if the sample is treated as a uniform discrete adsorber. Both these observations can be explained if the sample is considered to be acting as a column of finite length, despite the small amount of sample used. A number of new models have been developed to account for this behaviour (See above). Results from one of these models are shown in Figures 4 and 5. As shown above the new model reproduces the trend in amplitude well. There is also reasonable agreement with the phase lags measured experimentally but there appears to be a consistent deviation at high frequency (Figure 5). One parameter in the model is the fractional voidage in the column, the calculated values for this are generally unrealistically high. However, if the sample length is replaced by an effective length equal to the real length multiplied by a tortuosity factor, the quality of the fit is unaffected but a more reasonable physical picture of what is happening in the sample is produced, i.e. the sample is acting as a column (despite being only 1-2mm deep) because the gas path is tortuous.

Reliable fitting of the derived model to experimental data has proved difficult. Often multiple minima exist and the best method of fitting has been found to be Monte-Carlo methods. Various versions of the model have been tested based on 2 different treatments of

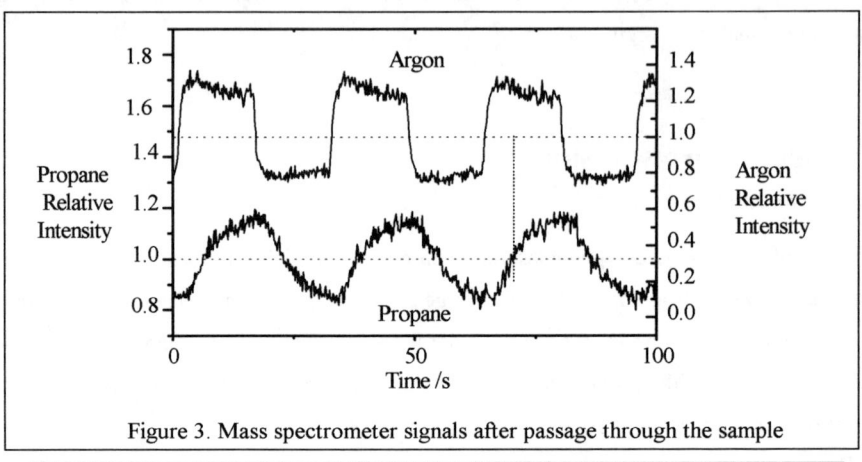

Figure 3. Mass spectrometer signals after passage through the sample

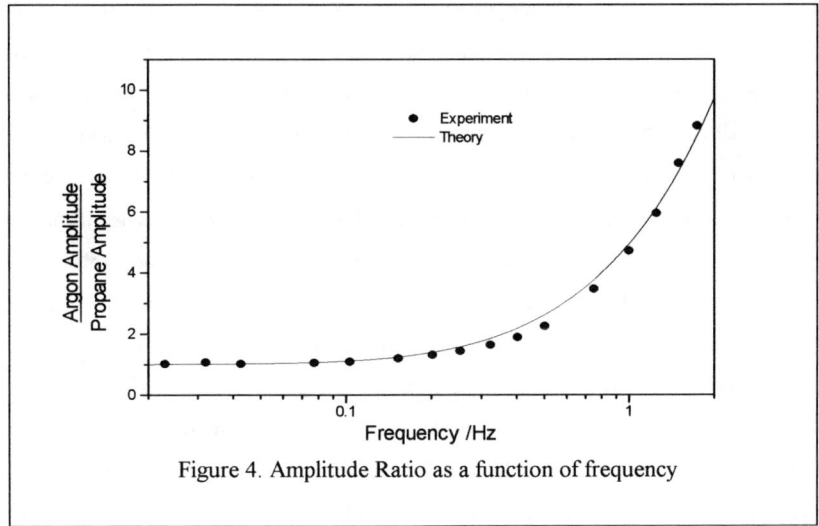

Figure 4. Amplitude Ratio as a function of frequency

the behaviour of the adsorbate in the particles following previous work by Do [9] and Ruthven [10]. Both models give very similar curves and neither are able to model the high frequency phase behaviour well. Inclusion of a film resistance only slightly improves high frequency fitting. The values of the diffusion coefficient produced by the models differ from each other but are both in reasonable agreement with previous macroscopic measurements of propane in silicalite-1 [7] (e.g. $3 \times 10^{-8} cm^{-2} s^{-1}$ at 348K). These values are lower by 2-3 orders of magnitude than PFG NMR values [7] which are themselves in good agreement with the most recent frequency response measurements under batch conditions [6]. However, due to the uncertainty in the fitting the numerical values found thus far should be treated with caution.

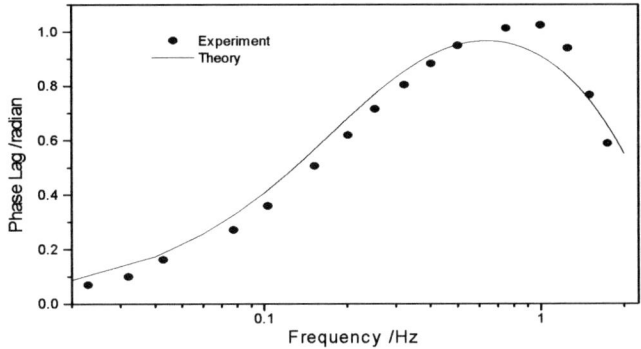

Figure 5. Phase lag relative to reference gas against frequency

CO on Na-Y

CO adsorption on Na-Y at 300-400K under flow conditions gives 2 peaks in the CO stretching region at 2163 cm^{-1} and 2120 cm^{-1} (Figure 6).

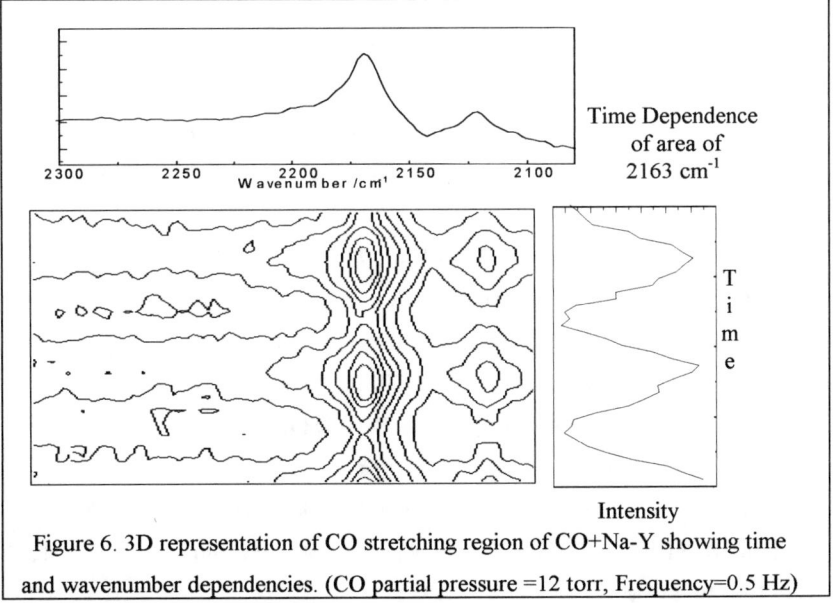

Figure 6. 3D representation of CO stretching region of CO+Na-Y showing time and wavenumber dependencies. (CO partial pressure =12 torr, Frequency=0.5 Hz)

In accordance with the literature, these are assigned to CO co-ordinated to Na^+ via the carbon and CO co-ordinated to Na^+ via the oxygen (or through oxygen and carbon) [11] respectively.

Fluctuations in both these peaks can be seen with good signal to noise and thus the dynamic response of the 2 species can be determined independently. Simultaneous to the oscillations in the DRIFT spectrum, partial pressure modulations can be observed in the mass spectrometer.

This part of the work is at an early stage and no quantitative results have been produced but the feasibility of simultaneous DRIFTS and mass spec. frequency response under flow conditions has clearly been demonstrated for the first time.

CONCLUSIONS

An integrated mass spec./DRIFTS frequency response reactor has been constructed and been shown to work well. Good quality frequency domain data is obtainable over a frequency range of 0.02 Hz to 3 Hz. With simultaneous measurement of the dynamic responses of both the gas and adsorbed phases this techniques is clearly of great potential for the study of heterogeneous catalysis.

REFERENCES

1. D.R. Coughanowr, Process Systems Analysis and Control, (McGraw-Hill, N.Y., 1991)
2. L.V.C. Rees and D. Shen, Gas Sep. and Purif., **7(2)** 83-89 (1993)
3. Gy. Oneystak, D. Shen and L.V.C. Rees, J.Chem.Soc. Farad. Trans., **92(2)** 307-315 (1996)
4. J.R. Schrieffer and J.H. Sinfelt, J.Phys.Chem., **94** 1047-1050 (1990)
5. Y.-E. Li, D. Willcox, R.D. Gonzalez, A.I.Ch.E.J. **35(3)**,. 423-429 (1989)
6. L. Song and L.V.C. Rees, Microporous Materials, **6** 363-374 (1996)
7. J. Kärger and D.M. Ruthven, Zeolites, **9** 267-281 (1989)
8. A. Zecchina and C. Otera Aréan, Chem.Soc.Rev., **25(3)** 187-197 (1996)
9. I.S. Park, M. Petkovska and D.D. Do, Chem.Eng.Sci., **53(4)** 833-843 (1998)
10. H.A. Boniface and D.M. Ruthven, Chem.Eng.Sci., **40(11)**, 2053-2061 (1985)
11. V. Bolis, B. Fubini, E. Garrone, E. Gianello and C. Morterra, Stud.Surf.Sci.Catal.,**48** 159-166 (1998)

ENTROPICALLY DETERMINED ADSORPTION PECULIARITIES STUDIED ON KFI BY TPD, MICROCALORIMETRY, ^{13}C CP MAS NMR AND FTIR

J. JÄNCHEN*, W.J.M. VAN WELL§, J.H.M.C. VAN WOLPUT§, H. STACH*

* ZeoSys GmbH (Zeolithsysteme), Haus 6.1, Rudower Chaussee 5, D-12484 Berlin-Adlershof, Germany
§ Schuit Institute of Catalysis, Laboratory of Inorganic Chemistry and Catalysis Eindhoven University of Technology, P.O. Box 513, 5600 MB Eindhoven, The Netherlands

ABSTRACT

The adsorption behaviour of some n-alkanes and methanol in KFI (ZK-5) was investigated by temperature programmed desorption, microcalorimetry, ^{13}C CP MAS NMR and FTIR. A two step desorption behaviour of n-pentane, n-hexane and n-heptane as well as a pronounced step in the adsorption isotherm of n-hexane can be correlated to an entropically determined adsorption behaviour in the ZK-5. Using the ^{13}C CP MAS NMR the sitings of the adsorbed molecules in K-ZK-5 are estimated. Whereas at low loadings small molecules such as propane and n-butane are adsorbed in both of the two types of cavities of the K-ZK-5, the longer n-pentane and n-hexane are preferentially adsorbed in the larger α-cages. At higher loadings (after a pronounced step in the isotherm but at constant heats of adsorption) the remaining part of the pore volume, including the γ-cages, are filled too. A two step desorption behaviour and a step in the isotherm of the polar methanol in H-ZK-5 can be related with the strong specific interaction of methanol and with the energetical heterogeneity of the acid sites of H-ZK-5.

INTRODUCTION

In the present paper we report about some unusual adsorption properties of ZK-5 (KFI). The KFI structure consist of a three dimensional network of 2 larger α- (diameter 11.6 Å) and 6 smaller γ-cages (6.6x10.8 Å) per unit cell. The cavities are connected with each other through 8 membered rings.[1,2] Many of those rings contain a cation in the K-form of ZK-5.[3] As a consequence of the difference in size of the two kinds of cavities and the cation positions, differences in the adsorption behaviour as function of the adsorbed amount for molecules with variation in the chain length can be expected. Further, energetic heterogeneity of the structure and the polarity of the sorbate molecules can cause discontinuities (steps) in the adsorption isotherms or a two stage desorption behaviour in the TPD. Thus, energetic as well as entropic reasons for this behaviour can be expected. Entropically determined adsorption behaviour, for instance, has been recently reported for n-paraffin adsorption in the MFI structure.[4,5] In this study we use temperature programmed desorption (TPD), microcalorimetry, ^{13}C CP MAS NMR and FTIR spectroscopy to get more insight into the adsorption properties of some n-paraffins and methanol in the KFI structure.

EXPERIMENTAL

The heats of adsorption of n-alkanes were measured at 303 K on a SETARAM microcalorimeter connected with a balance (TG-DSC 111). The pulse technique via sample loop and a defined sorbate concentration by a saturator in a nitrogen gas stream (of in total 0.9 l/h) was used to attain small stepwise increasing loadings of the sample. The heat curve of methanol was measured at 303 K on the SETARAM C80 calorimeter connected with a volumetric adsorption apparatus. Additional, adsorption isotherms of n-hexane and methanol were measured at 298 K with a common McBain quartz spring balance. Temperature programmed desorption was performed on the TG-DSC 111 equipment after saturation of the sample at 303 K with the corresponding adsorbate with a heating rate of 5 K/min to a temperature of 673 K in a nitrogen stream of 0.9 l/h.

The ^{13}C Cross Polarised (CP) Magic Angle Spinning (MAS) NMR measurements were carried out on a Bruker MSL 400 spectrometer at room temperature with a MAS spinning rate of 2.5 kHz. The contact time for the CP was 1 ms while the repetition time was 5 s. Defined adsorbed amounts on the samples (15 to 30 mg) were obtained volumetrically on a separate vacuum system using spinner fitting small glass tubes. The IR spectra of KFI were taken on a Bruker FTIR-spectrometer (IFS 113v) equipped with a vacuum cell. Self-supporting discs with a thickness of 7.5 mg/cm^2 were used with different adsorbed amounts of methanol.

The KFI sample with a Si/Al ratio of 3.3 was obtained from EXXON Chemical Europe and was used in the as synthesised potassium form and, after ion exchange, in the H-form. The ion exchange was carried out three times followed by careful calcination up to 670 K. The high micropore adsorption capacity of methanol in the H-ZK-5 of 0.28 cm^3/g (from isotherm measurements) compares well with the free pore volume of ZK-5 of 0.3 cm^3/g verifying good crystallinity of both samples. The calorimetric titration of the acidic sites (mainly H$^+$) gives the same concentration as Al in the lattice suggesting a very high ion exchange degree.

RESULTS AND DISCUSSIONS

Table 1 summarises the results evaluated from the temperature programmed desorption measurements. As can be seen from the table the maximum loadings, at the given relative

Adsorbate	a in mmol/g	a in cm³/g	Relative pressure	Temperature in °C	
methanol	3.95	0.16	0.15		140
n-pentane	1.42	0.165	0.15	95	167
n-hexane	1.15	0.15	0.15	58	167
n-heptane	0.64	0.094	0.15	48	163
methanol/H-ZK-5	6.16	0.25	0.15	112	218

Table 1 Maximum loadings (a) of different sorbate molecules in K-ZK-5 and temperatures in the TPD profiles at which differential mass loss is at maximum

pressure, for methanol and n-pentane are quite comparable suggesting that these molecules fill up the pore system in a comparable way. The amount of n-hexane is somewhat lower but for n-heptane the adsorbed amount is significant lower pointing to the fact that parts of the whole structure are not filled at the given conditions.

The second part of the table (last column) shows the temperatures at which the differential mass

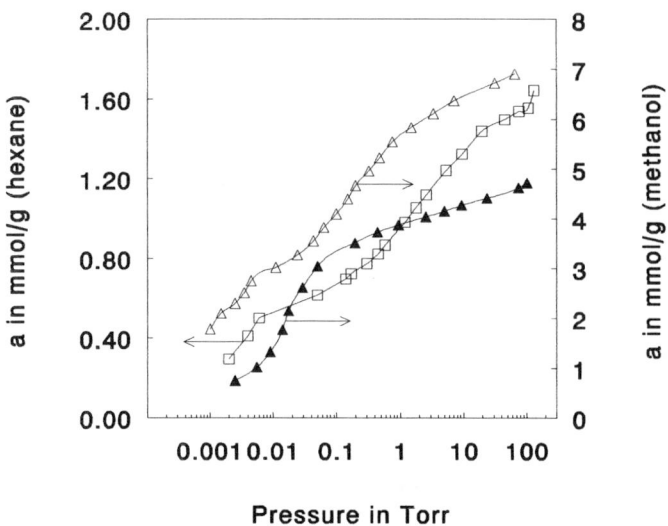

Figure 1. Adsorption isotherms of methanol in H-ZK-5 at 303 K (△), methanol in K-ZK-5 at 298 K (▲) and n-hexane in K-ZK-5 at 298 K (□)

loss is at maximum in the TPD profiles. A pronounced two-stage desorption is found for n-pentane, n-hexane, n-heptane and methanol/H-ZK-5 characterised in the table by two desorption maximum temperatures while the relatively small polar methanol on K-ZK-5 as well as propane and n-butane (not shown in the table) desorb in a single step. This is in good agreement with the shape of the adsorption isotherms of methanol (no step in K-ZK-5, step in H-ZK-5) and n-hexane (pronounced step) in Figure 1. The isotherms have been proved to be reversible. Thus, the adsorption behaviour of the n-paraffins may be structure related due to the two types of cavities in ZK-5 and the size of the sorbate molecules.

To prove this hypothesis ^{13}C CP MAS NMR spectra were measured of n-alkanes with different chain length (C3 to C6) and two different loadings of the KFI. Since the carbon chemical shift depends on the medium or environment where the molecule is located the sitings of the sorbate molecule can be studied.[6] The zeolite RHO was used as a reference since it has only α-cavities comparable in size with the α-cavities in KFI. The α-cavities in KFI are also accessible through 8-ring windows.

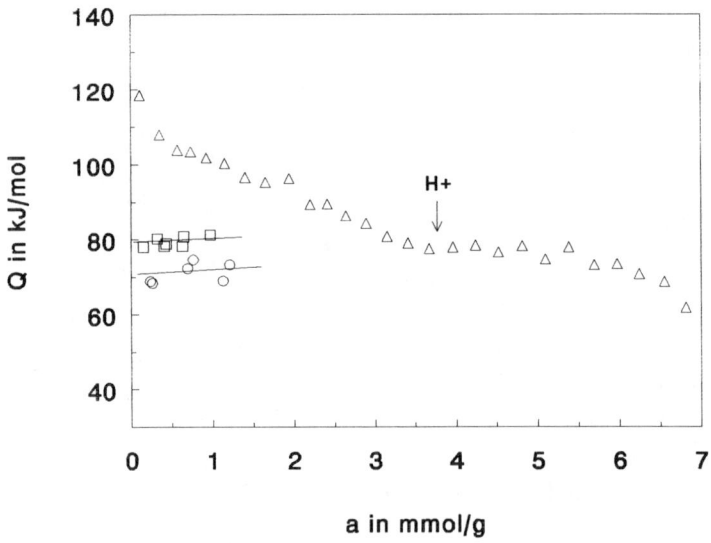

Figure 2. Differential molar heats of adsorption of methanol in H-ZK-5 (Δ), n-hexane in K-ZK-5 (□) and of n-pentane in K-ZK-5 (O).

In principle, one signal for the CH_3 and another signal for the CH_2 group(s) can be expected in the ^{13}C NMR of adsorbed propane and n-butane. A third signal must appear for n-pentane and n-hexane, due to the different surrounding of their inner CH_2 group(s). This was found for zeolite RHO upon adsorption of about 0.4 mmol/g of the mentioned paraffins: propane, CH_3 15.1 /CH_2 16.1 ppm; n-butane, 13.0 /25.3 ppm; n-pentane, CH_3 13.1 /CH_2 22.6 /CH_2 34.6 ppm, n-hexane: CH_3 12.7+13.1 /CH_2 22.4 /CH_2 31.3 ppm. Very comparable spectra with single signals for each C were found for n-pentane and n-hexane upon adsorption of 0.4 mmol/g in the K-ZK-5 (n-pentane: 14.0/23.0/35.9 ppm, n-hexane: 13.6/23.0/32.0 ppm). However, the increase of the adsorbed amount, for instance of hexane to 1.1 mmol/g (compare step in the isotherm in Figure. 1), results in a splitting or broadening of the signals due to adsorption in the smaller cages of the ZK-5: CH_3 14.4+18.8 /CH_2 23.4+25.5 /CH_2 32.6: broadening from 0.8 to 2.2 ppm. The same result was found for n-pentane. For n-butane two CH_2 signals and one broad CH_3 signal are found at all loadings in K-ZK-5. Since no such a splitting of the CH_2 signal was observed in RHO, we can conclude that n-butane is at all loadings distributed over both types of cages. The same conclusion holds for propane. Thus, propane and n-butane desorb in a single-step and n-pentane and n-hexane show a two-stage desorption TG profile.

However, from the heats of adsorption of n-C_5 and n-C_6 (Figure 2) one would not expect a two-stage desorption, since they are almost independent from the loading and do not show a discontinuity which is in good agreement with results for H-ZK-5.[7] Further, the decreasing peak temperature of the first peak with increasing chain length of the paraffins (Table 1) is unexpected too. Because of the increasing heats of adsorption with rising C-number (see Figure 2) one would expect otherwise. Both observations can be explained in terms of entropy determined adsorption taking into account basic thermodynamics given by

$$\Delta G = \Delta H - T \Delta S.$$

A great loss in entropy can over-compensate an increasingly negative heat of adsorption which results in a less negative free enthalpy of adsorption.

It should finally be noted that the occurrence of two different sets of NMR signals indicates that the exchange of molecules between the two types of cavities through the 8-ring windows in ZK-5 is slow. This confirms the existence of a large barrier (about 100 kJ/mol) for diffusion of linear butene molecules through the 8-ring window as calculated in ref. 8.

Different to the n-alkanes and the methanol (see ref. 9) in K-ZK-5 the heat curve of methanol in the H-ZK-5 shows a stepwise decrease. Thus, energetical heterogeneity contribute to the two-stage desorption profile and should be the reason for the step in the isotherm. As could be expected, the FTIR spectra confirm that the methanol is first strongly bonded on the acidic OH, preferably on the HF band (3610 cm^{-1}) and later on the weaker LF band (3570 cm^{-1}). The total number of H^+ (about 3.8 mmol/g) corresponds with the amount of stronger bonded molecules in the heat curve (see Figure. 2).

CONCLUSIONS

It can be concluded that strong specific interaction of methanol with the acidic OH and energetic heterogeneity of the H+ in H-ZK-5 leads to a two stage desorption pattern in the TG as well as to a step in the isotherm. The non-specific adsorption of n-hexane in K-ZK-5 results in the same TG pattern and shape of the isotherm but for entropic reasons.

Different to propane or n-butane, n-hexane adsorbs preferentially in the larger cavities with up to about 4 molecules per unit cell (0.6 mmol/g). The additional 6 molecules, to fill the entire pore volume (1.6 mmol/g), can only be adsorbed after increasing the equilibrium pressure (step in the isotherm) corresponding with high losses of entropy. Consequently, the desorption of these less strongly bonded molecules (less negative free enthalpy, ΔG) with rising temperature in the TG occurs first followed by desorption of the 4 stronger adsorbed molecules.

ACKNOWLEDGEMENT

The supply of the K-ZK-5 sample by J.P. Verduijn (EXXON Chemical Europe, Machelen) is gratefully acknowledged.

REFERENCES

1. W.M. Meier, D.H. Olson and Ch. Baerlocher, Atlas of Zeolite Structure Types, 4th ed. (Elsevier, London, 1996), pp. 118-119.
2. G.T. Kerr, Science **140**, 1412 (1966).
3. J.L. Lievens, J.P. Verduijn and W.J. Mortier, Zeolites **12**, 690 (1992).
4. W.J.M. van Well, J.P. Wolhuizen, B. Smit, J.H.C. van Hooff and R.A. van Santen, Angew. Chem. Int. Ed. Eng. **34**, 2543 (1995).
5. D.H. Olson and P.T. Reischman, Zeolites **16**, 434 (1996).
6. W.J.M. van Well, X. Cottin, J.W. de Haan, B. Smit, G. Nivarthy, J.A. Lercher, J.H.C. van Hooff and R.A. van Santen, J. Phys. Chem. B **102**, (1998)
7. F. Eder and J.A.Lercher, J. Phys. Chem. B **101** 1273 (1997).
8. F. Jousse, L. Leherte and D.P. Vercauteren, Mol. Simul. **17**, 175 (1996).
9. S.G. Izmailova, I.V. Karetina, S.S. Khvoshoev, M.A. Shubaeva, J. Colloid Interface Sci. **165**, 318 (1994).

CATALYTIC SIGNIFICANCE OF STRONG ACID SITES IN DEALUMINATED FAUJASITES AND MORDENITES

I.V. MISHIN, T.R. BRUEVA and G.I. KAPUSTIN

N.D.Zelinskii Institute of Organic Chemistry RAS, Russia; igo@ioc.ac.ru

ABSTRACT

By the removal of aluminium from zeolites, new very strong acid sites are generated that are characterised by the heats of adsorption of NH_3 higher than 115 kJ/mol for faujasites and higher than 140 kJ/mol for mordenites. The number of strong sites plotted against the mole fraction of the framework aluminum is described by the curves with a maximum. The position of the maximum corresponds to the framework composition with the 50% dealumination. Catalytic activity for the cracking of octanes and disproportionation of ethylbenzene follows the course of strong acidity with a maximum for the samples dealuminized to ~50%. The similarity of the curves shows that the topological approach to acidity is valid for mordenites.

INTRODUCTION

The topological model developed to explain the acidity of fajasites emphasises that the site isolation has a major effect on the appearance of strong acidity [1]. An earlier work [2] suggested that the initial dealumination of mordenite may remove a pair of neighboring aluminum atoms sharing one four-member ring to increase the fraction of «isolated» ($0NNN$) AlO_4 tetrahedra. A concomitant increase in the strength of acid sites was indicated later by microcalorimetric measurements [3]. However in a number of recent publications the applicability of this approach to mordenites is denied and the formation of strong acid sites on aluminum removal questioned [4,5]. In order to resolve the controversy around this problem, the knowledge of acid-site strength distribution in zeolites is needed. In this work, an attempt is made to use microcalorimetry to determine the distribution of acid sites in faujasites and mordenites with similar framework compositions. In addition to the acidity data, the reactivity of acid sites is supplemented by catalytic measurements.

EXPERIMENTAL

To prepare decationated samples, the NaY and NaM zeolites were equilibrated with a 1N solution of NH_4NO_3 followed by calcination at 500°C. High-silica zeolites were prepared by treating Na-faujasites with $SiCl_4$ vapor and refluxing small port Na-mordenites in solutions with varying concentration of hydrochloric acid. In this work, the decationated samples are denoted by symbols HY and HM, dealuminated zeolites are designated by DY and DM, while the Si/Al framework ratio is indicated by the subscript number. Thus the dealuminated mordenite with the Si/Al ratio of 10 will be designated by a symbol DM_{10}. Differential heats of adsorption were measured at 300°C by using a Calvet type microcalorimeter connected to the adsorption unit. Catalytic activity in the cracking of n-octane and isooctane (2,2,4-trimethylpentane) was evaluated in a micropulse unit at 400°C. All the mordenite samples gradually dropped in activity during the course of the reaction. Correspondingly, the apparent rate constants (kK), calculated from the initial conversions according to the Bassett and Habgood equation [6] were taken as a measure of activity. Ethylbenzene disproportionation was performed in a stainless steal tubular fixed-bed reactor at 150°C and atmospheric pressure. The rate of disproportionation served as a measure of activity. For the decationated mordenite HM_5 the reaction rate was calculated from the initial conversion, whereas for other samples the reaction rates were derived from the conversions measured at steady-state [7].

RESULTS

Acid-site strength distribution

The plots of differential heats of adsorption *vs* coverage for the hydrogen forms of zeolites are described by stepwise curves [8]. The profile of the curves is accounted for by the presence of sites with different strengths. By differentiating the curves one can derive the acidity spectra, which clearly show how modification of zeolites affects the number and strength of acid sites. Analysis of these data can be used to understand the effect of framework composition on the nature of acid sites

Y zeolites. The distribution of acid strength in faujasites is given in Table 1. Calorimetric measurements imply that all the sites adsorbing NH_3 with q>90 kJ/mol can be related to the acid sites. We were able to define three groups of sites with different acid strengths (with q=90-100 kJ/mol, q=100-115 kJ/mol and q=115-130 kJ/mol) and determine the corresponding number of acid sites for every group. The evidence for existence of different acid strengths was found in the careful analysis of the stepwise calorimetric curves.

The main portion of acid sites in the $HY_{2.4}$ zeolite adsorbs ammonia with the heats of 100-115 kJ/mol and stronger sites could not be identified in the sample. After dealumination, the total number of acid sites decreases. By comparing these data with the results obtained for the $DY_{6.5}$ faujasite it can be concluded that weak acid sites with q < 100 kJ/mol are preferentially removed. The aluminum deficient Y zeolites have, however, stronger (q>115 kJ/mol) sites than the parent zeolites. Based on the data for faujasites with Si/Al ratios of 17 and 42, the increase in the strength of acid sites can be seen up to a limiting value of N_{Al}= 25-30 atoms per u.c., after which an increase in acidity strength could not be observed. That means that with the advanced dealumination, only strong sites are extracted. Accordingly, the total number of acid sites decreases linearly with the reducing density of the framework aluminum, whereas the dependence of the number of the strong acid sites (q=115-130 kJ/mol) on the aluminum content is described by the curve with a maximum at Si/Al = 5-7, corresponding to ~50% dealumination (Figure 1).

Sample, Si/Al	Number of sites (mmol/g) with q (kJ/mol)			
	>90	90-100	100-115	115-130
$HY_{2.4}$	3	1.2	1.8	
$DY_{6.5}$	1.5	0.3	0.85	0.35
DY_{17}	0.3	0.15		0.15
DY_{42}	0.05			0.05

Table 1. Distribution of acid sites in dealuminated faujasites

Mordenites. Similar trends are found for evolution of acidity in mordenites. The acid sites in mordenites can be combined to give the groups with q = 90-115, 115-145, 145-165 and 165-180 kJ/mol. In the HM_5 sample, acidity is represented by sites with q = 90-145 kJ/mol, the total number of these sites makes up about 80% of the framework aluminum. Again, at the early stage of dealumination, weak acid

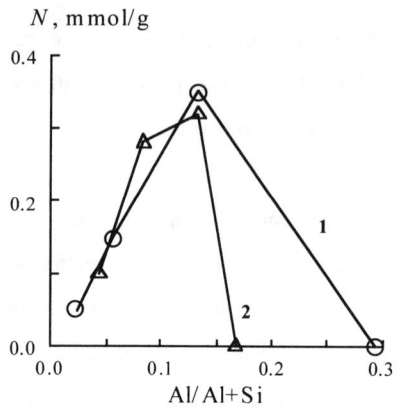

Figure 1.

The effect of the framework composition on the number of strong acid sites in the high-silica zeolites

(1 - faujasites; 2 - mordenites)

sites with q = 115-145 kJ/mol are readily removed, while new strong sites with q = 145-180 kJ/mol appear. Beyond a dealumination level of 50% (N_{Al}=3.5-4), an increase in the strength of acid sites can not be detected and the reduction in density of acid sites occurs at the expense of strong acidity. As for faujasites, removal of the second half of the framework aluminum does not generate new strong acid sites. Accordingly, the plot of the number of strong acid sites with q = 145-165 kJ/mol against the Al content is represented by the curve with a maximum corresponding to ~50% dealumination (Figure 1). Thus, a decrease in the total number of acid sites and appearance of super acid sites are the main features characterising the evolution of acidity with progressive dealumination.

Sample, Si/Al	Number of sites (mmol/g) with q (kJ/mol)				
	>90	90-115	115-145	145-165	>165
HM_5	1.85	0.24	1.14	-	-
$DM_{6.3}$	1.05	0.25	0.49	0.11	0.21
DM_{10}	0.8	0.12	0.4	0.12	0.16
DM_{27}	0.32		0.22	0.1	

Table 2: Distribution of acid sites in dealuminated mordenites

Acidity and catalytic activity. Figure 2 shows the relationship between the activity in the cracking of *n*-octane and the Al/Al+Si ratio in faujasites and mordenites. The size of *n*-octane molecule (0.47 nm) is essentially smaller than the dimensions of windows leading to the large cages of faujasite (0.75 nm) and to the 12MR mordenite channels. Accordingly, the rate of the cracking would depend, primarily, on the number and the strength of acid sites rather than on the rate of diffusion. Since the total number of acid sites in faujasites and mordenites over the composition range of Al/Al+Si < 0.15 is nearly the same, it can be therefore expected that the reaction rates for the unbranched linear molecules would be very sensitive to the strength of acid sites.

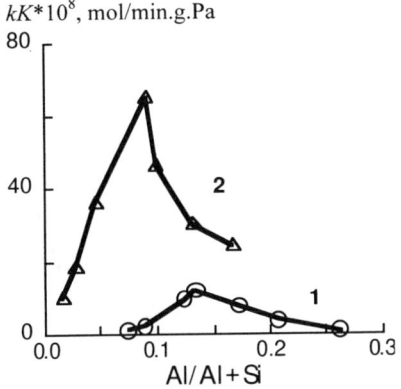

Figure 2.

Catalytic activity of high-silica faujasites (1) and mordenites (2) in the cracking of n-octane

Reaction studies demonstrate that mordenites crack n-octane 5-6 times faster than faujasites. The differences in the reaction rates between faujasites and mordenites can be explained by the presence of stronger sites in mordenites. This can be interpreted to mean that strong acid sites are enhanced in dealuminated mordenites compared with faujasites with the same aluminium content. In particular, the sites adsorbing NH_3 with the heats higher than 120 kJ/mol are absent in $HY_{2.4}$, whereas in the HM_5 mordenite they account for 70% of total acidity. In faujasites, the sites with q>120 kJ/mol can be observed only in dealuminated samples, while extraction of aluminum from the mordnite framework creates the sites with q>145 kJ/mol. Nearly 50% of acidity in the DM_{27} mordenite is represented by the sites with q>130 kJ/mol that can not be found in dealuminated faujasites. Enhanced reaction rates in the cracking of n-octane shown by mordenites can apparently be attributed to the presence of stronger acid sites.

As can be seen in Figure 2, the activity of mordenites and faujasites increases with rising degree of dealumination, passes through a maximum and then decreases upon further dealumination. Both curves follow the course of very strong acid sites and the maximum in the curves of the catalytic activity corresponds to the maximum of the strong acidity. The position of the maximum corresponds to the values of Al/Al+Si =0.15 for faujasites and 0.09 for mordenites. These values are close to the chemical composition of the frameworks containing the highest number of «isolated» aluminum atoms. The frameworks with these Si/Al ratios are typical for faujasites and mordenites with the extent of dealumination near 50%. The coincidence of the position of maximum on the curves relating strong acidity and activity to the framework composition implies an important catalytic role of acid sites associated with the «isolated» aluminum atoms.

The size of isooctane molecule (~0.7 nm) is comparable with the dimensions of the 12MR in mordenite. Consequently, faujasites are much more active in the cracking of i-octane than mordenites [8]. In agreement with the changes in acidity, progressive dealumination of Y zeolites results first in an increase and then in a decrease in catalytic activity. The crucial role of the aluminum removal is clearly demonstrated for zeolites treated with $SiCl_4$ (Figure 3). The cracking activity doubles, when the aluminum content in zeolites produced by direct synthesis decreases from 60 to 50 Al/u.c. For comparison, after extraction of 10 more Al atoms from the Y zeolite with $SiCl_4$, the value of kK increases by nearly one order of magnitude. The activity reaches a maximum when about 25 Al/u.c remains in the framework. This shows that the activity of the faujasites increases 8-10 times as the N_{Al} values decrease from 50-60 for the parent hydrogen forms to 20-25 Al per u.c. for the samples with a dealumination degree of 50%. With the further increase in the aluminum density, the catalytic activity declines. A similar trend can be observed for the variation of the strong acidity with progressive dealumination. On the other hand, the data in Figure 3 indicate that the catalytic activity over mordenites is limited by the accessibility of active sites rather than by their intrinsic acidity.

In Figure 4, the reaction rates for the disproportionation of ethylbenzene over faujasites and mordenites are plotted against Al/Al+Si ratios. The size of ethylbenzene molecule (0.67 nm) is only slightly less than the windows in the 12MR of mordenite. Correspondingly, faujasites show higher reaction rates than mordenites over a broad range of Al/Al+Si ratios. The activity of faujasites and mordenites in this reaction

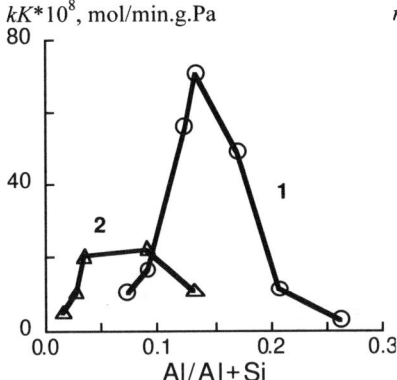
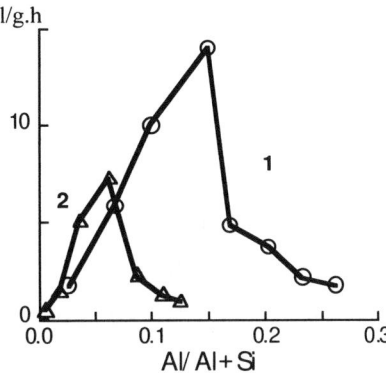

Figure 3. Catalytic activity of high-silica faujasites (1) and mordenites (2) in the cracking of isooctane

Figure 4. Catalytic activity of high-silica faujasites (1) and mordenites (2) in the ethylbenzene disproportionation

increases 3-4 times for the samples with a dealumination degree of 50% and then declines with the further decrease in the Al/Al+Si ratios. The results presented in Fig.4 illustrate the importance of the "isolated" Al atoms in forming the active surface of mordenites.

In a number of works a suggestion is made that the dramatic increase in the catalytic activity caused by dealumination can be related to elimination of extraneous material or/and to the increase in the micropore volume making diffusion of reacting molecules more facile. However earlier measurements of adsorption of benzene on dealuminated mordenites [3] indicate that when even a small amount not exceeding 20% of the framework aluminium atoms is removed by dealumination, the adsorption of benzene in micropores considerably increases. At small extents of dealumination, the amount of benzene adsorbed at low relative pressures increases 1.5 times indicating an essential increase in the micropore volume during dealumination. When the extent of dealumination increases, the adsorption of benzene in the micropores gradually decreases. At the same time the highest activity in hydrocarbon conversions is exhibited by zeolites dealuminated to 50%. Moreover, the drastic increase in catalytic activity of faujasites treated with $SiCl_4$ vapor can be attributed only to the increasing strength of acid sites since this procedure of dealumination remains the pore structure of Y zeolites intact.

CONCLUSIONS

The data outlined above demonstrate the catalytic significance of strong acid sites created by aluminium removal from zeolites. The impact of dealumination on the behaviour of topologically different zeolites is similar since the results obtained with hydrocarbon molecules varying in size and configuration are similar. The activity reaches a maximum when about 50% of the initial Al remains in the framework. For all three reactions a maximum in the catalyst activity occurs when the framework aluminium ions have no second nearest neighbour aluminum. This is in good agreement with the maximum in the concentration of strong acid sites. Consistent with the finding of this study and the work [9], it is proposed that the topological model of acidity is valid for the mordenite type zeolites. In addition, mordenites are much more active in the cracking of *n*-octane than faujasites over the whole range of composition. In the absence of steric effects on the reaction rates this observation illustrates an important role of strong acidity in the catalytic performance of zeolites.

REFERENCES

1. Beamont R., Barthomeuf D., *J.Catal.*,**30**, 228, (1973).
2. Klyachko A., Mishin I., *Neftekhimia*, **30**,339, (1990).
3. Mishin I., Brueva T., Kapustin G., *Kinet.Catal.*, **38**, 574, (1997).
4. Datong D., Pingshuan S., Quinghua J., *et al.*, *Zeolites*, **14**, 65, (1994).
5. Takahashi T; Kato M., Itabashi K., *J. Phys. Chem.*, **98**, 5742, (1994).
6. Bassett, Habgood, *J. Phys. Chem.*, **64**, 769, (1960)
7. Karge H., Ladebeck J., Sarbak Z., Hatada K., *Zeolites*, **2**, 94, (1982).
8. Mishin I., Klyachko A., Brueva T., *et al.,Kinet.Catal.*, **34**, 502, (1993).
9. Stach H., Janchen J., Jerschkewitz H.-G., *et al., J.Phys.Chem.*, **96**, 8480, (1992).

DISTRIBUTION OF ACID SITES IN ZEOLITES OF DIFFERENT TYPES BASED ON MICROCALORIMETRIC MEASUREMENTS

G.I KAPUSTIN, T.R. BRUEVA and I.V. MISHIN

N.D.Zelinskii Institute of Organic Chemistry RAS, Russia; gik@ioc.ac.ru

ABSTRACT

A Calvet type microcalorimeter connected to a volumetric system was used to measure the differential heats of adsorption of ammonia at 573K, and to evaluate the acidity of various zeolites (Y, Mor, ZSM-5, ERI) modified by decationization, dealumination, and ultrastabilization. The adsorption heats plotted against filling show stepwise curves indicating the energetic heterogeneity of zeolite surfaces. The strength of acid sites depends primarily on the chemical composition of zeolite frameworks and, in the case of mordenites and erionites on the framework topology. The strongest B-sites in Y and ZSM-5 associated with the isolated Al atoms adsorb ammonia with the heats of 120-130 kJ/mol, whereas in mordenites and erionites these sites are characterized by the differential heats that are ≈30 kJ/mol higher.

INTRODUCTION

The relation between the framework composition, topology and acidity of zeolites is important for understanding the intrinsic activity of zeolite catalysts. The technique of adsorption calorimetry contributes to our knowledge of the distribution of acid sites in zeolites and can provide an approximate basis for correlation between physical parameters and catalytic properties of zeolites.

EXPERIMENTAL

The isotherms and the heats of adsorption were measured at 573 K by a Calvet type microcalorimeter connected with a volumetric system. Before measurements all samples were evacuated for 40 h at 753 K. Adsorption of ammonia on zeolites is accompanied by a slow evolution of the heat. It was shown that this process caused by migration of the adsorbed molecules to the strongest sites, is observable for several days in the experiments conducted at 303 K whereas at 573 K it can be traced only for several hours. The variation of the adsorption heats with progressive filling represented by the stepwise curves can be related to the energetic heterogeneity of the adsorption sites.

RESULTS

Figure 1a shows the heats of adsorption of ammonia at 573 K on HY zeolite with Si/Al = 2.4 and on Y zeolites, dealuminated with $SiCl_4$ vapor. The parent Y zeolite adsorbs NH_3 with the initial heats < 112 kJ/mol. On this sample, the heats of adsorption of the main portion of NH_3 (1.8 mmol/g) are larger than 90 kJ/mol. After dealumination, the total number of acid sites decreases, so that not more than 0.05 mmol/g was adsorbed with the heats higher than 90 kJ/mol by the sample with a Si/Al ratio of 42. The DY zeolites are characterized by stronger (q > 112 kJ/mol) acid sites than the parent faujasite. An increase in the strength of acid sites can be observed until the framework composition reaches a limiting value of N_{Al} = 25 - 30 Al atoms per u.c., beyond which an increase in the acidity strength could not be more found. Accordingly, the dependence of the number of strong acid sites with q = 120- 140 kJ/mol on the aluminum content is described by a curve with the maximum at Si/Al = 5-7, corresponding to the 50% dealumination. In agreement with the changes in acidity, initial dealumination of Y zeolites results in the increasing activity in the cracking of isooctane while with the advanced aluminum removal catalytic activity decreases [1].

Earlier we have showed [2] that the heats of adsorption of ammonia on acid sites exceed 80-90 kJ/mol. These sites can be due to Brönsted (B) or Lewis (L) acidity. The results obtained on dehydroxylating H-MOR serve to clarify the chemical nature of acid sites. It might be argued that the q values of 110-160 kJ/mol can be ascribed to the adsorption on B-sites and those with q= 100 and 170-180 kJ/mol can be related to L-sites. This assignment was supported by IR-data [3]. It was concluded that the heat effects with q=170-180 kJ/mol accompany dissociative chemisorption of NH_3, on L-sites, whereas the heats of 100 kJ/mol characterize the adsorption on L-sites without dissociation. When NH_3 was adsorbed on B-sites the NH_4^+-ions were formed with heat effects of 110-160 kJ/mol

Appearance of acid sites with q =120-130 kJ/mol in DY zeolites can be readily explained using a simple model, which implies that the strongest acid sites are associated with the «isolated» AlO_4 tetrahedra. Based on the data shown in Figure 1b, the same acid sites are formed by ultrastabilization of Y zeolites.

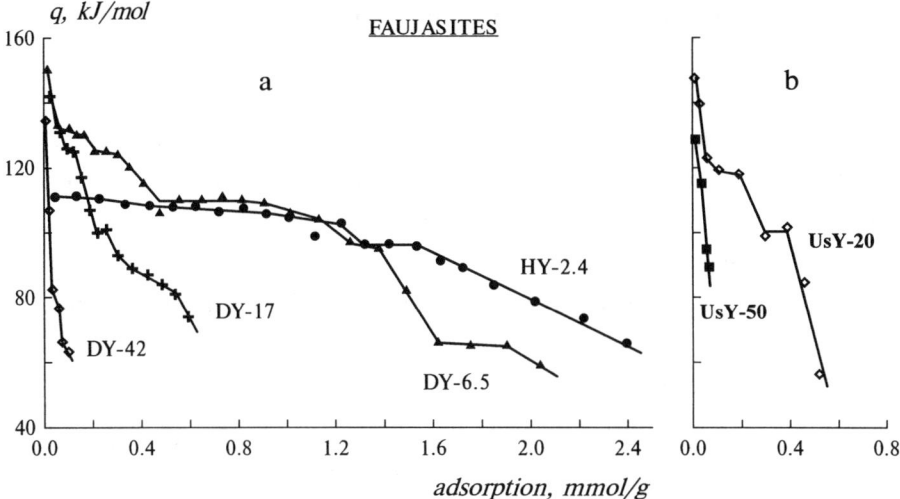

Figure 1. The heats of NH_3 adsorption on faujasites with different Si/Al ratios.

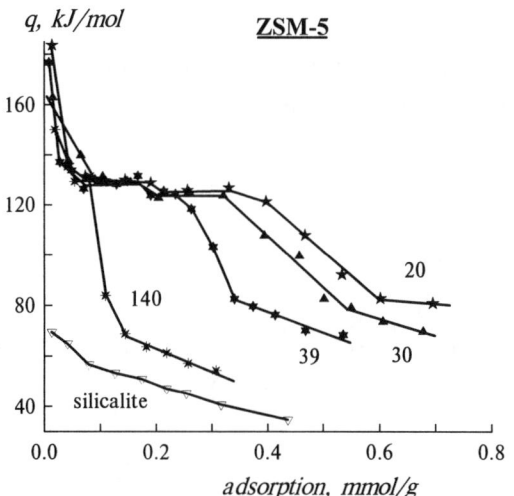

Figure 2. The heats of NH_3 adsorption on ZSM-5 zeolites of different Si/Al ratio.

ZSM-5 zeolites are synthesized with Si/Al > 14 and all the Al atoms should be located in the «isolated» AlO_4 tetrahedra. This assumption is supported by the data in Figure 2 that indicate that the sites of the same nature with q = 130 kJ/mol can be readily recognized in all ZSM-5 samples independent on the Si/Al ratio. High values of the initial heats imply the adsorption on L-sites. The heats on silicalite were measuered at 303 K.

The data in Figure 2 show a fairly good reproducibility of calorimetric measurements. Moreover, two different levels of heat values related to the adsorption on B-sites can be distinguished which we attribute to different sittings of NH_4^+-ions in a zeolite lattice.

From similarity between the heats of adsorption on Y and ZSM-5 zeolites it is tempting to conclude that the strength of acid sites is determined only by Si/Al ratio rather than by the zeolite topology. However, the data obtained for the adsorption of ammonia on MOR and ERI suggest important role of the framework structure in the interaction of ammonia with acid sites.

Thus, Figure 3a shows that the small port H-MOR with Si/Al = 6.5 has the acid sites with the heats of 120-140 kJ/mol. After dealumination with a HCl solution, the fraction of this acid sites decreases but the sites with q up to 160 kJ/mol appear. As in the case of FAU, an enhancement in the initial heats on MOR is related to the «isolated» Al atoms in the zeolite lattice. That indicates that MOR contains stronger acid sites than FAU and ZSM-5. Apparently, higher q values are caused by the adsorption in narrower 8-ring pockets of MOR. However, according to the earlier observation, [4], the B-sites in MOR with q = 150 kJ/mol are several times more active in the cracking of *n*-octane than acid sites in ZSM-5. It appears that higher heats of adsorption are determined by stronger acid sites of mordenite rather than by larger energies of the dispersion interactions inside the mordenite channels.

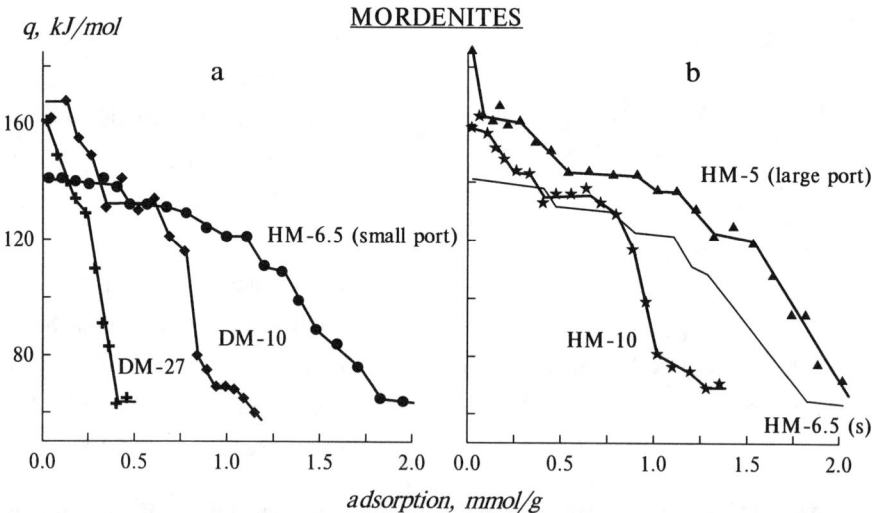

Figure 3. The heats of NH_3 adsorption on mordenites with different Si/Al ratios.

Recent work on configuration matrix indicates a possibility of arranging «isolated» Al atoms in the low silica mordenite with Si/Al=5 avoiding Al-O-Si-Al-bounds[5]. The data on the heats of adsorption of NH_3 on the H-form of a Norton MOR (large port) confirm such structure since the strong B-sites with q > 150 kJ/mol were also found in this sample. (Fig.3b). It is obvious that the conditions of synthesis significantly affect the Al distribution in the zeolite lattice. The same strong B-sites are present in the high-silica mordenite prepared by the direct synthesis with Si/Al = 10 (Fig. 3b). The presence of strong B-sites seems to be important for the understanding of adsorption properties and catalytic behavior of MOR, in particular, those of NH_4-forms [6]. The complete decomposition of NH_4^+-ions leading to formation of these sites occurs at temperatures higher than 550°C. If the temperature used for pretreatment of a NH_4-form is sufficiently lower, the residual ammonium ions located in the large channels of MOR can prevent the adsorption of aromatic hydrocarbons. Thus, the adsorption of C_6H_6 on the NH_4-form of a Norton zeolite at $p/p_s = 0.1$ achieves the maximum value of 0.083 cm³/g only after the sample was heated at 923 K, while values as low as 0.031 and 0.041 cm³/g were found for the samples calcined at 693 and 753 K, respectively. The sample of $NH_4MOR_{6.5}$ contains no such strong sites with q=150-160 kJ/mol, and the maximum adsorption of C_6H_6 is attained already after heating at 753K.

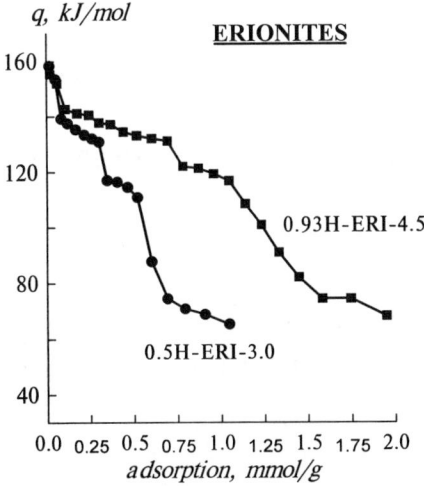

Figure 4. The heats of NH_3 adsorption on erionites

For erionite, the heats of adsorption of NH_3 were measured for two samples exchanged to 50 and 93%. As can be seen from Figure 4, the samples have three type of B-sites with q = 155, 130-140 and 115-120 kJ/mol. Thus ERI has the same strong acid sites as MOR. The number of the sites adsorbing NH_3 with q > 90 kJ/mol accounts for 30-50% of the total number of the sites vacated by cations.

The difference can be explained by the presence of weak nonacid sites in the hydrogen form of ERI, that adsorb NH_3 with lower heats. In dealuminated ERI the same strong acid sites as in MOR can therefore be expected

CONCLUSIONS

The strength of acid sites is determined not only by the Si/Al ratio but can be affected by the framework topology of zeolites. Ammonia appears to be a suitable test-molecule to study acidity of zeolites using adsorption calorimetry. The acid strength distribution obtained by this technique is in good agreement with the catalytic performance of zeolites.

ACKNOWLEDGMENT

The authors are indebted to Prof. Ariel Klyachko who has initiated the pioneering works on application of microcalorimetry to adsorption and catalysis in the Zelinskii Institute of Organic Chemistry in Moscow.

REFERENCES.

1. Mishin I.V., Klyachko A.L., Brueva T.R. et al., *Kinetics and Catalysis*, 1993, vol. 34, no. 3, p. 502
2. Rukhadze A.D., Kapustin G.I., Brueva T.R. et al., *Kinetics and Catalysis*, 1981, vol. 22, no. 2, p. 474.
3. Kapustin G.I., Kustov L.M., Glonti G.O., et al., *Kinetics and Catalysis*, 1985, vol. 26, no. 3, p. 706.
4. Klyachko A.L., Kapustin G.I., Brueva T.R., Rubinstein A.M.., *Zeolites*, 1987, vol. 7, p.119
5. Takahashi T.and Kato M., *J. Phys. Chem.*, 1994, vol. 98, p. 5742.
6. Mishin I.V., Brueva T.R., Kapustin G.I., et al., *Kinetics and Catalysis*, 1997, vol. 38, no. 3, p. 417

PRINCIPLE FOR GENERATION OF ACIDITY IN Y ZEOLITE FOUND BY AMMONIA TEMPERATURE-PROGRAMMED DESORPTION: STOICHIOMETRIC GENERATION OF ACID SITES WITH A CONSTANT STRENGTH BY ISOLATED FRAMEWORK Al ATOMS

HIROFUMI IGI, NAONOBU KATADA and MIKI NIWA

Department of Materials Science, Faculty of Engineering, Tottori University, Koyama-cho, Tottori 680-8552, Japan; katada@chem.tottori-u.ac.jp

ABSTRACT

Water vapor treatment method of temperature-programmed desorption (TPD) of ammonia was applied to Y zeolite in order to remove the unnecessary *l*- (lower temperature) peak and to measure the acidic property from the *h*- (higher temperature) peak. The determined acid amount was in agreement with the stoichiometric generation of acid site by isolated framework aluminum atom. The strength of acid site (adsorption heat of ammonia) ascribed to the framework aluminum was $ca.$ 110 kJ mol^{-1} with a few kilojoules per mole of distribution, but the significant influence of the extra-framework aluminum was observed.

INTRODUCTION

Y-type zeolite has widely been used as a solid-acid catalyst especially for the cracking of hydrocarbon in the refinery process of petroleum, and the analysis of acidic property, namely the number, strength and nature (Brønsted or Lewis) of the acid sites, must be important to interpret the catalytic functions of Y zeolite. Temperature-programmed desorption (TPD) of ammonia is a convenient method to quickly determine the number and strength of acid sites on zeolites, as reviewed [1]. However, this method has several problems which often mislead the interpretation.

TPD spectrum has usually two desorption peaks, named *l*- (lower) and *h*- (higher temperature) peaks, as shown below. The latter is ascribed to ammonia or ammonium cation adsorbed on the acid site, while the former is due to the weakly-held ammonia and unnecessary to analyze the acidic property [2]. The peaks seriously overlap in the case of such a weakly acidic zeolite as Y-type. Therefore, the acidic property of this most important zeolite in the petroleum refinery process has not been interpreted exactly from the TPD spectrum. The removal of *l*-peak is necessary to measure the acidic property.

Chester et al. [3, 4] and Bagnasco [5] reported that the introduction of water vapor into the carrier gas removed the *l*-peak. The introduced water vapor probably replaces the weakly-held ammonia.

We applied this principle to Y zeolite, and the replacement of the unnecessary species of ammonia by water molecule was confirmed. The obtained TPD spectra were first analyzed by a method to calculate the adsorption heat of ammonia and its distribution [6]. By these improved methods, the acidic property of Y zeolite was quantitatively determined with varying the compositions, and the principle of generation of acidity was clarified.

EXPERIMENTAL

The samples shown in Table 1 were used in the present study; the HNa-type samples were prepared from the Na-type by the ion-exchange in an ammonium nitrate solution at 353 K followed by calcination at 813 K for 4 hours in flowing nitrogen. As a comparison, H-mordenite was prepared from Na-mordenite (a reference catalyst JRC-Z-M15, Catalysis Society of Japan) under the same conditions.

The equipment and procedure of ammonia TPD were detailed in our review [1]. After evacuation at 773 K, zeolite (0.1 g) was exposed to ammonia (13.3 kPa) at 373 K for 30 minutes. After evacuation for 30 minutes, water vapor treatment was carried out, i.e., the sample was exposed to water vapor (ca. 3 kPa, vapor pressure at room temperature) at 373 K for 30 minutes, followed by evacuation for 30 minutes; the treatments were twice repeated in standard procedure. The adsorbed ammonia was desorbed in helium flow (0.044 mmol s^{-1}) under 13.3 kPa of the total pressure with 10 K min^{-1} of the heating rate from 373 K, and analyzed by a mass-spectrometer (ULVAC, UPM-ST-200P). The fragment with $m/e = 16$ was used to quantify ammonia.

Table 1: Y zeolite samples used.

No.	Note	Si / Al$_2$ molar ratio	Na / Al molar ratio	[Al] - [Na] mol kg^{-1}
1	JRC-Z-HY4.8*	4.8	0.01	4.84
2	Na-Y(5.0)**	5.0	1.0	0.00
3	ion-exchanged from 2	5.1	0.90	0.43
4	ion-exchanged from 2	5.0	0.66	1.53
5	ion-exchanged from 2	5.2	0.65	1.52
6	ion-exchanged from 2	5.1	0.61	1.72
7	ion-exchanged from 2	5.2	0.54	2.02
8	ion-exchanged from 2	4.7	0.38	2.96
9	ion-exchanged from 2	5.3	0.27	3.24
10	ion-exchanged from 2	5.0	0.24	3.53
11	ion-exchanged from 2	5.0	0.10	4.24
12	Na-Y(5.5)**	5.5	1.0	0.00
13	ion-exchanged from 12	6.0	0.81	0.74
14	ion-exchanged from 12	5.7	0.78	0.89
15	ion-exchanged from 12	5.5	0.76	0.99
16	JRC-Z-HY5.6*	5.6	0.26	3.16

*: Reference catalysts from Catalysis Society of Japan.
**: from Catalysts & Chemicals Ind.

In order to show the crystal structure, powder X-ray diffraction (XRD) was recorded by a Rigaku Miniflex-plus diffract meter with Cu-kα X-ray source (30 kV, 15 mA). Adsorption isotherm of nitrogen was measured at 77 K. Infrared (IR) spectrum was collected on the self-supporting disk with 10 mm of the diameter molded from *ca.* 7 mg of the zeolite in an *in-situ* cell by a JASCO FT/IR-5300 spectrometer.

RESULTS

Figure 1 (a) shows the TPD spectrum observed over H-mordenite after the adsorption of ammonia followed by evacuation at 373 K. The spectrum had well-separated *l*- and *h*-peaks, as previously shown [6]. After the adsorption of ammonia, the water vapor treatment was carried out; the zeolite was exposed to the water vapor for 30 minutes at the same temperature (373 K) and then evacuated for 30 minutes. After the water vapor treatment, the *l*-peak became quite small, as shown in Figure 1 (b). By 2 - 3 times repeating the introduction of water vapor and evacuation, the *l*-peak was almost completely removed, while the *h*-peak was not affected (c and d). Thus obtained profile could be fitted well with the curve (e) simulated based on the theoretical equation [6]. After the water vapor treatment, the fragment with $m/e = 18$, showing the desorption of water, was observed at 450 - 500 K, where the *l*-peak had been observed before the water vapor treatment. These findings suggest that the water vapor replaced the ammonia species corresponding to the *l*-peak.

On the Na-mordenite, only the *l*-peak was observed. The water vapor treatment completely removed the peak (spectrum not shown).

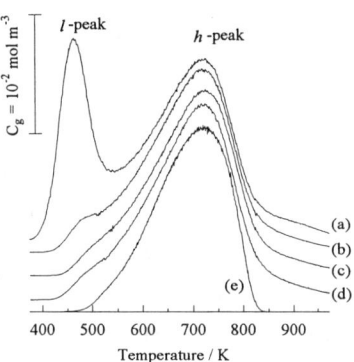

Figure 1. TPD spectrum over H-mordenite ($Si/Al_2 = 15.0$, $Na/Al = 0.04$) after the adsorption of ammonia (a), followed by the repetition of the introduction of water vapor and evacuation (b: 1, c: 2 and d: 3 times repetition).

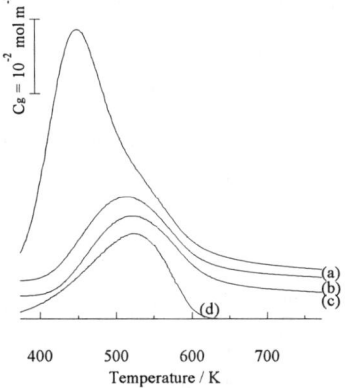

Figure 2. TPD spectrum over HNa-Y (sample 7) after adsorption of ammonia (a) followed by the introduction of water vapor and evacuation (b: 2 and c: 4 times repetition), and the simulated curve (d).

The *l*- and *h*-peaks seriously overlap in the case of Y zeolite unless the water vapor treatment was applied, as shown in Figure 2 (a). Figure 2 (b) shows that the twice repetition of the introduction of water vapor and evacuation reduced the peak in < 500 K of the temperature region. The resulted simple peak was unchanged by the further water vapor treatment (c).

The TPD profiles over the Y zeolite samples with various compositions were recorded as shown in Figure 3. The Na-Y zeolite (sample 2) showed no peak (a), and the peak appeared with the ion-exchange. From 0.9 (b) to 0.3 (d) of the Na / Al ratio, the peak became larger with the ion-exchange degree, but the complete ion-exchange made the peak small and broad (e).

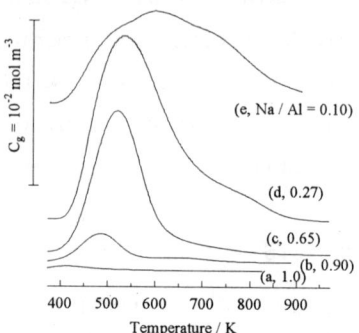

Figure 3. TPD spectra on Na-Y (a: sample 2), and HNa-Y (b: sample 3, c: sample 5, d: sample 9 and e: sample 11).

Figure 4 (a) shows the IR spectrum of the H-Y zeolite evacuated at 773 K. The isolated silanol and acidic hydroxyl groups were observed at 3745 and 3560 - 3640 cm^{-1}, respectively, and the skeletal vibrations were observed at 1860 (not shown) and 1630 cm^{-1}. After the adsorption of ammonia (b), the N-H stretching was observed at 2500 - 3500 cm^{-1}, and the H-N-H deformation bands of NH_4^+ cation (1450 cm^{-1}) and NH_3 molecule (1630 cm^{-1}, overlapping the skeletal vibration of zeolite) [7] were observed in place of the acidic hydroxyl groups. No change of the shape and intensity of the peak due to NH_4^+ at 1450 cm^{-1} was observed after the water vapor treatment at 373 K (c), showing that the water vapor could not replace the NH_4^+ cation adsorbed on the acid site. On the contrary, the NH_3 molecule (1630 cm^{-1}) was affected by the introduction of water vapor, and a new peak appeared at 1640 cm^{-1}. Although the deformation bands of H_2O and NH_3 are close,

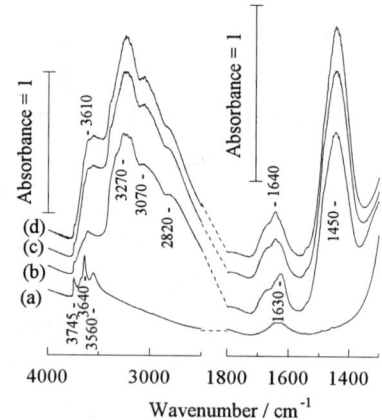

Figure 4. Infrared spectra of H-Y (sample 16) evacuated at 773 K for 1 hour (a), followed by adsorption of ammonia at 373 K (b), introduction of water vapor (c: once and d: twice repetition) at 373 K.

we suppose from the observed behavior that the generated peak is due to the H_2O molecule. Twice repetition of the water vapor treatment did not affect the spectrum furthermore (d).

The samples prepared in this study showed the XRD patterns characteristic to the faujasite (FAU) structure, but the intensity became small by the ion-exchange from sodium to proton, especially with less than 0.4 of the Na / Al ratio, as shown in Figure 5 ● and ▲. For some of these samples, the nitrogen adsorption isotherm was measured at 77 K. All of them showed typical I type isotherms, corresponding to the microporous struc-

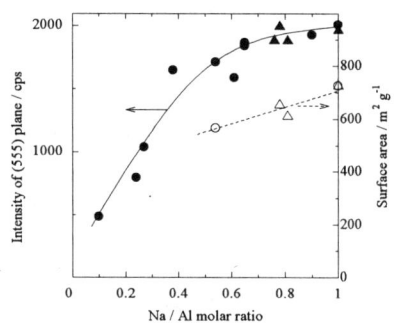

Figure 5. Intensity of X-ray diffraction of (555) plane (● and ▲) and Langmuir surface area (○ and △) against Na / Al molar ratio on Y zeolites ion-exchanged from Na-Y sample 2 (○ and ●) and sample 3 (△ and ▲).

ture. The ion-exchange decreased the Langmuir surface area also, as shown in Figure 5 ○ and △. In conclusion, the present samples with high ion-exchange ratios seem to possess partly collapsed structures containing amorphous silica-alumina and/or extra-framework aluminum.

DISCUSSION

Twice repetition of the introduction of water vapor over the acidic zeolites which had adsorbed ammonia completely removed the unnecessary l-peak, while the h-peak was not affected, as shown in Figures 1 and 2. Further repetition of the introduction of water vapor unchanged the peak shape. The Na-type zeolite without acidity showed no peak after the water vapor treatment. It is concluded that the water vapor deposition method under these conditions selectively removes the l-peak to extract the h-peak showing the acidic property.

Different two species of ammonia are suggested by these observations. One must be adsorbed on the acid site; this shows the h-peak in TPD spectrum and is hardly replaced by water. The another shows the l-peak, exists also on the Na-type zeolite, can be evacuated at such a low temperature as 373 K and is replaced by water. The IR spectroscopy showed the two species showing the absorptions at 1630 and 1450 cm^{-1}. The latter is ascribed to the NH_4^+ cation adsorbed on Brønsted acid site, presumably corresponding to the h-peak. The former has two possibilities; the NH_3 molecule coordinated to Lewis acid site or hydrogen-bonded [7]. Earl and

Lunsford et al. showed the presence of the NH_3 molecule hydrogen-bonded to the NH_4^+ cation, which had been adsorbed on the acid site of Y-zeolite [8]; in other words, an oligomer of ammonia should be formed on one acid site. On the other hand, Woolery and Chester et al. proposed that aluminum species attached to the zeolite framework with weak Lewis acidity showed the l-peak in TPD, and the ammonia coordinated to this species could be replaced by water [4]. The IR study demonstrated that the NH_4^+ cation was never affected by the water vapor treatment at 373 K, while the NH_3 molecule was replaced by the H_2O molecule, as shown in Figure 4.

Thus the efficiency of water vapor treatment

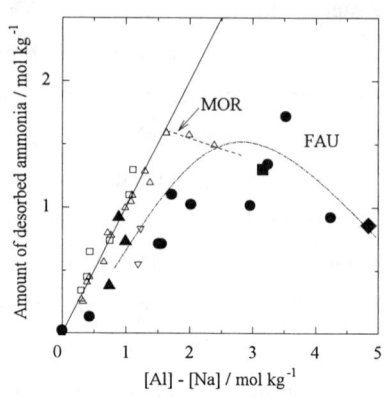

Figure 6. Plot of determined acid amount against [Al] - [Na] content over H- and HNa-type mordenite (△), ZSM-5 (□), β (◇) and Y zeolites (◆: sample 1, ●: ion-exchanged from sample 2, ▲: from sample 12 and ■: sample 16).

was confirmed, and the method was applied to the Y zeolite samples with various compositions. From the resulted peak area, the acid amount was determined as shown in Figure 6. In the low [Al] - [Na] region (< 2.5 mol kg^{-1}), the acid amount on Y zeolite was close or slightly lower than the [Al] - [Na] content, ca. 2/3 to 1 of [Al] - [Na]. This approximately agrees with the stoichiometric generation of acid site by isomorphous substitution of aluminum atom into the silicate matrix; one Al atom generates one acid site, and one Na atom neutralizes one acid site. However, this stoichiometry was not complete compared with in mordenite (△) and ZSM-5 (□), on which almost 1 : 1 relationship was observed between the acid amount and [Al] - [Na] [6]. On the present Y zeolite samples, the destruction of crystallite by increasing the ion-exchange ratio was suggested from the XRD and nitrogen adsorption isotherm, as shown in Figure 5. Probably the distorted part loses the acidity. Also on zeolite β (▽), the stoichiometry was not complete because of the low crystallinity [9].

The maximum acidity seems to exist at ca. 2.5 - 3 mol kg^{-1} of the [Al] - [Na] content. There are two possibilities to explain this phenomenon:
1) In the case of mordenite, similar phenomenon was observed; the acid amount increased with increasing the [Al] - [Na] content, showed the maximum at 1.6 mol kg^{-1}, and decreased again [6]. Secondary-neighboring aluminum atoms must be formed with > 2.5 and > 1.6 mol kg^{-1} of Al in

FAU and MOR structures, respectively [10], suggesting that such Al atoms, or at least a fraction of them, do not exhibit the acidity. The maximum acidity is also in good agreement with the catalytic activities for many kinds of reactions summarized by Barthomeuf [10]; it is noteworthy that the maximum activity shown by Barthomeuf is with respect to the [Al] content, but the present maximum acidity was obtained by varying mainly the [Na] content.

2) However, this can be simply due to the crystal destruction with increasing the ion-exchange ratio, as shown above. Calcination after the ion-exchange and/or hydration by humidity even at room temperature [11] can destroy the FAU crystal with high Al content and low Na/Al ratio.

Anyway, the volcano shape relationship between the acid amount and the [Na] - [Al] content, similar to the relationship often observed between the catalytic activity and [Al] content, was found.

In order to determine the acid strength, the curve-fitting method [6] was first combined with the water vapor-treated TPD profiles of Y zeolites. The fitting was not complete in many cases, because the experimental peak has a broad tail at 600 - 800 K, while the simulated curve suddenly dropped at *ca.* 600 K, as shown in Figure 2. We have found that the extra-framework trivalent cations, *i.e.*, aluminum [12] and gallium [13], showed a large tail, or in some cases a large peak at such a high temperature. It is therefore considered that the major part of the TPD spectrum, namely the part near the peak maximum, shows the properties of acid sites generated by the framework aluminum. The curve-fitting method generally gave the simulated curves fitted with the observed ones near the peak maxima, and the acid strength and its distribution was determined, as shown in Figure 7. The distribution of acid strength (adsorption heat) was only a few kilojoules per mole. The averaged acid strength was almost constant at 105 to 120 kJ mol^{-1}, in most cases *ca.* 110 kJ mol^{-1}, independently of the composition, namely independently of the acid amount. This is in good agreement with careful measurements of calorimetry [14] and quantum chemical calculation, showing that the faujasite structure

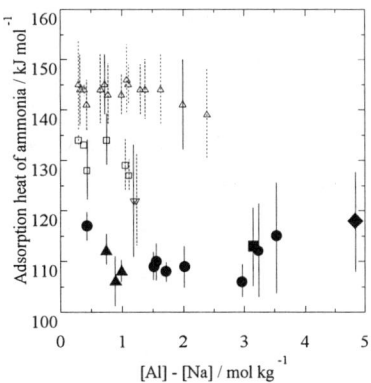

Figure 7. Determined acid strength (adsorption heat of ammonia). The symbol (see Figure 6) shows the averaged adsorption heat, and the vertical bar shows its distribution (standard deviation).

induces 113 kJ mol^{-1} of the adsorption heat of ammonia [15], while conventional analysis for TPD gave lower values [16]. The acid strength, not only the averaged value but also the strength of most of the acid sites ascribed to the framework aluminum, is concluded to be determined by the crystal structure as MOR (*ca.* 145 kJ mol^{-1}) > MFI (130) > BEA (120) > FAU (110). In addition with this simple principle, we have to emphasize that the Y zeolite has significant amount of different type of acid sites due to the extra-framework aluminum.

REFERENCES

1. M. Niwa and N. Katada, Catal. Surveys Jpn. **1,** 215 (1997).
2. M. Niwa, M. Iwamoto and K. Segawa, Bull. Chem. Soc. Jpn. **59,** 3735 (1986).
3. A.W. Chester, J.B. Higgins, G.H. Kuehl, J.L. Schlenker and G.L. Woolery, in Preprint of the 11th Meeting on the Reference Catalyst of Japan, (The Catalysis Society of Japan, Tokyo, 1987) p. 22.
4. G.L. Woolery, G.H. Kuehl, H.C. Timken and A.W. Chester, Zeolites **19,** 288 (1997).
5. G. Bagnasco, J. Catal. **159,** 249 (1996).
6. N. Katada, H. Igi, J.-H. Kim and M. Niwa, J. Phys. Chem., B **101,** 5969 (1997).
7. M.R. Basila and T.R. Kantner, J. Phys. Chem. **71,** 467 (1967).
8. W.L. Earl, P.O. Fritz, A.A.V. Gibson and J.H. Lunsford, J. Phys. Chem. **91,** 2091 (1987).
9. N. Katada, S. Iijima, H. Igi and M. Niwa, Stud. Surf. Sci. Catal. **105,** 1227 (1996).
10. D. Barthomeuf, J. Phys. Chem. **97,** 10092 (1993).
11. W.O. Haag, personal communication.
12. T. Kunieda, N. Katada and M. Niwa, 12th International Zeolite Conference, (1998) A23.
13. T. Miyamoto, N. Katada and M. Niwa, J. Phys. Chem., B in press.
14. G. Kapustin, T. Brueva and I. Mishin, 12th International Zeolite Conference, (1998) P53.
15. M. Brändle and J. Sauer, J. Am. Chem. Soc. **120,** 1556 (1998).
16. H.G. Karge, V. Dondur and J. Weitkamp, J. Phys. Chem. **95,** 283 (1991).

ADSORPTION, ACIDIC AND CATALYTIC PROPERTIES OF DECATIONIZED LOW-ALUMINA ZEOLITES OBTAINED THROUGH DIRECT SYNTHESIS

J. Ya. SMORODINSKAYA[*], Yu. I. AZIMOVA[*], M. I. LEVINBUK[#], M. Ya. MELNIKOV[*].

[*]Moscow State University, Department of Chemistry.
[#]CATACHEM Co., Moscow, Russia

ABSTRACT

Low-alumina Y zeolites (LAY zeolites) obtained by direct synthesis (molar SiO_2/Al_2O_3 ratios are 5.5-7.2) without dealumination treatment and adding templating agents were found through adsorption and XRD investigations to possess a stable zeolite framework in a wide range of their decationizing degree. The samples investigated were established by NMR to have no extraframework aluminum.

Acidic properties of LAY and LZ-210 zeolites were studied through the NH_3 thermodesorption within a temperature range from 20 to 600°C. The concentration of strong acidic centers on the LZ-210 zeolite is somewhat higher than on LAY zeolites. The latter, though, feature a higher content of strong acidic centers than conventional HY-zeolites.

Basic distinctions between LAY zeolite catalytic properties and those of conventional HY and ultrastable zeolites were established. To take an example, the activation energy of n-hexane cracking was found to be 29 kcal/mol at the beginning of the process and change to 19 kcal/mol at a roughly 20% conversion. The proportion of isomer products increases with the growth of the residual Na_2O content of LAY zeolites.

INTRODUCTION

Some physical and chemical properties of LAY zeolites [1] were reported in our publications [2,3] earlier. The regularities and rules determining the influence of decationizing degree variations of LAY zeolites on their adsorption, acidic and catalytic properties in the n-hexane cracking process are discussed in the present.

EXPERIMENTAL

Adsorption, acidic and catalytic properties of LAY and LZ-210 zeolites were investigated on a high-vacuum unit, which allowed adsorption measurements to be carried out through the volume absorption method and permitted kinetic measurements to be fulfilled with gas continuously circulating via a reactor with catalyst where n-hexane cracking proceeded. During the process, samples were taken from the gas phase and analyzed by chromatography at different n-hexane conversions. LAY zeolite samples with M=6.0 and a residual Na_2O content of 0.1 to 6.2 wt. %, LAY samples in a non-decationized form as well as LZ-210 (M=9, 0.25 wt. % of Na_2O) and HY (M=4.9, 0.25 wt. % of Na_2O) zeolite samples were studied.

RESULTS AND DISCUSSION

Physical and chemical properties of LAY zeolites with M=6 and a varying residual Na_2O percentage (0.1-6.2 wt. %) were characterized by NMR and IR methods [4]. The NMR study of LAY zeolite samples demonstrated the absence of extraframework aluminum. IR analyses of LAY zeolites revealed a higher thermal stability of acidic OH-groups as compared with the ultrastable LZ-210 zeolite.

This work is devoted to the investigation of adsorption, acidic and catalytic properties of LAY zeolites. H_2O, n-C_6H_{16} and N_2 adsorption over LAY and LZ-210 zeolites respectively was measured at 24°C and -196°C, which served as a standard test for the probable presence of the amorphous phase in decationized samples. Adsorption researches were controlled through XRD-analyses of corresponding samples. A 1.0 wt. % residual Na_2O content was determined as the lowest limit of LAY zeolite framework stability, below which the framework partially decomposed during decationizing.

Adsorption of N_2 at -196°C and C_6H_{14} at 24°C was measured on samples of LZ-210 and LAY zeolites with different Na_2O contents (Figure 1). The character of isotherms and the fact that they are identical independently from the degree of the sample decationizing testify to the framework stability during LAY zeolite decationizing and to the absence of a secondary porous structure in the samples under investigation, as differentiated from dealuminated zeolites of USY type.

The study of thermodesorption of ammonia chemisorbed at 20°C permitted acidic qualities of LAY zeolites with various residual Na_2O contents to be compared with those of LZ-210 and conventional HY zeolites (Figure 2). The LZ-210 zeolite features a higher content of acidic centers of different forces than LAY zeolites. However, the quantity of strong acidic centers from which ammonia could not be removed even after prolonged pumping-out at 600°C are rather similar for LAY zeolites and the LZ-210. In addition, LAY-zeolites, in contrast to the LZ-210, allow the concentration of such strong acidic centers to be changed. Besides, LAY zeolites have a higher content of strong acidic centers than conventional HY zeolites.

All the LAY and LZ-210 samples above-mentioned were put through n-hexane cracking experiments at 500°C and an initial n-hexane pressure of 60 mm Hg. Kinetic curves describing n-hexane consumption and product formation were determined. These curves permitted us to find out the initial rates for these processes and to establish the correlation between the yields and the n-hexane conversion for all the samples under investigation. Effective activation energies of n-hexane cracking and activation energies of C_3H_6 and C_3H_8 formation were determined at various conversions for one of the LAY samples (4 wt. % of Na_2O). The experiments performed over LAY samples with various residual Na_2O content allowed us to carry out a study of the influence of the Na_2O content on the cracking selectivity.

Chromatographic analyses have exposed the formation of C_1-C_5 alkanes with the following row of weight yields: C_3H_6> C_3H_8> C_2H_4> C_2H_6> i-C_4H_{10}> CH_4> n-C_4H_{10}> i-C_4H_8> i-C_5H_{12}> tr-C_4H_8, cis-C_4H_8, C_4H_8-1. The data concerning the initial reaction period are the most informative for the comparison of alkane and alkene yields, because products formed at the beginning of the process participate in secondary reactions at higher n-hexane conversions. For instance, alkenes proved to undergo H-transfer reactions to give alkanes.

The ratio of the yield of alkanes to that of alkenes at a n-hexane conversion of about 8% was determined to be equal to 0.47 over the LAY with 1.2 wt. % of Na_2O, 0.22 over the LAY with 4 wt. % of Na_2O and 0.20 over the LAY with 6.2 wt. % of Na_2O. The decrease in the alkane share with the growth in the residual Na_2O content in LAY zeolite at the beginning of the process can be demonstrated by the ratios of the yields of individual alkanes and alkenes. This is most obviously shown by the ratio of the yield of butanes to that of butenes (Figure 3). At conversions below 15%, these ratios are less than 1 and diminish with a residual Na_2O content increase for all the LAY samples. At higher conversions, the increase in the residual Na_2O content, on the contrary, leads to a more effective rise in these ratios due to butane formation in secondary H-transfer reactions.

The ratio of butane/butene yields over the LZ-210 exceeds 1 from the very beginning of the process, which surpasses the corresponding values for LAY samples at conversions up to 25-30%. The substantial divergence in the butane share at the beginning of the process for LAY and LZ-210 zeolites correlates with the difference in the alkane share for these zeolites. For instance, the ratio of alkane yield to that of alkenes at an 8% conversion is equal to 0.2 for LAY (4-6 wt. %) and 0.3 for the LZ-210. The 1.5-time difference in the ratios of the yields (alkanes/alkenes, butanes/butenes) at the beginning of the process implies a distinction between the cracking mechanisms over LAY zeolites and the ultrastable LZ-210 zeolite.

The comparison of data on selectivity and those on ammonia thermodesorption (Figure 2) shows that at the beginning of the cracking process the share of alkanes (butanes) is the highest over the samples with the biggest concentration of strong-acidic centers (the LZ-210 and LAY with 1.2 wt. % of Na_2O), and the proportion of alkenes (butenes) at the beginning of the process is higher over the samples with a lower concentration of strong-acidic centers and, correspondingly, a higher residual Na_2O content.

It was supposed that the tendency observed might be connected with the competition between various carbon-chain cracking mechanisms (protolytic and β-cracking ones) taking place over the samples with different acidic properties.

The increase in the residual Na_2O content in LAY zeolite samples (unlike conventional low-alumina zeolites) was established to result in a higher proportion of isomeric products in the process of n-hexane cracking (Figure 4). In the latter case the yield of isomeric products rises with the decrease in the residual Na_2O content. The share of isomeric products is higher over LAY zeolites than over the ultrastable LZ-210 zeolite (Figure 5).

It is noteworthy that the selectivity can be directed in a desirable trend by varying the residual Na_2O content. That is a substantial advantage of LAY zeolites over LZ-210.

Initial rates of n-hexane cracking were determined on the basis of kinetic curves describing n-hexane consumption at 500°C (Table 1).

The highest initial rate of n-hexane consumption over the LAY zeolite samples with a modulus of 6 was attained over a sample with a residual Na_2O content of 4 wt. %. This result implies an extremal character of the relationship between n-hexane cracking rate and the residual Na_2O content of LAY zeolite. These experimental data seem unusual and need further investigation, because conventional low-alumina zeolite samples demonstrate a monotonous dependence of the initial rate on the Na_2O content; the initial rate rises with an increase in the zeolite decationizing degree.

The effective activation energies of n-hexane cracking and product formation were established within the 450-550°C temperature range. A sample with a residual Na_2O content of 4 wt. % featuring the highest initial rate of cracking was used in these experiments. The values of the effective activation energy were determined both at the beginning of the process and at the conversion about 20% (Figure 6). The effective activation energy of n-hexane cracking at the beginning of the process (conversion approximating 0) proved to be equal to 29 kcal/mol.

The effective activation energy of cracking over conventional HY zeolites was estimated [5] at 50 kcal/mol. The cracking activation energy was discovered to change with the growth in conversion over LAY zeolite. The effective activation energy of cracking at a 20% conversion is equal to 19 kcal/mol. Conventional HY zeolites also demonstrated a decline in the activation energy at a higher conversion, but the cracking activation energy at a 20% conversion is equal to 30 kcal/mol [5]. The low values of cracking activation energy (about 17 kcal/mol) are usually connected with the initiation of cracking by L-centers [6]. But, as it was mentioned above, the NMR studies of LAY samples demonstrated the absence of extraframework aluminum, hence the low activation energy of n-hexane cracking over LAY zeolite is not connected with conducting this process on L-centers. The difference between activation energies over LAY zeolites and HY and LZ-210 ones testifies to varying cracking mechanisms over these zeolites.

Activation energies of the formation of individual products, determined in this work, are presented in Table 2.

As it is obvious from these data, the effective energy of C_3H_6 and C_3H_8 formation coincides with that of n-hexane cracking in a wide range of conversion, both at the beginning of the process and at the conversion about 20%.

The cleavage of C_6H_{14} molecule into C_3 fragments with C_3H_6 as the primary product was assumed to be the initial step in n-hexane cracking over LAY. On accumulating a sufficient amount of the main products, the change of the limiting stage takes place, which is reflected by the variations in the activation energy (Table 2, Figure 6).

CONCLUSIONS

1. LAY zeolites obtained via direct synthesis were established to possess a stable zeolite framework in a wide range of their decationizing degree. The LAY samples under investigation were proved by NMR to contain no extraframework aluminum [4].

2. It was discovered that the composition of LAY zeolites can be optimized in terms of their acidity, activity and selectivity in cracking by varying the residual Na_2O content. The share of isomeric cracking products depends antibately on the LAY zeolite decationizing degree.

3. The effective activation energy of n-hexane cracking over LAY zeolite was revealed to be equal to 29 kcal/mol at the beginning of the process and 19 kcal/mol at the conversion around 20%. This fact testifies to the change of the limiting stage at this level of n-hexane conversion. The effective activation energies are substantially lower than the ones over conventional HY zeolites at any conversion. The low values of effective cracking activation energy over LAY

zeolites are not connected with initiating the process by L-centers absent in the LAY zeolite samples.

4. The results obtained imply a difference in the cracking mechanisms over LAY zeolites and HY and LZ-210 ones obtained through standard techniques.

Catalyst	M	Na$_2$O wt. %	W, mmol/(min·g)
LZ-210	9	0.25	0.082*
LAY	6	1.2	0.103
LAY	6	4.0	0.147
LAY	6	6.2	0.069
HY	4.9	0.25	0.039

Table 1. Initial rates of *n*-hexane cracking over Y zeolites (T=500°C, $P_o^{C_6H_{14}}$=40 mm Hg).
*The reaction with a longer induction period

	Conversion 0%	Conversion 20%
C_6H_{14}	29	19
C_3H_6	31	20
C_3H_8	30	19

Table 2. The activation energy of *n*-hexane cracking and C_3H_6 and C_3H_8 formation, kcal/mol.

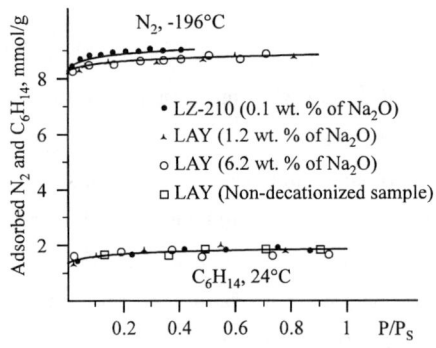

Figure 1. Adsorption isotherms of nitrogen at −196°C) and *n*-hexane (at 24°C) over LAY zeolite samples with various residual Na$_2$O content.

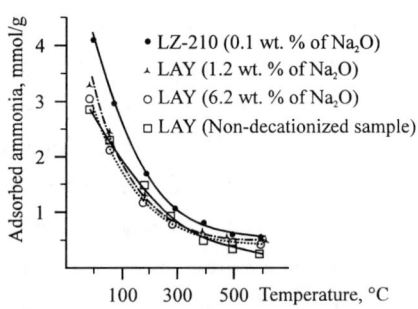

Figure 2. The quantity of chemisorbed ammonia *vs.* desorption temperature over differing types of decationized Y zeolites.

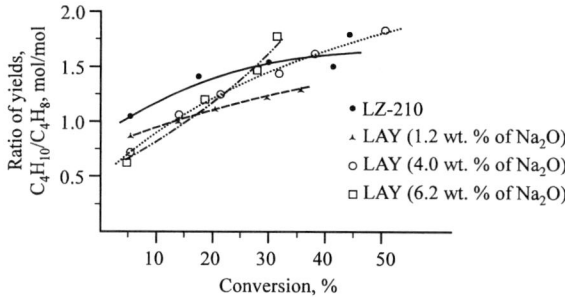

Figure 3. The ratio of butane/butene yields *vs.* conversion.

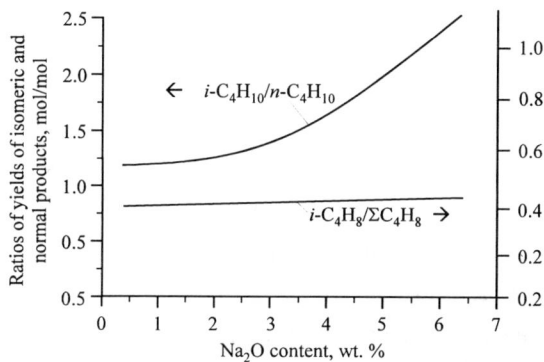

Figure 4. Yields of i-C_4H_{10} and i-C_4H_8 related to yields of corresponding normal products.

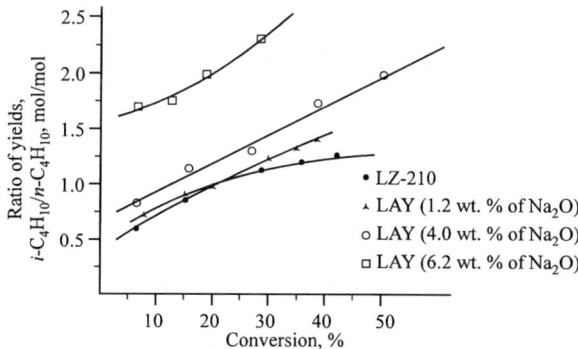

Figure 5. The ratio of i-butane yield to n-butane yield *vs.* n-hexane conversion.

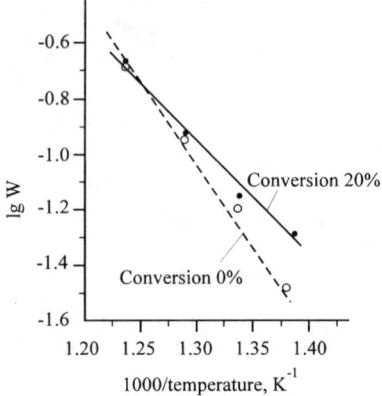

Figure 6. The determination of the effective energy of *n*-hexane cracking at the beginning of the process and at a 20% conversion.

REFERENCES

1. M.L.Pavlov, M.I.Levinvuk et al. Russian Patent #2090502; (September 20, 1997).
2. M.I.Levinvuk, V.K.Smirnov, M.L.Pavlon et al. Symp. on Adv. in FCC Conversion Cat. 211th ACS Nat. Meet. New Orleans, LA. 1996, p.419.
3. J.Smorodinskaya, Yu.Azimova, M.Levinbuk, et al. 3rd Europ.Congr. on Cat. EuropCat - 3. Crakow 1997, vol.2, p.813
4. M.I.Levinbuk, M.L.Pavlov, L.M.Kustov, T.V.Vasina, A.V.Kazakov, Yu.I.Azimova, J.Y.Smorodinskaya, J.P.Fraissard, Appl. Catalysis, **172** (1), to be published.
5. M.I.Levinbuk, Ph.D. Thesis. Moscow State Unitversity, 1980
6. R.J.Gorte, A.I.Biaglow, Symp. on Adv. in FCC Conversion Cat. 211-th ACS Nat. Meet. New Orleans, LA. 1996, p.381-384

STUDY OF THE ADSORPTION STATE OF PHENOL ON HY ZEOLITE BY INFRARED SPECTROSCOPY

XUAN-WEN LI, XIONG SU and XING-YUN LIU

Department of Chemistry, Peking University, Beijing, 100871, China;
xwli@ chemms.chem.pku.edu.cn; Fax 86-10-6275-1708

ABSTRACT

The adsorption of phenol on HY zeolite was studied employing infrared spectroscopy. With portion-wise addition of phenol vapor into the infrared cell, it was observed that the OH groups of phenol interact preferentially with hydroxyls of HY zeolite in the supercage and do not interact with the hydroxyls in the small cage significantly. The 3520 and 3470cm^{-1} bands, which appeared on further adsorption of phenol, are ascribed respectively to the OH stretching vibration of phenol dimer and trimer adsorbed on HY zeolite. These results suggest that phenol hydroxyls are attached to the acidic OH and framework oxygen of HY zeolite, while the aromatic ring of phenol are perpendicular to the HY surface in the adsorption state of phenol on HY zeolite.

INTRODUCTION

The alkylation of phenol by methanol and olefins is an important industrial reaction for producing intermediates of dyes, resins, antioxidants, stabilizers of polymers. Much research has been carried out on oxide and zeolite catalysts to improve the catalytic activity, especially the selectivity and to elucidate the reaction mechanism [1]. However, it is still unclear in certain cases whether the course of phenol alkylation is determined by shape selectivity or chem-selectivity [2]. Some workers [3,4] cited the adsorption state of phenol or anisole to describe the course of alkylation. But only a few references on phenol adsorption are available [2-5]. In order to elucidate the course of phenol alkylation, we carried out study of phenol adsorption on HY zeolite using infrared spectroscopy. The results are presented in this paper.

EXPERIMENTAL

HY zeolite was prepared by thermal decomposition of NH_4Y in vacuum at 400° C for 2h in an infrared cell equipped with CaF_2 windows. The NH_4Y has a very low sodium content (<0.2%). The adsorption of phenol was carried out *in situ* by portion-wise addition of phenol to the pretreated HY wafer in infrared cell at room temperature. The weight of HY wafer was 12mg (16mm in diameter). Each time 20ml of saturated phenol vapor was introduced into the IR cell at room temperature.

RESULTS AND DISCUSSION

1. Interaction of phenol with HY zeolite surface.

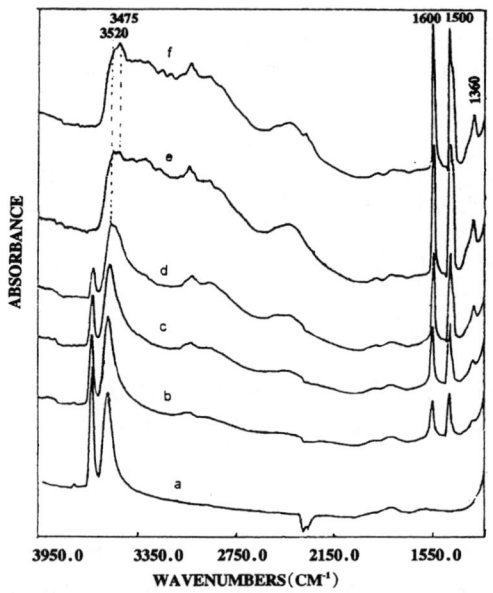

Figure 1. Infrared spectra of phenol adsorption on HY zeolite with portion-wise addition (a) HY, (b) to (f) introduced respectively 1, 2, 3, 4 and 6 portions of phenol.

The FTIR spectra of phenol adsorption on HY zeolite with increasing amount of phenol at room temperature are shown in Figure 1. Before the adsorption of phenol, three absorption bands of HY were observed at 3740, 3640 and 3550cm^{-1} which are related to silanols, hydroxyls in supercages and hydroxyls in sodalite cages respectively. After the first portion of phenol was introduced, the intensity of 3640cm^{-1} band decreased while there was no change at the 3550cm^{-1} band. With increasing amount of phenol adsorption, the 3640cm^{-1} band diminished progressively and disappeared completely after four portions of phenol were adsorbed. A shoulder at 3520cm^{-1} appeared after three portions of phenol were introduced into the IR cell and the intensity of 3520cm^{-1} band increased with increasing the amount of phenol adsorption. A 3475cm^{-1} band appeared after the fourth portion of phenol was adsorbed and became the maximum absorption band in the hydroxyl region at higher loading of phenol adsorption. These results suggest that the phenol interacts preferentially with the acidic hydroxyl in the supercages of HY and does not interact with the hydroxyl in the small cages significantly. The meaning of appearance of 3520cm^{-1} and 3475cm^{-1} bands will be discussed later.

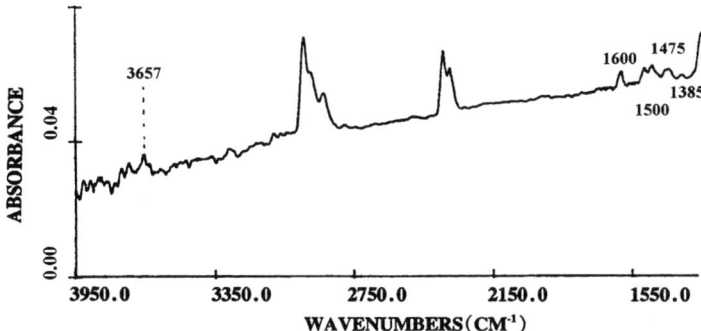

Figure 2.   Infrared spectra of phenol vapor in the infrared cell.

The FTIR spectra of saturated phenol vapor at room temperature when HY sample is absent in IR cell are presented in Figure 2. In the hydroxyl region, a band appeared at 3657cm^{-1} which is assigned to the hydroxyl absorption band of bare phenol. The absorption bands of aromatic ring vibration of phenol were observed at 1600cm^{-1}, 1500cm^{-1} and 1475cm^{-1} while the C-O-H in-plane bending vibration appeared at 1385cm^{-1}.

The results from Figure 1 demonstrate that the 3657 cm^{-1} band disappeared and the 1385 cm^{-1} band shifted to 1360cm^{-1} after adsorption of phenol on HY zeolite. All these changes reflect the interaction of hydroxyl of phenol with HY surface.

In the study of benzene adsorption on HY zeolite [7,8,9], it was found that if the protons of HY interact through hydrogen bonding with π-electrons of aromatic ring, the CH out-of-plane vibration and in-plane skeletal vibration of aromatic ring will shift. Our result in Figure

1 shows that no CH out-of-plane vibration was observed and the vibration bands of aromatic ring, such as 1600 cm^{-1}, 1500 cm^{-1} and 1475 cm^{-1} were the same after adsorption of phenol on HY as those in the liquid or vapor phases of phenol. Only their intensities increased with increasing phenol loading which reflect the augmentation of concentration of phenol adsorbed on HY surface. All these results demonstrate the absence of interaction of HY protons with aromatic rings of phenol in the adsorption process.

2. The adsorption state of phenol on HY zeolite.

In a previous work, Ebata et al. [6] studied the IR spectra of monomers, dimers and trimers of phenol. They observed the red shift of hydroxyl stretching vibration from 3657 cm^{-1} for bare phenol to 3654, 3530 cm^{-1} for phenol dimers and to 3441, 3449 cm^{-1} for phenol trimers. The 3530 cm^{-1} band was assigned to the hydroxyl stretching vibration of the proton donating phenol and 3654 cm^{-1} to that of the proton accepting phenol in phenol dimers.

In our study, the disappearance of 3640 cm^{-1} band of HY and 3657 cm^{-1} band of phenol associate with the strong interaction of acid hydroxyls of HY zeolite with oxygens of phenol OH groups at the beginning of adsorption. This leads to the delocalization of H from both hydroxyls of HY and phenol as shown in Figure 3. In this system, the protons may exist but would not exhibit infrared bands because they are highly delocalized as suggested by Barthomeuf [10].

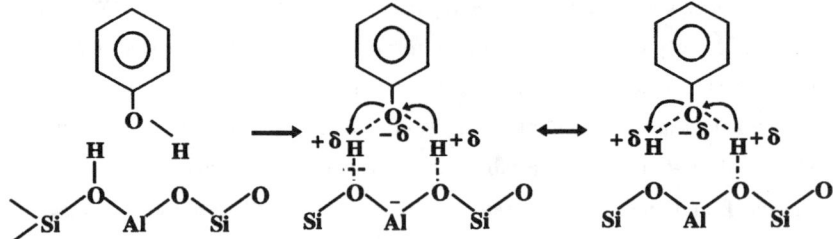

Figure 3.   Interaction of acidic hydroxyl of HY with phenol oxygen.

The appearance of 3520cm^{-1} and 3470cm^{-1} bands with increasing phenol adsorption may be related also to the strong interaction of protons of HY with phenol oxygen which leads phenol OH groups to have strong tendency for interacting through hydrogen bonding with both oxygen atoms of zeolite framework and another phenol. Thus, the dimer and trimer adsorption state could be formed with increasing the amount of adsorption as shown in Figure 4.

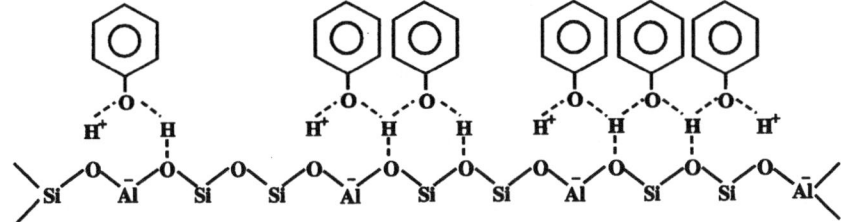

Figure 4. The adsorption state of phenol on HY zeolite.

The adsorption of phenol at 150°C on HY zeolite did not alter the main feature of the IR spectra as that obtained at the room temperature. After desorption of adsorbed phenol at 150°C under 1.3×10^{-2} Pa, three bands (3740, 3640, 3550 cm^{-1}) were restored almost completely, indicating the reversibility of phenol adsorption.

CONCLUSION

The phenol OH group interacts preferentially with hydroxyl group in the supercage of HY zeolite. The strong interaction of acidic hydroxyls of HY with oxygen of phenol would lead the OH groups of phenol to have strong tendency for interacting through H bonding with both oxygen atoms of zeolite framework and another phenol. This may cause the formation of monomer, dimer and trimer state of phenol adsorbed on HY zeolite. In the adsorption state, phenol hydroxyl groups attach to the surface of HY zeolite, while the aromatic ring is perpendicular to the surface.

ACKNOWLEDGMENT

This work was supported by the National Natural Science Foundation (No.29473088) and SINOPEC of China.

REFERENCES

1. R. F. Parton, J. M. Jacobs, D. R. Huybrechts and P. A. Jacobs Proc. Int. Symp. On Zeolites as catalysts, Sorbents and Detergent Builders, Wurzburg, FRG, 1988, Elsevier, Amsterdam, p.169 (1989).
2. Worfgang F. Hoederich, Guidlines for mastering the properties of molecular sieves. Edited by D. Barthomeuf et al. Plenum Press, New Yourk, p.320 (1990).
3. K. Tanabe, Proc. 6th Intern. Congr. Catalysis, p.863 (1976).
4. P. Beltrame, P. L. Beltrame, P. Carnitti, A. Castelli and L. Forni, Appl. Catal., **29**, 327 (1987).
5. I. Yu. Ponmarenko, E. A. Paukshtic, L. M. Koval, Zh. Fiz. Khim, **67**, 1726 (1993).
6. T. Ebata, T. Watanabe and N. Mikami, J. Phys. Chem. **99**, 5761 (1995).
7. A.De Mallmann, and D.barthomeuf. Proc. 7th Int. Zeolite Conf., Kodansha-Elsevier, Tokyo, p.609 (1986).
8. P. A. Jacobs, J. A. Martens, J. Weitkamp and H. K. Beyer, Faraday Disc. Chem. Soc., **72**, 353 (1981).
9. J. Datka, J. Chem. Soc., Faraday Trans. 1, **77**, 551 (1981).
10. D. Barthomeuf, in Stud. Surf. Sci. Catal. **5**, 56 (1980).

ESR STUDY OF NaY SUPPORTED Pd AND Pt IONS AND CLUSTERS

H. DU[§], R. KLEMT[§], F. SCHELL[$], J. WEITKAMP[$] and E. RODUNER[§]*

[§]Institute of Physical Chemistry, University of Stuttgart, D-70569, Germany
[$]Institute of Chemical Technology I, University of Stuttgart, D-70550, Germany

ABSTRACT

Pd/NaY and Pt/NaY were prepared by the aqueous ion-exchange method and investigated by ESR spectroscopy. Both, Pd^{2+} and Pd^{3+} were detected in oxidized Pd/NaY. Small intra-cavity clusters are formed upon H_2 reduction between 343 K and 623 K. Pd loadings in the range of 4wt% to 8wt% are favorable for cluster formation under mild reduction at 296 K. Platinum exists in a Pt^{2+} state in oxidized Pt/NaY. Upon reduction at 563 K, a sharp axial signal with $g_\perp=2.007$ and $g_\parallel=2.088$, possibly a positively charged diatomic cluster, Pt_2^+, is observed.

INTRODUCTION

Pd and Pt supported faujasites are important catalysts in petroleum and petrochemical industries [1]. The regular channels and cages provide perfect matrices for the dispersion of metal cations and metal particles, which have been extensively characterized by IR, XPS, SAXS, EXAFS and TEM, and by catalytic tests [2,3]. Palladium and platinum in their various oxidation states are highly active in hydrogenation, hydrocracking and dimerization reactions of small olefins [4]. In recent years, tiny metal clusters encaged in zeolites have been on special attention due to their higher catalytic activity towards hydrogenation [5] and hydrogenolysis [6,7], a different selectivity in the conversion of methylcyclopentane [8] and a stronger resistance to sulfur poisoning [9]. Encapsulated clusters show also enhanced stability towards sintering [10].

Electron spin resonance (ESR) has been proven to be valuable for the characterization of Pd ions in zeolites [11,12] but little attention has been paid on small 'electron-deficient' Pd clusters which are of high bifunctional activity [13]. We have not seen ESR reports on zeolite-supported Pt ions or clusters. The present study reports the ESR characterization of Pd/NaY and Pt/NaY.

EXPERIMENTAL

Ion-exchange of NaY zeolite with palladium was carried out at 343 K for 48 h by dropwise addition of 0.01M $[Pd(NH_3)_4]Cl_2$ solution to NaY slurry. The exchanged sample was filtered, washed with deionized water till Cl⁻ free and dried in air at 296 K. Calcination of Pd/NaY was conducted from room temperature to 773 K at 0.5 K min^{-1} in flowing O_2, and maintained at 773 K for 10 h. The oxidized sample was then reduced statically by H_2 at stepwise increasing temperature with six mol H_2 per mol Pd.

Platinum was introduced via exchange of NaY with 0.001 M [Pt(NH$_3$)$_4$]Cl$_2$ solution at 295 K for 24 h. The product was filtered, washed and dried at 353 K, then calcined at 563 K for 3 h at a heating rate of 0.5 K min^{-1} in flowing O_2. Reduction was performed from 343 K to 563 K at 7.6 K min^{-1} in flowing H_2 (300 ml min^{-1} g^{-1}), maintaining the final temperature for 6 h.

X-band ESR spectra were recorded on a Bruker EMX spectrometer in a temperature range of 10 K to 296 K. The samples were evacuated for 2 h at room temperature before measurement.

RESULTS AND DISCUSSION

NaY supported Pd ions and clusters

Oxidized sample: Exchange with [Pd(NH$_3$)$_4$]Cl$_2$ solution replaces some Na$^+$ by Pd(NH$_3$)$_4^{2+}$ ions. Subsequent calcination at 773 K in flowing O_2 destroys the ammine ligands and removes physisorbed water, leading to the formation of Pd^{2+}. Some of the Pd^{2+} may be further oxidized to Pd^{3+} in the presence of oxygen.

Fig.1 shows the ESR spectra of calcined Pd/NaY. No signal is observed at room temperature. At 171 K, an isotropic line with g=2.23 is attributed to Pd^{3+} with 4d^7 configuration [11,14]. Its width decreases from 95 G (171 K), to 81 G (158 K), and to 63 G (149 K). This behavior is typical for spin-lattice relaxation. One contribution should come from the spin-orbit interaction of Pd^{3+}. In Y zeolite, Pd^{3+} ions exist in a high-spin state because ligands like water, hydroxyl, and framework oxygen are too weak to produce low-spin Pd^{3+}. The presence of three unpaired d electrons in Pd^{3+} should lead to a relatively strong spin-orbit interaction. The color of the calcined sample is beige pink, showing the presence of diamagnetic Pd^{2+} with 4d^8 configuration. The oxidized Pd/NaY sample thus contains both divalent and trivalent Pd ions which presumably are located at the regular cationic positions in Y zeolite [1].

Reduced sample: Upon admission of H_2 at 296 K, the g=2.23 signal disappears immediately, and the beige pink color changes to ash gray, indicating that both Pd^{3+} and Pd^{2+} are easily reduced. At the same time, four anisotropic signals B, C, D and E with $g_\perp$= 2.01-2.12 and $g_\parallel$ = 2.17-3.30 appear (Fig.2b). They are assigned to monovalent Pd$^+$ ions [16,17]. In zeolite Y, cation positions are the well defined SI sites in hexagonal prisms, SI' and SII' sites in sodalite cages,

SII, SII*, SIII, and SIII' sites in supercages [18]. Pd ions at these positions may have different coordination symmetries and show different ESR signals. Most probably, the observed four ESR lines correspond to Pd^+ located at different cationic sites. Upon reduction at 343 K, the signals B, C, and D disappear nearly. They are further reduced to diamagnetic Pd metal at higher temperature. Meanwhile, signal E enhances, suggesting that the corresponding Pd^+ ions are located at less accessible positions like SI or SI' whose reduction needs higher temperature, or that some Pd^+ ions transform by migration to the Pd^+ corresponding to signal E.

A dominant signal F with $g_\perp = 2.30$ and $g_\parallel = 2.05$ appears after reduction at 343 K (Fig.2c) and much stronger at 423 K (Fig.2d), while signal E disappears completely. This may be due to the reduction of the corresponding Pd^+ to Pd^0 or the transformation of Pd^+ to Pd^0 species corresponding to signal F. In literature, this signal F was assigned to Pd^+[19]. To us, this seems unlikely due to the following facts: 1) The Pd^+ signal is characterized by its strong anisotropy with $g_\perp \ll g_\parallel$ [16], but F has $g_\perp \gg g_\parallel$. For d^9 transition metal ions, the reverse order occurs when

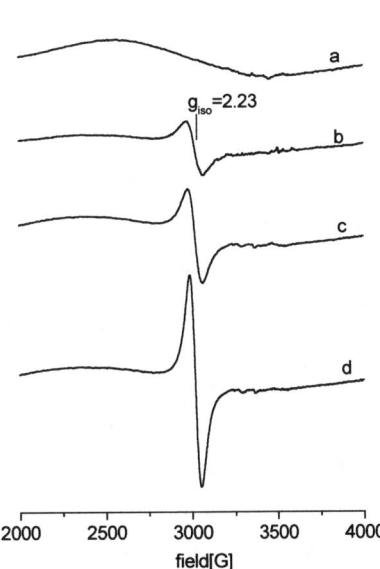

Fig.1 X-band ESR spectra of Pd/NaY oxidized at 773 K for 10 h under O_2 flow measured at (a) 296 K; (b)171 K; (c)158 K; (d)149 K.

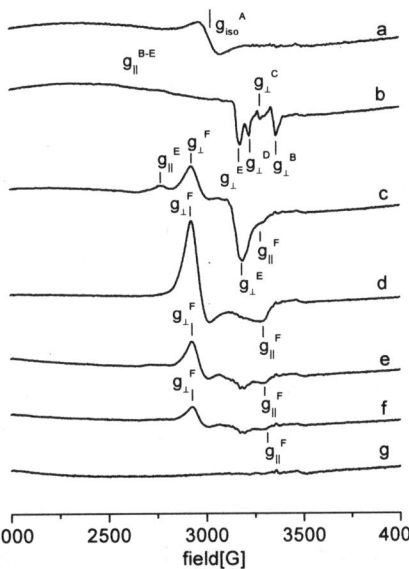

Fig.2 X-band ESR spectra of Pd/NaY (4wt%) at 180 K. (a) 10 h oxidation at 723 K; 1 h H_2 reduction at (b) 296 K; (c) 343 K; (d) 423 K; (e) 523 K; (f) 573 K; (g) 623 K. $g_{iso}^A = 2.23$; $g_\perp^B = 2.01$; $g_\perp^C = 2.05$; $g_\perp^D = 2.09$; $g_\perp^E = 2.12$; $g_\perp^F = 2.30$; $g_\parallel^{B-E} = 2.17$-3.30; $g_\parallel^E = 2.42$; $g_\parallel^F = 2.05$.

the unpaired spin resides in a $|3z^2-r^2>$ ground state [19]. For a d^9 ion, this is expected for trigonal bipyramidal, trigonally compressed tetrahedral, tetragonally compressed square pyramidal, or tetragonally compressed octahedral geometries [20]. The $g_{\|}$ values in the above cases are reported to be in the range of 0.740 to 2.004 [21]. However, signal F has $g_{\|}=$ 2.05, and it is too high for such d^9 cationic geometries in which the reverse order of g values is possible [19,20]. 2) Comparing with the typical Pd^+ signals shown in Fig.2b, the line width of signal F is much larger. Assuming that F arises from Pd^+, one possibility for the signal broadening is spin-spin interaction which becomes effective at high paramagnetic concentration. As long as Pd ions are located at cationic positions which are well defined and separated, spin-spin interaction seems unlikely. Since the Pd atom is diamagnetic the only remaining candidates are Pd clusters. However, in Na-MCM-22 zeolite or K-L zeolite small Pd clusters show an isotropic signal [22], different from the present signal F.

XPS, IR and EXAFS studies have proven that the small Pd clusters encaged in NaY are formed during H_2 reduction via the following procedures [13,23,24]:

$$n\,[\,2Pd^{2+} + 2O^{2-}_{wall}\,] + n\,H_2 \rightarrow 2Pd^0_n + 2n\,H\text{-}O_{wall}$$

$$Pd^0_n + z\,H\text{-}O_{wall} \rightarrow [\,Pd_n\text{-}H_z\,]^{z+} + z\,O^-_{wall}$$

The clusters are anchored by the nearby protons, and thus become 'electron-deficient'. These positively charged Pd particles are further stabilized by framework oxygen, and some of them may be paramagnetic. Similar tiny charged Na_4^{3+} and Ag_3^{2+} clusters encapsulated in zeolites have also been reported [25,26]. Such a charged cluster model is compatible with our ESR results: Pd clusters with different nuclearity or size could be responsible for the broadening of the signal F. The presence of the protons acting as strong ligands could cause an asymmetric coordination, leading to the anisotropy of the signal.

The assignment of the signal F to small charged clusters is further supported by the following facts: 1) Bergeret et al. investigated a similar Pd/NaY system by SAXS and found that 5-7Å Pd clusters, compatible with the supercage diameter, were formed in Y zeolite upon reduction at 423 K [27]. 2) SAXS and SEM studies showed that 20Å particles were formed outside the cavities of NaY at the expense of small Pd clusters inside the cages upon reduction above 500 K [27]. Indeed, the ESR spectrum (Fig.2e) after reduction of Pd/NaY at 523 K shows clearly the decrease of signal F, revealing the decrease of the number of paramagnetic clusters in the zeolite. After reduction at 573 K, signal F declines further (Fig.2f) and disappears around 623 K (Fig.2g), indicating the further migration out of the cavities. 3) From the enlarged ESR spectra of reduced Pd/NaY we see that after reduction at 423K signal F is overwhelmingly dominant (Fig.3a).

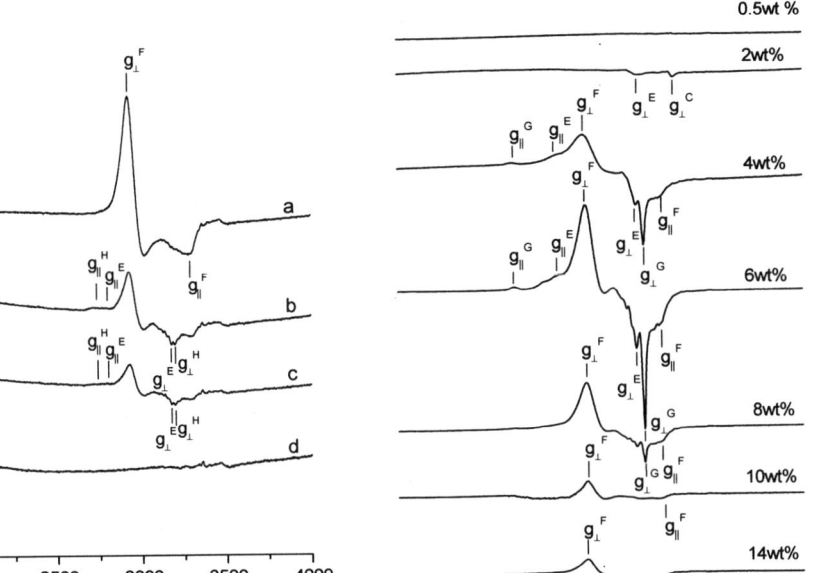

Fig.3 X-band ESR spectra of Pd/NaY (4wt%) oxidized at 773 K followed by H_2 reduction for 1 h at (a) 423 K; (b) 523 K; (c) 573 K; (d) 623 K.
g values: $g_\perp^E$=2.12; $g_\perp^F$=2.30; $g_\perp^H$=2.10; $g_\parallel^E$=2.42; $g_\parallel^F$=2.05; $g_\parallel^H$=2.48.

Fig.4 X-band ESR spectra of Pd/NaY oxidized at 623 K followed by H_2 reduction at 296 K for 8 days.
g values: $g_\perp^C$=2.05; $g_\perp^E$=2.12; $g_\perp^F$=2.30; $g_\perp^G$=2.09; $g_\parallel^E$=2.42; $g_\parallel^F$=2.05; $g_\parallel^G$=2.61.

However, upon reduction at the higher temperature of 523 K, weaker Pd^+ signals E with $g_\perp$=2.12 and $g_\parallel$ = 2.42 and H with $g_\perp$ = 2.10 and $g_\parallel$ = 2.48 appear (Fig. 3b). They are also present in the ESR spectrum of Pd/NaY reduced at 573 K (Fig.3c). This apparent back oxidation of Pd atom to Pd^+ in the presence of H_2 is somewhat unusual. 4) Fig.4 shows the ESR spectra of Pd/NaY with different Pd loadings reduced at 296K. No signal F is observed below 2wt% due to the insufficient palladium content for cluster formation under mild reduction. F appears at 4wt% loading, reaching the maximum around 6wt%, and decreases with the further increasing palladium content. Such a change of signal F with Pd loading agrees well with the general law that higher concentration favors cluster formation. Here, the further increase of Pd amount in zeolite is favorable for the formation of large ESR inactive Pd particles on the external surface of the zeolite at the cost of small Pd clusters, leading to the decrease of the signal intensity.

Fig.5a and Fig.5b show the ESR spectra of Pd/NaY reduced at 423 K and half a month later under static H_2, respectively. It can be seen that signal F is still present with the intensity being about two-thirds of that of the original line after half a month at 296 K. This means that most of the small clusters inside the zeolite do not migrate and coalesce at room temperature. Upon vacuum treatment at 423 K under 10^{-5} mbar (Fig.5c), signal F still remains with the intensity about one-third of that of the original line, revealing that the corresponding clusters are relatively stable. This is in agreement with the EXAFS result [28]. The reason for the cluster stability may be that the small encaged $[Pd_n-H_z]^{z+}$ clusters are chemically anchored by protons and further stabilized by framework oxygen in zeolite.

$[Pd_n-H_z]^{z+}$ encapsulated in zeolite is responsible for the high activity in the neopentane conversion reaction [7]. From the catalytic viewpoint, such active centers should exhibit medium strength stability. When $[Pd_n-H_z]^{z+}$ is unstable, it could decompose or convert to other species during raising temperature or catalytic reaction. When $[Pd_n-H_z]^{z+}$ is too stable, it cannot efficiently interact with reactant molecules and activate them. For $[Pd_n-H_z]^{z+}$ catalyzed reactions

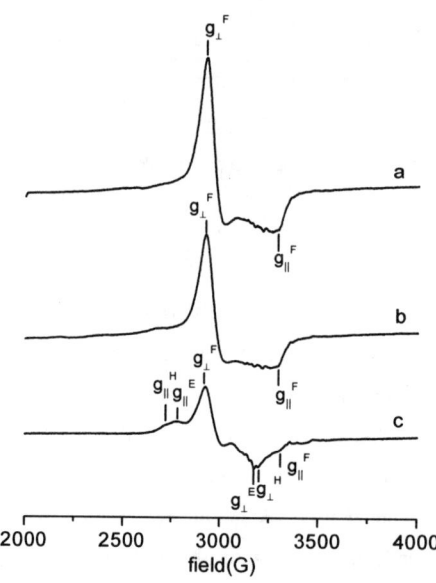

Fig.5 X-band ESR spectra of Pd/NaY at 177 K (a) 10 h oxidation at 773 K followed by 1 h H_2 reduction at 423 K; (b) 15 days later at 296 K after (a); (c) 1.5 h evacuation at 423 K after (a).
g values: $g_\perp^E$=2.12; $g_\perp^F$=2.30; $g_\perp^H$=2.10; $g_\parallel^E$=2.42; $g_\parallel^F$=2.05; $g_\parallel^H$=2.48.

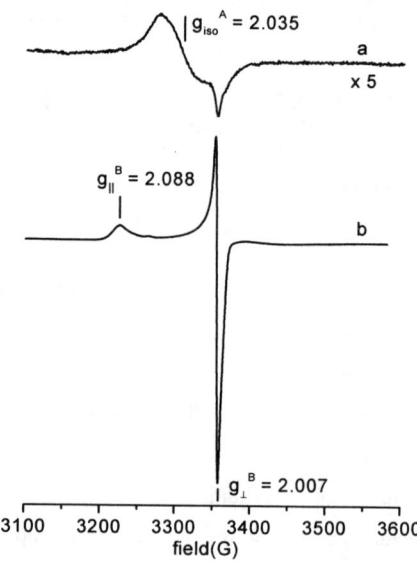

Fig.6 X-band ESR spectra of Pt/NaY with 1wt% Pt loading oxidized at 563 K under O_2 flow followed by reduction at 563 K under H_2 flow. (a) 288 K; (b) 10 K.

like neopentane and methylcyclopentane conversion [29], the reaction temperature is normally in the range of 423 K to 553 K. In this region, $[Pd_n-H_z]^{z+}$ fits the requirement of medium stability.

NaY supported Pt ions and clusters

No signal is observed with oxidized Pt/NaY, indicating that Pt exists in a state of $Pt^{2+}(5d^8)$, which is consistent with IR and SAXS results [30]. The ESR spectrum of the reduced Pt/NaY displays a strong isotropic signal A at g=2.035 superimposed on a sharp axial signal B with $g_\perp$= 2.007 and $g_\parallel$=2.088 (Fig.6a), which can be seen more clearly at 10 K (Fig.6b). Unlike the line B, A declines with decreasing temperature and becomes unobservable below 100 K. The combined intensities correspond to ca. 1% of the platinum loading.

The line shape of signal B looks like that of Pt^+ supported in a Kr-matrix except for the absence of a hyperfine structure and for much lower g-anisotropy in the zeolite [31]. The low g-anisotropy points to a larger s-orbital contribution at the cost of d-character of the unpaired electron. The increased s-character should lead to a large hyperfine interaction for the ^{195}Pt isotope (I = 1/2, natural abundance 33.8%). The reasons for the absence of a hyperfine structure could be either a large anisotropy leading to extremely broad lines, or just too large an isotropic value (1.2 T for atomic Pt which has $5d^96s^1$ configuration) rendering it ESR silent in the X-band. In view of this, a tentative candidate for the assignment of signal B is a diatomic cluster with single positive charge, Pt_2^+. Indeed, small clusters were reported to be formed in NaY after similar treatment [32]. At low temperature they are thought to be fixed to the zeolite lattice, but at higher temperature they could occupy a state in which they are sufficiently mobile to average the axial g-anisotropy, giving rise to the isotropic species A at g=2.035. This is supported by the opposite temperature dependence of signal A and signal B. Further characterization including W-band ESR and a study of oxidation behavior is in progress.

REFERENCES

1   R.A. Dalla-Betta and M. Boudart, in Proceedings of the 5th International Congress on Catalysis, Edited by H. Hightower (North Holland, Amsterdam, 1973) p.1329.
2   S.T. Homeyer, Z. Karpiński and W.M.H. Sachtler, Recl. Trav. Chim. Pays-Bas, **109**, 81 (1990).
3   V. Romannikov, K. Ione and L. Pedersen, J. Catal., **66**, 121 (1980).
4   A.L. Lapidus and K.M. Minachev, Neftehimiya, **18**, 212 (1978).
5   J. Bandiera, J. Chim. Phys., **77**, 303 (1980).

6. C. Naccache, N. Kaufherr, M. Dufaux, J. Bandiera and B. Imelik, in Molecular Sieves-II, Edited by J. Katzer, ACS (Washington DC, 1977) p. 538.
7. S.T. Homeyer, Z. Karpiński and W.M.H. Sachtler, J. Catal., **123**, 60 (1990).
8. X. Bai and W.M.H. Sachtler, J. Catal., **129**, 121, (1990).
9. T.M. Tri, J. Massardier, P. Gallezot and B. Imelik, in Catalysis by Zeolites, Edited by B. Imelik, (Elsevier, Amsterdam, 1980) p. 279.
10. B. Gate, Chem. Rev., **95**, 511 (1995).
11. A.K. Ghosh and L. Kevan, J. Phys. Chem., **92**, 4439 (1988).
12. C. Naccache, J.F. Dutel and M. Che, J. Catal., **29**, 179 (1973).
13. W.M.H. Sachtler and A.Y. Stakheev, Catal. Today, **12**, 283 (1992).
14. A.K. Ghosh and L. Kevan, J. Amer. Chem. Soc., **110**, 8044 (1988).
15. J. Michalik, H. Lee and L. Kevan, J. Phys. Chem., **89**, 4282 (1985).
16. J. Michalik, M. Heming and L. Kevan, J. Phys. Chem., **90**, 2132 (1986).
17. T. Saint-Pierre, X.H. Chen and L. Kevan, J. Phys. Chem., **97**, 932 (1993).
18. H. Klein, C. Kirschloch and H. Fuess, J. Phys. Chem., **98**, 12345 (1994).
19. A. Abragam and B. Bleaney, Electron Paramagnetic Resonance of Transition Metal Ions, (Oxford Press, London, 1970) p. 456.
20. J. Michalik, M. Narayana and L. Kevan, J. Phys. Chem., **89**, 4553 (1985).
21. M. Narayana and L. Kevan, J. Chem. Phys., **78**, 3573 (1983).
22. A.M. Prakash, T. Wasowicz and L. Kevan, J. Phys. Chem., B **101**, 1985 (1997).
23. A.Y. Stakheev and W.M.H. Sachtler, J. Chem. Soc., Faraday Trans., **87**, 3703 (1991).
24. L. Xu, Z. Zhang and W.M.H. Sachtler, J. Chem. Soc., Faraday Trans., **88**, 2291 (1992).
25. M.R. Harrison, P.P. Edwards, J. Klinowski, J.M. Thomas, D.C. Johnson and C.J. Page, J. Solid State Chem., **54**, 330 (1984).
26. L.. Gallen, W.J. Mortier, R.A. Schoonheydt and J.B. Uytterhoeven, J. Phys. Chem., **85**, 2783 (1981).
27. G. Bergeret, T.M. Tri and P. Gallezot, J. Phys. Chem., **87**, 1160 (1983).
28. K. Moller, D.C. Koningsberger and T. Bein, J. Phys. Chem., **93**, 6116 (1989).
29. Z. Zhang, B. Lerner, G. Lei and W.M.H. Sachtler, J. Catal., **140**, 481 (1993).
30. P. Gallezot, A. Alarcon-Diaz, J. Dalmon, A. Renoprez, B. Imelik, J. Catal., **39**, 334 (1975).
31. R.J. Van Zee and W. Weltner Jr., Chem. Phys. Lett., **266**, 403 (1997).
32. K.I. Pandya, S.M. Heald, J.A. Hriljac, L. Petrakis and J. Fraissard, J. Phys. Chem., **100**, 5070 (1996).

STUDY OF PYRIDINE AND AMMONIA SORPTION IN FAUJASITE AND MORDENITE ZEOLITES BY THE FREQUENCY-RESPONSE TECHNIQUE

DONGMIN SHEN[*,1], GY. ONYESTYÁK[2] and L. V. C. REES[1]

[1]Department of Chemistry, University of Edinburgh, West Mains Road, Edinburgh EH9 3JJ, UK
[2]Central Research Institute for Chemistry, Hungarian Academy of Sciences, Budapest, Hungary

ABSTRACT

The frequency response (FR) technique was applied to study the kinetics of pyridine and ammonia sorption processes and to characterise acidic sites in hydrogen and sodium forms of faujasite and mordenite zeolites. The pyridine FR spectra for HY reflected pyridine chemisorption, while the spectra for HM were determined by pyridine diffusion processes. For acidic site characterisation using the FR technique, ammonia was found to be a better probe molecule because it, having a small kinetic diameter, diffuses fast and is able to access most sorption sites, compared to pyridine. The ammonia FR spectra for different forms of Y and M zeolites were significantly different, characterising chemisorption processes on different acidic sites. The FR technique provides an alternative method in the aspect of sorption kinetics for studying the chemisorption of ammonia on acidic sites in zeolites.

INTRODUCTION

By the frequency response (FR) technique, a "rate spectrum" characterising kinetic phenomena of gas/solid surface can be obtained. Various rate processes which occur simultaneously, e.g. diffusion in micropores and macropores, adsorption/desorption on/from different sites, complex reactions, can be determined [1]. The FR method has been successfully applied to the studies of intracrystalline [2-4] and intercrystalline [5, 6] mass transfer kinetics in zeolites and their pellets. The applications of the FR method for the study of chemisorption of ammonia on various zeolitic catalysts have begun recently [7-9].

Pyridine and ammonia have been widely used as probe molecules for characterisation of zeolitic acidic sites. The difference in size and basity between these two molecules may provide

---

[*] Current and corresponding address: The BOC Group Technical Center, 100 Mountain Avenue, Murray Hill, New Jersey 07974, USA. Email Address: dongmin.shen@us.gtc.boc.com.

more information on acidic sites in zeolites and rate-controlling steps when chemisorption and/or diffusion processes are involved. The aim of this paper is to apply the FR technique for studying sorption kinetics of pyridine and ammonia in faujasite and mordenite zeolites exchanged with cations: NH_4^+ (or H^+) and Na^+ in order to determine mass transfer rate-controlling steps for the probe molecules and to characterise Brönsted and Lewis acidic sites present in the zeolites.

EXPERIMENTAL

The theories of the frequency response method have been comprehensively developed for various kinetic behaviours of a gas-surface system [1, 10, 11]. The experimental FR spectrum for a sorption controlled system with n sorption sites may be described by the in-phase, $K\delta_{in}$, and out-of-phase, $K\delta_{out}$, characteristic functions [1]:

$$\frac{P_B}{P_Z}\cos\phi_{Z-B} - 1 = K\delta_{in} = \sum_{i=1}^{n} K_i \frac{t_i^2}{t_i^2 + \omega^2} \tag{1}$$

$$\frac{P_B}{P_Z}\sin\phi_{Z-B} = K\delta_{out} = \sum_{i=1}^{n} K_i \frac{t_i \omega}{t_i^2 + \omega^2} \tag{2}$$

where

$$K_i = \left(\frac{\partial A_i}{\partial P}\right)_e \frac{RT}{V_e} \tag{3}$$

which is related to a gradient of a sorption isotherm arising from A_i, the amount sorbed on sorption site i; t_i the time constant for the sorption process of the sorbate on the site i; P_B and P_Z are the pressure responses to c. $\pm 1\%$ volume perturbations in the absence (B) and presence (Z) of adsorbent and ϕ_{Z-B} is the difference between phase lags and ω is the angular velocity of the wave generator. The in-phase function (δ_{in}) approaches K_i at low frequencies. The maximum of the out-of-phase function (δ_{out}) appears at a perturbation frequency of the resonance, which is dependent on kinetic time constants that characterise the sorption processes. The associated kinetic parameters of the FR spectra, K_i and t_i, can be determined by fitting the characteristic functions with an appropriate theoretical model. The adsorption and desorption rate constants may be determined from the pressure dependence of the time constant t_i values [1].

On the other hand, the characteristic functions for the *Fickian* diffusion of a single component in a spherical solid of a radius, r, subjected to a periodic, sinusoidal surface concentration modulation can be described by [1]:

$$\frac{P_B}{P_Z}\cos\phi_{Z-B} - 1 = K\delta_{in} = K\frac{3}{\eta}\left(\frac{\sinh\eta - \sin\eta}{\cosh\eta - \cos\eta}\right) \quad (4)$$

$$\frac{P_B}{P_Z}\sin\phi_{Z-B} = K\delta_{out} = K\frac{6}{\eta}\left(\frac{1}{2}\frac{(\sinh\eta + \sin\eta)}{(\cosh\eta - \cos h)} - \frac{1}{\eta}\right) \quad (5)$$

where $\eta = (2\omega r^2/D)^{1/2}$ and D is the intracrystalline diffusivity.

In the FR measurements, c. 50 mg of zeolite sample was placed into a sorption chamber and activated at 450 °C (or in some cases at a lower temperature) for 14 hours before carrying out FR experiments. The sorbate was then admitted to the sample and allowed to come to sorption equilibrium at a desired pressure and temperature. Measurements were carried out in the presence and absence of the zeolite sample to obtain the respective FR parameters.

NaY zeolite (Na_{55} Al_{54} Si_{138} O_{384}) was provided by Union Carbide in 1970. HY zeolite ($H_{53.5}$ $Na_{0.75}$ $Al_{54.3}$ Si_{138} O_{384}) was prepared with 1 M NH_4Cl solution after 15 times repeated ion-exchange to the NaY zeolite, and then calcinated at 400 °C. NaM zeolite ($Na_{8.36}$ $Al_{8.36}$ $Si_{39.64}$ O_{96}) and HM zeolite ($H_{5.5}$ $Na_{0.47}$ $Al_{6.33}$ $Si_{41.8}$ O_{96}) were received from Norton Corp., Mass., USA in 1970. The pyridine was from Aldrich with GC purity. The ammonia sorbate was from ARGO International with purity of 99.96%.

RESULTS AND DISCUSSION

Figure 1 shows the FR spectra of pyridine sorption on HY, HM, NaY and NaM zeolites. On HY zeolite the in-phase characteristic function intersected the out-of-phase curve at its maximum, indicating that the pyridine sorption on the acidic sites in the HY zeolite was the rate-controlling step rather than pyridine diffusion. Figure 2 shows the temperature and concentration dependences of pyridine sorption kinetic time constant in the HY zeolite, as determined from the FR spectra. The sorption kinetic time constant decreases with increasing temperature or equilibrium pressure. For the pyridine/HM system, the in-phase function curve joined the out-of-phase function curve at high frequencies. This kind of characteristic function curves, typical of a FR spectrum for a diffusion controlled process, indicates that pyridine diffusion in the HM zeolite was the rate-controlling process. Figure 3 shows the concentration dependence of the measured diffusion coefficients for pyridine in the HM zeolite along with FTIR data measured at a very low pressure [12, 13]. The pyridine diffusivity in HM measured by the FR method at relative high equilibrium pressures (0.25 ~ 2 torr) is about three orders of magnitude faster than those

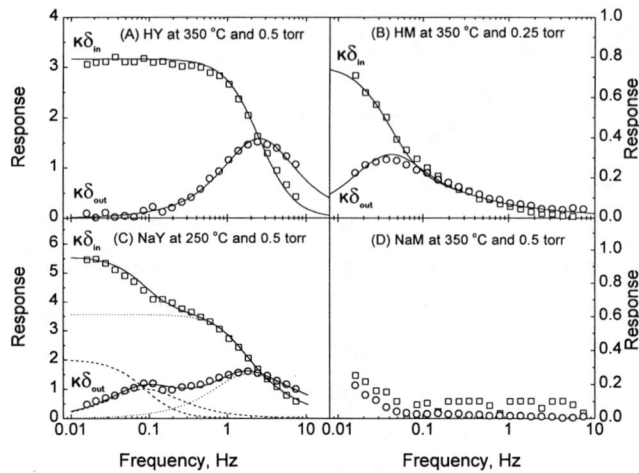

Figure 1: FR spectra of the in-phase (□) and out-of-phase (O) characteristic functions for pyridine in HY and HM zeolites. Lines are fits by the sorption model for HY and the diffusion model for HM.

measured by the FTIR method on the same sample at an extremely low pressure range from 0 to 0.02 torr [12, 13]. This discrepancy may be due to the difference in the pyridine concentrations and/or the different techniques used. An early NMR study of water and ammonia self-diffusion in NaX zeolite showed that the diffusion coefficients of water and ammonia were significantly influenced by the sorbate concentration [14], e.g. of Type III diffusion process [15]. The diffusivity of ammonia at higher concentrations is much faster than those at lower concentrations because strong interactions between the basic molecules and acidic sorption sites pose energetic hindrance to the intracrystalline transport at much lower concentrations, whereas at higher concentrations, the molecular mobility may increase significantly due to progressive saturation of the strongest sorption sites and increased mobility of the cations.

It is of great interest to note that the sorption kinetics of pyridine in the HY and HM zeolites are dramatically different. The reason for these phenomena may be due to the difference in their framework structures and pore sizes, which strongly affect the rate-controlling steps for pyridine in these two types of zeolites. In the three-dimensional faujasite zeolite, which has large supercages and wide pore openings (7.4 Å), diffusion processes are normally very fast. Therefore, the rate-controlling steps for pyridine in the HY zeolite is most probably determined by the chemisorption processes on the acidic sites. On the other hand, pyridine diffusion in the mordenite zeolite is the dominant process due to smaller "one-dimensional" channel structure of mordenite with respect to pyridine, which cannot access the "side-pockets" of the mordenite zeolite. Similar phenomenon was also observed from the FR study of ammonia sorption in NaA and KA zeolites [16]. The rate-controlling steps for NH$_3$ sorption in NaA zeolite, were determined by sorption processes, whereas in KA zeolite, ammonia diffusion was dominant.

On NaY, the FR spectrum for pyridine became more complicated and two response peaks were observed. This is due to different sorption processes occurred on different sorption sites present in the sodium form of zeolite Y. Compared with HM zeolite, the FR spectrum for pyridine on NaM appeared at a very low frequency outside the frequency window, indicating that pyridine diffusion in the Na$^+$ form of mordenite was significantly slower than that in the H$^+$ form of the mordenite zeolite. These FR data have shown that the pyridine molecule can only be used to characterise sorption sites in large pore zeolites, but not in small pore zeolites where pyridine diffusion processes are the dominant steps.

The FR spectra of NH$_3$ on the same HY, HM, NaY and NaM zeolites are plotted in Figures 4 and 5. Significant differences in the FR spectra were observed on these zeolites when NH$_3$ was used as a probe molecule.

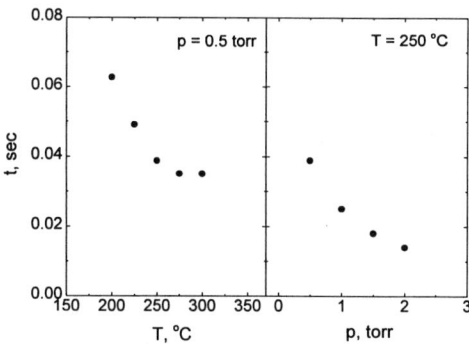

Figure 2: Temperature and concentration dependences of pyridine sorption time constant in HY zeolite

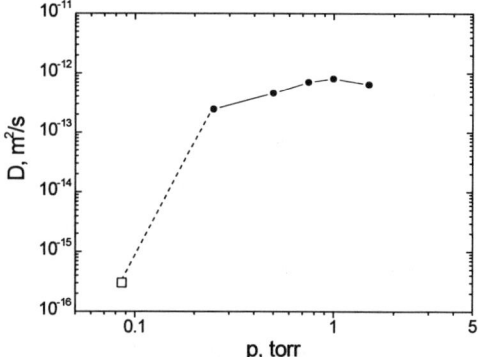

Figure 3: Concentration dependence of pyridine diffusivity in HM zeolite at 350 °C measured by FR (●) and FTIR (□) [12, 13] methods.

A wide temperature range was applied in order to distinguish different sorption processes, characterising the acid-base interactions between NH$_3$ and various sorption sites in the zeolites investigated. The most expressive spectra were given at 100, 250 and 450 °C for comparisons. The NH$_3$ responses were generally more complicated compared with those for pyridine because the small ammonia molecules were able to interact closely with different sites present in these zeolites. The complexity may be associated with different sorption rates for NH$_3$ sorption on different acidic sites. The profiles of the characteristic functions indicate that NH₃ chemisorption was the dominant rate-controlling process. None of the NH₃ response spectra reflected a diffusion-controlled process. In the cases where more than one response peak was observed, the FR spectra could be fitted by the multi-sites sorption model. NH₃ has shown great advantages as a basic probe molecule because of its small kinetic diameter, compared to pyridine.

The changes in shapes and intensities of the NH₃ FR spectra with increasing temperature are significantly different between the sodium form and hydrogen form of the faujasite and mordenite zeolites. As the temperature increased, the intensity of the response curves on both Na⁺ form and H⁺ form of the zeolites was reduced due to the decrease in the amount of NH₃ adsorbed and different forms of the chemisorbed species associated with different strengths of the sorption sites. At 450 °C, no FR responses of NH₃ sorption were observed on the NaY and NaM zeolites, but the FR responses were still strong on the HY and HM zeolites, indicating that at low temperatures the FR spectra may reflect both physical sorption and chemisorption processes in the zeolites, whereas at high temperatures the FR responses, e.g. for HY and HM, characterise the acidic sites in the H⁺ forms of zeolites.

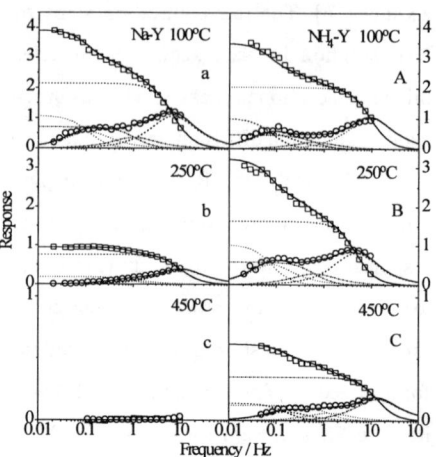

Figure 4: NH₃ FR spectra in NaY (abc) and HY (ABC) at 1 torr. Lines are fits by the sorption model.

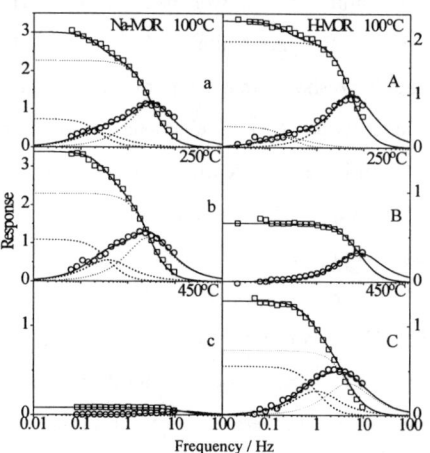

Figure 5: NH₃ FR spectra in NaM (abc) and HM (ABC) at 1 torr. Lines are fits by the sorption model.

These results clearly demonstrate the need for studying FR spectra over a wide temperature range in order to differentiate "physical' and a variety of chemisorption sorption processes in different forms of zeolites.

Figure 6 shows the temperature dependence of overall intensities of the response signals for NH_3 on Y and M zeolites. On the Na^+ form of Y and M zeolites, the intensity of response reduced rapidly as the temperature increased, whereas on the H^+ form of the zeolites, a minimum and a maximum were observed. This phenomenon may reflect different NH_3 sorption processes, e.g. physical sorption processes and/or chemisorption on Brönsted and/or Lewis acidic sites. The large NH_4^+ ions can act as Lewis acid sites and can adsorb NH_3 through their lone electron pairs. This reaction is weaker than NH_3 protonation reaction i.e. the interaction be-

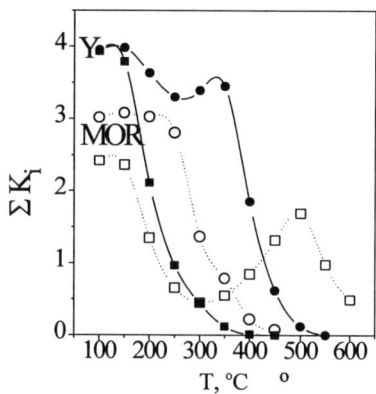

Figure 6: Influence of temperature on intensity of NH_3 responses in NaM (O), HM (□), NaY (■) and HY (●).

tween NH_3 and a strongly acidic surface hydroxyl proton. At low temperatures, NH_3 can form ensembles of mono-, bi-, tri-, etc. ammoniates and thus the response at low temperatures is less affected by the strength of the Brönsted acidic sites. Below approx. 300 °C and/or high NH_3 pressures, strong acidic sites are saturated, and therefore, NH_3 sorption rate spectra are mainly determined by physical sorption processes and/or NH_3 chemisorption on weaker Lewis acid sites. At high temperatures, NH_3 molecules can only be adsorbed on strong Brönsted acidic sites, and this leads to appearance of response peaks for HY and HM at high temperatures, but not for NaY and NaM. Furthermore, the maximum FR response for ammonia on HY zeolite appears at the high temperature of ~500 °C, whereas on HM zeolite it occurs at 350 °C. This may characterise that the acidic sites in the HY zeolite are much stronger than those in the HM zeolite.

CONCLUSIONS

The frequency-response technique is a useful tool capable of distinguishing and determining the rate-controlling steps of pyridine and ammonia sorption processes in various forms of faujasite and mordenite zeolites. The framework structures, pore sizes and cation forms strongly affect pyridine uptakes in faujasite and mordenite zeolites. The chemisorption of pyridine on acidic sites was the rate-controlling step in the faujasite zeolite, whereas pyridine diffusion was the dominant process in the mordenite zeolite.

For acidic site characterisation using the FR technique, ammonia was found to be a better probe molecule because it, having a small kinetic diameter, diffuses fast and is able to access most sorption sites, compared to pyridine. The ammonia FR spectra, specially at high temperatures, can be used to characterise chemisorption on different sorption sites present in different cation forms of zeolites. Further study is needed to develop the FR technique as a sorption kinetics method for studying the chemisorption of ammonia on acidic sites in zeolites.

REFERENCES

1. Y. Yasuda, Heterog. Chem. Rev. **1,** 103 (1994).
2. N.G. van Begin and L. V. C. Rees, Zeolites: Facts, Figures, Future, edited by P. A. Jacobs and R. A. van Santen, (Elsevier, Amsterdam, 1986), p915.
3. D. Shen and L.V.C. Rees, J.C.S. Faraday Trans., **90,** 3011 (1994).
4. D. Shen and L.V.C. Rees, J.C.S. Faraday Trans., **90,** 3017 (1994).
5. Gy. Onyestyak, D. Shen and L.V.C. Rees, J.C.S. Faraday Trans., **91,** 1399 (1995).
6. Gy. Onyestyak, D. Shen and L.V.C. Rees, Microporous Materials, **5,** 279 (1996).
7. Gy. Onyestyak, D. Shen and L. V. C. Rees, in Proc. ZEOCAT'95 Catalysis by Microporous Materials, Edited by H.K. Beyer, H.G. Karge, I. Kiricsi and J. B. Nagy, (Elsevier, Amsterdam, 1995), pp116-123.
8. Gy. Onyestyak, D. Shen and L.V.C. Rees, J.C.S. Faraday Trans., **92,** 307 (1996).
9. Gy. Onyestyak, J. Valyon and L. V. C. Rees in Proc. 11th Int. Zeolite Conf. Progress in zeolite and microporous materials, Edited by H. Chon, S. Ihm, and Y. Uh, (Elsevier, Amsterdam, 1996), pp703-710.
10. R.G. Jordi and D.D. Do, Chem. Eng. Sci., **48,** 1103 (1993).
11. L.M. Sun, and V. Bourdin, Chem. Eng. Sci., **48,** 3783 (1993).
12. H.G. Karge and K. Klose, Berichte der Bunsen-Gesellschaft, **79,** 454 (1975).
13. H. Bludau, H. G. Karge and W. Niessen, Microporous Mesoporous Materials, in press.
14. C. Förste, A. Germanus, J. Kärger, A. Möbius, M. Bülow, S.P. Zhdanov and N.N. Feoktistova, IsotopenPraxis, **25,** 48 (1989).
15. J. Kärger and H. Pfeifer, Zeolites, **7,** 90 (1987).
16. J. Valyon, Gy. Onyestyak and L.V.C. Rees, in Proceedings of the 6th International Conference on Fundamental of Adsorption, (France, 1998), in press.

EFFECT OF ACIDO-BASICITY OF BETA ZEOLITES ON THE CONVERSION OF CHLOROMETHANE TO HYDROCARBONS AS STUDIED BY FTIR AND TPD-MS

B.-L. SU[1*], D. JAUMAIN[1], K. NGALULA[1] and M. BRIEND[2]

[1]Laboratoire de Chimie des Matériaux Inorganiques, Université de Namur, 61 rue de Bruxelles, 5000 Namur, Belgium; Fax: (32) 81 724530 and email: bao-lain.su@fundp.ac.be and [2]Laboratoire de Réactivité de Surface, Université P. et M. Curie, 4 Place Jussieu, 75252, Paris, France

ABSTRACT

Adsorption and direct catalytic conversion of chloromethane to higher hydrocarbons on a series of acidic and basic Beta zeolites have been studied using FTIR spectroscopy. The products formed from the conversion have been identified using TPD-mass spectroscopy. The present study shows that both acidic and basic Beta zeolites are very active for the conversion of CH_3Cl. Higher hydrocarbons such as aromatics, alkenes and alkanes are formed. The elimination of HCl is the first step of the conversion both on acidic and basic zeolites. However, the oxygen atoms, being the intrinsic basic sites, are the active sites in cationic Beta zeolite while the protons of the acidic Beta zeolites catalyse the conversion of CH_3Cl. The different active sites lead to different reaction mechanisms. It is found that on NaBeta, the removal of HCl is an intermolecular reaction, giving more linear products, whereas on HBeta, the elimination of HCl is an intramolecular reaction, giving surface methoxy intermediate and further surface carbene by deprotonation as observed in the conversion of methanol. The formation of the aromatics and branched alkenes and alkanes is favored on protonated Beta zeolite.

INTRODUCTION

Much attention has been devoted to efficient catalytic conversion of methane to higher hydrocarbons after the oil crisis in 1970s. Recently, the conversion of methane via chloromethane to higher olefins has shown several advantages over other routes for chemical utilization of the abundant natural gas [1]. This direct conversion presents some other obvious importances. Chlorinated volatile organic compounds (VOC) are often the industrial hazardous solvent wastes. Their catalytic destruction, or if possible, their conversion to useful organic compounds like higher hydrocarbons will be an interesting matter for environment protection [2]. Direct conversion of CH_3Cl to higher hydrocarbons has recently been studied on HZSM-5 and on faujasites exchanged with alkali cations and some divalent metal cations [1, 3]. Beta zeolite since its synthesis has widely been used as adsorbent and as catalyst in different processes of hydrocarbon compounds. The aim of the present work is to study the effect of acido-basic properties of Beta zeolites on the adsorption and conversion of CH_3Cl and to try to elucidate the reaction mechanism in order to develop the new catalysts with advanced performances.

EXPERIMENTAL

Materials

The starting Beta zeolite (Si/Al= 17.3) was synthesized in alkaline medium in the presence of tetraethylammonium hydroxide. HBeta zeolite was obtained from a calcination of the as-synthesized Beta sample in nitrogen and then in oxygen at 773K. NaBeta was prepared by ion exchange from organic template-free HBeta zeolite. The ion exchange procedure was repeated two times to get fully Na^+-exchanged Beta zeolites.

Infrared studies

Self-supported zeolite wafers (15 mg/cm^2) were prepared with a pressure of 5 t/cm^2 and calcined in a Pyrex IR cell at 723K in oxygen and then in vacuum. The adsorption of known and increasing amounts of CH_3Cl on the wafers was performed [4]. The catalytic conversion of CH_3Cl was carried out in-situ at 673K during 15 min. After reaction, the IR cell containing the sample was quickly cooled to RT and maintained for 1 h for equilibration and IR spectra were then recorded. These were followed by a desorption at RT for 30 min to study the irreversible adsorption of products and reactant. The crystallinity of catalysts after reaction was checked using the following procedure: 1g of zeolite powder was loaded in a reactor and pretreated as described above. The temperature of the samples was then adjusted to 673K. A flow of pure CH_3Cl gas was introduced during 1 h. The crystallinity of the catalysts was then examined using XRD.

Identification of products using TPD-MS

30 mg of sample was loaded in a reactor. The sample was pretreated as described above.

After cooling the reactor to room temperature, a known amount of CH_3Cl was introduced into the reactor. After equilibration, the temperature of the sample containing adsorbed CH_3Cl was raised rapidly to 673 K and this temperature was maintained for 15 minutes. The reactor was then cooled to room temperature and linked to a TPD-vacuum line. The desorbed products from the sample were directly analysed by Mass Spectrometer.

RESULTS AND DISCUSSION

The interaction of CH_3Cl with catalysts at room temperature and the conversion of CH_3Cl at 673K on acidic and basic Beta zeolites have been carried out at various of CH_3Cl loadings. The crystallinity of the catalysts was checked after reaction using XRD and showed unchanged. For the sake of brevity, only IR spectra of zeolites saturated with CH_3Cl are shown here.

Adsorption and conversion of CH_3Cl on Na-Beta zeolite

Figure 1 reports the IR absorbance spectra of species adsorbed on NaBeta zeolite after introduction of 20 torrs of chloromethane (Fig. 1b), after reaction at 673K for 15 min (Fig. 1c) and after desorption at room temperature during 30 minutes (Fig. 1d). The zeolite phase alone is also given (Fig. 1a) for comparison. Only one sharp peak at 3745 cm^{-1} and a small band at 3674 cm^{-1}, corresponding to the external silanols and the extra-framework Al-OH species, respectively, are detected after pretreatment (Fig. 1a). This indicates that the ion exchange was complete. However, the presence of the small peak at 3674 cm^{-1} suggests the slight dealumination of the zeolite during the pretreatment. Upon adsorption of increasing and known amounts of CH_3Cl, four peaks at 2963, 2860, 1442 and 1349 cm^{-1}, assigned to the asymmetric and symmetric stretching and bending vibrations of CH_3 groups, appear. The peak of silanols decrease progressively in intensity and a new broad band centred at 3621 cm^{-1} attributed to the interaction of CH_3Cl with silanols is generated. Even in the presence of a high pressure of CH_3Cl in the cell (Fig. 1b), the silanol peak does not disappear completely, implying that not all the silanols can interact with CH_3Cl.

After heating NaBeta zeolite saturated with CH_3Cl at 673K during 15 minutes, important changes take place (Fig. 1c). The intensity of the peaks at 2963, 2860, 1442 and 1349 cm^{-1}, assigned to the vibrations of CH_3 groups of chloromethane decreases strongly and a series of new peaks appears. The wavenumbers and their assignments are given in Table 1. Higher hydrocarbons such as aromatics (C=C at 1504 cm^{-1}), alkenes and alkanes (C=C at 1627, -CH_3 at 2960 cm^{-1} and -CH_2- at 2933, 1465, 1387 and 1375 cm^{-1}) are formed. A small peak at 1530 cm^{-1}, corresponding to C=C vibration of coke is also observed, indicating the formation of some coke on this zeolite. The zeolite wafer becomes indeed slightly black after reaction, confirming the formation of coke.

As can be seen from Fig. 1c, the OH region becomes more complicated after reaction. Four new broad peaks are observed at 3651, 3558, 3479 and 3219 cm^{-1}.

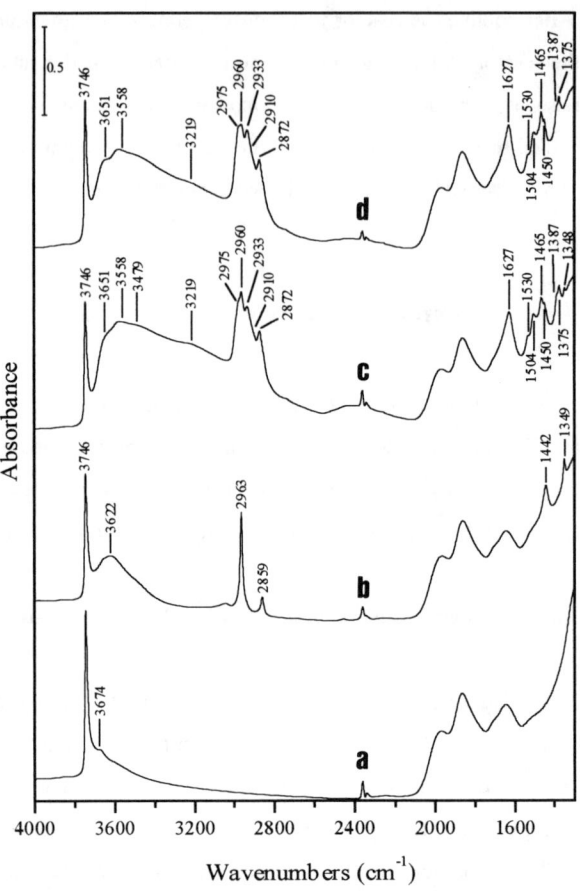

Figure 1, Adsorption, conversion of chloromethane and desorption of formed products on NaBeta; a: after pretreatment at 723K; b: adsorption of CH_3Cl (saturation condition); c: after reaction at 673K and d: after evacuation at room temperature during 30 min

On the basis of the previously reported results on the interaction of aromatics, alkenes and alkanes with protonated zeolites, the broad bands observed at 3219 and 3558 cm^{-1} should correspond to the interaction of hydrocarbons, containing π electrons such as aromatics and alkenes, with bridging framework OH groups and the external silanols, respectively. The two other peaks at 3479 and 3651 cm^{-1} can be assigned to the interaction of alkanes with bridging framework OH groups and silanols, respectively. The above observation proves also the formation of higher hydrocarbons such as aromatics, alkenes and alkanes from CH_3Cl. However, as described above,

Table 1, Wavenumbers and assignments of IR peaks of the products formed on two Beta zeolites

Wavenumbers (cm^{-1})		Assignments	Wavenumbers (cm^{-1})		Assignments
on NaBeta	on HBeta		on NaBeta	on HBeta	
2975	-	(C-H) of =CH$_2$	1504	1504	(C-C) of aromatics
2960	2960	ν(C-H)$_{as}$ of CH$_3$	-	1471	(C-C) of benzene
2933	2935	ν(C-H)$_{as}$ of -CH$_2$-	1465	1465	δ(C-H) of -CH$_2$-
2906	2910	-	1450	1450	δ(C-H)$_{as}$ of CH$_3$
2872	2872	ν(C-H)$_s$ of CH$_3$	1387	1388	(CH$_2$) deformation
2710	2710	-	1375	1375	(CH$_2$) deformation
1627	1627	(C=C) of alkenes	-	1367	-
-	1606	(C=C) of alkenes	1348	1348	δ(C-H)$_s$ of CH$_3$
1530	1535	(C=C) of coke			

no bridging framework OH groups on fully Na$^+$-exchanged Beta zeolite areobserved after pretreatment. The presence of these new broad bands corresponding to the bridging framework OH groups interacting with the formed products indicates that there is a generation of some bridging framework OH groups after conversion of CH$_3$Cl at 673K. These OH groups might be generated by some compounds containing hydrogen atoms, formed during the reaction and these compounds should be enough acid to interact with the oxygen atoms of NaBeta zeolite to give OH groups. It was reported that the adsorption of HCl on dehydrated NaY zeolite at room temperature [5] can generate some OH groups due to the interaction of highly polar HCl molecules with oxygen atoms of the NaY zeolite framework and these generated OH groups have similar behaviour to those of HY zeolite. However, these created OH groups can easily be removed from the zeolite by an evacuation at 423 K and the adsorption of HCl on NaY creating these OH groups does not affect the crystallinity of zeolite. The above results lead us to suggest that the compound containing protons, being able to interact with framework oxygen atoms to form OH groups, should be HCl, a product of CH$_3$Cl conversion. In a previous study on the interaction of n-propyl chloride with cationic zeolites, the creation of OH groups on the studied zeolites has also been attributed to the interaction of HCl molecules formed by decomposition of n-propyl chloride with surface oxygen atoms [6].

All above observations suggest that NaBeta zeolite dehydrohalogenates the CH$_3$Cl molecules splitting off HCl molecules. Since our study is carried out in a closed IR cell, the formed HCl molecules stay in the cell after reaction. When the cell is cooled to room temperature, HCl in the gas phase will interact with zeolite. The proton part of HCl molecules then in-situ reacts with surface oxygen atoms and attaches itself to the oxygen atoms creating OH groups. The chloride part will associate with the surface cations. The formed OH groups will thus interact with the formed products. Desorption at room temperature of products adsorbed on NaBeta zeolite (Fig. 1d) does not result in a significant modification, implying that the formed products adsorb strongly on NaBeta zeolite compared to CH_3Cl since a same desorption can remove all the CH_3Cl adsorbed on this zeolite (spectra are not shown here).

Adsorption and conversion of CH_3Cl on H-Beta zeolite

Two peaks at 3612 and 3746 cm^{-1}, corresponding to the bridging framework OH groups and silanols, respectively, are present after pretreatment of H-Beta (Fig. 2a). Two small peaks at 3782 and 3668 cm^{-1}, previously attributed to the Al-OH species near to one or more Si-OH groups generated when Al leaves the framework and the extra-framework, are also detected. With introduction of CH_3Cl (Fig. 2b), four peaks at 2963, 2862, 1442 and 1349 cm^{-1}, assigned to vibrations of CH_3 groups, appear. Moreover, the peak at 3612 cm^{-1} disappears and simultaneously a new broad band at around 3223 cm^{-1} appears. This broad band corresponds to the interaction of bridging OH group with CH_3Cl and shifted towards low wavenumber. After complete disappearance of the peak at 3612 cm^{-1}, the intensity of the silanol groups decreases and a new small broad band is generated at 3602 cm^{-1}. This band is attributed to the silanol groups interacting with CH_3Cl and shifted towards low wavenumber. In the presence of a high pressure of CH_3Cl in the cell (Fig. 2b), silanols are still present, indicating that not all the silanols can interact with CH_3Cl molecules.

After reaction at 673K, important modifications are observed in the regions of 4000-2800 and 1700-1200 cm^{-1} (Fig. 2c). The peaks at 2963, 2862, 1442 and 1349 cm^{-1}, corresponding to vibrations of CH_3 groups disappear completely, implying the complete conversion of CH_3Cl at this temperature. The wavenumbers of new peaks and their possible assignments are listed in Table 1. It can be seen that higher hydrocarbons such as aromatics, alkenes and alkanes are formed. The intensity of the peaks in the region of 4000-3000 cm^{-1} is highly increased after reaction. Three important broad bands at 3652, 3510, 3204 cm^{-1} instead of only two at 3602 and 3223 cm^{-1} before reaction are detected. The important increase in intensity for the peaks generated by the interaction of OH groups with the formed products indicates clearly the creation of the new OH groups by the conversion of CH_3Cl. As observed in NaBeta zeolite, HCl molecules are formed as a product of the conversion of CH_3Cl. HCl molecules then in-situ reacts with surface oxygen atoms and attaches themselves to the oxygen atoms creating supplementary bridging framework OH groups. The formed OH groups will interact with the formed products. That is why the intensity of the broad bands observed in the region of 4000-3000 cm^{-1} is increased. After desorption of the products adsorbed on HBeta zeolite at room temperature (Fig. 2d), the intensity of all the peaks decreases, implying the removal of some products which adsorb weakly on HBeta zeolite.

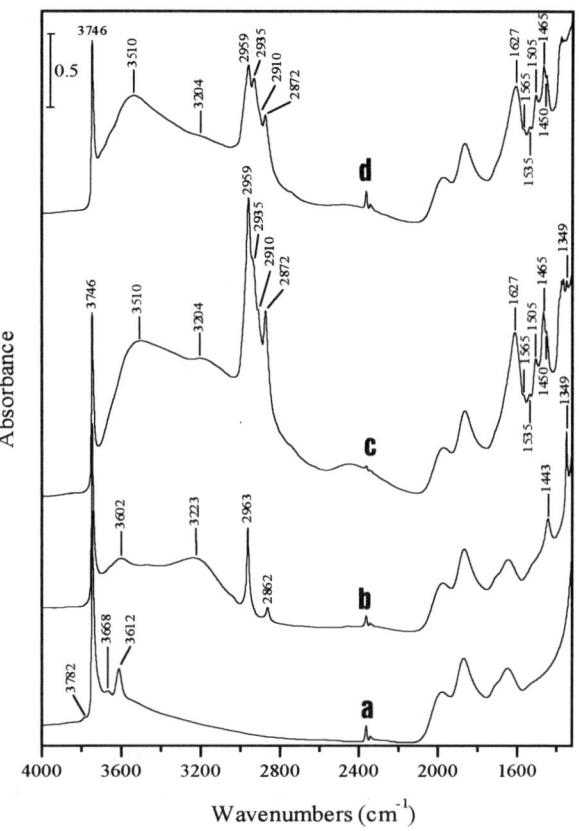

Figure 2, Adsorption, conversion of chloromethane and desorption of formed products on HBeta; a: after pretreatment at 723K; b: adsorption of CH_3Cl (saturation condition); c: after reaction at 673K and d: after evacuation at room temperature during 30 min

Identification of products using TPD-Mass Spectrometer

CH_3^+, Cl, HCl, $CH_3\text{-}CH_2^+$, $CH_2\text{=}CH_2$, $CH_2\text{=}CH^+$, $CH_2\text{=}CH\text{-}CH_2^+$, $CH_2\text{=}CH\text{-}CH_3$, $CH_3\text{-}CH_2\text{-}CH_2^+$, $C_4H_9^+$ and C_4H_{10} are detected both on HBeta and NaBeta zeolites. $C_5H_{11}^+$ is observed only on NaBeta while $C_6H_4^+$, $C_6H_5^+$ and $C_6H_5\text{-}CH_2^+$ are detected on HBeta. Aromatics observed by IR on NaBeta are not detected here probably due to the small quantity formed. Both IR and Mass Spectrometer demonstrate that the formation of aromatics on NaBeta is less important than on HBeta zeolite. It is very important to note that a CH_4 signal is not observed, indicating that methane is not a product, at least not a main product of the chloromethane conversion.

Comparison of the CH$_3$Cl conversion on NaBeta and HBeta

The ratio of the intensity of -CH$_3$ groups at 2960 cm^{-1} and -CH$_2$- groups at 2934 cm^{-1} on NaBeta and HBeta after reaction (Figs. 1c and 2c) is very different and is quite higher on HBeta than that on NaBeta. This means that the products formed on HBeta are more branched than those on NeBeta. This is in agreement with what we observed by catalytic test. The reaction mechanism on these two zeolites with different acido-base characters are probably different. The elimination of the HCl from CH$_3$Cl molecules is observed to be the first step both for acidic and basic Beta zeolites. However, on NaBeta, the removal of HCl is an intermolecular reaction, giving more linear products, while on HBeta, the elimination is an intramolecular reaction, giving surface methoxy intermediate and further surface carbene by deprotonation as observed in the conversion of methanol. However, more work will be done to confirm the above observations.

CONCLUSION

Both acid and basic Beta zeolites are very active for the conversion of CH$_3$Cl. However, for sodium zeolites, the oxygen atoms, being the intrinsic basic sites, are the active sites and zeolites with low Si/Al ratio have high catalytic activity. By contrast, for protonic zeolites, the protons are the active sites. Sodium zeolites give more linear products while HBeta produces more branched hydrocarbons and even some aromatics. On NaBeta, the removal of HCl is an intermolecular reaction while on HBeta, the elimination is an intramolecular reaction, giving surface methoxy intermediate and further surface carbene by deprotonation.

REFERENCES

1. X. R. Xia, Y. L. Bi, T. H. Wu and K. J. Zhen, Catal. Lett., **33**, 75, (1995)
2. F. Solymosi, J. Rasko, E. Papp, A. Oszko and T. Bansagi, Appl. Catal., A, **131**, 55 (1995)
3. D. K. Murry, J. W. Chang and J. F. Haw, J. Am. Chem. Soc., **115**, 4732 (1993)
4. B. L. Su and V. Norberg, Zeolites, **19**, 65 (1997)
5. A. de Mallmann, PhD thesis, Université de Paris, France, 1989
6. C. L. Angell and M. V. Howell, J. Phys. Chem., **74**, 2737 (1970)

CHLOROMETHANE AS PROBE MOLECULE TO CHARACTERIZE THE BRÖNSTED ACIDITY OF ZEOLITES: AN IN-SITU FTIR STUDY

B.-L. SU* and D. JAUMAIN

Laboratoire de Chimie des Matériaux Inorganiques, Université de Namur, 61 Rue de Bruxelles, B-5000 Namur, Belgium; Fax: (32) 81 72 45 30 and E-mail: bao-lian.su@fundp.ac.be

ABSTRACT

A series of protonated zeolites with different structures and Si/Al ratios such as HY, H(Na)EMT, HMOR, HBeta, HZSM-5 has been studied by means of in-situ FTIR spectroscopy using chloromethane as probe molecule. The negatively charged chlorine atom of CH_3Cl molecule interact with the hydroxyls of acidic zeolites through hydrogen bonding, i.e., $H_3C-Cl^{\delta-}\cdots H^{\delta+}-OZ$, causing the shift of the hydroxyl infrared peaks towards the lower wavenumbers and the extent of the shift supplies the important information on the strength of Brönsted acid sites of zeolites. The acid strength ranks in the following order for the studied zeolites: HBeta (Si/Al=12.5) > HMOR (9.5) ~ HZSM-5 (21.7) > HY (2.4) ~ H(Na)EMT (3.6), being in agreement with what observed in the conversion of propene and chloromethane as test reaction using these zeolites as catalysts.

INTRODUCTION

Because of its importance in industrial catalytic applications, the Brönsted acidity of zeolites, which is related to the framework hydroxyl group Si-OH-Al, has widely been investigated with experimental and quantum methods [1-9]. The spectroscopic study using probe molecules has been shown to be an important way to determine the nature and the number of acid sites in zeolites. In the recent years, the different probe molecules have greatly enlarged the amount of available information on the zeolite acidity. Two main types of probes have been considered [9]. First, the probe molecules such as NH_3, pyridine, amine, etc., form a chemical bond with the protons of hydroxyls, this gives access to the concentration of acid sites of zeolites. On the other hand, the aromatics, olefins, CO, H_2S, etc., which can interact with the protonic sites through hydrogen bonding, can be used to acquire information on the strength of protonic sites and on the accessibility of the Brönsted acid sites to the probe molecules. However, the role, the nature and the strength of acid sites in solid catalysts have been the object of a dispute which is still far from being settled. The knowledge of what is the required acid concentration and strength for a defined reaction catalysed by acid solids is primordial in the modification of conventional catalysts and in the design of new catalysts with advanced performance. Although different probe molecules have been developed, searching a judicious molecular probe is still of important interest to get the desired information. The criteria for the selection of probe molecules has recently been summarised in an excellent review by H. Knözinger [9].

Recently, different small halogenated hydrocarbon molecules have been studied (mostly ethane and ethene derivatives) employing IR and NMR as analytical tools for the measurement of shifts in wavenumbers and chemical shifts of the signals originating from acidic bridging OH groups of zeolites [10, 11]. However, monitoring the effect of chloromethane adsorption as a measure of acidity of zeolites is less reported [12]. Since the high electronegative chlorine atom in chloromethane molecule is slight negatively charged, it can interact with the hydroxyls through strong hydrogen bonding. The modifications of chloromethane and the extent of its interaction with the acid sites (Figure 1) contains information not only on the strength of the accessible protonic sites but also on the number of these sites. Here, we report the first results of our study on the acid strength of zeolites.

Figure 1, Schematic representation of the interaction of CH_3Cl with hydroxyls of zeolites.

EXPERIMENTAL

HY, H(Na)EMT, HBeta and HZSM-5 were prepared by ion exchange from their Na form as described in refs. [5-7]. HMOR is a commercial product provided by Toyosoda Company. The chemical composition and characteristics of these zeolites are listed in Table 1.

Self-supported zeolite wafer ($15mg/cm^2$) prepared with a pressure of 5 tons/cm^2 was first calcined in a Pyrex IR cell with two NaCl windows at 723K in a flow of dry oxygen overnight and followed by evacuation for 6-8h at same temperature. After cooling to room temperature, the spectrum of zeolite wafer alone was recorded as reference using a Perkin-Elmer Fourier Transform Infrared Spectrometer Spectrum 2000. The adsorption of known and increasing amount of chloromethane was then performed on wafer as described in refs. [5-7]. After 1 h equilibration, the IR spectra of zeolite adsorbed chloromethane were recorded. All the peak values are obtained by deconvolution of the IR spectra using Winspec program. The desorption experiments have been performed at room temperature during 30 minutes.

RESULTS AND DISCUSSION

Adsorption of chloromethane on HY zeolite

Figure 2 reports the changes in the IR absorbance spectra of hydroxyls of HY zeolite in the range of 4000-2800 cm^{-1} upon adsorption of increasing and known amounts of chloromethane. Three peaks at 3745, 3640 and 3544 cm^{-1} are observed on HY zeolite after pretreatment (Fig. 2a). The hydroxyls which give rise to around 3741 cm^{-1} band have previously been attributed to the external silanols. Two other peaks at 3640 and 3544 cm^{-1} arise from the bridging framework Si-OH-Al groups located in the large cages and those in the sodalite cages and hexagonal prisms, respectively. When first molecules of chloromethane (4 molecules per unit cell) are

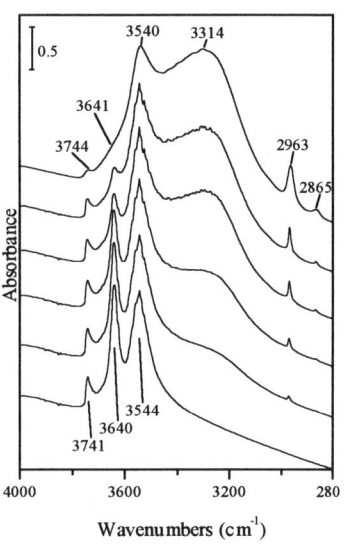

Figure 2, Changes in the IR absorbance of the hydroxyls of HY upon adsorption of increasing amounts of CH$_3$Cl

introduced into the cell (Fig. 2b), two bands at 2963 and 2865 cm^{-1}, corresponding to the CH$_3$ symetric and asymetric stretching vibrations appear immediately, indicating the adsorption of chloromethane on HY zeolite. It is also observed that the intensity of the peak at 3640 cm^{-1} assigned to the bridging framework Si-OH-Al groups located in the large cages decreases and a new broad band in the range of 3400-3200 cm^{-1} appears simultaneouly. The bridged Si-OH-Al groups interacting with CH$_3$Cl are shifted to a lower value and generate a broad band centred at 3314 cm^{-1}. This is in accordance with the observations previously reported [5-7]. With increasing the amount of chloromethane introduced, the intensity of the broad band centred at 3314 cm^{-1} increases while that of the peak at 3640 cm^{-1} decreases. When a high pressure of chloromethane is introduced into the cell, the peak at 3640 cm^{-1} disappears completely and the broad band is the

Table 1. Chemical composition and characteristics of the studied zeolites and extent of the shift, Δv_{OH} (cm^{-1}), of different hydroxyls in all the studied zeolite upon interaction with CH$_3$Cl

Zeolite	Chemical composition	Si/Al ratio	Characteristics	Δv_{OH} (cm^{-1})		
				Si-OH-Al	extra-framework	terminal silanols
HY	$H_{58}Al_{58}Si_{134}O_{384}$	2.4	large pore	325*	-	100
H(Na)EMT	$Na_{3.5}H_{17.5}Al_{21}Si_{75}O_{192}$	3.6	large pore	322*	-	106
HMOR	$H_4Al_4Si_{44}O_{96}$	9.5	large pore	415	241	132
HBeta	$H_{4.7}Al_{4.7}Si_{59}O_{128}$	12.5	large pore	427	-	128
HZSM-5	$Na_{0.4}H_{3.6}Al_4Si_{92}O_{192}$	21.7	medium pore	412	-	132

*Only the hydroxyls located in the large cages interact with CH$_3$Cl and the hydroxyls at 3544 cm^{-1} for HY and at 3553 cm^{-1} for H(Na)EMT are inaccessible for CH$_3$Cl and unaffected.

most intense. The silanol peak decreases also in intensity and a new shoulder at 3641 cm^{-1} is generated, indicating that part of silanol hydroxyls can also interact with CH$_3$Cl. However, the peak at 3544 cm^{-1} is not affected upon adsorption of CH$_3$Cl even in the presence of a high pressure of CH$_3$Cl, indicating the inaccessibility of this OH group. In fact, the peak at 3544 cm^{-1} corresponds to the bridged Si-OH-Al located in the hexagonal prisms and the sodalite cages. Chloromethane molecule cannot penetrate into these small cavities due to its volume.

Adsorption of chloromethane on H(Na)EMT zeolite

Same experiments have been made on H(Na)EMT zeolite and the spectra of hydroxyls of H(Na)EMT zeolite in the range of 4000-2800 cm^{-1} upon adsorption of increasing and known amounts of CH$_3$Cl are depicted in Figure 3. Three peaks at 3745, 3636 and 3553 cm^{-1} are observed on HEMT zeolite after pretreatment (Fig. 3a). The hydroxyls at 3745 cm^{-1} have previously been attributed to the external silanols. Two other peaks at 3636 and 3553 cm^{-1} arise from

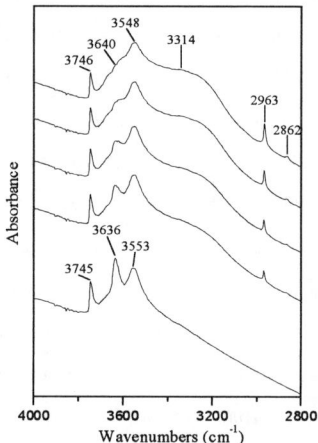

Figure 3, Changes in the IR absorbance of the hydroxyls of H(Na)EMT upon adsorption of increasing amounts of CH$_3$Cl

the bridging framework Si-OH-Al groups located in the large cages and in the sodalite cages and hexagonal prisms, respectively. EMT is the hexagonal analogue of faujasite. Two large cages (hypo- and hypercages) are present in EMT and different from the supercages in faujasite. Furthermore, the sodalite cages and hexagonal prisms in EMT structure are not equivalent to those occurring in faujasites. The wavenumber of Si-OH-Al groups located in the large cages (3636 cm^{-1}) and in the small cavities (3553 cm^{-1}) in EMT structure is therefore different from those in faujasite (3640 and 3544 cm^{-1}, respectively). The adsorption of increasing and known amounts of CH_3Cl gives the same observation as in the case of HY zeolite. The bridged Si-OH-Al groups located in the large cages are shifted to low wavenumber and give a broad band centred at 3314 cm^{-1} due to their interaction with CH_3Cl. The OH groups located in the small cavities remain unaffected, only the wavenumber changes slightly upon adsorption of CH_3Cl, implying their inaccessibility for chloromethane molecules. The intensity of the peak belonging to silanols decreases slightly, indicating interaction of part of silanols with CH_3Cl.

Adsorption of chloromethane on HMOR zeolite

In HMOR, three peaks at 3746, 3658 and 3608 cm^{-1} are detected after pretreatment and assigned to the silanols, the hydroxyls attached to the extra-framework Al species and the bridging framework Si-OH-Al groups, respectively. Upon adsorption of CH_3Cl (Figure 4), a broad band centred at 3194 cm^{-1} and a broad shoulder at 3417 cm^{-1} appear at the expense of the peaks at 3608 and 3658 cm^{-1}, respectively. At higher chloromethane loadings, the peak of silanols decreases also in intensity. This indicates that all three OH groups present in H-Mordenite can interact with CH_3Cl.

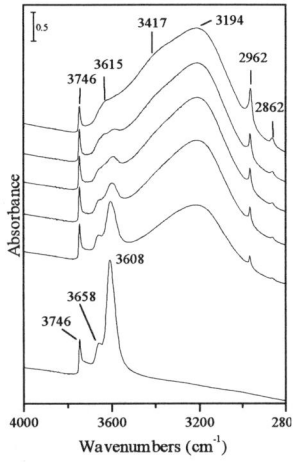

Figure 4, Changes in the IR absorbance of the hydroxyls of HMOR upon adsorption of increasing amounts of CH_3Cl

Adsorption of chloromethane on HZSM-5 zeolite

In HZSM-5, besides the silanols at 3745 cm^{-1}, a small peak at 3724 cm^{-1}, a sharp peak at 3612 cm^{-1} and a broad band in the range of 3745-2900 cm^{-1} are recorded after pretreatment (Figure 5a) and attributed respectively to the isolated internal silanols, the bridging framework Si-

OH-Al groups and the hydrogen bond formed by the internal silanols. Upon adsorption of CH_3Cl, the peak at 3611 cm^{-1} decreases progressively in intensity and a broad band centred at 3201 cm^{-1}, steming from the interaction of the bridging OH groups with CH_3Cl, grows in intensity (Figure 5). In the presence of a high pressure of chloromethane in the cell, the peak at 3611 cm^{-1} disappears completely and the silanols decrease in intensity and a broad band centred at 3610 cm^{-1}, originating from the interaction of silanols with CH_3Cl, is observed. This part of results indicates that both silanols and bridging framework OH groups of HZSM-5 can interact with chloromethane. The changes of the peak at 3724 cm^{-1}, upon adsorption of CH_3Cl, cannot be evaluated due to its low intensity.

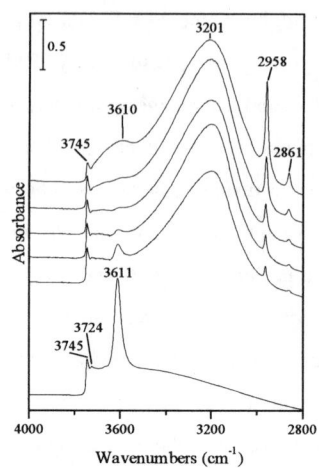

Figure 5, Changes in the IR absorbance of the hydroxyls of HZSM-5 upon adsorption of increasing amounts of CH_3Cl

Adsorption of chloromethane on HBeta zeolite

In HBeta, other than the external silanols at 3745 cm-1, a small peak at 3781 cm-1 and a sharp peak at 3610 cm-1, corresponding to the Al-OH species near to one or more Si-OH generated when Al leaves the framework and to the bridging framework Si-OH-Al species, respectively. Upon adsorption of CH_3Cl, the complete disappearance of the bridging framework hydroxyls is accompanied by the generation of a broad band at low wavenumber, 3183 cm-1 (Figure 6). The peak at 3745 cm-1 in the presence of a high pressure of CH_3Cl disappears completely, giving an intense broad band at 3614 cm-1. This means that all the silanols in HBeta can interact with CH_3Cl. This is not observed in other zeolites studied here. The high intensity of the band at 3614 cm-1 is due to the large number of silanols present in HBeta zeolite.

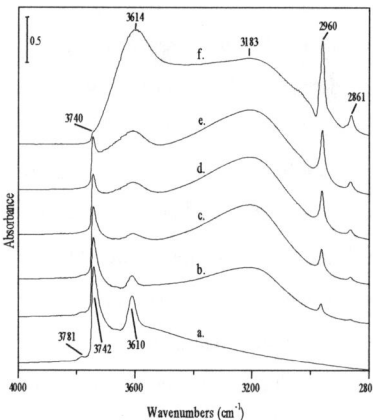

Figure 6, Changes in the IR absorbance of the hydroxyls of HBeta upon adsorption of increasing amounts of CH_3Cl

DISCUSSION

It was well known that OH groups interacting with adsorbed bases can generate a new band shifted to lower wavenumber value. The extent of the shift being characteristic of the strength of the Brönsted acidic sites [2, 5-7]. This method is applied in the present study and the extent of the shift for all the studied zeolites is calculated and listed in Table 1. It is clear that OH groups associated to the framework Al are more acid than other hydroxyls, and the bridging framework hydroxyls of HBeta gives the highest shift value, signifying that the framework hydroxyls of this zeolite have the highest acid strength in all the accessible hydroxyls of studied zeolites. The acid strength ranks in the following order for the studied zeolites: HBeta (Si/Al=12.5) > HMOR (9.5) ~ HZSM-5 (21.7) > HY (2.4) ~ H(Na)EMT (3.6). This is not in the same sense as that of Si/Al ratio, indicating the strong effect of zeolite structure on the acid strength although Eicher et al. [13] evidenced on the basis of quantum chemical results that the effect of the composition (Si/Al ratio) is much more important on acidity of zeolites. The previous work using benzene and ammonia as probes reported that a highly exchanged HEMT is more acid than HY [5-7], and that HBeta (Si/Al = 14 and 22) is less acid than a HZSM-5 (Si/Al ratio = 35) [2]. The presence of some Na^+ ions in H(Na)EMT zeolite, the low Si/Al ratio of our present HZSM-5 zeolite and using different molecular probes result probably in the difference in the order of acid strength between the results obtained from the present study and those previously reported. Nevertheless, the higher acidity of the present HBeta zeolite compared to our present HZSM-5 is confirmed in our more recent catalytic study in the conversion of CH_3Cl and propene using these two zeolites as catalysts [14, 15]. The higher acidity of HMOR compared to HZSM-5 was previously observed by Ghosh, et al. [16] in the propene reactions. The extra-framework hydroxyls in HMOR gives a shift value around 241 cm^{-1}, indicating the medium acid strength of these groups. The extent shift value of the silanols in all the studied zeolites varies from 100 to 132 cm^{-1}. It is evident that silanols are less acid in comparison with the framework and extra-framework hydroxyls. Hence, the obtained acid strength order using chloromethane as probe is convincing and reflects the real acidity of zeolites. However, the verification of the present results *via* application of independent techniques such as microcalorimetry, TPD, ^1H MAS NMR and suitable test reaction will be made. The desorption experiments have shown that an evacuation of 30 min at RT is needed to remove a great part of adsorbed CH_3Cl on these zeolites, implying a medium interaction strength between CH_3Cl and the hydroxyls of these zeolites.

CONCLUSION

The acid strength of a series of protonated zeolites has been evalauted using CH_3Cl as porobe molecule and ranks in the following order: HBeta (Si/Al=12.5) > HMOR (9.5) ~ HZSM-5 (21.7) > HY (2.4) ~ H(Na)EMT (3.6), being in accordance with what we observed in the conversion of propene and CH_3Cl as test reaction. The present work evidences that CH_3Cl should be an efficient molecular probe in determination of the strength and the accessibility of the Brönsted acid sites of zeolites. The obtained information is very useful not only for quantum calculation, in understanding the role and the nature of acid sites in zeolites, but also in the classification of acid strength of protonated zeolites.

REFERENCES

1. G. J. Kramer and R. A. van Santen, J. Am. Chem. Soc., **115**, 2887 (1993)
2. S. G. Hedge, R. Kumar, R. N. Bhat and P. Ratnasamy, Zeolites, **9**, 231 (1989)
3. M. Jia, H. Lechert and H. Förster, Zeolites, **12**, 32 (1992)
4. M. A. Makarovo, A. E. Wilson, B. J. van Lient, C. M. A. M. Mesters, A. W. de Winter and C. Williams, J. Catal., **172**, 170 (1997)
5. B. L. Su and D. Barthomeuf, J. Catal., **139**, 81 (1993); Zeolites, **13**, 626 (1993)
6. B. L. Su and V. Norberg, Zeolites, **19**, 655 (1997)
7. B. L. Su, J. M. Manoli, C. Potvin and D. Barthomeuf, J. Chem. Soc., Faraday Trans, **89**, 875 (1993)
8. M. Hanger, Catal. Rev., **39**, 345 (1997)
9. H. Knözinger, in Handbook of Heterogeneous Catalysis, Edited by. G. Ertl, H. Knözinger and J. Weitkamp, VCH, Weinheim, Vol. **2**, 707 (1997)
10. E. Brunner, J. Kärger, M. Koch and B. Standte, Stud. Sur. Sci. Catal., **105**, 463 (1997)
11. H. Sachsenröder, E. Brunner, M. Koch, H. Pfeifer and B. Staudte, Microporous Mater., **6**, 341 (1996)
12. U. Zscherpal, B. Staudte and H. Jobic, Z. Physik. Chem., **194**, 31 (1996)
13. U. Eichler, M. Brändle and J. Sauer, J. Phys. Chem., **B101**, 10035 (1997)
14. B. L. Su, D. Jaumain, K. Ngalula and M. Briend, to be published in present proceedings,
15. S. Siffert, L. Gaillard and B. L. Su, to be published in present proceedings,
16. A. K. Ghosh and R. A. Kydd, J. Catal., **100**, 185 (1986)

PECULIARITIES OF BRØNSTED ACID SITES IN FER-TYPE ZEOLITES

J. WEITKAMP[a*], M. BREUNINGER[a], H.G. KARGE[b] and M. HUNGER[a]

[a]Institute of Chemical Technology I, University of Stuttgart, D-70550 Stuttgart, Germany;
fax: +49/711/685-4065, e-mail: jens.weitkamp@po.uni-stuttgart.de
[b]Fritz Haber Institute of the Max Planck Society, D-14195 Berlin, Germany

ABSTRACT

Peculiarities concerning the location of Brønsted acid sites in a FER-type zeolite HZSM-35 were found by FTIR spectroscopy after adsorption of pyridine and ethylbenzene and by disproportionation of ethylbenzene. The total adsorption capacity for pyridine was compared with those of zeolites HZSM-5, HZSM-22 and HZK-5. From the results, it is inferred that ca. 75 % of the acidic OH groups in the zeolite HZSM-35 are located at sites with size constraints, e.g., in 8-ring pores. About 25 % of the acidic OH groups were found to be accessible for pyridine and are, therefore, located in 10-ring pores and/or on the outer surface of the zeolite particles. The high activity of the zeolite HZSM-35 in ethylbenzene disproportionation indicates a significant number of acidic hydroxyl groups on the outer surface of the crystallites. It is proposed that this location of Brønsted acid sites in zeolite HZSM-35 plays a role in the selective isomerization of 1-butene to isobutene over this material and the variation of the isobutene selectivity with time-on-stream.

INTRODUCTION

The conversion of 1-butene to isobutene needed for the production of MTBE has been considered in recent years [1]. The FER-type zeolite HZSM-35 is a shape selective catalyst suitable for skeletal isomerization of butenes. In catalytic investigations, this material has shown a higher selectivity to isobutene than other 10-ring zeolites such as HZSM-5 and Theta-1 [2]. According to Mériaudeau et al. [3], a monomolecular reaction mechanism involving bridging OH groups as catalytically active sites is responsible for the high selectivity to isobutene. In recent studies, the density and strength of Brønsted acid sites [4, 5] and the location of coke deposits and hydroxyl groups in FER-type zeolites were investigated by catalytic reactions [6, 7] and FTIR spectroscopy [7, 8]. Fripiat et al. [9] concluded from Hartree-Fock calculations that the positions T2 and T4 are the two preferred aluminum sites in the framework of FER-type zeolites. These sites are located in 6-membered rings forming the walls of 8-ring pores. Hydroxyl groups bound to T4 sites point into these 8-ring pores. Applying ^{27}Al MQMAS NMR and ^{29}Si MAS NMR spectroscopy, Sarv et al. [10] found that aluminum atoms are incorporated into the framework of FER-type zeolites at positions T2 + T3 and T1 + T4 + T5 in a ratio of 4 : 5. This indicates that acidic bridging OH groups reside at a variety of framework sites in this zeolitic material. In the present work, FTIR and ^{29}Si MAS NMR spectroscopy, adsorption experiments and catalytic investigations were carried out to clarify the location of OH groups inside the pores and on the outer surface of the ferrierite crystals.

EXPERIMENTAL

Zeolites ZSM-35 (bulk n_{Si}/n_{Al} = 13.0, particle size 1.0 μm) and ZSM-22 (bulk n_{Si}/n_{Al} = 40, particle size 1.0 μm) were synthesized using pyrrolidine [11] and 1,6-diaminohexane [12] as templating agents, respectively. Zeolites ZSM-5 (bulk n_{Si}/n_{Al} = 23, particle size 2.0 μm) and ZK-5 (bulk n_{Si}/n_{Al} = 3.5, particle size 2.0 μm) were synthesized without template ([13] and [14], respectively). After calcination, the zeolites were threefold ion-exchanged with a 1 N aqueous solution of NH_4NO_3 leading to an exchange degree of $\geq$ 98 %. X-ray diffraction patterns were collected on a Siemens D5000 instrument, and SEM studies were made on a Cambridge Instruments CAM SCAN44. For FTIR spectroscopy, self-supporting wafers (ca. 10 mg cm^{-2}) were prepared and activated in vacuum (p < 10^{-6} mbar) at 773 K. For the cell and procedure compare Ref. [15]. Pyridine vapor (p = 3.4 mbar) was introduced at 393 K for 30 min and then desorbed at the same temperature for 1 h. Adsorption of ethylbenzene (p = 2.0 mbar) was performed at temperatures up to 523 K. The FTIR spectra were recorded on a Perkin-Elmer 1760X spectrometer. ^{27}Al and ^{29}Si MAS NMR spectra were recorded on a Bruker MSL 400 NMR spectrometer at resonance frequencies of 104.3 MHz and 79.5 MHz and with sample spinning rates of 10.0 kHz and 3.5 kHz, respectively. The total adsorption capacity of the materials for pyridine was measured under flow conditions at 323 K. Disproportionation of ethylbenzene was performed in a flow-type apparatus with a fixed bed reactor at atmospheric pressure and with on-line capillary gas chromatographic analysis of the reaction products.

RESULTS AND DISCUSSION

The primary objective of the present work was a comparison between the location of hydroxyl groups in zeolites HZSM-35 (10-ring ↔ 8-ring pores) and HZSM-5 (10-ring ↔ 10-ring pores). Therefore, zeolites HZSM-5 and HZSM-35 with a similar morphology and particle size (2.0 μm and 1.0 μm, respectively) were used. Powder X-ray diffraction proved that the calcined materials and the H-forms were highly crystalline. The ^{27}Al MAS NMR spectra of the calcined and rehydrated samples showed only weak signals (< 2 %) of extra-framework aluminum atoms

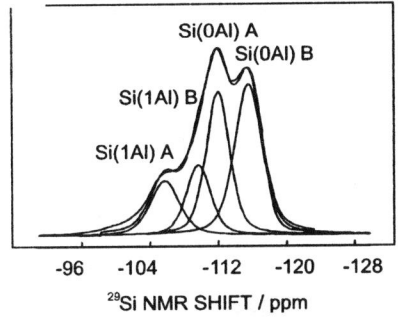
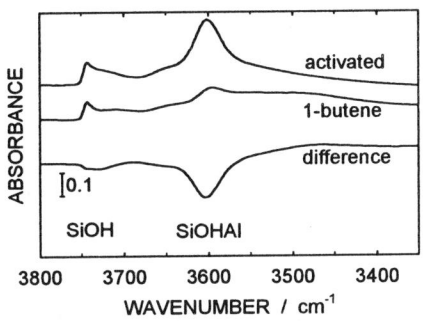

Figure 1. ^{29}Si MAS NMR spectrum of zeolite HZSM-35 and its simulation assuming two groups A and B of Si(nAl) signals.

Figure 2. FTIR spectrum of zeolite HZSM-35 before (top) and after adsorption of 1-butene (middle); difference of both spectra (bottom).

at about 0 ppm which indicates that most of the aluminum atoms were incorporated in the zeolite framework. In such a case, the n_{Si}/n_{Al} ratio obtained by ^{29}Si MAS NMR spectroscopy is equal to that determined by ICP-AES (*vide supra*). The ^{29}Si MAS NMR spectrum of zeolite HZSM-35 (Fig. 1) was simulated using two groups A and B of Si(nAl) signals (n = 0, 1) which differ in their chemical shifts by about 4 ppm. The framework n_{Si}/n_{Al} ratio of 13.1 calculated in this way agrees well with the bulk n_{Si}/n_{Al} ratio of 13.0 obtained by ICP-AES. According to Sarv et al. [10], the two groups A and B of Si(nAl) signals are due to silicon atoms at crystallographically non-equivalent T sites with significantly different local geometries. Group A was attributed to silicon atoms at T2 and T3 positions and group B to silicon atoms at T1, T4 and T5 positions. In the ^{29}Si MAS NMR spectrum shown in Figure 1, the intensity ratio of the signals of groups A and B is about 1 : 1 (accuracy of 5 %). This indicates an incorporation of aluminum atoms at sites (T2 + T3) and sites (T1 + T4 + T5) with about equal probability. Based on theoretical studies, Fripiat et al. [9] found a preferential incorporation of aluminum atoms in the framework of FER-type zeolites at sites T2 and T4. According to these findings, formation of hydroxyl groups at different bridging oxygen atoms must be expected. In ideal FER-type crystals, 33.3 % of the bridging oxygen atoms are located in both 10-ring and 8-ring pores and have the largest space around them, while 27.8 % of them are pointing into 10-ring pores [6]. The remaining 38.9 % are bridging oxygen atoms which contribute to 8-, 6- and 5-membered rings [6]. FTIR spectroscopy and adsorption of probe molecules is a suitable method for the investigation of the accessibility of OH groups in zeolites. The FTIR spectrum of the calcined zeolite HZSM-35 (Fig. 2, top) consists of bands with peaks at 3740 cm^{-1} and 3601 cm^{-1} due to silanol and bridging OH groups, respectively. Broad humps at about 3720 cm^{-1} and 3660 cm^{-1} indicate a small amount of internal silanol groups and hydroxyl groups at extra-framework aluminum species. After adsorption of 1-butene (Fig. 2, middle), the strong band at 3601 cm^{-1} is significantly decreased and a broad band remains at about 3595 cm^{-1}. The bridging OH groups which are influenced by 1-butene cause a negative band at 3604 cm^{-1} (Fig. 2, bottom). Based on an estimation of the bathochromic shift of the stretching vibrations of hydroxyl protons in zeolite pores with different sizes, Zholobenko et al. [8] decomposed the FTIR spectrum of a ferrierite with n_{Si}/n_{Al} = 6.3 into four bands at 3609 cm^{-1} (ca. 20 %), 3601 cm^{-1} (ca. 50 %), 3587 cm^{-1} and 3565 cm^{-1}. These bands were attributed to bridging OH groups pointing into 10-ring pores, 8-ring pores, 8-membered oxygen rings and 6-membered oxygen rings [8], respectively. According to this assignment, the strong band at 3601 cm^{-1} (Fig. 2, top) is mainly caused by bridging OH groups in 8-ring pores which are accessible for 1-butene (see Fig. 2, middle). The wavenumber of 3604 cm^{-1}, found for the band of OH groups interacting with 1-butene (Fig. 2, bottom), is due to an overlap of stretching vibrations of hydroxyl groups located in 8-ring (3601 cm^{-1}) and 10-ring pores (3609 cm^{-1}).

Previous investigations have shown that pyridine is a suitable probe molecule allowing the study of the accessibility of acidic bridging OH groups in 10-ring zeolites (see, e.g., Ref. [16]). In the following FTIR investigations, ethylbenzene has been adsorbed as well, since its catalytic disproportionation is used in the present paper for the characterization of the Brønsted acid sites in zeolites HZSM-35 and HZSM-5. The FTIR spectrum of zeolite HZSM-5 (Fig. 3, bottom) consists of bands at 3610 cm^{-1} and 3740 cm^{-1} due to bridging OH groups in the 10-ring pores and silanol groups at framework defects, respectively. Upon adsorption of pyridine (Fig. 3, top) and ethylbenzene (Fig. 3, middle), the band at 3610 cm^{-1} is strongly reduced. This indicates that most

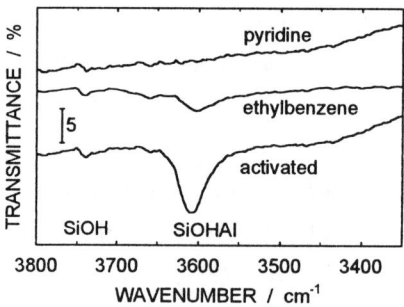

Figure 3. FTIR spectra of zeolite HZSM-5, activated (bottom) and after adsorption of pyridine (top) and ethylbenzene (middle).

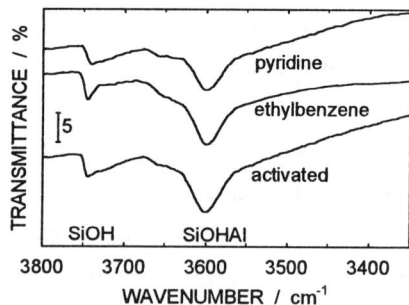

Figure 4. FTIR spectra of zeolite HZSM-35, activated (bottom) and after adsorption of pyridine (top) and ethylbenzene (middle).

of the Brønsted acid sites in the MFI-type material are accessible for pyridine and ethylbenzene, and it is a straightforward conclusion that the conversion of ethylbenzene over zeolite HZSM-5 is catalyzed by Brønsted acid sites located inside the 10-ring pores. In contrast, FTIR spectroscopic investigations of pyridine adsorption on zeolite HZSM-35 (Fig. 4, top) revealed that most of the bridging OH groups at 3601 cm^{-1} are not accessible for the probe molecules which supports an assignment of this band to bridging OH groups located in 8-ring pores (0.35 nm × 0.48 nm). Adsorption of ethylbenzene on zeolite HZSM-35 (Fig. 4, middle) also affects only a fraction of the hydroxyl groups giving rise to the band at 3601 cm^{-1}. The determination of the amount of acidic OH groups in zeolites HZSM-5 and HZSM-35 interacting with pyridine was performed by a quantitative evaluation of the 1700 cm^{-1} to 1400 cm^{-1} region of the FTIR spectra (Fig. 5). The weak intensity of the bands at 1455 cm^{-1} due to pyridine molecules coordinatively bound at Lewis sites (LS) agrees well with the small amount of extra-framework aluminum atoms found by ^{27}Al MAS NMR spectroscopy (*vide supra*). Integration of the band at 1545 cm^{-1} and using the extinction coefficients given in Ref. [17] led to an amount of Brønsted acid sites (BS) interacting with pyridine of 0.68 mmol/g in HZSM-5 and of 0.30 mmol/g in HZSM-35. For zeolite HZSM-5, the above-mentioned amount of Brønsted acid sites corresponds to the amount of framework aluminum atoms (0.70 mmol/g). On the other hand, in zeolite HZSM-35 only ca. 25 % of the framework aluminum atoms (1.25 mmol/g) contribute to bridging OH groups which are accessible for pyridine.

To explore whether pyridine molecules are able to enter 10-ring pores with a size like that of zeolite HZSM-35, the adsorption of pyridine was studied on various zeolites. By passing a flow of nitrogen loaded with pyridine over the calcined powder materials (Fig. 6), total adsorption

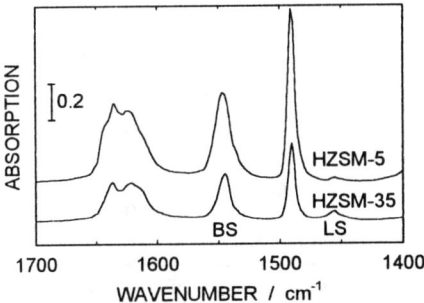

Figure 5. 1700 cm^{-1} to 1400 cm^{-1} region of the FTIR spectrum of zeolites HZSM-5 and HZSM-35 after adsorption of pyridine.

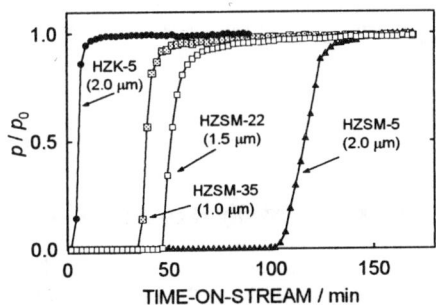

Figure 6. Adsorption of pyridine on zeolites HZK-5, HZSM-35, HZSM-22 and HZSM-5 under continuous flow conditions.

capacities of 1.7 mmol/g, 0.8 mmol/g and 0.6 mmol/g were determined for HZSM-5 (0.53 nm × 0.56 nm ↔ 0.51 nm × 0.55 nm), HZSM-22 (0.55 nm × 0.44 nm) and HZSM-35 (0.54 nm × 0.42 nm), respectively. For HZK-5, which possesses 8-ring pores (0.39 nm × 0.39 nm) only, an adsorption capacity for pyridine on its outer surface of less than 0.1 mmol/g was found. These values strongly suggest that pyridine molecules do enter the 10-ring pores of zeolites HZSM-5, HZSM-22 and HZSM-35. Nevertheless, as shown in Figure 4, most of the Brønsted acid sites in zeolite HZSM-35 causing the band at 3601 cm^{-1} are not affected by these molecules which supports the assumption that the majority of these hydroxyl groups is located in the 8-ring pores.

In the disproportionation of ethylbenzene, zeolites HZSM-5 and HZSM-35 showed comparable activities (Figs. 7 and 8). After a time-on-stream of 50 h, the conversions of ethylbenzene on HZSM-5 and HZSM-35 were 2.9 % and 2.6 %, respectively, even though a much larger amount of framework aluminum exists in HZSM-35. The distribution of the diethylbenzene isomers $n_{1,2\text{-DE-Bz}} : n_{1,3\text{-DE-Bz}} : n_{1,4\text{-DE-Bz}}$ was 0 % : 60 % : 40 % over HZSM-5 and 1 % : 67 % : 32 % over HZSM-35 which is not typical for medium-pore zeolites [18]. Since FTIR spectroscopy yields no interaction of bridging OH groups in zeolite HZSM-35 with ethylbenzene, the question concerning the nature and location of Brønsted acid sites catalyzing the disproportionation of these reactant molecules arises. Ethylbenzene disproportionation can occur on Brønsted acid sites located in the 10-ring pores and/or on the outer surface of the zeolite particles. Upon preadsorption of the bulky base collidine which leads to a selective poisoning of the sites on the outer surface, ethylbenzene conversion over zeolite HZSM-5 decreased from 2.9 % (Fig. 7) to 1.7 % (Fig. 9). Interestingly, there was a significant change in the distribution of

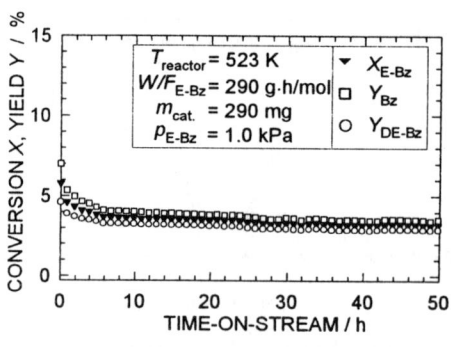

Figure 7. Disproportionation of ethylbenzene on zeolite HZSM-5.

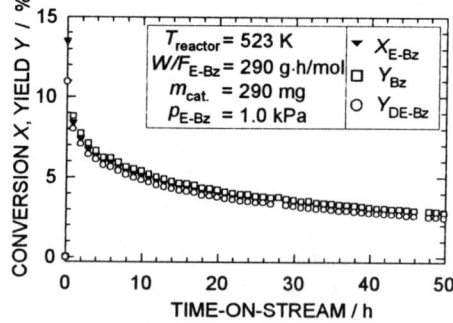

Figure 8. Disproportionation of ethylbenzene on zeolite HZSM-35

diethylbenzene isomers (Fig. 10). With increasing time-on-stream, the 1,4-diethylbenzene isomer is preferentially formed ($n_{1,2\text{-DE-Bz}} : n_{1,3\text{-DE-Bz}} : n_{1,4\text{-DE-Bz}} = 0 : 11 : 89$ after 50 h). This indicates a significant effect of Brønsted acid sites on the outer surface of the HZSM-5 crystallites on the distribution of diethylbenzene isomers. After preadsorption of collidine on zeolite HZSM-35, no catalytic activity at all was found any more. Hence, all of the Brønsted acid sites of zeolite HZSM-35, which are active in the disproportionation of ethylbenzene must be located on the outer surface. No shape selectivity can be expected in catalysis on these sites, and this explains the finding that the distribution of the diethylbenzene isomers on the unpoisoned HZSM-35 catalyst (*vide supra*) resembles the one found on Y-type zeolites [19].

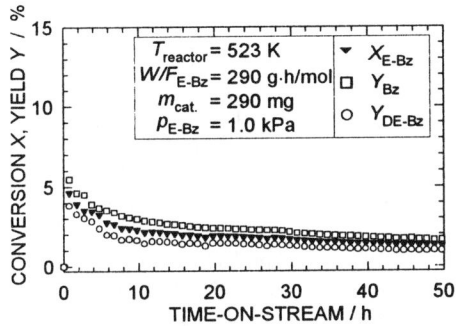

Figure 9. Disproportionation of ethylbenzene on zeolite HZSM-5 after preadsorption of collidine.

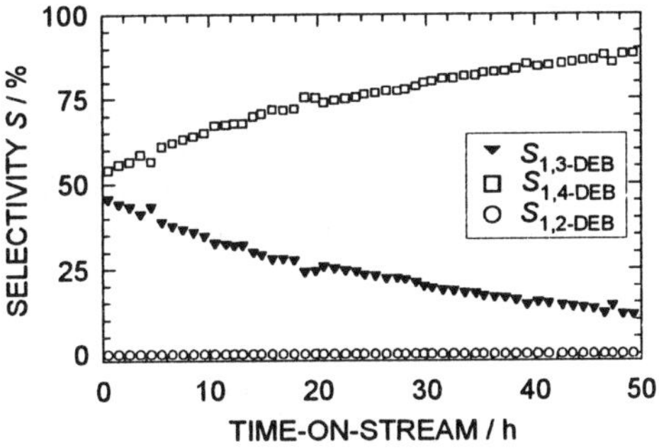

Figure 10. Isomer distribution obtained by disproportionation of ethylbenzene on zeolite HZSM-5 after preadsorption of collidine.

CONCLUSIONS

Spectroscopic and catalytic investigations performed on the FER-type zeolite HZSM-35 indicate that ca. 75 % of the Brønsted acid sites are located at sites with strong size constraints, presumably in 8-ring pores. About 25 % of the acidic OH groups in zeolite HZSM-35 are located in 10-ring pores and on the outer surface of the crystals. The Brønsted acid sites on the outer surface of the HZSM-35 particles cause an ethylbenzene conversion activity comparable to that of OH groups inside the 10-ring pores of zeolite HZSM-5. In agreement with these results, Monte Carlo calculations carried out by Seo et al. [5] suggest a preferential location of single 1-butene molecules in the 8-ring pores of FER-type zeolites. This is consistent with a monomolecular reaction mechanism for the isomerization of 1-butene to isobutene and explains the unique selectivity of FER-type zeolites for isobutene. Based on these facts, the initial increase of the isomerization activity and the strong decrease of the disproportionation activity observed for

FER-type catalysts as a function of time-on-stream [7], are best interpreted in terms of an initial deactivation of Brønsted acid sites located in 10-ring pores and on the outer surface of the ferrierite crystals. At longer reaction times, the decrease of the accessibility of the Brønsted acid sites located in 8-ring pores would result in a deactivation of the FER-type catalysts.

ACKNOWLEDGMENTS

Support of this work by Deutsche Forschungsgemeinschaft, Max-Buchner-Forschungsstiftung and Fonds der Chemischen Industrie is gratefully acknowledged.

REFERENCES

1. P. Mériaudeau, T. Vu, H. Le Ngoc and C. Naccache, in Progress in Zeolite and Microporous Materials, Edited by H. Chon, S.K. Ihm and Y.S. Uh, Studies in Surface Science and Catalysis, Vol. 105 (Elsevier, Amsterdam, 1997) p. 1373.
2. H.H. Mooiweer, K.P. de Jong, B. Kraushaar-Czarnetzki, W.H.J. Stork and B.C.H. Krutzen, in Zeolites and Related Microporous Materials: State of the Art 1994, Edited by J. Weitkamp, H.G. Karge, H. Pfeifer and W. Hölderich, Studies in Surface Science and Catalysis, Vol. 84 (Elsevier, Amsterdam, 1994) p. 2327.
3. P. Mériaudeau, R. Bacaud, N.H. Le and A.T. Vu, J. Mol. Catal. A **110**, L177 (1996).
4. R.J. Pellet, D.G. Casey, H.-M. Huang, R.V. Kessler, E.J. Kuhlmann, C.-L. O'Young, R.A. Sawicki and J.R. Ugolini, J. Catal. **157**, 423 (1995).
5. G. Seo, H.S. Jeong, J.M. Lee and B.J. Ahn, in Progress in Zeolite and Microporous Materials, Edited by H. Chon, S.K. Ihm and Y.S. Uh, Studies in Surface Science and Catalysis, Vol. 105 (Elsevier, Amsterdam, 1997) p. 1431.
6. W.-Q. Xu, Y.-G. Yin, S.L. Suib and C.-L. O'Young, J. Phys. Chem. **99**, 758 (1995).
7. P. Andy, N.S. Gnep, M. Guisnet, E. Benazzi and C. Travers, J. Catal. **173**, 322 (1998).
8. V.L. Zholobenko, D.B. Lukyanov, J. Dwyer and W.J. Smith, J. Phys. Chem. B **102**, 2715 (1998).
9. J.G. Fripiat, P. Galet, J. Delhalle, J.M. André, J.B. Nagy and E.G. Derouane, J. Phys. Chem. **89**, 1932 (1985).
10. P. Sarv, B. Wichterlová and J. Cejka, J. Phys. Chem. B **102**, 1372 (1998).
11. S.J. Rane, C.V.V. Sayanarayana and D. K. Chakrabarty, Appl. Catal. **69**, 177 (1991).
12. S. Ernst, J. Weitkamp, J.A. Martens and P.A. Jacobs, Appl. Catal. **48**, 137 (1991).
13. S. Ernst and J. Weitkamp, Chem.-Ing.-Tech. **63**, 748 (1991).
14. H.E. Robson, US Patent 3,720,753 (13.03.1973).
15. H.G. Karge and W. Niessen, Catal. Today **8**, 451 (1991).
16. J.C. Védrine, A. Auroux, V. Bolis, P. Dejaifve, C. Naccache, P. Wierzchowski, E.C. Derouane, J.B. Nagy, J.P. Gilson, J.H.C. van Hooff, J.P. van den Berg and J. Wolthuizen, J. Catal. **59**, 248 (1979).
17. M.A. Makarova, K. Karim and J. Dwyer, Microporous Mater. **4**, 243 (1995).
18. J. Weitkamp, S. Ernst, P.A. Jacobs and H.G. Karge, Erdöl, Kohle-Petrochem. **39**, 13 (1986).
19. H.G. Karge, S. Ernst, M. Weihe, U. Weiß and J. Weitkamp in Zeolites and Related Microporous Materials: State of the Art 1994, Edited by J. Weitkamp, H.G. Karge, H. Pfeifer and W. Hölderich, Studies in Surface Science and Catalysis, Vol. 84 (Elsevier, Amsterdam, 1994) p. 1805.

BASIC AND ACIDIC SITES IN Cs/Na FAUJASITES: AN IR STUDY

E. GARRONE*, P. MARTURANO*, B. ONIDA[†], M. LASPÉRAS[§] and F. Di RENZO[§]

*Dipartimento di Chimica Inorganica, Fisica e dei Materiali, Università di Torino, Italy
[†] Dipartimento di Scienza dei Materiali e Ingegneria Chimica, Politecnico di Torino, Italy
[§] Ecole National Superieure de Chimie de Montpellier, Montpellier, France

ABSTRACT

The substitution of Na^+ for Cs^+ in X and Y zeolites in the range 40 to 95% has been studied by means of IR spectroscopy of adsorbed CO_2, pyrrole and methylacetylene. Data have been interpreted with the help of similar results on the much simpler systems alkali-cation ZSM5 zeolites. CO_2 adsorbs in a linear form on Cs^+ cations only: Na^+ ions seem to be no more available at the supercages; the adsorbed amount being dictated by the available space. Carbonate are formed, though less stable than on Na^+ samples: the usual rule relating the degree of symmetry with basicity of the oxygen atom does not hold strictly. Methylacetylene interacts both with the basic oxygen through the acidic proton, and the cation via the triple bond. This fact does not allow absolute determinations of the basicity: use of pyrrole still results the best available method.

INTRODUCTION

Basic zeolites are promising substitutes for strong liquid bases in catalysis. Faujasitic zeolites are usually considered, with low Al/Si ratio, i.e. X and Y types. Their basicity depends markedly on the alkaline cation, in that Cs-substituted samples are much more basic than Na-substituted ones. The strength of basic oxygen is usually measured in the IR by the interaction with acidic molecules like pyrrole, whose N-H stretching mode engages in H-bonding and therefore undergoes a bathochromic effect [1]. The role of cations as Lewis acidic sites has been studied to a much lesser detail.

The substitution Na^+ for Cs^+ is never complete, and a pure Cs-faujasite is not yet available. In the present paper, we study the process of Na^+/Cs^+ exchange comparing five samples, i.e. pure NaX and NaY with one Cs-NaY and three Cs-NaX, one of which a nearly 100% Cs sample. The presence of Lewis sites has been studied by CO adsorption, as nearly customary in adsorption studies: the data are, however, exceedingly complex and will be reported elsewhere. Selective information on both Lewis acidity and basicity has been gained through the adsorption of CO_2, which may act either as a probe of acid/base pairs by the formation of carbonates, and on Lewis acidity alone by the formation of linear adducts. Basicity has been study through the adsorption of pyrrole, and also of weaker acids like substituted acetylenes, as suggested by Kustov et al. [2].

EXPERIMENTAL

Samples were NaX (13X, Aldrich-Chemie) and NaY (Linde SK-40, Alfa), as such and twice exchanged in water with Cs acetate at room temperature, giving zeolite CsX42 (exchanged 42%, $Na_{50}Cs_{36}Al_{86}Si_{106}O_{384}$) and CsNaY exchanged 60% ($H_5Na_{17}Cs_{33}Al_{55}Si_{137}O_{384}$) [3]. CsX72 (exchanged 72%) and CsX94 (exchanged 94%). Samples were normally outgassed at 600 °C for activation. Spectra were obtained on a Perkin-Elmer 1760-X spectrophotometer equipped with a MCT cryodetector, at a resolution of 2 cm^{-1}.

RESULTS AND DISCUSSION

Adsorption of carbon dioxide

Figure 1 compares the room temperature spectra of the three CsX samples under 0.5 mbar CO_2 in the region around the asymmetric stretch of the free molecule (2349 cm^{-1}). A single band is observed at 2349 cm^{-1}: the interaction is weak and reversible, as shown by the limited shift from that frequency. Note that with NaY and NaX the corresponding band falls at 2349 cm^{-1}: the interaction is stronger and intensities very large, so that features become evident which are usually negligible, i.e. the ^{13}C satellite band at 2276 cm^{-1}, and combination bands with soft intermolecular modes at 2407 and 2450 cm^{-1}.

The intensity in Figure 1 is observed to increase when passing from CsX42 to CsX72, whereas no further increase is seen when moving to CsX94.

No other component is seen. The conclusion is that CO_2 does not adsorb on Na^+ but only on Cs^+ ions. When the exchange is 42%, Na^+ cations are no more available to interaction. The presence of Cs^+ cations changes the electric field within the zeolites: as a consequence the CO_2 band is observed to shift somewhat to lower frequency.

The same spectra as in section *a* of Figure 1 are shown in a different spectral region in section *b*. On NaX (lower curve) asymmetric carbonates are formed. When Cs is present, the bands get closer to each other. The intensities are seen first to decrease, when passing from Na to Cs containing samples, then to increase again. Carbonates are not formed on Na^+ O^- pairs, as Na^+ ions seem not to be available at the supercages. The increased basicity, caused by the presence of

Cs⁺ ions and documented by the results below, is in agreement with the more symmetric nature of the carbonates, according to a criterion proposed by Davydov [3]. This does not, however, apply to the comparison NaY/NaX (figures not reported), so showing that the difference in frequency between the asymmetric and symmetric stretching modes of carbonates assumed to be an inverse measure of basicity does not apply in general to zeolitic systems.

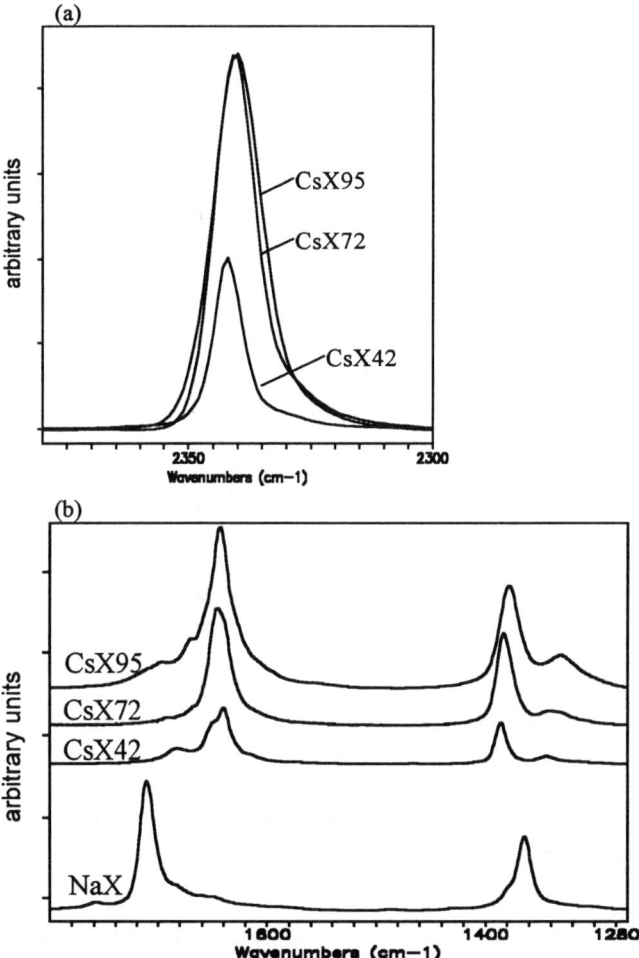

Figure 1. Adsorption of 0.5 mbar CO_2 on X samples. a) region of ν(asymm) of the free molecule; b) carbonate region.

Note that TPD measurement of the amount of CO_2 held by Cs containing systems [4] shows that these are less stable than on Na containing systems.

Adsorption of pyrrole and methylacetylene

Figures describing the adsorption of pyrrole are not reported for sake of brevity: the corresponding results are summarised in Figure 3. Figure 2 shows instead the adsorption of 10 mbar methylacetylene in the CH stretching region on NaX and NaY. The intense bands around 3250 cm^{-1} are due to the C-H stretch of the acetylenic moiety, shifted to lower frequency.

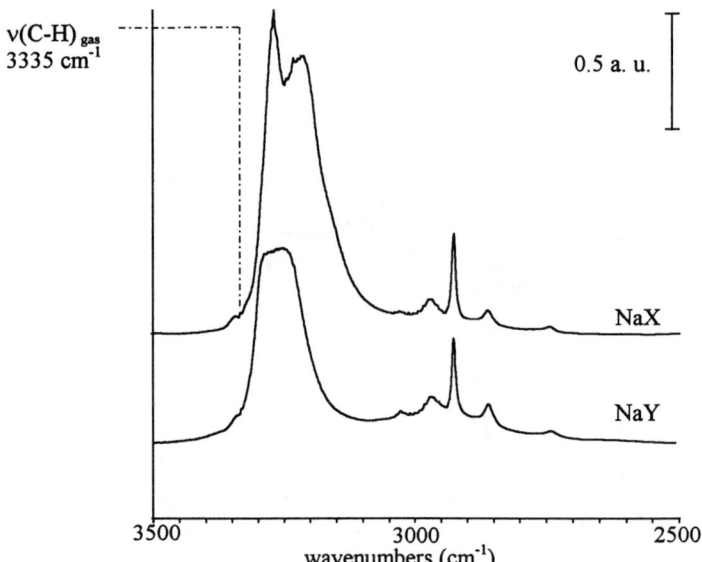

Figure 2. Adsorption of methylacetylene on unexchanged samples. Vertical broken line: reference for the gas-phase molecule.

Actually in the NaX sample, two such bands are seen, of which that less shifted is narrower. The same happens with Cs substituted samples: narrow bands (having widths typical of free molecular species) are superimposed to the broad band of the H-bonded C-H group. We interpret such bands as due to molecules mainly interacting with the Na$^+$ cation through the triple bond, and to a lesser (or nil) extent with the basic oxygen. Interaction of a positive centre with the triple bond does decrease the frequency of the C-H stretch in acetylenes [5]. Such bands interfere to some extent with the band due to H-bonded species, as in the NaY spectrum, where a high-frequency component in the band envelope around 350 is barely seen.

On the other hand, it is probable that a methylacetylene molecule, H-bonded to a basic oxygen, also interacts with a cation, if account is taken of their density in X or Y zeolites.

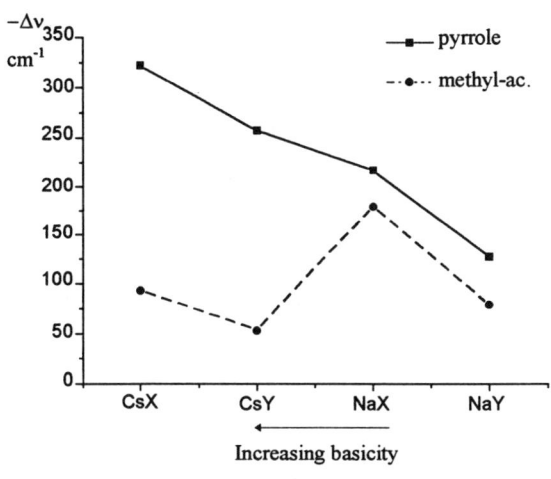

Figure 3

The comparison of basicity as measured through pyrrole adsorption and through methylacetylene adsorption (Figure 4), shows that, whereas pyrrole yields the trend expected on the basis of catalytic behaviour, i.e. CsX > CsY > NaX > NaY, methylacetilene undergoes much larger shifts with NaX and NaY than with CsX and CsY, because of the polarising role of the hard cation Na^+, which the molecule also binds to. The correct order is found for the couples NaX > NaY and CsX > CsY, when the effect of the cation is the same, and therefore a change in basicity is indeed measured.

CONCLUSIONS

Both linearly adsorbed carbon dioxide and carbonate species indicate that Na^+ cations are no more available to interaction in systems with 42% Cs or more. This does not probably mean that Na^+ cations are only present in sodalite units or hexagonal prisms. Indeed, spectra not shown of CO adsorbed at nominal 77 K seem to show the residual presence of Na^+ cations. The most plausible interpretation is that, because of their size, Cs^+ cations sterically inhibit interaction with less accessible Na^+ cations. Methylacetylene interacts both with the basic oxygen through the acidic proton, and the cation via the triple bond, whereas with pyrrole, if any such double interaction occurs, the contribution of H-bonding is prevailing. As a consequence, even though pyrrole bands are broad and ill-defined, this molecule still affords the best probe for basicity: the spectroscopic features of carbonates, instead, do not constitute a proper guideline in zeolitic systems.

REFERENCES

1. D. Barthomeuf in "Zeolite Microporous Solids", Kluver (1992), p. 193
2. Kustov *et al*. Proceeding of EuropacatII, (1995), p.395.
3. A. A. Davydov, M. L. Shepot'ko e A. A. Budneva, Kinetics and Catalysis, **35** 272 (1994).
4. M. Laspéras *et al.*, Microp. Mater. **7** 61 (1996).
5. P. Ugliengo *et al*. J. Phys. Chem. **100** 3632 (1996).

NATURE OF Zn SITES IN Zn-MFI: A FTIR INVESTIGATION

S. VALANGE[(1)], Z. GABELICA[(1)], B. ONIDA[(2)] and E. GARRONE[(3)]*

[(1)] ENSCMulhouse, Université de Haute Alsace, Laboratoire de Matériaux Mineraux, Mulhouse Cedex, France

[(2)] Dipartimento di Scienza dei Materiali e Ingegneria Chimica, Politecnico di Torino, Italy

[(3)] Dipartimento di Chimica Inorganica, Fisica e dei Materiali, Università di Torino, Torino, Italy

ABSTRACT

The structural state of Zn^{2+} ions in Zn-containing MFI synthesized by the "alkylamine route" has been investigated by FTIR. A medium-to-strong band at 3640 cm^{-1} is attributed to OH groups linked to isolated Zn^{2+} ions. While these species do not behave as true Brønsted acids (absence of ammonium species in the interaction with ammonia), the extent of the shift with CO shows that they are more acidic than silanol groups in silicalite and OH groups on the surface of ZnO. Exposed Zn^{2+} species also interact with CO: FTIR data suggest that the Zn species are highly dispersed within the MFI matrix, possibly partly anchored to the zeolite framework.

INTRODUCTION

Zn-containing MFI zeolites, essentially used as bifunctional catalysts for aromatization reactions [1,2], are currently prepared by indirect methods such as impregnation [3], ion exchange [2,4], or by mechanical admixing of ZnO to the zeolite. Hagen et al. [4] seem to be the only team having attempted to prepare Zn-MFI through direct synthesis by using classical procedures in basic media. As a part of an extensive research aiming to prepare MFI metallosilicate catalysts through less conventional routes yielding more homogeneous metal zeolite admixtures, we have succeeded to synthesise Zn-MFI by mobilizing through complexation with short-chain alkylamines various Zn salts in the gel precursor prior to hydrothermal crystallization [5]. FTIR spectroscopy, a key tool to investigate local structural changes in zeolites induced by e.g. heteroatom insertions, was used to probe the nature of the Zn^{2+} ions.

EXPERIMENTAL

Zn-containing MFI was prepared by direct crystallization following the "methylamine route". The advantage of synthesizing MFI metallosilicate zeolites using short chain amines as mineralizing-complexing species was reported in more detail elsewhere [5]. A solution containing Zn nitrate ($Zn(NO_3)_2 \cdot 9H_2O$, Fluka) pre-dissolved in a small amount of water, was added dropwise to an aqueous solution containing TPABr (Merck, P. A.) pre-dissolved in methylamine (CH_3NH_2 40 % aq., pure, Fluka). The resulting clear solution (pH abput 12.5) was stirred for 1h at ambient temperature before addition of solid silica (Areosil-200 from Degussa). The final gel of composition:

0.94 SiO_2: 0.02 $Zn(NO_3)_2 \cdot 9H_2O$: 2 methylamine: 0.25 TPABr: 37 H_2O

was stirred for another 2h until complete homogeneization. It was heated in a Teflon stainless steel autoclave at 185 °C, in static conditions, for 11 days. The as-synthesized solid was filtered, washed thoroughly with cold water and dried 90 °C for 12 hours. A reference silicalite was synthesized by using the same method. Atomic absorption was used to measure the overall Zn content, while the Zn gradient concentration from the crystal surface to a depth of about 1 micron was evaluated by Rutherford Backscattering Spectroscopy (RBS) [6].

IR spectra were recorded on a Perkin-Elmer 1760-X spectrophotometer equipped with MCT cryodetector, at a resolution of 2 cm^{-1} (about 100 scans). The samples were slowly outgassed up to 750 K, then further oxidized at this temperature in O_2. Being very difficult to prepare samples as self-supporting wafers for IR measurements, thin films of powder were deposited on a Si disc. For the adsorption of CO at low temperature, a cell was used allowing the cooling of the sample at the liquid nitrogen temperature (nominal 77 K).

RESULTS AND DISCUSSION

Sample characterization

The as-synthesized sample was pure MFI (XRD, not reported), with no amorphous phase present (SEM data). The Zn-MFI crystallites exhibit a uniform size and morphology (elongated prisms of 45 x 10 x 5 μm). The average amount for Zn per unit cell containing 96 T (T = Si + Zn) was 1.8 ± 0.1, as determined by atomic absorption. RBS data confirm that the methylamine route readily leads to a uniform metallic distribution in MFI at least in the first outer shell of the crystallites.

OH species

The OH stretching region of silicalite and of the Zn-MFI sample, both outgassed at 750 K, are reported in Figure 1 (curve a and b, respectively). Peaks at about 3730, 3690 cm^{-1} and a broad absorption in the range 3500-3200 cm^{-1} are observed in both spectra. The peak at 3730 cm^{-1} is due to terminal Si-OH, present in lattice defects (OH nests) in the zeolite structure, to which corresponds the broad band below 3500 cm^{-1}, due to H-bonded partners. A weak component at 3745 cm^{-1}, related to isolated SiOH groups at the external surfaces, contributes to this absorption. The assignment of the band at 3690 cm^{-1} is less straightforward, as this band was not observed on silicalite so far. It appears to be related to internal silanols in peculiar environment, probably caused by the particular synthesis route used.

The presence of Zn in MFI generates a new OH species (medium-strong band at 3640 cm^{-1}, marked with an asterisk in the figure). As the IR modes of isolated OH groups on ZnO appear in the range 3675-3620 cm^{-1} [7], the band at 3640 cm^{-1} is ascribed to isolated Zn-OH species. The thermal stability of this species is lower than that of the SiOH species, as seen by the dramatic decrease of the intensity of this band after treatment at 850 K in vacuo.

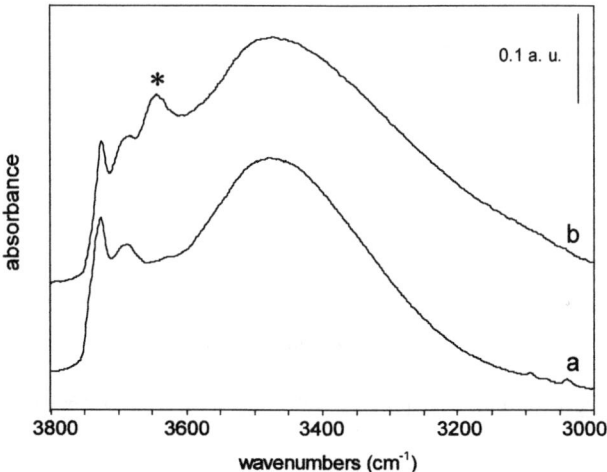

Figure 1. FTIR spectra of silicalite (a) and of Zn-MFI (b) outgassed at 750 K.

Adsorption of ammonia

Experiments were carried out on both silicalite and Zn-MFI and the corresponding spectra are reported in figures 2a and 2b, respectively. In the case of silicalite, no ammonium species are formed for sure, since it is well known that silanols cannot give any proton transfer to ammonia. In figure 2a, two bands appear at about 1630 and 1480 cm^{-1}, both being reversible upon outgassing at room temperature. The first band is due to the asymmetric bending mode of molecular ammonia interacting via H-bonding with silanols. The second band is tentatively assigned to a combination mode of the symmetric bending mode of H-bonded NH_3 (~950 cm^{-1}) and an intermolecular SiOH---NH_3 mode (around 260 cm^{-1}) [8]

The same bands are also observed in the FTIR spectra of Zn-MFI (figure 2b). Upon outgassing at room temperature, the band at about 1630 cm^{-1} shows an irreversible component in the 1620-1615 cm^{-1} range, ascribed to molecular ammonia strongly coordinated to Lewis acidic centres due to the presence of zinc in Zn-MFI.

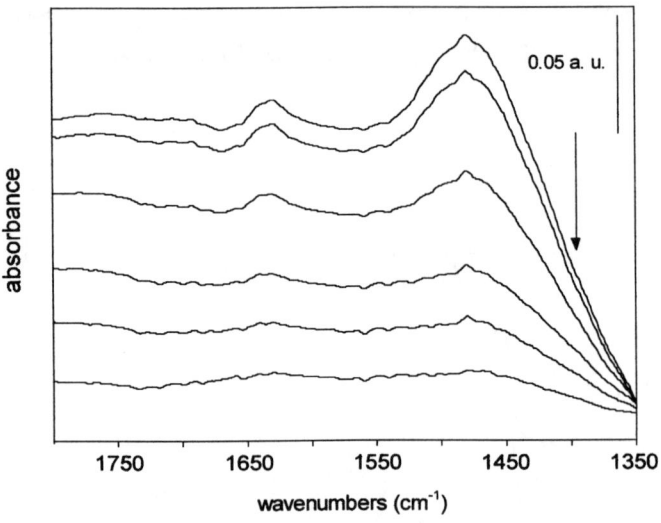

Figure 2a. FTIR spectra of silicalite progressively outgassed at room temperature after adsorption of ammonia.

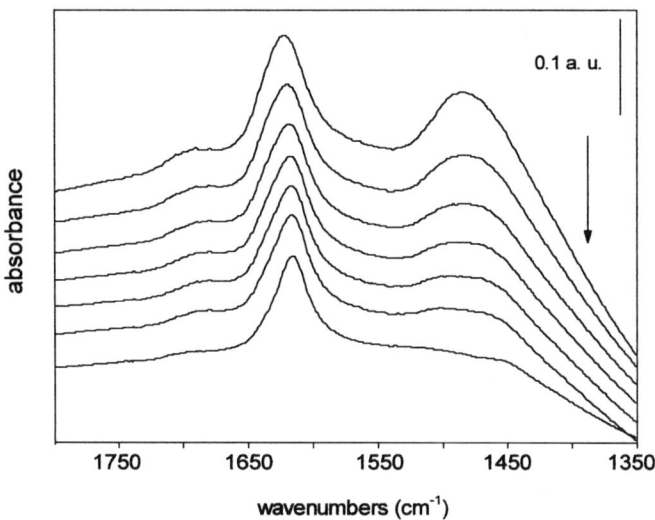

Figure 2b. FTIR spectra of Zn-MFI progressively outgassed at room temperature after adsorption of ammonia.

Adsorption of CO

The spectrum in the OH stretching region of Zn-MFI recorded at 77 K (nominal) under 1 torr of CO is reported in Figure 3 and compared to that of the bare sample. The only band decreasing under these circumstances is that at 3640 cm^{-1}: simultaneously a broad band increases at about 3450 cm^{-1} (stretching mode of ZnOH species engaged in H-bonding with CO). At higher pressures, the other bands also decrease (spectra not reported). The shift observed for the ZnOH species is of about 190 cm^{-1}. This value is much larger than the shift observed for the most acidic ZnOH species present on ZnO, (about 45 cm^{-1} [9]), and definitely larger than the shift observed for the most acidic silanols in silicalite (about 110 cm^{-1}).

We therefore conclude that ZnOH species in Zn-MFI are more acidic than silanols in silicalite and definitely far more acidic than the OH species present on the surface of ZnO.

Figure 4 shows some other interesting features in the CO stretching region.

Upon introduction of CO, a band appears at about 2200 cm^{-1}. Upon increasing the CO amount, this band increases and shifts somewhat to lower frequencies. It characterized species related to the presence of Lewis acid sites, already detected by adsorption of ammonia and consisting of exposed Zn species that can interact with CO. CO adsorbed on pure ZnO, gives rise to bands at and below 2190 cm^{-1} [9]. This suggests that Lewis sites in Zn-MFI are stronger than those on ZnO and confirms that the structural state of Zn in MFI is definitely different than that on the surface of bulk ZnO.

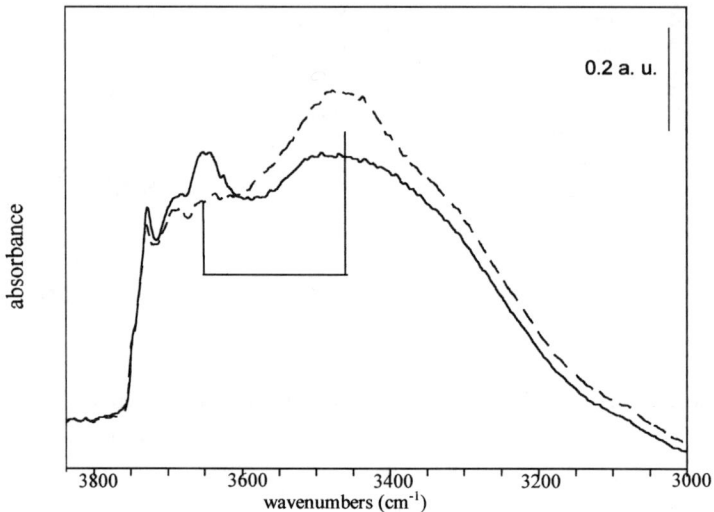

Figure 3. FTIR spectra of Zn-MFI outgassed at 750 K (solid curve) and after adsorption of 1 torr of CO (broken curve).

At increasing coverage of CO, an intense band at 2165 cm^{-1} increases. It is ascribed to CO interacting with ZnOH species through H-bonding, the corresponding OH stretching frequency falling at 3450 cm^{-1}. Note the higher frequency of the 2165 cm^{-1} band with respect to the band at 2156 cm^{-1} related to CO in interaction with silicalite silanols [6] (in the spectra of figure 4 such band contributes to the tail of the 2165 cm^{-1} band on the low-frequency side). This observation is again in line with the stronger acidity of the ZnOH groups in MFI.

The band which increases at about 2140 cm^{-1} is due to CO physisorbed inside the channels of the zeolite.

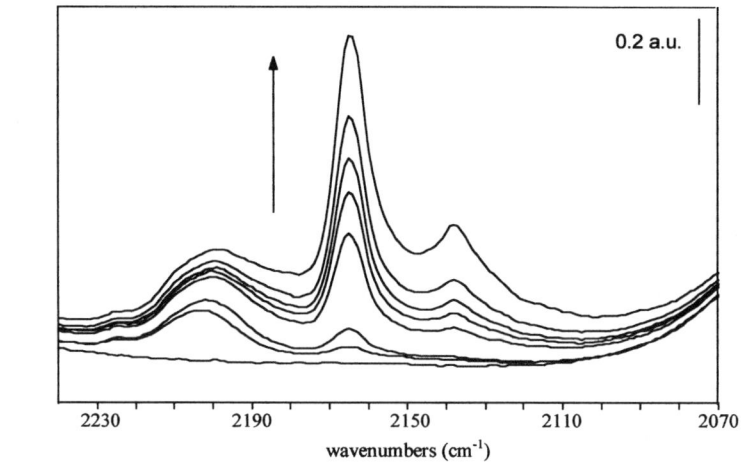

Figure 4. FTIR spectra related to the adsorption at 77 K of increasing amounts of CO on Zn-MFI, outgassed at 750 K.

CONCLUSIONS

Evidence has been given that Zn species are highly dispersed within the MFI matrix. They give rise to both new Brønsted and Lewis acidities. For a possible model describing the location of Zn^{2+} ions in the Zn-MFI structure, although further investigations are needed, our data reported are compatible with the idea that Zn^{2+} ions achieve their coordination partly with OH groups and are partly anchored to the oxygens of the zeolite framework.

ACKNOWLEDGEMENTS

ECC is acknowledged for Human Capital and Mobility contract and for a grant to one of the authors (Sabine Valange).

REFERENCES

1. M. Guisnet, N. S. Gnep, H. Vasques and F. R. Ribeiro, Stud. Surf. Sci. Catal., **69** 321 (1991).
2. A. Hagen and F. Roessner, Stud. Surf. Sci. Catal., **98** 182 (1995).

3. R. J. McIntosh and D. Seddon, Appl. Catal., **6** 307 (1983).
4. A. Hagen, K. H. Hallmeier, C. Hennig, R. Szargan, T. Inui and F. Roessner, Stud. Surf. Sci. Catal. **94** 195 (1995).
5. Z. Gabelica and S. Valange, Res. Chem. Intermediates, **24** 227 (1998).
6. Z. Gabelica, S. Valange, M. Jacobs and G. Demortier, abstract proposed at this meeting.
7. A. Zecchina, S. Bordiga, G. Spoto, L. Marchese, G. Petrini, G. Leofanti and M. Padovan, J. Phys. Chem., **96** 4991 (1992).
8. J. Sauer, P. Ugliengo, E. Garrone and V. R. Saunders, *Chem. Rev.* **94** 2095 (1994).
9. D. Scarano, G. Spoto, S. Bordiga, A. Zecchina and C. Lamberti, Surface Science, **276** 281 (1992).

COMPARATIVE STUDY OF n-HEPTANE HYDROCRACKING OVER Pt HEMT AND Pt HFAU CATALYSTS

A. BERREGHIS, P. MAGNOUX and M. GUISNET

UMR 6503, Laboratoire de Catalyse en Chimie Organique, Université de Poitiers
40, Avenue du Recteur Pineau 86022 Poitiers Cedex France
Fax : 00 33 5 49 45 37 79
E mail : michel.guisnet@cri.univ-poitiers.fr

ABSTRACT

The catalytic properties of two series of Pt catalysts (Pt content from 0 to 1.5 wt%) prepared from HFAU and HEMT samples with similar acid site densities were compared in n-heptane hydrocracking at 250°C. For identical values of the number of accessible platinum atoms, the Pt HEMT samples are more active than the Pt HFAU samples, the latter being more stable and selective for the formation of isomers, in particular of monobranched isomers. These differences in activity, stability and selectivity may be related to the higher strength of the acid sites of HEMT samples.

INTRODUCTION

Zeolite catalysts having hydrogenating and acid functions are widely used in hydroisomerization and hydrocracking of petroleum fractions [1, 2]. For n-alkane hydrocracking, it has been shown that the balance between the two functions determine to a large extent the activity, stability and selectivity of bifunctional Pt zeolite catalysts [3, 4]. nPt/nA, the ratio of the number of accessible Pt atoms and of accessible acid sites was chosen in order to represent this balance.

In this work, the catalytic properties of two series of Pt zeolites : Pt HFAU and Pt HEMT with platinum content from 0 to 1.5 wt% are compared in n-heptane transformation for identical values of nPt/nA. EMT which is the hexagonal analogue of FAU has been recently synthesized by using crown ethers as templating agents [5]. The EMT zeolite contains equal amounts of two types of large cages : hypercages (13 x 14 Å) slightly greater than FAU supercages with two circular apertures similar to those of FAU (7.2 Å) and three elliptical apertures (7.4 x 6.9 Å) and hypocages (7 x 13 Å) accessible through three elliptical apertures. The properties of EMT zeolites are significantly different from those of FAU zeolites. In particular hypocages were found to be responsible for transition-state shape selectivity in various reactions [6-9] and the acidity of HEMT samples was found to be stronger than that of HFAU samples [9-11]. It will be shown that the large differences found between the two series of Pt zeolites are mainly related to the stronger acidity of HEMT.

EXPERIMENTAL

The parent NaEMT sample (Si/Al=3.8), synthesized at the Laboratoire des Matériaux Minéraux (Mulhouse) [5] was completely exchanged by ammonium ions according to the procedure described by Dougnier et al [12]. The ammonium sample was then submitted to a hydrothermal treatment at 450°C for 30 minutes, leading to a HEMT sample (HEMT5.5nw) with a framework Si/Al ratio of 5.5 (unit cell formula $Na_{0.04}H_{14.66}Al_{14.7}Si_{81.3}O_{192}$, with 6.7 extraframework aluminium species). Most of these extraframework species were eliminated by acid washing at 25°C with a 0.01N HCl solution without modification of the framework Si/Al ratio (HEMT5.5w). The HFAU sample (HFAU4.5 : $Na_{0.8}H_{34.1}Al_{34.9}Si_{157.1}O_{384}$, 25 extraframework aluminium species) results from the calcination of an ultrastable NH_4Y zeolite (LZY 82 from Union Carbide) at 500°C under dry air flow. The acidity and the porosity of these samples were previously characterized [13, 14]. Series of Pt zeolites (0 to 1.5 wt%) were prepared from HFAU4.5 and HEMT5.5w (as well as a 1.0wt% Pt HEMT5.5nw) by ion exchange with $Pt(NH_3)_4Cl_2$ followed by calcination at 300°C and reduction at 500°C under the conditions previously reported [4]. The metal dispersions (70 to

100%) were estimated by TEM and from the activity of the Pt samples for toluene hydrogenation. n-Heptane transformation was carried out in a flow reactor at 250°C under atmospheric pressure with a hydrogen/hydrocarbon molar ratio of 9. The reaction products were analyzed on line by GC using a squalane capillary column.

RESULTS AND DISCUSSION

The catalytic properties of the Pt HEMT and Pt HFAU samples will be compared for identical values of the balance between the hydrogenating and the acid functions. nPt/nA where nPt is the number of accessible platinum atoms and nA the number of acid sites retaining pyridine adsorbed at 150°C (determined by IR spectroscopy), was considered as representative of this balance. It should be emphasized that the nA values are very close for the three samples: HFAU4.5 : $0.900 \ 10^{-3}$ mol.g^{-1}, HEMT5.5nw : $0.775 \ 10^{-3}$ mol.g^{-1}; HEMT5.5w : $0.907 \ 10^{-3}$ mol.g^{-1}.

Activity and stability

The initial activity (A_0) and stability were determined for conversions lower than 10%. Figure 1 shows that for both series of catalysts, A_0 initially increases with nPt/nA and then levels off to a plateau for a certain value of this ratio (0.013 with Pt HFAU4.5 and 0.006 with Pt HEMT5.5w). This behaviour is that as expected from the classical bifunctional mechanism : at low platinum contents the limiting step is the dehydrogenation of n-heptane on the platinum sites whereas at high platinum contents the limiting step is the transformation of alkene intermediates on the acid sites. However, there is a large difference in the activity of the two series of Pt zeolites. At the plateau the activity of the Pt HEMT5.5w samples is 2.5 times greater than that of the Pt HFAU4.5 samples, which means that the acid sites of HEMT5.5w are 2.5 times more active than those of HFAU4.5. The difference is still more pronounced with Pt HEMT5.5nw whose acid sites are 7.5 times more active than those of HFAU4.5. This confirms i) that the acid sites of HEMT are stronger than those of HFAU [9, 11] and

ii) that the removal, by washing, of extraframework species from HEMT5.5nw causes a decrease in acid strength. Very strong protonic sites possibly resulting from the interaction of extraframework aluminium species with the framework hydroxyl groups are indeed present in HEMT5.5nw.

For both series of catalysts, the stability represented by the ratio of the final activity A_f (after 3 hours of reaction) to the initial activity A_0, increases as nPt/nA increases. However for identical values of nPt/nA, Pt HFAU are more stable than Pt HEMT catalysts (Fig 2). This indicates that the stronger the acid sites the lower the stability. This is expected, for the formation of coke is favoured by the acid strength [15]. In agreement with this negative effect of the acid strength on the catalyst stability, Pt HEMT5.5nw is less stable than Pt HEMT5.5w.

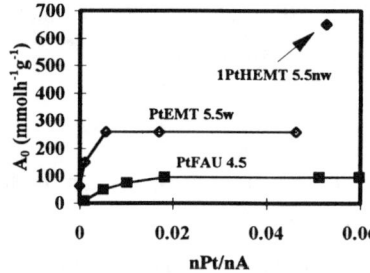

Figure 1: Initial activity (A_0) of Pt HFAU and Pt HEMT catalysts versus nPt/nA ratio

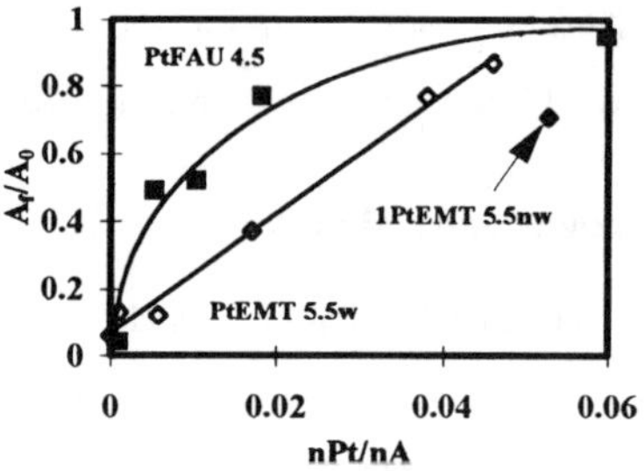

Figure 2: Ratio of the final and initial activities (A_f/A_0) of the Pt HFAU and Pt HEMT catalysts versus nPt/nA ratio

Product distribution

Identical products are formed on all catalysts. These products can be classified into three categories :
- M : Monobranched isomers (mainly 2 and 3 methylhexanes).
- B : Multibranched isomers (mainly 2,2; 2,3 and 2,4 dimethylpentanes).
- C : Cracked n-heptane (mainly isobutane and propane).

With all the samples, the product distribution was determined over a large range of conversions(1-70%) and the reaction scheme was established. For Pt HFAU4.5 samples with nPt/nA < 0.0103 and Pt HEMT5.5w samples with nPt/nA < 0.017 M, B and C are formed directly from n-heptane with an apparent parallel scheme.

$$nC_7 \rightleftarrows \begin{array}{c} M \\ B \\ C \end{array}$$

For the other samples, C is a secondary product, the apparent reaction scheme being as follows :

$$nC_7 \rightleftarrows (M, B) \longrightarrow C$$

The ideal reaction scheme :

$$nC_7 \rightleftarrows M \rightleftarrows B \longrightarrow C$$

is not observed for any sample, which indicates that olefinic intermediates can undergo more than one reaction during their diffusion between two Pt crystallites [4]. The selectivity of the catalyst samples can be compared by plotting the I/C (isomerized/cracked n-heptane) and the M/B ratio at the same conversion (10% in figure 3) as a function of nPt/nA.

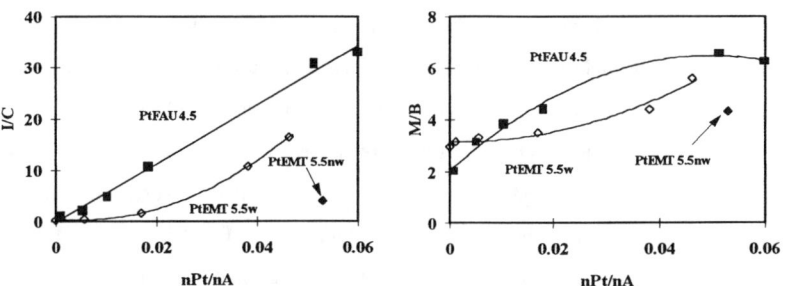

Figure 3 : Isomerized/Cracked n-heptane ratio (I/C) and Monobranched/Multibranched isomer ratio (M/B) on Pt HFAU and Pt HEMT catalysts versus nPt/nA ratio

Figures 3 show that lower values of I/C and M/B are observed with Pt HEMT5.5w than with Pt HFAU4.5. This can be expected for the active sites of HEMT5.5w are stronger than those of HFAU4.5, hence more active. Therefore for the same value of nPt/nA, the number of reactions undergone by the olefinic intermediates is always greater in the case of Pt HEMT5.5w. In agreement with this effect of the acid strength, the values found with Pt HEMT5.5nw sample are lower than with Pt HEMT5.5w.

Monobranched isomer distributions depend very little on the catalyst and conversion. Methylhexanes constitute over 96% of monobranched isomers, 3-ethylpentane content being at the best equal to 70% of its equilibrium value. 2- and 3-methylhexanes are in a ratio close to equilibrium. The distribution of multibranched isomers changes with conversion (Table 1) ; the 2,3- and 2,4- dimethylpentanes which predominate see their quantities decrease in favour of 2,2- and 3,3- dimethylpentanes and 2,2,3- trimethylbutane. All the dimethylpentanes seem to be formed initially on all the catalysts. 2,2,3- trimethylbutane, however, is not a primary reaction product for values of nPt/nA equal to or higher than 0.0103 with Pt HFAU4.5 samples and than 0.017 with Pt HEMT5.5w samples. This indicates that with these latter samples n-heptenes resulting from n- heptane dehydrogenation can undergo only two successive isomerization steps during their diffusion between two platinum crystallites. It should be remarked that while the multibranched isomer distributions at zero conversion are close on Pt HEMT and Pt HFAU samples, it is not the case at 50% conversion : with PtHEMT samples (e.g. table 1) the percentages of 2,3- dimethylpentane and especially of 2,4- dimethylpentane are more significant than on Pt HFAU samples. This can be attributed either to the higher activity of Pt HEMT samples or to limitations in the hypocages of HEMT of interisomerization of dibranched isomers.

On the other hand the distribution of cracking products is practically independent on the catalyst and on the conversion. Approximatively 98% of these products are constituted by propane and butanes, isobutane being largely predominant (iso/n > 30).

This confirms that whatever the apparent reaction scheme the light products mainly result from cracking of dibranched alkene intermediates [16]. The 2% of C_5-C_6 (iso/n ≈ 2) which are observed result from a dimerization-cracking pathway.

Catalysts	Conversion (%)	2,2-DMC$_5$ (%)	2,3-DMC$_5$ (%)	2,4-DMC$_5$ (%)	3,3-DMC$_5$ (%)	2,2,3-TMC$_4$ (%)
Pt FAU	50	22.9	33.4	25.0	15.6	3.1
	0	14.4	46.3	32.5	6.8	0
Pt EMTw	50	21.0	35.1	29.9	11.0	3.0
	0	14.1	47.5	33.4	5.0	0

Table 1 : Distribution of multibranched isomers at 50% conversion and at 0% conversion (extrapolated values) for 1PtHFAU4.5 (nPt/nA=0.0103) and for 1.5PtHEMT5.5w (nPt/nA=0.017).

CONCLUSION

The effect of the balance between the hydrogenating and acid functions on the catalytic properties of Pt HFAU and Pt HEMT samples is that as expected from the classical bifunctional mechanism. Large differences are observed between the activity, stability and selectivity of Pt HFAU and Pt HEMT samples with similar hydrogenating activities and acid site densities. All these differences can be related to the greater strength of the acid sites of HEMT samples. The resulting much higher activity of EMT catalysts could lead to the development of new hydrocracking catalysts provided that a much cheaper method for EMT preparation is developed.

Acknowledgement : Thanks are due to Dr. Kessler from the « Laboratoire des Matériaux Minéraux » de Mulhouse for the gift of an EMT sample.

REFERENCES

1- A.P. Bolton in Zeolite Chemistry and Catalysis, Edited by J.A. Rabo, (ACS Monograph 171, American Chemical Society, Washington D.C., 1976) pp. 714-779.
2- M. Guisnet and G. Perot in Zeolites : Science and Technology, Edited by F.R. Ribeiro, A.E. Rodrigues, L.D. Rollmann, C. Naccache, (NATO ASI Series E 80, Martinus Nijhoff Publishers, The Hague, 1984) pp. 387-420.
3- J.A. Martens, P.A. Jacobs and J. Weitkamp, Appl. Catal. **20**, 239 (1986).
4- M. Guisnet, F. Alvarez, G. Giannetto and G. Perot. Catal. Today, **1**, 415 (1987).
5- F. Delprato, L. Delmotte, J.L. Guth and L. Huve, Zeolites, **90**, 546 (1990).
6- M. Stöcker, H. Mostad and T. Rorvik, Catal. Letters **28**, 203 (1994).
7- J.A. Martens and P.A. Jacobs, J. Mol. Catal **78**, L47 (1993).
8- E.J.P. Feijen, J.A. Martens, P.A. Jacobs in 11th International Congress on Catalysis. 40th Anniversary, Edited by J.W. Hightower et al. , Stud. Surf. Sci. Catal, Elsevier, Amsterdam, vol 101, 721 (1996).
9- S. Morin, A. Berreghis, P. Ayrault, N.S. Gnep and M. Guisnet. J. Chem. Soc., Faraday Trans., **93** (17) (1997) 3269.
10- G.A. Doka Nassionou, P. Magnoux and M. Guisnet., Microporous and Mesoporous Materials, Accepted for publication.
11- B.L. Su and D. Barthomeuf, Zeolites **13**, 626 (1993).
12- F. Dougnier, J. Patarin, J.L. Guth and D. Anglerot, Zeolites **12**, 160 (1992).
13- A. Berreghis, S. Morin, P. Magnoux, M. Guisnet, V. le Chanu, H. Kessler, J. Chim Phys **93**, 1525 (1996).
14- S. Morin, P. Ayrault, N.S. Gnep and M. Guisnet., Appl. Catal, **166**, 281 (1998).
15- M. Guisnet and P. Magnoux, Catalysis Today **36**, 477 (1997).
16- G. E. Giannetto, G.R. Perot and M.R. Guisnet. Ind. Eng. Chem. Prod. Res. Dev. **25**, 481 (1986).

AROMATIZATION OF 1,3-BUTADIENE ON BASIC ZEOLITES IN THE VAPOR PHASE

J. ACKERMANN, E. KLEMM and G. EMIG

Institute of Technical Chemistry I, Friedrich-Alexander-University of Erlangen-Nürnberg
Egerlandstraße 3, D-91058 Erlangen, Germany
phone: (+49-9131) 857420, fax: (+49-9131) 857421, email: emig@tc.uni-erlangen.de

ABSTRACT

Vapor phase aromatization of 1,3-butadiene on basic zeolites has been investigated as a possible useful way for the depletion of growing butadiene surplus. Different alkaline and earth alkaline metal exchanged FAU-type zeolites with basic properties were tested as potential catalysts for the production of ethylbenzene. Additional intrazeolitic sodium oxide in zeolite NaX showed beneficial effects on selectivity to ethylbenzene. The different active sites were identified by using DRIFT-spectroscopy and adding CO_2 to the feed. Furthermore, variation of reaction conditions (temperature, modified residence time) was studied on zeolite NaX to optimize yield.

INTRODUCTION

Currently, more than 80 % of the world's 1,3-butadiene is extracted from crude C_4-effluents as a by-product from naphtha-based ethylene and propylene plants. The growth rate of butadiene production in ethylene plants is estimated to be 28 % greater than its demand, which has resulted in a global butadiene surplus. Therefore, new routes for the upgrading of butadiene seem to be profitable [1].

One possibility could be the aromatization of 1,3-butadiene to ethylbenzene. In the first step, 1,3-butadiene undergoes cyclodimerization via a so called Diels-Alder reaction ([4+2]-cycloaddition) to produce 4-vinylcyclohexene. Ethylbenzene is then built by isomerization and dehydrogenation, respectively. The overall reaction scheme is as follows:

1,3-Butadiene → (Diels-Alder Reaction, [4+2]-Cycloaddition) → 4-Vinylcyclohexene → (Isomerization/Dehydrogenation, $-H_2$) → Ethylbenzene

As known from literature, zeolites are useful catalysts for the first reaction step, since the Diels-Alder reaction is catalyzed by Lewis acid sites (extra-framework cations) [2]. Another advantage of zeolites in the Diels-Alder reaction is the restricted transition state shape selectivity resulting from their well defined microporous structure. Within the second step of the consecutive reaction, Lewis basic sites (framework oxygen) should form the catalytic active species [3]. Keeping these facts in mind, the intention of our work is to investigate the aromatization of 1,3-butadiene to ethylbenzene on zeolites with basic properties.

EXPERIMENTAL

Catalysts were prepared by conventional ion-exchange of the sodium form of FAU-type zeolites (Y: $n_{Si}/n_{Al} = 2,5$ from Grace, X: $n_{Si}/n_{Al} = 1,3$ from Fluka) according to their ion-exchange isotherms [4,5]. Following the method of HATHAWAY AND DAVIS [6], samples were threefold exchanged with 0,1 N solutions of alkaline and earth alkaline metal chlorides at 75 °C for 3 hours. The samples were filtered and washed chlorine free after each exchange. Finally, the catalysts were dried at 100 °C in air.

According to LASPÉRAS ET AL. [7], sodium-added zeolite NaX was prepared by impregnation with an aqueous solution of sodium acetate of appropriate concentration and subsequent calcination at 550 °C (heating rate: 1 °C/min) for 1 h. Loadings of 1 (NaX+1Na) to 7 (NaX+7Na) additional sodium atoms per unit cell (corresponding to 6,9 – 34,2 wt.-% Na_2O) were obtained by this method.

The catalytic activity was determined by measuring the butadiene conversion in an isothermal integral fixed-bed reactor operated at atmospheric pressure (wall material: heat-resistant steel, length: 340 mm, inner diameter: 20 mm). The catalysts were activated at 450 °C for 1,5 h in nitrogen. Catalyst screening was carried out under the following standard reaction conditions: T = 400 °C, W/F = 800 (g·min)/mol, $m_{cat.}$ = 2 g, $p_{nitrogen}:p_{butadiene}$ = 4:1, $d_{particle}$ = 1,1 - 2 mm.

To obtain the selectivity and yield of each reaction product, analysis was performed by on-line gas chromatography (HP 5890 series II) using a FID-detector with a HP-Innowax column.

DRIFTS measurements were carried out using a FT-IR spectrometer (Perkin-Elmer Paragon 1000) equipped with a MCT detector, a "praying mantis" optical attachment and a high-temperature reaction cell (Harrick Scientific HVC-DR2). Before taking spectra at 450 °C with a resolution of 4 cm^{-1} and an accumulation of 100 scans, samples were first dried under flowing nitrogen at 450 °C for 0,5 h.

RESULTS AND DISCUSSION

Catalyst screening

Distribution of the main products (ethylbenzene and 4-vinylcyclohexene) and catalytic activity of the different alkaline metal exchanged zeolites are depicted in Figure 1 and Figure 2 (conversion-selectivity-plots including lines of constant yield). More basic X-type zeolites (this property is well known from [8]) show equal or higher conversion of butadiene and higher total yield of valuables than Y-type zeolites. The completion of the consecutive reaction seems to be favored on zeolites X, whereas on less basic zeolites Y the reaction mainly remains on the level of the intermediate 4-vinylcyclohexene. The Diels-Alder product 4-vinylcyclohexene is best produced on CsY (Y_{4-vch} = 15,4 %). Li-exchanged samples as catalysts with the highest Lewis acidity in the series of alkaline metal cations [9] show high activity but no selectivity to valuables and are therefore not presented in the figures below.

Among earth alkaline metal exchanged samples only BaX with the highest supposed basicity [10] shows appropriate yield of ethylbenzene (Figure 2). Although all other samples show high conversion (X ≈ 100 %), selectivity to desired products is low ($S_{tot.}$ < 0,5 %), regardless of the type of zeolite.

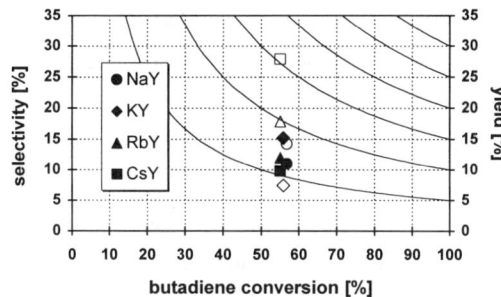

Figure 1: Catalytic activity of ion-exchanged zeolites Y
(after 15 min time-on-stream)

empty symbols: 4-vinylcyclohexene
filled symbols: ethylbenzene

Figure 2: Catalytic activity of ion-exchanged zeolites X
(after 15 min time-on-stream)

empty symbols: 4-vinylcyclohexene
filled symbols: ethylbenzene

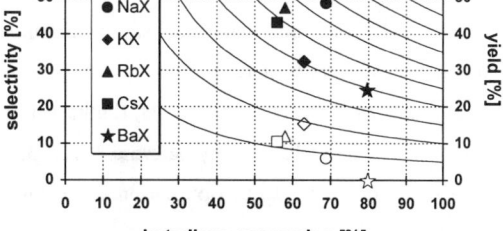

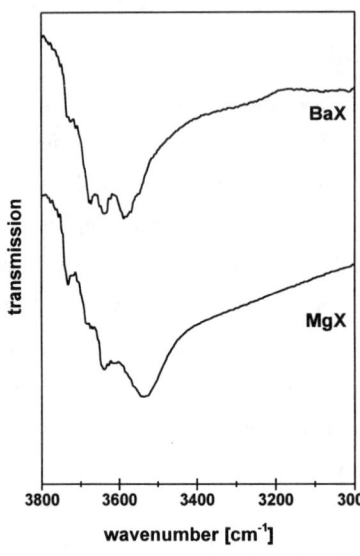

Figure 3: DRIFT-spectra of two selected earth alkaline metal exchanged zeolites in the ν(OH) region (shown in transmission mode)

- relevance of shown IR-bands [12]:

~ 3560 cm^{-1}	Si-OH-Al low frequency	acidic
~ 3650 cm^{-1}	Si-OH-Al high frequency	acidic
~ 3670 cm^{-1}	Me(OH)$^+$	basic
~ 3740 cm^{-1}	Si-OH	non-acidic

- catalytic activity:

catalyst	conversion [%]	total selectivity [%]
BaX	79,4	24,0
MgX	99,3	0,5
HX	79,4	1,5

DRIFT-spectra of two selected earth alkaline metal exchanged samples - BaX and MgX - are given in Figure 3, as well as data about their catalytic activity. Both catalysts show a band at ~ 3670 cm^{-1} resulting from Brønsted basic Me(OH)$^+$ species. These species are formed during dissociation of coordinated water molecules, when the divalent cation Me(H$_2$O)$^{2+}$ becomes localized during dehydration of the zeolite [10,11]. As a consequence of the dissociation, protons are formed and react with other exchange sites to bridged Si-OH-Al groups (characterized by a low frequency band at ~ 3560 cm^{-1} and a high frequency band at ~ 3650 cm^{-1}), which can be ascribed to Brønsted acidity.

In comparison with the H-form, Mg-exchanged zeolite X shows the same catalytic behavior: high activity but very low selectivity. Obviously, Brønsted acid sites inevitably cause the formation of carbonium ion intermediates [13] from butadiene so that cracking products and coke deposits are favorably formed on these catalysts. On BaX, selectivity to ethylbenzene is higher but does not reach the results of alkaline metal exchanged samples. This might be explained by the fact that its basicity resulting both from O^{2-} and from Me(OH)$^+$ partly compensates Brønsted acidity.

Summarizing these experiments, highest total yield of valuables is achieved on NaX ($Y_{tot.}$ = 37,4 %), yield of ethylbenzene being 33,3 %. The combination of Lewis acidity and Lewis basicity seems to be optimal on this catalyst. In any case, it is crucial to avoid Brønsted acidity.

Influence of additional sodium oxide impregnation

The influence of additional basic active sites was investigated on zeolite NaX. Preliminary experiments showed that of all alkaline metal oxides intrazeolitic sodium oxide had the strongest influence on the catalytic behavior. As shown in Figure 4, butadiene conversion drops with increasing sodium oxide loading because of the diminishing microporevolume. However, selectivity to ethylbenzene rises from 48 % (NaX) to approximately 62 % (NaX+4Na and NaX+5Na) and then only slightly decreases (Figure 5). This growth in selectivity is not caused by the decreasing conversion, which could be shown by measurements on a constant level of conversion.

Impregnation leads to a greater number of basic sites. Thus, more 4-vinylcyclohexene can selectively react to ethylbenzene, so that less 4-vinylcyclohexene is transformed into coke via side reactions (e.g. the Diels-Alder reaction between 4-vinylcyclohexene and butadiene).

Finally, there is an optimum amount of additional sodium oxide. At a loading of about 4 additional sodium atoms per unit cell (corresponding to 23 wt.-% sodium oxide), total yield of valuables is acceptable at still high selectivity. Compared to the variation of basicity by ion-exchange, impregnation has a stronger effect on the production of ethylbenzene.

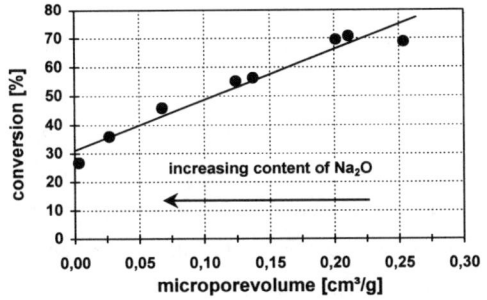

Figure 4: Dependence of butadiene conversion over sodium oxide loaded zeolite NaX on microporevolume

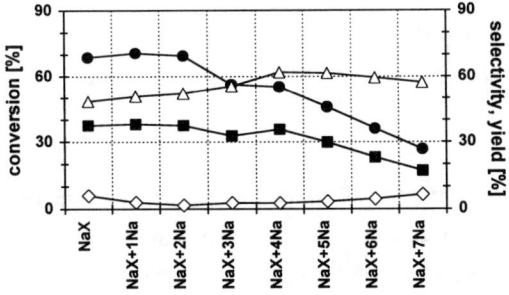

Figure 5: Effect of additional intra-zeolitic sodium oxide in zeolite NaX on conversion, selectivities and total yield

● butadiene conversion
◇ selectivity 4-vinylcyclohexene
△ selectivity ethylbenzene
■ total yield

Influence of carbon dioxide contact during reaction

Because of the electrophilic (acidic) nature of CO_2, its adsorption onto active basic sites (e.g. O^{2-}) leads to the formation of carbonate species and subsequently can block base-catalyzed activity [14,15]. Due to its parallel existing basic nature, CO_2 can also chemisorb on Lewis acid sites (cations) by ion-dipole interaction [16]. To inhibit one or both reaction steps and obtain more information about involved active sites, CO_2 was added to the feed stream (20 vol.-%) during the reaction. The effect on catalytic activity of two alkaline ion-exchanged X-type zeolites with different basicity is shown in Table 1.

Table 1: Effect of CO_2 contact on catalytic activity of X-type zeolites with different basicity

catalyst	NaX		CsX	
concentration [%] CO_2	0	20	0	20
conversion [%] butadiene	67,8	63,6	56,0	49,1
selectivity [%] 4-vinylcyclohexene	4,7	0,3	10,5	24,4
selectivity [%] ethylbenzene	48,7	20,2	43,1	15,3
yield [%] total	36,2	13,0	30,0	19,5

For both samples butadiene conversion is reduced because of the competing adsorption of CO_2. On less basic zeolite NaX, selectivity to 4-vinylcyclohexene as well as selectivity to ethylbenzene decreases. This behavior can be explained by a nearly equal adsorption of CO_2 on the sodium cations and the framework oxygen [17], resulting in an inhibition of both reaction steps. However, more basic zeolite CsX shows decreasing selectivity to ethylbenzene, but increasing selectivity to 4-vinylcyclohexene. Here, CO_2 strongly interacts with the framework oxygen, whereas only weak interaction with the cesium cations takes place [17]. For this reason, only the second reaction step is impeded, i.e. less produced 4-vinylcyclohexene is transformed into ethylbenzene. According to these results, basic sites must take part in the second step of the investigated reaction.

Influence of reaction conditions

Variation of reaction temperature is illustrated in Figure 6 for zeolite NaX. While butadiene conversion increases continuously, yield of ethylbenzene reaches a maximum at 400 °C. At higher temperatures yields decrease, as coke formation and the building of by-products due to cracking is favored. In the temperature range below 360 °C ethylbenzene formation is kinetically inhibited and the intermediate 4-vinylcyclohexene is preferably built.

Figure 7 illustrates the influence of modified residence time (W/F) on catalyst performance for zeolite NaX. Butadiene conversion as well as total yield of valuables rise continuously with increasing residence time. Yield of 4-vinylcyclohexene first increases and then only slightly decreases with increasing W/F ratio, while yield of ethylbenzene increases steadily. This indicates the typical behavior of a consecutive reaction, in which ethylbenzene is built via the intermediate 4-vinylcyclohexene. Over the whole range of varied modified residence time total selectivity to valuables remains almost constant.

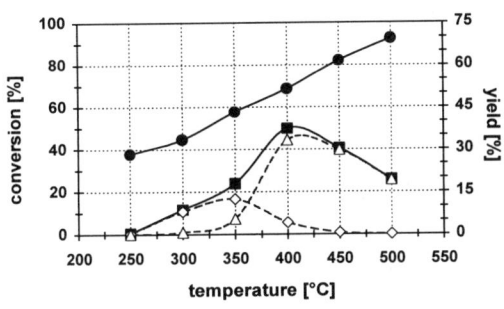

Figure 6: Influence of temperature on butadiene conversion and product yields over zeolite NaX (W/F = 800 (g·min)/mol)

- ● butadiene conversion
- ◇ yield 4-vinylcyclohexene
- △ yield ethylbenzene
- ■ total yield

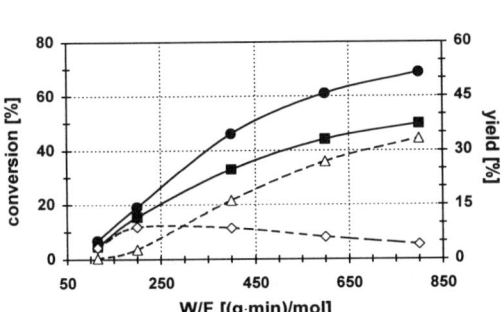

Figure 7: Influence of modified residence time (W/F) on butadiene conversion and product yields over zeolite NaX (T = 400 °C)

- ● butadiene conversion
- ◇ yield 4-vinylcyclohexene
- △ yield ethylbenzene
- ■ total yield

CONCLUSIONS

Using bifunctional FAU-type zeolites with Lewis acid and Lewis basic properties, ethylbenzene can be built from 1,3-butadiene via the intermediate 4-vinylcyclohexene in a one-step vapor phase process.

While earth alkaline metal exchanged zeolites are less selective catalysts, alkaline metal exchanged zeolites, especially more basic X-type zeolites, show acceptable yields. Best results were obtained on zeolite NaX. On this catalyst, yield of ethylbenzene can be optimized by varying reaction conditions (temperature, modified residence time). Furthermore, additional intrazeolitic sodium oxide enhances selectivity to ethylbenzene. In comparison to the variation of basicity by ion-exchange, impregnation has a more beneficial effect on the formation of ethylbenzene.

Concerning active sites, a complex correlation between Lewis basicity (framework oxygen) and Lewis acidity (extra-framework cations) seems to exist for this consecutive reaction. The influence of basic sites in the production of ethylbenzene from 4-vinylcyclohexene could be shown by adding CO_2 during the reaction. DRIFT-spectroscopic studies demonstrated, that Brønsted acidity definitely has to be avoided.

REFERENCES

[1] M.L. Morgan; DGMK Tagungsbericht **9705** (1997), 9-16
[2] R.M. Dessau; J. Chem. Soc., Chem. Commun. (1986), 1167-1168
[3] H. Suzuka, H. Hattori; J. Mol. Catal. **63** (1990), 371-385
[4] H.S. Sherry; J. Phys. Chem. **70** (1966) 4, 1158-1168
[5] H.S. Sherry; J. Phys. Chem. **72** (1968) 12, 4086-4094
[6] P.E. Hathaway, M.E. Davis; J. Catal. **116** (1989), 263-278
[7] M. Laspéras, H. Cambon, D. Brunel, I. Rodriguez, P. Geneste; Microporous Mater. **1** (1993), 343-351
[8] D. Barthomeuf; J. Phys. Chem. **88** (1984) 1, 42-45
[9] J.W. Ward; J. Catal. **10** (1968), 34-46
[10] J. Xie, M. Huang, S. Kaliaguine; Catal. Lett. **29** (1994), 281-291
[11] J.W. Ward; in Zeolite Chemistry and Catalysis (J. Rabo, ed.), ACS Monograph **171** (1976), 118-284
[12] L.M. Kustov, V.Y. Borovkov, V.B. Kazansky; J. Catal. **72** (1981), 149-159
[13] P.A. Jacobs; Carboniogenic Activity of Zeolites (Elsevier, Amsterdam, 1977)
[14] J.C. Lavalley; Catal. Today **27** (1996), 377-401
[15] F. Yagi, H. Tsuji, H. Hattori; Microporous Mater. **9** (1997), 237-245
[16] P.A. Jacobs, F.H. van Cauwelaert, E.F. Vansant; J. Chem. Soc., Farad. Trans. **69** (1973), 2130-2139
[17] S. Huber; PhD thesis, Ludwig-Maximilians-University of Munich (1997)

CONTROLLED REMOVAL OF EXTRA-FRAMEWORK ALUMINUM SPECIES IN USY ZEOLITE

E. BENAZZI, J. LYNCH, A. GOLA, S. LACOMBE, C. MARCILLY

Institut Français du Pétrole, 1 et 4 Avenue de Bois-Préau, F-92852 Rueil-Malmaison Cedex, fax : (33) 1 47 52 60 55, e-mail : sylvie.lacombe@ifp.fr

ABSTRACT

A HY zeolite was first dealuminated by steaming. The extra-framework species (EFAL) were then removed by treatments of various strengths with either nitric acid, or ammonium hexafluorosilicate (HFS) or ethylenediaminetetraacetic acid (EDTA). It was found that the HFS and EDTA treatments allowed controlled removal of EFAL without significantly affecting the framework composition, whereas a HNO_3 treatment of medium strength contributed to the dealumination of the framework. As a consequence, the formation of mesopores was favoured when using HNO_3. A loss of micro and mesoporosity was observed on the HFS treated sample, which was attributed to the formation of a silicic layer on the cristallite surfaces.

INTRODUCTION

Zeolite Y is used as the main component in FCC and hydrocracking catalysts. In both cases, the zeolite is modified in order to adapt its properties to the catalytic application. The dealumination consists generally first in a hydrothermal treatment in order to dealuminate the zeolite framework, then an acid leaching in order to solubilize the extra-framework species (EFAL) created during the steaming. The dealumination steps generate a secondary pore system, whose pore diameters are in the range 10-40 nm [1,2]. The latter is known to affect the catalytic properties of the zeolite. Consequently, an understanding of the factors controlling its formation is of primary importance. This study consisted in investigating the influence of three different post-steaming treatments [3,4] on the zeolite chemical composition and textural properties, in particular on the characteristics of the created mesopores.

EXPERIMENTAL

A commercial NaY (mol. Si/Al 2.8) zeolite supplied by Tosoh Corporation was transformed by ionic exchanges in a NH_4NO_3 solution to the NH_4-Y form containing less than 500 ppm wt Na (dry basis). It was then dealuminated by steaming. The zeolite was first heated in a dry air flow. The water vapour was introduced at 473 K and the dry air flow was stopped. The steaming was continued until 923 K and left at 923 K for 4 hours, under 100% steam. The temperature was then reduced to 673 K and water injection stopped. The zeolite was allowed to dry at this temperature under dry air flow. The steamed sample thus obtained was called S6. The S6 sample was treated according to three different procedures : i) acid leached with HNO_3 solutions (5 ml/g - 0.1 to 3 N) at 373 K during 2 hours, filtered and rinsed with distilled water, then dried overnight at 393 K ; ii) treated by aqueous solutions of $EDTAH_2Na_2$ (20 ml/g - 0.1 to 0.4 M) at 373 K during 4 hours, filtered and rinsed with distilled water, then dried overnight at 393 K ; the sodium eventually introduced in the zeolite was then exchanged with ammonium nitrate ; iii) treated with aqueous 0.4 M solutions of $(NH_4)_2SiF_6$ at 353 K during 2 hours. In this case the steamed zeolite was first put in suspension in an ammonium acetate solution heated at 353 K (10 ml/g) such that the molar ratio of ammonium acetate to further added $(NH_4)_2SiF_6$ equalled 15. A various volume of $(NH_4)_2SiF_6$ 0.4 M was then slowly added and the mixture was stirred for 4 hours. The zeolite was filtered and rinsed with distilled water at 353 K, then dried overnight at 393 K.

The global aluminum content was determined by XRF. The aluminum framework composition was obtained by averaging the values obtained by IR spectroscopy, ^{29}Si-MAS-NMR and XRD. The EFAL content was obtained as the difference between the total amount of aluminum atoms and the amount of framework aluminum atoms (FAL) per unit cell. ^{27}Al-MAS-NMR analyses were performed on fully hydrated samples, using a Brucker MSL400 instrument. The textural properties of the samples were determined after calcination at 773 K under helium for 12 hours. This treatment aimed at removing the physisorbed water and should not effect the pore structure. The used techniques were : i) *Nitrogen adsorption :* the nitrogen adsorption-desorption isotherms were recorded at 77K with a Φ-SORB apparatus (Vinci-Technologies). The total sorbed volume (V_{pT}), including adsorption in micropores, mesopores and on the external surface, was obtained from the amount of nitrogen adsorbed at the relative pressure $P/P_0=0.95$. The micropore volume ($V_{\mu m}$) was determined using the Dubinin theory of pore volume filling. The difference between the total sorbed volume and the micropore volume gave a reasonable estimation of the mesopore volume (V_{mes}). The distribution of the mesopores was calculated from the adsorption branch using the Broekhoff and de Boer method. ii) *Mercury porosimetry* : mercury porosimetry measurements were peformed using a Autopore III micromeritics apparatus.

RESULTS AND DISCUSSION

Influence of the post-steaming treatments on the zeolite composition

In table 1 are summarised the physico-chemical characteristics of the original NH_4-Y form, of the steamed and post-steaming modified samples.

Sample		XRD cristallinity (%)	Si/Al		Al/uc		S_{BET} (m^2/g)
			Global	Frw.	FAL	EFAL	
NH_4Y		100	2.8	2.8	51	0	930
S6		83	2.8	11	16	35	730
	[HNO3] (M)						
A_1	0.1	94	5.9	11	16	12	870
A_3	0.5	94	6.2	13	14	13	886
A_4	1.0	94	6.8	14	13	13	854
A_5	1.5	85	6.9	14	13	11	912
A_6	1.9	75	10.8	17	11	5	nd
A_8	3	64	43	47	4	0.4	nd
	HFS/Al (mol/mol)						
H_2	0.22	75	3.2	10.0	17	28	704
H_3	0.32	76	5.3	10.0	17	13	739
H_4	0.42	80	6.0	9.1	19	8	758
H_5	0.64	89	7.5	11	16	7	813
H_6	0.8	80	9.4	11.8	15	4	669
H_7	1.0	68	14.0	15.7	12	1	628
	EDTA/Al (mol/mol)						
E_1	0.46	94	3.0	9.1	19	29	nd
E_2	0.56	92	3.5	9.7	18	25	785
E_3	0.60	86	4.7	10.3	17	17	850
E_4	0.91	93	7.8	11.8	15	7	850
E_5	2.05	91	8.5	9.7	18	2	895
E_6	2.09	91	10.0	11.0	16	1	900

Table 1: Effect of steaming and post-steaming treatments of increasing strength, on the composition and BET surface area of the Y zeolite. *nd= not determined*

Concerning the effect of the nitric acid treatments, it is observed that a great part of the EFAL aluminum species, about 23 Al per unit cell (Al/uc) out of a total of 35, are easily removed with low concentrated nitric acid treatment (0.1 M). This suggests that a great part of the removed EFAL are constituted of cationic species and weakly polymerised aluminum species. When increasing the severity of the acid treatment up to 1.5 M, the EFAL species content remains constant, about 12 Al/uc, whereas the framework is slightly dealuminated. When the nitric acid concentration is higher than 1.5 M, the treatment leads concomitantly to the elimination of the remaining EFAL species and to the dealumination of the framework (Figure 1). When the post-steaming treatments are operated using either $(NH_4)_2SiF_6$ or $EDTAH_2Na_2$ aqueous solutions, it is observed that unlike the nitric acid solutions, these reactants allow a selective and controlled removal of the EFAL without dealuminating the framework, except when the $(NH_4)_2SiF_6/Al$ or $EDTAH_2Na_2/Al$ ratios are very high. The aluminum framework content is about 16-17 Al/uc and the amount of EFAL varies from 35 (steamed sample) to less than 2 for the higher reactants concentrations. So the use of $(NH_4)_2SiF_6$ or $EDTAH_2Na_2$ aqueous solutions, with appropriate concentrations, allows the preparation of samples with controlled amount of EFAL species for a given framework composition.

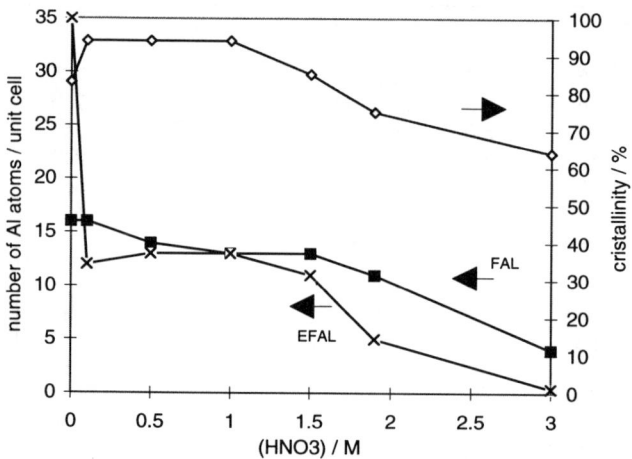

Figure 1. Number of framework (FAL), non framework (EFAL) aluminum atoms and cristallinity as a function of the strength of the post-steaming HNO_3 treatment.

The nature of the extra-framework species was investigated using ^{27}Al-MAS-NMR This technique provides useful information although part of the aluminum atoms may not be visible. The analysis was performed on the steamed sample (S6) and after treatment with the three dealuminating agents. Some relevant spectra are shown on Figure 2.

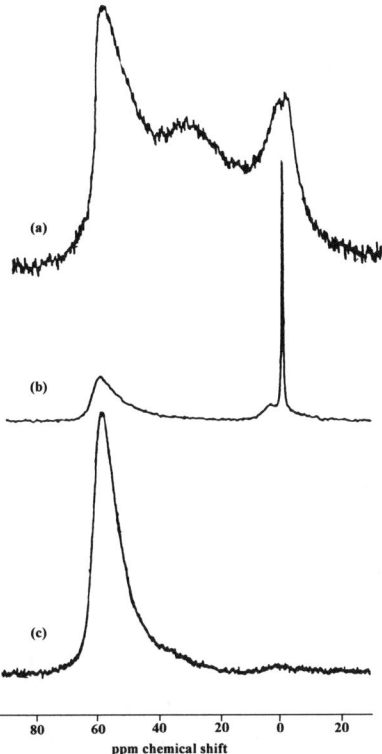

Figure 2. ^{27}Al-MAS-NMR spectra of (a) steamed sample S6, (b) steamed and treated with nitric acid sample A5, (c) steamed and treated with $(NH_4)_2SiF_6$ sample H5.

The ^{27}Al MAS-NMR on the steamed sample presents three peaks located at 60, 30 et 0 ppm (see spectrum (a) on Figure 2). They are attributed to Al IV, Al V et Al VI respectively. The Al V and Al VI species correspond to EFAL species. The three post-steaming dealuminating agents contribute to the fast disappearance of the pentacoordinated aluminum Al V. On the other hand some differences are observed on the octahedral aluminum peak according to the dealuminating agent. After the nitric acid treatment, the peak of the octahedral aluminum Al VI is enlarged and a very narrow and intense peak is observed (see spectrum (b) on Figure 2). The latter corresponds to aluminum atoms in a highly symmetrical environment and is therefore attributed to $Al(H_2O)_6^{3+}$. The Al IV is highly asymmetric, suggesting that part of the EFAL species are in a tetrahedral configuration. The spectra obtained after $EDTAH_2Na_2$ and $(NH_4)_2SiF_6$ treatment are very similar and differ from the one obtained with nitric acid. The non symmetry of the peak at 60 ppm is not as pronounced as in the case of nitric acid treatment. Furthermore no narrow and intense peak is observed at 0 ppm, but a large peak which disappears when the amount of EFAL decreases to around 12 per unit cell. A possible explanation is therefore that the $Al(H_2O)_6^{3+}$ species observed after nitric acid treatment are created during the framework dealumination. It is also possible that they were present but invisible in the steamed sample. As a conclusion the dealuminating agent acts not only on the zeolite composition but also on the chemical state of the left EFAL species. The nitric acid treatment induces a different chemical state of the aluminum atoms than the two other agents, which seem to act on the solid in a more similar way.

Influence of the post-steaming treatments on the textural properties of the zeolite

When performing the nitric acid leaching and $EDTAH_2Na_2$ treatments, the materials show an increase of the surface area compared to the steamed sample when the severity of the treatment increases (see Table 1). This result is in agreement with the fact that the elimination of the EFAL species created during the steaming allows the nitrogen molecules to access the microporosity. The initial surface area of 930 m^2/g (sample NH_4Y) is nevertheless not recovered. This has to be correlated with the decrease of cristallinity, attesting of a partial destruction of the zeolite and the formation of an amorphous phase. Surprisingly the treatments with $(NH_4)_2SiF_6$ aqueous solutions of increasing strength induce a decrease of the surface area.

In Table 2 and Figure 3 are given, for some selected samples, the microporous, mesoporous and total porous volumes. The values on the mesopores obtained both by N_2 adsorption and Hg porosimetry were coherent and are here not differenciated.

Samples	V_{pT} (ml/g)	$V_{\mu m}$ (ml/g)	V_{mes} (ml/g)	$\emptyset_{mésopores}$ range (nm)	Mesopores central value (nm)
NH_4-Y	0.393	0.356	0.037	--	--
S6	0.376	0.300	0.076	4 - 25	9
A_1	0.474	0.362	0.112	2 - 40	15
E_4	0.482	0.355	0.127	4 - 30	16
H_2	0.343	0.265	0.078	4 - 30	12
H_6	0.324	0.293	0.031	4 - 25	11

Table 2 : Effect of the steaming and post-steaming treatments on the porosity of the Y zeolite.

In the case of the NH_4-Y sample, a significant mesoporous volume is measured. It is attributed to the pores which are formed by the crystallites packing. When operating the steaming, the total sorbed volume and the microporous volumes both decrease, whereas the mesoporous volume is strongly increased. These results indicate a partial destruction of the framework of the Y zeolite and the formation of mesopores.

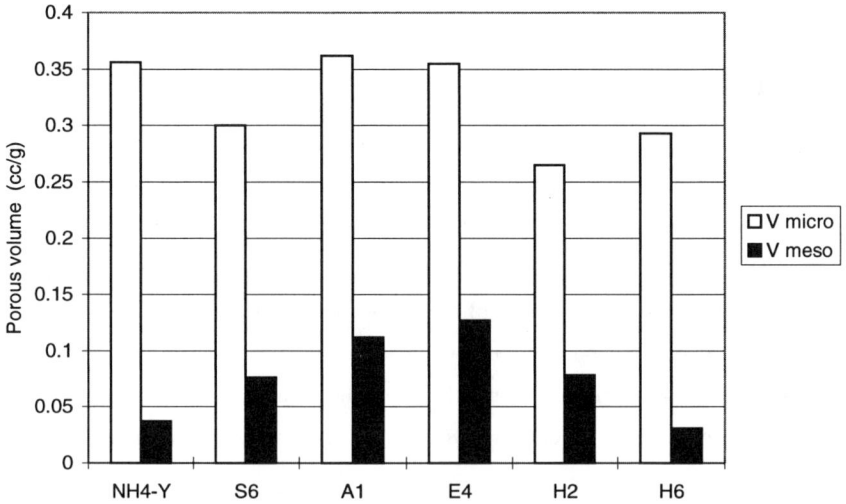

Figure 3. Microporous and mesoporous volumes on Y zeolite as a function of the dealumination treatment.

Concerning the effect of the post-steaming treatment, it is interesting to compare first the A_1 and E_4 samples, containing respectively 12 and 7 EFAL per unit cell. They both present higher mesoporous volumes than the steamed sample, which is expected, but also higher microporous volumes. The HNO_3 and $EDTAH_2Na_2$ treatments probably free some micropores which were plugged by EFAL species. Furthermore, although the A_1 sample is less dealuminated, A_1 and E_4 present similar microporous and mesoporous volumes. The mesopore sizes are also identical. This suggests that the nitric acid treatment, even when it is weak, favours the creation of mesopores. It is likely related to the fact that the zeolite framework is very rapidly dealuminated at the same time as the elimination of the EFAL species, so that the mesopores exist even when large amounts of EFAL are still present. Conversely, when dealuminating with EDTA, it seems to be necessary to remove the majority of the EFAL to obtain a significantly larger mesoporosity than in the steamed sample.

When performing a weak treatment with HFS after steaming (sample H_2), some microporosity is lost compared to the steamed sample and there is no gain in mesoporosity. The nearly complete elimination of the EFAL species by HFS (sample H_6) even induces a lost in mesoporosity. This can be explained by the formation of a silicic layer on the cristallites surface, which was confirmed by XPS analysis. It is also coherent with the already observed loss in surface area.

CONCLUSION

The use of different dealuminating agents after steaming of a Y zeolite allows control of the chemical composition of the zeolite but also its porosity, both parameters being important in catalytic applications.

1) The use of $(NH_4)_2SiF_6$ or $EDTAH_2Na_2$ aqueous solutions with appropriate concentrations allows the preparation of samples with controlled amount of EFAL species for a given framework composition. On the contrary, the treatment with HNO_3 induces dealumination of the framework even at low concentration. In parallel the chemical state of the left aluminum species is different when using nitric acid, compared to the two other dealuminating agents.

2) HNO_3 and $EDTAH_2Na_2$ treatments induce a gain in meso- and microporosity. The formation of mesopores is favoured when using HNO_3. The $(NH_4)_2SiF_6$ treatment induces a loss in micro- and mesoporosity, which is attributed to the formation of a silicic layer on the cristallite surface.

The influence of the nature of the extra-framework aluminum species on the catalytic activity has been investigated using as model reaction the n-decane hydrocracking. The results will be published elsewhere.

REFERENCES

1. J. Lynch, F. Raatz and P. Dufresne, Zeolites **7**, 333 (1987)
2. C. Choi-Feng, J.B. Hall, B.J. Higgins and R.A. Beyerlein, J. Catal. **140**, 395 (1993).
3. G.W. Skeels and D.W. Breck, Proceedings of the 6th IZC, Reno, 87 (1984).
4. A. Corma, V . Fornès, F. Rey, Appl. Catal. **59**, 267 (1990).
5. G.T. Kerr, J. Phys. Chem. **72**, 2594 (1968).

REINSERTION OF NONFRAMEWORK ALUMINUMS IN DEALUMINATED HZSM-5 ZEOLITE BY ACID TREATMENT

T. SANO, R. TADENUMA, Y. UNO, Z.B. WANG and K. SOGA

School of Materials Science, Japan Advanced Institute of Science and Technology
Tatsunokuchi, Ishikawa 923-1292, Japan; t-sano@jaist.ac.jp

ABSTRACT

Reinsertion of nonframework aluminums in dealuminated HZSM-5 zeolites by HCl and NaOH solution treatments was investigated by means of nitrogen adsorption, XRD, SEM, ^{27}Al MAS NMR, FT-IR and ICP analyses. In the case of HCl treatment, it was found that no structural degradation takes place and that a part of nonframework aluminums in dealuminated HZSM-5 zeolites are reinserted into the framework during the acid treatment. On the other hand, in the case of NaOH treatment, it was demonstrated that the possibility of the reinsertion of nonframework aluminums is low. A plausible realumination mechanism by acid treatment was also discussed.

INTRODUCTION

Recently, realumination as well as dealumination of zeolites has received considerable attention from a standpoint of controlling physicochemical properties of zeolites such as thermal stability and catalytic, sorptive and ion-exchange abilities. There are several papers concerning the realumination of dealuminated zeolites by treatment with alkali and ammonium nitrate solutions, suggesting the reinsertion of nonframework aluminums into the framework [1-9]. In the case of alkali treatment however, the possibility of removal of silicon atoms from the zeolite framework or formation of extraframework aluminosilicate on crystal surface is pointed out by several researchers [10-12].

We are also interested in the reversibility of dealumination-realumination process of zeolite from a standpoint of regeneration of zeolite catalysts. Based on the result obtained from a systematic study of dealumination behavior of HZSM-5 zeolite by thermal and hydrothermal treatments, namely the dealumination via hydrolysis of Si-O-Al bond in the zeolite framework is catalyzed by an acid (proton) that moves freely in the zeolite pores [13], we have studied the realumination by acid treatment. Very recently, the possibility of the reinsertion of nonframework aluminums in dealuminated HZSM-5 zeolites by treatment with 2 M HCl was pointed out [14].

In this paper, in order to get more information concerning the realumination of dealuminated HZSM-5 zeolites by acid and alkali treatments, the more systematic study was conducted.

EXPERIMENTAL

HZSM-5 zeolite was prepared following the procedure previously described [13]. The identification of the zeolite was achieved by X-ray diffraction (Rigaku RINT 2000). The total SiO_2/Al_2O_3 ratio of 71 was determined by X-ray fluorescence (XRF, Philips PW-2400). The framework SiO_2/Al_2O_3 ratio of 78 was determined by ^{27}Al MAS NMR (Varian VXP-400) at 104.3 MHz with an ca. 4.5 kHz spinning speed and 1.73 μs pulses for 4,000 scans using a calibration curve, which was produced measuring ^{27}Al MAS NMR spectra of HZSM-5 zeolites with various SiO_2/Al_2O_3 ratios [13].

Dealumination of the HZSM-5 zeolite was carried out by thermal and hydrothermal treatments at 600°C using a muffle furnace and a quartz reactor tube, respectively. The water vapor pressure of hydrothermal treatment was 10 kPa, generated using nitrogen as a diluent gas. The total gas flow rate (nitrogen + water vapor) was kept constant (0.531 mol h^{-1}).

Acid or alkali treatment was carried out as follows: 1 g of dealuminated HZSM-5 zeolite was treated with 100 ml of 2 M HCl at 100°C or 0.05-0.2 M NaOH at 80°C, stirring for a given time. The product was filtered off, washed thoroughly with deionized water, dried at 100°C and calcined at 380°C for 8 h.

Characterization of dealuminated and realuminated HZSM-5 zeolites was also achieved by diffuse reflectance FT-IR (JEOL JIR-7000), nitrogen adsorption (Bel Japan Belsorp 28SA), and scanning electron micrograph (SEM, Hitachi S-4000). The FT-IR spectra were taken at 4 cm^{-1} resolution for 500 scans at room temperature. The powdered zeolite was placed in a thin-walled ampule and then evacuated to ca. 10^{-5} torr at 400°C for 2 h. Before nitrogen adsorption measurements at -196°C, the powdered zeolites (ca. 0.1 g) was evacuated at 400°C for 12 h.

RESULTS AND DISCUSSION

Acid treatment

Figure 1 shows nitrogen adsorption isotherms of HZSM-5 zeolites before and after HCl treatment. No difference was observed for the nitrogen adsorption isotherms before and after HCl treatment, indicating no structural degradation. This was also confirmed from the fact that X-ray diffraction diagrams of HZSM-5 zeolites treated with HCl treatment show no peaks other

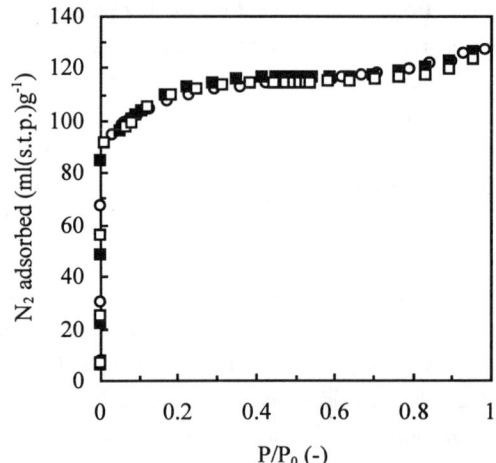

Figure 1. Nitrogen adsorption isotherms of HZSM-5 zeolites of (○) parent, (■) after thermal treatment at 600°C for 48 h and (□) after treatment of (■) with 2 M HCl at 100°C for 120 h.

than those corresponding to ZSM-5 zeolite and the intensities of peaks observed were almost same as those of the parent zeolite (Figure 2). Figure 3 shows the scanning electron micrographs of HZSM-5 zeolites before and after HCl treatment. Although the broken pieces of zeolite crystal are attached on crystal surfaces as shown in the SEM image of zeolite after HCl treatment, no crystal dissolution seems to occur during the acid treatment.

Figure 4 (A) shows FT-IR spectra of the parent HZSM-5 (total SiO_2/Al_2O_3 ratio: 71), the dealuminated HZSM-5 prepared by thermal treatment at 600°C for 48 h (total SiO_2/Al_2O_3 ratio: 72) and the realuminated HZSM-5 after HCl treatment (total SiO_2/Al_2O_3 ratio: 78). The decrease in the peak intensity at 3605 cm^{-1}

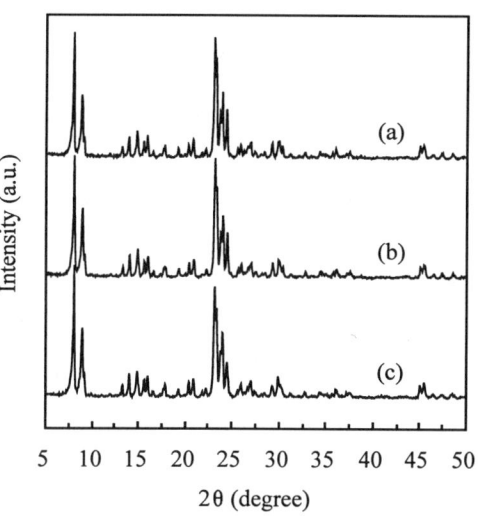

Figure 2. XRD patterns of HZSM-5 zeolites of (a) parent, (b) after thermal treatment at 600°C for 48 h and (c) after treatment of (b) with 2 M HCl at 100°C for 120 h.

assigned to an acidic bridged OH of Si(OH)Al was observed in the spectrum of dealuminated HZSM-5 zeolite, while the clear increase in the peak intensity was observed in the spectrum of realuminated zeolite. This result suggests the reinsertion of nonframework aluminums in the dealuminated zeolite into the framework. The slight decrease in the peak intensity at ca. 3500cm^{-1} assigned to hydrogen bounding adjacent hydroxyl groups is possibly corresponding to that in the number of silanol group by the condensation reaction.

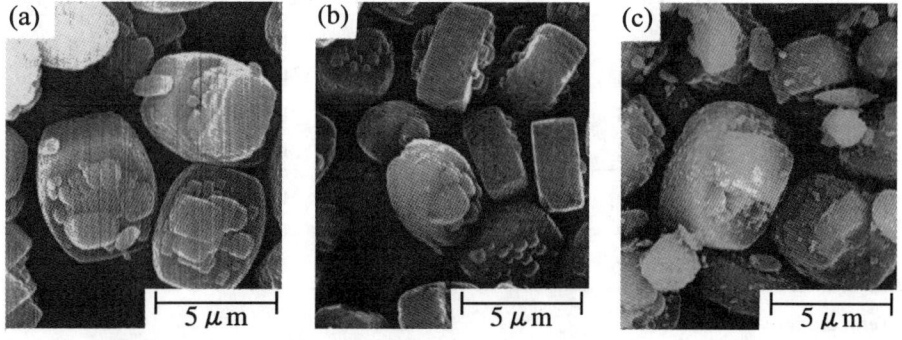

Figure 3. SEM images of HZSM-5 zeolites of (a) parent, (b) after thermal treatment (600°C, 48 h) and (c) after treatment of 2 M HCl (100°C, 120 h).

To clarify this possibility of reinsertion, the ^{27}Al MAS NMR spectra of these samples were measured. The peak intensity at 53 ppm was normalized based on 1 g of zeolite. As shown in Figure 4 (B), the peak intensity of the dealuminated HZSM-5 zeolite definitely increased by HCl treatment. The calculated framework SiO_2/Al_2O_3 ratios of the dealuminated and the realuminated

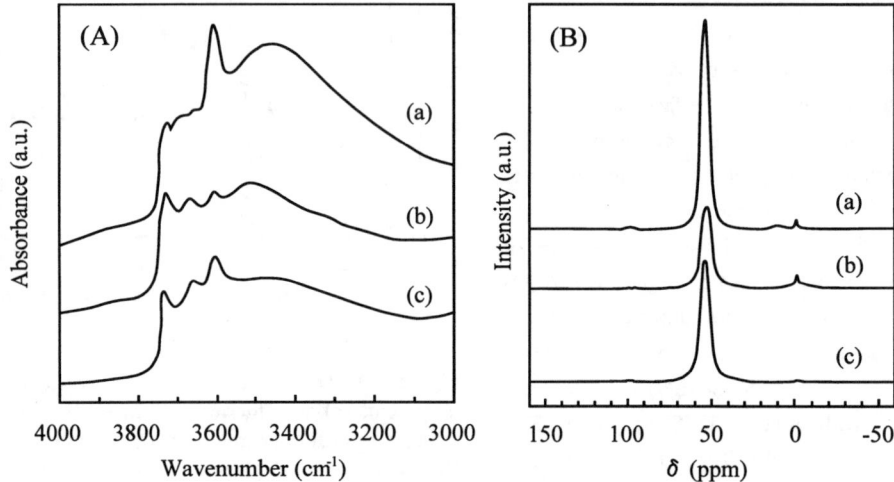

Figure 4. FT-IR and ^{27}Al MAS NMR spectra of HZSM-5 zeolites of (a) parent, (b) after thermal treatment at 600°C for 48 h and (c) after treatment of (b) with 2 M HCl at 100°C for 120 h.

Table 1: Degrees of elution of silicon and aluminum atoms from dealuminated zeolites during HCl treatment.

Run No.	Dealumination		Realumination			Degree of elution [c]	
	Surrounding gas	Time (h)	Solution	Temp. (°C)	Time (h)	Si (%)	Al (%)
1	Air [a]	12	2 M HCl	100	120	2.08	8.84
2	Air	24	2 M HCl	100	120	1.33	9.48
3	Air	48	2 M HCl	100	120	0.989	9.36
4	Air	72	2 M HCl	100	120	1.73	15.2
5	Steam [b]	3	2 M HCl	100	120	1.65	7.93
6	Steam	12	2 M HCl	100	120	1.47	8.13
7	Steam	72	2 M HCl	100	120	2.44	14.6

a) Temperature: 600°C.
b) Temperature: 600°C, P(H$_2$O): 10 kPa.
c) Determined by ICP.

HZSM-5 zeolites were 173 and 140, respectively. The total SiO_2/Al_2O_3 ratio (78) of the realuminated zeolite was higher than that of the parent zeolite (71), namely the dissolution of nonframework aluminums slightly occurred during HCl treatment. These results indicate that a part of nonframework aluminums in dealuminated HZSM-5 zeolite are reinserted into the framework. However, it is very reasonable to consider that the increase in the peak intensities of ^{27}Al MAS NMR and FT-IR spectra is attributable to the removal of the framework silicon atoms [10-12]. To elucidate this, therefore, silicon and aluminum concentrations in liquid phase of HCl treatment were analyzed by inductively coupled argon emission spectroscopy (ICP, Seiko SPS7700). The degrees of elution of silicon and aluminum atoms from the dealuminated zeolite during HCl treatment are summarized in Table 1. It is obvious that the removal of framework silicon atoms hardly takes place during acid treatment. The high stability of HZSM-5 zeolite in

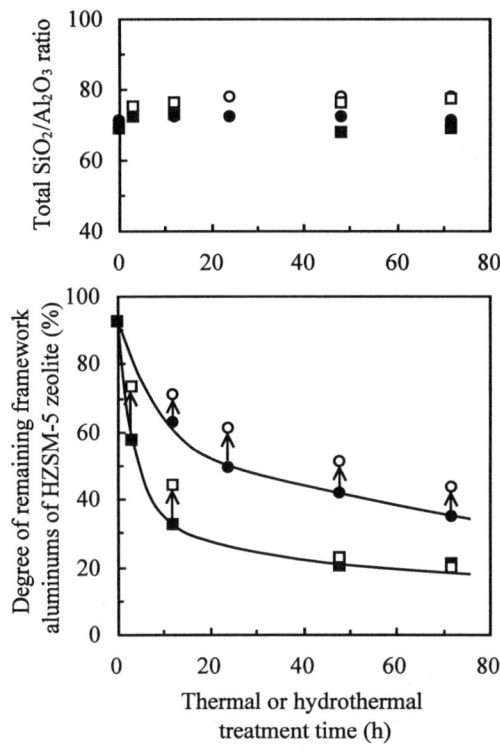

Figure 5. Total SiO_2/Al_2O_3 ratios and degrees of remaining framework aluminums of dealuminated and realuminated HZSM-5 zeolites. (●) thermal treatment at 600°C, (■) hydrothermal treatment at 600°C and $P(H_2O)$ of 10 kPa, (○,□) after treatment of (●,■) with 2 M HCl at 100 °C for 120 h.

the acid solution is also reported by Kooyman et al. [15]. Therefore, we can conclude that the realumination of dealuminated zeolite occurs during the acid treatment.

To get more information about the realumination, the realumination of several HZSM-5 zeolites with different degrees of dealumination was carried out using 2 M HCl at 100°C for 120 h. The dealuminated HZSM-5 zeolites with different degrees of dealumination were prepared varying the thermal treatment time. Figure 5 shows the total SiO_2/Al_2O_3 ratios and the degrees of remaining framework aluminums of dealuminated and corresponding realuminated HZSM-5 zeolites. The total SiO_2/Al_2O_3 ratios of realuminated HZSM-5 zeolites were slightly higher than those of dealuminated zeolites. All framework aluminum atom concentrations of dealuminated HZSM-5 zeolite increased by HCl treatment.

Dealuminated HZSM-5 zeolites were also prepared by hydrothermal treatment at 600°C under the water vapor pressure of 10 kPa and then treated with HCl and the result is also illustrated in Figure 5. The reinsertion of nonframework aluminums into the framework was also observed for the dealuminated HZSM-5 zeolites obtained at the hydrothermal treatment time below 15 h, whereas no reinsertion for the dealuminated zeolites with the treatment time more than 40 h.

Alkali treatment

In order to compare with the acid treatment, the dealuminated HZSM-5 zeolites were also

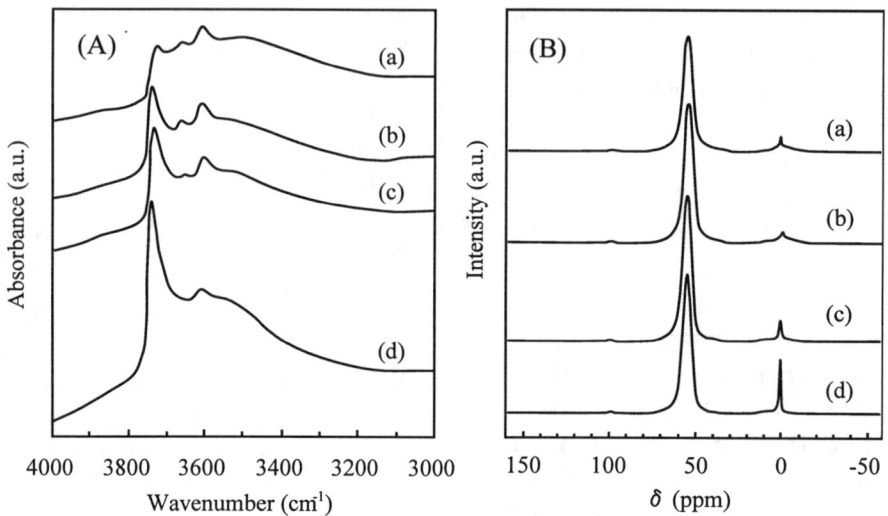

Figure 6. FT-IR and ^{27}Al MAS NMR spectra of HZSM-5 zeolites of (a) after thermal dealumination at 600°C for 48 h, (b) after treatment of (a) at 80°C for 0.5 h with NaOH of 0.05 M, (c) 0.1 M and (d) 0.2 M.

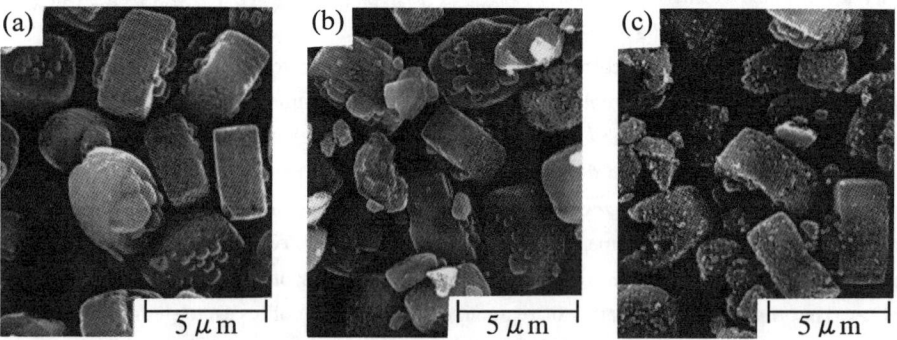

Figure 7. SEM images of ZSM-5 zeolites of (a) after thermal treatment (600°C, 48 h), (b) after alkali treatment of (a) at 80°C for 0.5 h with 0.1 M NaOH and (c) 0.2 M NaOH.

treated with 0.05-0.2 M NaOH aqueous solution at 80°C for 0.5 h. The slight increase in the peak intensities corresponding to the framework aluminums was confirmed by FT-IR and ^{27}Al MAS NMR spectra (Figure 6). This phenomenon was consistent with that described in literature [1-9]. However, the dissolution of crystal surface was observed in SEM images as shown in Figure 7, indicating the extraction of framework silicon atoms. This was also confirmed from ICP analysis of liquid phase of NaOH treatment (Table 2). Significant quantities of silicon atoms were detected in spite of very short treatment time compared with the acid treatment.

Table 2: Degrees elution of silicon and aluminum atoms from dealuminated zeolites during NaOH treatment.

Dealumination		Alkali treatment			Degree of elution [b)]	
Surrounding gas	Time (h)	Conc. (M)	Temp. (°C)	Time (h)	Si (%)	Al (%)
Air [a)]	48	0.05	80	0.5	25.8	0.234
Air	48	0.1	80	0.5	44.7	1.04
Air	48	0.2	80	0.5	78.2	7.92

a) Temperature: 600°C.
b) Determined by ICP.

Although we can not absolutely exclude the reinsertion of nonframework aluminums into the zeolite framework by alkali treatment, the possibility of realumination seems to be low, namely, the increase in the SiO_2/Al_2O_3 ratios of dealuminated HZSM-5 zeolites with alkali treatment is mainly due to the removal of framework silicon atoms owing to the hydrolysis of Si-O-Si bonds.

Mechanism of realumination

It is well known that the nonframework aluminum is not uniform and exists in different states [16]. Taking into account the fact that the reinsertion of framework aluminums hardly takes place on dealuminated zeolites prepared by the long hydrothermal treatment, in which the complete extraction of aluminum atoms from the framework seems to proceed, we have speculated the realumination mechanism by acid treatment as follows (Figure 8). The realumination proceeds

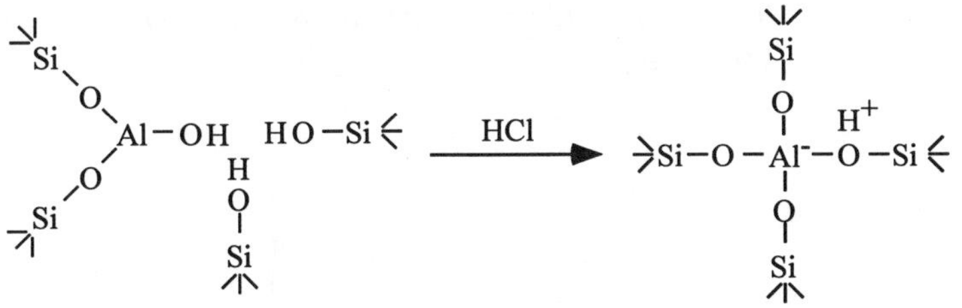

Figure 8. Plausible mechanism of the realumination of HZSM-5 zeolite by acid treatment.

mainly through the reinsertion of nonframework aluminum species where the aluminum atom is connected to the zeolite framework only by one or two remaining chemical bonds.

CONCLUSIONS

It was concluded that a part of nonframework aluminums in dealuminated HZSM-5 zeolite are reinserted into the framework by acid treatment. It seems that the nonframework aluminum species removed from the framework completely, probably Al_2O_3 and AlOOH, can not reinsert into the framework. As the degree of reinsertion of framework aluminums into the zeolite framework is not so high, a further study is needed to achieve the high degree of reinsertion.

REFERENCES

1. D.W. Breck and G.W. Skeels, Proceeding of the 5th International Zeolite Conference, Edited by L.V.C. Rees (Heyden, London, 1980) p. 335.
2. X. Lin, J. Klinowski and J.M. Thomas, J. Chem. Soc., Chem. Commun. 582 (1986).
3. H. Hamdan, B. Sulikowski and J. Klinowski, J. Phys. Chem. **93**, 350 (1989).
4. Z. Zhang, X. Liu, Y. Xu and R. Xu, Zeolites **11**, 232 (1991).
5. J. Datka, B. Sulikowski and B.Gil, J. Phys. Chem. **100**, 11242 (1996).
6. V. Calsavara, E. Falabella Sousa-Aguiar and N.R.C. Fernandes Machado, Zeolites **17**, 340 (1996).
7. D. Lin, S. Bao and Q. Xu, Zeolites **18**, 162 (1997).
8. C. Yang and Q. Xu, J. Chem. Soc., Faraday Trans. **93**, 1675 (1997).
9. G.L. Woolery, G.H. Kuehl, H.C. Timken, A.W. Chester and J.C. Vartuli, Zeolites **19**, 288 (1997).
10. G. Engelhardt and U. Lohse, J. Catal. **88**, 513 (1984).
11. L. Aouali, J. Jeanjean, A. Dereign, P. Tougne and D. Delafosse, Zeolites **8**, 517 (1988).
12. W. Lutz, W. Gessner, R. Bertram, I. Pitsch and R. Fricke, Microporous Mater. **12**, 131 (1997).
13. T. Sano, H. Ikeya, T. Kasuno, Z.B. Wang, Y. Kawakami, K. Soga, Zeolites **19**, 80 (1997).
14. T. Sano, R. Tadenuma, Z.B. Wang and K. Soga, J. Chem. Soc., Chem. Commun. 1945 (1997).
15. P.J. Kooyman, P. van der Waal and H. van Bekkum, Zeolites **18**, 50 (1997).
16. E. Loeffler, U. Lohse, Ch. Peuker, G. Oehlmann, L.M. Kustov, V.L. Zholobenko and V.B. Kazansky, Zeolites **10**, 266 (1990).

CO AND NO ADSORPTION ON THE COPPER EXCHANGED SAPO-34 MOLECULAR SIEVE CATALYST

DEEPAK B. AKOLEKAR AND SURESH K BHARGAVA

Department of Applied Chemistry, RMIT University, City campus, Melbourne, Vic.3001, Australia; D.AKOLEKAR@rmit.edu.au

ABSTRACT

This paper discusses the CO and NO adsorption on the well characterised Cu-exchanged SAPO catalyst studied by IR spectroscopy. The *in-situ* reactions of CO, CO and NO on the Cu-SAPO-34 were studied using an FT-IR quartz cell. Adsorption of CO over the Cu sites of the catalyst leads to formation of mono- and dicarbonyl species. Introduction of NO on the preadsorbed CO Cu-catalyst indicated the presence of Cu(I)CO, Cu(I)(CO)$_2$, Cu(II)-NO and Cu(II)-NO$_2$ species. Effect of the pressure, temperature and evacuation on the CO and NO species have been studied.

INTRODUCTION

Noble metals such as Rh, Pt and Pd are used as catalysts for the important reactions between CO and NO in automobile exhausts. Transition metal oxides - zeolite system have attracted much interest as alternative low cost materials for the direct conversion of NO and CO[1,2]. Among various types of Cu exchanged low and high silica zeolites developed for the NOx decomposition, only the Cu-ZSM-5 exhibits high NO decomposition activity with limitations. Aluminophosphate molecular sieves are the new generation materials with mild to strong acidity, catalytic activity, ion-exchange capacity and framework flexibility[2-7]. Recently, metal exchanged new generation SAPO-n molecular sieves have been employed for the NOx conversion[8]. These materials have shown promising results as compared with the zeolites in nitric oxide conversion reactions[8,9].

We have reported the FTIR investigations of the NO interaction with Cu-SAPO-34[4] and NO adsorption results are compared with those of Cu-ZSM-5 zeolite. So far, IR studies on single (CO) and two (CO, NO) component adsorption on the Cu-SAPO catalyst are not reported. This work attempts to investigate the type of species formed after CO and NO adsorption at different pressures and temperatures on the Cu-SAPO catalysts.

EXPERIMENTAL

Catalyst preparation, characterization and FTIR measurements:

A gel of molar composition 2.5TEAOH 1Al$_2$O$_3$ 0.82 P$_2$O$_5$ 0.42SiO$_2$ 51H$_2$O was hydrothermally crystallised in a 200 cm^3 Teflon autoclave initially at 423K for 48h and at 473K for 58h. The crystallization product was thoroughly processed and calcined at 793K for 16h. The calcined material was ion-exchanged with Cu in a 0.014M copper acetate solution at 348K for 4h, processed, ion-exchanged three times and finally heated in the presence of nitrogen at 793K for 5h. Details of the SAPO preparation, chemicals and characterization already have been reported[4]. *In-situ* FTIR studies were performed on self-supported wafers (~19mg, *thickness* 10.0mg.cm^{-2}) using an Perkin Elmer FT-IR 1725X spectrometer. IR spectra were obtained at 2 cm^{-1} resolution. All the sample wafers were activated in vacuum (< 10^{-4}Torr) at 623K for 16h. The spectrum of dehydrated sample was used as a background from which the adsorbed spectrum was subtracted. High purity CO and mixture of N$_2$ (95%) and NO (5%) gases (supplied by Linde and BOC, Australia) were used.

RESULTS AND DISCUSSION

The characteristics of the copper exchanged silico-aluminophosphate catalyst prepared from the SAPO-34 material are presented in Table 1. The copper exchanged catalyst contained 19 wt.% copper, 30wt.% aluminium; 26wt.% phosphorous and 25 wt% silicon.

Table 1 Properties of the Cu-SAPO-34 catalysts

Cu-CASPO-34	12.0 Cu 44.4 Al 33.0 P 10.6 Si
Crystal shape	Irregular
Crystal size, μm	<3
N$_2$ sorption capacity, Cu-SAPO-34 mmol.g^{-1}	3.40
Micropore volume of Cu-SAPO-34, cm^3.g^{-1}	0.14
Cu-SAPO-34: XPS analysis	Cu$_{2p3/2}$3.2, Si$_{2p}$2.40, Al$_{2p}$23.10, P$_{2p}$ 13.0, O$_{1s}$
Surface composition (atomic %)	58.3.
Binding energy (eV)/FWHM (eV)[a]	Cu$_{2p3/2}$ 935.2(2.6), Cu$_{2p3/2}$933.2, Si$_{2p}$ 103 (2.2), Al$_{2p}$ 75.1(2.6), P$_{2p}$ 134.8 (2.6), O$_{1s}$ 532.5(2.8)
[a]Referenced to C$_{1s}$ = 285eV	

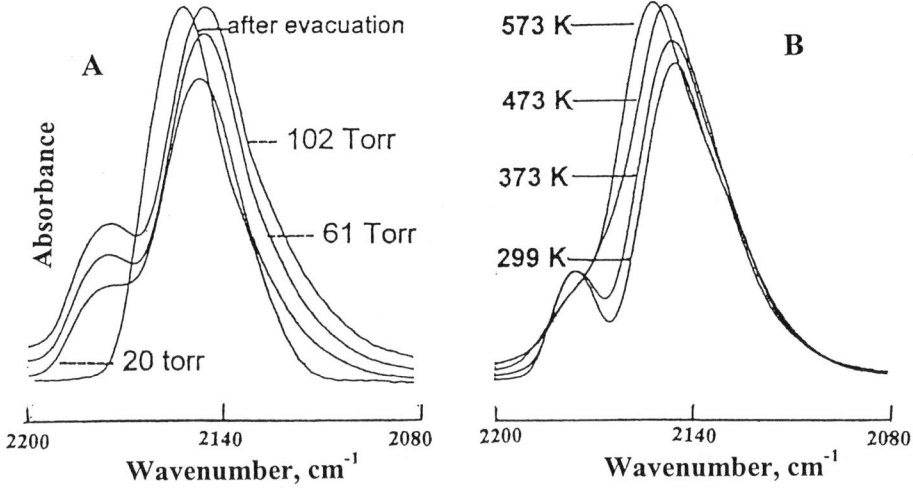

Figure 1: IR spectra of CO adsorbed on the Cu-SAPO-34 at (A) different pressures and (B) at different temperatures.

Adsorption of CO (20 Torr) on the Cu-SAPO-34 leads to appearance of two main bands at 2175, and 2147 cm^{-1} (Figure 1A). The deconvolution of spectrum(CO, 20 Torr) shows the presence of third band at 2134 cm^{-1}. The observed bands at 2175 and 2147 cm^{-1} are assigned to dicarbonyls related with the isolated Cu$^+$ sites while the associated Cu$^+$ cation sites are indicated by a CO band at 2134 cm^{-1} [10]. The band at 2134 cm^{-1} is characteristic of high loaded copper SAPO sample in which CO adsorbs only on associated Cu$^+$. Similar behaviour and the adsorbed CO stretching frequencies for CuO-Silica and high Cu loaded ZSM-5 samples are reported[10]. The increase of CO equilibrium pressure to 106 Torr (at high surface coverage) leads to a shift of the 2147 cm^{-1} band to 2145 cm^{-1} and an increase in the concentration of dicarbonyl species as evidenced by increase in the intensity of the 2175 and 2147 cm^{-1} bands. After CO adsorption at 106 Torr, the evacuation experiment was carried out in order to explore the behaviour of copper and carbon monoxide species on the Cu-SAPO-34 catalyst surface. After evacuation of CO gas at 296 K (vacuum <10^{-4} Torr, 50 minutes), the bands at 2175 and 2145 disappeared, whereas significant increase in the intensity of the new band at 2152 cm^{-1} and the shift of the 2137 cm^{-1} band to 2142 cm^{-1} was observed. The observed band at 2152 cm^{-1} is resistant toward evacuation and indicates that the copper (I) sites strongly hold CO (i.e. CO molecules are chemisorbed on the Cu$^+$ sites). In the previous CO adsorption studies on the Cu-ZSM-5, it was observed that the evacuation decreases the concentration of di-carbonyl species and increases the concentration of mono-carbonyl species.

In case of CO adsorbed on Cu-ZSM-5, Peiplu et al.[11] reported that with increase in the CO equilibrium pressure, dicarbonyl formation increases while monocarbonyl species formation decreases. On this basis, we have assigned the bands at 2175 and 2145 cm^{-1} to the v_s(CO) and v_{as}(CO) of Cu$^+$(CO)$_2$ (dicarbonyls) type species whereas the 2152 cm^{-1} band has been attributed to v(CO) of Cu$^+$ -CO monocarbonyls formed on the same Cu$^+$ sites of the SAPO catalyst. The trend is similar to that observed over the copper exchanged high silica zeolite catalyst[5,6]. The reaction of CO with the Cu sites of the SAPO catalyst can be explained as follows:

$$Cu^+ \xrightarrow{CO} Cu^+(CO) \ [2152 \ cm^{-1}] \underset{-CO}{\overset{+CO}{\leftrightarrow}} Cu^+(CO)_2 \ [2175 \ and \ 2145 \ cm^{-1}]$$

Figure 1B shows the decomposition of adsorbed CO species at different temperatures and constant pressure (102 Torr) over the dehydrated Cu-SAPO-34 catalyst. The adsorption of CO at 299K resulted in the formation mono- and di-carbonyl species. After the heating of CO adsorbed material at 373, 473, 573 K for 50 minutes changes in the band frequencies and the intensities were noticed. It is noticeable that until 373 K, dicarbonyl species are not significantly affected. The band position of monocarbonyl species were effected by increasing the sample temperature. However, above 473 K, the transformation of dicarbonyls to monocarbonyl species occurs which is evident from the increase in the intensity of band at 2152 cm^{-1} (monocarbonyl species) and decrease in the intensity of band at 2175 cm^{-1} (dicarbonyl species). This shows that at higher temperatures (>373 K), the concentration of monocarbonyls increases at the expense of dicarbonyl species over the Cu-SAPO-34.

Figure 2 shows the spectrum of NO adsorption over the Cu-SAPO-34 (temperature 296 K; NO pressure 21 Torr). Assignment of the type of NOx species to the bands[4] are as follows:

Cu(I)-NO, 1806 cm^{-1}; Cu(II)-NO, 1900 cm^{-1}; Cu(I)-(NO)$_2$(asym), 1732 cm^{-1} and Cu(II)-NO$_2$, 1627 cm^{-1}.

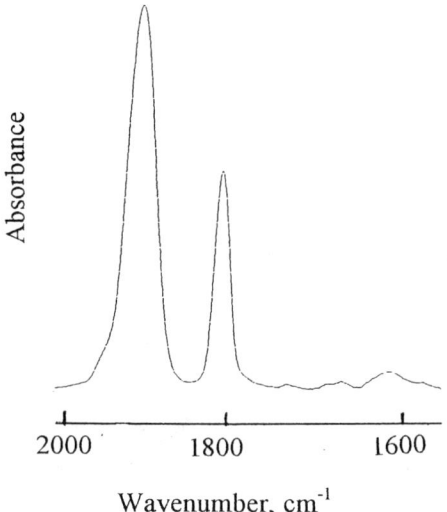

Figure 2 FTIR spectrum of NO adsorbed on the Cu-SAPO-34

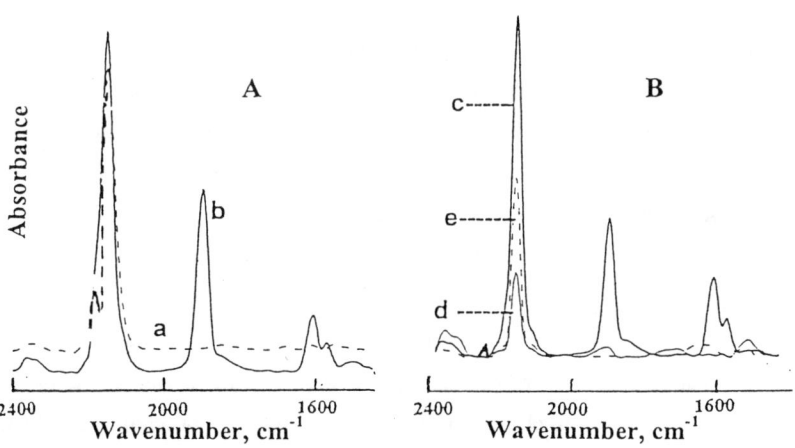

Figure 3 CO and NO adsorption on the Cu-SAPO-34 [A] (a) CO adsorption, (b) introduction of NO after CO adsorption at 299 K; [B] CO and NO adsorption at (c) 373 K, (d)613 K, (e)after evacuation.

In the CO and NO adsorption experiments (Figure 3A), first CO (12 Torr) was adsorbed on the catalyst at 299K giving rise to absorption bands at related with dicarbonyl and monocarbonyl species and then NO gas (60 Torr) was introduced in the chamber. Introduction of the NO gave strong and weak bands at 1898 and 1618 cm^{-1}, respectively. These bands were caused by the Cu(II)-NO and Cu(II)-NO$_2$ species. No Cu(I)-NO and Cu(I)-(NO)$_2$ species were observed because the Cu(I) sites were preoccupied by CO. However, only NO adsorption shows the presence of NO species related to Cu(I) and Cu(II)[5]. NO introduction did not affect the concentration of CO species on the copper catalyst surface.

The decomposition of adsorbed CO and NO at different temperatures over dehydrated Cu-SAPO-34 catalyst is shown in Figure 3B. After heating the CO and NO adsorbed material at 373, 473 and 613K for 40 minutes, the peak intensities of CO and NO species decreases, with significant decrease occurring only > 473K. As compared to the CO species, the concentration of NO species was strongly affected > 473K. The Cu(II)-NO$_2$ species was completely destroyed above 613K. The reaction of CO and NO on the Cu-SAPO-34 above 473K leads to formation of isocyanate (NCO) species (2240 cm^{-1}). The concentration of NCO species increases with the reaction temperature.

CONCLUSIONS

Adsorption of CO on the Cu-SAPO-34 shows the presence of associated and isolated Cu$^+$ sites. Increase in the CO pressure leads to the higher formation of dicarbonyl species and vice versa. After evacuation only monocarbonyls are detected which are stable and resistant toward evacuation. The monocarbonyls species are chemisorbed. The isolated Cu$^+$ ions possess two vacancies which determine their ability to form dicarbonyls. Increase in the sample temperature shows decrease in the concentration of dicarbonyl species.

Adsorption of NO on the CO preadsorbed Cu-catalyst exhibits the presence of mostly Cu(II)NO species and did not displace the CO species. The sample temperature shows strong effect on the CO and NO species and on the formation of isocyanate species. The ratio of CO and NO species indicated the strong changes of the surface complexes concentration. Evacuation of the NO and CO adsorbed sample leads to removal of NO species and suggest that the CO species are more strongly adsorbed than the NO species.

REFERENCES

1. Iwamoto, M; "Future Opportunities in Catalytic and Separation Techn.", Ed.M.Misono et al." Elsevier, Amsterdam ,1990, p.121.
2. Iwamoto, M; Stud.Surf.Sci.Catal., 1994, 84,1395.
3. Wilson, S.T. and Flanigen, E.M, ACS Symp.Ser., 1988, 398, 329.
4. Akolekar, D.B., J.Catal., 1993, 144, 148.
5. Akolekar, D.B., J.Chem.Soc., Faraday Trans., 1994, 90(7), 1041.
6. Akolekar, D.B., Huang, M., and Kalaiaguine, S., Zeolites, 1994, 14, 519.
7. Akolekar, D.B., and Howe, R.F., J.Chem.Soc., Faraday Trans., 1997, 93(17), 3263.
8. Ishihara, T; Kagawa, M; Hadama, F; Takita, Y; Stud.Surf.Sci.Catal., 199 84,1493.
9. Akolekar, D.B; Bhargava, S.K; Foger, K; JCS, Faraday Trans., 1998, 94(1), 155.
10. Hadjiivanov,K.I; Kantcheva, M.M; Klissurski,D.G; JCS, Faraday Trans., 1996, 92(22), 4595.
11. Pieplu,T; Poignant, F; Vallet,A.; Saussey, J; Lalalley, J.C; Stud.Surf.Sci.Catal. 1995, 96, 619.

AROMATIZATION OF n-HEPTANE ON THE MODIFIED MFI ZEOLITES

N. BILBA[a], I. ASAFTEI[a], GH. IOFCEA[b] and N. NAUM[a]

[a] Faculty of Chemistry, "Al.I.Cuza" University of Iasi, 6600 Iasi, ROMANIA; nbilba@ch.tuiasi.ro
[b] S.C. CAROM S.A. Onesti, 5450 Onesti, ROMANIA

ABSTRACT

The aromatization of n-heptane on the MFI zeolites modified by both isomorphous substitution with Ga and ion exchange with Zn (II), Ni (II) and Ag (I) was studied by the chromatographic pulse method in the temperature range of 673 - 823 K at atmospheric pressure. The catalysts studied, i.e. HGaSil-1 (R = 216, 100, 76), HGaAlZSM5 (wt. % Ga_2O_3 + Al_2O_3 1.52 +3.52, 1.65 + 3.31, 1.74 + 3.24), HZSM5 (R = 33, 9), Zn-HZSM5 (wt. % Zn, 0.69, 1.32, 1.51), Ni-HZSM5 (wt. % Ni, 0.57, 1,09, 1.34) and Ag-HZSM5 (wt. % Ag, 0.95, 1.51, 1,90) having different acid strength distribution exhibit a conversion depending on the temperature (except Zn-HZSM5) and a yield of aromatics between 21.1 - 43.3, wt.%, at 823 K.

The metal actions and the acidic properties of zeolites have an important effect on the aromatization of n-heptane.

INTRODUCTION

Among the known zeolite networks, that of the MFI (ZSM5) type in H-form has the greatest interest due to its unique systems of channels and acidity as well as its appreciable thermal and hydrothermal stability [1]. One of the catalytic processes in which the modified HZSM5 catalysts have proved their superiority is that of conversion of light paraffins (C_2-C_7) and olefins ($C_3^=$-$C_4^=$) into C_6-C_9 aromatic compounds (Cyclar, M-2 Forming, Aroforming) [2]. Due to the shape-selective properties of the HZSM5 framework, mainly mononuclear aromatic hydrocarbons are formed (BTX). The best results were obtained on the HZSM5 catalysts modified with gallium or zinc [3-7].

The present work deals with the results obtained by investigation of aromatization of n-heptane on HZSM5 zeolite catalysts isomorphously substituted with gallium, in the forms HGaSil-1 and HGaAlZSM5, in comparison with those of the HZSM5 type ion-exchanged with Zn (II), Ni (II) and Ag (I).

EXPERIMENTAL

Synthesis

MFI metal silicates with Ga and with Ga and Al were synthesized under hydrothermal conditions with tetrapropylammonium bromide (TPABr) as the organic structure-directing agent in the zeolite synthesis mixture following well known literature procedures. The starting materials were: sodium silicate solution (29.63 % SiO_2, 9.55 % Na_2O and 60.8 % H_2O), gallium nitrate $Ga(NO_3)_3 \cdot 9\ H_2O$ (Aldrich), aluminum sulphate $Al_2(SO_4)_3 \cdot 18\ H_2O$ (Aldrich), concentrated sulphuric acid (Riedel - de Haën), TPABr (Aldrich) and distilled water.

The parent Na-ZSM5 was synthesized with ethylene glycol as the organic molecule.

Crystallization of the homogeneous gels took place over 24 h at autogenous pressure and 453 K in 0.150 L teflon-lined autoclaves with intermittent stirring. The synthesis products were filtered, washed repeatedly with distilled water, dried at 383 K in air for 6 h and calcined at 823 K in air for 6 h in order to remove the organic agent.

Chemical composition of the gels and of the calcined crystalline zeolites are given in Table 1.

Table 1: Chemical composition of the gels and of the corresponding calcined zeolites.

Sample	GaSil-1			GaAlZSM5			ZSM5
a. Chemical composition of gels, molar ratio							
SiO_2/Ga_2O_3	276.920	125.670	83.760	-	-	-	-
$SiO_2/Ga_2O_3+Al_2O_3$	-	-	-	48.23	51.06	54.60	-
SiO_2/Al_2O_3	-	-	-	-	-	-	60.68
$(TPA)_2O/SiO_2$	0.216	0.216	0.216	0.216	0.216	0.216	-
EG/SiO_2	-	-	-	-	-	-	0.76
HO^-/SiO_2	0.166	0.166	0.166	0.166	0.166	0.166	0.164
H_2O/SiO_2	31.14	31.14	31.14	28.04	28.04	28.04	38.14
b. Chemical composition of calcined zeolites, wt. %							
SiO_2	97.34	96.03	94.86	92.32	92.48	92.47	92.70
Ga_2O_3	1.39	2.98	3.86	1.52	1.65	1.74	-
Al_2O_3	-	-	-	3.52	3.31	3.24	4.54
Na_2O	0.46	0.99	1.28	2.64	2.56	2.55	2.76
R	216	100	76	36.0	37.3	37.5	33.9

The calcined zeolites were converted into the H-forms by three successive ion exchanges with 1 M solution of NH_4NO_3 at 353 K for 6 h (5 ml 1 M solution per gram zeolite), followed by drying and air calcining at 823 K for 6 h.

From the HZSM5 sample, by ion exchange with 0.1 M solutions of $Zn(NO_3)_2$, $Ni(NO_3)_2$ and $AgNO_3$, the MeH-forms with different contents of metal were prepared.

Characterization

Powder XRD patterns of the samples were recorded on a TUR-M 62 (HZG 3) diffractometer by using Ni - filtered Cu K_α radiation with a scanning speed of 0.5 ° θ per minute. Infrared spectra of the framework vibrations were recorded on a Perkin-Elmer 1700 spectrometer. The SEM photographs were obtained on a TESLA B300 electron microscope. Acidic properties of the catalysts were measured by TPD method of pre-adsorbed NH_3. BET surface area was obtained through the N_2 - adsorption on SORPTOMATIC - CARLO ERBA. Adsorption measurements of water and n-hexane were carried out on a self - constructed McBain type gravimetric unit. The chemical analyses were performed by a combination of wet chemical, spectrophotometric (SPEKOL) and atomic absorption methods. Catalytic tests on aromatization of n-heptane were carried out in a pulse microreactor coupled with a GCH gas chromatograph. Aromatics were analyzed on a SE30 / Chromosorb P (60-80 mesh) silanized column.

RESULTS AND DISCUSSION

The diffractograms proper to synthesized zeolites, in calcined state, are typical of pure ZSM5 (MFI) [8,9] with respect to the positions and relative intensities of the observed peaks. Figure 1 shows the XRD powder patterns of the H-forms of ZSM5 (33.9), GaAlZSM5 (37.3) and GaSil-1 (216). The isomorphous substitution of Ga for Al results in the expansion of the unit cell due to larger bond length Ga-O in comparison with Al-O, the whole XRD pattern moving to increased "d" values. Another observation is the decreasing of the relative intensity of the (101) and (200) reflections in the pattern of GaSil-1 (216).

Tetrahedrical inserting of Ga in the zeolite network was proved on the basis of the IR spectra [10] presented in Figure 2. The absorption band of asymmetric vibrations at 1100 cm^{-1} [10], specific to ZSM5 (Si - O - Si) is shifted to lower wavenumbers, 1098.1 cm^{-1} for GaAlZSM5 (1.74% Ga_2O_3), 1092.3 cm^{-1} for GaAlZSM5 (1.52% Ga_2O_3) and 1090.1 cm^{-1} for GaSil-1 (216)(Si-O-Ga).

The SEM images of GaSil-1 (216), and ZSM5 (33.9) shown in Figure 3 reveal that the crystals have a different morphology, from spherical, composed of much smaller microcrystallites to twinned.

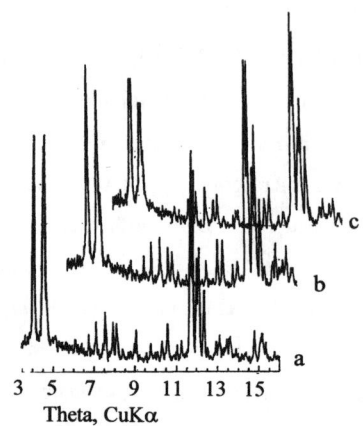
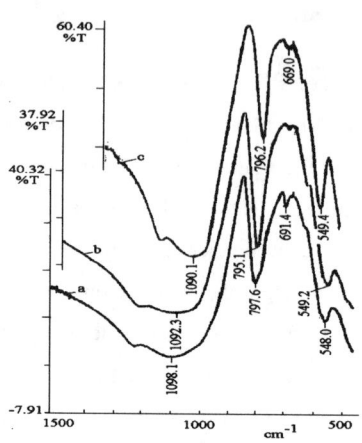

Figure 1. XRD powder patterns of the (a) ZSM5 (33.9), (b) GaAlZSM5 (37.3) and (c) GaSil-1 (216).

Figure 2. I.r. framework vibration spectra of (a) GaAlZSM5 (1.74% Ga_2O_3), (b) GaAlZSM5 (1.52% Ga_2O_3) and (c) GaSil-1 (216).

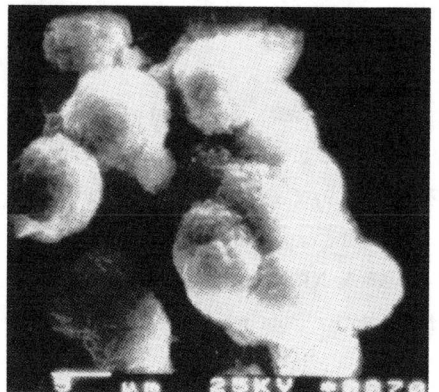

Figure 3. Scanning electron micrographs of (a) GaSil-1 (216) and (b) ZSM5 (33.9).

The acidity of the tested MFI zeolites (TPD - ammonia), due to the bridging Si-OH-Al, Si-OH-Ga and Si-OMe-Al groups changes in the acid site density and strength (Table 2).

The values of the BET specific surface area are between 290 and 310 m^2/g and the equilibrium sorption capacities for water (polar) and n-hexane (nonpolar) are 11.3 wt.% and 11.0 wt.% for HZSM5(33.9), 5.2 % and 9.9 % for HGaSil-1 (216), 10.4 % and 9.3 % for HGaSil-1 (76), 10.2% and 8.8% for HGaAlZSM5 (1.74 % Ga_2O_3 + 3.24 % Al_2O_3), 10.96 % and 9.2 % for HGaAlZSM5 (1.52 % Ga_2O_3 + 3.52 % Al_2O_3).

Table 2. The acid strength distribution of modified MFI zeolites.

Catalyst	Acidity, mmol NH_3/g weak and medium sites 353-673 K	strong sites 673-973 K
HZSM5 (33.9)	0.676	0.298
HGaSil-1 (216)	0.490	0.137
HGaAlZSM5 (1.74% Ga_2O_3 + 3.24% Al_2O_3)	0.555	0.229
HGaAlZSM5 (1.52% Ga_2O_3 + 3.52% Al_2O_3)	0.604	0.196
Zn-HZSM5 (1.32 % Zn)	0.620	0.135
Ni-HZSM5 (1.34 % Ni)	0.645	0.212
Ag-HZSM5 (1.90 % Ag)	0.661	0.115

Catalytic studies

The catalytic properties of the modified MFI zeolites were tested in acid-catalyzed n-heptane aromatization. The catalysts were pressed, crushed and sorted into grains smaller than 0.147 mm and than 0.1 g, were packed into a microreactor and heated under a stream of N_2 at 823 K for 4 h (with exception of Zn-HZSM5 sample which was activated at 723 K for 4 h). The catalytic activity was investigated in the temperature range of 673-823 K with 0.2 ml n-heptane pulse in N_2 flow as carrier and the variations of conversion vs. temperature on HGaSil-1, HGaAlZSM5, HZSM5 and Me-HZSM5 catalysts are presented in Figure 4.

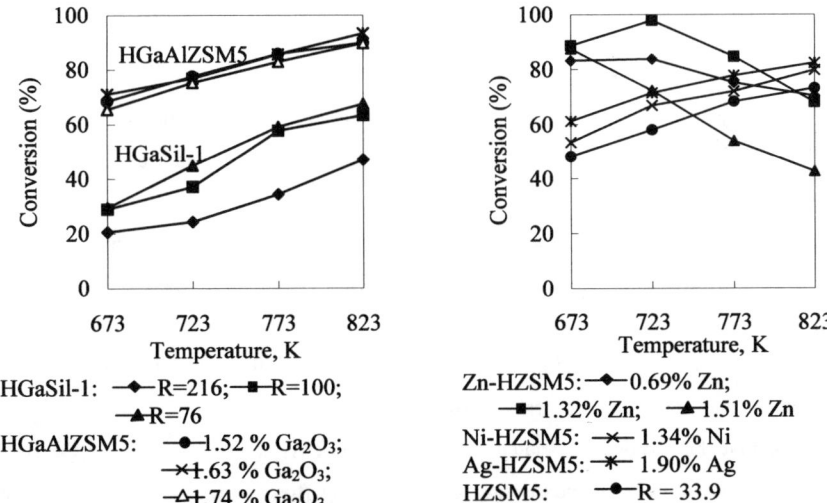

HGaSil-1: ◆ R=216; ■ R=100; ▲ R=76
HGaAlZSM5: ● 1.52 % Ga_2O_3; × 1.63 % Ga_2O_3; △ 1.74 % Ga_2O_3
Zn-HZSM5: ◆ 0.69% Zn; ■ 1.32% Zn; ▲ 1.51% Zn
Ni-HZSM5: × 1.34% Ni
Ag-HZSM5: * 1.90% Ag
HZSM5: ● R = 33.9

Figure 4. Conversion of n-heptane vs. temperature on a series of the MFI catalysts.

For each temperature, four successive pulses were injected and the average values of conversion and aromatic yield were calculated.

Catalytic conversion of n-heptane increases as temperature increases except the Zn-HZSM5 samples which decrease, probably due to displacement of Zn from active sites. The highest conversion of n-heptane (~90 %) at 823 K was recorded on HGaAlZSM5 and the lowest (~40 %) on HGaSil-1 (216), following the series:

HGaAlZSM5 > Ag-HZSM5 > Ni-HZSM5 > HZSM5 ⩾ Zn-HZSM5 > HGaSil-1 (216)

The highest conversion of n-heptane at 673 K was recorded on Zn-HZSM5 (~ 85 %) catalysts and it follows the series:

Zn-HZSM5 > HGaAlZSM5 > Ag-HZSM5 > Ni-HZSM5 > HZSM5 > HGaSil-1 (216)

The yield in aromatics vs. temperature is given in Figure 5.

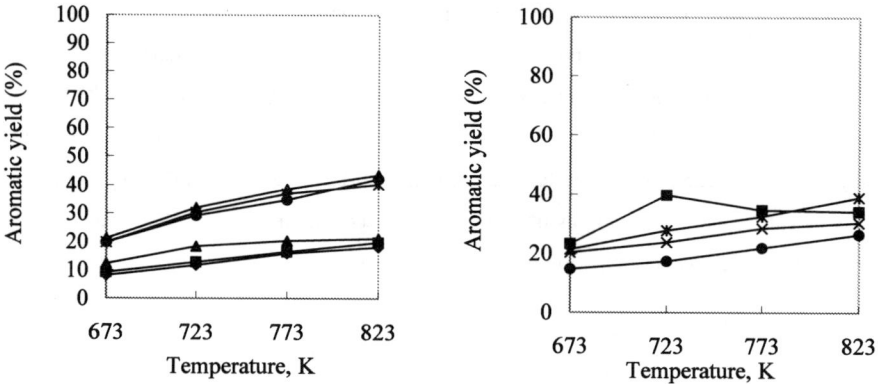

Figure 5. Aromatic yields vs. working temperature, key as Figure 4.

The distribution of aromatic hydrocarbons in liquid fraction depends on the catalyst and the working temperature as shown in Figure 6.

The main component of liquid fraction is toluene, its concentration being controlled by the catalyst and the working temperature. As long as the working temperature increases, an increase of benzene and C_8-aromatics concentrations takes place, probably due to the disproportionation reaction as well as toluene dealkylation. The formation of toluene illustrates that cyclization may occur by a mechanism involving 1,6 - ring closure.

It is stated that the protons associated to the framework Al are responsible for the strong Brönsted acidity of zeolites. Ga possessing chemical properties similar to those of Al, but different atomic size and electronic properties can be incorporated into zeolitic framework. The Ga framework generates Brönsted acidity too, but the strength of acid sites is less than of the Al framework (Table 2). With the increase of Ga content in the HGaSil-1 catalysts an increase of

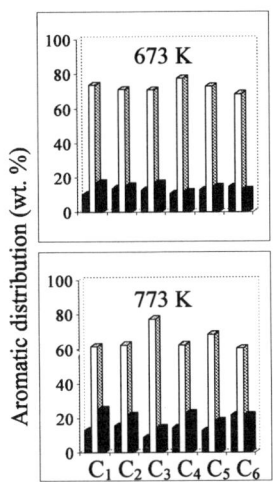
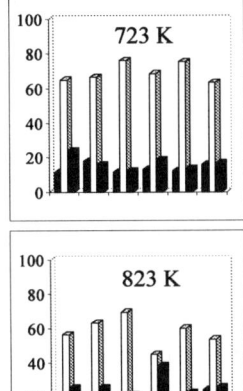

Figure 6. Aromatics distribution (wt %) on conversion of n-heptane vs. temperature over MFI catalysts : HGaSil-1 (216) (C_1); HGaAlZSM5(1.74% Ga_2O_3) (C_2); Zn-HZSM5(C_3); Ag-HZSM5(C_4); Ni-HZSM5 (C_5); H-ZSM5 (C_6) (⊟ benzene,☐ toluene,■ C_8-arom)

conversion of n-heptane takes place, but the aromatic yield is shortly changed (Figure 4, 5). An increased activity and selectivity for aromatics is obtained when in the framework silicate are present both Al and Ga, and when Ga may be in nonframework either. The results are comparable with those reported on HGa[AlZSM5] obtained by ion exchange [11]. The conversion of n-heptane and the aromatic yield are higher on acidic HZSM5 than on HGaSil-1.

It is accepted that the aromatization of n-heptane over monofunctional (H-form) and bifunctional catalysts take place by direct dehydrogenation and cyclization and by hydride transfer which lead to the cracking of n-heptane followed by oligomerization, cyclization and dehydrogenation. The gallium catalyses, the dehydrogenation steps and H^+- acidic sites of zeolite are responsible for the other aromatization reaction sequences which involve hydride transfer and deprotonation of carbenium ions. The activity and selectivity for n-heptane aromatization on HGaAlZSM5 at 823 K is higher than that on the other studied catalysts.

The incorporation of Zn (II) into cationic positions of HZSM5 zeolite affects the acidity and the large attraction of Zn for hydrogen is responsible for higher conversion (~ 90 %) and aromatic yield (~ 40 %) for n-heptane aromatization. Because the conversion values increase during the four pulses, the reduction of Zn(II) to Zn^o is possible, the reduced form having a greater dehydrogenating activity. Catalytical conversion of n-heptane on Zn-HZSM5 decreases at temperature higher than 723 K because of sublimation or migration from the active sites. The conversion of n-heptane on Ag-HZSM5 at 823 K is similar to that on HGaAlZSM5 but the aromatic distribution is different. The increase of temperature determines an increase of xylenes concentration. Ni - ion exchange ZSM5 was less effective in the aromatization of n-heptane than Ag-, Zn- and HGaAlZSM5 but much more effective than HGaSil-1 and HZSM5.

CONCLUSIONS

In the n-heptane aromatization on modified MFI zeolites, the conversion and the aromatic yield depend on the metal cation incorporated by ionic exchange and the T atoms from framework incorporated by isomorphic substitution where they serve as active sites for dehydrogenation and for olefin interconversion and less for cracking.

The conversion of n-heptane within temperature range 673-773 K is in the following series:

Zn-HZSM5 > HGaAlZSM5 > AgHZSM5 > NiHZSM5 > HZSM5 > HGaSil-1

After 773 K, the activity of ZnHZSM5 decreases, and at 873 K becomes similar to HZSM5.

The yield in aromatics within temperature range 673-773 K is in the following series:

HGaAlZSM5 ≥ Zn-HZSM5 > Ag-HZSM5 > Ni-HZSM5 > HZSM5 > HGaSil-1.

REFERENCES

1. N.Y. Chen, W.E. Garwood and F.G. Dwyer, Shape Selective Catalysis in Industrial Applications, Marcel Dekker, Inc., Eds,. New York, 1989
2. M. Guisnet, N.S. Gnep and F. Alario, Appl. Catal. A, **89**, 1 (1992)
3. H. Kitagawa, Y. Sendoda and Y. Ono, J. Catal., **101**, 12 (1986)
4. N. S. Gnep, J.J. Doyemet and M. Guisnet, J. Mol. Catal., **45**, 281 (1988)
5. Liu Xinsheng and J. Klinowski, J. Phys. Chem., **96**, 3403 (1992)
6. Y. Ono, H. Kitagawa and Y. Sendoda, JCS Far. I, **83**, 2913 (1987)
7. Y. Ono, Catal. Rev. Sci. Eng., **34**, 179 (1992)
8. E.L. Wu, S.L. Lawton, D.H. Olson, A.C. Rohrman, Jr. and G.T. Kokotailo, J. Phys. Chem., **83**, 2777 (1979)
9. R. von Ballmoos and J.B. Higgins, Zeolites, **10**, 444S (1990)
10. G. Giannetto, R. Monque, J.A. Pérez, J. Papa and L. Garcia, Zeolites, **13**, 557 (1993)
11. G. Giannetto, J.A. Pérez, R. Sciamanna, L. Garcia, R. Galiasso and R. Monque, in E.G. Derouane, F. Lemos, A. Naccache and F. Ramoa Ribeiro, eds., Zeolite Microporous Solids: Synthesis, Structure and Reactivity, NATO ASI C 352, Kluwer, Dordrecht, 1992.

PREDICTING EXTRAFRAMEWORK CATION POSITIONS IN ZEOLITES: ENERGY MINIMIZATION AND (N,V,T) MONTE CARLO SIMULATIONS IN LILSX(X)

C. F. MELLOT†,* and A. K. CHEETHAM‡

†Institut Lavoisier, UMR CNRS 173, Université de Versailles St-Quentin, 45, avenue des Etats-Unis, Versailles Cédex 78035, FRANCE; mellot@chimie.uvsq.fr
‡Materials Research Laboratory, University of California, Santa Barbara, CA 93106, USA.

Abstract
A knowledge of extraframework cation positions can yield crucial insight into the adsorption and catalytic properties of zeolites. However, diffraction methods are not always able to provide this information with sufficient reliability. Li-containing zeolites, for example, which are very important in air separation, are particularly difficult to characterize experimentally because of the small X-Ray and neutron scattering powers of lithium. We describe here how simulation tools can be used to predict the supercage cation positions in zeolite lithium X and LSX. In a first step, energy minimizations (zero K) have been performed in LiLSX and the impact of the framework relaxation on the placement of supercage Li ions investigated. Significant rearrangement of cation positions is observed when framework relaxation is taken into account, leading to a better agreement with a recent powder neutron diffraction study. In a second step, we have investigated a new methodology using canonical Monte Carlo simulations for exploring the location of extraframework cations in zeolites. In the case of LiX, (N,V,T) Monte Carlo simulations are shown to be especially appropriate for exploring the disordered placement of supercage Li ions.

Introduction

A substantial number of diffraction studies have examined the location of extraframework cations in faujasite type zeolites, including hydrated and dehydrated forms of zeolites X and Y [1]. However, in the case of monovalent cations forms of zeolite X, a significant proportion of the cations are often undetected in X-Ray studies. In hydrated NaX [2], KX [3], TlX[4], for example, the proportion of undetected cations is almost 50%, leading Baur [5] to conclude that many of the water molecules and exchangeable cations in faujasite zeolites move freely through the aluminosilicate framework. The dehydrated forms of zeolites find utility in number of commercial separation processes, including air separation [6-8] and hydrocarbon separations [9], and in comprehensive determination of the cation positions in such materials is an important prerequisite to understanding their properties [10,11].

Even in the prototypic system, NaX, where the numbers of cations in the SI, SI' and SII sites appears to be well-established, the precise location of the remaining cations has been the subject of much controversy. For example, a recent single crystal X-ray study of NaX [12] describes three additional positions for these Na ions, a result that contrasts with several other studies, including an earlier single crystal structure [13] in which only one extra site was detected. In a recent paper [14], we have demonstrated that computer modeling techniques could help to resolve uncertainties in the location of extraframework cations in NaX. In the present work, we extend our modeling work to the important case of the zeolite LiLSX, which is an excellent sorbent for non-cryogenic air separation [15]. First, we use packing and energy minimization (zero K) procedures to explore the most favorable cation positions. We also compare our findings with a recent powder neutron diffraction study of LiLSX [16]. Second, we investigate how canonical (N,V,T) Monte Carlo simulations can be used for exploring the placement of extraframework cations in zeolites. We present a case history of the LiX structure related the placement of supercage Li ions at room temperature. The interest of using the (N,V,T) Monte Carlo methodology for studying partially disordered structures is outlined.

COMPUTATIONAL METHODOLOGY

Packing and energy minimization

Our purpose in this computational study was to probe the most favorable positions of the supercage cations in the dehydrated LiLSX structure. We utilized a combination of packing and energy minimization procedures, which has recently been used to study the Li-ABW [17], 4-A [18], ETS-10 [18] and NaX [14] structures. The use of random generation of non-framework positions in the zeolite is of special interest since it circumvents the pre-assignement of SIII and SIII' sites occupancies. Our simulations were carried on a LiLSX structure, $Li_{96}Si_{96}Al_{96}O_{384}$, with the strict alternation of SiO_4 and $[AlO_4]^-$ tetrahedra. The starting host structure consisted of a unit-cell made of the framework, 32 Li ions in sites I' and 32 Li ions in sites II, using the coordinates from recent neutron diffraction data on LiLSX, obtained at room temperature and refined in the Fd3 space group [16]. In the first step of the simulation, the 32 missing Li cations were inserted randomly into the host structure (using the packing procedure of the Catalysis package, MSI). Dummy atoms assigned with high repulsive Lennard-Jones parameters were placed at the centers of the sodalite cages so that insertions of cations happened exclusively in the supercages. This whole procedure was reiterated to generate a set of 15 initial structures. These were then subjected to structure optimization with the Discover_3 program and the cvff_aug forcefield [19]. The zeolite was presumed to be semi ionic with atoms carrying the following partial charges Si(+2.4), Al(+1.4), O(-1.2), Li(+1). The total non-bonded energy included a coulombic term calculated using an Ewald summation, and repulsive-dispersive term described with a Lennard-Jones potential and calculated with a short-range cut-off of 12.5 Å. All simulations were performed in the triclinic P1 space group. Two different sets of energy minimizations were performed: (i) the first set of initial structures were subjected to partial energy optimization, keeping the framework and the Li ions in sites SI' and SII fixed and allowing the 32 inserted Li ions to relax. (ii) a second set of minimized structures was obtained by fully-minimizing the structures of the first set, i.e. allowing both the framework and the cations to relax.

Canonical (N,V,T) Monte Carlo simulations

We have investigated a new methodology using (N,V,T) Monte Carlo simulations for probing the location of extraframework cations in zeolites [20]. Typically, energy minimizations are equivalent to zero K calculations and yield minimized structures that can only be compared with low temperature diffraction studies. More over, the combination of packing and energy minimizations give access to a limited number of trials and optimized structures, failing at giving the statistical and averaged information required in such systems.

In this section, our purpose was to use (N,V,T) Monte Carlo simulations to capture the equilibrium distribution of supercage Li cations in the LiX (Si:Al=1.2) zeolite, considering extraframework cations as guest particles and the zeolite structure as the host. In this work, the host structure consisted of an incomplete and non-neutral unit-cell, $[Li_{64}Si_{106}Al_{86}O_{384}]^{-22}$, made of the framework, 32 Li ions in sites SI' and 32 Li ions in sites SII. The host structure was built in a highly symmetric (Fd3), non-relaxed geometry and kept rigid during the course of the Monte Carlo run. The (N,V,T) Monte Carlo simulation itself was then performed in the triclinic P1 space group. The number of guest Li ions, N, was fixed to 22 so as to match the required charge balance and zeolite composition. Dummy atoms assigned with high repulsive Lennard-Jones parameters were placed at the centers of the sodalite cages so that insertions of cations happened exclusively in the supercages. The (N,V,T) simulations were performed at 300 K. Interactions between the inserted Li ions and the host were described similarly than in the previous section, using the Sorption software (MSI) and the same cvff_aug forcefield parameters for describing Li/host interactions. The average energies were obtained over 10^6 iterations, after an equilibrium period of 10^5 steps, with a short range summation taken up to cutoff radius of 12.5 Å and an Ewald summation regarding the electrostatic term. Further details are given in ref. [20].

RESULTS AND DISCUSSION

Packing and Energy minimizations in LiLSX

According to [16], the most probable sites for the location of the supercage Li ions in LiLSX are those in the vicinity of the four membered rings of oxygen atoms in the supercage. The specific sites of interest are the SIII site, which is above the 4-ring window, and the SIII' site, which is on the edge of the 4-ring window, i.e. in the plane of the 12-ring window (Figure 1). Unlike the SI, SI' and SII sites, which are on special positions, these low symmetry sites with their low occupancy factors are difficult to pinpoint by diffraction method.

The partially minimized structures of LiLSX were analyzed in terms of the local framework environment of each of the 32 packed Li ions. The most stable structure is shown in Figure 2a and has nearly all supercage Li ions located at the edges of the 4-rings in SIII' sites, as we found previously in our simulation of the NaLSX [14]. The cations are coordinated by two oxygens O(1) and O(4), at average distances of ~2 Å. Further inspection reveals that the Li ions are exclusively located in SIII' sites facing AlO$_4$ tetrahedra, as expected on electrostatic grounds. These simulations are in partial agreement with the recent neutron diffraction study [16], which finds approximatly half of the cations in this SIII' position at 300 K, and somewhat more in the low temperature orthorombic structure. The remainder are observed in the SIII sites [16].

In view of the residual discrepancies between the simulations and the experimental results, we have explored the way in which our predictions vary if we allow the host framework to relax completely. In that purpose, the above partially minimized structures were subjected to full optimization, computing the total non-bonded energy in a similar fashion than in the partial optimization runs and using the cvff_aug forcefield. Under these conditions, we find that the balance between the SIII and the SIII' sites is inverted, with approximatly 75 % of the supercage ions now occupying SIII sites (Figure 2b). Such a redistribution of Li ions from SIII' sites into SIII sites is correlated with distorsions of the framework, especially of the 4-ring windows. Such a distorsion of the 4-ring window permits a close coordination between two O(4) oxygens and the Li ion at a distance of ~2 Å, which could not occur in a non-relaxed structure. Besides a better agreement with both the low temperature and room temperature diffraction results of LiLSX [16], our results show that framework relaxation has an important impact on the location of the supercage Li ions and especially on the balance between sites SIII and SIII' occupancies.

(N,V,T) Monte Carlo simulations in LiX

The calculated value for the internal energy for the insertion of 22 Li ions in $[Li_{64}Si_{106}Al_{86}O_{384}]^{-22}$ is of -312.4 kJ mol^{-1}, with a contribution of +41 kJ mol^{-1} and –350 kJ mol^{-1} for the short-range and electrostatic terms, respectively. As shown by the predominance of the electrostatic term to the total energy, the distribution of the 22 Li ions over the supercages is obviously driven by a minimization of long range electrostatic interactions. With the aim of unambigously characterize the placement of the supercage Li ions, their room temperature equilibrated distribution was analyzed in terms of their framework and intercationic environments. In that respect, the radial distribution functions (RDFs) involving the inserted Li ions are of special interest (Figure 3). Figure 3-a shows the RDFs related to each oxygen of the framework, i.e. Li⋯O(1), Li⋯O(2), Li⋯O(3), Li⋯O(4). The striking feature in the RDFs is that all

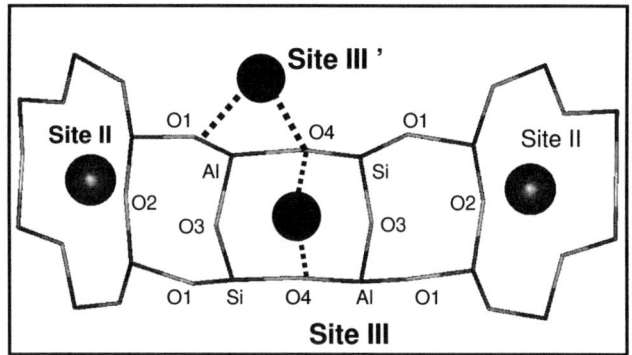

Figure 1. Location of SIII and SIII' cation sites in LiLSX.

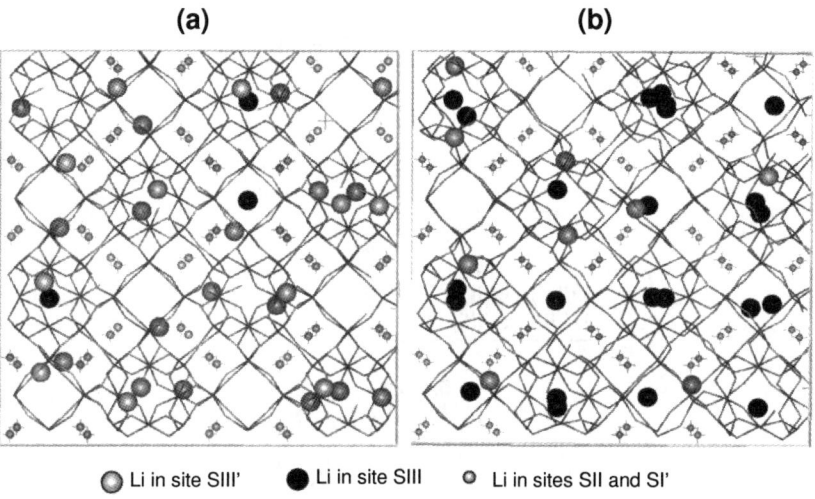

Figure 2. Distribution of the supercage Li ions over sites SIII'(white) and sites SIII (black) in the LiLSX unit-cell. *(a)* after partial minimization of the structure (only supercage Li ions were allowed to relax, while the framework, SI' and SII sites are fixed. *(b)* after full optimization of the structure. (Li ions in sites SI' and SII are represented with small white spheres for clarity).

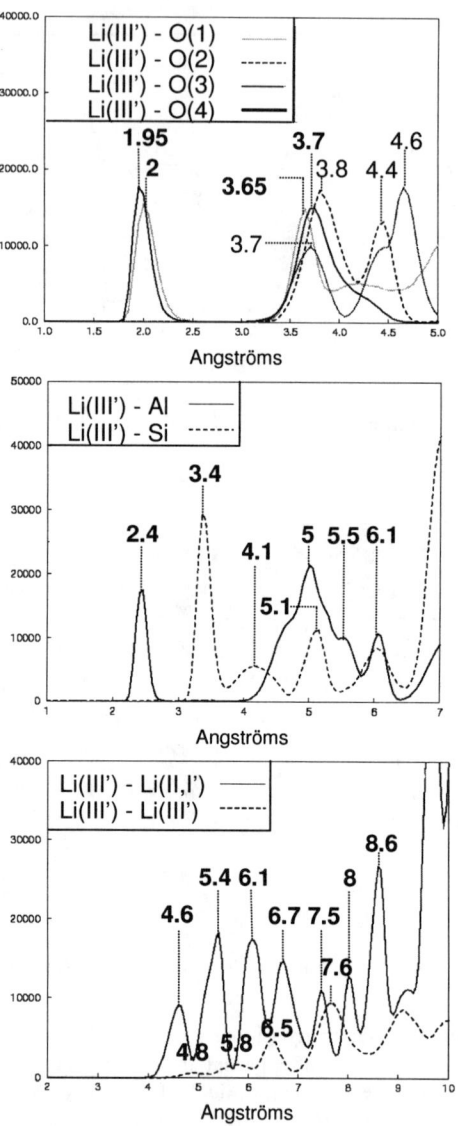

Figure 3. Radial Distribution Functions related to the supercage Li ions in the LiX structure from the (N,V,T) Monte Carlo simulations.

inserted Li ions are coordinated to O(1) and O(4) exclusively, at 2 Å and 1.95 Å respectively, while no peak in the Li⋯O(3) or Li⋯O(2) RDFs is observed within van der Waals distances. This is however in contrast with the diffraction findings in [20]. The exclusive coordination of Li ions to O(1) and O(4) is the signature of the location of Li ions in SIII' sites (see Figure 1). In Figure 3-b, the peak at 2.4 Å in the Li⋯Al's RDF shows that Li ions are preferentially located in SIII' sites facing AlO_4 tetraedra, leaving the nearest Si atoms at distances of 3.4 Å. The respective features of Li⋯Al and Li⋯Si RDFs corroborate the location of Li ions in sites SIII', since the location of any Li ions in site SIII would give a single Li⋯Si RDF peak at ~3 Å. The occupancy of sites SIII' facing AlO_4 tetraedra is similar to our previous study of the NaX structure [14], and is understandable on the basis of the local compensation of the anionic charge of the framework. Regarding the relative placement of Li cations towards one another, the analysis of the Li⋯Li RDFs is of special interest. Figure 3-c shows the RDFs related to Li(III')⋯Li(II,I') and Li(III')⋯Li(III') distances, leading to specific and discrete intercationic distances. These Li⋯Li RDFs give the clear evidence that the placement of the supercage cations is the result of a cooperative and disordered placement of ions with respect to one another.

In view of the significant impact of framework relaxation shown by our energy minimizations on the LiLSX structure, (N,V,T) Monte Carlo simulations with a relaxed host geometry for the zeolite host have been investigated and modifications in the supercage Li ions distribution observed [20].

CONCLUSION

Energy minimizations on the LiLSX have shown that the distribution of Li ions over sites SIII' and SIII is significantly modified when framework relaxation is taken into account. More importantly, we have shown that the use of canonical (N,V,T) Monte Carlo simulations can give valuable insights into the placement of ions in zeolites. We show that (N,V,T) simulations are especially adapted for exploring partially disordered structures such as the LiX and LiLSX zeolites, through the analysis of the radial distribution functions related to Li ions. The experimental and simulation work on LiLSX(LiX), together with the previous history of such cation sites, points to the likelihood that the energies of the SIII and SIII' sites must in fact be very similar. This conclusion raises the possibility that the supercage cations might even be redistributed in the presence of sorbates such as nitrogen. Our simulations may also carry implications with respect to the use of LiLSX for air separation. Unlike the diffraction measurements, which yield only an average picture of the structure, the simulations predict the short range ordering, too. For example, our predicted structures contain many pairs of SIII/SIII' cations at separations in the range 4.5 to 6 Å. If, as suggested in. [21], N_2 molecules are tightly bound between such pairs, their presence in the structure may have an important role in air separation.

ACKNOWLEDGMENTS

This work was supported by the MRL Program of the National Science Foundation under Award No DMR96-32716. CFM acknowledges the French Ministère des Affaires Etrangères for a Lavoisier fellowship, and AKC thanks the Fondation de l'Ecole Normale Supérieure and the Région de l'Ile de France for a Chaire Internationale de Recherche, Blaise Pascal.

REFERENCES

1. W. J. Mortier, in Compilation of Extra-Framework Sites in Zeolites; Butterworth Scientific Ltd.: Guildford, U.K. (1982).
2. D. H. Olson, J. Phys. Chem. **74**, 2758 (1970).
3. W. J. Mortier, H. J. Bosmans, J. Phys.Chem. **75**, 3327 (1971).
4. J. J. de Boer, I. E. Maxwell, J. Phys. Chem. **78**, 2395 (1974).
5. W. H. Baur, Am. Mineral. **49**, 697 (1964).
6. S. Sircar, Sep. Sci. Technol. **23**, 2379 (1988).
7. C. G. Coe, in Gas Separation Technology ; Elsevier: Amsterdam (1989).
8. M. S. Baksh, E. S. Kikkinides, R. T. Yang, Sep. Sci. Technol. **27**, 277 (1992).
9. D. M. Ruthven, in Principles of Adsorption and Adsorption Processes;Wiley: New York, 1984.
10. G.Vitale, L. M. Bull, R. E. Morris, A. K. Cheetham, B. H. Toby, C. G. Coe, J. E. MacDougall, J. Phys. Chem. **99**, 16087 (1995).
11. C. Mellot, J. Lignières, Molecular Simulation **18**, 349 (1997).
12. D. H. Olson, Zeolites **15**, 439 (1995).
13. T. Hseu, Ph.D. Thesis, University of Washington (1972).
14. G. Vitale, C. F.Mellot, L. M. Bull, A. K. Cheetham, J. Phys. Chem. **23**, 4559 (1997).
15. T. R. Gaffney, Current Opinion in Solid State and Mat. Sc. **1**, 69 (1996).
16. J. Plévert, F. Di Renzo, F. Fajula, G. Chiari, J. Phys. Chem. **101**, 10340 (1997).
17. J. Newsam, C. Freeman, A. Gorman, B. Vessal, Chem. Comm. **16**, 1945 (1996).
18. M.E. Grillo, J. Carrazza, J. Phys. Chem. **100**, 12261 (1996).
19. Catalysis 4.0 Software Suite and Discover, MSI: San Diego, CA.
20. C. F. Mellot, A. K. Cheetham, submitted for publication.
21. I. Papai, A. Goursot, F. Fajula, D. Plee, J. Weber, J. Phys. Chem. **99**, 12925 (1995).

REACTIVITY OF NO ON Co^{2+}/Co^{3+} REDOX SITES IN CoAPO-18. FTIR AND UV-Vis-NIR STUDIES

E. GIANOTTI, L. MARCHESE*, G. MARTRA and S. COLUCCIA

Dipartimento di Chimica IFM, University of Turin, via P. Giuria, 7, I-10125 Torino, Italy; marchese@ch.unito.it

ABSTRACT

NO activation on Co^{2+}/Co^{3+} centres of CoAPO-18 catalysts is studied by means of FTIR and diffuse reflectance UV-Vis-NIR spectroscopies both at 298 and 85K. Two families of Co^{2+} sites are found in the CoAPO-18 structure. A) *Ions in framework $[Co^{2+}(OH)P]$* (associated with Brønsted acid sites) adsorb NO to produce dinitrosyls absorbing at 1903 and 1834 cm^{-1}, the OH group is displaced and forced to H-bond to the zeolitic framework. These dinitrosyl complexes are reactive; in fact, Co^{2+} in $[Co^{2+}(OH)P]$ sites is oxidized to Co^{3+} and N_2O is formed. B) *Structural defect Co^{2+}* (Lewis acid sites) stabilize dinitrosyls, absorbing at 1813 and 1900 cm^{-1}; these complexes are formed slowly through activated processes which probably involve the breaking of some Co-O bonds of the zeolitic framework. Interestingly, we found that the two kinds of dinitrosyl complexes have different reactivity in presence of oxygen.

INTRODUCTION

CoAPO-18 is a microporous acid catalyst for the conversion of methanol to light olefins. This occurs on Brønsted hydroxyl groups (OH bridged between Co^{2+} and P of the AEI framework) [1,2]. Combined XANES, EXAFS [2], FTIR [2-4] and DR UV-Vis [4] studies revealed that the Brønsted acidity is associated with Co^{2+}/Co^{3+} redox couples: in fact, when Co^{2+} framework ions are oxidized to Co^{3+} (in oxygen at 823K), the Brønsted acidity disappears. Such behavior also reveals the potential redox properties of the catalyst. In this work we investigate the structural characteristics of the cobalt centres by their interactions with NO.

NO adsorption on Co^{2+}-exchanged zeolites leads mainly to the formation of dinitrosyls which react in the presence of O_2 to form NO_2 species at temperatures higher than 473K [5-6].

---

* To whom correspondence should be addressed

Interestingly, this reaction occurs rapidly on CoAPO-18 even at room temperature. Differences in the reactivity of the various types of Co^{2+} centers present in CoAPO-18 catalysts [3-4,7] give further information on the structures of the active sites.

EXPERIMENTAL

Synthesis and activation of CoAPO-18 catalysts are reported in ref. [3]. N,N-diisopropylethylamine (DIPE) was used as the organic template and a gel with a composition of $0.08CoO:Al_2O_3:P_2O_5:0.16HAc:1.7DIPE:50H_2O$ was crystallized in a Teflon-lined autoclave at 180°C for 10 days. FTIR spectra (resolution of 4 cm^{-1}) were collected on pelletized samples using a Bruker IFS88 spectrometer, and UV-Vis-NIR diffuse reflectance (DR) experiments were performed on a Perkin Elmer (Lambda 19) spectrometer equipped with an integrating sphere attachment. The IR spectra of adsorbed NO are shown in absorbance scale, having subtracted the background due to the bare sample.

RESULTS AND DISCUSSION

Figure 1 illustrates the FTIR spectra of NO (low dosage, p<0.02 mbar) adsorbed at 85K on CoAPO-18 reduced catalyst. Notice that, upon adding doses of adsorbates, a "negative" band grows in progressively at 3571 cm^{-1}. This negative band is due to the stretching mode of OH groups in [Co-O(H)-P] acid sites [4] which are perturbed by the addition of NO. The disappearance of this band is accompanied by the appearance of a new much broader band at 3104 cm^{-1}, indicating that the vibrations of the hydroxyl groups shift to lower frequency. The overall phenomenon and the size of the band shift ($\Delta v= 467$ cm^{-1}) resemble those observed upon CO adsorption [4]. It was demonstrated that a ligand displacement occurred at 298K, forcing the OH group towards the frame of the material. We propose that nitric oxide modifies the Brønsted acid centers by a similar displacement process [4] as shown in the scheme 1. The spectra in the 2300-1600 cm^{-1} region (figure 1B) reveal that NO adsorption produces two bands at 1903 and 1834 cm^{-1} whose intensity grows with NO pressure (≤ 0.02 mbar). These bands are assigned to the symmetric and asymmetric stretching modes of dinitrosyl species adsorbed on Co^{2+} ions [5-6]. The adsorption mechanism is depicted in scheme 1: two NO molecules interact with Co^{2+} ions and force the OH to H-bond with a structural oxygen atom. <u>All Brønsted sites are involved in this process</u>. Beside the bands at 1903 and 1834 cm^{-1}, a very sharp peak at 1869 cm^{-1} assigned to liquid-like NO is also observed [8].

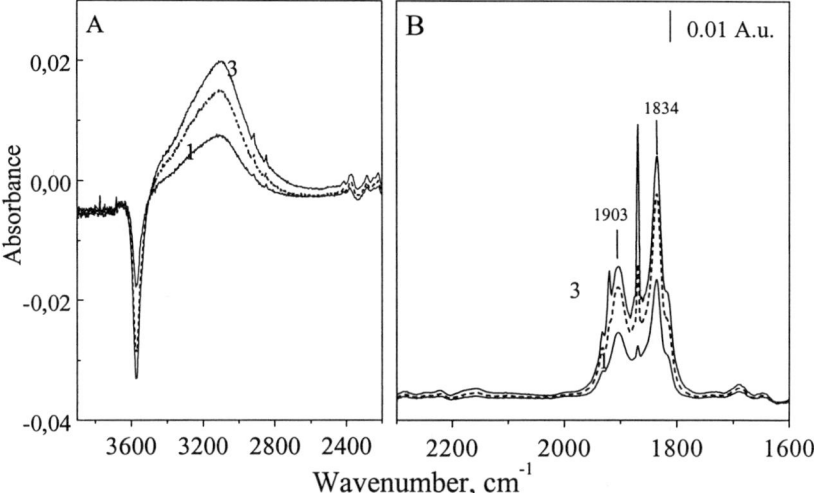

Fig. 1. FTIR spectra of NO adsorbed (p_{max} = 0.02 mbar) at 85K on reduced CoAPO-18 (increasing pressures from spectrum 1 to 3).

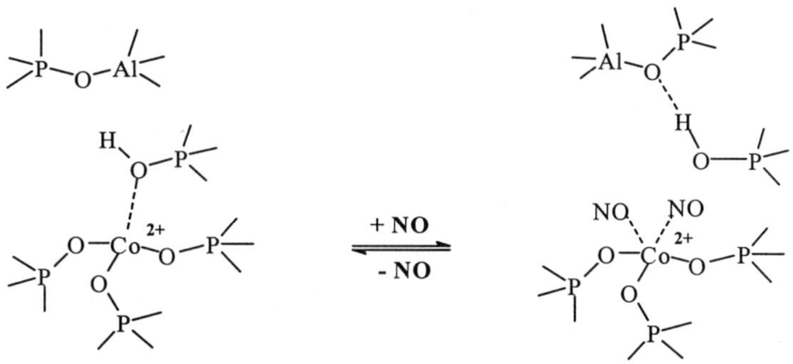

Scheme 1

At higher NO dosages (figure 2, curve b; p=1 mbar), new and very intense absorptions appear in the 2000-1750 cm^{-1} region, due to (NO)$_x$ clusters [8], together with a much weaker band at 2237 cm^{-1}, due to N$_2$O species [8] adsorbed on Co^{2+} ions. This was confirmed by adsorbing N$_2$O on CoAPO-18. *By evacuating and warming* (from 85 to 153K) the sample (figure 2, curve c), a doublet was observed with bands at 1900 and 1813 cm^{-1} also assignable to dinitrosyls on Co^{2+} ions, together with another absorption at 2160 cm^{-1} due to NO$_2^+$ species [5,6].

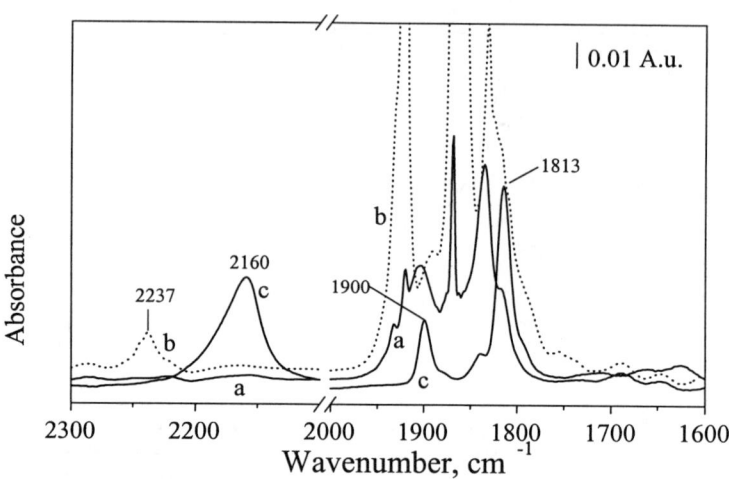

Fig. 2. FTIR spectra of NO adsorbed at 85K on reduced CoAPO-18: a) 0.02 mbar, b) 1 mbar. Spectrum (c) was obtained by warming up to 153K (in vacuo).

These data, supported by the results of adsorbing NO at room temperature (not shown for brevity) indicate that the dinitrosyl complexes at 1900-1813 cm^{-1} (Fig. 2c) have a different origin from those at 1903-1834 cm^{-1} (Fig. 2a): in fact, whilst the latter ones are observed immediately after the admission of small doses of NO, the former are produced slowly, either at temperature higher than 153K or in presence of high NO coverages. It is proposed that the dinitrosyls at 1900-1813 cm^{-1} are formed on defect Lewis acid Co^{2+} sites (those which are not related with Co(OH)P groups [3-4,7]) through an activated mechanism which may involve the breaking of some framework Co-O bonds. These dinitrosyls are irreversible even at 298K and are not reactive in presence of O$_2$. Conversely, the oxidation to NO$_2$ occurs rapidly in excess of NO and in presence of O$_2$, even at 298K.

Diffuse reflectance (DR) UV-Vis-NIR spectroscopy (figure 3) was used to obtain direct information on the oxidation state of cobalt ions before (curve a) and after (curve b) NO adsorption at RT.

The absorptions in the visible (20000-15000 cm^{-1}) and in the near IR (10000-4000 cm^{-1}) are attributable to $^4T_1(P)\leftarrow{}^4A_2(F)$ and $^4T_1(F)\leftarrow{}^4A_2(F)$ ligand field transitions tetrahedral Co^{2+} ions [4,9].

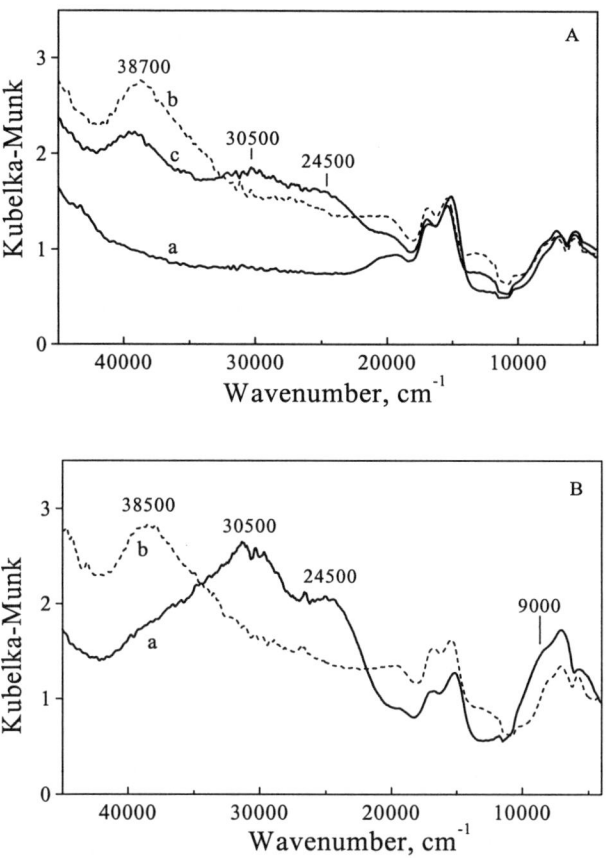

Fig. 3. Section A: DR UV-Vis-NIR spectra of reduced CoAPO-18 before (a) and after (b) NO adsorption at 298K; (c) was obtained by evacuating the sample for 2 hours at 298K. Section B: DR UV-Vis-NIR spectra of oxidized CoAPO-18 before (a) and after (b) NO adsorption.

These bands decrease somewhat in intensity after NO admission and new absorptions in the 45000-20000 cm^{-1} appear, dominated by a very intense band at about 38700 cm^{-1}. This may be assigned to ligand to metal charge transfer (LMCT) transitions of NO complexes adsorbed on Co^{3+} on the basis of the results obtained by adsorbing NO directly on thermally oxidized CoAPO-18 catalyst (figure 3B).

In fact, in the case of the oxidized catalyst, the electronic transitions at 30500 and 24500 cm^{-1} (charge transfer bands) and at around 9000 cm^{-1} ($^5T_2 \leftarrow {^5E}$) [4], associated with tetrahedral Co^{3+} centers (Fig. 3B, curve a) are depleted by the adsorption of NO (curve b), and the band at

38500 cm^{-1} is formed. When the excess of NO is removed by evacuating the CoAPO-18 catalyst (Fig. 3A, curve c), the 38700 cm^{-1} band decreases and, contemporarely, the doublet at 30500 and 24500 cm^{-1} increases. These results suggest that a fraction of Co^{2+} ions is irreversibly oxidized to Co^{3+} in the presence of NO at RT.

CONCLUSIONS

The use of NO as molecular probe has revealed the presence of two kinds of dinitrosyl groups adsorbed on Co^{2+} centres with different structural environments. FTIR and diffuse reflectance UV-Vis-NIR studies have indicated that a fraction of Co^{2+} sites are oxidizable to Co^{3+}. We tentatively propose that such Co^{2+} ions are those associated with Brønsted acid sites [Co^{2+}(OH)P].

REFERENCES

1. J. Chen and J.M. Thomas, J. Chem. Soc., Chem. Commun. (1994), 603.
2. J.M. Thomas, G.N. Greaves, G. Sankar, P.A. Wright, J. Chen, A.J. Dent and L. Marchese, Angew. Chem. Int. Ed. Engl. **33**, 1871 (1994).
3. L .Marchese, J. Chen, J.M. Thomas, S. Coluccia and A. Zecchina, J. Phys. Chem. **98**, 13350 (1994).
4. L. Marchese, G. Martra, N. Damilano, S. Coluccia and J.M. Thomas, Stud. Surf. Sci. Catal. **101**, 861 (1996).
5. Y. Li, T.L. Slager and J.N. Armor, J. Catal. **150**, 388 (1994).
6. A.W. Aylor, L.J. Lobree, J.A. Reimer and A.T. Bell, Stud. Surf. Sci. Catal. **101**, 661 (1996).
7. P.A. Barret, G. Sankar, C.R.A. Catlow, and J.M. Thomas, J. Phys. Chem. **100**, 8977 (1996).
8. J. Lanee, J.R. Ohlsen, Progress Inorg. Chem. **27**, 465 (1980).
9. B.N. Figgis, "Introduction to Ligand Field", Wiley-Interscience, New York, 1966, p. 239.

N_2O DECOMPOSITION OVER [Fe]-ZSM-22 ZEOLITES

L. MATACHOWSKI, M. KASTURE, T. MACHEJ and M. DEREWIŃSKI

Institute of Catalysis and Surface Chemistry Polish Academy of Sciences, Cracow, Poland;
ncderewi@cyf-kr.edu.pl

ABSTRACT

Decomposition of N_2O in the presence of excess oxygen over [Fe]-ZSM-22 zeolites with a different iron content was studied. The increase in zeolite activity in proportion to iron content was observed. The samples with Si/Fe ratio below 40 exhibit high activity in N_2O decomposition in the temperature range 400-600^{0}C. The long term calcination of H form of iron rich sample [Fe]-ZSM-22(29) has a limited effect on its catalytic activity indicating that the state of the iron species after such a treatment does not change considerably. The drop in activity observed for the sodium form and significant increase of activity observed for iron exchanged sample, in comparison to the activity of H form of the zeolite [Fe]-ZSM-22(29), points to the dominating role of the non-framework iron species in N_2O decomposition.

INTRODUCTION

Nitrous oxide is a compound contributing to the global warming effect and is presumably involved in the stratospheric ozone depletion. The development of catalytic processes for the control of N_2O emission seems to be possible with the use of transition metal exchanged ZSM-5 zeolites, which are active in the catalytic decomposition of N_2O into nitrogen and oxygen at relatively low temperatures [1, 2]. Among transition metals, Cu and Co are the most active in the above reaction.

Oxygen has an inhibiting effect on the rate of N_2O decomposition. However for iron-exchanged zeolites (Fe-MOR, Fe-ZSM-5) the absence of oxygen inhibition was observed [3,4] and for ferrisilicate [Fe]-ZSM-5 even a positive effect of oxygen was reported [5].

In the present study the catalytic properties of a series of ZSM-22-type zeolites isomorphously substituted with iron was investigated in the decomposition of N_2O in the presence of excess oxygen.

EXPERIMENTAL

Syntheses of the framework iron-containing ZSM-22 zeolites with 1-ethyl pyridinium bromide as a template were carried out using the procedure outlined in Ref. [6]. Four samples with different Si/Fe ratios, ranging from 29 to 81, were obtained. XRD and SEM data showed that all samples were highly crystalline and with no unreacted or amorphous phase present. The samples were designated as [Fe]-ZSM-22(X) where X stands for the Si/Fe ratio. All zeolites prepared were white, indicating that no bulky iron oxide was present on the surface of the crystals. The incorporation of iron into the framework of the [Fe]-ZSM-22 was proved by the changes in the unit cell volume and the shift in the position of the lattice vibrations bands (IR) as well as with Mössbauer and ESR spectroscopies.

As-synthesized zeolites were calcined at $500^{\circ}C$ for 16 h in dry air to decompose the organic template. H forms, designated as H[Fe]-ZSM-22(X), were prepared by exchange with 0.1M NH_4NO_3 solution at $60^{\circ}C$ for 24 h and then deammoniation at $480^{\circ}C$ for 3 h.

The BET surface area was measured both for calcined and H forms of zeolites, in order to determine whether the applied procedure of deammoniation does not result in the deironation of samples. High specific BET surface areas of H[Fe]-ZSM-22(29) and H[Fe]-ZSM-22(37) samples in comparison to those measured for calcined preparations (Table) indicates that during NH_4 decomposition only limited iron extraction from the lattice takes place and no bulky, extraframework Fe_xO_y species are present in the channels. This is confirmed by examination of the IR spectra of H form of the preparations (in the region of vibration of acidic OH groups). The presence of the band at 3623 cm^{-1} attributed to SiOHFe sites confirmed that at least part of iron is still present in the framework positions.

Table
BET surface area of calcined and H forms of [Fe]-ZSM-22(X) samples

Sample	BET surface area [$m^2 \cdot g^{-1}$]	
	calcined $500^{\circ}C$, 16h	H form (calc.$480^{\circ}C$, 3h)
[Fe]-ZSM-22(29)	247	219
[Fe]-ZSM-22(37)	226	206

Sodium and iron exchanged samples were obtained by ion exchange with $NaNO_3$ and $Fe(NO_3)_3$ respectively. The [Fe]-ZSM-22(29) sample calcined at $500^{\circ}C$ was treated with diluted HCl to remove the extra framework species from the zeolite. Subsequently ion-exchange with 0.1M $NaNO_3$ solution ($60^{\circ}C$, 24 h) was performed to introduce Na$^+$ ions into the cationic positions. The obtained sodium form was used as a starting material to prepare iron exchanged sample designated as Fe[Fe]-ZSM-22(29). The treatment with 0.1M Fe(NO$_3$)$_3$ solution, at 60°C for 24 h, was repeated twice.

Catalytic tests were carried out in a fixed-bed flow microreactor in the temperature range 400 - 600°C. A mixture of ca 2 vol. % N$_2$O, 2 vol. % O$_2$ and nitrogen was passed over 0.5 ml of catalyst at a space velocity of 4000 h^{-1}. The effluent gases were analyzed by means of gas chromatography. The activity of the samples was checked at increasing temperatures from 400°C to 600°C after activation at 400°C.

RESULTS AND DISCUSSION

Fig.1 presents the N$_2$O conversion as a function of the reaction temperature for the samples with different iron contents. The data clearly show that the activity of the samples is directly

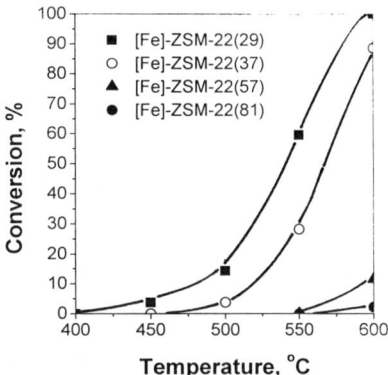

Fig. 1. Conversion of N$_2$O as a function function of the reaction temperature.

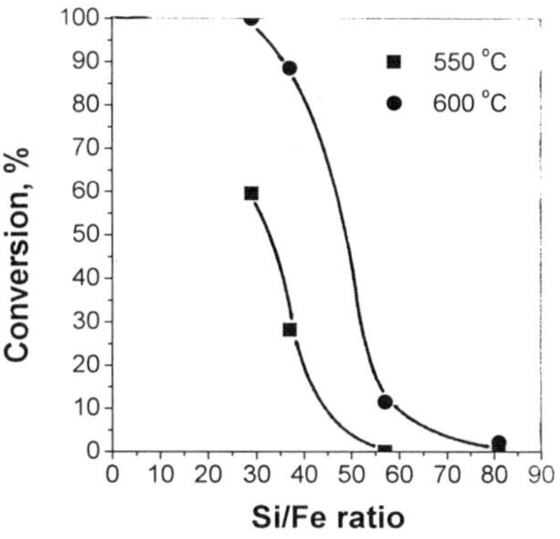

Fig. 2. Conversion of N$_2$O as a function of the Si/Fe molar ratio.

related to the amount of iron present in the zeolite. The samples with Si/Fe ratio higher than 50 exhibit poor activity whereas the iron rich samples i.e. those with Si/Fe ratios of 37 and 29, show a considerably higher activity in the nitrous oxide decomposition. This is better seen in Fig. 2 where the dependence of the N$_2$O conversion at 550°C and 600°C on iron content expressed as the

Si/Fe molar ratio is presented. Although the framework Fe^{3+} ions are quite stable under reducing conditions, the removal of iron atoms from the framework positions occurs upon calcination at high temperature in the presence of O_2/air. It takes place during the thermal decomposition of NH_4^+ cations. In order to check the thermal stability of the iron species in the reaction conditions, the activity of the Fe-rich sample i.e. [Fe]-ZSM-22(29) was determined upon increasing and then decreasing temperature. The same N_2O conversion level obtained in both cases indicates that the state of the iron species present in H form of preparations remains practically unchanged during the catalytic test. Additionally it was found that prolonged calcination at 600^0C (7h) did not influence significantly the activity of the sample. The conversion of N_2O on such a sample increased by ca 3% over the whole studied temperature range indicating that even prolonged high temperature treatment did not considerably increase the number of sites which are involved in the catalytic process. This can be explained by the stabilizing effect of extralattice iron, formed during the high temperature decomposition of NH_4^+ cations, on the remaining tetrahedrally coordinated framework iron. Therefore the H form of our zeolites contains always both H^+ and Fe^{3+} on extralattice positions.

The activity of zeolite samples was compared to the activity of 2 wt. % Fe_2O_3 supported on γ-Al_2O_3 and amorphous SiO_2. The N_2O conversion levels in the latter cases were very small (below the value obtained for the poorest iron sample i.e. [Fe]-ZSM-22(81)). This suggests that iron oxide species are not responsible for the catalytic properties in the N_2O decomposition to N_2 and O_2.

The data on N_2O decomposition obtained for the ZSM-5-type zeolite isomorphously substituted with iron [5] revealed two activity temperature ranges. The framework iron species are active at lower temperatures (below 400^0C) whereas extraframework Fe^{3+} introduced as the framework charge countercations are responsible for the activity at higher temperatures. The zeolites under study are active at higher temperatures (>400°C) which suggests that in the case of ZSM-22 only extra framework iron species are active in the N_2O decomposition.

In order to check which type of iron species

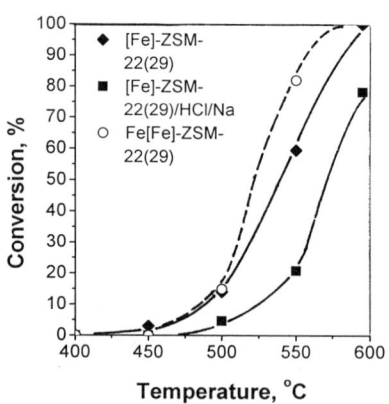

Fig. 3. The effect of sodium and ion exchange with $Fe(NO_3)_3$ on the activity of [Fe]-ZSM-22(29) zeolite.

(framework or extra framework) is responsible for the catalytic activity of the samples in question, the catalytic activities of sodium [Fe]-ZSM-22(29)/HCl/Na and iron exchanged Fe[Fe]-ZSM-22(29) samples were compared to that of H form [Fe]-ZSM-22(29) (Figure 3). It can be seen that the activity of the sodium-modified sample is much lower than that observed for H form which can points to the important role of the extra framework iron species. However, another explanation is also possible. The sodium cations introduced into the zeolite may decrease the accessibility of the framework Fe to the N_2O molecules. To explain this, the catalytic activity of the sample in which both sodium and H were replaced by iron cations, was checked. The significant increase in the activity in N_2O decomposition observed for iron exchanged sample Fe[Fe]-ZSM-22(29) (Figure 3) clearly shows that extraframework iron species introduced to the zeolite during the ion-exchange with iron nitrate solution or formed during the thermal decomposition of NH_4^+ cations at 480^0C, are active sites involved in decomposition of nitrous oxide. The high catalytic activity observed for samples [Fe]-ZSM-22(29) and [Fe]-ZSM-22(37) can be explain by higher deironation level of those iron rich preparations, during thermal treatment prior the catalytic tests.

REFERENCES

1. Y. Li and J.N. Armor, Appl. Catal. B, **1**, L21 (1992).
2. Y. Li and J.N. Armor, Appl. Catal. B, **3**, 55 (1993).
3. J. Leglise, J.O. Petunchi, W.K. Hall, J.Catal., **86**, 382 (1984).
4. G.I. Panov, V.I. Sobolev, A.S. Kharitonov, J.Mol.Catal., **61**, 85 (1990).
5. Y.A. Chang, J.G. Mc Carty and Y.L. Zhang, Catal. Lett. **34**, 163 (1995).
6. R. Kumar, P. Ratnasamy, J.Catal., **116**, 440 (1989).

STRUCTURE AND ACTIVITY OF CERIUM-PROMOTED AG-ZSM-5 FOR THE SELECTIVE CATALYTIC REDUCTION OF NITRIC OXIDE WITH METHANE

ZHIJIANG LI and MARIA FLYTZANI-STEPHANOPOULOS

Department of Chemical Engineering, Tufts University, Medford, MA 02155, U.S.A.
Fax: (617)-627-3991, E-mail: zli1@ emerald.tufts.edu, mstephanopoulos@infonet.tufts.edu

ABSTRACT

Cerium-promoted silver ion-exchanged ZSM-5 was found very active and selective for the catalytic reduction of nitric oxide by methane in the presence of oxygen. It was found that silver sites at high Ag exchange level (>50%) are more active for the methane oxidation, while those at low Ag loading are more effective for the SCR reaction. Characterization of the catalyst by HRTEM/EDS, STEM/EDS and UV-VIS (DRS) has shown that the performance of the catalyst is closely related to the dispersion and oxidation states of silver. Ag is well dispersed in the zeolite channels as Ag^+ ions at low loadings, while at high Ag loadings, metallic Ag particles are also found on the zeolite surface. The dispersed silver state provides sites for the SCR reaction, i.e. the formation of N_2, while Ag particles are active for CH_4 combustion.

INTRODUCTION

The selective catalytic reduction (SCR) of NOx with methane in oxygen-rich exhaust gas is a promising new NOx reduction technology, which would benefit both the transportation and power generation sectors (CNG vehicles, gas utilities, etc.) To-date, several single metal ion-exchanged zeolites have been shown active for this application[1-3]. However, the NO reduction rates over these catalysts are too low for commercialization. Recently, we have found that cerium-promoted Ag-ZSM-5 is an active and stable catalyst for the SCR of NO with CH_4[4]. Previous reports on the SCR activity of silver-supported on alumina and zirconia catalysts were limited to hydrocarbons other than methane[5-8]. The activity of these catalysts was reported to strongly depend on the dispersion and oxidation state of silver[6,7]. In this paper we report on the structure of silver in Ce-Ag-ZSM-5 and its effects on the catalyst activity for the SCR of NO with CH_4.

EXPERIMENTAL

Ce-promoted Ag-ZSM-5 was prepared by ion-exchange of Na-ZSM-5 (Davison, Si/Al=13.8) in dilute aqueous solutions of the metal nitrates. Cerium was exchanged first (80°C, 2 h), followed by washing, drying (110°C, 10-12 hr) and heat treatment in air(500°C, 2 hr). Exchange of Ag took place at RT, 24 h in the dark. Elemental analysis of the samples was performed by ICP. The % exchange is based on Al:Ag=1:1 and Al:Ce=3:1.

Parametric and kinetic studies were conducted in a quartz, fixed-bed microreactor equipped with a HP 5890 GC/TCD. A 5A molecular sieve column was used to separate O_2, N_2, CH_4, NO, and CO. The typical experimental conditions used were a gas mixture of 0.5%NO, 0.5%CH_4, 2.5%O_2 in He at atmospheric pressure, a temperature range of 350-650°C and a GHSV of 7,500 h^{-1}(STP) in parametric tests or 30,000-180,000 h^{-1} (STP) in kinetic studies. Selected catalyst samples were characterized by surface and bulk analytical techniques including HRTEM, STEM/EDS and UV-VIS diffuse reflectance spectroscopy (DRS).

Results and Discussion

Figure 1 shows the conversion vs. temperature plots for NO→N_2 and CH_4→CO_2 for single Ag- and Ce-ZSM-5 and promoted Ce-Ag-ZSM-5 catalysts. The NO conversion to N_2 over the unpromoted Ag-ZSM-5 is very low (<30%) at the test conditions. Ce-ZSM-5 is also a poor catalyst for the SCR of NO by CH_4. However, a significant promotion effect was found over cerium modified Ag-ZSM-5. The rate of NO reduction to N_2 over Ce-Ag-ZSM-5 is comparable to that reported for Co-ZSM-5 catalyst at temperatures <500°C and higher than that at higher temperatures[1]. It should be pointed out that the addition of cerium does not improve the catalyst activity for CH_4 combustion as evidenced by the similar profiles of CH_4 conversion vs. temperature over the promoted and unpromoted Ag-ZSM-5 samples in Figure 1. Therefore, the presence of cerium enhances only the catalytic activity for SCR, not for CH_4 oxidation. In other words, cerium increases the selectivity of Ag-ZSM-5.

The activity of Ce-Ag-ZSM-5 was found to depend on both cerium and silver loadings. SCR experiments conducted over Ce-Ag-ZSM-5 catalysts with various cerium exchange levels from 0 to 54% (corresponding to 2.6wt% Ce), showed that approximately 20-30% cerium ion exchange (1-1.5wt%) could provide the maximum promotion to Ag-ZSM-5. Excess cerium exchange is not necessary. However, as shown in Figure 2, a linear relationship exists between NO conversion to N_2 and Ag^+ exchange level up to 50-60% Ag exchange (5.5-6.5 wt%) at three temperatures of 450, 500 and 550°C. Above 60% silver exchange, no further effect of Ag loading is observed. This means that not all the Ag^+ sites are equivalent in terms of their activity for the SCR of NO with CH_4. Within the range of 0-60% Ag exchange, each Ag site seems to have approximately the same activity as indicated by the linear dependence of the NO conversion on Ag exchange level. However, as Ag loading was increased further from 60 to 80%, the NO conversion did not increase, suggesting that some silver sites at high exchange level are less active than those at low loadings.

To examine the effect of Ag loading on the activity of the Ce-Ag-ZSM-5 catalyst for the CH_4 oxidation by O_2, the reaction of CH_4+O_2 (CH_4 combustion) was conducted over samples with various silver exchange levels from 0 to 78% (corresponding to 8.5wt% Ag) and a fixed cerium exchange level of 21-24% (~1% wt% Ce). The results are shown in Figure 3 in terms of CH_4 conversion vs. Ag exchange level at the temperature of 450, 500 and 550°C. Clearly, a non-linear relationship between CH_4 conversion and Ag loading is observed at these temperatures. The increase of CH_4 conversion with Ag exchange level is faster at higher exchange levels. This means that at higher silver loadings, the state of silver is more active for CH_4 combustion. This observation is contrasted to the correlation of NO conversion to N_2 with Ag loading shown in Figure 2. Therefore, it is apparent that different silver sites are formed at higher exchange levels. These sites are less active for the SCR reaction, but very active for CH_4 combustion.

The dispersion and oxidation state of silver in low and high silver-containing Ce-Ag-ZSM-5 samples were characterized by HRTEM/EDS, STEM/EDS and UV-VIS (DRS). Figures 4(a) and (b) show the HRTEM images of Ce(21)-Ag(42)-ZSM-5 and Ce(24)-Ag(78)-ZSM-5, respectively. The EDS X-ray emission spectra (c) and (d) were taken in the areas A and B, respectively, of the sample Ce(24)-Ag(78)-ZSM-5 (Figure 4(b)). On Ce(21)-Ag(42)-ZSM-5, only highly dispersed clusters of 1-3 nm size were observed, while on Ce(24)-Ag(78)-ZSM-5 large particles of 8-12 nm

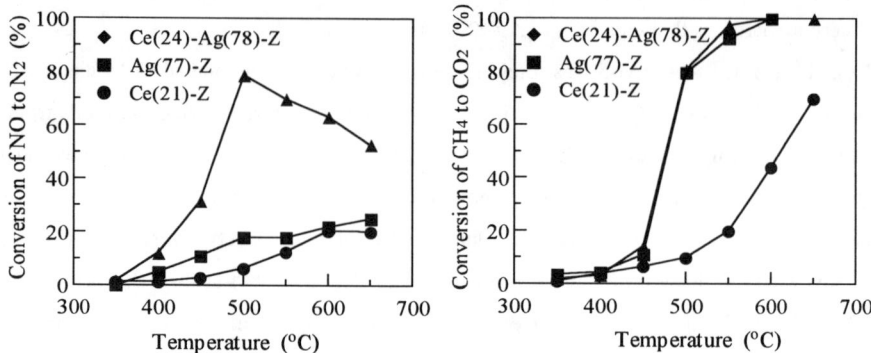

Figure 1 Promotion effect of Ce on Ag-ZSM-5 for the SCR of NO with CH_4. Feed gas: 0.5%NO-0.5%CH_4-2.5%O_2-bal. He, S.V.=7,500 1/h.

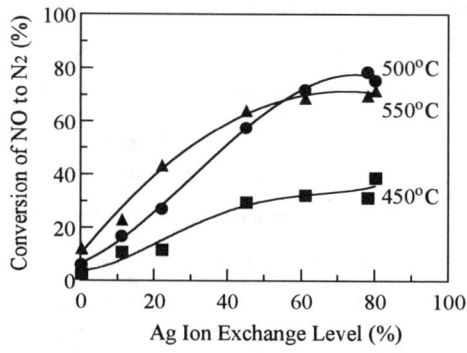

Figure 2 Conversion of NO to N_2 as a function of Ag loading in the SCR of NO by CH_4 over Ce-Ag-ZSM-5. Feed gas: 0.5%NO-0.5%CH_4-2.5%O_2-bal. He, S.V.=7,500 1/h, Ce=21-24%.

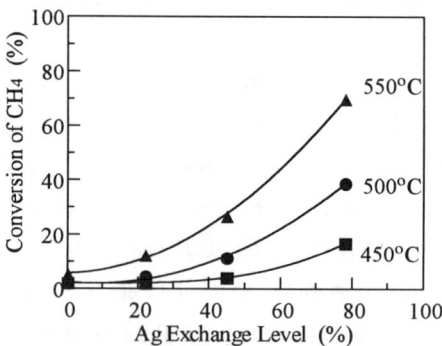

Figure 3 CH_4 conversion as a function of Ag loading in the CH_4 combustion over Ce-Ag-ZSM-5 catalysts. Ce=21-24%. Feed gas: 0.5%CH_4-2.5%O_2-bal. He, S.V.=7,500 1/h.

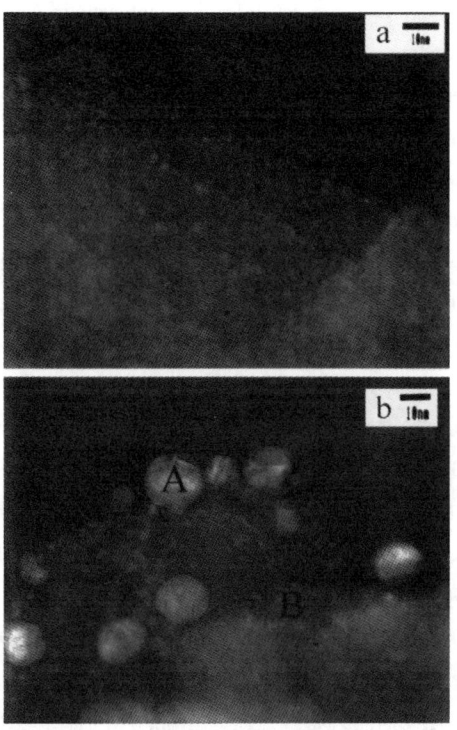
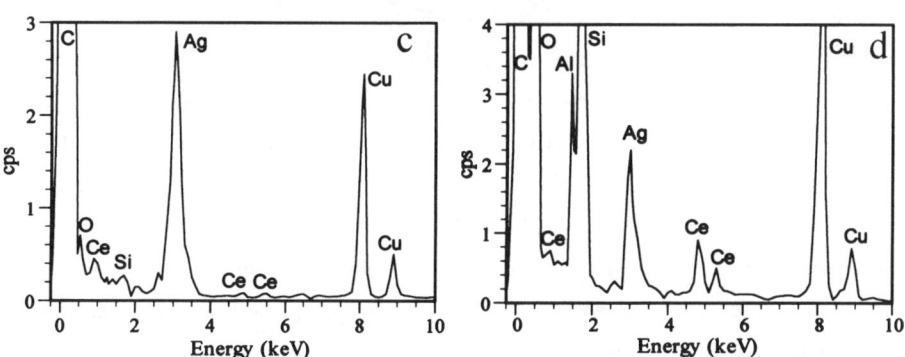

Figure 4 HRTEM of air-calcined Ce(21)-Ag(42)-ZSM-5(a) and Ce(24)-Ag(78)-ZSM-5(b) and EDS X-ray emission spectra from areas A(c) and B(d) in image (b).

size were also seen on the catalyst surface. The EDS spectrum (c) which was taken on a large particle identified it as Ag. Note that in spectrum (c), the intensity of Ce, Si and O X-ray bands is substantially lower compared to that of Ag. (The carbon and copper bands are from carbon coating and the copper grid, respectively.) Therefore, the large particles are likely in metallic silver form. However, in area B where large silver particles are absent, strong Ag as well as Ce X-ray emissions were obtained in addition to Al and Si emissions, as shown by the EDS spectrum (d). In this case, the band corresponding to silver may be caused by the emission of silver ions embedded in the zeolite matrix. Therefore, the 1-3 nm clusters seen in the two HRTEM images (Figure 4(a) and (b)) are cerium species, most likely in the 4+ oxidation state as determined by UV-VIS (DRS) (see below).

Formation of silver particles in the high Ag-containing sample was further confirmed by STEM/EDS analysis. It was found that the distribution of Ag closely matches that of Al in the low Ag-containing sample Ce(21)-Ag(42)-ZSM-5, suggesting high dispersion of silver at this Ag exchange level(<50%). However, in the high Ag-containing sample Ce(24)-Ag(78)-ZSM-5, agglomeration of silver was observed, although the distribution of Al is still uniform.

The oxidation states of silver and cerium were examined by UV-VIS (DRS). The spectra for four catalyst compositions are shown in Figure 5. Isolated Ag^+ ions were identified by three strong bands at 196, 212 and 222 nm caused by the electronic transition of $4d^{10} \rightarrow 4d^9 5s^1$[9]. These three bands grew as Ag loading was increased from 0 in Ce(21)-ZSM-5 to 78% in Ce(24)-Ag(78)-ZSM-5. But the Ag^0 phase could not be distinguished because of the overlapping of its band with the charge transfer band of $Ce^{4+} \leftarrow O^{2-}$ in CeO_2 clusters at 294 nm. However, metallic silver was identified in Ag(77)-ZSM-5 by UV-VIS (data not shown in Figure 5). The CeO_2 clusters are several nanometers in size according to [10], consistent with our HRTEM observations. In addition to the CeO_2 phase, isolated Ce^{3+} species were also suggested by the two UV-VIS bands at 210 and 260 nm in Ce(21)-ZSM-5.

Summarizing the Ce-Ag-ZSM-5 characterization, we found that Ag is well dispersed in the zeolite channels at low loadings, while at high Ag loadings, Ag particles are also found on the zeolite surface. By recalling the different catalytic properties of silver at low and high Ag loadings for the SCR and CH_4 oxidation reactions, it can be concluded that the dispersed state of silver is highly active and selective for the SCR reaction, while the silver nanoparticles are non-selective and catalyze the CH_4 oxidation by O_2. Nanocrystalline (5-10 nm) silver particles on zirconia have been found to be very active for the complete oxidation of methane in recent work in our lab[11].

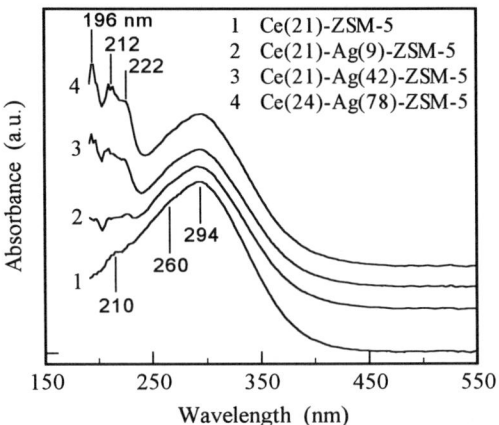

Figure 5 UV-VIS DR spectra of fresh Ce- and Ce-Ag-ZSM-5 catalysts. The spectra were taken in air and recorded after subtracting the spectrum of parent Na-ZSM-5. All samples were heat-treated at 500°C for 2 h in air.

In a recent paper, Bethke and Kung [6] have explained the low activity of a 6wt% Ag/Al_2O_3 for the SCR reaction of NO with C_3H_6 along similar terms. Thus, metallic particles of silver in this material catalyzed the combustion of propylene rather than the SCR reaction. The opposite was true for a low-content 2wt% Ag/Al_2O_3 which contained silver in oxidized form.

CONCLUSION

Cerium-promoted Ag-ZSM-5 was found to be very active and selective for the SCR of NO by CH_4 in the presence of excess oxygen. Different correlations between the catalyst activity and Ag exchange levels were obtained for the SCR reaction and the CH_4 oxidation by O_2. Silver sites at high Ag exchange level (>50%) are more active for the methane oxidation, while those at low Ag loading are more effective for the SCR reaction. The different catalytic properties of silver are determined by its dispersion and oxidation state. Isolated Ag^+ ions and metallic silver particles of 8-12 nm size on the zeolite surface were observed. The dispersed silver state provides sites for the

SCR reaction, i.e. the formation of N_2, while metallic Ag particles are active for the CH_4 combustion. Cerium exists in the catalyst in two oxidation states: 3+ and 4+, primarily in the form of CeO_2 clusters on the zeolite surface.

References
[1] Y. Li and Armor, J.N., Appl. Catal. B **1**, 239(1993); **5**, L257(1995).
[2] K Yogo and E. Kikuchi, Stud. Sur. Sci. Catal. **84**, 1547(1994).
[3] Y. Nishizaka and M. Misono, Chem. Lett., 1295(1993).
[4] Z. Li and M. Flytzani-Stephanopoulos, Appl. Catal. A **165**, 15(1997).
[5] T. Miyadera, Appl. Catal. B **2**, 199(1993).
[6] K.A. Bethke and H.H. Kung, J. Catal. 172, **93**(1997).
[7] T.E. Hoost, R.J. Kudla, K.M. Collins and M.S. Chattha, Appl. Catal. B **13**, 59(1997).
[8] M. Haneda, Y. Kintaichi, M. Inaba, and H. Hamada, Bull. Chem. Soc. Jpn. **70**, 499(1997).
[9] J. Texter, J.J. Hastrelter, and J.L. Hall, J. Phys. Chem. **87**, 4690 (1983).
[10] A. Bensalem, J.C. Muller and F. Bozon-Verduraz, J. Chem. Soc., Far Trans. **88**, 153(1992).
[11] L. Kundakovic and M. Flytzani-Stephanopoulos, J. Catal. (accepted).

CeO_2-H-ZSM-5 COMPOSITES—A BIFUNCTIONAL SYSTEM FOR THE SELECTIVE CATALYTIC REDUCTION OF NO BY METHANE

T. LIESE, D. RUTENBECK[1] AND W. GRÜNERT

Lehrstuhl für Technische Chemie, Ruhr-Universität Bochum, D-44780 Bochum, Germany,
FAX +49 234 7094115, email w.gruenert@techem.ruhr-uni-bochum.de

[1]Institut für Technische Chemie, Universität Leipzig, D-04103 Leipzig, Germany

ABSTRACT

A composite catalyst system that provides promising activities for the SCR of NO with methane but fails with propene was prepared from CeO_2 and NH_4-ZSM-5. It exhibits a bifunctional synergy between CeO_2 and the sites in the zeolite, which allows its application as a mechanical mixture. The zeolite sites engaged in the bifunctional interaction are the acidic protons. NO_2 formation is one of the functions of the CeO_2, but the synergy between the cooperating sites is not mediated by gas-phase NO_2 transport. CeO_2/H-ZSM-5 composites offer a high flexibility for further modification.

INTRODUCTION

In recent years great effort has been taken to develop the SCR of NO_x with hydrocarbon reductants (SCR-HC), mostly for application with mobile sources (diesel, lean-burn engines). This development may also provide opportunities for the current SCR technology with stationary sources, where the reductant ammonia might be replaced by cheap hydrocarbons usually employed as raw material for NH_3 production (methane, liquid petrol gases). Up to now, the best results were reported for zeolites (MFI, MOR) doped with Cu, Co, Ga, In, Ce, Pt and Pd [1-3]. With the reductant CH_4 (SCR-methane), lower activities are generally found, and some dopants (Cu, Ce), which provide high activities with higher hydrocarbons show poor activities with methane. Over Cu-ZSM-5, e.g., CH_4 is mostly oxidised to CO_x and H_2O, while Ce-ZSM-5 which is promising with propene, is inferior with methane [1, 4, 5].

We report here the catalytic properties of catalysts containing CeO_2 precipitated onto the external surface of H-ZSM-5 ("CeO_2-Zeolite-Composites"). In contrast to known Ce-ZSM-5 catalysts [4, 5] they exhibit promising activity with the reductant methane but fail with propene. Investigations aiming at the elucidation of the catalytic functions provided by these composites will be presented.

EXPERIMENTAL

Catalysts combining cerium and ZSM-5 were prepared by exchange (Ce/Na(or H)-ZSM-5), by precipitation of cerium oxide onto the external surface of ZSM-5 (CeO$_2$/H-ZSM-5, several batches, see below) and as physical mixtures between ZSM-5 and CeO$_2$ (CeO$_2$//Na (or H)-ZSM-5).

Ce/Na-ZSM-5 was obtained by exchanging Na-ZSM-5 (Si/Al≈13, Chemiewerk Bad Köstritz, Germany) with a cerium nitrate solution (20 mmole/l) at 353 K for 16 hours. Ce/H-ZSM-5 was prepared in the same way using the NH$_4^+$ form of the zeolite. The Ce contents of Ce/Na and Ce/H-ZSM-5 were 0.70 and 0.57 wt-%, respectively.

The preparation of the CeO$_2$/H-ZSM-5 composites (batches I and III) was performed by adding rapidly aqueous NH$_3$ (25 %) to a suspension of the zeolite in a cerium nitrate solution (12.9 mmole/l). The zeolites used were a H-ZSM-5 provided by Süd-Chemie, Germany (Si/Al ≈ 14; → batch I) and a NH$_4$-ZSM-5 obtained from CK Bitterfeld, Germany (Si/Al≈ 19, → batch III). Batch II was made by adding aqueous NH$_3$ (25 %) to a suspension of NH$_4$-ZSM-5 (Si/Al≈13, *vide supra*) in a cerium sulphate solution (20 mmole/l) after the suspension

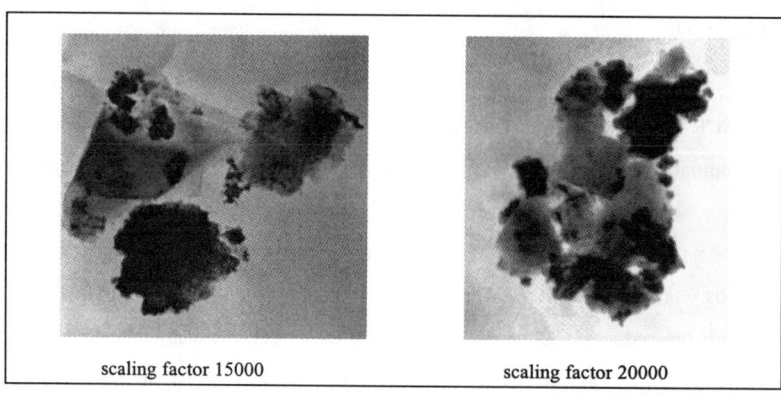

scaling factor 15000   scaling factor 20000

FIG 1. TEM pictures of CeO$_2$/H-ZSM-5(II). The dark particles were identified as CeO$_2$ by EDX, no Ce (above the detection limits of EDX) was detected in the (light) zeolite particles

had been stirred at room temperature for 16 h. The cerium content of the composites was 8.2 wt-% (I and III), and 14.5 wt-% (II). The preparation of more involved composite materials is detailed below. TEM of typical composite samples shows the CeO$_2$ as particles of up to 100 nm size deposited on the external surface of the zeolite grains (Figure 1).

Physical mixtures were prepared by mixing zeolite and CeO_2 (from calcination of $Ce(NO_3)_3 \cdot 6\,H_2O$) in different mesh fractions in order to allow the subsequent separation of the components. Their Ce content was 17.5 wt-%. They are distinguished from precipitated preparations by '//' separating CeO_2 and the zeolite code.

The selective catalytic reduction of NO with methane was studied in a flow reactor using a gas mixture containing 1000 ppm NO, 1000 ppm CH_4 or C_3H_6, and 2 % O_2 in He. The NO conversion, X_{NO}, will be reported for a space velocity of 10,000 h^{-1} scanning the temperature range from 873 K down to 673 K (for propylene - 573 K) in steps of 50 K (standard reaction conditions, only reduction product observed - N_2). The products were analysed by a combination of gas-chromatographic (O_2, N_2, CH_4), mass-spectrometric (NO, CH_4) and non-dispersive IR techniques (CO, CO_2, N_2O). Prior to each run, the samples were calcined in flowing He at 873 K for 2 h. In some experiments, the CeO_2 and H-ZSM-5 components were not mixed but applied in layers.

The NO oxidation activity of the catalyst components was studied with a mixture of 1000 ppm NO and 2 % O_2 in He. The space velocity (referred to the amount of the component - CeO_2 or zeolite - in the catalysts) was five times higher than in the SCR runs. The ratio of NO_2 formed to the sum of NO_2 and NO ("γ") was used as a measure for the NO oxidation activity. The temperature was decreased stepwise from 773 to 673 K. NO and NO_2 were analysed by an IR detector.

RESULTS

The figures 2 and 3 report NO conversions obtained with the CeO_2/H-ZSM-5 composites and with Ce-exchanged ZSM-5 using methane or propene as reductants. Figure 2 compares the composite CeO_2/H-ZSM-5 (batch I) with pure H-ZSM-5 (reductant CH_4). Over the latter, X_{NO} increases with the reaction temperature up to 18 % NO conversion at 873 K. The composite shows higher NO conversions at all temperatures, with a peak conversion of 64 % at 873 K.

In figure 3, the behaviour of Ce-exchanged samples and a CeO_2/H-ZSM-5 composite with respect to the reductants methane and propene is compared. Curve 1 describes the NO conversion with propene over Ce/Na-ZSM-5. The trend, with a peak conversion of 51 %, agrees well with the literature [4, 6]. Ce/H-ZSM-5 was inferior with propene ($X_{NO} \leq 10$ %) due to coking, and did not perform well with methane either (curve 2). On the contrary, an apreciab-

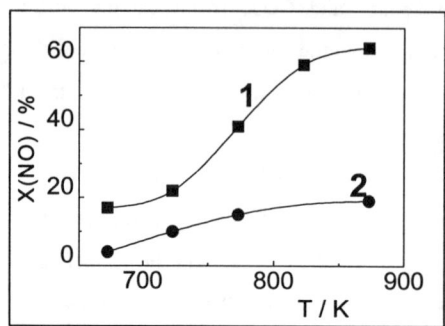

FIG 2. NO conversion vs. reaction temperature, reductant methane.
1 - CeO$_2$/H-ZSM-5 (I), 2 - H-ZSM-5

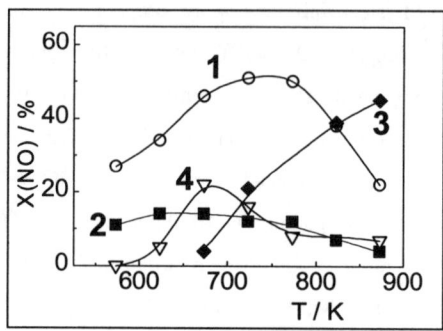

FIG 3. Comparison of exchanged and precipitated systems.
1 - Ce/Na-ZSM-5 (C$_3$H$_6$), 2 - Ce/H-ZSM-5 (CH$_4$), 3 - CeO$_2$/H-ZSM-5 (II) (CH$_4$), 4 - CeO$_2$/H-ZSM-5 (II) (C$_3$H$_6$)

le NO conversion was found with CeO$_2$/H-ZSM-5 (batch II) with the reductant methane, while propene was again inferior with this acidic catalyst due to coking (curve 4).

Figure 4 shows NO conversions measured with a physical mixture of CeO$_2$ and H-ZSM-5 and with the zeolite recovered from the mixture after the run, and compares it with the behaviour of pure H-ZSM-5. The mixture CeO$_2$//H-ZSM-5 shows the same increasing conversion as the composite (curve 1). The overall activity is somewhat lower, but it is higher than that of H-ZSM-5 at all temperatures. After catalysis, the recovered zeolite shows the same SCR activity as the pure H-ZSM-5 (curves 2 and 3). For pure CeO$_2$ and the CeO$_2$//Na-ZSM-5 mixture, the NO conversions observed were negligible ($\leq$ 3 %). The zeolite component separated from the mixture after the SCR run was also analysed by ICP-AES and XPS [7]. A small amount of Ce could be detected by both techniques (ICP: 0.1 wt.-%).

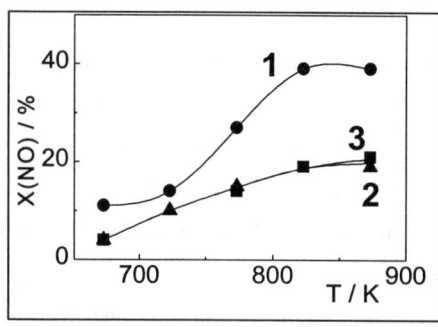

FIG 4. SCR-methane with a physical mixture (reductant methane).
1- CeO$_2$//H-ZSM-5 mixture; 2 - fresh H-ZSM-5; 3 - H-ZSM-5, recovered from CeO$_2$//H-ZSM-5 after catalysis

The SCR reaction proceeds under conditions where solid state ion exchange of

sample	protons 10^{20}/ g	extra-framework aluminium 10^{20}/ g
CeO$_2$/H-ZSM-5 before reaction[1]	2.80	0.35
CeO$_2$/H-ZSM-5 after reaction	1.65	0.36

Table 1: Proton and extra-framework aluminium concentration before and after reaction as determined by ^{1}H and ^{27}Al-NMR.

[1]sample dehydrated at 673 K; proton concentration of H-ZSM-5 - $3.60 \cdot 10^{20}$ / g

cerium into H-ZSM-5 can be assumed to proceed. In view of the difficulties to detect intra-zeolite Ce^{n+} coexisting with extra-zeolite CeO$_2$, the expected consumption of protons by this process was studied by ^{1}H-NMR[a] combined with ^{27}Al-NMR [7]. A significant loss of protons was found by comparing samples before and after reaction, while ^{27}Al-NMR indicated only minor dealumination (cf. Table 1). This implies that solid-state ion-exchange proceeds to a certain extent during the reaction without, however, consuming all acidic protons of the sample. The role of intra-zeolite Ce^{n+} was further studied by preparing CeO$_2$ - Ce/H-ZSM-5 systems (employing Ce/H-ZSM-5 prepared by ion exchange) both as mechanical mixtures and as composites. In both cases, the system containing exchanged Ce in the zeolite was inferior to that prepared with the pure H-form. Moreover, we prepared a composite by precipitating the CeO$_2$ onto Na-ZSM-5 by aqueous NH$_3$, which most likely leads to some exchange of Na$^+$ by NH$_4^+$. Such composite is active in the SCR giving peak NO conversions of about 50 % under the standard reaction conditions. However when the sample was calcined at 873 K to simulate the reaction, and the residual acidity (introduced by the use of aqueous NH$_3$, *vide supra*) was poisoned by re-exchange with NaNO$_3$ solution, the sample became completely inactive. It was made sure by blank experiments that Ce on ion-exchange positions is not re-exchangeable by Na under the conditions applied.

In the SCR of NO by methane, the oxidation of NO to NO$_2$ is often considered a step of the catalytic reaction cycle [6, 8, 9]. This was also confirmed for our system by experiments

---

[a] cooperation by Dr. Brunner (Leipzig University) gratefully acknowledged

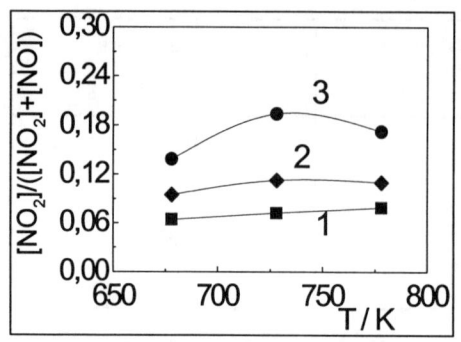
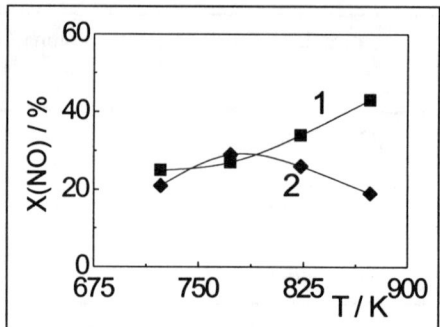

FIG 5. NO oxidation activity *vs.* temperature of 1 - H-ZSM-5 (Si/Al ≈ 14; Süd-Chemie, Germany), 2 - H-ZSM-5 (Si/Al ≈ 14, Alsi Penta, Germany), 3 - CeO_2

FIG 6. NO conversion *vs.* reaction temperature, with 1 - CeO_2//H-ZSM-5 (Si/Al ≈ 14, Süd-Chemie, Germany), 2 - CeO_2//H-ZSM-5 (Si/Al ≈ 14, Alsi Penta, Germany)

with NO replaced by NO_2 (in presence and absence of oxygen) [7]. To assign the NO oxidation activity to one of the composite components, this reaction was studied with pure CeO_2 and H-ZSM-5 of different origin. Figure 5 shows the results with CeO_2 and two H-ZSM-5 samples of Si/Al≈14. With CeO_2, equilibrium was achieved already at 673 K ($\gamma = 0.2$).

With H-ZSM-5, the NO conversion was far from equilibrium even at 773 K ($\gamma = 0.05$). Figure 6 shows the SCR activity obtained with the physical mixtures of the same H-ZSM-5 materials with CeO_2. It can be seen that the zeolite with the better NO oxidation activity is an inferior component for the composite SCR catalyst. This conclusion is supported by similar comparison of the zeolite NO oxidation activity and the SCR activity of the corresponding physical mixture with CeO_2 for other zeolites as exemplified in Table 2.

zeolite[1]	γ at 773 K	$X_{NO, max}$ (in physical mixture with CeO_2 at 873 K) / %
H-BEA (1)	0.15	2
H-MOR (2)	0.05	25
H-ZSM-5 (3)	0.12	5
Na/H-Y (4)	0.05	5

[1] sources (1) - Süd-Chemie, (2) and (4) - Chemiewerk Bad Köstritz, (3) - Elf Acquitaine

Table 2: Comparison of NO oxidation activity of several zeolites and SCR activity of their physical mixtures with CeO_2.

DISCUSSION

Composites of CeO_2 and H-ZSM-5 are a rare example of SCR catalysts performing better with the reductant methane than with propene. They are different from Ce-exchanged ZSM-5, which is a good catalyst with propene, but poor with methane [4-6]. In addition, SCR with propene over Ce-ZSM-5 requires the removal of acidic protons by alkali while the zeolite acidity of CeO_2/H-ZSM-5 is essential for the SCR-methane.

CeO_2/H-ZSM-5 composites are bifunctional systems: They exhibit a higher activity as compared to the individual components even when applied as mechanical mixtures. The synergy disappears as the mixture is separated. This result implies that the cerium detected in the zeolite component after the SCR run (probably introduced by solid-state ion exchange or by non-ideal sieving) is not responsible for the catalytic effect by its own.

Our NMR work shows that during the catalytic reaction with composites some Ce migrates into the zeolite by solid-state ion exchange. This intra-zeolite Ce is, however, not of benefit for the SCR with methane, even in the presence of extra-zeolite CeO_2, since CeO_2 - Ce/H-ZSM-5 systems (composites, mixtures) performed inferior to the corresponding CeO_2 - H-ZSM-5 system. This is a clear difference to the behaviour of mechanical mixtures of CeO_2/Ce-ZSM-5 in the SCR with propene where Yokoyama and Misono [10,11] reported a beneficial effect of CeO_2 admixture. The contradiction can be only explained by assuming that different mechanisms are operative with propene and methane.

In the CeO_2/H-ZSM-5 composite, catalytic functions are distributed between CeO_2 and a site in the zeolite. The experiments cited above exclude Ce^{n+} on exchange positions as the intrazeolite cooperative site, and the complete disappearance of any activity upon poisoning of protons with Na^+ shows that the zeolite site engaged in the SCR reaction is, indeed, the acidic proton or a site being formed from it during the reaction.

The distribution of the catalytic functions between CeO_2 and H-ZSM-5 is not known so far except for the NO oxidation. Our experiments concerning this step (Fig. 5, 6) show that NO_2 is formed mostly on the cerium oxide. CeO_2 is much more active than the zeolites, and there is no correlation between the NO oxidation activity of the zeolites and their performance as a composite component applied for the SCR reaction (Fig. 5, 6, Table 2).

The NO_2 might be considered to mediate the bifunctional interaction via transport through the gas phase as often assumed in the literature [9, 10]. We have examined this hypothesis for our system by arranging the components in layers (CeO_2 in front of, in between, and after the zeolite bed). The results, which will be reported in more detail in a forthcoming paper, show that such an arrangement gives no increase of the SCR activity beyond the level of H-ZSM-5 but enhances only the total oxidation. Hence, the bifunctional interaction operates only over a very short distance and is mediated by a species different from NO_2.

CONCLUSIONS

Composites prepared by precipitating CeO_2 onto NH_4-ZSM-5 exhibit promising activities in the SCR-methane. Their mode of action is different from that of Ce-exchanged ZSM-5 known to catalyze the SCR of NO with propene. The composites are bifunctional catalysts, they may be applied as mechanical mixtures. The catalytic functions are distributed between extra-zeolite CeO_2 and acidic zeolite protons. NO_2 is mainly formed over CeO_2, but the synergy between the cooperating sites is not mediated by gas-phase NO_2 transport. The composites offer high flexibility for further modification.

REFERENCES

[1] M. Shelef, Chem. Rev. **95**, 209 (1995).
[2] J. N. Armor, Catal. Today **26**, 147 (1995).
[3] R. Burch, P. J. Millington, Catal. Today **26**, 185 (1995).
[4] M. Misono, K. Kondo, Chem. Lett. 1001 (1991).
[5] Y. Nishizaka, M. Misono, Chem. Lett. 1296 (1993).
[6] C. Yokoyama, M. Misono, Catal. Today **22**, 59 (1994).
[7] T. Liese, W. Grünert, unpublished results
[8] E. Kikuchi, K. Yogo, Catal. Today **22**, 73 (1994)
[9] Z. Li, M. Flytzani-Stephanopoulos, Appl. Catal. A **165**, 15 (1997).
[10] C. Yokoyama, M. Misono, Catal. Lett. **29**, 1 (1994).
[11] M. Misono, Y. Hirao, C. Yokoyama, Catal. Today **38**, 157 (1997).

We acknowledge gratefully financial support by the Deutsche Forschungsgemeinschaft.

IN-SITU SYNTHESIS OF ZEOLITES ON CORDIERITE AND THEIR CATALYTIC BEHAVIOR IN DECOMPOSITION OF NO

N. J. GUAN*, X. L. SHAN*, K. ZHANG[§], D. S. WANG*, S.H. XIANG*

*ICM, Chem. Inst., Nankai University, Tianjin, P.R.China; guanj@public1.tpt.tj.cn
[§]State key lab. of C_1 Chem., Tech. of Tianjin University, Tianjin, P.R.China;

ABSTRACT

ZSM-5, Beta and metallosilicate with ZSM-11 structure containing cobalt have been synthesized on a honeycomb ceramic support (cordierite, $2MgO \cdot 2Al_2O_3 \cdot 5SiO_2$) by hydrothermal crystallization under static conditions. XRD and IR characterized the monolithic samples for their crystalline nature and framework vibration. These new synthesized materials were proved to be zeolites crystallized directly on the support. Furthermore, they have all shown catalytic activity in the reaction of NO decomposition. This offers a potential way to produce exhaust gas converters.

INTRODUCTION

In the last few years, researchers worldwide are aggressively searching for a catalyst, which will reduce NO_X emissions in oxygen-rich exhaust. Many of the research papers reported that certain zeolite catalysts are able to reduce or decompose NO_X in the presence of oxygen [1-3]. Catalytic removal of NO by decomposition is simpler than by chemical reduction, because no reducing agent is required and the reaction is independent of exhaust gas composition. Iwamoto et al's breakthrough discovery in the mid 1980's of a Cu/ZSM-5 catalyst for the direct decomposition of NO_X into N_2 and O_2 has resulted in a storm of research work focused on characterizing and understanding the performance of Cu/ZSM-5 [4-5]. But there is a significant limitation for commercial use due to its poor durability in the real exhaust conditions. In order to improve the activity and durability of this catalyst, we have recently developed a new method of in-situ crystallization of zeolites on honeycomb catalyst support. In recent years, zeolite/ceramic composite membranes become more important because of zeolites have uniform pore size, high porosity, huge specific surface area and selective adsorption and catalytic properties [6-7]. The honeycomb structure cordierite is a traditional automotive catalyst substrate ($2MgO \cdot 2Al_2O_3 \cdot 5SiO_2$). And the zeolites are usually washcoated on the substrate to provide enough specific surface to carry metallic active component. In our work, zeolites/cordierite monolithic catalyst for NO decomposition was prepared by in situ synthesis. This technology is supposed to superior to the typical "washcoat" method for the well-distributed zeolite layer and possible durability resulted from the interaction between zeolite and support. Otherwise, the metallosilicate molecular sieve in-situ synthesized on the support has developed the traditional impregnation or ion-exchange method for getting metal-modified catalysts.

EXPERIMENTAL

Catalyst preparation

A whole square-celled extruded honeycomb substrate (cordierite) with 400cells/inch^2 was broken into small pieces(2cm×1cm×0.5cm). The hydrothermal synthesis of zeolite was carried out in 100ml stainless-steel autoclaves with Teflon sleeves. The cordierite pieces were placed vertically in Teflon holder and were immersed in different mixtures according to the zeolite type wanted. (1) To get ZSM-5, the mixture is composed of tetrapropyl ammonium bromide (TPABr), Na_2SiO_3, $Al_2(SO_4)_3$, H_2SO_4 and distilled water. The molar composition of the gel was: $6Na_2O$: Al_2O_3: $44SiO_2$: $1840H_2O$: $5TPABr$. The crystallization was carried out at 443K without agitation for 24h. (2) To get zeolite Beta, tetraethyl ammonium hydroxide($(TEA)_2O$), silica sol, sodium aluminate and distilled water were added. The molar composition of the gel was: $3.6Na_2O$: Al_2O_3: $60SiO_2$: $200H_2O$: $6(TEA)_2O$. The crystallization was carried out at 423K without agitation for 60h. (3)To synthesize Co-ZSM-11, $Co(acac)_2(H_2O)_2$, tetraethyl orthosilicate(TEOS), tetrabutyl ammonium hydroxide (TBAOH) and distilled water were used. The molar composition of the gel was: Si/Co=40, TBAOH/Si=0.25, H_2O/Si=25. Consequently, the homogeneous gel was heated to 333K and stirred continuously for 2h. The crystallization was carried out at 448K without agitation for 50h. After crystallization, the autoclaves were taken out and quenched with cold water. And the substrates were washed thoroughly by ultrasonic wave generator. The precursors of zeolites/cordierite were dried at 373K for 12h and calcined at 773K for 3h to remove the templates and converted to the H^+ form zeolites by exchanging with 0.5M HCl. Cu-ZSM-5 was prepared by impregnation of zeolites/cordierite with copper nitrate. Then the catalysts were washed, dried, and calcined at 773K for 4h.

Catalyst characterization

Crystallinity of the "as-synthesized materials" were estimated on XRD patterns, recorded on a D/max-rA diffractometer using CuKα radiation. Framework i.r. spectra were recorded on a Perkin-Elmer 684 spectrometer and the samples were ground with KBr.

Catalyst evaluations

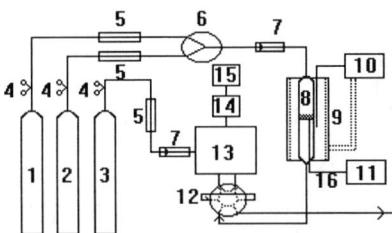

Fig.1 Apparatus used for catalytic test

1.Cylinder gas of Ar(99.9%); 2.Cylinder gas of 3%NO+97%He; 3.Cylinder gas of He(99.9%); 4.Compression release valve; 5.Dryer; 6.3-way valve; 7.Flowrator; 8.Reactor; 9.Heating muff; 10.YCC-126 temperature controller; 11.SK-1410A temperature indicator; 12.6-way valve; 13. SP-502 Gas chromatography; 14.TCD detector; 15.C-R3A Shimadzu data processor; 16.Electricthermo-couple.

Figure1 shows the apparatus used for catalytic test. 0.1~0.3g of catalyst(30-40 mesh, in case of zeolite/cordierite, cordierite included) was used in a fixed-bed flow reactor. The reaction temperature was varied in 523~773K, and the space velocity(GHSV) was varied in 5000~15000h^{-1}.

RESULTS AND DISCUSSION

The X-ray diffraction profiles were given in *Figure 2*. Compared with (A), the new peaks at 2θ=23~24.5° observed in (B), were in good agreement with typical XRD-pattern of ZSM-5[8]. A new peak at 2θ=22° observed in (C), was in agreement with typical XRD-pattern of Beta[8]. This indicated that ZSM-5 and Beta have successfully in-situ synthesized on the cordierite. A new peak at 15.6° appeared in (B) and (C) at the same time. It was supposed that this peak came from the change of the surface property of cordierite in presence of strong alkaline. Two new peaks at 2θ=8°and 23° observed in (D), were partly in agreement with typical XRD-pattern of ZSM-11[8]. This result couldn't fully prove that ZSM-11 has synthesized on the cordierite, so an IR study of

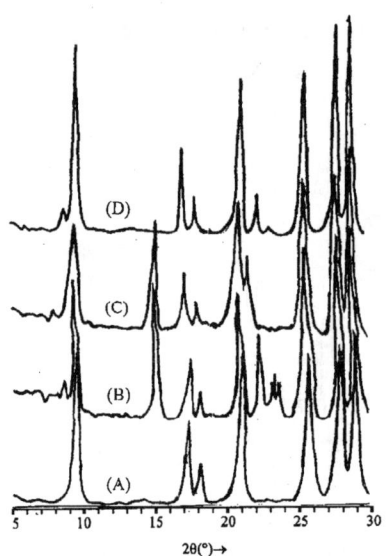

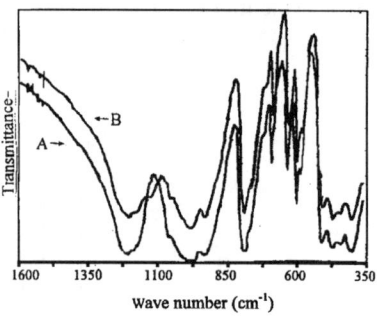

Fig.3 IR spectra of (A) cordierite; (B) as-synthesized sample Co-ZSM-11/cordierite

Fig.2 X-ray diffraction patterns of (A)cordierite; (B)as-synthesized sample ZSM-5/cordierite; (C)as-synthesized sample β/cordierite; (D)as-synthesized sample Co-ZSM-11/cordierite

the framework absorption of as-synthesized Co-ZSM-11/cordierite samples shown in *Figure 3* has given more evidence. Compared with the i. r. spectra of cordierite, new absorption near 1100cm^{-1} and 560cm^{-1} were present in the spectra of

Co-ZSM-11/cordierite. According to earlier investigators[9], typically framework absorption of ZSM-11 were near 1230cm^{-1}, 1100cm^{-1}, 560cm^{-1}, and 455cm^{-1}, and the typical absorption of non-crystalline SiO_2 were near 1100cm^{-1} and 475cm^{-1}. From the *Figure 3*, there was no absorption of 475cm^{-1}, so the non-crystalline SiO_2 was not present, and the absorption of 1100cm^{-1} was belonging to ZSM-11. The absorption of 560cm^{-1} was caused by double five-member ring blocks. Furthermore, the absorption of cobalt oxides was not observed, this indicated that cobaltic ions were highly dispersed. These results provide us the evidence in determining the ZSM-11 in-situ synthesized on the cordierite.

Figure 4 showed the temperature influenced on the activity of NO decomposition. In this case, Co-Beta/cordierite, Cu-ZSM-5/cordierite, Co-ZSM-11/cordierite were used with a space velocity of 10,000 h^{-1} and the temperature was varied from 523~773K. From the results, Co-Beta/cordierite exhibited the highest NO decomposition efficiency. Besides, with increase in temperature from 523~773K, the conversion of NO to N_2 on Co-Beta/cordierite and Co-ZSM-11/cordierite were not obviously increased, while the activity of Cu-ZSM-5/cordierite increased with the increase of temperature. The dependencies of the catalytic activities of Co-Beta/cordierite, Cu-ZSM-5/cordierite, Co-ZSM-11/cordierite on GHSV were shown in *Figure 5*. The reaction temperature was at 623K, and the GHSV were changed from 5,000~15,000h^{-1}. The results indicated that with increase in GHSV, the activity of all catalysts decreased rapidly. At GHSV<900h^{-1}, Co-Beta/cordierite showed the highest activity, but when GHSV>900h^{-1}, Cu-ZSM-5/cordierite was the most active one. It is easy to understand that the conversion of NO decreased in such a small amount of zeolite charge and so large GHSV(short contact time).

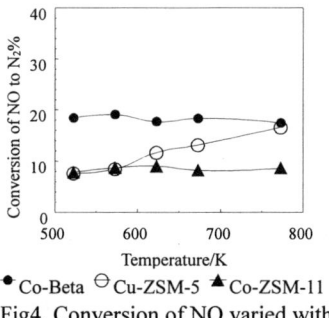

Fig4. Conversion of NO varied with reaction temperature

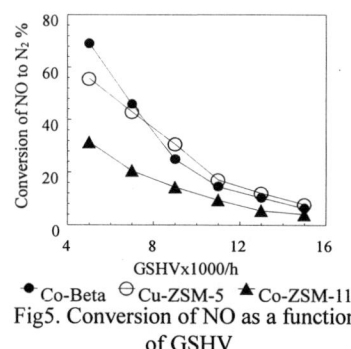

Fig5. Conversion of NO as a function of GSHV

CONCLUSION

ZSM-5, Beta and Co-ZSM-11 have been successfully in-situ synthesized on the cordierite. These materials have shown activities in the reaction of NO decomposition, and the further research work is being done.

REFERENCE

1. M. D. Amiridis, T. J. Zhang and R. J.Farrauto, Appl.Catal.(B), **10**, 25 (1996).
2. Masakazu Iwamoto, Catal. Today, **29**, 29 (1996).
3. Takeshi Tabata, Mikako Kokitsu, Hhirofumi Ohhtsuka, et.al, Catal. Today, **22**, 147 (1994).
4. Yun-feng Chang and Jon G. McCarty, J. Catal., **165**, 1 (1997).
5. Yuejin Li and John N.Armor, Appl. Catal. **76**, L1-L8 (1991).
6. T. Sano, Y. Kiyozumi, M. Kawamura, F. Mizukami et al. Zeolites **11**, 842 (1991).
7. Eduard R. Geus, Marcel J. den Exter and Herman van Bekkum J. Chem. Soc. Faraday trans **88**(20), 3101 (1992).
8. H.van Koningsveld, H.van Bekkum and J.C.Jansen, Acta Cryst., B**43**, 127(1987).
9. Gisele Coudurier, Claude Nacceache and Jacques C. Vedrine, J. Chem. Soc., Chem. Commun., 1434 mnjhuy76(1982).

PREPARATION, CHARACTERIZATION AND DeNO$_x$ ACTIVITY OF (Pt-Co)ZSM-5 AND (Pt-Cu)ZSM-5 ZEOLITE CATALYSTS

A. TAMÁSI[1], I. KIRICSI[1], Z. KÓNYA[1], Z. SCHAY[2], J. HALÁSZ[1] and L. GUCZI[2]*

[1]Applied Chemistry Department, József Attila University, Rerrich B. tér 1, H-6720 Szeged, Hungary; kiricsi@chem.u-szeged.hu
[2]Department of Surface Chemistry and Catalysis, Institute of Isotope and Surface Chemistry. Chemical Research Center, Hungarian Academy of Sciences, P. O. Box 77, H-1525, Budapest, Hungary; guczi@alpha0.iki.kfki.hu

ABSTRACT

(Pt-Co)ZSM-5 and (Pt-Cu)ZSM-5 bimetallic catalysts were prepared by ion exchange of Co^{2+} and Pt^{2+} as well as Cu^{2+} and Pt^{2+} ions into NaZSM-5 zeolite. Reduction of the samples was performed either by a stream of H$_2$ or by NaBH$_4$ in solution. The samples reduced by H$_2$ exhibited Brønsted acidity indicated by its high activity in 1-butene isomerization. On the other hand, the samples reduced in NaBH$_4$ solution possessed negligible Brønsted acidity with no isomerization activity. It was unambiguously proven that in the NO decomposition to form N$_2$ and O$_2$ both the metal function and the Brønsted acidity play a decisive role for the catalytic behavior of the sample reduced in H$_2$.

INTRODUCTION

Investigations on zeolite catalysts containing two transition or noble metal ions or bimetallic phases have widened in the last decade. These materials proved to be active and selective catalysts in various transformations such as methane activation [1], CO hydrogenation [2] and NO decomposition or reduction [3]. Among the zeolite supported bimetallic materials, the platinum-cobalt and platinum-copper have not been studied systematically yet. In the decomposition on NO$_x$ it has been suggested, using indirect evidence [4, 5], that in addition to the metal redox system, acid sites also responsible for the reaction producing nitrogen and oxygen. In the underlying work we report the synthesis of these systems and their characterization by means of various instrumental analytical methods. The catalytic activity of the samples has been compared in 1-butene isomerization and in NO decomposition.

---

* Corresponding author

EXPERIMENTAL

Preparation of catalysts

Catalysts were prepared by ion exchange of NaZSM-5 zeolite (Si/Al = 40) in a solution of 0.1 M $CoCl_2$ or $CuCl_2$ with stirring for 24 h at 333 K to prepare the corresponding CoZSM-5 and CuZSM-5 samples. The samples were washed chlorine-free and dried at 373 K. Having determined the degree of ion exchange, an equivalent amount of $[Pt(NH_3)]_4Cl_2$ complex was dissolved in water and the solution of the complex was added drop-wise to the suspension containing either CoZSM-5, or CuZSM-5 under constant stirring. After 4 h the samples were filtered, washed and dried at 373 K. The metal loading of the samples was found to be 0.1 mmol/g for Pt and Cu and 0.067 mmol/g for Pt and Co. One portion of the samples was suspended in water and the metal ions were reduced by 0.1 M $NaBH_4$ added drop-wise, stirred for 8 h, filtered and dried. The second portion of the samples was degassed in N_2 and reduced in a stream of H_2 at 573 K for 15 h.

Characterization of samples

After dissolving the Co^{2+} or Cu^{2+} ions in 1:1 HCl the degrees of ion exchange were determined by complexometry. The analysis of the solution after ion exchange gave an independent estimation for the ion loading. ^{23}Na-MAS-NMR spectroscopy (BRUKER 400-MHz equipment) was applied to determine the concentration and the coordination of sodium ions at 105.8 MHz and a 1.0 μs ($\Theta=\pi/12$) pulse was used with a repetition time of 0.1 s. Crystallinity of the samples was checked by means of DRON-3 X-ray diffractometer. Thermal behavior of the samples was investigated by means of a Derivatograph-Q equipment. BET area of the specimens was measured a volumetric adsorption apparatus. XPS spectra were measured in a Kratos XSAM-800 cpi ESCA equipment with an Al K_α X-ray source applied for excitation.

Infrared spectroscopy (KBr technique) was used to check the generation of framework vacancies upon various treatments. Self-supporting wafer technique was employed for acidity measurements using pyridine as a probe molecule and to monitor the NO decomposition by i.r. technique. After degassing the wafer at 673 K, 1.33 kPa pyridine was adsorbed at 473 K for 1 h followed by evacuation at the same temperature for 1 h. After cooling the sample to room temperature spectra of the adsorbed pyridine were taken. For NO adsorption and decomposition measurements 1.33 kPa of NO was adsorbed on the pretreated wafer and the products of decomposition was followed at room temperature.

NO decomposition was studied in a plug flow apparatus using mass spectrometer to analyze the reaction products. Catalytic activity of the samples created by acid sites was characterized by double bond isomerization of 1-butene carried out in a circulation batch reactor. Generally 50 mg of catalyst was placed into the reactor; the reaction was performed at 423 K with GC analysis.

RESULTS AND DISCUSSION

Characteristics of the samples prepared are summarized in Table 1.

Table 1.

	Reduced with	Weight loss at 1200 K, (m%)	BET area m^2/g	Brønsted acidity A_{1545}/mg	Lewis acidity A_{1450}/mg
(Pt-Cu)ZSM-5	H_2	5.38	328	0.18	0.16
	$NaBH_4$	5.19	313	0.04	0.58
(Pt-Co)ZSM-5	H_2	8.87	328	0.16	0.60
	$NaBH_4$	8.03	292	0.07	0.53

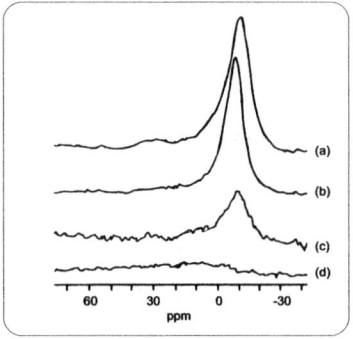

Figure 1. ^{23}Na NMR spectra of bimetallic samples reduced by H_2 and by $NaBH_4$

(a) (Pt-Co)ZSM-5/ $NaBH_4$
(b) (Pt-Cu)ZSM-5/ $NaBH_4$
(c) (Pt-Co)ZSM-5/H_2
(d) (Pt-Cu)ZSM-5/H_2

In Figure 1 the ^{23}Na NMR spectra of the (Pt-Co)- and (Pt-Cu)ZSM-5 samples reduced by H_2 or $NaBH_4$ are presented. The difference in the sodium content (compare the intensities by the resonance at -5 ppm) among the catalysts reduced by H_2 or $NaBH_4$ is obvious. In the latter case the loss of sodium is negligible since the Pt^{2+} ions are replaced by Na^+-ions, i.e. no Brønsted acid sites are generated after treatment with $NaBH_4$. This is confirmed by pyridine adsorption because in the samples reduced in H_2 much more Brønsted acidic sites are present.

IR spectra taken on the range of framework vibration of zeolites (400-1400 cm^{-1}) showed no new bands around 930 cm^{-1} which had been assigned to framework defects.

X-ray photoelectron spectroscopy was employed to measure the surface concentration of the various components using the intensities of Cu 2p, Co 2p, Si 2p and Pt 4f peaks. The surface concentrations of Pt, Cu and Co were determined and the Cu/Si, Co/Si and Pt/Si surface atomic ratios indicated the presence of these components in the samples after reduction both by hydrogen and NaBH$_4$. The surface concentrations are presented in Table 2 and the values are compared with the nominal compositions.

Pt 4f$_{7/2}$, Co 2p$_{3/2}$ and Cu 2p$_{3/2}$ binding energies were used to determine the oxidation state of the various metals after reduction. The Pt 4f$_{7/2}$ binding energy is 314.2 and 314.1 eV after reduction in H$_2$ and NaBH$_4$, respectively, while the respective Cu 2p$_{3/2}$ binding energies are 932.3 and 931.9 eV. These values are indicative that regardless of the mode of reduction, both metals are in zero oxidation state. On the contrary, in the (Pt-Co)ZSM-5 only Pt is reduced by both H$_2$ and NaBH$_4$ (B.E. are 314.8 and 315.1 eV, respectively), whereas cobalt remains in cobalt oxide state (B.E. are 781.3 and 782.4 eV, respectively). It is, therefore, suggested that bimetallic catalysts can be found only in the (Pt-Cu)ZSM-5 sample, whereas only Pt is reduced in the (Pt-Co)ZSM-5 sample. Generally, the metals are enriched on the surface but in the case of Cu/Pt the ratio does not change.

Table 2. Surface concentration of Co, Cu and Pt calculated from the Co 2p, Si 2p, Cu 2p and Pt 4f ratios measured by XPS

Samples reduced by	Co/Si bulk	Co/Si surf.	Cu/Si bulk	Cu/Si surf.	Pt/Si bulk	Pt/Si surf.
(Pt-Co)ZSM-5	0.008				0.008	
Hydrogen		0.025				0.015
NaBH$_4$		0.047				0.008
(Pt-Cu)ZSM-5			0.012		0.012	
Hydrogen				0.012		0.012
NaBH$_4$				0.042		0.037

Isomerization of 1-butene to cis/trans 2-butenes takes place over acid sites. As can be seen in Figure 2 on the (Pt-Cu)ZSM-5 samples reduced by H$_2$ (curve (c)) appeared to be very active in the double bond isomerization of 1-butene (see the initial slope of the 1-butene consumption), while that reduced by NaBH$_4$ there is proved to be much less active with a reminiscent Brønsted sites characteristic of the zeolite itself (curve (a)). When the sample reduced by NaBH$_4$ was treated with H$_2$ at 573 K no further reduction of Pt^{2+} or Cu^{2+} took place since the rate of 1-butene transformation did not change significantly (see curve (b)). The results are similar to (Pt-Co)ZSM-5 samples.

Decomposition of the 2 vol.% NO in argon at a flow rate of 20 cm^3 min^{-1} over 100 mg of catalyst samples was carried out at different temperatures. Since no N_2O and NO_2 were observed, in Figure 3. the logarithm of the QMS signal of the oxygen formed being indicative of the NO into N_2 and O_2 conversion, is plotted versus temperature. A continuous increase in the activity of the samples reduced in hydrogen was established with raising reaction temperature in contrast to what was observed on CuZSM-5 sample [6]. In the case of the samples reduced with NaBH$_4$ no activity in NO decomposition was observed.

Both platinum and the bimetallic samples maintain their activity at high temperature (the maximum NO conversion was about 50 %). The most active sample is PtZSM-5 followed by (Pt-Cu)ZSM-5. Since here only metallic components are present, the redox mechanism is not operative [7].

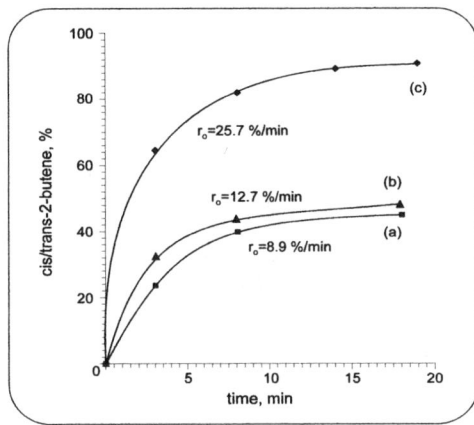

Figure 2. Kinetic curves of 1-butene transformation over (Pt-Cu)ZSM-5 reduced by hydrogen (c), NaBH$_4$, (a) and by NaBH$_4$ followed by hydrogen treatment (b).

However, the activity of the samples depends not only upon metal phase, but the presence of acid sites is prerequisite. It was proven not only by the inactivity of NaBH$_4$ treated samples, but by the treatment of the catalyst reduced by H$_2$ with NaCl. Since the Brønsted sites are partially removed after this treatment shown by isomerization of 1-butene, the decreased activity in NO decomposition points to the cooperative function of the acid and metallic sites. Insertion of copper into the platinum containing zeolite sample decreases the number of platinum sites exposed to the surface due to the copper enrichment on the surface of the bimetallic particles, thus their activity is lower than the one measured for PtZSM-5 sample. Change in the platinum sites exposed to the surface is supported by the result of the O$_2$-H$_2$ titration. It is demonstrated that the amount of Pt exposed to the surface decreases in the sequence of PtZSM-5 > (Pt-Co)ZSM-5 > (Pt-Cu)ZSM-5 the values being 14, 9, 3 µmole/g_{sample}, respectively. In the used samples the decrease in surface platinum content from PtZSM-5 to (Pt-Cu)ZSM-5 is from a value of 5.3 µmol/g_{sample} to about 1.3 µmol/g_{sample}, respectively. Additional treatment with NaCl does not diminish further the metal sites, but presumably decreases the number of acid sites. This is in line with the result observed on bifunctional PdZSM-5 catalyst for NO decomposition in the presence of hydrocarbons [8].

On the other hand, the (Pt-Co)ZSM-5 behaves in similar manner as PtZSM-5. Although we may not exclude formation of the Pt-Co bimetallic particles as was suggested in other NaY zeolite system [9] and the usual platinum enrichment on the platinum-cobalt alloy [10], but here it is most probable that most of the cobalt is not reduced indicated by XPS measurements. Consequently, only the Pt sites active both in PtZSM-5 and (Pt-Co)ZSM-5 samples and the similarity in the deNOx activity is not surprising. Similar finding was observed on the (Pd-Fe)ZSM-5 catalysts [11] because the mixing of the two different metals are more difficult in ZSM-5 zeolite than in NaY supercage.

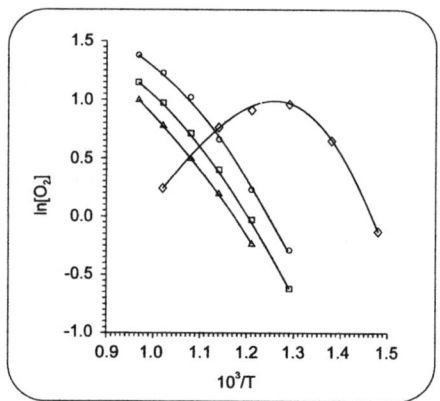

Figure 3. Arrhenius plot of NO decomposition over PtZSM-5 (o), (Pt-Cu)ZSM-5 (□), (Pt-Cu)ZSM-5 + NaCl treatment (Δ), CuZSM-5 (◊) from ref. 4.

The fact that no activity in the NO decomposition is developed after reduction by NaBH$_4$ provides further evidence that the metal and acid sites are together responsible for the catalytic activity. The continuous increase in activity at high temperature is the key issue on the Pt-containing zeolite based deNO$_x$ catalysts since on CuZSM-5 at similar temperature range the deNO$_x$ activity passes through a maximum [4]. Here we suggested that the decrease in activity at higher temperature was due to the decomposition of the NO$_2$ containing surface intermediates which had formed on acid sites. However, the present results point out that the role of the Brønsted sites are much more complex than it has been expected [12].

In order to confirm this catalytic feature, in situ IR experiments were carried out. As Figure 4 shows the intensities of the i.r. bands at 2125 and 1800 cm^{-1} decrease while those at 1620 and 1500 cm^{-1} increase with contact time at room temperature. No bands at 2240 and 2290 cm^{-1} were observed which could be assigned to the presence of adsorbed N_2O and indeed no N_2O was found among the products. From this it follows that the surface transformations are fast hindering the observation this intermediate. Bands at 1500 and 1620 cm^{-1} reveal the formation of NO_2 as surface product. However, in kinetic experiments performed at higher temperatures extensive oxygen formation was measured which proves that NO_2 must be also only an intermediate on this catalyst.

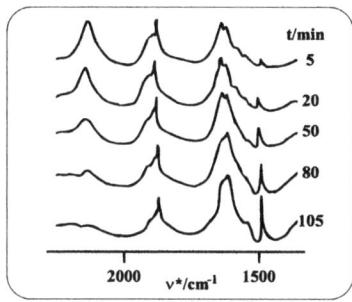

Figure 4. IR spectra of NO adsorbed on (Pt-Cu)ZSM-5 reduced by H_2 at room temperature

The ^{23}Na NMR spectra of the samples reduced in different ways showed only a small amount of sodium in the catalysts reduced with H_2. On these samples the acidic character of the OH groups was evidenced by IR spectroscopy. This finding was supplemented also by transformation of 1-butene.

The usual zeolite based deNO$_x$ catalysts such as CuZSM-5 does not have metallic phase, thus, the reaction mechanism should be different on our samples. The major difference is the declined activity of the CuZSM-5 sample at high temperature and a considerably large gap is developed above 970 K as illustrated in Figure 3.

CONCLUSIONS

Reduction of Pt and Co or Cu ions introduced in ZSM-5 zeolite by H_2 or NaBH$_4$ produced samples of different character. Reduction by H_2 of the samples resulted in Brønsted acidity and acidic activity, whereas the specimens reduced by NaBH$_4$ had no Brønsted acidity.

The concentration of metal components on the outer surface of the samples varied upon different reduction procedure. For (Pt-Cu)ZSM-5 sample the considerable increase of Pt was found after reduction with NaBH$_4$, while this treatment did not affect the state of platinum in the cobalt containing sample. Enhanced surface concentration of cobalt and copper was observed for samples reduced by NaBH$_4$.

Continuous increase of catalytic activity with reaction temperature was found for (Pt-Cu)ZSM-5 reduced by H$_2$ in the decomposition of NO, however the samples reduced with NaBH$_4$ were inactive in this reaction. This feature may be explained by the simultaneous presence of bimetallic component and the Brønsted acid sites.

ACKNOWLEDGEMENTS

The work was supported by the National Research and Science Fund (Grant No. TO-22117) and Hungarian Ministry of Education (Grant No. FKFP 1067).

REFERENCES

[1] L. Guczi, K.V. Sarma and L. Borkó, Catal. Lett. **39**, 43 (1996)

[2] G. Lu, T. Hoffer and L. Guczi, Appl. Catal. **93**, 61 (1992)

[3] L. Gutierres, A. Ribotta and J. Petunchi, Stud. Surf. Sci. Catal., Edited by J. W. Hightower, W. N. Delgass, E. Iglesia and A. T. Bell, (Elsevier Sci. Publ. Co., Amsterdam, 1996) **Vol. 101**, pp. 631,

[4] Z. Schay, H. Knözinger, L. Guczi and G. Pál-Borbély, Appl. Catal. B. in press

[5] J. Sárkány, J. L. d'Itri and W. M. H. Sachtler, Catal. Lett. **16**, 241 (1992)

[6] Z. Schay and L. Guczi, Catal. Today **17**, 175 (1993)

[7] H. Hamada, Y. Kuwahara, Y. Kindaichi and T. Ito, National Meeting Chem. Soc. Japan, p. 344 (1988)

[8] M. Misono, Y. Hirao and C. Yokoyama, Catal. Today **38**, 157 (1997)

[9] G. Lu, T. Hoffer and L. Guczi, Catal. Lett. **14**, 207 (1992)

[10] U. Bardi, B. C. Beard and P. N. Ross, J. Catal. **124**, 22 (1990)

[11] K. Lázár, B. M. Choudary and L. Guczi, Solid State Ionics **32/33**, 1000 (1989)

[12] A. Fritz and V. Pitchon, Appl. Catal. B. **13**, 1 (1997)

HYDROCONVERSION OF HEPTANE AND OCTANE OVER BIFUNCTIONAL ZEOLITES; INFLUENCE OF STRUCTURE AND METAL DISTRIBUTION ON ACTIVITY AND SELECTIVITY

A. JENTYS, A. LUGSTEIN, G. KINGER and H. VINEK

Vienna University of Technology, Institute for Physical Chemistry
Getreidemarkt 9, A-1060 Wien, Austria.
http://www.physchem.tuwien.ac.at/catalysis/

ABSTRACT

Hydroisomerization and hydrocracking of n-heptane and n-octane over Ni-containing MFI, mordenite and beta zeolites were investigated at 533 K and 20 bar total pressure. The activity increases in the sequence NiHBEA ≈ NiHMOR << NiHMFI. The selectivity for isomerization of n-octane was the highest on NiHBEA and the lowest on NiHMFI. These three zeolites have acid sites of equal strength, but a different pore structure and, therefore, we conclude that adsorption and diffusion processes are a main factor determining the activity and selectivity of these catalysts.

INTRODUCTION

Hydroconversion of light naphtha is a commercially important process that leads to an increase in the gasoline pool octane. In this reaction, hydrocarbons are isomerized and/or cracked in the presence of hydrogen over bifunctional catalysts, which combine the acid/base chemistry of zeolites with the hydrogenating/dehydrogenating activity of metals. The activity, selectivity and stability of these bifunctional catalysts can be favorably influenced by the use of a suitable combination between the acidic and the metallic components [1-3]. Typically, noble metals (such as Pt or Pd) or transition metals (such as Ni, Co, Mo and W, often applied in their sulfided form) supported on zeolites, amorphous silica and alumina are used [4-8].

The purpose of the work presented was to compare the activities and selectivities for n-heptane and n-octane hydroconversion on various Ni-containing zeolites and to evaluate the role of the pore geometry upon the activity and the selectivity of bifunctional catalysts.

EXPERIMENTAL

The starting materials were H-ZSM5 (MFI), H-Mordenite (MOR) and H-Beta (BEA) zeolites with Si/Al ratios of 26, 7.2 and 35, respectively. Ni was introduced into H-ZSM5 by ion exchange carried out at a pH of 6, 8 and 10 and into H-Mordenite and H-Beta by ion exchange at pH 8 using a 0.52 molar aqueous solution of $NiCl_2$. The composition of the zeolites was determined by energy dispersive X-ray emission (EDX). The crystallinity and stability of the samples after ion exchange and reduction was verified by XRD and BET measurements. The concentration of acid sites was determined by temperature programmed desorption of ammonia (NH_3-TPD). The degree of Ni exchange and the concentration of Brønsted acid sites after reduction were calculated from the decrease of the SiOHAl stretching vibration band (3610 cm^{-1}) investigated by IR-spectroscopy [9]. The strength and the accessibility of the acid sites were probed by adsorption of benzene and n-octane, followed by IR-spectroscopy.

n-Heptane and n-octane (Fluka 99.9%) conversion using a H_2/n-alkane molar ratio of 71 was carried out in a fixed bed downstream reactor at total pressure of 20 bar and a temperature of 533 K. Before the kinetic experiments, the catalysts were reduced in hydrogen (40 ml/min) at atmospheric pressure for 2 h at 773 K for ZSM5, Beta and at 803 K for Mordenite. After the reduction Ni was found to be in the metallic state [10].

RESULTS AND DISCUSSION

Structural properties of the zeolites

The BET-surface area, the concentration of Brønsted acid sites (BS) after reduction of the catalysts, the metal content, the metal to acid sites ratio (Ni/BS) of the reduced catalysts and the shift of the IR-band at 3610 cm^{-1} ($\Delta\nu_{OH}$), after adsorption of benzene (benzene partial pressure 10^{-3} mbar), are summarized in Table 1.

	BET [m^2/g]	Acid sites (BS) [mmol/g]	Ni [wt%]	Ni/BS	Δv_{OH}* [cm^{-1}]
NiHMFI (pH6)	356	0.29	0.7	0.2	350
NiHMFI (pH8)	348	0.29	10.9	3.4	350
NiHMFI (pH10)	355	0.29	25.3	9.4	350
NiHMOR	422	1.80	8.3	0.5	348
NiHBEA	477	0.57	8.3	1.6	346

Table 1: Physicochemical properties of the catalysts
* measured on H-MFI, H-MOR and H-BEA

Note that segregation during the reduction led to a decrease of metal dispersion with increasing metal content.

After reduction of NiHMFI, practically the same concentration of Brønsted acid sites was present, independently of the pH value at which the ion exchange was carried out. For the sample prepared at pH 6 only protons from the Brønsted acid sites were exchanged against Ni, while at higher pH values Ni(OH)$_2$ was additionally deposited on the external surface. For the NiHMFI sample exchanged at pH 10 a Ni(OH)$_2$ phase could be detected by XRD.

Hoang et al. [11] reported that in Ni exchanged ZSM5 only 75 % of the Ni atoms could be reduced to the metallic state and 25 % remained in the ionic state, which are located in "hidden positions" of ZSM5. We found that 76 % of the initially present Brønsted acid sites were present after the reduction in hydrogen, which is in good agreement with the value given by Hoang et al.

IR-spectroscopy indicated that all Brønsted acid sites on ZSM5 and Beta were accessible for n-heptane and n-octane, whereas only one third of these sites was accessible on Mordenite. The number of Brønsted acid sites, determined from NH$_3$-TPD, increased in the sequence:

ZSM5 < Beta < Mordenite

whereas the strength of the accessible acid sites, determined from Δv_{OH} after adsorption of benzene followed by IR-spectroscopy, was similar (see Table 1).

The space available for the molecules studied, i.e., NH$_3$ (kinetic diameter 2.6 Å), n-octane (kinetic diameter 4.3 Å), 2,5-dimethylhexane (kinetic diameter 5.0 Å) and 2,2,4-trimethylpentane (kinetic diameter 6.3 Å) in ZSM5, Mordenite and Beta is compared in Figure 1 [12]. In perfect agreement with the IR experiments it can be seen, that the whole pore system of Mordenite (including the side pockets) is accessible to ammonia, whereas only a part of it is accessible to bulkier molecules, such as n-octane and its isomers. Molecules that are too bulky to enter the pore system of the zeolites can only react on its external surface.

It can be visualized that ample space is available in the intersection of the two channels in ZSM5 for a skeleton isomerization of C8 molecules, however, the resulting multibranched molecules cannot diffuse out of the pores and consequently will be cracked.

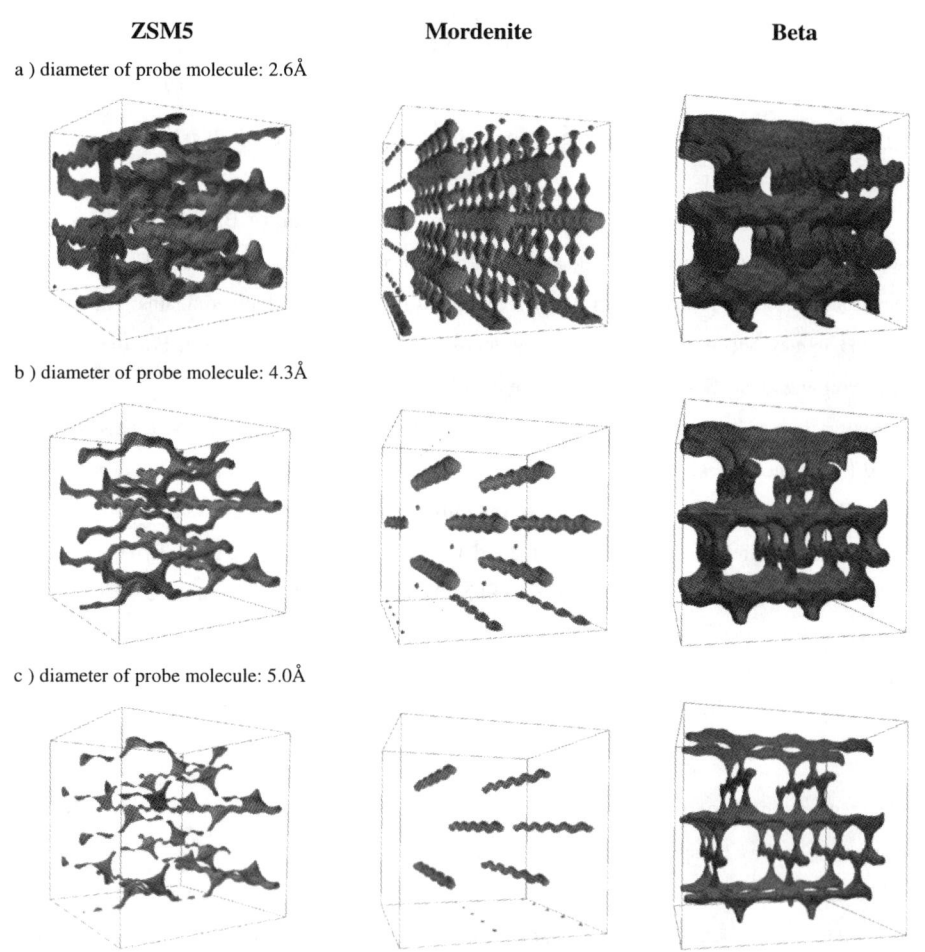

Figure 1 (*continued on next page*): Computer modeling of pore space for different probe molecules

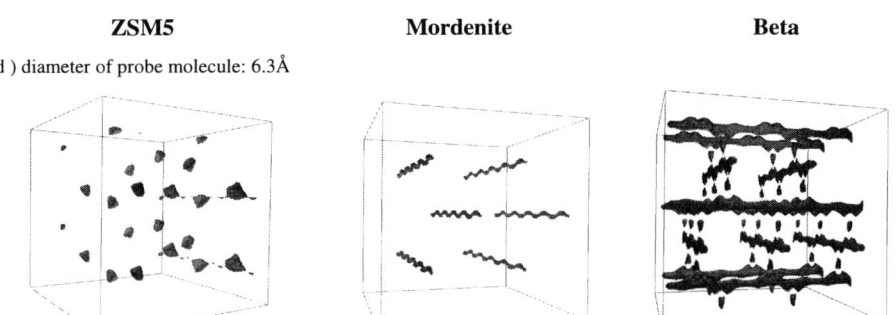

d) diameter of probe molecule: 6.3Å

Figure 1 (cont.): Computer modeling of pore space for different probe molecules

Activities and Selectivities

The activity and selectivity of the catalysts studied for the n-heptane and n-octane conversion at 20 bar and 533 K at a conversion of 10 % are shown in Table 2. For NiHMFI the highest activity and the lowest selectivity to isomerization was observed, while NiHBEA was the most selective catalyst for isomerization in this series of materials.

Catalyst	n-octane		Catalyst	n-heptane	
	Activity [mol·g^{-1}·s^{-1}]	Isomer-selectivity [%]		Activity [mol·g^{-1}·s^{-1}]	Isomer-selectivity [%]
NiHMFI(pH8)	1.3·10^{-5}	10.0	NiHMFI(pH6)	9.1·10^{-7}	5.0
NiHMOR	6.4·10^{-6}	55.1	NiHMFI(pH8)	6.7·10^{-6}	21.5
NiHBEA	5.1·10^{-6}	96.8	NiHMFI(pH10)	6.7·10^{-7}	20.4

Table 2: Catalytic activity of the catalysts

The selectivity for cracking and isomerization of the three types of zeolites investigated is compared in Figure 2. Under the reaction conditions studied the iso/n-C_4 ratio on NiHMFI (pH8) and on NiHMFI (pH10) in the conversion of n-heptane was 73 and 53, respectively. Although the observed selectivity for isomerization was only small on NiHMFI, isomerization products must be formed, which can be deduced from the high iso/n ratio of the cracked products.

Ideal bifunctional behavior in hydroconversion of n-alkanes is achieved, when the acidic and the metallic function of the catalyst are in the right balance. In this case, only products resulting from isomerization and saturated alkanes, resulting from cracking reactions, can be found.

In the series of Ni exchanged MFI zeolites, the NiHMFI (pH 6) sample was the least active. In the product distribution over this catalyst alkanes and alkenes were detected, therefore, the concentration of nickel is not sufficient to establish a balance between metallic and acidic sites. On the NiHMFI (pH 10) sample, a large amount of nickel is located on the external surface, which restrict the accessibility of the active sites and this reduces the activity of this catalyst. The most active catalyst tested was NiHMFI (pH 8) sample, which revealed an ideal bifunctional behavior.

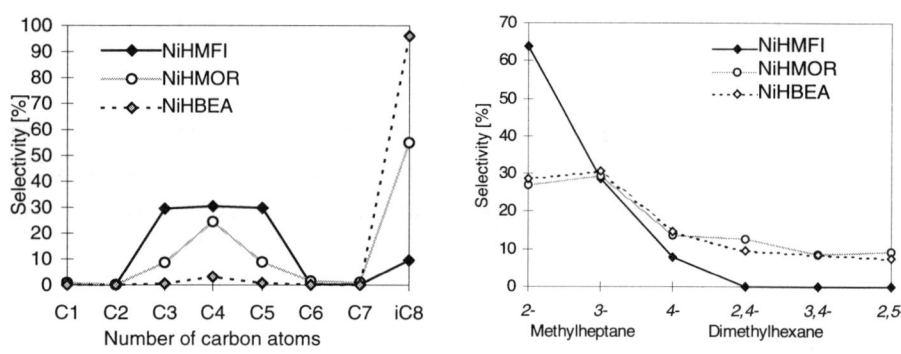

Figure 2: Isomerization selectivity

Mordenite exhibits a higher concentration of acid sites, but the main part of these sites was not accessible for n-heptane and n-octane (sites located in the side pockets of Mordenite [13]). Zeolite Beta has pores with a larger diameter (7.6 x 6.4 Å and 5.5 x 5.5 Å) compared to ZSM5 (5.3 x 5.6 Å and 5.1 x 5.5 Å) and although all acid sites were accessible for the reactants, the activity was smaller on Beta compared with ZSM5. These results can be explained using the dependence of the heat of adsorption on the ratio between the size of the hydrocarbons and the pore dimensions of the zeolite as described by Derouane [14], van Santen [15] and Lercher et al. [16]. The diameter of n-octane and n-heptane in ZSM5 should be close to this optimum ratio, because assuming the same true activation energy for the reaction, the higher heat of adsorption for the reactants should result in a lower apparent activation energy compared to the zeolites with larger pores and, therefore, in a higher reaction rate. Note that for zeolite Beta the pore dimensions and the concentration of accessible acid sites were similar to Mordenite (6.5 x 7.0 Å and 2.6 x 5.7 Å) and, therefore, a similar activity in cracking and isomerization was observed, although the selectivity to i-C8 was smaller on NiHMOR compared with NiHBEA.

CONCLUSIONS

Ni containing ZSM5 zeolites are more active for the hydroconversion of n-alkanes compared to Ni containing Mordenite and Beta, although these catalysts posses a higher concentration of accessible acid sites. We attribute this to the dimensions of the ZSM5 pores, which better fit to the size of n-heptane and n-octane and, therefore, favor their catalytic conversion.

The close distance between acidic and metallic sites on NiHMFI additionally favors a high iso/n-butane ratio and a high concentrations of heptane isomers, while for NiHMOR and NiHBEA catalysts a high selectivity to isomerization was found. In the hydroconversion of n-octane over ZSM5, having a 10-membered ring system, cracking to propane and pentanes was preferred over the formation of butanes (with a low iso/n alkane ratio), whereas on Mordenite and Beta higher concentrations of C4-products than of C3- and C5-products (with a high iso/n alkane ratio) were found.

ACKNOWLEDGMENTS

The work was supported by the "Fonds zur Förderung der Wissenschaftlichen Forschung" under project P11479 CHE. We thank Klaus Kleestorfer for the simulation of the pore accessibility.

REFERENCES

1   M. Guisnet, C. Thomazeau, J.L. Lemberton and S. Mignard, J. Catal. **151,** 102 (1995).
2   M. Guisnet and V. Fouche, Appl. Catal., **71**, 283 (1991)
3   A. Corma, M.I. Vasquez, A. Escardino, Ind. Eng. Chem. Res., **26**, 1495 (1987)
4   Z. Zhan, I. Manninger, Z. Paál and D. Barhomeuf, J. Catal., **147**, 333 (1994)
5   C.W. Ward, Stud. Surf. Sci. Catal., **16**, 587 (1983)
6   P.B. Weisz, Adv. Catal., **13**, 137 (1962)
7   J.A. Martens, M. Tielen and P.A. Jacobs, Catal. Today, **1**, 435 (1987)
8   M. Guisnet and G. Perot, "Zeolites: Science and Technology", Nato Series 80, 397 (1984)
9   A. Jentys, A. Lugstein and H. Vinek, Zeolites, **18**, 391 (1997)
10 A. Jentys, A. Lugstein and H. Vinek, Faraday Trans., **93** (9), 1837 (1997)
11 D. L. Hoang, H. Berndt, H. Miessner, E. Schreiner, H. Lieske, Appl. Catal., **114**, 295 (1994)
12 T.F. Nagy, S.D. Mahanti and J.L. Dye, Zeolites, **19**, 57 (1997)
13 J.A. Lercher, Ch. Gründling and G. Eder-Mirth, Catalysis Today **27,** 353 (1996)
14 E.G. Derouane, J. Catal., **100**, 541 (1986)
15 R.A. van Santen, J. Mol. Catal., **115**, 405 (1997)
16 J.A. Lercher and F. Eder, J. Phys. Chem. B, **101**, 1273 (1997)

COORDINATION CHEMISTRY OF TITANIUM IN TITANOSILICATE MOLECULAR SIEVES STUDIED BY ELECTRON SPIN RESONANCE AND ELECTRON SPIN ECHO MODULATION SPECTROSCOPY

A. M. PRAKASH and L. KEVAN*

Department of Chemistry, University of Houston, Houston, Texas 77204, USA; prakash@bayou.uh.edu; kevan@uh.edu

ABSTRACT

Electron spin resonance (ESR) and electron spin echo modulation (ESEM) spectroscopy are used to study the coordination of titanium in various titanosilicate molecular sieves. The titanosilicate molecular sieves TS-1, TiMCM-41, ETS-4 and ETS-10 have been synthesized hydrothermally. These materials were thermally activated, γ-irradiated at 77 K and then examined by ESR. An axially symmetric ESR signal with g values $g_{\parallel}$ = 1.970 and $g_{\perp}$ = 1.919 is observed for Ti(III) centers in TS-1 and TiMCM-41. On the other hand, ESR signals characterized by reversed g values are observed for Ti(III) centers (g_1 = 1.944, g_2 = 1.916 and g_3 = 1.862 for ETS-10 and $g_{\parallel}$ = 1.923 and $g_{\perp}$ = 1.862 for ETS-4) in ETS molecular sieves. The observed ESR parameters of Ti(III) can be explained on the basis tetrahedral coordination of titanium in TS-1 and TiMCM-41, and octahedral coordination in ETS-10 and ETS-4 materials. Titanium in ETS-10 is in a more distorted octahedral environment than in ETS-4. Tetrahedral titanium in TS-1 readily coordinates with adsorbate molecules as shown by a new ESR signal observed for Ti(III) after adsorbing D_2O, CH_3OD and C_2D_4. This is further confirmed by strong ^2D ESEM modulation observed in TS-1 after adsorption of D_2O, CH_3OD and C_2D_4. In contrast, octahedral titanium in ETS-10 does not coordinate with adsorbate molecules. No significant ^2D ESE modulation is observed in ETS-10 suggesting no direct insertion of ligands into the first coordination sphere of titanium in this material.

INTRODUCTION

Titanosilicate molecular sieves have been shown to be efficient catalysts for a wide range of selective oxidation of alkenes and alkanes [1]. The coordination environment of titanium plays a major role in the catalytic properties of these materials. Molecular sieves in which titanium enters into the framework in both tetrahedral and octahedral coordination have been reported. TS-1, TS-2, Ti-β and TiMCM-41 are examples in which titanium is believed to be in tetrahedral coordination, while ETS-4 and ETS-10 belong to a class in which framework titanium is in six coordination.

Electron spin resonance (ESR) spectroscopy has been widely used to characterize several transition metal incorporated molecular sieves. ESR characterization of Ti compounds, are limited due to the fact that the most common and stable oxidation state of titanium in these materials is tetravalent which is ESR silent. However, trivalent titanium having a $3d^1$ electron configuration has characteristic ESR signals depending on its coordination geometry. ESR studies on the nature of Ti in TS-1, TiMCM-41 and TAPO-5 have been reported recently[2-4]. Convincing evidence for substitution of Ti into the framework of various materials has been provided by both ESR and ESEM spectroscopy. In the present work, we demonstrate that electron spin resonance and electron spin echo modulation spectroscopy can be effectively used to identify the coordination environment of titanium in various titanosilicate molecular sieves.

EXPERIMENTAL

Synthesis. Titanosilicate molecular sieves TS-1, TiMCM-41, ETS-10 and ETS-4 were prepared following standard procedures reported in literature [5,8]. Tetrapropylammonium hydroxide and cetyltrimethylammonium hydroxide/chloride were used as organic templates for TS-1 and TiMCM-41 syntheses, respectively. No organic template was employed in the syntheses of ETS-10 and ETS-4. As-made samples of ETS-10 and ETS-4 were used for ESR and ESEM studies. TS-1 and TiMCM-41 were used in calcined form prepared by heating as-synthesized samples in oxygen at 823 K for 16 h to remove the organic species. The proton form HETS-10 was prepared by exchanging ETS-10 with NH_4Cl followed by heating at 673 K for 16 h.

Sample Treatment and Measurements. Powder X-ray diffraction (XRD) patterns were recorded on a Siemens D5000 X-ray diffractometer using CuKα radiation. Chemical analyses of the samples were carried out by electron microprobe analysis. In order to study titanosilicate molecular sieves by ESR these materials have to be reduced by suitable reduction methods. Methods such as high temperature thermal treatment, treatment with H_2 or CO at moderate to high temperature and γ-ray irradiation are generally employed to reduce Ti(IV) to Ti(III). Samples were loaded into 3 mm o. d. × 2 mm i. d. Suprasil quartz tubes and evacuated to a final pressure of 10^{-4} Torr at 295 K overnight. The samples were then heated to 673 K and kept at this temperature for 16 h. They were then contacted with 1 atm of O_2 at 673 K for 16 h followed by evacuation at the same temperature (activation). Activated samples were sealed and immersed in liquid nitrogen and then exposed to γ-radiation from a ^{60}Co source at 77 K to a total dose of 1.6 Mrad at a dose rate of 0.20 Mrad h^{-1}. In order to prepare Ti(III) complexes with various adsorbates the activated samples were exposed to the room temperature vapor pressure of D_2O and CH_3OD, and 100 Torr of C_2D_4 These samples with adsorbates were sealed and kept at room temperature for 24 h before exposing to γ-radiation at 77 K to a total dose of 1.6 Mrad. Details of ESR and ESEM measurements routinely employed in our lab are described elsewhere [4,9].

RESULTS AND DISCUSSION

Crystallinity and phase purity of all the titanosilicate samples are confirmed by X-ray powder diffraction. Substitution of titanium into the MFI and MCM-41 frameworks is also confirmed by XRD and other characterization methods. Microprobe analysis gives the following Si/Ti ratio for the various titanosilicates:

TS-1: 66;   TiMCM-41: 69;   ETS-10: 5;   ETS-4: 2.8.

The observed Si/Ti ratio in TS-1 corresponds to about 1.5 titanium ions per unit cell. A maximum limit of 2.3 Ti per unit cell is reported in the literature for TS-1 without any TiO_2 phase [1]. Since the observed value in our sample is well below the maximum limit, we assume that all titanium in our samples are in framework sites. Similarly we have shown earlier the absence of any TiO_2 phase in TiMCM-41 having Si/Ti = 69 [3]. The Si/Ti ratios observed in both ETS-10 and ETS-4 are the same as reported earlier [8].

Both hydrated and activated samples of various titanosilicates do not show any ESR signal for Ti(III) ions. Thus we assume titanium exists as Ti(IV). Isolated Ti(III) ions, however, can be generated in all titanosilicates by γ-ray irradiation at 77 K of activated samples. Figure 1 shows ESR spectra at 77 K after γ-ray irradiation of activated titanosilicate samples. The observed spectra are characterized by signals from radiation induced hole centers known as V centers and from Ti(III) centers. The ESR spectra of TS-1 and TiMCM-41 show a strong orthorhombic signal at g ≈ 2 from V centers and an axial signal with $g_\parallel$ = 1.970 and $g_\perp$ = 1.919 from Ti(III) ions. As shown in

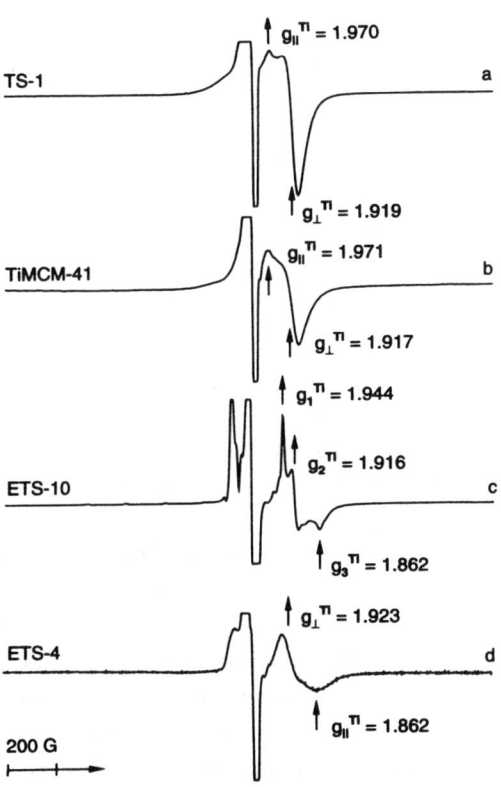

Fig.1. ESR spectra at 77 K of (a) TS-1, (b)TiMCM-41, (c) ETS-10 and (d)ETS-4 after γ-irradiation at 77 K of activated samples.

Figure 1, the V centers observed in ETS-10 and ETS-4 are more complex than in TS-1 or TiMCM-41. The Ti(III) ions observed in ETS-10 and -4 are different from those of TS-1 or TiMCM-41. A rhombic signal with $g_1 = 1.944$, $g_2 = 1.916$ and $g_3 = 1.891$ in ETS-10 and an axial signal with $g_\perp = 1.923$ and $g_\parallel = 1.862$ in ETS-4 are observed for the Ti(III) ions. Our assignment of Ti(III) in various titanosilicates is based on two observations. Ti(III) centers in several titanium containing compounds are reported to have ESR in the same region. For example, in alkali titanates, more than one Ti(III) center is observed after γ-irradiation. A broad signal with $g_\perp = 1.975$ and $g_\parallel = 1.890$ and a sharp signal with $g_\perp = 1.990$ and $g_\parallel = 1.981$

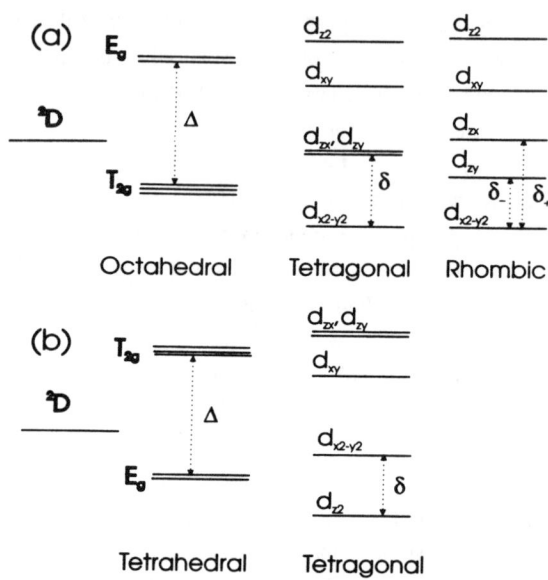

Fig.2 Energy levels of a $3d^1$ ion in (a) octahedral and (b) tetrahedral cubic fields

occur. They are identified as Ti(III) centers formed in TiO_6 units having one and two non-bridging oxygens, respectively [10]. Secondly, when silicalite-1 or SiMCM-41 is γ-irradiated under the same conditions only a signal due to V centers as in TS-1 or TiMCM-41 is observed [3].

The observed g components of the Ti(III) ion in various titanosilicates can be correlated to the specific crystalline field experienced by this ion. If titanium in TS-1 and TiMCM-41 occupies a framework site, tetrahedral site symmetry is expected for Ti(III) in dehydrated samples. On the other hand, in ETS-4 and ETS-10, the framework titanium is in octahedral coordination as suggested by several structural studies [7]. The ground state of Ti(III) with a $3d^1$ configuration is ^2D. When this ion is subjected to a perfect cubic crystalline field from tetrahedral or octahedral coordination, its five-fold degeneracy is lifted into a doublet and a triplet. In a tetrahedral field, the doublet lies lower in energy, while in octahedral coordination the triplet has the lower energy [11] as in Figure 2. An additional trigonal or tetragonal distortion is necessary to lift the degeneracy of the low lying doublet (tetrahedral) or triplet (octahedral) and is responsible for the g anisotropy and the deviation of g_{avg} from the free electron value $g_e = 2.0023$. In a tetragonally distorted tetrahedral field, the g values calculated to first order are [11] as follows.

$g_\parallel \approx g_e$ and $g_\perp \approx g_e - 6\lambda/\Delta$ (tetragonal compression)

$g_\parallel \approx g_e - 8\lambda/\Delta$ and $g_\perp \approx g_e - 2\lambda/\Delta$ (tetragonal elongation)

Here, λ is the spin orbit coupling constant and Δ is the energy splitting between the degenerate triplet and doublet levels in a cubic tetrahedral field (Figure 2b). From the ordering of the g values observed for Ti(III) in TS-1 and TiMCM-41 after γ-irradiation, the most likely situation is one of tetragonal compression. However, there are wide variations in the values reported for the ESR parameters of Ti(III) centers in various compounds.

It has been observed [12] that in a trigonal field, Ti(III) in octahedral coordination generally yields a spectrum for which the value of $g_\parallel$ is larger than $g_\perp$. But this is not the case observed in both ETS-4 and ETS-10. The observed ESR signals of Ti(III) in ETS-4 and ETS-10 can be explained on the basis of tetragonal or rhombic distortions to the octahedral crystal field of titanium similar to that observed for several $3d^1$ ions in a rutile structure [13]. The anisotropy of the ESR signal of Ti(III) arises by distortion of the cubic field due to displacement of one or both of axial oxygens or displacement of one or more of the four planar oxygens. The first possibility will give the required tetragonal field. The second possibility produces a crystal field of symmetry lower than tetragonal. The anisotropy can also arise due to distortion caused by strong interaction from nearby cations such as Na^+ or K^+. Thus, upon assuming tetragonally distorted octahedral symmetry for Ti(III) ion in ETS-4, the g values calculated to the first order are [11] $g_\perp \approx g_e - 8\lambda/\Delta$ and $g_\parallel \approx g_e - 2\lambda/\delta$. Using the values for $\lambda = 154$ cm^{-1}, and the observed g values, the calculated values of energy splitting are $\Delta = 15{,}700$ cm^{-1} and $\delta = 2700$ cm^{-1}.

The fact that Ti(III) in ETS-10 has a rhombic ESR signal with three g components indicates that the crystal symmetry is lower than tetragonal. Thus the second possibility of distortions caused by displacement of planar oxygens or by interactions from nearby ions seems to be more likely in this case. As result of this distortion, the degeneracy of the ground state triplet level is lifted. Assuming a D_{2h} site symmetry for Ti(III) ion in ETS-10, provided the spin-orbit interaction is not too high, the g values calculated to the first order are [13] $g_1 = g_e - 8\lambda/\Delta$; $g_2 = g_e - 2\lambda/\delta_+$; $g_3 = g_e - 2\lambda/\delta_-$. Here, Δ, δ_+ and δ_- are respectively, the energy separations of the upper doublet e_g (d_z, d_{xy}), and the middle lying d_{xz} and d_{yz} states from the ground state d_{x-y} (Figure 2a). Substituting the experimentally observed g values into the above expressions, we obtain $\Delta = 23{,}000$ cm^{-1}, $\delta_+ = 3800$ cm^{-1} and $\delta_- = 2700$ cm^{-1}. The octahedral field splitting Δ is still predominant, and additional separations of about 1000 wave numbers among the lower triplet component are enough to explain the observed g tensor. Highly distorted TiO_6 units have been also suggested in ETS-10 by EXAFS and XANES measurements aided by lattice energy minimization calculations [14].

The difference in coordination of titanium in TS-1 and TiMCM-41 compared to ETS-4 and ETS-10 is further reflected by their behaviors towards adsorption of various adsorbates. Upon adso adsorption of molecules such as D$_2$O, CH$_3$OD and C$_2$D$_4$ on TS-1 and TiMCM-41, new ESR signals are observed for Ti(III) suggesting significant modification of the crystal field around this ion by interaction from these molecules. For example Figure 3 shows the ESR spectra of various titanosilicates after adsorption of D$_2$O. Both TS-1 and TiMCM-41 show new Ti(III) species. On the other hand, no significant change in the ESR signal for Ti(III) is observed in ETS-10 after adsorption of these molecules. This suggests that there is no direct coordination or strong interaction between Ti(III) and these molecules. This assumption is further supported by the ESEM measurements on these samples.

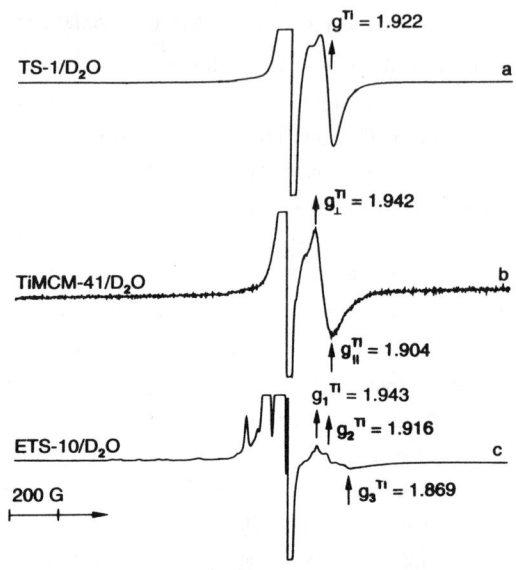

Fig.3. ESR spectra at 77 K of (a) TS-1 and (b) TiMCM-41 and (c) ETS-10 after adsorption of D$_2$O at 295 K and γ-irradiation at 77 K.

Strong deuterium modulations are observed in TS-1 after adsorption of various adsorbates. For instance, Figure 4a shows the deuterium modulation observed in TS-1 after D$_2$O adsorption. Similarly strong modulations are observed after adsorption of CH$_3$OD and C$_2$D$_4$. In contrast

Table 1 ^{2}D ESEM simulation parameters of TS-1 and ETS-10 after adsorption of various adsorbates.

Adsorbate	TS-1			ETS-10		
	No.of nuclei	Distance (Å)	A(MHz)	No.of nuclei	Distance(Å)	A(MHz)
[a]D$_2$O	1	3.0	0.30			
	1	3.5	0.04			
CH$_3$OD	1	2.8	0.14	1	4.0	0.05
C$_2$D$_4$	4	4.5	0.12	4	5.0	0.05

[a]In ETS-10 after adsorption of D$_2$O the echo was very weak and no significant deuterium modulation was observed.

either no or weak deuterium modulations are observed in ETS-10 after adsorption of the same adsorbates. In the specific case of D_2O no modulation is observed in ETS-10. As shown in Figure 4b adsorption of CH_3OD gives a very weak modulation. Similarly a weak modulation is also observed for C_2D_4. Simulations of the ^2D ESEM patterns observed in both TS-1 and ETS-10 after adsorption of various adsorbates are given in Table 1. The simulation parameters for TS-1 suggest a direct insertion of the ligands into the first coordination sphere of Ti(III). In the case of D_2O adsorption the simulation showing two deuterium nuclei at 3.0 Å and 3.5 Å seems consistent with the cleavage of Ti-O-Si bonds and the formation of TiOD and SiOD groups. Similarly with methanol, the simulation parameters for one deuterium interacting at 2.8 Å is consistent with coordination of one CH_3OD molecule to the Ti(III) ion. For ethylene, the four deuterium nuclei at 4.5 Å are consistent with one ethylene molecule coordinating with Ti(III) by π coordination.

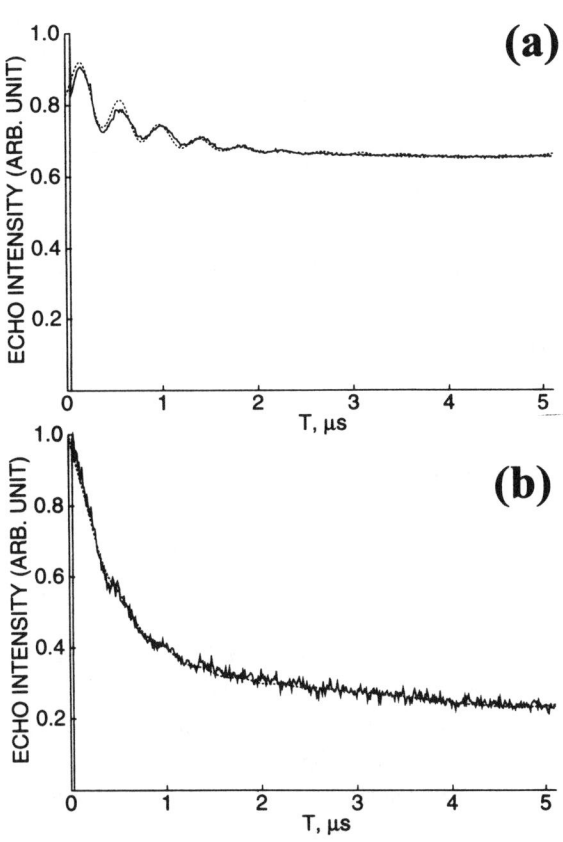

Fig.4. Experimental (—) and simulated (...) three pulse ^2D ESEM spectrum at 4.8 K of (a)TS-1 after adsorption of D_2O and (b) ETS-10 after adsorption of CH_3OD and subsequent γ-irradiation at 77 K.

The distances observed in ETS-10 are too long for any direct coordination between these olecules and Ti(III) ion. This is supportive evidence that titanium in ETS-10 is in octahedral coordination and that additional coordination with adsorbate molecules is unlikely.

The conclusions from these ESR and ESEM studies on various titanosilicate molecular sieves are significant for their possible catalytic implications. TS-1 and TiMCM-41 behave as Lewis acids due to coordinatively unsaturated Ti(IV) ions. Adsorption of various adsorbates significantly changes the ESR signal for Ti(III) suggesting direct insertion of the ligands into the first coordination sphere of titanium to increase the coordination number. This is confirmed by the ESEM results on these materials. ESEM indicates only one adsorbate molecule coordinating with Ti(III). This is consistent with EXAFS results within experimental error [15]. One adsorbate is also consistent with models proposed for catalytic test reactions. The specific coordination and high dispersion of titanium in these materials explains their high activity in various oxidation reactions. Although Ti(IV) in ETS-10 and ETS-4 can be reduced to Ti(III), direct insertion of ligands into the first coordination sphere of Ti(IV) is not observed. Such behavior of ETS-10 and ETS-4 may restrict their use as catalysts for oxidation reactions.

Acknowledgments. This research was supported by the National Science Foundation, the Robert A. Welch Foundation and the University of Houston Energy Laboratory

REFERENCES

1. B. Notari, Adv. Catal. **41,** 253 (1996).
2. A. Tuel , J. Diab, P. Gelin, M. Dufaux, J.-F. Dutel and Y. B. Tarit, J. Mol. Catal. **63,** 95 (1990).
3. A. M. Prakash, H. -M. Sung-Suh and L. Kevan, J. Phys. Chem. B **102,** 857 (1998).
4. A. M. Prakash, V. Kurshev and L. Kevan, J. Phys. Chem. B **101,** 9794 (1997).
5. A. J. H. P. van der Pol and J. H. C. van Hooff, Appl. Catal. A **92,** 93 (1992).
6. M. D. Alba, Z. Luan and J. Klinowski, J. Phys. Chem. **100,** 2179 (1996).
7. M. W. Anderson, O. Terasaki, T. Ohsuna, A. Phillippou, S. P. Mackay, A. Ferreira, J. Rocha and S. Lidin, Nature **367,** 347 (1994).
8. S. M. Kuznicki, K. A. Thrush, F. M. Allen, M. S. Levine, M. M. Hamil, D. T. Hayhurst, M. Mansour, in Synthesis of Microporous Materials, Edited by M. L. Occelli and H. Robson (Van Nostrand Reinhold, New York , 1992) p 427.
9. M. W. Anderson and L. Kevan, J. Chem. Soc., Faraday Trans. **83,** 3505 (1987).
10. Y. M. Kim and P. J. Bray, J. Chem. Phys. **53,** 71 (1970).
11. J. A. Weil, J. R. Bolton and J. E. Wertz, Electron Paramagnetic Resonance, Elementary Theory and Practical Applications (Wiley, New York, 1994), Chapter 8, pp.213-238.
12. H. M. Gladney and J. D. Swalen, J. Chem. Phys. **42,** 1999 (1965).
13. P. H. Kasai, Phys. Letts. 7, 5 (1963).
14. G. Sankar, R. G. Bell, J. M. Thomas, M. W. Anderson, P. A. Wright and J. Rocha, J. Phys. Chem. **100,** 449 (1996).
15. G. N. Vayssilov, Catal. Rev.- Sci. Eng. **39,** 209 (1997).

PALLADIUM SPECIES IN Co/Pd/H-ZSM-5 CATALYSTS FOR CH$_4$-SCR OF NOx

M. OGURA, M. HAYASHI and E. KIKUCHI

Department of Applied Chemistry, Waseda University, Shinjuku, Tokyo, Japan; ekikuchi@mn.waseda.ac.jp

ABSTRACT

The active state of palladium for NO reduction with methane (CH$_4$-SCR) was investigated by comparing the catalytic activity of Pd/H-ZSM-5 with those of Co/Pd/H-ZSM-5 and PdO/SiO$_2$. High catalytic activity for CH$_4$-SCR was given by Pd/H-ZSM-5 and much higher activity by Co/Pd/H-ZSM-5 in the temperature range of 300-500 °C. PdO/SiO$_2$ catalyzed the reaction between NO$_2$ and CH$_4$ in the absence of oxygen, which retarded the reaction by consuming CH$_4$ for combustion. CH$_4$ combustion occurred on either zeolite-supported or silica-supported catalyst, while NO preferentially retarded the combustion on Pd/H-ZSM-5. NO was found to be chemisorbed on the palladium sites in zeolite, while it was hardly chemisorbed on PdO/SiO$_2$. XPS and NaCl titration showed that the palladium species in zeolite is Pd^{2+}, on which NO is chemisorbed resulting in the high catalytic performance for CH$_4$-SCR.

INTRODUCTION

Selective catalytic reduction of nitric oxide by use of methane (CH$_4$-SCR) is one of the most promising way to reduce NOx from the lean-burn combustion system utilizing natural gas. Many investigations concerning CH$_4$-SCR have been reported [1-6], among which we have firstly reported that the palladium containing catalysts is most attractive because of its high catalytic performance even in the presence of water vapor [5]. Moreover, Co/Pd/H-ZSM-5 has been found to show high catalytic performance for removal of low concentration NOx in the moist atmosphere [5]. The cobalt and palladium on ZSM-5 play a bifunctional role to catalyze NO oxidation and NO$_2$ reduction, respectively.

Recent reports [5-9] indicate that the active state of palladium effectively catalyzing CH$_4$-SCR is highly dispersed, although the "dispersed palladium" is now in dispute to be located as Pd^{2+} or PdO. The aim of this paper is to discuss the active state of palladium comparing catalytic activities of various kinds of supported palladium and the chemisorptive properties for NO.

EXPERIMENTAL

Catalysts supported on ZSM-5 were prepared according to the description detailed in our previous work [5]. NH_4-ZSM-5 (SiO_2/Al_2O_3 molar ratio of 40.4) supplied by Tosoh Co. was ion-exchanged by stirring in a solution of $Pd(NH_3)_4Cl_2$ to give a Pd/H-ZSM-5 catalyst with a 0.4 wt% palladium loading. Co/Pd/H-ZSM-5 catalyst containing 1 wt% cobalt with 0.4 wt% palladium was prepared by ion exchanging Co/NH_4-ZSM-5 for 2 h using a solution of $Pd(NH_3)_4Cl_2$. These catalysts were dried at 110 °C overnight and were changed into the H-type of ZSM-5 by heat treatment at 500 °C prior to reaction. Silica, alumina, and silica-alumina having 0.4 wt% palladium were prepared by impregnation of the supports in palladium nitrate solutions. The chemical composition of catalysts was determined by means of ICP.

Reduction of NOx was carried out in a fixed-bed flow reactor, by feeding a mixture of 100 ppm NOx, 2000 ppm CH_4, 10% O_2, and 10% H_2O in He balance at a rate of 100 $cm^3 \cdot min^{-1}$ to 0.1 g catalyst. Effluent gases were analyzed by means of gas chromatography and chemiluminescence NOx analysis. N_2O was not detected in the reaction conditions employed. Catalysts were pretreated in a 20% O_2/He stream at 500 °C for 1 h prior to reaction. Catalytic activities were evaluated by the level of NOx conversion to N_2. The catalytic activity for CH_4 combustion was investigated using the same reaction conditions as those for NOx reduction without feeding NO.

Chemisorption of NO was conducted by means of NO-TPD measurements. A catalyst sample (0.1 g) was normally pretreated in flowing air at 500 °C for 1h, and NO was adsorbed from a stream of 1000 ppm NO in He at 50 °C for 1h, followed by evacuation for 30 min. Then, the sample was heated at a ramping rate of 10 °C/min. The desorbed NO was detected with an on-line quadrupole mass spectroscopy (TPD-AT-1; BEL Japan, Inc.).

The surface state of palladium species was studied by X-ray photoelectron spectroscopy (XPS). Spectra of Pd 3d were recorded with JPS90MX (JEOL). Monochromated Mg Kα radiation (1254 eV) was used for the measurements. The sample was pressed into a disk, set into the XPS sample holder, and outgassed to 1.0 x 10^{-8} Torr for more than 1 h before each measurement. The 1s level of adventitious carbon (atmospheric hydrocarbons and surface carbon oxides) was taken as the internal reference, with the peak position at 284.6 eV.

The amount of Pd^{2+} cation was determined by NaCl titration, which is an ion-exchange method of palladium with a 0.02M NaCl solution: 0.1 g of a sample after calcination at 500 °C was stirred in the NaCl solution at 80 °C for 1 h, filtered, and the concentration of palladium in the filtrate was determined by ICP.

RESULTS AND DISCUSSION

Figure 1 shows the catalytic activities of palladium catalysts loaded on H-ZSM-5 and silica for $NO\text{-}CH_4\text{-}O_2$, $NO_2\text{-}CH_4$, and $CH_4\text{-}O_2$ reactions. As noted in the previous paper [5], Pd/H-ZSM-5 has the catalytic activity for NOx reduction with CH_4 in the presence of water vapor, and that for CH_4 combustion. It is noteworthy that PdO/SiO_2, among the oxide-supported palladium catalysts tested in this study, showed catalytic activities for NO_2 reduction comparable to Pd/H-ZSM-5. The activities of the oxide-supported catalysts for CH_4 combustion and NO reduction in the presence of oxygen were comparable to and less than that of Pd/H-ZSM-5, respectively. Interestingly, the catalytic activity of Pd/H-ZSM-5 for CH_4 combustion was depressed by the coexistence of NO, while those of oxide-supported PdO were hardly affected by NOx. These results show that there is a difference between H-ZSM-5 and the oxide-supports on the interaction between NO and palladium species on them. Loughran and Resasco showed that PdO/SiO_2 has a catalytic activity for NO reduction with CH_4 when the catalyst is mixed with a catalytic component having acidity, such as H-ZSM-5 or SO_4^{2-}/ZrO_2 [4]. They speculated that the palladium on SiO_2 is the active catalytic species for $NO_2\text{-}CH_4$ reaction, although it requires NO oxidation sites as claimed by Nishizaka and Misono [3].

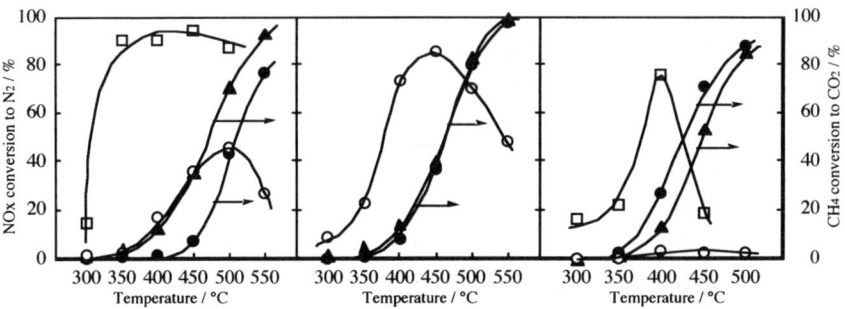

Figure 1 Catalytic activities of Pd/H-ZSM-5(a), Co/Pd/H-ZSM-5(b), and PdO/SiO_2(c) for NOx reduction with CH_4 and CH_4 combustion with O_2 in the presence of 10% H_2O.
Catalyst weight, 0.1 g.   ○ ●: $NO\text{-}CH_4\text{-}O_2$; □: $NO_2\text{-}CH_4$; ▲: $CH_4\text{-}O_2$.

The effect of O_2 concentration on the catalytic activities of Pd/H-ZSM-5 and PdO/SiO$_2$ for NO$_2$-CH$_4$ reaction was investigated. NO$_2$ conversion on PdO/SiO$_2$ sharply decreased with low concentrations of O$_2$, while CH$_4$ conversion increased with increasing concentration of O$_2$, so that the selectivity of CH$_4$ for NO$_2$ reduction was remarkably lowered by O$_2$. Pd/H-ZSM-5, on the other hand, moderately catalyzed NO$_2$ reduction even in the presence of high concentration O$_2$. As the effect of oxygen was reversible, the retardation of NO$_2$-CH$_4$ reaction by O$_2$ is not due to the destruction of the active palladium species or zeolite structure, and it might be attributed to the competitive chemisorption of NO and O$_2$ on the active site.

These results indicate that palladium is located on SiO$_2$ in a quite different state from palladium on zeolite, although NO$_2$-CH$_4$ reaction equally occurred on either state of catalysts. Figure 2 shows the temperature programmed desorption of NO chemisorbed on palladium. Palladium in H-ZSM-5 adsorbed NO in a molar ratio of NO to palladium to be close to unity, while palladium on SiO$_2$ hardly adsorbed NO. Most of NO (about 90%) chemisorbed on Pd/H-ZSM-5 desorbed at around 500 °C. The same result was obtained on Co/Pd/H-ZSM-5. Strongly chemisorbed NO seems to be the reason for retardation of the catalytic activity for CH$_4$ combustion and the high selectivity for NO reduction with CH$_4$.

Figure 3 shows the XPS results on Pd 3d in Pd/H-ZSM-5 and Pd/SiO$_2$ after the *ex-situ* calcination at 500 °C, the same pretreatment temperature for activity tests. As has been shown in our previous work [5], palladium on H-ZSM-5 is highly dispersed in the zeolite pores so that it could not appreciably be detected by XPS. The intense peaks assigned to PdO are evident for PdO/SiO$_2$. It is well-known that PdO has high catalytic activity for CH$_4$ combustion; therefore, it can be concluded that the state of palladium on SiO$_2$ gives the low selectivity for CH$_4$-SCR.

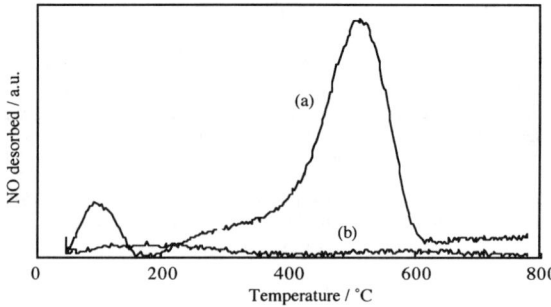

Figure 2 Temperature programmed desorption of NO from Pd/H-ZSM-5(a) and PdO/SiO$_2$(b).

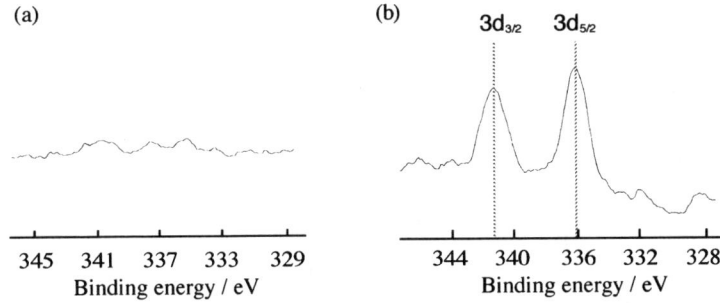

Figure 3   XPS spectra of Pd 3d on Pd/H-ZSM-5(a) and PdO/SiO$_2$(b).

Sample	total amount of Pd	eluted Pd^{2+} in NaCl solution	Pd^{2+}/total Pd
	mmol/g-cat	mmol/g-cat	%
Pd/H-ZSM-5		0.037	98
Co/Pd/H-ZSM-5	0.038	0.038	100
PdO/SiO$_2$		0.0	0.0

Table 1   NaCl titration for quantitative analysis of Pd^{2+}.

Table 1 summarizes the results of NaCl titration on calcined Co/Pd/H-ZSM-5, Pd/H-ZSM-5, and Pd/SiO$_2$. Palladium in Co/Pd/H-ZSM-5 and Pd/H-ZSM-5 was almost wholly eluted into the solution, showing that they are Pd^{2+} cations whether Co coexists or not. On the other hand, PdO on SiO$_2$ cannot be dissolved into the NaCl solution. From these studies of XPS and NO-TPD, it can be concluded that Pd^{2+} in zeolite structure has an important role to chemisorb NO and to reduce NO (or NO$_2$) with CH$_4$ even in the presence of excess amount of O$_2$ and H$_2$O.

CONCLUSIONS

The palladium species active for NOx reduction with CH$_4$ were investigated by comparing the catalytic activities of palladium on H-ZSM-5 with those on oxide supports. Among the oxide-supported palladium catalysts, PdO/SiO$_2$ showed comparable catalytic activity with Pd/H-ZSM-5 for NO_2 reduction by CH_4 in the absence of oxygen. However, oxygen in the reactant significantly retarded NO_2-CH_4 reaction on PdO/SiO_2, while not so much on Pd/H-ZSM-5. It is interesting to note that, on Pd/H-ZSM-5, the CH_4 conversion by O_2 decreased by NO to a large extent. PdO/SiO_2 showed similar CH_4 conversion even though NO coexisted or not. From the measurement of temperature programmed desorption of NO from these catalysts, it is apparently observed that palladium species on H-ZSM-5 chemisorbed NO, while those on SiO_2 hardly. XPS spectra indicated that the palladium species on SiO_2 are located as PdO. Pd^{2+} was eluted from calcined Pd/H-ZSM-5 and Co/Pd/H-ZSM-5 by ion exchange with NaCl solution; therefore, palladium species on zeolite were mainly located in a cationic state. From the results obtained in this chapter, Pd^{2+} is an important species to chemisorb NO, probably leading to show catalytic activity for NO-CH_4-O_2 reaction.

REFERENCES

[1] Y. Li and J. N. Armor, *Appl. Catal.*, **B1**, L31 (1992).
[2] K. Yogo, M. Ihara, I. Terasaki and E. Kikuchi, *Catal. Lett.*, **17**, 303 (1993).
[3] Y. Nishizaka and M. Misono, *Chem. Lett.*, **1993**, 1295.
[4] C. J. Loughran and D. E. Resasco, *Appl. Catal.*, **B7**, 113 (1995).
[5] M. Ogura, Y. Sugiura, M. Hayashi and E. Kikuchi, *Catal. Lett.*, **42**, 185 (1996).
[6] M. Suzuki and M. Niwa, *Chem. Lett.*, **1996**, 275.
[7] A. W. Aylor, L. J. Lobree, J. A. Reimer, and A. T. Bell, *J. Catal.*, **172**, 453 (1997).
[8] B. J. Adelman and W. M. H. Sachtler, *Appl. Catal.*, **B14**, 1 (1997).
[9] A. Ali, W. Alvarez, C. J. Loughran, and D. E. Resasco, *Appl. Catal.*, **B14**, 13 (1997).

CRYSTAL STRUCTURE OF A BENZENE SORPTION COMPLEX OF DEHYDRATED FULLY Mn(II)-EXCHANGED ZEOLITE X

Y. KIM[a], A. N. KIM[a], Y.W. HAN[b] and K. SEFF[c]

[a]Department of Chemistry, Pusan National University, Pusan 609-735, Korea; ykim@hyowon.pusan.ac.kr
[b]Department of Science Education, Pusan National University of Education, Pusan 607-071, Korea
[c]Department of Chemistry, University of Hawaii, 2545 The Mall, Honolulu, Hawaii 96822, U. S. A.; kseff@gold.chem.hawaii.edu

ABSTRACT

The crystal structure of a benzene sorption complex of dehydrated Mn^{2+}-exchanged zeolite X, $Mn_{46}Si_{100}Al_{92}O_{384} \cdot 40C_6H_6$ (a = 24.645(4) Å), has been determined by single-crystal X-ray diffraction techniques in the cubic space group $Fd\bar{3}$ at 21 °C. The structure was refined to the final error indices R_1 = 0.064 and R_w = 0.066 with 523 reflections for which $I > 3\,\sigma(I)$. Mn^{2+} ions are located at three different sites: 16 Mn^{2+} ions fill site I at the centers of the double six-rings, 4 are at site II, almost in the planes of single six-oxygen rings, and the remaining 26, also at site II, extend $ca.$ 1.22(2) Å into the supercage from their six-ring planes. The benzene molecules are found at two distinct sites within the supercages. Twenty-six benzenes lie on threefold axes in the large cavities where each interacts facially with one of the latter site–II Mn^{2+} ions; the remaining fourteen benzenes are found in the planes of the 12-rings where each is stabilized by 18 interactions (each weakly electrostatic and van der Waals) with framework oxygens.

INTRODUCTION

The structure of zeolite Na-X, which is isomorphous with the mineral faujasite, is well established by X-ray diffraction [1,2]. The system of benzene sorbed on synthetic faujasite-type zeolite has drawn significant attention in recent years. The location of benzene in sodium zeolite Y has been studied by powder neutron diffraction [3,4]. At room temperature, the benzene molecules are "largely delocalized within the supercages" of zeolite Y, whereas at 4 K the benzene molecules are found at two distinct sites: one centered on and normal to the threefold axes of the unit cell near site II, and the other centered in the plane of the 12-ring window between adjacent supercages. Using high-speed X-ray powder diffraction methods, the influence of temperature on the sorption of benzene in K^+, Ca^{2+}, and Sr^{2+} exchanged Y-type zeolites was investigated [5]. The ability of benzene to alter the cation distribution increases rapidly with the cation-benzene interaction energy. Using ^{2}H NMR, Cheetham et al. [6] found that C_6D_6 sorbed in Ca-X (one molecule per supercage) is bound strongly at site II and interacts facially with Ca^{2+}; this benzene molecule rotates rapidly about its sixfold axis in the temperature range 260-330 K. The activation energy for this rotation is 4.9 kJmol^{-1}. The precise nature of the cation-benzene ring interaction is unclear, but charge-quadrupolar terms are believed to be important [6].

This work was done to locate the sorbed C_6H_6 molecules, to determine the cation shifts upon sorption, and to observe cation-sorbate interactions in a transition-metal exchanged zeolite.

EXPERIMENTAL

Large single crystals of sodium zeolite X, stoichiometry $Na_{92}Si_{100}Al_{92}O_{384}$, were prepared in St. Petersburg, Russia [7]. One of these, a colorless octahedron about 0.20 mm in cross-section, was lodged in a fine Pyrex capillary. An exchange solution of 0.05 M $Mn(NO_3)_2$ was allowed to flow past the crystal at a velocity of 10 mm/s for 3 days. The crystal was dehydrated at 380 °C and 2 x 10^{-6} Torr for 44 h. It was then treated with ca. 92.2 Torr of zeolitically dried benzene for 2 days at 22(1) °C. The resulting colorless crystal, still in this benzene atmosphere, was sealed in its capillary by torch.

The cubic space group $Fd\bar{3}$ was used. Molybdenum K α radiation (weighted mean λ = 0.71073 Å) was used for all experiments. The unit cell constant at 21(1) °C, determined by least-squares refinement of 25 intense reflections for which $14° < 2\theta < 22°$, is $a = 24.645(4)$ Å. All unique reflections in the positive octant of an F-centered unit cell for which $2\theta < 60°$, $l > h$, and $k > h$ were recorded at 21(1) °C. Of the 1400 reflections examined, only the 523 reflections for which $I > 3\sigma(I)$ were used in subsequent structure determination.

STRUCTURE DETERMINATION

Full-matrix least-squares refinement was initiated by using the atomic parameters of the framework atoms [Si, Al, O(1), O(2), O(3) and O(4)] in Mn_{46}-X.30C_2H_4 [8]. Isotropic refinement of the framework atoms converged to an unweighted R_1 index $\Sigma |F_o - |F_c||/\Sigma F_o$ of 0.36 and a weighted R_w index $(\Sigma w(F_o - |F_c|)^2/\Sigma wF_o^2)^{1/2}$ of 0.42.

From a subsequent difference Fourier function, the Mn^{2+} ions at Mn(1) and Mn(2) were located, and isotropic refinement of framework atoms, Mn(1), and Mn(2) converged to $R_1 = 0.109$ and $R_w = 0.166$. From subsequent difference Fourier functions, the carbon atoms of the benzene molecules were located at the general positions C(1), C(2), and C(3). Anisotropic refinement of the framework atoms and Mn^{2+} ions, and isotropic refinement of the carbon atoms, converged to $R_1 = 0.072$ and $R_w = 0.077$. The thermal ellipsoid of Mn(2) became very elongated in subsequent refinements and was therefore split into Mn(2) at (0.23, 0.23, 0.23) and Mn(3) at (0.21, 0.21, 0.21). Anisotropic refinement of the framework atoms and Mn^{2+} ions at Mn(1) and Mn(2) converged to $R_1 = 0.063$ and $R_w = 0.065$; the C atoms and Mn^{2+} ions at Mn(3) varied isotropically.

Table 1. [a]Positional, Thermal, and Occupancy Parameters

Atom	Cation Site	x	y	z	U_{eq}[b] or U_{iso}[c]	[d]Occupancy varied	fixed
Si		-523(1)	1226(2)	345(1)	106(14)[b]		96.0
Al		-536(1)	375(1)	1215(2)	78(15)[b]		96.0
O(1)		-1096(3)	14(4)	1053(3)	236(54)[b]		96.0
O(2)		-21(3)	14(4)	1483(3)	106(39)[b]		96.0
O(3)		-303(3)	632(4)	594(4)	121(43)[b]		96.0
O(4)		-645(3)	832(4)	1691(3)	183(48)[b]		96.0
Mn(1)	I	0	0	0	150(11)[b]	15.4(2)	16.0
Mn(2)	II	2303(11)	2303(11)	2303(11)	194(10)[b]	25.4(3)	26.0
Mn(3)	II	2068(10)	2068(10)	2068(10)	817(161)[c]	4.2(3)	4.0
C(1)		2543(13)	3318(13)	2938(13)	1158(130)[c]	76.2(9)	78.0
C(2)		2559(13)	2883(14)	3317(13)	1148(125)[c]	76.2(9)	78.0
C(3)		4755(14)	4796(15)	5472(11)	1176(127)[c]	87.4(31)	84.0
H(1)[e]		2283	3600	2969			78.0
H(2)[e]		2275	2850	3576			78.0
H(3)[e]		4607	4676	5807			84.0

[a]Positional and thermal parameters are given × 10^4. Numbers in parentheses are the esd's in the units of the least significant digit given for the corresponding parameter. [b]U_{eq} is defined as one-third of the trace of the orthogonalized U_{ij} tensor. [c]U_{iso} (Å^2 × 10^4). [d]Occupancy factors are given as the number of atoms or ions per unit cell. [e]The positions of the hydrogen atoms were calculated with C-H distances of 0.95 Å.

The occupancy numbers at Mn(1), Mn(2), Mn(3), C(1), C(2) and C(3) were fixed as shown in Table 1 by the assumption of stoichiometry and the requirement of electrical neutrality. With C-H bond lengths of 0.95 Å, the positions of the hydrogen atoms H(1), H(2), and H(3), were calculated by the computer system, MolEN [9].

The final error indices converged to $R_1 = 0.064$ and $R_w = 0.066$. Atomic scattering factors for Si, Al, O⁻, C, and Mn^{2+} were used [10]. All scattering factors were modified to account for anomalous dispersion [11]. The final structural parameters and selected interatomic distances and angles are presented in Tables 1 and 2, respectively.

Table 2. Selected Interatomic Distances (Å) and Angles (deg)[a].

Si-O(1)	1.611(10)	Al-O(1)	1.689(11)
Si-O(2)	1.667(10)[b]	Al-O(2)	1.723(10)[b]
Si-O(3)	1.673(9)[b]	Al-O(3)	1.753(10)[b]
Si-O(4)	1.611(10)	Al-O(4)	1.648(10)
Mn(1)-O(3)	2.263(8)	C(1)-C(2)	1.43(5)
Mn(2)-O(2)	2.155(8)	C(1)-C(2)'	1.25(5)
Mn(2)-O(4)	3.007(9)	C(3)-C(3)	1.40(5)
Mn(3)-O(2)	2.13(2)	C(1)-H(1)	0.95[b]
Mn(3)-O(4)	2.93(3)	C(2)-H(2)	0.95[b]
		C(3)-H(3)	0.95[b]
Mn(2)-C(1)	3.01(3)	C(1)-C(2)-C(1)'	122(3)
Mn(2)-C(2)	2.94(3)	C(2)-C(1)-C(2)'	118(3)
		C(3)-C(3)'-C(3)"	120(3)
O(3)-Mn(1)-O(3)	89.8(3)		
O(2)-Mn(2)-O(2)	113.4(4)	H(1)-O(2)	3.16
O(2)-Mn(3)-O(2)	117.1(4)	H(2)-O(1)	3.52
		H(2)-O(2)	3.41
		H(3)-O(1)	2.99
		H(3)-O(4)	2.93

[a]Numbers in parentheses are estimated standard deviations in units of the least significant digit given for the corresponding value. [b]Hydrogen positions were calculated.

DISCUSSION

In this structure, the Mn^{2+} ions are located at three different crystallographic sites. The 16 Mn^{2+} ions at Mn(1) fill site I at the centers of the double six-rings. The octahedral Mn(1)-O(3) distance, 2.263(8) Å, is a little longer than the sum of the corresponding ionic radii, 0.80 +

1.32 = 2.12 Å [12]. The 30 Mn^{2+} ions are located at two different site II's in the supercage: 26 at Mn(2) each coordinate at 2.155(8) Å to three O(2) framework oxygens and also coordinate facially with a benzene molecule; the remaining four at Mn(3) do not coordinate to benzene molecules but coordinate trigonally to three O(2) framework oxygens at 2.13(2) Å (see Figure 1). The O(2)-Mn(2)-O(2) angle is 113.4(4)° ; O(2)-Mn(3)-O(2) is 117.1(4)°, closer to trigonal planar (see Table 2).

Crystallographically there are two types of benzene molecules. The first type is located deep inside the supercage with 156 carbon atoms (26 molecules of C_6H_6) at C(1) and C(2). These benzenes coordinate facially with Mn^{2+} ions at site II (Mn^{2+}-benzene center = 2.67 Å) (Figures 1 and 2). Although these benzenes have high thermal motion, their six-rings appear relatively normal (C(1)-C(2) = 1.43(5) and C(2)-C(1)' = 1.25(5) Å; C(1)-C(2)-C(1)' = 122(3)° and C(2)-C(1)-C(2)' = 118(3)°) (see Table 2 and Figures 1 and 2).

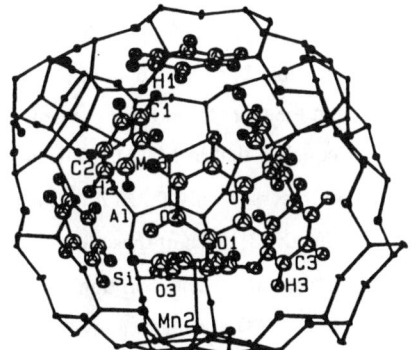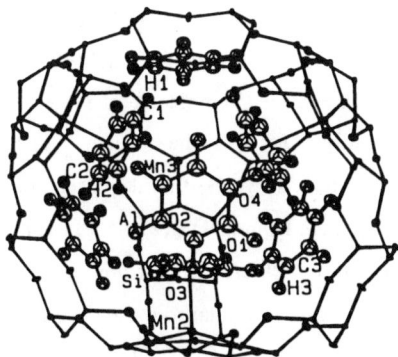

Figure 1. Stereoview of a supercage. Shown are three Mn^{2+} ions at Mn(2) each coordinated to a benzene molecule, one Mn^{2+} ion at Mn(3), and four benzene molecules at 12-ring centers. Ellipsoids of 20% probability are used.

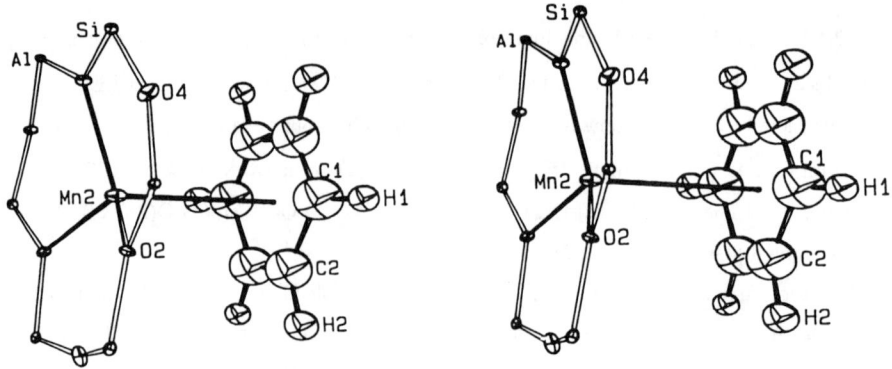

Figure 2. Stereoview of a Mn(C$_6$H$_6$)$^{2+}$ complex on the inner surface of a supercage is shown. Ellipsoids of 20% probability are used. Twenty-six Mn^{2+} ions per unit cell at Mn(3) coordinate to benzene molecules as shown.

The sorption of benzene causes Mn(2) to move 1.22 - 0.14 = 1.08 Å deeper into the supercage, as compared to its position in dehydrated Mn$_{46}$-X [8], from the plane of the O(2)'s to which it is bound. In this way, these Mn^{2+} ions are able to coordinate more octahedrally to benzene(1) (considering benzene to be tridentate). The deviation of these Mn^{2+} ions from the six-ring plane at O(2) into the supercage is larger than those in Mn$_{46}$-X·30CO [13] and Mn$_{46}$-X·30C$_2$H$_4$ [8]. This and the Mn(2)-O(2) bond distances (see Table 2) suggest that, in binding to the benzene(1) molecule, the interaction between Mn(2) and the O(2) framework oxygens is slightly reduced. The occupancy at benzene(1), 26/32, is equal to that at Mn(2). The four Mn^{2+} ions at Mn(3) that do not coordinate to benzene(1) are recessed only 0.22 Å into the supercage from their O(2) plane; compared to dehydrated Mn$_{46}$-X [8], each of these Mn^{2+} ions deviates 0.08 Å more from the plane of its three O(2) oxygens. The sorption site at benzene(1) is similar to those observed for benzene sorbed in Na-Y [3] at a loading level of 2.6 molecules, and in Ca-X [6] at a loading level of one molecule per supercage.

The second type of benzene molecule (about 14 per unit cell) are located at the centers of the 12-ring windows at positions of symmetry $\bar{3}$; their occupancy is 14/16. Each molecule can be generated crystallographically from the C(3) position by symmetry (C(3)-C(3)' = 1.40(5) Å and C(3)-C(3)'-C(3)" = 120(3)°). Each hydrogen of this type of benzene interacts with three 12-ring oxygens. Each molecule therefore has 18 such interactions. The 12 oxygen atoms of the 12-ring provide a close-fitting environment for this kind of benzene molecule (see Figure 3), where its position is stabilized by multiple van der Waals forces and electrostatic interactions.

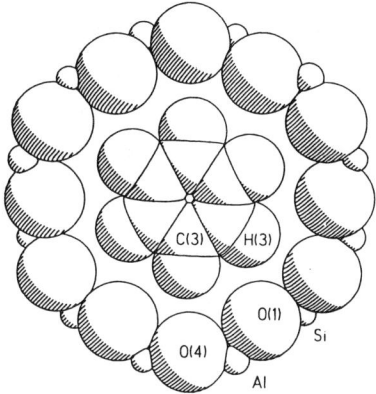

Figure 3. Van der Waals view of a 12-ring containing a benzene(2) molecule. Fourteen of the 16 12-rings per unit cell are so occupied. The van der Waals radii used for oxygen and hydrogen are 1.40 and 1.20 Å, respectively. The positions plotted are those found crystallographically except for those of the hydrogen atoms. Note that each H(3) atom is near one O(4) and two O(1) framework oxygens; altogether then each benzene(2) molecule has 18 such H...O interactions. The fit is remarkable. The view is along (111); the center is at (0.5, 0.5, 0.5); the symmetry is $\bar{3}$.

Benzene(2) molecules are usually found only at high benzene coverages [3]. The observation of simultaneous occupation at the 12-ring window site is in excellent agreement with the results of a neutron diffraction study [3] where it was found that benzenes occupy two sites, one near site II, and the other in the 12-ring window.

The shortest C-C distance (4.06(8) Å) between the benzene(1) and benzene(2) molecules is less than the correspomding distance (4.40(6) Å) between the benzene(1) molecules themselves. The distances between H(3) of the benzene(2) molecule and H(1) of benzene(1) are 2.5 and 2.7 Å. The H(2)-H(3) distances are 2.3 and 2.8 Å.

Altogether 40 benzene molecules are sorbed per unit cell. On average, 5 benzene molecules have been sorbed per supercage, 3.25 near site II and 1.75 in 12-rings. This can be compared to 28 sorbed in Ca_{46}-X at 21(1) °C at a vapor pressure of 48 Torr, with 3.5 benzene molecules per supercage, 2.75 near site II and 0.75 in 12-rings [14].

SUMMARY

Mn^{2+} ions occupy sites I and II: 16 fill sites I and 30 nearly fill sites II. There are two kinds of site II Mn^{2+} ions: 26 coordinate to benzene molecules and 4 do not. There are also two kinds of benzene molecules. The first kind (26 benzene molecules) coordinate facially to Mn^{2+} ions in the supercage. The second kind (14 molecules) nearly fill the 16 12-ring sites per unit cell. Each of these 14 benzene molecules is held in place by 18 van der Waals and weakly electrostatic interactions.

ACKNOWLEDGEMENT. This work was supported in part by the Korean Science and Engineering Foundation (Grant No. 971-0305-027-2). Mr. Shenyan Zhen prepared Figure 3.

REFERENCES

1. D.H. Olson, J. Phys. Chem. **74**, 2758 (1970).
2. D.H. Olson, Zeolites **15**, 439 (1995).
3. A.N. Fitch, H. Jobic and A. Renouprez, J. Phys. Chem. **90**, 1311 (1986).
4. A.N. Fitch, H. Jobic and A. Renouprez, J. Chem. Soc., Chem. Commun. 284 (1985).
5. J.J.L. van Dum, W.J. Mortier and J.B. Uytterhoeven, Zeolites **5**, 1257 (1985).
6. G. Vitale, L.M. Bull, R.E. Morris and A.K. Cheetham, J. Phys. Chem. **99**, 16087 (1995).
7. V.N. Bogomolov and V.P. Petranovskii, Zeolites **6**, 418 (1986).
8. S.B. Jang, Y.H. Yeom, M.S. Jeong, Y. Kim and K. Seff, J. Phys. Chem. B **101**, 9041 (1997).
9. Calculations were performed with Structure Determination Package programs, MolEN, (Enraf-Nonius, Netherlands, 1990).
10. International Tables for X-ray Crystallography, Vol.IV, (Kynoch Press, Birmingham, England, 1974) pp. 73-87.
11. Reference 10, pp. 149-150.
12. Handbook of Chemistry and Physics, 70th Ed. (The Chemical Rubber Co., Cleveland, Ohio, 1989/1990) p. F-187.
13. M.N. Bae, Y. Kim and K. Seff, submitted to Micropor. Mesopor. Mater. (1997).
14. Y.H. Yeom, A.N. Kim, Y. Kim and K. Seff, J. Phys. Chem. B **102**, in press.

CATION EXCHANGED ZEOLITES ZSM-5 FOR THE HYDROXYLATION OF BENZENE WITH NITROUS OXIDE

S. KOWALAK[§], K. NOWIŃSKA[§], M. ŚWIĘCICKA[§], M. SOPA[§], A. JANKOWSKA[§], G. EMIG*, E. KLEMM*, A. REITZMANN*

[§] A. Mickiewicz University, Faculty of Chemistry, Poznań, Poland;
skowalak@main.amu.edu.pl
*University of Erlangen-Nuremberg, Institute of Technical Chemistry I, Erlangen, Germany;
Elias.Klemm@rzmail.uni-erlangen.de

ABSTRACT

Iron exchanged zeolites ZSM-5 as well as some other transition metal modification of these zeolites show a considerable catalytic activity and selectivity for the direct hydroxylation of benzene with N_2O towards phenol. The best catalytic performance was obtained after calcination at 900 °C. The Co^{2+} and Cr^{3+} modified zeolites calcined at this temperature showed a noticeable activity too. The iron salt impregnated $AlPO_4$-5 or silicalite-1 were inactive for this reaction, whereas the silicalite treated with $FeCl_3$ vapors showed some activity. It can be assumed that iron-oxide cationic species combined with a negatively charged lattice of the MFI structure are responsible for the catalytic activity. Such species can be generated upon calcination of either Fe-ZSM-5 or ferrisilicalite.

INTRODUCTION

Direct oxidation of benzene to phenol has recently attracted the attention of many research groups. Although the cumene process has been known for 50 years (90% of total phenol production), this three step process exhibits some economic drawbacks, e.g. separation of reactive intermediates or corrosive reagents. The most important disadvantage results from a predicted surplus of the main by-product acetone in the year 2000 [1]. Attempts to oxidize benzene directly to phenol by means of dioxygen did not show much success [2, 5]. By replacing oxygen with nitrous oxide and using MFI type zeolites containing framework substituted iron (ferrisilicalites), considerable catalytic activities and selectivities were achieved in the direct hydroxylation of benzene to phenol [5, 6, 8]. It is believed that a substantial number of iron is released from the framework positions of the ferrisilicalite already on heating at 500 °C [7]. The resulting extra lattice iron species – so called α-sites – are made responsible for the decomposition of N_2O [3] as well as for the subsequent selective oxidation of benzene towards phenol [5, 6]. The nature of the α-sites is not known in detail, however, oxide-like clusters are considered as the main components [4].

The simplest way to introduce extra-framework species into zeolites is a conventional ion-exchange procedure. Such a procedure was applied in this study to prepare Fe(III)-ZSM-5 as well as Co^{2+} and Cr^{3+} modifications [9]. Alternatives in producing extra-framework species are liquid impregnation and chemical vapor deposition (CVD). These two methods have been applied for a modification of nonacidic molecular sieves such as silicalite-1 and $AlPO_4$-5 in order to find out which nature of the extraframework species is necessary. The thermal treatment of all modified zeolites plays an crucial role in generation of active sites.

The resulting samples were examined in hydroxylation of benzene to phenol, studying the catalytic activity and product selectivity. Different characterization methods were applied to obtain more information about the reaction mechanism and the active sites.

EXPERIMENTAL

Zeolite Na-ZSM-5 (Degussa, Si/Al ~ 15) was the principal substrate for the ion exchange which has already been described in detail [9]. Calcination was carried out in air for two hours, temperature was varied between 500 °C and 1000 °C. The amount of transition metal introduced was determined by XRF as following: Fe 0.5 wt.%, Co 0.6 wt.%, Cr 0.3 wt.%. Silicalite-1 (from UOP) and $AlPO_4$-5 (synthesized in our laboratories) were modified with iron salts by means of impregnation with aqueous solutions of $Fe(NO_3)_3$, $FeCl_3$, $Fe_2(SO_4)_3$. The amount of iron introduced by this way was about 1 wt.%. Thermal treatment of silicalite-1 with $FeCl_3$ vapors was conducted in nitrogen stream at 500 °C. The resulting product was washed with water, then dried and calcined at 900 °C. The iron content was 1.5 wt.%. The modified zeolite samples were characterized by means of XRD, FTIR (IFS 113v Bruker.), XPS, TPD of ammonia (Altamira), micropore volume (ASAP 1000).

The hydoxylation of benzene was performed in a fully automated laboratory setup including a plug-flow reactor [8] (see Table 1).

Parameter	Range
Temperature [°C]	350 – 450
W/F [g·min/mol]	90
Benzene Concentration [%]	0 – 12%
Nitrous Oxide Concentration [%]	0 – 25%

Table 1: Reaction conditions for benzene hydroxylation

RESULTS AND DISCUSSION

As will be discussed later the calcination temperature is crucial for the catalytic performance. Therefore, the influence of calcination on the properties of the zeolites is studied first.

Calc. Temperature	Microporevolume [cm^3/g]	Fe^{3+} (outer surface) [mol %]
550	0.15	3.7
700	0.15	2.5
900	0.11	1.0
1000	0.02	1.0

Table 2: Influence of calcination temperature on micropore volume and Fe^{3+} content at the outer surface (XPS) for the Fe-ZSM-5 prepared by ion exchange.

As can be seen from Table 2, the ion exchange in liquid phase results in a higher iron concentration at the outer surface of the crystallites measured by XPS. This can be explained by a hindrance of the bulky solvated Fe^{3+}- ion to diffuse into the pore system [13]. With increasing calcination temperature, the iron concentration at the outer surface decreases due to a solid ion exchange procedure.

Up to 700 °C no crystallinity loss was observed by XRD and sorption measurements (see Table 2). The sample calcined at 900°C showed no loss in crystallinity by XRD but a small loss of pore volume by sorption measurements indicating the formation of bulkier extra-framework species (see Table 2). Calcination at 1000 °C results in a substantial of crystallinity (XRD) as well as in pore volume.

ESR spectra of the iron cation modified zeolites could support this explanation [9]. Calcination results in a decrease of the **g** value signal at 2.0 and a significant increase in the intensity of the 4.3 **g** value signal. Usually the **g** value signal at 2.0 is attributed to octahedral iron species in extraframework position [10]. In the present case it probably results from hydrated Fe^{3+} clusters because the sample was not evacuated and measured at room temperature. The signal observed at 4.3 can be attributed to tetrahedral-coordinated Fe^{3+} which can be located in the framework (ferrisilicalite) or outside the framework as isolated cations in the zeolite channels (e.g. FeO$^+$) [9, 11]. But, as described in [11], an insertion of Fe^{3+} into the lattice seems to be unlikely. Thus, it can be assumed that after liquid ion exchange and calcination in air at low temperatures, still a significant amount of Fe^{3+} is present at the outer crystallite surface. During calcination at high temperatures in air Fe^{3+} ions diffuse to the cationic positions into the zeolite channels according to a solid ion exchange generating isolated cations like FeO$^+$.

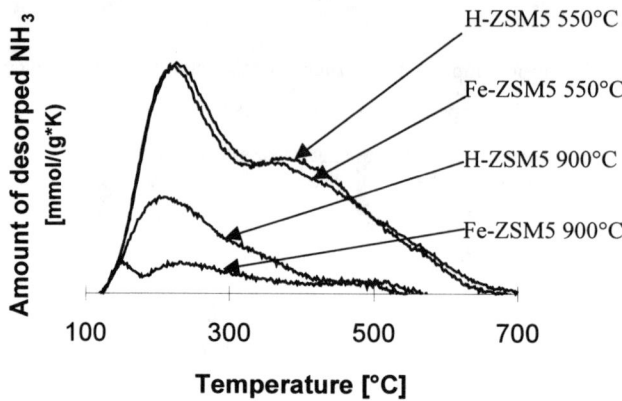

Figure 1: Influence of calcination temperature on acid properties

The TPD spectrum of ammonia (Fig. 1) shows that the iron exchanged sample calcined at 550°C possesses almost the same Brønsted acidity (high temperature peak at 400 °C) as the H-ZSM-5. It can be assumed that high acidity of Fe-ZSM-5 resulted from protons introduced upon ion-exchange from $Fe(NO_3)_3$ solution (pH~4) and from hydrolysis of Fe^{3+} cations. Part of the cations ($Fe(OH)_2^+$, $Fe^{3+}x6H_2O$) remained on the surface of zeolite. Additionally, some contribution of ammonia adsorption on Fe^{3+} species could responsible for this behaviour. After calcination at 900 °C, the amount of adsorbed ammonia is much lower in the case of the Fe-ZSM5 than in the case of H- form. This is consistent with the solid state ion exchange between Fe cations and protons occuring at this temperature.

The mid-IR spectra of zeolites ZSM-5 modified with Fe^{3+}, Co^{2+}, Cr^{3+} did not differ noticeably from those of parent Na ZSM-5 [9]. In the case of silicalite treated with $FeCl_3$ vapors, new distinctive bands are seen at 1050, 900 and 780 cm^{-1} (Fig. 2). Such bands were reported for ferrisilicalite [7, 10] which means that iron could be introduced into the lattice generating negative charges. So additional iron could be stabilized at the corresponding cationic sites.

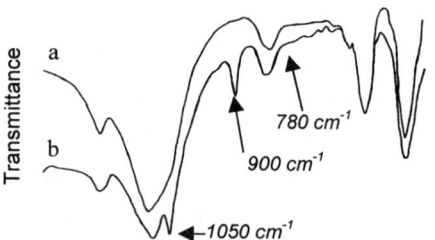

Figure 2: IR spectra of the framework vibration
a:silicalite, b:$FeCl_3$ treated silicalite

In an earlier work [9] we found that transition metal modified ZSM-5 showed substantial activity for N_2O decomposition. Activity of Co-ZSM-5 was very high and temperature indispensable to N_2O decomposition was lower than that of the other cation modifications, regardless the calcination temperature. The Fe-ZSM-5 samples calcined at extreme temperatures (500 and 900°C) showed lower activity than those heated between 600°C and 800°C.

The catalytic tests for the hydroxylation of benzene by nitrous oxide show a different activity order than that for the N_2O decomposition. Fe-ZSM5 has been found much more active than Co-ZSM-5 – particularlly, after calcination at 900 °C (see Fig. 3 and Fig. 4). Furthermore, it can be deduced from Figure 3 that Fe-ZSM-5 is the most active, but not the most selective catalyst. The rather unexpected activity of Na-ZSM-5 could be explained by some iron impurities (0.08 wt% Fe_2O_3). But this also supports the opinion that Brønsted protons are not necessary for the reaction [6]. The low activity of Na, Co and Cr modifications calcined at 550°C is consistent with findings of PANOV who believes that only iron can form the active complexes enable to decompose N_2O [3, 4] and subsequently to hydroxylize benzene [5, 6]. Calcination at 900 °C changes the level and the order of activity of the samples, emphasizing the superiority of iron modification.

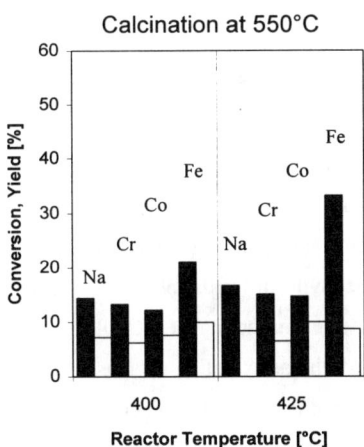 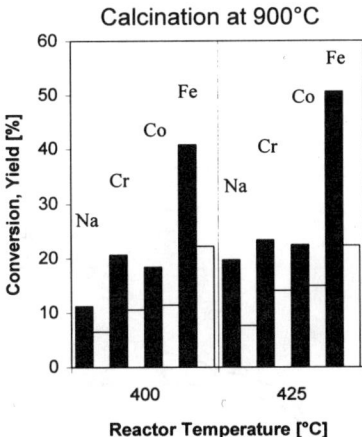

Figure 3: Influence of different cations on the catalytic activity in benzene hydroxylation dependent on calcination temperature ((a: 550°C, b: 900°C): ■ conversion of benzene, ☐ yield of phenol)

While the Na-ZSM-5 which was not practically affected by calcination at 900 °C, activity and selectivity in phenol production increased in the case of chromium– and cobalt modifications after 900°C calcination. Taking in account that the decrease of the micropore volume (0.15

cm^3/g → 0.1 cm^3/g) and of crystallinity was in the the same range for all samples, these metals also seem to be able to enhance activity. Activity of Fe-ZSM-5 as well as selectivity towards phenol exhibits maximum values at a calcination temperature of 900°C (Fig. 4) for all reaction temperatures measured. Calcination at 1000°C reduces activity and selectivity drastically probably due to the loss in crystallinity, as previously described.

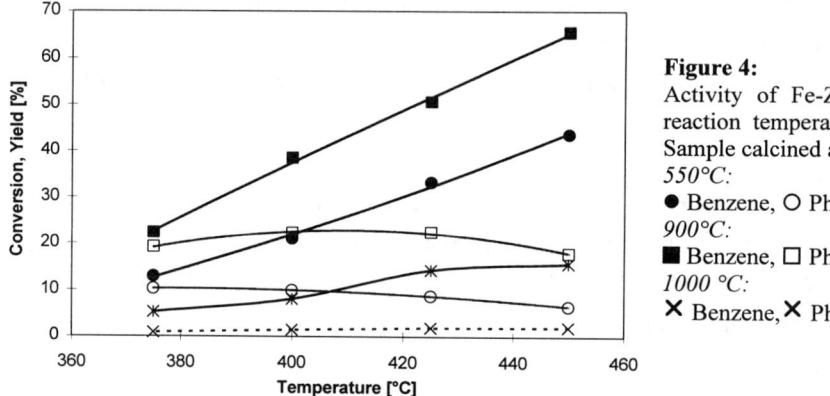

Figure 4:
Activity of Fe-ZSM-5 vs. reaction temperature. Sample calcined at
550°C:
● Benzene, ○ Phenol
900°C:
■ Benzene, □ Phenol
1000 °C:
✕ Benzene, ✕ Phenol

The phenol yield and benzene conversion are very similar at 375°C, but selectivity decreases above 400°C. The high chosen ratio of nitrous oxide to benzene of 6.25 in the first experiments also favours total oxidation; thus, the N$_2$O/C$_6$H$_6$ ratio in the feed was reduced step by step (Fig 5). Conversion of benzene decreased gradually, the phenol yield remained almost constant down to a ratio of 2:1. Selectivity towards phenol increased and the production of CO$_2$ decreased. Also, the deactivation decreases with a lower nitrous oxide content. Figure 6 demonstrates that at a N$_2$O/C$_6$H$_6$ ratio of 1 up to a time on stream of 180 minutes, the catalyst exhibits about 85% of its initial phenol yield. Selectivity was still increasing.

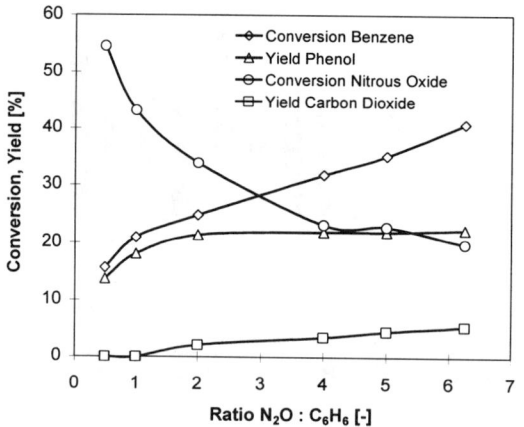

Figure 5: Influence of N$_2$O/C$_6$H$_6$ ratio on the catalytic performance in phenol production. Reaction temperature: 400 °C

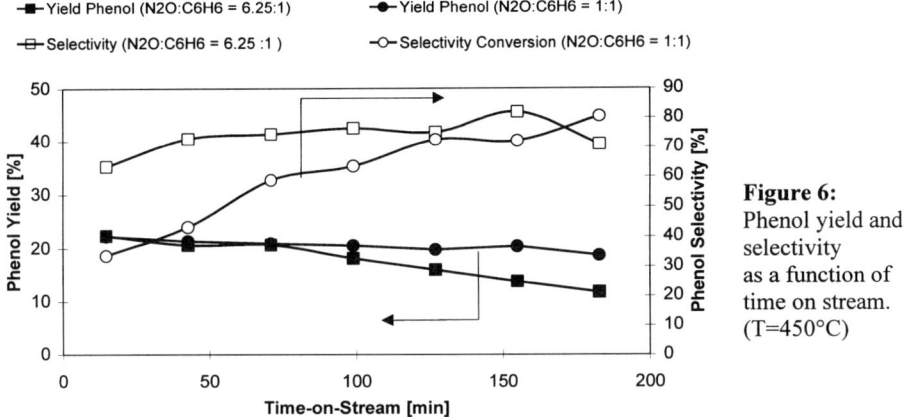

Figure 6:
Phenol yield and selectivity as a function of time on stream. (T=450°C)

Silicalite and AlPO$_4$-5 impregnated with various iron salts (FeCl$_3$, Fe(NO$_3$)$_3$, Fe$_2$(SO$_4$)$_3$) did not show any remarkable activity in phenol production, although the zeolite contained significant amounts of iron. However, the silicalite sample modified with FeCl$_3$ vapor showed some activity (about 1 to 2 % conversion and 0.5 % phenol yield) and no CO$_2$ was produced compared to the untreated silicalite which exhibited only total oxidation.

CONCLUSIONS

Although the nature of the iron species responsible for the catalytic activity remains still not quite clear, it seems very likely that there are positively charged moieties combined with the framework by means of ionic bonds playes the role of the active sites. It seems that such species can be generated either by removal of the framework iron from ferrisilicalite and aluminoferrisilicalite, respectively, or by thermal transformation of iron cations introduced into zeolite by ordinary ion-exchange.

ACKNOWLEDGEMENTS

The authors wish to thank TEMPUS program supporting the Polish German cooperation. We are grateful for the financial support for parts of the work from the Deutsche Forschungsgemeinschaft (DFG). A. Reitzmann wish to thank Max Buchner Stiftung of Dechema for a scholarship.

REFERENCES

[1]   A. Budzinski, Chem. Ind. **5,** 13 (1996).
[2]   G.I. Golodets, Heterogeneous Catalytic Reactions Involving Molecular Oxygen, Elsevier, Amsterdam, 1983.
[3]   V.I. Sobolev, G.I. Panov, A.S. Kharitonov, V.N. Romannikov, A.M. Volodin, K.G. Ione, J. Catal. **139**, 435 (1993).
[4]   M.J. Filatov, A.G. Pelmenschikov, G.M. Zhidomirov, New Frontiers in Catalysis, Ed. by L. Guczi (Elsevier, 1993) p. 312.
[5]   G.I. Panov, A.S. Kharitonov, V.I. Sobolev, Appl. Catal. A **98**, 1 (1993); A.S. Kharitonov, G.A. Sheveleva, G.I. Panov, V.I. Sobolev, Ye.A. Paukshtis, V.N. Romannikov, ibid. **98**, 33 (1993).
[6]   V.I. Sobolev, K.A. Dubkov, E.A. Paukshtis, L.V. Pirutko, M.A. Rodkin, A.S. Kharitonov, G.I. Panov, Appl. Catal. A. **141**, 185 (1996).
[7]   R. Szostak, V. Nair, T.L. Thomas, J. Chem. Soc., Farad. Trans. I **83**, 487 (1987).
[8]   A. Reitzmann, H. Friedrich, E. Klemm, M. Häefele, G. Emig, Proceedings of 3rd Polish-German Zeolite Colloquium, Edited by M. Rozwadowski (Nicholas Copernicus University Press, Torun, 1998) p. 239.
[9]   K. Nowinska, S. Kowalak, M. Sopa, M. Swiecicka, Proceedings of 3rd Polish-German Zeolite Colloquium, Edited by M. Rozwadowski (Nicholas Copernicus University Press, Torun, 1998) p. 215.
[10] S.Kaliguine, J.B. Nagy, Z. Gabelica, Keynotes in Energy Related Catalysis, Edited by S. Kaliguine (Elsevier, Amsterdam, 1988) p. 381, R.B. Borade, Zeolites **7**, 389 (1987).
[11] A.V. Kucherov, A.A. Slinkin, Kinetika i Kataliz **28** (5), 1199 (1987); Zeolites **8**, 110 (1988).
[12] M. Häfele, A. Reitzmann, D. Roppelt, G. Emig, Appl. Catal. A **150**, 153 (1997).
[13] L.V. Pirutko, A.S. Kharitonov, V.I. Bukhtiyarov, G.I. Panov, Kinetika i Kataliz **38** (1), 88 (1997).

MODIFICATION OF ZEOLITES BY Mo WITH THE USE OF CHEMICAL TRANSPORT REACTION

A.V. KUCHEROV and A.A. SLINKIN

Zelinsky Institute of Organic Chemistry, RAS, Moscow, Russia; avk@ioc.ac.ru

Abstract

New way of modification of zeolites is discussed being based on the use of the active gas-phase species formed *in-situ* upon thermal treatment of the mixture [H-zeolite + MoO_3] with an air flow containing CCl_4. Chemical transport reaction, with formation of reactive and mobile oxychloride fragments in the zeolitic bed, provides effective dissipation of the oxide phase and migration of active species into zeolitic channels at temperatures as low as 150 - 200°C. Peculiarities of introduction of Mo^{5+}-ions into H-forms of several zeolites (ZSM-5, Beta, ferrierite, USY) differing in the channel size/structure are monitored by ESR.

INTRODUCTION

Atomic-scale engineering of catalytic functions of isolated redox sites located in confined environments of zeolitic voids ('biomimetic' systems, filled zeolites) is of great interest. However, a conventional ion exchange from solutions is of limited usefulness for introduction of many hydrated multi-charged ions into small cavities of different zeolites. In some cases, a solid-state reaction between zeolites and different compounds can be used for introduction of one or several transition metal ions into cationic positions of zeolites [1,2] but for oxides this method usually requires too high temperature of calcination. Use of chlorides permits to reduce drastically the temperature of the solid-state exchange [1-5]. In some cases, the gas phase nature influences noticeably the process of ion migration upon calcination of zeolite/oxide mixtures [2, 6-8].

The aim of our work is to study how the presence of CCl_4 molecules in the gas phase enhances the possibility of "solid-state" introduction of Mo^{5+} ions in different zeolites due to the change in cationic fragments mobility by *in-situ* formation of oxychloride gas-phase species.

EXPERIMENTAL

Molybdenum-containing samples, with 2.0 wt% of MoO_3, were prepared by mechanical mixing of precalcined MoO_3 with H-forms of zeolites (ZSM-5, Si/Al = 15; beta, Si/Al = 8; ferrerite, Si/Al = 4; USY, Si/Al = 5; crystallinity of all the zeolites > 90%) in a mortar, by pestle. Powders were pressed without binder and crushed into 5-10 mm pieces. A reference sample, 2%MoO_3/H-ZSM-5, was prepared by incipient wetness impregnation (0.8 cm^3 per 1 g of zeolite) of H-ZSM-5 by water solution of ammonium molybdate with subsequent drying at ~120°C. Two samples, with 1% and 20 wt% of MoO_3 supported on amorphous Al_2O_3 (S $\cong$ 120 m^2/g), were prepared by the same method for reference.

Prepared granulated samples were placed in a quartz reactor tube, and calcined at 500°C in an air flow for 2 h. At 20°C the air flow was switched to the bubbler filled with CCl_4, reactor temperature was raised during 1 h to a given value (between 150 and 200°C), the sample was calcined in the stream of air saturated with CCl_4 vapor for 2 h, cooled to 20°C, and blown out with dry air for 2-3 h.

The ESR spectra, at 200°, 20° and -196°C, were obtained in the X-band (λ = 3.2 cm) on a reflecting type spectrometer equipped with a heater permitting the sample heating up to 400°C in the cavity. The ESR signal from DPPH (g = 2.0036) was used as a standard. The Origin 3.5 program for Windows was used for the treatment (baseline correction, double integration, and subtraction) of the recorded spectra. The ESR signals of Mo^{5+} were registered in the field region from 2500 to 3500 G (one scan with a sweep time of ~3.5 min).

The samples were crushed into 0.1 - 0.2 mm pieces, and 25 - 50 mg of a sample was placed in a glass ampoule with an inner diameter of 3 mm for ESR measurements. Two modes of ESR study were used:

1.*Comparison of different samples pretreated in identical conditions.* The samples pretreated in an [air + CCl_4] flow were placed in a glass ampoule, evacuated to 10^{-2} Torr, and sealed off. To provide the maximum accuracy of the ESR measurements, taken at 20° or -196°C, the packing height of the ampoule was 25 mm in all cases, with the center of the sample positioned in the center of the ESR cavity.

2.*In-situ ESR monitoring of the system transformation.* The [zeolite+oxide] mixture was precalcined at 450°C on air, placed in the glass ampoule positioned in the ESR cavity heater, and sealed to a static adsorption system permitting evacuation of the sample to 10^{-6} Torr and ingress of different gases. This system permits *in-situ* ESR monitoring of transformation of the starting sample upon heating at temperatures up to 400°C in atmosphere of different gases (O_2, H_2, CCl_4, C_2H_4).

RESULTS AND DISCUSSION

In a blank experiment interaction of pure H-ZSM-5 with an [air + CCl_4] flow was studied at 300°C. A rather weak ESR spectrum appears as a result of this treatment. The signal formed, with g_{xx} = 2.017; g_{yy} = 2.010; g_{zz} = 2.004, depends markedly on the O_2 pressure and can be unambiguously assigned to the signal from O_2^- radical species stabilized on the defect sites of the zeolitic lattice. So, the rupture of some stressed framework bonds in zeolite takes place upon the sample treatment in [air + CCl_4] mixture. However, the number of the defect sites formed comprises less than 1% from the number of the lattice Al^{3+} ions in the sample. Thus, only minor dealumination of the zeolite occurs as a result of the sample treatment even at 300°C.

Molybdenum introduction into zeolites

(a) Calcination of oxide samples.

The solid-state interaction, at T ≤ 600°C, between mordenite or ZSM-5 and MoO_3, in air or in vacuum, did not result in an ESR signal from Mo(V) ions. The same is true of the other zeolites studied in present work. Reduction of both starting and air-calcined mixtures, by H_2 at 400°C, produces identical weak Mo(V) ESR signals due to reduction of the surface of the MoO_3 phase (2% wt.) added to zeolite. Therefore, no disintegration of bulk MoO_3 admixture takes place upon oxidative calcination. It could be assumed that Mo(VI) ions cannot enter into zeolitic channels despite the rather high volatility of MoO_3.

From the other hand, a solid-state interaction of $MoCl_5$ in vacuum with H-ZSM-5 at 150°C resulted in effective dispersion/stabilization of isolated Mo(V) ions in the zeolite. Mo(V) bonding

(b) CCl₄-assisted interaction between molibdena and zeolites.

An effectiveness of the CCl_4-assisted process depends noticeably on conditions of treatment : volatile Mo-compounds formed upon too severe heating of the sample could provide sublimation of Mo out from the hot reactor zone. At given temperature, this carrying of Mo away from the sample is much more pronounced in an air stream (as compared with static treatment used in our *in-situ* ESR measurements). For MoO_3/zeolite mixtures the upper limit of calcination temperature is 200°C at the air flow of 50 - 100 cm³/min.

Starting MoO_3/zeolite mixtures demonstrate no ESR signals. Treatment of different samples at 200°C in the [air + CCl_4] flow (50 cm³/min, 2 h) results in appearance of narrow symmetric ESR spectra typical of interacting Mo^{5+} ions in molibdenum blues (Fig. 1).

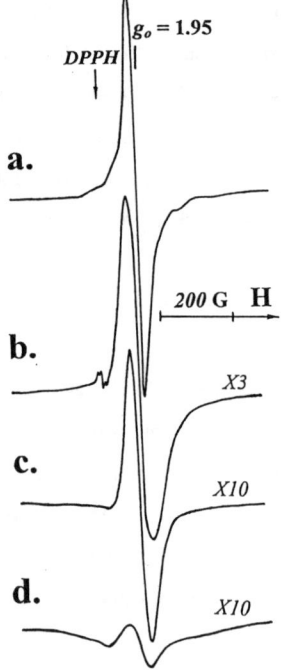

Fig. 1. ESR spectra, obtained at 20°C, of 2%MoO_3/zeolite mixtures treated at 200°C for 2 h in the [air+CCl_4] flow of 50 cc/min :
(a) – H-beta;
(b) – H-ZSM-5;
(c) – ultrastable H-Y;
(d) – H-ferrierite.

In spite of identical treatment conditions, the result obtained depends drastically on the type of the zeolite: for both H-ZSM-5 and H-Beta strong ESR signals from Mo^{5+} ions are obtained (Fig. 1a,b), a moderate ESR spectrum is seen for USY (Fig. 1c), and only a weak Mo^{5+} spectrum is observed for H-ferrierite (Fig. 1c). Samples differ also in color change caused by treatment. Although all starting mixtures are white, Mo/ZSM-5 and Mo/Beta take on a quite intense blue color after treatment (suggesting formation of disperse phases of 'molybdenum blues'), whereas the two other ones retain its white color. Thus, the process of MoO_3 dissipation takes place in condition chosen but stabilization of mobile Mo-species depends strongly on the type of the zeolite. A negligible stabilization of isolated Mo^{5+} ions by H-ferrierite (a two-orders below those typical of H-ZSM-5 or H-Beta) can be related with an essential difference in kinetic diameters of the zeolites studied (0.8 nm for Y, 0.6 nm for ZSM-5 and Beta, and 0.4 nm only for ferrierite). One can assume that the difference in the pore sizes becomes critical just for migration of oxychloride Mo-species into a narrow pore structure of ferrierite, and a weak ESR signal is associated with Mo^{5+} ions stabilized on the outer zeolitic surface only.

A noticeable difference in ability of USY and H-ZSM-5 (or H-Beta) to stabilize Mo^{5+} upon identical treatment cannot be explained by the geometrical factor: the effective size of entrances in H-Y exceeds noticeably the diameter of channels in H-ZSM-5 or H-Beta. Therefore, a rather low concentration of isolated Mo^{5+} ions stabilized by USY is not related with steric hindrance for migrating species. Formation of noticeable portion of clustered Mo^{5+} species is most likely to occur inside large pores of the zeolite Y.

The crucial role of the support nature/structure in effectiveness of the CCl_4-assisted Mo dispersion is confirmed upon the comparative study of MoO_3/alumina samples. Only a weak effect is observed for the sample with 20% of MoO_3, and no stabilization of Mo^{5+} species is found in the 1%MoO_3/Al_2O_3 treated with [air + CCl_4] at 200°C. It is likely that the difference observed is representative of sharp distinctions in nature of the support surfaces: in contrast to the zeolites used, Al_2O_3 surface contains negligible amount of protonic (Broensted) acid sites.

The two zeolitic systems provide the most effective CCl_4-assisted filling of channels with Mo^{5+}-species : H-ZSM-5 and H-Beta. These two samples demonstrate intense (~10^{20} spin/g) and narrow ESR lines from Mo^{5+} ions. It seems that just these two zeolites, with intermediate channel size, both allow a quite easy migration of reactive Mo-species to cationic positions inside channels and provide preferential fixation of Mo^{5+}-species in restricted inner voids. It is difficult to imagine, for high-silica zeolite with a large distance between neighbor lattice Al atoms, a site capable of polyvalent cation coordination without additional ligands. Therefore, we suggest that isolated complex species [$(MoO_2)^+$, $(MoOCl_2)^+$, or $(MoCl_4)^+$], and not single Mo^{5+} ions, are filling the zeolitic channels.

Properties of Mo-containing zeolites

The Mo(V) state stabilized by the ZSM-5 matrix is quite stable at room temperature. However, Mo(V) oxidation state is not tolerant to oxidative calcination, and the sample treatment in air at 200-300°C results in total disappearance of the Mo^{5+}-ESR signal pointing to a quantitative oxidation of the Mo(V).

Reduction of the oxidized sample by H_2 (~10 Torr) at 400°C leads to appearance of a new Mo^{5+} ESR spectrum, shown in Fig. 2.

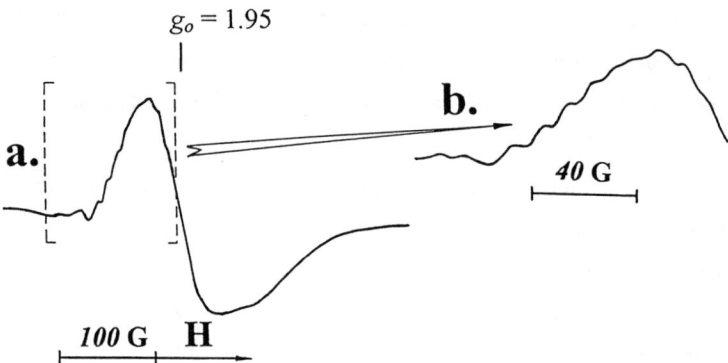

Fig. 2. ESR spectra, taken at 20°C, of MoO_3/H-ZSM-5 treated at 200°C in the [air + CCl_4] flow, oxidized at 400°C in air for 1 h, and reduced by H_2 (~10 Torr) at 400°C for 2 h.

This less intense and asymmetric ESR-line from isolated Mo^{5+} ions ($g_\perp$ = 1.955, $g_\parallel$ = 1.895) demonstrates a very informative peculiarity : an additional superhyperfine splitting, with $\Delta H \cong$ 7÷8 G, can be distinguished (Fig. 2a). The formation of the extra structure may be only due to an interaction of Mo^{5+} unpaired electron with an outershell ^{27}Al (nuclear spin 5/2). Therefore, in spite of non-perfect resolution of lines, just presence of the shfs of this type gives an evidence of cationic location of the Mo^{5+} specie, with the lattice Al^{3+} of the zeolite positioned in the second coordinative sphere.

Reduction of the CCl_4-treated Mo/H-ZSM-5 by C_2H_4 (~10 Torr) at 400°C results also in formation of the less intense and slightly asymmetric ESR-line from isolated Mo^{5+} ions resembling the spectrum of the sample reduced by H_2 but shf-splitting is absent in this case. The lack of the shfs could be explained by the fact that residual CCl_4 is not removed before the reduction, and further treatment with hydrocarbon molecules is accompanied by formation of the "coke" residue preventing electronic interaction of the Mo^{5+} cations with the lattice Al^{3+} ions. Deposition of the coke of the sample is confirmed by appearance of the narrow ESR line with $g \cong 2.002$.

In-situ ESR monitoring of Mo^{5+}/ZSM-5 and Mo^{5+}/Beta in flowing gas mixtures

The Mo^{5+}-ESR signal is not altered at 500°C in He flow save for an intensity decrease due to the Curie-Weiss law. When treated with a flow of [1% vol H_2 + He] the ESR signal associated with the isolated Mo^{5+} ions did not change the shape in the temperature range from 20 to 500°C.

Switch to [0.4%NO + He] flow at 20°C is accompanied by fast appearance of a new signal from paramagnetic species, with $g_1 = 2.041$ and $g_2 = 2.011$. The intensity of this signal reaches a maximum in the first 2-4 min in stream, and a monotone drop of this signal is observed during the next 1 h. At the same time, the signal from isolated Mo^{5+} ions, with $g_\perp = 1.956$ and $g_\parallel = 1.895$, disappears gradually. Back switch to pure He does not restore the starting Mo^{5+}-ESR signal. Therefore, irreversible change of the paramagnetic Mo(V) sites occurs upon interaction of Mo/ZSM-5 with NO at 20°C. One can assume that intermediate formation of paramagnetic complex specie (NO^{2-}) takes place, with subsequent quenching of the two unpaired spins and oxidation of Mo(V) to Mo(IV). It seems that oxidation of the Mo(V) state at room temperature by NO occurs noticeably easier than oxidation by the oxygen of air. At temperatures >200°C the complete oxidation of Mo(V) by the [0.4%NO + He] flow occurs very fast, and no formation of intermediates can be monitored.

The ESR-signal from Mo^{5+} ions, taken *in-situ* at 500°C in a flow cell [9], does not decrease in a flow of [He+H_2+NO] at different stoichiometries, going from a stoichiometric ratio, H_2/NO, of ~7 to 1.2. Going to stronger oxidizing conditions results in noticeable drop of this signal but the signal decrease is completely reversible upon back switch to the reducing gas mixture. Therefore, at 500°C in the reaction mixture of NO and H_2 the step of the catalytic site reduction is fast, and the dynamic equilibrium of the redox reaction $Mo^{4+} \leftrightarrow Mo^{5+}$ seems to be strongly shifted to Mo^{5+}.

CONCLUSIONS

Modification of zeolites with use of active gas-phase species formed *in-situ* upon thermal treatment of the mixture [H-zeolite + modifier oxide] with an [air + CCl_4] flow can be treated as a promising way of cation introduction. The presence of CCl_4 molecules enhances the possibility of "solid-state" introduction of policharged ions in zeolites due to a sharp increase in cationic fragments mobility by *in-situ* formation of active species. Chemical transport reaction, with formation of reactive and mobile oxychloride fragments in zeolitic bed, provides effective dissipation of oxide phase and migration of active species into zeolitic channels in mild conditions.

ACKNOWLEDGMENT

This work was supported by Russian Foundation for Basic Research (Grant 98-03-32010a) .

REFERENCES

1. H.G. Karge and H.K. Beyer, Stud. Surf. Sci. Catal., **69**, 43 (1991).
2. A.A. Slinkin and A.V. Kucherov, Catal. Today, **36**, 485 (1997).
3. J.A. Rabo and P.H. Kasai, Progr. Solid State Chem., **9**, 1 (1975).
4. A. Clearfield, C.H. Saldarriaga, and R.C. Buckley, Proc. 3-rd Int. Conf. Molec. Sieves, Sept. 3-7, 1973, Zurich, Switzerland, (Recent Progress Reports, J.B.Uytterhoeven, (Ed.) Univ. Leuven Press, 1973) Paper 130, p. 241.
5. H.G. Karge, Y. Zhang, and H.K. Beyer, Catal. Letters, **12**, 147 (1992).
6. B. Wichterlova, S. Beran, S. Bednarova, K. Nedomova, L. Dudikova, and P. Jiru, Stud. Surf. Sci. Catal., **37**, 199 (1987).
7. S. Beran, B. Wichterlova and H.G. Karge, J. Chem. Soc., Faraday Trans.1, **86**, 3033 (1990).
8. Zhu, Jian-hua, Cuihua Xuebao (Chines J. Catal.), **14**, 294 (1993).
9. A.V. Kucherov, J.L. Gerlock, H.-W. Jen, and M. Shelef, J. Phys. Chem., **98**, 4892 (1994).

CATALYTIC DECOMPOSITION OF N_2O OVER NaY-SUPPORTED AND USY-SUPPORTED Rh CATALYSTS

KOICHI YUZAKI [a], TAKAYOSHI YARIMIZU [a], KENJI AOYAGI [a], SHIN-ICHI ITO [a], TAKAYUKI SATO [a], SHIGENOBU HAYASHI [b] AND KIMIO KUNIMORI [a]

[a] Institute of Materials Science, University of Tsukuba, Tsukuba, Ibaraki 305-8573, Japan
[b] National Institute of Materials and Chemical Research, 1-1 Higashi, Tsukuba, Ibaraki 305-8565, Japan

Abstract

The catalytic decomposition of nitrous oxide to nitrogen and oxygen has been studied over zeolite (NaY, USY) supported rhodium catalysts. Rh/USY catalysts were more active than Rh/NaY catalysts, and the activity of N_2O decomposition was increased by the precalcination at 700°C of the zeolite support (NaY, USY). The precalcination effect was more significant for Rh/NaY catalysts. The results were discussed in terms of characterization of the catalysts using H_2 and CO chemisorption and ^{29}Si and ^{27}Al NMR.

Introduction

Nitrous oxide (N_2O) is a noteworthy environmental pollutant because it contributes to the catalytic destruction of stratospheric ozone and is a greenhouse gas. The development of an active catalyst for N_2O decomposition will be necessary to abate N_2O emission into the atmosphere. So far, Li and Armor [1] reported that Rh (or Ru)/ZSM-5 catalysts were very active for N_2O decomposition. Oi et al. [2, 3] studied the activities of Rh catalysts supported on various oxides, and reported that Rh/ZnO was the most active catalyst. Recently, we reported the activities (and TOF values) of N_2O decomposition over Al_2O_3-supported and zeolite-supported Rh catalysts [4], and found that Rh/Al_2O_3 and Rh/USY (ultrastable Y zeolite) prepared from $Rh(NO_3)_3$ were comparable to or more active than the other catalysts in the literature [1-3]. In this work, we have studied the effect of the precalcination of zeolite supports on the catalytic activity. All the supports were examined by ^{29}Si and ^{27}Al high-resolution solid-state NMR technique with magic-angle spinning. The BET surface areas and pore volumes were also measured.

Experimental

The catalyst supports used are NaY (SK-40, GL SCIENCE) and USY (TOSO, $Si/Al_2 = 14.6$). Rh/NaY and Rh/USY catalysts were prepared by incipient wetness impregnation with aqueous solutions of $RhCl_3$ or $Rh(NO_3)_3$, followed by air calcination at 600°C for 3h. Before the impregnation, the NaY and USY supports were precalcined in air for 3h at 600°C or 700°C (designated as NaY(600), USY(600), NaY(700) and USY(700)). The Rh loading was 2wt% for all the catalysts studied. The N_2O decomposition reactions were carried out on a 50 mg catalyst sample with a total flow rate of 50 cm^3/min using a micro-catalytic reactor [5]. N_2O concentration was 5000 ppm in helium. Unless otherwise stated, the catalysts were pretreated in H_2 at 500°C before use. The reaction ($N_2O \rightarrow N_2 + 1/2\ O_2$) was catalytic (i.e., no occurrence of the non-catalytic N_2O reaction: $N_2O \rightarrow N_2 + O(a)$ [6]) because the product $N_2 : O_2$ ratio was 2 : 1, and the activity did not decrease during the reaction for more than 8h. The H_2 and CO chemisorption measurements were performed by a conventional volumetric adsorption apparatus [7]. The total adsorption of H_2 (H/Rh), as well as irreversible adsorption of CO (CO/Rh), was measured after H_2 reduction at 500°C [7]. The TOF values (molecules of N_2O converted per number of surface Rh per second) were calculated by assuming that the adsorption stoichiometry (H/surface Rh, CO/surface Rh) is unity. The BET surface areas and pore volumes of zeolite supports were measured by a Gas Sorption Analyzer (Coulter Omunisorp Series). NMR measurements were carried out by a Bruker MSL400 spectrometer with a static magnetic field strength 9.4 T [8]. Larmor frequencies were 79.50 and 104.26 MHz for ^{29}Si and ^{27}Al, respectively. The samples were spinning about an axis inclined at the magic angle (54.7°) to the static field. The sample temperature was ambient. ^{29}Si and ^{27}Al spectra were obtained using the single pulse sequence under MAS conditions at a spinning rate of 4 and 10 kHz, respectively.

Results and Discussion

Figure 1 shows the activities of N_2O decomposition over zeolite (NaY, USY) supported Rh catalysts. The Rh/USY(NO_3) catalysts prepared from Rh(NO_3)$_3$ were more active than the other catalysts. The USY-supported Rh catalysts showed 100% conversion of N_2O (to nitrogen and

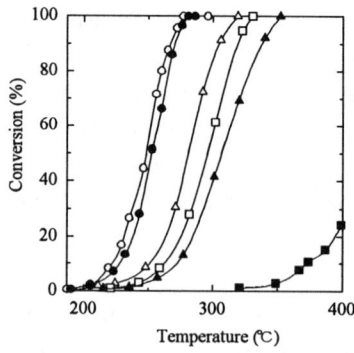

Figure 1. N_2O conversion on Rh/zeolite catalysts as a function of catalyst temperature. ○: USY(700)(NO_3); ●: USY(600)(NO_3); △: NaY(700)(NO_3); ▲: NaY(600)(NO_3); □: NaY(700)(Cl); ■: NaY(600)(Cl). NO_3: Rh(NO_3)$_3$; Cl: RhCl$_3$.

oxygen) at the temperatures above about 280°C. The catalysts showed almost the same activity in the run after O_2 oxidation at 500°C. The activity of the Rh/NaY(600)(Cl) prepared from $RhCl_3$ was much lower than that of the Rh/NaY(700)(Cl). The lower activity of the Rh/NaY(600)(Cl) may be caused by residual Cl in the catalyst, because the activity was increased by washing with hot water (80°C) [4]. On the other hand, the activity of the Rh/NaY(700)(Cl) did not change by washing and Cl was hardly detected. In order to avoid the poisoning effect of Cl, we compared the activities of the catalysts prepared from $Rh(NO_3)_3$. As shown in Fig.1, the precalcination at 700°C of the NaY zeolite support resulted in higher activity (i.e., Rh/NaY(700)(NO_3) > Rh/NaY (600)(NO_3)). The activity of the Rh/USY(700) was also slightly higher than that of the Rh/USY(600). The BET surface area and the micropore volume were slightly decreased (from 765 to 729 m^2/g and from 0.329 to 0.315 cc/g) by the calcination at 700°C for the NaY, but almost the same (ca. 690 m^2/g and 0.260 cc/g) for the USY.

Table 1 summarizes the H_2 and CO chemisorption values, and the TOF values of the Rh/zeolite catalysts. The Rh dispersion (H/Rh or CO/Rh) of the Rh/NaY(700)(NO_3) was significantly higher than that of the Rh/NaY(600)(NO_3), which may be in good accordance with the higher activity of the former. A little higher value (0.62) of the Rh/USY(700) may be obtained as a result of hydrogen spillover onto the support. Judging from the CO/Rh values, the Rh dispersion may be similar between the two Rh/USY catalysts, but the Rh/USY(700) showed the highest TOF value.

Figures 2 and 3 show ^{29}Si and ^{27}Al NMR spectra of NaY and USY, respectively. The ^{29}Si spectra of NaY have four peaks assigned to Si(3Al), Si(2Al), Si(1Al) and Si(0Al), while those of USY show only one peak attributed to Si(0Al). Only the peak corresponding to four-coordinated Al is observed in ^{27}Al spectra of NaY, whereas three peaks are observed in those of USY, which are ascribed to Al atoms coordinated by four, five and six oxygens. The five- and six-coordinated Al ions correspond to extraframework species. The framework Si/Al ratios are estimated from the relative intensities, $I_{Si(nAl)}$, of the Si(nAl) signals in the ^{29}Si spectra using the following formula,

$$Si/Al = \sum_{n=0}^{4} I_{Si(nAl)} / \sum_{n=0}^{4} 0.25 n I_{Si(nAl)}$$

Table 1 Comparison of the Rh dispersion and the TOF values of Rh/zeolite catalysts (5000 ppm)

Catalyst	H/Rh	CO/Rh	TOF (x 10^{-3} s^{-1}) [a]	TOF (x 10^{-3} s^{-1}) [b]	T_{100} (°C) [c]
Rh/USY (700)(NO_3)	0.62	0.47	25.7	33.3	277
Rh/USY (600)(NO_3)	0.54	0.46	17.4	20.4	281
Rh/NaY (700)(NO_3)	0.52	0.37	3.77	5.25	319
Rh/NaY (600)(NO_3)	0.27	0.19	2.24	3.18	352
Rh/NaY (700)(Cl)	0.22	0.17	4.09	5.52	331
Rh/NaY (600)(Cl)	0.11	0.03	0.03	0.07	≫400 [d]

a) At 250°C, based on the H/Rh value. b) At 250°C, based on the CO/Rh value. c) The temperature at which 100% N_2O conversion was achieved. d) The activity was very low.

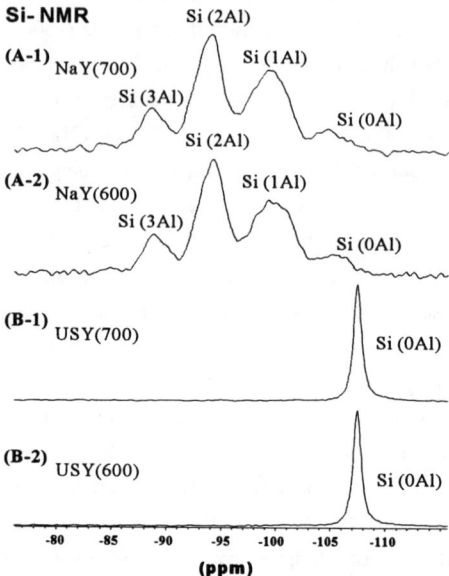

Figure 2. ^{29}Si-NMR spectra of NaY zeolite (A) and USY zeolite (B).

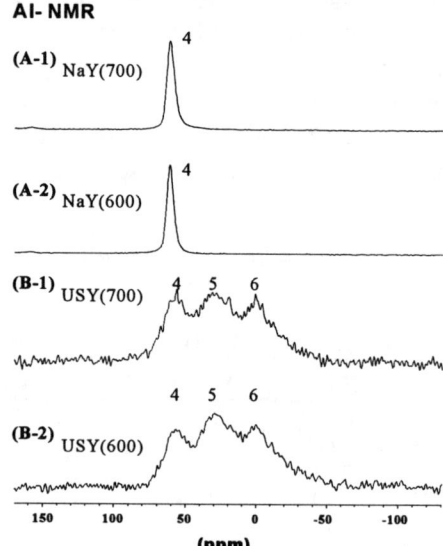

Figure 3. ^{27}Al-NMR spectra of NaY zeolite (A) and USY zeolite (B). Numbers on the peak indicate the coordination number.

the results of which are shown in Table 2 for NaY. Although the difference between NaY(600) and NaY(700) is quite small, dealumination from the framework proceeded slightly by the higher-temperature calcination. As for USY, the structural change is not clear, though subtle changes are observed in ^{29}Si and ^{27}Al spectra.

The origin of the precalcination effect on the catalytic reaction is not clear at the present stage. The higher-temperature precalcination may prevent a structural change during the following calcination and reduction treatments (e.g., encapsulation of Rh particles due to pore blockages). In practice, the H/Rh and CO/Rh values of the Rh/NaY(700)(NO$_3$) was higher than those of the Rh/NaY(600)(NO$_3$), in which part of Rh particles may be encapsulated due to the possible pore blockages. Another possibility may be that the extraframework Al plays an important role in the higher activity. For NaY, dealumination from the framework proceeded slightly by the higher-temperature (700°C) calcination (Table 2). For USY, the existence of the extraframework Al (both the five- and six-coordinated Al species) was evident from the ^{27}Al NMR studies. In practice, the Rh/USY catalysts were more active than the Rh/NaY catalysts. Furthermore, the precalcination may affect the properties of the extraframework ion species, resulting in different metal-support interactions.

Table 2 Si/Al ratio of the NaY zeolite samples estimated from the relative intensities of the Si(nAl) signals in the ^{29}Si NMR spectra

Sample	Si/Al
NaY (600)	2.43
NaY (700)	2.45

Conclusions

The catalytic decomposition of N$_2$O (to N$_2$ and 1/2 O$_2$) has been studied over zeolite (NaY, USY) supported Rh catalysts. Rh/USY catalysts were more active than Rh/NaY catalysts. The precalcination of the NaY zeolite support at 700°C resulted in higher activity (i.e., the Rh/NaY(700) was more active than the Rh/NaY(600)). The activity of the Rh/USY(700) was also slightly higher than that of the Rh/USY(600). The origin of the precalcination effect is not clear. For the NaY support the BET surface area and the micropore volume were slightly decreased, and dealumination from the framework proceeded slightly by the calcination at 700°C, but the structure change is not clear for the USY support. It is suggested that the extraframework Al may play a role in the higher activity of N$_2$O decomposition.

Acknowledgement

This work was partly supported by Iketani Science and Technology Foundation.

References

1. Y. Li and J.N. Armor, Appl. Catal. B **1**, 21 (1992).
2. J. Oi, A. Obuchi, A. Ogata, H. Yagita, G.R. Bamwenda and K. Mizuno, Chem. Lett. 453 (1995).
3. J. Oi, A. Obuchi, G. R. Bamwenda, A. Ogata, H. Yagita, S. Kushiyama and K. Mizuno, Appl. Catal. B **12**, 277 (1997).
4. K. Yuzaki, T. Yarimizu, S. Ito and K. Kunimori, Catal. Lett. **47**, 173 (1997).
5. K. Kunimori, K. Yuzaki, T. Yarimizu, M. Seino and S. Ito, Studies in Surf. Sci. and Catal. **105**, 2083 (1997).
6. Y. Li and M. Bowker, Surf. Sci. **348**, 67 (1996).
7. K. Kunimori, T. Uchijima, M. Yamada, H. Matsumoto, T. Hattori and Y. Murakami, Appl. Catal. **4**, 67 (1982).
8. M. Satozawa, K. Kunimori and S. Hayashi, Bull. Chem. Soc. Jpn. **70**, 97 (1997).

TOTAL CATALYTIC OXIDATION OF ACETIC ACID BY H_2O_2 OVER TRANSITION METAL-EXCHANGED NaY ZEOLITES

S. LÉVESQUE, Y. YANG, F. LARACHI, A. SAYARI

Department of Chemical Engineering & CERPIC– Université Laval, Ste-Foy, Québec, Canada, G1K 7P4

ABSTRACT

During the wet oxidation of contaminated wastewaters, the destruction of low molecular weight carboxylic acid intermediates such as acetic, glyoxalic, and oxalic acids is often the rate-controlling step. Oxidation of acetic acid, the most recalcitrant intermediate, requires compelling treatment severity. Heterogeneous catalytic wet oxidation of model acetic acid aqueous solutions was conducted under mild conditions (below water normal boiling point) using hydrogen peroxide over various transition metal-exchanged NaY zeolites. Treatment of Cu^{2+}-NaY with oxalic acid [OA] led to a catalyst, Cu^{2+}–NaY [OA], with significantly improved properties in terms of total organic carbon (TOC) removal efficiency and catalyst stability against leaching. This catalyst outperformed homogeneous Cu^{2+} by a factor of 2-2.5 times. Continuous feeding of H_2O_2 reduced its undesirable decomposition. Improvement of the TOC-degradation performance by Cu^{2+}–NaY [OA] were tentatively attributed to the removal of sodium and possibly aluminum in the zeolite.

INTRODUCTION

Advanced oxidation processes for wastewater treatment designate a number of aqueous phase oxidation technologies based primarily on the intermediacy and immense oxidizing potential of the hydroxyl free radical (OH$^\bullet$) in the oxidative destruction of organic pollutants. The hydroxyl free radicals can be generated by various catalytic (homogeneous and heterogeneous) wet oxidation routes where hydrogen peroxide is used as an oxidizer. In the Fenton's reaction, the reagent system (Fe^{2+}/Fe^{3+}–H_2O_2) is used to unleash OH$^\bullet$ by the redox reaction between H_2O_2 and Fe^{2+}. Addition of trace amounts of Cu^{2+}, Mn^{2+} or Co^{2+} enhances the efficacy of Fenton's route [1]. The major weakness of these systems is the tight pH control to prevent precipitation of $Fe(OH)_3$, which occurs at pH > 5 [2]. Homogeneous catalysts consisting of a mixture of Cu^{2+}, Mn^{2+}, Fe^{2+} used in conjunction with H_2O_2 showed synergistic effects and high destruction levels under mild conditions [3,4]. However, without proper pH adjustment, precipitation of the catalyst may also occur. In summary, even though all the Fenton-like catalysts offer a cost-effective source of hydroxyl radicals and involve easy-to-handle reagents, additional steps are required both before and after the oxidation step for pH adjustment and catalyst precipitation/separation.

To circumvent the above drawbacks, attempts are being undertaken to explore the feasibility of heterogeneously catalyzed Fenton-like routes for the destruction of the water-dissolved organic pollutants [5,6]. This strategy combines the advantages of the Fenton-like systems and those of heterogeneously catalyzed wet oxidation. This paper is a preliminary account on the wet oxidation of acetic acid by hydrogen peroxide over new heterogeneous catalysts based on Cu^{2+}, Fe^{2+}, Fe^{3+}, Mn^{2+}, and Ce^{3+} exchanged NaY zeolites. There are two main reasons behind this selection. First, the destruction of low molecular weight carboxylic acid intermediates such as acetic acid is often a rate-controlling step in the overall oxidation process. Because of their resistance to oxidation, they tend to build up in the solution.. It is thus imperative to develop a heterogeneous catalyst able to efficiently degrade acetic acid. Second, some cations (e.g. Cu^{2+}, Fe^{2+}) were found to be excellent homogeneous catalysts. Since zeolites can play the role of a «solid solvent» in which the exchanged cations are stabilized within the pore structure, while they can move relatively freely as in solution, it is anticipated that some of these catalysts combine the efficiency of homogeneous cations and the convenience of heterogeneous catalysts. In addition, these catalysts must exhibit marginal leaching of active cations, and minimal H_2O_2 parasitic decomposition.

EXPERIMENTAL

Table 1 lists the various catalysts used in the screening tests. They were prepared by ion exchange using dilute solutions of appropriate salts. For multimetallic systems, metal cations were exchanged either individually or simultaneously.

Approximately 100 mL of a 2 g/L acetic acid (AA) aqueous solution (TOC = 800 ppm) was treated batchwise at 363 K and 1 atm in a 500 mL glass reactor equipped with a magnetic stirrer and a condenser. After the solution was heated up to the preset temperature, both catalyst and H_2O_2 (33 wt%) were added. No attempt was made to control pH during the course of the reaction.

The relative amount of H_2O_2 used will be referred to as R which represents the ratio of the actual amount of H_2O_2 to the amount required to oxidize stoichiometrically acetic acid into carbon dioxide and water. To assess the effect of the method of H_2O_2 addition on the extent of the parasitic decomposition, the peroxide was either injected at once (R = 2.6) or fed continuously (R = 1.5) over a period of 60 to 120 min.

The TOC content of solutions before and after reaction was analyzed using a Shimadzu 5050 instrument. The metal content in fresh (Table 1) and used catalysts was determined by neutron activation. The metal leaching-off was calculated accordingly. Residual H_2O_2 during wet oxidation was titrated iodometrically.

RESULTS AND DISCUSSION

A series of samples were tested with the objective of finding heterogeneous catalysts with the following properties: i) high activity in terms of TOC reduction, ii) no leaching of active ingredients to benefit from the advantages of heterogeneously catalyzed reactions, and to eliminate the additional pollution by the release of soluble metal compounds, iii) high H_2O_2 conversion with little or no decomposition.

Only fresh catalysts were employed in the screening tests. Table 2 shows TOC conversions along with the concentration of metals that leached into the solution after 60 minutes of reaction at 363 K. In these experiments, H_2O_2 was added at once, and the initial catalyst loading was 5 g/L. Only marginal TOC removal occurred in blank experiments (1st and 2nd entries). As for the catalysts, they exhibited three types of behaviors: (i) efficient TOC depletion but excessive leaching: Fe^{2+}–NaY, Fe^{3+}–NaY, Cu^{2+}:Mn^{2+}–NaY, Cu^{2+}:Ce^{3+}–NaY, Fe^{2+}:Cu^{2+}:Mn^{2+}–NaY; (ii) low TOC reduction regardless of their stability against leaching: Mn^{2+}–NaY, Ce^{3+}–NaY, Mn^{2+}:Ce^{3+}–NaY; and (iii) acceptable TOC conversion and excellent stability to leaching: Cu^{2+}–NaY. The latter catalyst was therefore chosen for further investigation.

Using equal copper loadings, homogeneous Cu^{2+} (Table 3, 2nd entry) was found to be slightly better than fresh Cu^{2+}–NaY (5th entry). With the former catalyst, less residual H_2O_2 remained, but more H_2O_2 decomposed. In the absence of organic substrate, both catalysts fully decomposed H_2O_2 after one hour.

The stability of Cu^{2+}–NaY was evaluated by running three wet oxidation cycles using the same catalyst under identical conditions (catalyst loading = 5g/L, 363 K, R = 2.6, initial TOC = 800 ppm). When the catalyst was reused directly, Cu^{2+}–NaY [U] activity remained almost constant, and almost no leaching took place. Interestingly, the catalyst activity was boosted significantly by calcination at 673 K in air for 6 hours (Cu^{2+}–NaY [UC]) and the TOC conversion was superior to that obtained over homogeneous Cu^{2+}. Recalcination and reuse of the catalyst in the third cycle (Cu^{2+}–NaY [UCUC]) did not affect its performance. The Al:Si ratio remained constant, while Na:Si decreased steadily indicating that Na^+ cations were gradually exchanged. After the third cycle, Cu^{2+} species began to leach as inferred from decreasing Cu:Si ratio.

The strong enhancement of catalyst performance despite decreasing copper content is most intriguing. At first, oxalic acid (OA) which is a possible reaction intermediate, was thought to affect the catalyst via dealumination [7]. To support this contention, a fresh catalyst was treated in a 4 g/L OA solution at the reaction temperature (363 K) for one hour, and tested for wet oxidation of AA. Table 3 shows that, under otherwise identical conditions, a remarkable improvement of the TOC removal performance was attained while Cu^{2+} exchanged ions did not leach. It is also seen that during treatment with OA, the Na:Si ratio decreased more than the Al:Si. In light of all these data, it may be tentatively concluded that the improvement in TOC abatement is related to the removal of Na^+. Additional work is underway to unravel this problem.

To evaluate the effect of the method of H_2O_2 addition on the TOC conversion and H_2O_2 decomposition, the oxidant was fed continuously over 1 or 2 hrs. Data are shown in Fig. 1 as a function of the number of moles of H_2O_2 added in the continuous mode (R = 1.5) and the pulse injection mode (R = 2.6). As seen, less decomposition (51 vs. 80 %) occurred when the continuous mode was used. Moreover, although less H_2O_2 was used, for a 2 hr injection duration, the TOC conversion reached 63 %, corresponding to an improvement of 2.5 times over the homogeneous Cu^{2+}.

REFERENCES

1. Mitsubishi Heavy Ind. K.K., Japan patent, AN 78-73285A (1978).
2. L. Plant, M. Jeff Chem. Eng., EE16 (September 1994).

3. M. Falcon et al., Rev. Sci. Eau **6**,411 (1993).
4. M. Falcon et al., Environ. Technol. **16,** 501 (1995).
5. K. Fajerwerg, H. Debellefontaine, Applied Catal. B: Environ. **10**, L229 (1996).
6. K. Fajerwerg et al., Wat. Sci. Tech. **35**, 103 (1997).
7. M.R. Apelian et al., J. Phys. Chem. **100**, 16577 (1996).

Catalyst	wt-% of active metal
Fe^{2+}-NaY	3.3
Fe^{3+}-NaY	2.8
Cu^{2+}-NaY	2.9
Mn^{2+}-NaY	3.3
Ce^{3+}-NaY	3.5
$Fe^{2+}:Mn^{2+}$-NaY	2.7, 2.3
$Cu^{2+}:Mn^{2+}$-NaY	2.7, 2.3
$Cu^{2+}:Ce^{3+}$-NaY	2.6, 4.4
$Mn^{2+}:Ce^{3+}$-NaY	3.0, 3.6
$Fe^{2+}:Cu^{2+}:Mn^{2+}$-NaY	2.5, 2.0, 1.3

Table 1 Tested catalysts.

Catalyst	% TOC removal	[Metal] in the solution (ppm)
no catalyst	10	–
NaY	5	–
Fe^{2+}-NaY	31.5	(28)
Fe^{3+}-NaY	33	(15)
Cu^{2+}-NaY	18	(<3)
Mn^{2+}-NaY	6	(35)
Ce^{3+}-NaY	6	–
$Fe^{2+}:Mn^{2+}$-NaY	15.5	(24), (32)
$Cu^{2+}:Mn^{2+}$-NaY	34	(27), (20)
$Cu^{2+}:Ce^{3+}$-NaY	26	(41), (–)
$Mn^{2+}:Ce^{3+}$-NaY	5	(35), (33)
$Fe^{2+}:Cu^{2+}:Mn^{2+}$-NaY	38	(9), (32), (20)

Table 2 Removal of acetic acid by various heterogeneous catalysts.

Catalyst	%TOC removal	H_2O_2 % residual	H_2O_2 % decomp.	Al:Si	Na:Si	Cu:Si	Na+2Cu :Si
Cu^{2+} (homo.)	no AA	0	100	–	–	–	–
Cu^{2+} (homo.)	26	6	84	–	–	–	–
Cu^{2+}-NaY	no AA	0	100	–	–	–	–
NaY	–	–	–	0.41	0.41	0.00	0.41
Cu^{2+}-NaY	18	19	74	0.41	0.29	0.06	0.41
Cu^{2+}-NaY [U]	16	29	66	0.42	0.14	0.06	0.26
Cu^{2+}-NaY [UC]	40	9	74	–	–	–	–
Cu^{2+}-NaY [UCUC]	40	15	70	0.40	0.08	0.03	0.15
Cu^{2+}-NaY [OA]	56	0	79	0.30	0.12	0.06	0.24

Table 3 Oxidation of acetic acid by Cu^{2+} and leaching-off of Na, Cu, Al from NaY

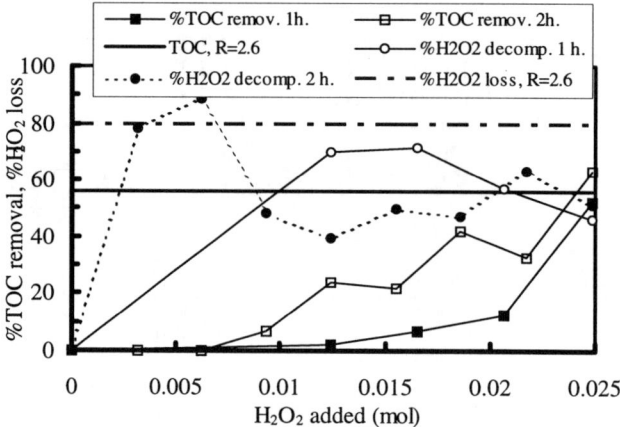

Figure 1 Oxidation of acetic acid over Cu^{2+}–NaY [OA] in the continuous mode.

REDUCTION OF IRON IONS TO THE METALLIC STATE IN X AND Y ZEOLITES BY SODIUM AZIDE

H.K. BEYER[§], G. ONYESTYÁK[§], B.J. JÖNSSON[†], K. MATUSEK[‡] and K. LÁZÁR[‡*]

[§] Institute of Chemistry, H-1525 Budapest, P.O.B. 17, Hungary; beyer@cric.chemres.hu
[†] Department of Condensed Matter Physics, Royal Institute of Technology, S-10044 Stockholm, Sweden; jj@kth.se
[‡] Institute of Isotope and Surface Chemistry, H-1525 Budapest, P.O.B. 77, Hungary; lazar@iserv.iki.kfki.hu, fax: +36-1-395-9075

ABSTRACT

Reduction of extra-framework iron ions into the zerovalent state with sodium vapour generated by thermal decomposition of sodium azide in mixtures with Fe-X, Fe-Y and La modified Fe-Y zeolites is reported. The process was studied by thermogravimetry, X-ray diffraction, magnetic susceptibility, Mössbauer spectroscopy and chemisorption measurements. Only a part of iron ions was reduced to the zerovalent state in each sample. A certain recrystallisation of the framework takes place in Fe-X, to a smaller extent in Fe-Y, while the Fe,La-Y is the most stable. Accordingly, metallic iron is detected in two size ranges in Fe-X and Fe-Y; metallic particles of 4 - 6 nm and small clusters about 1 nm in size. In the Fe,La-Y sample zerovalent iron was observed only in form of the smaller clusters.

INTRODUCTION

The preparation of metallic particles inside zeolite cages is of distinguished interest: the stabilised highly dispersed metallic phase may exhibit particular properties, e.g., in catalysis or as a nanocrystalline system. Among various elements, possibilities for preparation of iron clusters were also studied by direct reduction of extra-framework iron ions using the strong reducing agent, sodium vapour. In an early study a high temperature (1066 K) treatment was applied resulting in the formation of relatively large metallic iron particles partially located outside of the cages [1]. Application of lower temperature and longer treatments (673 K, 5 - 48 h) resulted in formation of metallic particles inside the cages in a significant portion [2].

In the present communication, the direct in situ generation of metallic sodium in a mixture of sodium azide with iron zeolites is reported. Principally, the sodium vapour formed upon thermal decomposition of sodium azide (2 $NaN_3 \rightarrow$ 2 Na^0 + 3 N_2) is utilized for an immediate reduction of the extra-framework iron ions. In fact, the decomposition of NaN_3 in faujasite is a more complex process and may result in the formation of Na_4^{3+}, Na_6^{5+} and Na_x^0 clusters as well [3].

---

[*] To whom correspondence should be addressed

Various iron zeolites (Fe-X, Fe-Y, Fe,La-Y) were subjected to this reduction process. X and Y zeolites were selected to compare frameworks of different Si/Al moduli. Incorporation of La^{3+} was performed to fill up the sodalite cages [4] prior to exchange of Fe^{2+} in order to confine the reduction of iron to proceed preferably in the supercages. The process was studied by thermogravimetry, the solid products were characterized by X-ray diffraction, AC magnetic susceptibility, Mössbauer spectroscopy, TPR and chemisorption measurements.

EXPERIMENTAL

Fe-X and Fe-Y were prepared by aqueous ion exchange in 0.1 N $FeSO_4$ solutions under nitrogen atmosphere, and partial exchange of sodium for iron was performed. The samples were dried on air. Their compositions were $Na_{38.5}Fe_{16.9}[Al_{81.5}Si_{110.5}O_{384}] \cdot 191\ H_2O$ for Fe-X and $Na_{18.1}Fe_{16.3}[Al_{54}Si_{138}O_{384}] \cdot 255\ H_2O$ for Fe-Y, respectively.

Fe,La-Y was prepared by a two-step exchange procedure. First La was introduced to the sodalite cages by the recipe of Lee and Rees [4] with ion exchange from a 0.7 N solution of $LaCl_3$, followed by a heat treatment for 5 h at a 813 K. In the second step iron was introduced into this zeolite from 0.1 N solution of $FeSO_4$. The composition of the final product was:
$La_{11.1}Na_{6.7}Fe_{7.0}[Al_{53.4}Si_{137.1}O_{384}] \cdot 238\ H_2O$.

The adsorbed water was removed from the zeolite samples by overnight drying at 623 K. Then the zeolite was mixed in an agate mortar with NaN_3 under argon in a dry box. NaN_3 was used in excess (30 % at Fe-X and Y, and 50 % at Fe,La-Y) of the stoichiometric Na/Fe ratio of 3. Then a pellet was pressed and placed into the Mössbauer cell, as fast as possible (ca. 5 min) in order to minimize the extent of readsorption of water. Further, the mixture was evacuated (10^{-1} Pa) at 550 K in the measuring cell to remove the traces of the adsorbed water remained. The decomposition of NaN_3 and the simultaneous reduction of the iron ions were performed by treating the sample in situ for 30 min at 800 K in a flow of purified nitrogen.

The reduction process was followed by thermogravimetry and differential thermal analysis using a MOM 1600 derivatograph. The solid products were characterized by different methods:
- X-ray diffractograms were recorded with a Philips PW 1200 powder diffractometer equipped with a graphite monochromator using K_α radiation of copper.
- AC susceptibility measurements were performed in a high sensitivity (10^{-7} emu/g) three-coil two-position mutual inductance bridge [5]. The susceptibility of samples was measured at different frequencies (9, 180 and 890 Hz) and in 1, 10 and 45 Oe fields in the 4 - 300 K range.
- Mössbauer spectra were recorded in an in situ cell by a conventional constant acceleration spectrometer.
- Temperature programmed reduction measurements were performed on 300 mg sample (0.3 mmol iron) in 10 % H_2/Ar mixture (20 ml min^{-1} flow rate, 20 °C min^{-1} ramp). Hydrogen - oxygen titrations were also carried out at various temperatures.

It should be noted that the Mössbauer measurements were performed in situ, the other ones "ex situ", i.e., oxidation of the formed metallic particles might have taken place when the samples were exposed to air when transferring them from one apparatus to the other.

RESULTS AND DISCUSSION

Thermogravimetry and thermal analysis

Two steps were observed on the TG records. The first one (at ca. 450 K) is accompanied by an endothermal DTA peak and is due to the removal of adsorbed water. The second one (at ca. 700 K) is associated with a strong exothermic effect, and corresponds to the decomposition of sodium azide. However, the weight losses observed upon the decomposition of azide were generally smaller (with ca. 20 %) than expected from the stoichiometry. This phenomenon is attributed to the contact-induced ion exchange of the protons of even very weak Brønsted acidic sites present in the parent zeolite, or the protons generated by hydrolytic scission of water for the sodium ions of azide and to the subsequent escape of the formed volatile HN_3.

X-ray diffraction

The diffractogram of the treated X zeolite revealed a partial recrystallization: peaks of non-specified aluminium silicate phase(s) were observed with typical reflections at $2\Theta = 21.1$ and 34.8 degree. In the Y zeolite samples the original crystallinity was essentially preserved. Reflections of metallic iron crystals (or iron oxides) were not observed in any diffractogram.

AC susceptibility

The dependences of the in-phase AC susceptibilities on the temperature at various fields are shown for an $FeX + NaN_3$ sample in Figure 1. A characteristic break appears at 120 K: the exponential shape changes to a modest increase above this temperature. The recorded curve is most probably a sum of the contributions of oxidic nanoparticles: similar behaviour was observed on oxidized iron (diameter: ca. 6 nm) with a break around 100 K [5]. The frequency dependence of the signal was modest: the superparamagnetic feature was less pronounced.

Mössbauer spectroscopy

Characteristic 77 K in situ Mössbauer spectra of iron-zeolite samples mixed with NaN_3 and decomposed at 800 K in nitrogen are shown in Figure 2. Spectra of X and Y samples are composed from the sextet of metallic α-iron and various doublets of Fe^{2+} (for the assignments see eg. [6]). In additon, a singlet component is also present (IS ≈ 0.0 mm/s) which can also be attributed to zerovalent iron. In the Fe,La-Y sample the same components are found except the sextet of metallic α-iron. Fe^{3+} component (indicative of the presence of oxides) was not detected in any sample (Table 1).

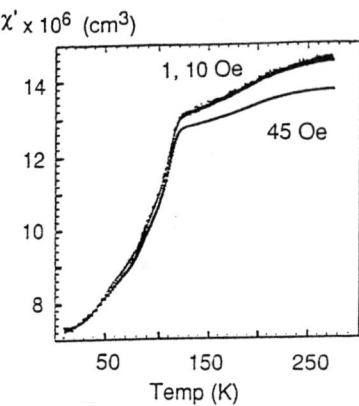

Figure 1. Temperature dependence of the in-phase AC susceptibility of the Fe-X + NaN$_3$ sample recorded at 181 Hz in various fields

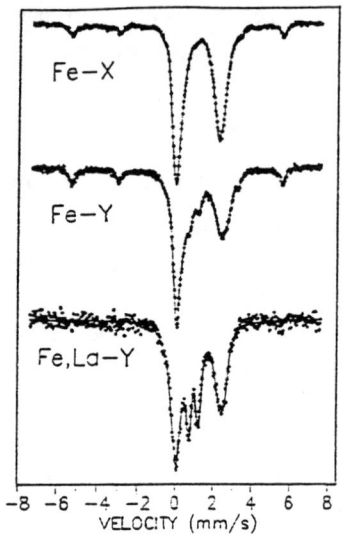

Figure 2. 77 K Mössbauer spectra of reduced zeolites.

As for the zerovalent iron two types can be distinguished by the characteristic size. The size of particles contributing with a sextet in the spectra can be estimated to be in the 4 - 6 nm range. The upper limit can be estimated from the lack of metallic iron reflections in the X-ray diffractograms; the lower size value is the threshold value necessary for the appearance of magnetic ordering in the spectra (ca. 4 nm at 77 K [7]). As for the size of clusters of the second type iron (IS $\approx$ 0.0 mm/s, without magnetic splitting), the contribution should originate from particles smaller than 4 nm. Since the proportion of this contribution is increasing with the stability of the framework it can be assumed that these zerovalent iron clusters are located in the non-destroyed supercages of the lattice (diameter 1.2 nm).

It is worth mentioning that the stabilising effect of La is also demonstrated by the lack of the sextet of iron, i.e. no larger voids were formed in which 4 - 6 nm metallic particles could be accumulated (contrasting the cases of the Fe-X and Fe-Y samples). Further, certain IS and QS values for the respective Fe^{2+} components are different in the three samples. The spectrum of FeX + NaN$_3$ probably contains contributions from the recrystallized aluminosilicate while the IS = 0.96, QS = 0.5 mm/s component in the spectra of Y samples probably originates from iron ions located in the characteristic trigonal prism sites of faujasites with distorted symmetry at the entrance of the sodalite cages [8].

	Fe-X + NaN$_3$			Fe-Y + NaN$_3$			Fe,La-Y + NaN$_3$		
	IS	QS	RI	IS	QS	RI	IS	QS	RI
Fe^{0,a}	0.10		10	0.10		12			
Fe^{0,b}	0.02		9	0.03		18	-0.17		15
Fe^{2+}				0.96	0.52	12	0.97	0.49	28
Fe^{2+}	1.01	2.44	24	1.08	2.33	35	1.20	2.14	37
Fe^{2+}	1.13	1.75	28						
Fe^{2+}	1.20	2.54	28				1.30	2.59	20
Fe^{2+}				1.48	2.28	22			

Table 1: Spectral components in the 77 K in situ Mössbauer spectra.
IS: isomer shift, related to α-Fe, mm/s; QS: quadrupole splitting, mm/s; RI: relative intensity, %
Notes: [a] magnetic component, with 34 Tesla internal magnetic field (sextet)
[b] non-magnetic component (singlet)

Temperature programmed reduction and chemisorption

A reduced Fe-Y zeolite sample (300 µmol total iron) was measured in the TPR apparatus. The exposure to air could not be excluded during the sample transfer. In order to restore a reduced state a treatment was applied first by heating the sample up to 720 K in 10% H$_2$/Ar (with 48 µmol H$_2$ spent). A subsequent room temperature oxygen titration consumed 0.48 µmol oxygen. After a repeated TPR treatment (14 µmol spent H$_2$) the oxygen titration was performed at a slightly elevated temperature, at 390 K (to remove the formed and adsorbed water); the O$_2$ consumption increased to 5 µmol. Upon the repetition of the reduction-oxydation cycle the oxygen consumption was practically the same as previously, i.e. this amount can be considered as the easily exchangeable proportion.

The combination of these results with those of the Mössbauer measurements allows to draw a rough estimation of the accessibility of the zerovalent iron to gas molecules (dispersion). Assuming that the iron ions do not take part in the 300 - 390 K oxidation-reduction processes and considering that ca. 1/3 of the iron was found in zerovalent state in the Mössbauer spectra, the 5 µmol oxygen uptake corresponds to ca. 1/10 of the total amount of zerovalent iron. Thus, as for the Fe-Y sample, only this portion is easily accessible to gas molecules, while the remaining part is probably occluded into the recrystallized aluminosilicate or the access to them is blocked by various extra-framework Na,Al-oxides formed in the zeolite pores.

CONCLUSIONS

The presented results clearly demonstrate the possibility for the creation of zerovalent iron from extraframework iron ions via reduction with sodium vapour generated by thermal decomposition of sodium azide. The process proceeds only partially, i.e., only a smaller part of iron is reduced to the zerovalent state.

The reduction process is accompanied with a simultaneous partial recrystallization and formation of non-porous aluminosilicates. The extent of recrystallization is decreasing in the Fe-X > Fe-Y > Fe,La-Y order. This order reflects also the stabilising effect of La insertion into sodalite cages. Amount and size of the zerovalent iron particles are in correspondence with the degree of recrystallization: in the Fe-X and Fe-Y samples metallic particles in the 4 - 6 nm size range were also detected. These particles can only be located in large voids of the structure formed upon partial destruction and recrystallisation. Simultaneous presence of smaller size iron clusters was also demonstrated, the proportion of this contribution increased with increasing stability of the framework (Fe-X < Fe-Y < Fe,La-Y). Thus, a characteristic size of about 1 nm and location of these clusters in the supercages is suggested.

The accessibility of metallic particles to gaseous molecules is restricted in the Fe-Y sample as revealed by TPR and oxygen titration measurements: ca. 1/10 of the zerovalent iron is available in mild chemisorption and redox processes.

ACKNOWLEDGMENTS

The financial supports provided by the Commission of the European Communities in the frame of Chemistry Action COST D5 (PECO project No. CIPE CT 92 6107) and by the Hungarian National Scientific Research Fund (OTKA T021131) are thankfully acknowledged.

REFERENCES

1. J.B. Lee, J. Catal., **68**, 27 (1981).
2. F. Schmidt, W. Gunsser, J. Adolph, in Molecular Sieves II. (Edited by J.R. Katzer), A.C.S. Symp. Series, Vol. **40**, (Am. Chem. Soc., Washington, 1977) p. 291.
3. M. Brock, C. Edwards, H. Förster, M. Schröder, Stud. Surf. Sci. Catal., **84**, 1515 (1994).
4. E.F.T. Lee, L.V.C. Rees, Zeolites, **7**, 143 (1987).
5. B.J. Jönsson, T. Turkki, V. Ström, M.S. El-Shall, K.V. Rao, J. Appl. Phys., **79**, 1 (1996).
6. R.G. Burns, Hyperfine Interactions, **91**, 739 (1994).
7. B.S. Clausen, H. Topsøe, S. Mørup, Appl. Catal., **48**, 327 (1989).
8. Z. Gao, L.V.C. Rees, Zeolites, **2**, 79 (1982).

DEHYDROISOMERIZATION OF N-BUTANE OVER BIFUCNTIONAL CATALYSTS

G. D. PIRNGRUBER, K. SESHAN AND J. A. LERCHER

University of Twente, Faulty of Chemical Technology, P. O. Box 217, 7500 AE Enschede, The Netherlands; J.A.Lercher@ct.utwente.nl

ABSTRACT

The activity and selectivity of Pt-ZSM5 catalysts with varying Si/Al-ratios for dehydroisomerization of n-butane was investigated. At 830 K and 1.8 bar total pressure isobutene yields of up to 13% were achieved. The reaction proceeds *via* a classical bifunctional mechanism. Conversion and yield of dehydrogenated products mainly depend on the metal loading, while the ratio isobutene/butene only depends on the concentration of acid sites. The main routes of byproduct formation are hydrogenolysis of butane over the metal sites and dimerization/cracking of butene over the acid sites.

INTRODUCTION

In industrial practice the conversion of n-butane to isobutene is carried out in a two-step process. The use of a bifunctional catalyst, which combines isomerization and dehydrogenation, offers the possibility to perform a one-step conversion of n-butane to isobutene. Several types of materials have been tested for the dehydroisomerization of butane, e.g., Pt-MOR [1], Ga-LTL [2], Pt-zincosilicate [3]. The highest isobutene yield was obtained with a Pt/Re-{B}-ZSM11 [4]. In the present contribution we want to focus on the mechanism of the reaction and especially the source of byproducts which should lead to design-parameters for improved catalysts.

EXPERIMENTAL

The parent zeolite ZSM5 with SiO_2/Al_2O_3 ratios between 80 and 480 were supplied by ZEOLYST Int. In the following text the SiO_2/Al_2O_3 ratio will be denoted in brackets at the end of the sample name. Pt was incorporated by liquid state ion exchange with a very diluted solution of $Pt(NH_3)_4(OH)_2$ and ammonia. The samples were calcined in air at 723 K and reduced in H_2 for 2 h at 773 K [5].

For the catalytic tests 10 to 100 mg of the catalyst (particle size 300 to 600 µm) were mixed with about 100 mg quartz and filled into quartz tube with an inner diameter of 4 mm. The samples were reduced *in situ* at 830 K for 1h in a mixture of H_2/Ar (18/82). The reaction conditions were 830 K and 1.8 bar, the feed composition 10% n-butane, 20% H_2, rest Ar. The reaction was followed for at least 2.5 h. Usually, only a slight deactivation of the catalysts was observed. For comparison between different experiments the values after 100 min time on stream were used, since all catalysts were in quasi steady state or only marginally deactivating at this stage. Conversion and yields are reported on a carbon basis.

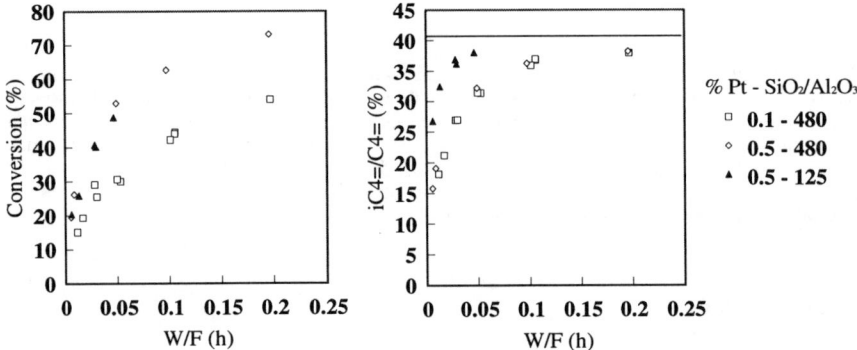

Figure 1 a) Conversion of n-butane over Pt-ZSM5 with different Pt-loadings and SiO_2/Al_2O_3 ratios. b) Ratio isobutene/butene. Standard reaction conditions.

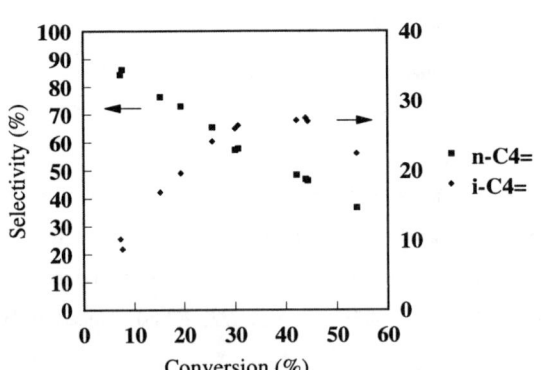

Figure 2 Selectivity to n–butene and isobutene as a function of conversion. 0.1 Pt-ZSM5 (480). Standard conditions.

Results and Discussion

The reaction sequence

The conversion of n-butane increased with metal loading, but did not depend on the concentration of acid sites (see Fig. 1a). A Pt-free ZSM5 gave virtually zero conversion.

The ratio isobutene/butene, on the other hand, which can be used as a measure for the butene isomerization activity, did not depend on the metal loading, but only on the concentration of acid sites (see Fig.1b).

The relation between selectivity and conversion (Fig.2) shows that n-butene was a primary and isobutene a secondary reaction product. Thus, we concluded that the dehydroisomerization proceeds *via* a classical bifunctional mechanism where n-butane is first dehydrogenated over the metal sites and then isomerized to isobutene over the acid sites (see scheme 1). The metal loading governed the activity of the catalyst, the acid site concentration determined the degree of isomerization.

$$n\text{-}C_4H_{10} \underset{-H_2}{\overset{Pt}{\rightleftarrows}} n\text{-}C_4H_8 \overset{H^+}{\rightleftarrows} i\text{-}C_4H_8 \qquad \text{Scheme 1}$$

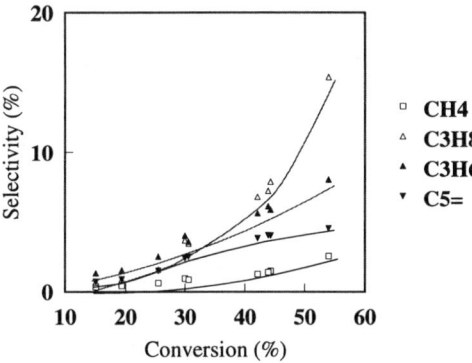

Figure 3 Selectivity to the main byproducts as a function of n-butane conversion. 0.1 % Pt ZSM5 (480), measured under standard conditions.

Byproduct formation

Thermodynamics limits the possible yields of butene and isobutene to 53% and 22%, respectively, under the chosen reaction conditions. The optimum yields achieved with Pt-ZSM5 were 39% butene and 13% isobutene. At high conversions side reactions consumed a significant fraction of butene to unwanted byproducts, mainly propane, propene and pentenes, followed by smaller amounts of methane, ethane and isobutane (see Fig. 3). Several routes of byproduct formation are conceivable, i.e., (i) cracking of butane over the acid sites, (ii) hydrogenolysis/isomerization over the metal and (iii) butene cracking over the acid sites.

As indicated by Fig. 3 most of the byproduct formation at high conversions results from secondary reactions, i.e. the contribution of direct cracking of and hydrogenolysis of butane (both are primary reactions) is small. The most important route of byproduct formation is dimerization of butenes followed by cracking, leading to propene and pentenes [6]. Propane is concluded to be formed by hydrogenation of propene over the metal sites.

Isomerization of butene and dimerization/cracking are competing reactions over the acid sites. In order to obtain information on these secondary reactions of the butenes, the conversion of butene over ZSM5 (480) and Pt-ZSM5 (480) was studied (feed 7.5% 1-$C_4=$, 20% H_2, 830 K, 1.8 bar). Some representative results are given in Table 1. Over the parent ZSM5 the intrinsic selectivity to isomerization was relatively high (about 85 % at zero conversion, see Fig. 4). As isomerization approached equilibrium conversion, the selectivity changed in favor of the byproducts propene, pentene, ethene and hexene, formed by dimerization followed by cracking. Ethane or propane were not formed.

Over 0.1% Pt-ZSM5 the rates of isomerization and of byproduct formation were drastically higher compared to the parent material. Large concentrations of alkanes (propane, ethane and methane) were found in the byproducts. The rates of ethane and propane formation (Fig. 5) increased significantly above 40% conversion. For a Pt-free ZSM5 dimerization/cracking began to be favoured over isomerization at these conversions, which indicated that ethane and propane indeed resulted from hydrogenation of the alkenes that were formed by cracking, rather than from hydrogenolysis.

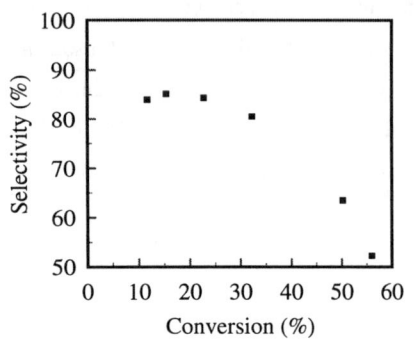

Figure 4 Selectivity to isobutene as a function of conversion of 1-butene over ZSM5 (480). Double bond isomerization was not regarded as conversion.

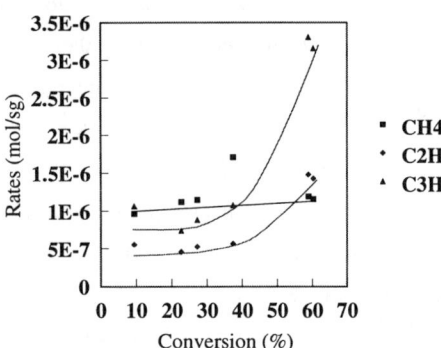

Figure 5 Rates of methane, ethane and propane formation as a function of n-butene conversion in the reaction of 1-butene over 0.1 % Pt ZSM5 (480).

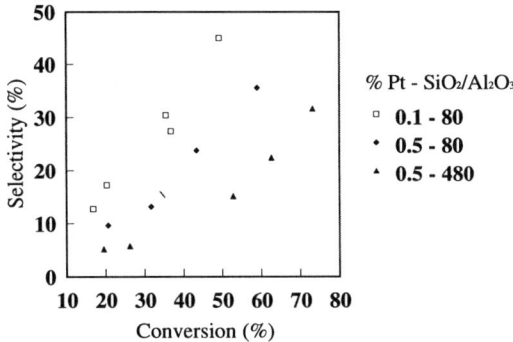

Figure 7 Selectivity to secondary cracking products (counted as sum of propane, propene and pentenes) in the dehydroisomerization of n-butane over Pt-ZSM5

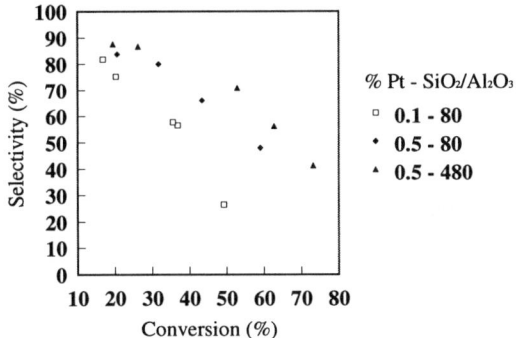

Figure 6 Selectivity to butenes in the dehydroisomerization of n-butane over Pt-ZSM5

Pt (%)	0	0.1	0.1
WHSV	34.2	36.2	80.6
Conversion (%)	22.8	37.4	22.7
Yields (%)			
C2H6	0.01	0.16	0.06
C3H8	0.00	0.45	0.14
C3H6	1.55	3.05	1.49
C5=	1.35	1.68	0.82
n-C4	0.16	6.13	3.59
i-C4=	19.20	23.58	14.89

Table 1 Conversion of n-butene and yield of the main products in the reaction of 1-butene over Pt-ZSM5 (480).

The influence of metal loading and acid site concentration on selectivity

Fig. 6 and 7 show the selectivity to dehydrogenation, i.e. to butenes, and to secondary cracking products (counted as the sum of propane, propene and pentenes) for catalysts of different metal loadings and acid site concentrations in the dehydroisomerization of n-butane. With increasing ratio of metal to acid sites the selectivity to dehydrogenation increased while the selectivity to secondary cracking decreased. This observation can be explained by the fact that in general the formation of secondary cracking products increased drastically (see Fig. 8) when the isomerization of n– to isobutene approached equilibrium (the equilibrium ratio is 41 % at the applied reaction conditions). Catalysts with a high concentration of acid sites approached isomerization equilibrium faster (see Fig. 1b), leading to a drop in selectivity to dehydrogenation already at low levels of n-butane conversion.

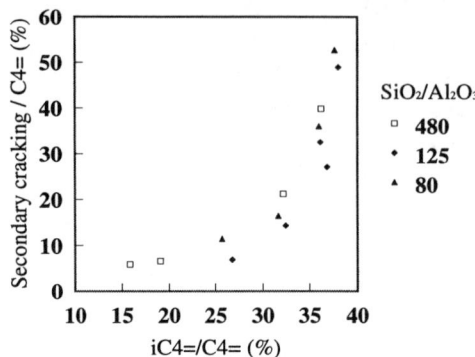

Figure 8 Selectivity to secondary cracking vs. isomerization in the dehydroisomerization of n-butane over 0.5% Pt-ZSM5

From this we conclude that catalysts with a high metal loading and a low acid site concentration are desirable for the dehydroisomerization of n-butane. They allow to approach the dehydrogenation equilibrium and, thus, achieve higher yields of butene, before the selectivity becomes limited due to the approach of the isomerization equilibrium.

CONCLUSIONS

The dehydroisomerization of butane proceeds *via* a bifunctional mechanism. n-Butane is dehydrogenated over the metal and then isomerized over the acid sites. The activity (i.e., the conversion) is governed by the metal loading. The concentration of acid sites determines how much of the primarily formed n-butene is converted to isobutene. However, as butene isomerization approaches equilibrium the selectivity to cracking products (propene, pentene, ethene) increases. These cracking products can be hydrogenated over the metal sites (if thermodynamics allows it, which is the case for ethene and propene). Both effects, butene cracking and subsequent hydrogenation, contribute to the high formation of byproducts at high contact times and limit the yield of isobutene that can be achieved.

In order to improve the performance of the catalyst it is necessary to improve the selectivity of butene isomerization *vs.* secondary cracking. While decreasing the acid site concentration allows the system to go further in the direction of the dehydrogenation equilibrium, before becoming limited by secondary cracking, it does not change the intrinsic selectivity of isomerization vs. cracking of ZSM5 (see reference [6] and Fig. 8). Thus, other isomerization catalysts and staged bed designs will be explored as alternatives. The key is to find isomerization catalysts that permit to approach selectively the thermodynamic limits of n-butene conversion.

ACKNOWLEDGEMENTS

This work was performed under the auspices of NIOK, the Netherlands Institute for Catalysis Research (Lab report No. UT 98-1-06). IOP Catalysis (IKA 94023) is gratefully acknowledged for financial support. GDP is thankful to KNCV and NWO for supplementary grants.

REFERENCES

1. M.M. de Agudelo, T. Romero, J. Guaregua, M. Gonzalez, US 5.416.052, 1995.
2. A.J. Kolombos, C.D. Telford, D. Young, EP 0.042.252, 1981.
3. V.K. Shum, US 4.962.266, 1990.
4. O' Young, J.E. Browne, J.F. Matteo, R.A. Sawicki, J. Hazen, US 5.198.597, 1993.
5. A.C.M. van den Broek, J.van Grondelle, R.A. van Santen, J. Catal. 167 (1997), 417.
6. J. Houzvicka, R. Klik, L. Kubelkova, V. Ponec, Appl. Catal. A 150 (1997), 101.

WATER AND SULFUR RESISTANT PT-BASED ZEOLITE CATALYST FOR NOx REDUCTION

S. E. MAISULS, S. FEAST, K. SESHAN, J. G. VAN OMMEN and J. A. LERCHER

UNIVERSITY OF TWENTE, Faculty of Chemical Technology, Laboratory for Catalytic Processes and Materials, P.O. Box 217, 7500 AE Enschede, The Netherlands.

ABSTRACT

The selective catalytic reduction of NO by propene in the presence of excess oxygen has been studied. Pt containing catalysts were found to be particulary active producing, however, large amounts of N_2O. Bimetallic Co-Pt/ZSM-5 catalysts with low Pt contents (0.1 wt %) combine high stability and activity of Pt catalysts with the high N_2 selectivity of monometallic Co catalysts. In addition these catalysts showed high water and sulfur tolerance above 620 K. Based on the experimental kinetic data and the results of *in- situ* i.r. studies, the reaction pathways over this catalyst are discussed.

INTRODUCTION

Spurred by increasingly stringent environmental legislation, the interest in developing new catalytic processes for the reduction of NOx under conditions of excess oxygen, needed, e.g., for stationary sources, diesel engines and lean burn Otto engines is currently very high [1,2]. Most likely this requires a shift from ammonia to hydrocarbons as reductant. While a large number of catalysts based on zeolites was proposed and tested (e.g., Cu/ZSM-5 [3,4], Co/ZSM-5 [5], Co/FER [6], Pt/ZSM-5 [7]), sensitivity towards water and sulfur compounds has been found to be prohibitive to a breakthrough. Reversibly and irreversibly, both molecules retard the rate at which NOx is reduced and may under some instances completely destroy the zeolite. Recently, a breakthrough in that respect was shown by the groups of Hall [8] and Sachtler [9] based on Fe loaded zeolites. The drawback of these materials was, however, that stable activity was only attained at temperatures above 400 °C. In the present communication we describe an alternative zeolite based catalyst, Co-Pt/ZSM-5, that is sulfur and water tolerant and considerably extend the limit of operation towards lower temperatures. The roles of Pt and Co for the elementary steps of NOx reduction and the implications for catalyst design are discussed.

EXPERIMENTAL

The catalysts were based on ZSM-5 and the metals were introduced *via* solid state ion exchange. In a typical procedure NH_4^+-ZSM-5 (5g) was thoroughly ground with $CoCl_2 \cdot 6H_2O$ (0.59g) and $PtCl_4$ (0.01g), The resulting mixture was heated to 500 °C at 2 K/min in He and was held at this temperature for 12 hrs. During the heating cycle it was maintained for three hours near the melting point (80 °C) and decomposition point (105°C) of $Co(Cl)_2$ and then at 365 °C for 3 hours allowing $PtCl_4$ (melting and decomposition point 370 °C) to decompose slowly. For catalyst testing, the catalyst was activated *in situ* by heating to 500 °C for 1hr in flowing He. The typical reactant gas mixture consisted of 1000 ppm NO, 1000 to 2250 ppm C_3H_6, 5% O_2 balancing with He to one bar. A total gas flow of 100 ml/min was passed through the catalyst bed (0.2 g). The resulting GHSV was 15000 hr^{-1}, based on the apparent bulk density of the catalyst bed (0.5 g/cm^3). The products were analyzed simultaneously by gas chromatography (Varian 3700, N_2, N_2O, O_2, C_3H_6, CO, CO_2 analysis) and a chemiluminescence NO-NO_2-NO_x analyzer (Thermo Environmental Instr., Model 42C, NO, NO_2 analysis). The I.r. measurements were performed *in situ* on a BRUKER IFS 88 FTIR spectrometer (resolution of 4 cm^{-1}) in a continuous gas flow mode using the transmission-absorption technique. The i.r. cell was equipped with a heatable sample holder and CaF_2 windows. The catalyst (Co-Pt/ZSM-5 with 2.7 wt % Co and 0.1 wt % Pt) was pressed into self supporting 2 mg wafer and activated in flowing He at 500°C for one hour. Subsequently, the catalyst was brought to the required temperature and the reactant gases were passed over the catalyst.

RESULTS AND DISCUSSION

The chemical composition of the catalysts and the most important kinetic results are compiled in Tables 1 and 2. The catalysts have been prepared with similar Co and Pt loading to allow straightforward comparison. Table 1 shows the influence of the Co concentration on the catalytic behavior of Co/ZSM5 catalysts. Increasing the Co content improved the activity of the catalysts. This is reflected in the rates of NO and C_3H_6 conversion. High Co content (5%) however, lead to lower selectivity towards N_2. This was the result of increased unselective C_3H_6 combustion. Under these conditions oxidation of NO to NO_2 was favored. EXAFS measurements suggest that the excess of Co on this catalyst (Co/Al mol ratio of 2.10) is present as Co oxide clusters. These oxide clusters are known to promote propene combustion [10]. Co/ZSM-5 with 2.8 wt % Co conc. was a good compromise between the high activity and high selectivity to N_2. Thus, this catalyst was chosen for further optimization.

Table 1. Chemical composition and kinetic results of the Co loaded ZSM-5 catalysts

Catalyst	Co [wt%]	Co/Al [Ratio]	Conversion[1] [mol s^{-1} g^{-1}] NO	Selectivity[2] [mol%]			Conversion[1] [mol s^{-1} g^{-1}] C_3H_6
				N_2	N_2O	NO_2	
H-ZSM-5	--	--	0.08E-07	100[3]	0	0	0.06E-06
Co/ZSM-5	0.7	0.3	0.18E-07	96	4	0	0.08E-06
Co/ZSM-5	2.8	1.2	1.28E-07	85	15	0	0.63E-06
Co/ZSM-5	5.0	2.1	6.84E-07	60	13	27	8.87E-06

ZSM-5 Si/Al ratio = 38.5; Reaction conditions: $T = 350$ °C, 1000ppm NO, 1000ppm C_3H_6 and 5% O_2 in He, [1] Contact time 0.02 s, [2] At 30% NO conversion (by varying the contact time, [3] at 2 % NO conversion. (product NO_2 is observed only when C_3H_6 conversion reaches 100%).

All catalyst investigated showed that NO conversions and N_2 yields were closely related to the propene conversion (see Fig 1.). Thus, more active catalysts should produce N_2 at lower temperatures. Aiming at improving the activity of these Co catalysts, bimetallic Pt-Co catalyst were prepared. The influence of the addition of Pt on the catalytic behavior of Co/ZSM5 catalysts is documented in Table 2. Co/ZSM-5 showed the lowest activity and the highest selectivity to N_2. Pt/ZSM-5 on the contrary showed (as reported previously [11]) high activity, but low selectivity to N_2. Addition of Pt improved the activity of Co based catalysts at the expense of increased selectivity to N_2O. Note that the catalytic properties observed are not a simple linear combination of the activity

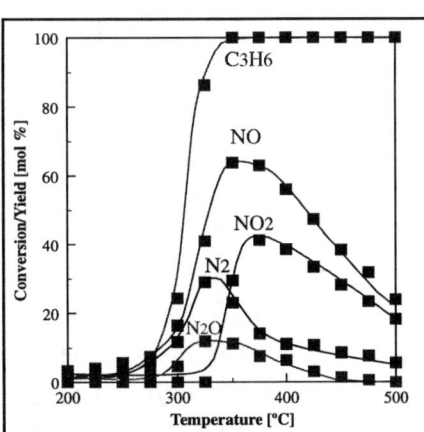

Figure 1 Influence of the temperature on conversions and yields over Co-Pt/ZSM-5 Conditions: 1000 ppm NO, 1000 ppm C_3H_6, 5% O_2 in He at 15000 hr^{-1}.

of the two metals, but indicate a more complex behavior. Especially Pt concentrations above 0.1 wt % increased the tendency to form N_2O. Thus, the Co-Pt/ZSM-5 material with 0.1 wt% Pt was chosen for detailed studies.

Table 2. Chemical composition and kinetic results of the catalysts studied

Catalyst	Co [wt%]	Pt [wt%]	Conversion[1] [mol s^{-1} g^{-1}] NO	Selectivity[2] [mol%]			Conversion[1] [mol s^{-1} g^{-1}] C_3H_6
				N_2	N_2O	NO_2	
H-ZSM-5	--	--	0.08E-07	100[3]	0	0	0.06E-06
Pt/ZSM-5	--	0.1	3.52E-07	45	55	0	1.03E-06
Co/ZSM-5	2.8	--	1.28E-07	85	15	0	0.63E-06
Co-Pt/ZSM-5	2.8	0.1	2.36E-07	80	20	0	1.04E-06

ZSM-5 Si/Al ratio = 38.5; Reaction conditions: At 350 °C, 1000ppm NO, 1000ppm C_3H_6 and 5% O_2 in He, [1] Contact time 0.02 s, [2] At 30% NO conversion (by varying the contact time, [3] at 2 % NO conversion. (product NO_2 is observed only when C_3H_6 conversion reaches 100%).

Fig 2. shows the time on stream behavior of this catalyst. Under the conditions employed the catalyst showed good stability, high activity and high selectivity to N_2. The presence of water (up to 6%) did not influence the NO conversion and the catalyst stability. Additionally, water improved the selectivity to N_2 at the expense of NO_2 (from 48 % N_2, 9 % N_2O and 43% NO_2 without water to 71 % N_2, 8 % N_2O and 21 % NO_2 in the presence of water), at 82% NO conversion. At a higher space velocity (19500 hr^{-1}) addition of water completely suppressed NO_2 formation without affecting the NO conversion (from 68 % N_2, 14 % N_2O and 18% NO_2 to 90 % N_2, 9 % N_2O and 0 % NO_2, at 62% NO conversion). The effect of water addition was completely reversible indicating that the catalyst does not undergo alterations in

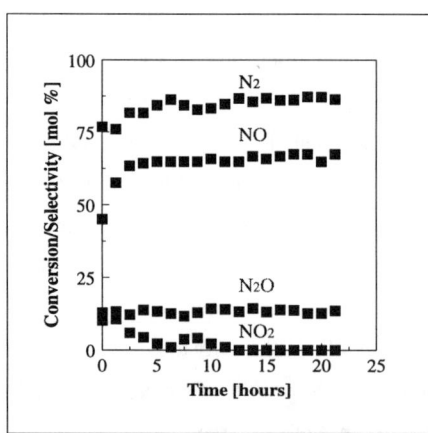

Figure 2 Time on stream behavior of Co-Pt/ZSM-5. Conditions: 300 °C, GHSV=5250 hr^{-1}, NO = 1000 ppm, C3H6 = 2250 ppm, O_2 = 5% in He. At ~ 100 % C_3H_6 conversion

presence of water vapor at 350°C. Under these conditions presence of SO_2 (200 ppm) did also not affect the stability or activity (NO conversion).

MECHANISTIC STUDIES

In situ EXAFS and XANES of Co-Pt/ZSM-5 catalysts indicate that Co is present in Co^{2+} and Co^{3+} oxidation state. Pt is present as highly dispersed metallic particles. In the fresh Co-Pt/ZSM-5 catalyst Co^{2+} assumes 50 % of the Brønsted acid sites of the parent H-ZSM-5 indicating that it is present as $Co(OH)^+$ or $CoCl^+$. Formation of oxidic Co^{2+}/Co^{3+} clusters during reaction suggests that the concentration of Brønsted acid sites is even higher in the working catalyst.

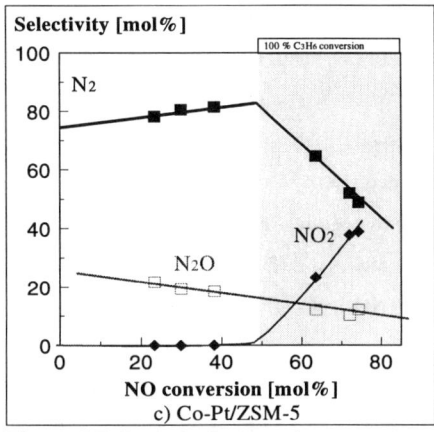

Figure 3 Influence of conversion on selectivity. Conditions: 350 °C, 1000 ppm NO, 2200 ppm C_3H_6, 5% O_2 in He.

Fig 3 shows the variation of selectivities with the conversion for Co-Pt/ZSM-5 at 350 °C. It can be deduced from the figure that N_2 and N_2O are primary products and that some interconversion of N_2O to N_2 occurs at higher NO conversions. When C_3H_6 conversion reaches 100 % the yields of N_2 reaches its maximum value. NO_2 formation is observed beyond this point. At still higher NO conversions N_2 selectivity decreases rapidly at the expenses of NO_2, i.e., the reduction of NO by C_3H_6 replaced by the NO oxidation by oxygen. This latter reaction is catalyzed by Co and Pt as deduced from separate experiments.

When passing NO alone over Co-Pt/ZSM-5, N_2 or N_2O were not detected in the gas phase. An NO/O_2 mixture produced exclusively NO_2. When NO and C_3H_6 were co-fed, N_2 and N_2O were observed as products. The same products were also observed when NO was passed over the catalyst previously exposed to propene. When NO, C_3H_6 and O_2 were passed simultaneously significant selective reduction of NO to N_2 took place.

The surface adsorbed species formed under the reaction conditions over the Co-Pt/ZSM-5 catalyst were studied by *in- situ* i.r.spectroscopy. At room temperature, when only NO was passed over the catalysts, i.r. bands attributed to the following N containing species were detected: Co-(NO) (1935 cm^{-1}), Co-(NO)$_2$ (1896, 1813 cm^{-1}), Co-(NO$_2$) (1633 cm^{-1}), Co-(NO$_3$) (1601, 1580,1540 cm^{-1}). The -(NO$_2$) and -(NO$_3$) species seemed to be the most abundant. At reaction temperatures all the peaks except that of Co-(NO) were still present. Similar results were obtained when NO and O$_2$ were co-fed, only the band intensities of the -(NO$_2$) and -(NO$_3$) species were stronger. When NO, C$_3$H$_6$ and O$_2$ were present, at reaction temperatures (see Fig 4), bands related to -NCO (2260, 2230 cm^{-1}) and to a smaller extent -NC(2160 cm^{-1}) were observed. A broad band resulting from propene (also observed when only propene is passed over the catalyst) centered around 1590 cm^{-1} makes it difficult to verify the presence of Co-(NO$_2$), Co-(NO$_3$) type species on the surface. These results are similar to an earlier report by Aylor *et al.*[12] using CH$_4$ as reductant over Co/ZSM-5. They suggested on the basis of their i.r. results a mechanism involving the formation of nitro-(NO$_2$) type species which are reduced by CH$_4$ to result in a complex (C-H-N-O) type species which further react *via* -CN type species, with NO or NO$_2$ to give N$_2$. The presence of Co-(NO$_2$), Co-(NO$_3$) of the *in situ* infrared study discussed above indicated that NO$_2$ could be an intermediate in the formation of N$_2$. To investigate this possibility experiments comparing the reactivity of NO and NO$_2$ were performed. It was found that the rates of N$_2$ formation were doubled when NO$_2$ was used as reactant. This results again show the role of NO$_2$ as an intermediate in the selective catalytic reduction of NO to N$_2$. The presence of Pt in Co-Pt/ZSM-5 causes enhanced N$_2$O formation. Our results indicate that most of the N$_2$O is formed over Pt and a small amount over Co. Experiments with N$_2$O show that Co-Pt/ZSM-5 and Co-/ZSM-5 are both active in the decomposition of N$_2$O. Therefore part of the N$_2$O produced over Pt is decomposed over Co (see Table 2 comparing Co-Pt/ZSM-5 and Pt/ZSM-5) to give N$_2$. This explains the interconversion of N$_2$O to N$_2$ at higher conversions. The role of the Brønsted acid sites, if any is not clear at the moment, under reaction conditions these sites seem to be covered with propene oligomers or coke.

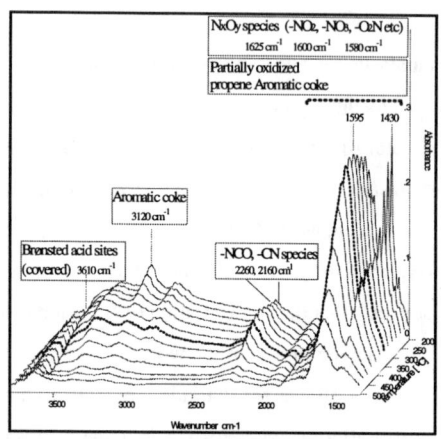

Figure 4 I.R spectra of the surface species formed under reaction conditions over Co-Pt/ZSM-5.

CONCLUSIONS

Addition of Pt improved the activity of Co/ZSM-5, however low Pt contents are necessary to maintain high N_2 selectivities. A Co-Pt/ZSM-5 catalyst with 0.1% Pt content is shown to be active, selective and stable under reaction conditions even in the presence of water and SO_2. In this Catalyst, Pt is present as well dispersed metal particles and Co as Co exchanged at the acid sites and as oxidic clusters. The results of the *in situ* infrared study toguether with the kinetic experiments using NO_2 as reactant indicate that the reaction mechanism over Co-Pt/ZSM-5 involve the formation of NO_2 adsorbed species as intermediates in the formation of N_2.

ACKNOWLEDGMENTS

This project was supported by the EU, in the framework of the R and D programme in the field of the environment, sub-programme: Technologies for protecting and rehabilitating the environment, Contract no. EVSV-CT-94-0535. This work was performed under the auspices of NIOK, the Netherlands Institute for Catalysis.(Report number UT-98-1-02).

REFERENCES

1. M. Iwamoto, Catal. Today, **29**, 29 (1996).
2. *Catal. Today,* **26** 107 (1995) and references therein.
3. M. Iwamoto, *Proc. Catal. Technol. for the Removal of Nitrogen Monoxide, Tokyo* (1990) 17.
4. W. Held, A. Konig, T. Richter and L. Puppe, *SAE Paper* 900496 (1990).
5. Y. Li and J.N. Armor., *Appl. Catal.*, B1, L31 (1992).
6. Y. Li and J.N. Armor., *Appl. Catal.*, B1, L1 (1993).
7. A.Obuchi, A.Ohi, M.Nakamura, A.Ogata, K.Mizuno and H.Ohuchi, *Appl.Catal.,* B2,71(1993).
8. X. Feng and W.K. Hall, *Catal. Lett* 4, 45-46 (1996).
9. H.-Y. Chen, W.M.H. Sachtler, *Catalysis letters*, Vol. 50 (Issue 3-4), 1998, 125-130 (6)
10. E. Finocchio, G. Busca, V. Lorenzelli and V.S. Escribano, *J. Chem. Soc., Faraday Trans.,***92**(9), 1587-1593 (1996).
11. H. Hirabayashi, H. Yahiro, N. Mizuno, and M. Iwamoto, *Chem. Lett*, 2235 (1992).
12. A.W.Aylor, L.J. Lobree, J.A. Reimer, and A. T. Bell, Studies in Surface Science and Catalysis, Vol 101,661 (1996).

FAUJASITE Y CONFINED NI(II)-TETRAKIS(N-METHYL-4-PYRIDYL)-PORPHYRIN AS HYDROGENATION CATALYST

B. -Z. ZHAN[a], P. A. JACOBS[b] and X. -Y. LI[a*]

[a] Department of Chemistry, The Hong Kong University of Science and Technology, Clear Water Bay, Kowloon, Hong Kong, P. R. China; chzhan/chxyli@ust.hk
[b] Centrum voor Oppervlaktechemie en Katalyse, K.U. Leuven, 92, Kardinaal Merciertaan, B-3001, Heverlee (Leuven), Belgium; Pierre.Jacobs@agr.kuleuven.ac.be

ABSTRACT

Ni(II)-tetrakis-(N-methyl-4-pyridyl)-porphyrin (NiTMPyP) was successfully encapsulated inside the supercages of faujasite Y (NaY) by using our developed "build-bottle-around-ship" method. X-ray powder diffraction (XRD) and X-ray induced fluorescence (XIF) analysis indicate that the zeolite framework around the guest molecules is NaY with a Si/Al ratio of 1.6. Several techniques were employed to characterize the synthesized host-guest composite. No evidence was observed for any chemical modification on NiTMPyP (such as decomposition or demetallization) in the process of hydrothermal synthesis. The synthesized NiTMPyP-in-NaY composite, denoted as NiTMPyP@NaY, exhibits high activity in the hydrogenation of cyclooctadiene (COD). The product is selective to cyclooctene (COE) with a K_1/K_2 value of over 9. The shape-selectivity imposed by the zeolite framework was clearly observed in the competitive catalysis over our catalyst. The hydrogenation rate for COE is about 33 times higher than that for the larger-sized cyclododecene (CODD).

INTRODUCTION

In contrast to the extensive study on the oxidation reactions of alkenes and alkanes over "ship-in-a-bottle" catalysts, the hydrogenation reaction catalyzed by zeolite confined transition metal complexes are scarce [1-3]. This has been mainly due to the difficulty in the encapsulation of an appropriate hydrogenation catalyst inside the cages/channels of zeolite. For the purpose of catalytic hydrogenation, only Pd(II) and Ni(II) complexes of a flexible ligand salen (bis-salicylidene-ethylenediamine) were successfully occluded in the supercages of faujasite X and Y by using "assemble-ship-inside-bottle" method [1-3]. We have recently developed a novel, efficient and quantitative "build-bottle-around-ship" method to efficiently encapsulate

---

*To whom correspondence should be addressed.

metalloporphyrins with cationic periphral groups inside the supercages of NaY with high purity and controllable loading [4]. We report here the occlusion of NiTMPyP inside the supercages of NaY by using this method. The synthesized host-guest composite, denoted as NiTMPyP@NaY, were systematically characterized by such techniques as XRD, micro Resonance Raman spectroscopy (µRRS), diffuse-reflectance-spectroscopy (DRS), XIF, scanning electron microscopy (SEM), nitrogen adsorption, and thermogravimetrical analysis (TGA). Our synthesized catalyst displays a high activity and selectivity in the hydrogenation of certain alkenes.

EXPERIMENTAL

Synthesis

NaY was synthesized by conventional hydrothermal method. The aluminosilicate gel was prepared by freshly mixing silicate with aluminate solution containing 4.6 g silica, 6.2 g NaOH, 3.2 g $NaAlO_2$ and 70 ml H_2O. The gel was then crystallized at 95 ± 2 °C under static and autogenous condition in a stainless steel bomb (250 ml) for 48 hours. Solid products were recovered by filtration, completely washed with distilled water and then dried at 60 °C for 24 hours.

NiTMPyP@NaY was synthesized by using our newly developed "build-bottle-around-ship" method [4]. Typically, 240mg of $NiTMPyPCl_4$ was mixed into an aluminosilicate gel as described above for the synthesis NaY and crystallized under the same condition as that for the synthesis of NaY. After natural cooling to room temperature, the solid products were recovered by filtration through a Buechner funnel. The surface adsorbed complexes were removed by a thorough extraction with distilled water and organic solvents, repetitively. The removed NiTMPyP complexes can be easily re-collected. Therefore, the loss of the NiTMPyP catalyst in the crystallization-purification process is negligibly small. For the comparison purpose, we have also synthesized NiTMPyP adsorbed on the external surfaces of NaY crystals. This system, denoted as NiTMPyP/NaY, was prepared by the conventional impregnation method using methanol as the solvent. Since a planar NiTMPyP has a diameter of ~18 Å, which is significantly larger than the diameter of a channel window (~7 Å) on the surface of NaY, it is therefore expected that NiTMPyP can only be adsorbed on the external surface of NaY after the impregnation process.

Characterization

TGA was conducted on a Setaram TGA 92 thermogravimetrical analyzer/differential thermal analyzer. XIF was performed on a Philips Sequential X-ray spectrometer (Model PW 2400) with a standard Rh X-ray tube. XRD was carried out on a Philips PW 1830 system equipped with a PW 1710 diffractometer control. DRS was recorded in the UV-VIS region with a Cary 5 instrument. Micromeritics ASAP 2000 was used for the measurement of nitrogen adsorption and surface areas at liquid nitrogen temperature. Prior to the adsorption measurement, the samples were activated at 423 K for 4 h in high vacuum (~8 x 10^{16} Pa). RRS with 413.1 nm excitation was performed on a SPEX 1403 double monochrometer equipped with a Kr^+ ion laser (Coherent Inova K400) and a photomultiplier detector and an Olympus BH2-UMA microscope. All GC analyses were performed on a HP GC equipped with a 50m CP Sil-5 (dimethylpoly siloxane) capillary column (Chrompack, 0.32 mm internal diameter, 1.2 mm film thickness) and a FID-detector. The reaction products were analyzed by GC-MS on a GC-8000 GC and FISONS MD800 Mass-Spectrometer in CI^+ (=positive chemical ionization) or EI mode with a 60 m DB-5MS column (J & W, 0.32 mm internal diameter, 1mm film thickness).

Catalysis

The hydrogenation reaction in this work was carried out in a home made stainless steel autoclave with a reaction volume of 10 ml. For the hydrogenation of *cis*-, *cis*-1,5-cyclooctadiene(COD), 3.0 mmol COD was mixed with 4.0g *n*-hexane as the solvent in a 10ml autoclave with a stir bar. 0.3g NiTMPyP@NaY was added as the catalyst. The H_2 pressure was 3 mPa at room temperature.

RESULTS AND DISCUSSION

Characterization of Synthesized NiTMPyP@NaY Catalyst

Figure 1 displays the XRD patterns of our synthesized plain NaY and NiTMPyP@NaY samples. The XRD patterns are in excellent agreement with the theoretical calculation for a faujasite type zeolite [5] in both peak positions and relative intentities. All observed diffraction peaks can be indexed according to the simulated results [5]. However, certain peaks in the XRD pattern of NiTMPyP@NaY show slightly downshift in 2 theta or the broadening of the peak shape, in comparison with their counterparts in pure NaY. This suggests that there occurred a slight distortion of the zeolitic cages (unit cells) as a consequence of the occlusion of the relatively larger-sized NiTMPyP. XIF data gives a Si/Al ratio of 1.6 for the synthesized NiTMPyP@NaY composite.

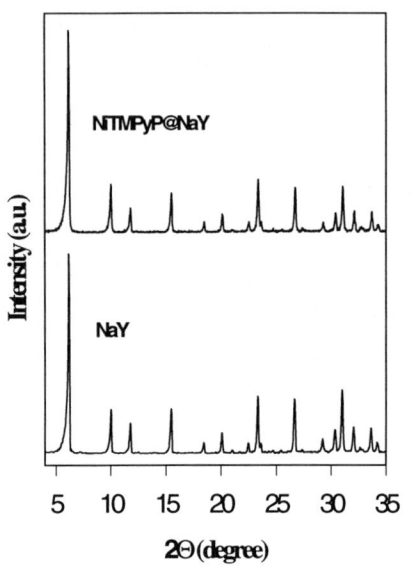

Figure 1. XRD patterns of NaY (lower) and NiTMPyP@ NaY (upper).

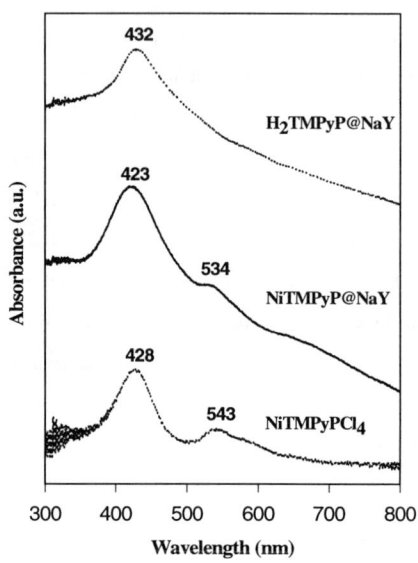

Figure 2. DRS spectra of NiTMPyPCl$_4$, NiTMPyP@NaY and H2NiTMPyP@NaY.

Figure 2 displays the DRS spectra of pure NiTMPyP guest, NiTMPyP@NaY and H$_2$TMPyP@NaY, respectively. The Soret and Q bands of NiTMPyP are retained well in shape and relative intensity when confined into the zeolitic framework. For the occluded NiTMPyP, Soret and Q bands are observed at 423 and 534 nm, respectively, somewhat blueshifted comparing to the 428 and 543 nm in NiTMPyPCl$_4$. The free base H$_2$TMPyP, when entrapped inside the supercages of NaY, gives a Soret band at 432 nm, which is significantly different from NiTMPyP@NaY, suggesting that no demetalization occurred during the hydrothermal synthesis.

The TGA of NaY, NiTMPyP/NaY and NiTMPyP@NaY samples all show endothermic weight losses at temperatures below 250 °C, attributable to the desorption of water in zeolite. In contrast to that of the plain NaY, the TGA of both NiTMPyP@NaY and NiTMPyP/NaY exhibit an additional exothermic weight loss at ~ 400 °C, readily attributable to the decomposition or burning of NiTMPyP in the air flow. The decomposition temperature is 407°C for NiTMPyP/NaY sample. The TGA of NiTMPyP@NaY displays a shoulder at lower temperature of 395 °C. This decrease in the decomposition temperature of the entrapped NiP can be ascribed to the distortion of porphyrin molecules inside the smaller zeolitic cages. This distortion perturbs the well-delocalized π skeleton of the macrocycle, and therefore destabilizes the entrapped molecule. This result provides direct evidence that the NiTMPyP cations are indeed entrapped in the cavities of NaY in NiTMPyP@NaY sample. From the observed exothermic weight loss, the loading of NiTMPyP in NiTMPyP@NaY composite was calculated to be 0.90 wt.%, which is further confirmed by an elemental analysis.

Nitrogen adsorption experiments indicate that the adsorption of 0.90 wt.% NiTMPyP on the external surface of NaY by impregnation method leads to a reduction of 8.0% in pore volume from 206.0 to 189.6 cc/g at a P/P_0 of 0.6. However, the nitrogen accessible volume in NiTMPyP@NaY decreased to 178.4 cc/g, about 13.4 % less than that in the plain NaY, indicating that the entrapped guest molecules block certain zeolitic pores/channels. This difference between NiTMPyP/NaY and NiTMPyP@NaY in their nitrogen adsorption volumes provides another piece of evidence that NiTMPyP is indeed encapsulated in the supercages of NaY by our synthetic method.

RRS with different laser excitations proven to be a very powerful technique in studying either heme structure in various hemeproteins or transition metal complexes confined in zeolites [6,7]. Figure 3 displays the RR spectra of NiTMPyP@NaY, pure NiTMPyPCl$_4$ and pure NaY acquired with a laser excitation at 413.1 nm, in close resonance with the Soret band (423 nm) of NiTMPyP cation. NaY zeolite itself is a poor light scatter, allowing a wide spectral window for the study of the occluded guest molecules. Figure 3 clearly shows that the basic RR features of homogeneous NiTMPyPCl$_4$ are all retained in the spectrum of NiTMPyP@NaY, indicating that neither decomposition nor demetallization occurred during the process of hydrothermal crystallization. However, differences between the two spectra, including peak shifts and band-broadening, were clearly observable. Comparing with the spectrum of NiTMPyPCl$_4$, pyridinium related RR bands in the region of 600 ~ 800 cm^{-1} become much broader and blue-shifted in the spectrum of the NiTMPyP@NaY. The ν_{22}(Pyr half-ring)$_{asym}$ band upshifted from 1000 to 1006 cm^{-1} when NiTMPyP cation was confined into the zeolite cages. However, the ν_9 band downshifted to 1088 cm^{-1} from 1094 cm^{-1} after encapsulation. The macrocyclic skeleton stretching bands in the region of 1100 to 1600 cm^{-1}, including $\nu_2(C_\beta C_\beta)$, $\nu_3(C_\alpha C_m)_{sym}$ and $\nu_4(C_\alpha C_\beta)$, show pronounced change of their relative intensities, with negligible shift in their frequencies. These changes provide additional evidences for the assertion that the entrapped NiTMPyP was forced to distort its macrocycle in order to fit its larger size into the rigid and relatively smaller supercage of zeolite, a conclusion in good agreement with the results obtained from XRD, TGA and DRS.

Catalytic Hydrogenation

In the hydrogenation of *cis*-, *cis*-1,5-cyclooctadiene (COD), NiTMPyP@NaY exhibits much higher activity than either NaY blank or NiY catalyst, which was prepared by ion-

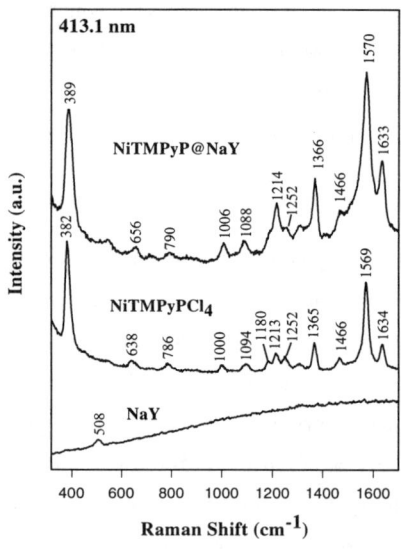

Figure 3. RR spectra of NaY, NiTMPyPCl$_4$, and NiTMPyP@NaY.

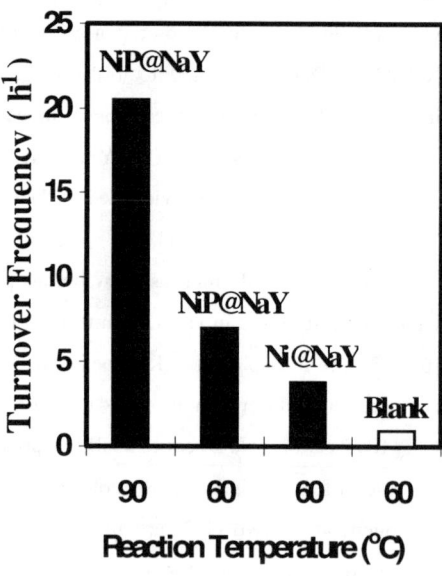

Figure 4. Hydrogenation of COD over NaY, Ni@NaY, NiP@NaY (60 °C) and NiP@NaY (90 °C).

exchange of NaY in NiAc solution (Figure 4). This shows that the complexation of Ni^{2+} with porphyrin ligand indeed enhance strongly its catalytic property. The formation of 1,4-diene via isomerization reaction was observed over both NiTMPyP@NaY and NiY catalysts, even though the latter shows relatively higher isomerization activity. Figure 4 also shows that the hydrogenation activity of NiTMPyP@NaY depends sensitively on the reaction temperature. The hydrogenation rate (turnover frequency or TOF) increases significantly with the elevation of reaction temperature. The TOF value increased from 7.0 to 20.5 h^{-1} when the temperature was elevated from 60 to 90 °C. The temperature effect can be accounted for partially by the diffusion effect arising from the space constraint imposed by zeolitic channel, where the reactants and products had to diffuse in and out of the active sites.

One of the key properties of zeolite catalysts is their shape-selectivity in the competitive reactions. This property was clearly observed in our synthesized NiTMPyP@NaY catalyst. The competitive reactions were studied over NiTMPyP@NaY on a model system, cyclooctene(COE) and cyclododecene (CODD) mixed in a molar ratio of 1:1. The C_8/C_{12} ratio for the hydrogenation can be over 35. The slight conversion of CODD may also be attributed to the reactor itself, as we did observe a slight activity in the blank hydrogenation of diene (Figure 4). The favorable conversion of COE over CODD can be understood from the shape-selectivity of zeolitic framework. The smaller sized COE can diffuse easily into the supercages of NiTMPyP@NaY through the 12MR channel (~ 7Å in diameter), while the bulkier CODD diffuses into this channel with a much lower rate than that of COE. Although the shape-selectivity is a common phenomenon in the zeolite related catalysis, our work is the first on the competitive hydrogenation catalyzed by a zeolite-confined metallocomplex. This shape-selectivity further demonstrates that the NiTMPyP cation is indeed encapsulated inside the supercage of NaY by our developed "build-bottle-around-ship" method.

CONCLUSIONS

NiTMPyP tetracations are successfully encapsulated inside the supercages of NaY by using our developed "build-bottle-around-ship" method [4]. The synthesized host-guest catalyst was fully characterized and it displays a high activity and selectivity in the catalysis of hydrogenation of alkenes.

ACKNOWLEGMENTS

We acknowledge the RGC-HK and HKUST for the financial support. We also like to thank Dr. P.P Knops-Gerrits, Dr. D. E. De Vos and Ms. M. Tielen for their kind help.

REFERENCES

1. S. Kowalak, R. C. Weiss, and Jr., K. J. Balkus, J. Chem. Soc., Chem. Comm., 57-58 (1991).
2. D. Chatterjee, H. C. Bajaj, A. Das, and K. Bhatt, J. Mole. Catal., **92**, 235-238 (1994).
3. D. E. De Vos and P. A. Jacobs, Stud. Surf. Sci. and Catal., 615-622 (1993).
4. B.-Z. Zhan and X.-Y. Li, J. Chem. Soc.: Chem. Commun., 349-350 (1998).
5. M. M. J. Treacy, J. B. Higgins and R. von Ballmoos, Collection of Simulated XRD Powder Patterns for Zeolites, 3rd edn. (Elsevier, 1996).
6. T. G. Spiro and X.-Y. Li in Biol. Appl. of Raman Spectrosc (Wiley & Sons, New York, 1988), Vol. 3, Chapt. 1.
7. B.-Z. Zhan and X.-Y. Li, Stud. Surf. Sci. Catal., **105A**, 615-622 (1997) and references cited therein.

NITRILE-TO-AMIDE HYDROLYSIS CATALYZED BY FAUJASITE-Y CONFINED CHROMIUM(III) PORPHYRIN

B. -Z. ZHAN[a], P. A. JACOBS[b] and X. -Y. Li[a] *

[a] Department of Chemistry, The Hong Kong University of Science and Technology, Clear Water Bay, Kowloon, Hong Kong, China; chzhan/chxyli@ust.hk
[b] Centrum voor Oppervlaktechemie en Katalyse, K.U. Leuven, 92, Kardinaal Merciertaan, B-3001, Heverlee (Leuven) Belgium; Pierre.Jacobs@agr.kuleuven.ac.be

ABSTRACT

We report a novel, efficient and rational " build-bottle-around-ship" method to construct the nanocages of faujasite-Y(NaY) around catalytically active Cr(III) porphyrin cations (CrP). Our method introduces an electrostatic recognition mechanism between the inherently negatively-charged building blocks of zeolite (the host) and the positively-charged peripheral substituents on the CrP (the guest), making the process of host-crystallization/guest-nanoinclusion being anchored by the guest molecules themselves. Our method allows the occlusion of pure metalloporphyrins (MP) in NaY with a variable loading concentration. The successful encapsulation of MP inside the supercages of NaY by our method was fully demonstrated by various methods of physical characterization, which indicate certain distortion but not decomposition or demetallation of the entrapped CrP during hydrothermal crystallization. Our catalyst, denoted as CrP@NaY, displays an efficient and selective nitrile-to-amide hydrolysis activity as shown by its catalysis of the hydrolysis of acetonitrile to acetamide.

INTRODUCTION

There has been extensive interest in developing viable catalysts for the controlled hydrolysis of nitriles to their corresponding amides with high activity and selectivity under mild conditions [1]. In nature, this goal is achieved by nitrile hydratase (Nhase) enzymes [2], which were shown to contain either a non-heme iron or a non-corrinoid cobalt at their catalytic sites [2]. Whole cells containing NHase have been used as catalysts for nitrile to amide transformations [3-4]. Synthetically, certain platinum and rhodium complexes have been shown to catalyze the hydrolysis of nitrile to amide [5], often with low turnover numbers (TON). For

---

* To whom correspondence should be addressed.

example, a TON of 150 can be achieved with Rh(OH)(CO)(PPh$_3$)$_2$ as the catalyst for the hydrolysis of acetonitrile to acetamide [5]. It was also shown that hydrolytic nitrile-to-amide conversion can also be achieved in a quite low yield in aqueous/ethanol solvent in the presence of hydrogen peroxide (H$_2$O$_2$) with alkali (OH$^-$) [5]. Encapsulation of catalytic transition metal complexes (TMCs) inside the supercages of zeolites, often referred to as the " ship-in-a-bottle" system, has been considered as a promising approach to improve the efficacy, selectivity and stability of the catalysts, and has been widely explored in the oxidation of alkanes and alkenes [6-10]. We report here the nanoinclusion of a carefully chosen CrP inside the supercages of NaY by using our developed " build-bottle-around-ship" method [11]. Our method is novel in that it introduces an electrostatic recognition mechanism between the host (zeolite) and the guest (CrP) during the nanoinclusion process. This was achieved by taking advantage of the negatively-charged (AlO$_2$)$^-$ sites inherently on the building blocks of the zeolite framework and the positively-charged substituents, e.g. *N*-methyl-4-pyridinium, synthetically attached to the porphyrin skeleton. The catalyst synthesized, denoted as CrP@NaY, has been fully characterized by various techniques. Our catalyst displays an excellent catalytic activity and selectivity in the hydrolysis of acetonitrile to acetamide in the presence of H$_2$O$_2$.

EXPERIMENTAL

NaY was synthesized by crystallizing the freshly prepared aluminosilicate gel, containing 4.6g silica, 6.2g NaOH, 3.2g NaAlO$_2$ and 60ml H$_2$O, at 95±2 °C for 48 hours. The zeolite encapsulated CrP [17], denoted as CrP@NaY, was synthesized by adding 800mg of CrP complex to the above amount of aluminosilicate gel. The gel was crystallized at 95±2 °C under static and autogenous conditions in a stainless steel bomb (250ml) for 48 hours. After naturally cooling to room temperature, the solid products were recovered by filtration through a Buechner funnel. Surface adsorbed complexes were removed by thorough and sequential extraction with distilled water, methanol, pyridine(2%)/methanol, and methanol again, respectively. The removed complexes can be fully recovered for the next round of synthesis. The crystals were then dried at 60 °C for 24 hours. For the purpose of comparison, CrP impregnated on the external surfaces of NaY, denoted as CrP/NaY, was prepared by an impregnation method in which the pure CrP was mixed thoroughly with NaY in methanol, the suspension was then refluxed at 60 °C for 10 hours, followed by the removal of solvent. The powder obtained was then dried at 60 °C for 24 hours. Since the selected CrP molecules have a diameter of approximately 1.8 nm, which is significantly larger than the diameter of a channel window of ~ 0.7 nm on NaY, it is reasonable to assume that CrP will be unable to diffuse into the supercages of NaY, and can only be adsorbed on the external surface.

Thermogravimetric analysis (TGA) was conducted on a Setaram TGA 92 thermo-gravimetrical analyzer/differential thermal analyzer. The temperature was scanned from 50 to 900 °C with a rate of 10 °C per minute. X-ray induced fluorescence analysis (XIF) was performed on a Philips Sequential X-ray spectrometer (Model PW 2400) equipped with a standard Rh X-ray tube. XRD was carried out on a Philips PW 1830 diffractometer equipped with a PW1710 control. Micro-Resonance Raman scattering (μRRS) was conducted on a 1403 SPEX double monochrometer equipped with an argon ion laser, an Olympus BH2-UMA microscope, and a photomultiplier tube on a Renishaw-3000 Ramanscope equipped with a He/Ne laser. Diffuse reflectance spectroscopy (DRS) was recorded in the range of 200 ~ 2500 nm with a Cary 5 instrument, a type II diffuse reflectance attachment, and an Eastman Kodak White Reflectance standard. The hydrolysis products were analyzed and quantified by GC-MS (Finnigan TSQ 7000) and GC (HP 5890 S2$^+$).

Hydrolysis of acetonitrile was conducted at room temperature under aerobic conditions in a glass tube with a rubber stopper. 20 mmol acetonitrile was mixed with 0.5g H_2O_2 (30 wt.%) and 0.2 g CrP@NaY or NaY(the blank) as the catalyst.

RESULTS AND DISCUSSION

Characterization of CrP@NaY

The XRD patterns of NaY and CrP@NaY with different loadings are displayed in Figure-1. The observed XRD patterns are in excellent agreement with that calculated for a faujasite model [12]. However, the effect of encapsulation of CrP in the supercages of NaY can indeed be reflected in XRD patterns: certain diffraction peaks of CrP@NaYsamples shift down in 2θ and have a broader bandshape than their counterparts in the XRD pattern of pure NaY. This can be more clearly seen in the inset of Figure 1 which displays an enlarged view of [111] diffractions of NaY and CrP@NaY. The broader and down-shifted peaks in the XRD patterns of CrP@NaY samples suggest that certain unit cells or supercages in the host-guest crystals are slightly enlarged or/and distorted as a consequence of the CrP occlusion. This is somewhat expected since the diameter of a CrP molecule as a whole (~ 1.8 nm) is slightly larger than the diameter of a supercage (~ 1.3 nm) in NaY. XIF analysis revealed a Si/Al ratio of 1.4 for NaY and 1.6 for CrP@NaY, respectively, confirming that the host lattices in CrP@NaY crystals are indeed NaY type.

Figure 2 shows the TGA profiles of CrP/NaY and CrP@NaY, respectively. An endothermic weight loss was observed at temperatures below 250 °C in both samples, which is readily attributed to the desorption of water since it also appears in the TGA of NaY. Comparing with NaY, both CrP/NaY and CrP@NaY exhibit an additional exothermic weight loss over 300 °C. This is attributed to the decomposition or burning of CrP in air flow. The decomposition temperature of CrP in CrP/NaY is 342 °C, much lower than that of 369 °C observed in CrP@NaY. This stabilization effect of the host matrix on the guest molecule provides a piece of direct evidence that CrP cations are indeed entrapped inside the supercages of NaY by our "build-bottle-around-ship" method. From the observed exothermic weight loss, the loading of CrP in CrP@NaY is calculated to be 3.2 wt.%, in good agreement with our result obtained from elemental analysis. This loading transforms into a concentration of one CrP molecule per eighteen supercages of NaY host. Of course, the loading concentration of CrP in NaY can be controlled in our method by simply varying the amount of pure CrP added in aluminosilicate gel. For example, everything else remained the same, the loading concentration of 0.89% was achieved when 240 mg of CrP was added to the aluminosilicate gel mentioned above.

The DRS spectrum of CrP@NaY, measured in the UV-Vis-NIR region, displays a strong Soret band at 446 nm and two relatively weak Q bands at 566 and 608 nm, respectively, which are not much different from their counterparts at 447, 567 and 606 nm, respectively, in the DRS spectrum of the pure CrP sample, suggesting that the effect of the host matrix on the electronic structure of the guest is quite small. The preservation of the characteristic DRS spectrum of CrP inside NaY also indicates that the entrapped CrP is neither demetalated nor chemically modified/destroyed.

RRS has proven to be a very powerful technique in studying heme structure and dynamics in various hemeproteins and in zeolitic host-guest systems [13,14]. Figure 3 displays the RR spectra of NaY, CrP, CrP/NaY and CrP@NaY, respectively, acquired with laser excitation at 457.9 nm, in close resonance with the Soret band of CrP complex. NaY itself, like SiAlO type zeolites, is a poor light scatterer. Its Raman spectrum is very simple and well-defined. Only one major scattering peak was clearly observed at 508 cm^{-1}, corresponding to the bending character of its T-O-T skeleton [14]. As expected, the RR spectrum of CrP/NaY is basically the same as that of a homogeneous pure CrP sample. The RR spectrum of CrP@NaY also displays all the basic features observed in that of CrP and CrP@NaY, indicating that no chemical modification of the entrapped CrP, such as demetalation, dimerization or autooxidation, occurred during the process of hydrothermal crystallization [15,16]. Certain changes are however noticeable in the RR spectrum of CrP@NaY in comparison with that of CrP: (a) most of the observed RR bands of CrP@NaY are broader and sometimes split in comparison with their counterparts in CrP and

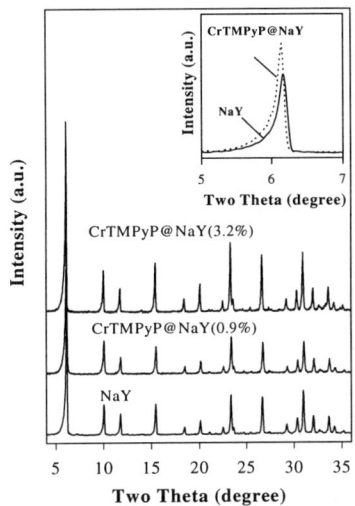

Figure 1: XRD patterns of NaY and CrP@NaY.

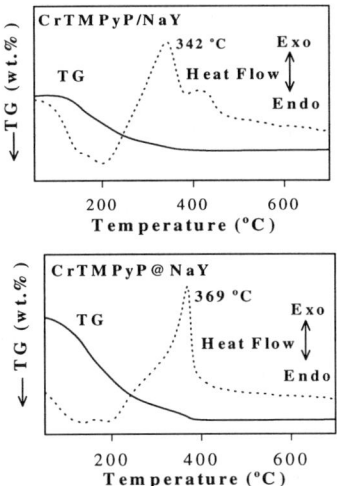

Figure 2: TGA profiles of CrP/NaY and CrP@NaY.

CrP/NaY; (b) $v_1(C_m$-Py) downshifts by about 8 cm^{-1} to 1245 cm^{-1}, implying weakening of the C_m-Py bond in the entrapped CrP, which is presumably caused by a ruffling distortion at methine bridges; such distortion is enforced by the four methyl-pyridinium groups stretching toward the four channels tetrahedrally connected to a supercage of NaY, allowing an entrapped CrP to fit itself into a relatively rigid and smaller supercage; (c) the tetrahedral ruffling of the entrapped CrP is also reflected by the band broadening/splitting in the 600 ~ 800 cm^{-1} region. For example, the 724 cm^{-1} band in CrP splits into two peaks in CrP@NaY at 714 and 738 cm^{-1}, respectively; (d) the v_8(Cr-N) band shifts down by about 4 cm^{-1} to 394 cm^{-1}, indicating the decrease of the bond strength between Cr^{3+} and the porphyrin skeleton; (e) CrP@NaY shows the characteristic pyridinium ring stretching mode ϕ_4(Py) at 1634 cm^{-1}, 6 cm^{-1} lower than its counterpart in CrP, indicating the change of counter anion from chloride to zeolitic framework.

Catalytic Hydrolysis

The result of hydrolysis of acetonitrile over NaY and CrP@NaY at room temperature are given in Figure 4. Over CrP@NaY catalyst, 19.4% of acetonitrile was converted to acetamide with over 99% selectivity in 24 hours, in comparison with < 0.2% conversion obtained over NaY in the blank experiment. This enormous difference in catalytic activity between CrP@NaY and NaY demonstrates that the entrapped CrP is indeed the active site for the catalytic hydrolysis. The turnover number(TON) for each entrapped CrP is about 5930 in 24 hours. This huge TON over CrP@NaY is a benefit from the host framework of NaY, which can effectively protect the guest molecules from dimerization and/or autooxidation under the oxidative conditions. The catalytic hydrolysis is via a Cr-OOH intermediate, as confirmed by *in situ* Raman spectroscopy which shows a band at 920 cm^{-1}, readily assigned to the v_{O-O} stretching in Cr(III)-OOH, and is higher than the v_{O-O} of H$_2$O$_2$ at 877 cm^{-1}. The remarkable activity and selectivity of our CrP@NaY catalyst under mild condition makes it an excellent catalyst for either industrial or benchtop transformation of nitriles to amides.

CONCLUSIONS

We have introduced a new " build-bottle-around-ship" method for the encapsulation of pure catalytic metalloporphyrins in NaY. The method takes advantages of the inherent negatively charged AlO$_2^-$ sites in zeolite precursors, and demonstrated the necessity to have positively charged guest molecules as the anchoring seeds in the hydrothermal crystallization process. Well-designed CrP containing peripheral cationic groups were successfully encapsulated into NaY via this method. The loading of the guest catalyst can be varied in a controllable manner. The successful encapsulation of guest molecules inside the supercages of zeoilte Y was demonstrated by various physical characterization methods, including XRD,

TGA, DRS and RR spectroscopies. The synthesized CrP@NaY displays an excellent activity and selectivity in catalyzing the hydrolysis of acetonitrile to acetamide.

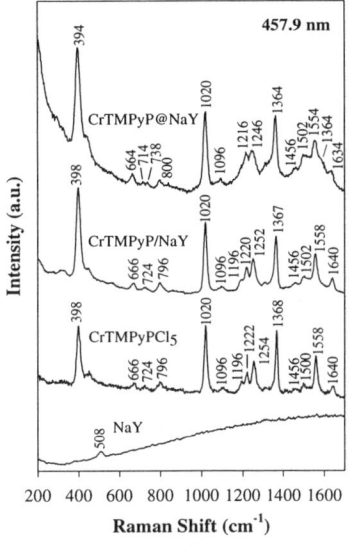

Figure 3: RR spectra of NaY, CrP, CrP/NaY and CrP@NaY.

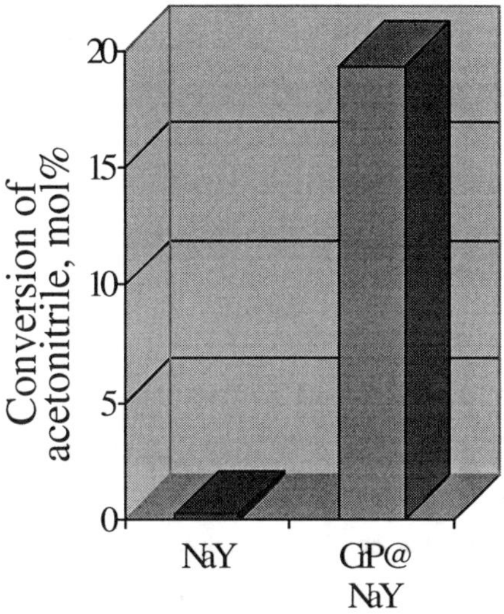

Figure 4: Hydrolysis of acetonitrile over NaY and CrP@NaY at room temperature.

ACKNOWLEDGMENT

We thank RGC-HK and HKUST for the financial support of this project.

REFERENCES

1. M. Kobayashi, T. Nagasawa, H, Yamada, Trends in Biotechnol., **11**, 402-408 (1992).
2. S. Nagashima et al., Nature Struct. Biol., **5**, 347-351 (1998).
3. J. Crosby, et al, J. Chem. Soc. Perkin Trans I, 1679-1687 (1994).
4. N. J. Turner, M. A. Cohen, J. Sawden in Preparative Biotransformations: Whole Cell and Isolated Enzymes in Organic Synthesis (Wiley & Sons, 1992).
5. B. R. Brown, in The Organic Chemistry of Aliphatic Nitrogen Compounds(Oxford, 1994), p.28.

6. B. V. Romanovsky and A. G. Gabrielov, J. Mole. Catal., **74**, 293-303 (1992), and references cited therein.
7. N. Herron, and W. E. Farneth, Adv. Mater., **8**, 959-968 (1996) and references cited therein.
8. Jr., K. J. Balkus, M. Eissa, and R. Levado, J. Am. Chem. Soc., **117**, 10753-10754 (1995) and references cited therein.
9. P. P. Knops-Gerrits, D. D. De Vos, F. Thibault-Starzyk, and P. A. Jacobs, Nature, **369**, 543-546 (1994).
10. C. Bowers, and P. K. Dutta, J. Catal., **122**, 271-279 (1990).
11. B. Z. Zhan, X.-Y. Li, J. Chem. Soc.: Chem. Commun., 349-350 (1998).
12. M. M. J. Treacy, J. B. Higgins, and R. von Ballmoos, Collection of Simulated XRD Powder Patterns for Zeolites, 3rd edn. (Elsevier, 1996).
13. T. G. Spiro and X.-Y Li, in Biological Applications of Raman Spectroscopy (Wiley, 1988), Vol. 3. Chapt. 1.
14. B.-Z. Zhan and X.-Y. Li, Stud. Surf. Sci. Catal., **105A**, 615- 622 (1997).
15. X.-Y. Li, et. al, J. Phys. Chem., **94**, 31-47 (1990).
16. N. Blom, J. Odo, K. Nakamoto, D. P. Strommen, J. Phys. Chem., **90**, 2847-2852 (1986).
17. Cr(III) porphyrin derivative used in this paper, including the synthesis, characterization and catalysis, is Cr(III) *tetra*-(*N*-methyl-4-pyridium)porphyrin chloride salt, denoted as the CrTMPyP. Another Cr(III) porphyrin carrying peripheral cations is Cr(III) *tetra*(4-*N,N,N*-trimethylanilinium)porphyrin chloride salt, which has also been encapsulated successfully in NaY using the same method [11].

RESTRICTED TRANSITION STATE AT PORE MOUTH CATALYSIS IN THE SELECTIVE HYDROISOMERISATION OF NORMAL AND METHYL BRANCHED C_8 PARAFFINS OVER MONODIMENSIONAL 10-RING MOLECULAR SIEVES.

P.MÉRIAUDEAU, VU. A. TUAN, G. SAPALY, VU. T. NGHIEM AND C. NACCACHE

Institut de Recherches sur la Catalyse/CNRS, 2 Avenue Albert Einstein - 69626 Villeurbanne Cedex, FRANCE

ABSTRACT

Hydroconversion of n-octane, 2-methyl heptane and 2,2,4 trimethylpentane over monodimensional 10-membered ring molecular sieves supporting Pt-Pd has been studied. 2,2,4 TMP is almost exclusively hydrocracked into iso-butane. The molecule is too bulky to enter the pores and reacts exclusively on the external surface. n-octane and 2-Me C_7 react inside the pore at a short distance from the surface. The high selectivity towards isomerisation is attributed to transition state shape selectivity at the pore mouth. Monobranched isomers, principally 2-Me C_7 and 3-Me C_7 are formed inside the pores. The isomerisation of 2-Me C_7 produced a considerably higher amount of n-octane than that predicted, smaller are the pore sizes smaller is the ration 3-Me C_7/nC_8, from the isomerisation of 2-Me C_7. It is suggested that the limited space available inside the 10-membered ring channels produces a larger constraint on 1,2 methyl shift intermediate than on the protonated cyclopropane intermediate.

INTRODUCTION

The hydroconversion of n-paraffins occurs through several consecutive steps, which leads to the isomerisation into mono and multibranched paraffins, followed by cracking of these intermediates. The rate of β-scission increases with the degree of branching, monobranched alkyl carbenium ions being hardly susceptible to cracking. It results that high selectivity towards skeletal isomerisation occurs when multibranching reactions are restricted. Methyl branching reaction proceeds through the protonation of the olefin formed by dehydrogenation of the alkane on the metallic site, followed by cyclization of the resulting sec-carbenium ion into corner protonated alkyl cylclypropyl carbenium ion and subsequent ring opening of the CPCP intermediate. Hence monobranched, dibranched and multibranched isomers are formed consecutively, the multibranched carbenium ion being very sensitive to cracking by β-scission.

Based on these considerations high yields of mono and dibranched methyl alkanes formed during the hydroconversion of n-alkanes on monodirectional medium pore molecular sieves, ZSM-22 and 23, SAPO-11, SAPO-31, SAPO-41 were interpreted as resulting from the circumscribed space available in the channels to accomodate multibranched alkylcarbenium ions. Based on isomer product distribution observed it was concluded that the monobranching reaction of n-alkanes occurs within the pores at the pore mouth [1-2], in the pore mouths and on the external surface [3,4], within the pores diffusional limitation occurring [5-6]. In the present work comparative studies on the isomerisation of n-octane and 2-methylheptane over unidimensional medium pore molecular sieves are reported. The aim was to investigate the effects of small change in the dimension of the 10-membered ring on the isomer product distribution. Our interest was also to determine how far the concept of restricted transition state catalysis at the pore mouths is valid in these systems.

EXPERIMENTAL

Materials: 10-membered ring SAPO-11, SAPO-41, ZSM-22, ZSM-23 were synthesized following the patent literature and according to recipes described in ref. [1-6]. For comparison monodimensional 12-membered ring SAPO-5 was also prepared. The Si/Al ratio in ZSM-22 and ZSM-23 as determined by chemical analysis amounts to 55. The grain sizes are in the range 0.1-0.5 µm for the SAPO' samples (SAPO-5 grain size is 1-2 µm). ZSM-22 and ZSM-23 are in the form of thin needles of 1 µm length.

The Bronsted acidity of these samples was determined by TPD of adsorbed ammonia, and by the relative intensity of the IR band around 3630 cm^{-1} due to acidic OH groups in SAPO' samples and around 3600 cm^{-1} for ZSM-22 and ZSM-23 also attributed to acid OH. The data will be detailed in a forthcoming paper [2]. The conclusions of these studies, which are worthwhile to mention, are that SAPO' samples have almost equivalent Bronsted acidity, while the acidity of ZSM-22 and ZSM-23 is higher in strength.

The bifunctional catalysts were obtained by impregnating the H-form molecular sieves with Pt and Pd tetra-amine chloride solution successively. The contents of Pt and Pd were adjusted such that the metal loading on each sample were 0.6 wt % Pt and 1.2 wt % Pd. Samples were calcined in air at 673 K and reduced in H_2 at 773K prior to the reaction. The metal dispersion as determined by H_2 adsorption was around 50-60 %.

Catalytic reactions

The hydroconversion of n-octane, 2-Me heptane and 2,2,4 trimethylpentane was carried out in a fixed bed microreactor at atmospheric pressure and at 523 K. The mole ratio H_2/HC was 60. The space velocity, WHSV, was adjusted for each sample such that the conversion never exceeded 10 %. WHSV was varied in the range 0.2-10 h^{-1}. The relative rate of the C_8 conversions was derived from the product αxWHSV (α = conversion). Reactant and products were analyzed on line by gas chromatography using a PONA capillary column for all the C_8 isomers and a second chromatograph equipped with an unibed 3 S column was used for the analysis of the cracked products.

RESULTS

At such low conversion, less than 10 %, the amount of cracked products was relatively low even for large pore SAPO-5 catalyst. The distribution of the cracked products from the reaction of n-octane showed for all the catalysts a symmetrical curve centered around C_4 alkanes. Furthermore the amounts of C_1 and C_2 alkanes which resulted generally from hydrogenolysis reaction on the metallic sites were low, less than 3 % of the total cracked products. These observations indicated that true bifunctional catalysts are operating.

In table 1 are given the relative reaction rates expressed by the product αxWHSV for respectively n-C_8, 2-Me C_7 and 2,2,4 TMP reactions over the catalysts based on SAPO and on ZSM-type.

Table 1: Hydroconversion of C_8 isomers : comparison of individual rates (α x WHSV) at 250°C. $\alpha < 10\%$

Reactant	SAPO-5	SAPO-11	SAPO-41	ZSM-22	ZSM-23
n-C_8	27.5	162	129	245	99
2Me-C_7	69	215	189	253	68
2,2,4 TMP	92.7	29.3	32.9	14.8	12.4

SAPO-11 and SAPO-41 which showed almost the same Bronsted acidity and grain size dimensions in the same range 0.1-0.5 μm, and pore size very similar 0.63 x 0.39 nm for SAPO-11, 0.7 x 0.43 nm for SAPO-41, did not differ much in their relative activity for the hydroconversion of the C_8 alkanes. Similar trend was shown when ZSM-22 and ZSM-23 were compared. However comparison between ZSM-type and SAPO-type molecular sieves is not straightforward unless careful determination of their respective Bronsted acidity (number and strength) has been achieved. In addition it is not clear why SAPO-5 with large pores, was less active than medium pore SAPO-catalysts unless the Bronsted acidity was weak.

Nevertheless the results in table 1 show that on SAPO-5 where the C_8 isomers can diffuse in the intracrystalline voids the order of decreasing reactivity from the multibranched 2,2,4 TMP to the monobranched and linear C_8 was obeyed. 2,2,4 TMP gave almost exclusively iso-butane. By contrast medium pore molecular sieves where the bulky 2,2,4 TMP cannot diffuse in the intracrystalline voids showed poor activity for the reaction of 2,2,4 TMP. Only the external grain surface was thus concerned for the hydroconversion of 2,2,4 TMP.

In table 2 are reported the selectivities towards hydrocracking and towards hydroisomerisation when n-octane reacted over the catalysts. In table 2 are also listed the relative percentage of 2-Me heptane (2-Me C_7) and 3 methylheptane (3-MeC_7) in the monomethylheptane isomers formed, as well the ratio 2,5 dMeC$_6$/2,4 dMeC$_6$.

Table 2: n-octane conversion at 250°C H_2/HC = 60. Product distribution at conversions < 10 %

Samples	SAPO-5	SAPO-11	SAPO-41	ZSM-22	ZSM-23
Cracking selectivity %	20	5.4	1.1	11	12
Isomerization selectivity %	80	94.6	98.9	89	88
2-MeC$_7$ %	30	56.9	56.4	63	56
3-MeC$_7$ %	34	38	39.4	23	26
2,5 dMeC$_6$/2,4 dMeC$_6$	0.88	1.8	2	4	2.3

Table 2 shows, as expected that monodimensional medium pore molecular sieves are very selective for hydroisomerisation reaction of long chain alkanes. Large pores SAPO-5, where multibranched isomers can be formed inside the intracrystalline voids, was less selective for isomerisation since the rate of cracking for multibranched isomers is very fast as compared to the rate of cracking of monobranched isomers. Within the experimental conditions used in this work it appears from table 2 that the selectivity for hydroisomerisation of n-octane is slightly lower over alumino-silicates than over SAPO-11, SAPO-41. It is no clear at present whether this is due to the higher acid strength of ZSM-22 and –23 as compared to SAPO-11 and SAPO-41 or to the smaller dimensions of the 10-membered ring in ZSM-22 and ZSM-23. Both parameters will contribute to longer residence time of the intermediate carbenium ions in the channels, thus to a higher probability for β-scission this will cause enhanced cracking reaction.

The hydroconversion of 2-Me C_7 over medium pore molecular sieves have shown, similarly to n-octane reaction, that at low conversion (less than 10 %) hydroisomerisation reaction has occurred almost exclusively. The cracking reaction represented less than 3-4 % of the total conversion. Also branching rearrangement of 2-MeC$_7$ into dimethylhexanes represented a small fraction of the total conversion. The experimental results indicated that the major reactions which have occurred over these 10-membered ring molecular sieves were :
i) rearrangement with no change of the degree of chain branching, such rearrangement occurred by 1,2 methyl shift, 2-Me C_7 producing 3-MeC$_7$, ii) rearrangement with decrease of the degree of chain branching 2-MeC$_7$ into n-C$_8$ such rearrangement involves PCP intermediate.

$$n\text{-}C_8 \xleftarrow{k_2} 2\text{-MeC}_7 \xrightarrow{k_1} 3\text{-MeC}_7$$
$$\quad\quad\quad\text{PCP} \quad\quad\quad\quad CH_3 \text{ shift}$$

Table 3 gives the ratio of 3-MeC$_7$/nC$_8$ resulting from the reaction of 2-Me C_7 at 523 K over the molecular sieve based catalysts. The conversion was kept low, less than 10 % to avoid further secondary reactions.

Table 3. Reaction of 2-MeC$_7$ at 250°C : Molecular ratio 3-MeC$_7$/nC$_8$ at conversion < 10%.

Catalyst	SAPO-5	SAPO-41	SAPO-11	ZSM-22	ZSM-23
Pore size nm	0.73x0.73	0.7x0.43	0.63x0.39	0.54x0.44	0.52x0.45
3MeC$_7$/nC$_8$	8	2.26	1.84	1.19	0.76

The product distribution resulting from the reaction of 2,2,4 TMP over the samples, indicates that essentially hydrocracking reaction has occurred. The selectivity for cracking, even at conversion less than 10 %, was in the range 60-80 %. Within the cracked products the selectivity to iC$_4$ was always higher than 98 %. In addition to the cracking reaction, isomerisation by 1,2 methyl shift has occurred producing essentially 2,3,4 TMP isomer.

DISCUSSION

The hydroconversion of C$_8$ paraffins follows a bifunctional mechanism : the olefin formed by dehydrogenation of the alkane on the metal is protonated and forms sec-octyl carbenium ion. These key intermediates experience carbon chain rearrangements via protonated cyclopropyl mechanism (branching rearrangement) and/or via 1,2 methyl shift mechanism (non-branching rearrangement).

The rate of cracking by β-scission increases with the degree of branching [7,8] such that trimethylpentyl intermediates are cracked considerably faster than dimethyl or monomethyl C_8 isomers. The key feature of the medium pore molecular sieves studied is the limited void space available within the channels, which renders impossible the formation of trimethyl pentyl ions within the intracrystalline void. It resulted that β-scission is considerably hampered and occurs rapidly only from the multibranched isomers formed on the external grain surface. The high selectivity showed by these catalysts for hydroisomerisation of long chain alkanes is thus explained.

It is well known [7] that rearrangement of the carbon chain without branching by methyl shift is faster than rearrangement where degree of branching increases or decreases (PCP intermediate). Furthermore cracking by β-scission of multibranched carbon chain where the starting and the resulting carbenium ions are both tertiary is faster than carbon chain rearrangements. Hence in non-constraint molecular sieves the hydroconversion rates of C_8 isomers must follow the order TMP > Me-C7 > n-C_8. The results shown in table 1 for SAPO-5, large pore molecular sieve, are in agreement with the previsions. The narrow 10-membered ring channel molecular sieves do not obey these rules. 2,2,4 TMP is indeed too bulky to enter the 10-ring channels and thus reacts only with the Bronsted sites at the grain surface [1,5]. n-octane and 2-methyl heptane by contrast can penetrate the pores, the molecules being totally or partly inside the channels. Hence the number of Bronsted sites available for the reaction and reached by the reactants will be by far much larger than when 2,2,4 TMP is considered. It results that the hydroconversion of n-octane or 2-Me heptane will be higher than that of 2,2,4 TMP in agreement with the results given in table 1.

One key question remains still open and debated in the literature : the alkylcarbenium ion with branched carbon chain is located partly the tail inside the channel and the branched ramification outside [4] or as suggested in [2] the ions are entirely in the channels at the pore mouth at a short distance from the grain surface.

Within large pore SAPO-5, n-octyl and 2-Me heptyl ions are formed inside the pores. The isomerisation of these intermediates is not hampered by space constraint, it results that the conversion of 2-Me C_7^+ which involves principally 1,2 methyl shift will be faster than that of n-C_8^+ which involves PCP intermediate in agreement with the results shown in table 1. Similarly if only the tails of n-C_8^+ and of 2-Me C_7^+ penetrate into the pore and the remaining branching carbon chain outside the pore, one should expect that the amount of Bronsted sites accessible to the reactants is very identical and again one expects that 2-Me C_7 will react faster than n-C_8 over small pore molecular sieves in contrast to the results shown in table 1. Based on these considerations we suggest than n-octene and 2-Me heptene formed at the metal surface penetrate into the medium pores of SAPO-11, and others, to experience monobranching and 1,2 methyl shift or debranching reactions respectively. The penetration of 2-Me C_7 into the micropores of SAPO-11, and others, involves diffusional transportation at a short distance from the pore mouth which decreases the true isomerisation rate of the monobranched alkane. We concluded that the rearrangements of the alkyl carbenium ions formed inside the pores occur also inside the pores. In the narrow channels of the 10-membered ring molecular sieves, the constraint exerted by the limited space hampered the formation of the PCP ring far from the terminal carbon chain (the approximative size of cyclopropane ring is 0.43 nm). Hence the most favoured alkylcyclopropyl ion would be the one with the ring at terminal position. Ring opening of these PCP intermediates would favour the preferential formation of 2-methyl branched and to less extent 3-Me branched isomers in agreement with the results shown in table 2. In conclusion the preferential of 2-Me C_7 from n-C_8 beyond the thermodynamic value (2-Me C_7/3-Me C_7 = 0.79) and beyond the predicted statistical distribution using equal weights of PCP intermediates (2-Me C_7/3-Me C_7 = 0.32) is due to restricted transition state selectivity at the pore mouth.

The final point of interest is shown by table 3. It has been indicated that with monomethylbranched carbenium ions migration of methyl group by 1,2 methyl shift occurs faster than the PCP debranching (or branching) of methyl group. Hence over systems producing no constraint on the reaction the preferential isomerisation of 2-MeC_7 is to form 3MeC_7 rather than n-C_8 or dMeC_6. Table 3 shows indeed such trend. However as the constraint due to the pore dimensions increases the ratio of 3-MeC_7/nC_8 decreases and for the smallest pores, ZSM-23, debranching becomes preferential.

Such behavior nicely reinforces the concept of transition state shape selectivity at the pore mouth. The approximative size of cyclopropane ring is 0.43 nm, while monomethylalkanes have the dimension of the molecule around 0.5 nm. The PCP intermediates where the ring is at terminal position will be favoured with respect to the monomethyl carbenium ion involved in the 1,2 methyl shift reaction. Hence methyl debranching by PCP will be favoured, and 2-MeC_7 would form preferentially n-C_8 as the constraint increases. To be accommodated in the molecular sieve channels, the terminal PCP intermediate should have the PCP cycle considerably distorted, favouring ring opening into-C_8.

Apparently the distortion of the PCP cycle must be easier than that of the methylbranched carbenium ion ; 1,2 methyl shift along the carbon chain is thus hindered.

In conclusion this work demonstrates that restricted transition state at the pore mouth is responsible for the high efficiency of monodimensional 10-membered ring molecular sieves. The high selectivity towards 2-MeC_7 is due to the preferential formation of end-PCP intermediate which can easily distort to fit with the sizes of the channel.

REFERENCES

1 - S.I Miller, Microporous Mater., 2, 439, (1994).

2 - P. Mériaudeau, V.A. Tuan, F. Lefebvre, V.T. Nghiem and C. Naccache, Microporous Mater., (1998).

3 - S.J. Ernst, J. Weitkamp, J.A. Martens and P. Jacobs, Appl. Catal., 48, 148, (1989).

4 - J.A. Martens, R. Parton, L. Uytterhoeven, P.A. Jacobs, Appl. Catal. 76, 95, (1991).

5 - P.A. Jacobs, J.A. Martens, J. Weitkamp and H.K. Beyer, Farad. Discussion Chem. Soc., 72, 353, (1981).

6 - P.Mériaudeau, V.A. Tuan, V.T. Nghiem, S.Y. Lai, L.N. Hung and C. Naccache, J. Catal., 169, 55, (1997).

7 - D.M. Brouwer, in "Chemistry and Chemical Engineering of Catalytic Processes" (R. Prins, G.C.A. Schuit, Eds.), NATO ASI Series E n° 39, p. 137 (1980).

8 - J. Weitkamp, P.A. Jacobs, and J.A. Martens, Appl. Catal. 8, 123, (1983).

CATALYTIC CRACKING OF PALM OIL TO HYDROCARBON LIQUID FUELS OVER VARIOUS ZEOLITE CATALYSTS: OPTIMIZATION STUDIES

S. BHATIA[*], NOOR ASMAWATI MOHD ZABIDI AND F. TWAIQ

[*]School of Chemical Engineering, Universiti Sains Malaysia, Perak Branch Campus, 31750 Tronoh, Perak, Malaysia; Fax: (605) 3677055; kkbhatia@kimia.eng.usm.my

ABSTRACT

The catalytic cracking of palm oil to liquid hydrocarbon fuels was studied over HZSM-5, zeolite β and USY zeolite catalysts in a fixed bed micro-reactor. The reactor was operated at atmospheric pressure, a temperature range of 350 – 450°C and weight hourly space velocity (WHSV) of 1 – 4 hr^{-1}. Organic liquid product (OLP) rich in hydrocarbons, light hydrocarbon gases and water were the major products obtained from the cracking reaction. Conversion of palm oil to an organic liquid product (OLP) varied significantly with reaction temperature and WHSV. HZSM-5 catalyst showed high shape selectivity for aromatics (75 – 94 wt. %) in the gasoline obtained, whereas zeolite β and USY gave 10 – 56 wt. % and 15 – 44 wt. % respectively. Statistical Design of Experiment (DOE) approach was used to assess the effect of reaction temperature, WHSV and catalyst pore size, on the gasoline yield and conversion of palm oil. Response surface methodology was utilized to obtain the optimum values of reaction parameters for maximum gasoline yield and conversion.

INTRODUCTION

Palm is widely grown in Malaysia, 90 % of palm oil production is used for food production and the remaining 10 % in oleochemicals applications. The Malaysian palm oil triglyceride composition is (C44: 0.07, C46: 1.18, C48: 8.08, C50: 39.88, C52: 38.77, C54: 11.35, C56: 0.59 wt. %) [1].

Currently, some researchers are concentrating on developing alternative and renewable sources of liquid fuels that are "environmental friendly". Catalytic conversion of various types of fatty oils like palm, canola, tall and soybean oils has been reported [2,3]. The shape selective zeolite catalyst have been used for catalytic cracking of canola oil and 60 – 95 wt. % of canola oil was converted to hydrocarbons in gasoline boiling range, light gases and water [3]. The choice and characteristic of the catalyst played an important role in the conversion of bio-oil to organic liquid product (OLP) rich in hydrocarbons and selectivity of particular hydrocarbon products [3].

The objectives of this study were to maximize the amount of the organic liquid product and its hydrocarbon contents as well as to optimize the selectivity for gasoline range hydrocarbons from the catalytic cracking of palm oil. The effects of important operating variables, such as reaction temperature, space velocity and type of zeolite cracking catalyst were studied using Statistical Design of Experiment (DOE) [4,5].

EXPERIMENTAL

Experiments were conducted in a stainless steel micro-reactor (155mm×10mm I. D.) fitted with a thermocouple in the center of the catalyst bed. One gram of calcined catalyst loaded over 0.2g quartz wool was placed in the reactor. The reactor was heated by electrical vertical tube furnace. Argon gas was allowed to flow at 0.2 l/hr through the system for 1 hr before feeding the palm oil in the reactor. The palm oil was fed by a syringe infusion pump (supplied from Cole-Parmer model E-74900-05) and was preheated in a horizontal tube furnace before entering the reactor.

The products were cooled by circulated water at 40°C in a condenser. The liquid products were collected in a glass sampler trap and the gas products separated in the trap were collected in a Teflon sampling bag. The aqueous phase of the liquid product was separated using a micro-syringe prior to distillation of the liquid products in a micro-distillation unit in order to estimate the quantity of the residual oil.

The gas products and the distillate (OLP) were analyzed in a gas chromatograph (HP model 5890 series II) using a flame ionization detector (FID) and a Porapak Q (2m length×6mm I. D.) column. The composition of organic liquid product was analyzed over a glass capillary column (50m length, 0.2mm I. D.) containing petrocol DH 50.2 phase with 0.5 microns film thickness. The spent catalyst was washed with hexane to determine the remained residual oil over the catalyst. The coke formed during the cracking reaction was determined using a thermal gravimetric analyzer (TGA).

RESULTS AND DISCUSSION

The performance of three zeolite catalysts having different pore sizes was studied in terms of conversion, yield of gasoline and coke formation. Among the three zeolite catalysts, HZSM-

5 with pore size of 5.5Å with 10 ring pore system gave highest conversion about 99 wt. % and maximum gasoline yield at low WHSV of 1 hr^{-1}. The conversion remains almost constant with the change in reaction temperature (350 – 450°C) at WHSV of 1 hr^{-1}. Zeolite β has pore size of 5.7 × 7.5Å with interconnecting 12 ring pore system and chiral pore sections. In the presence of zeolite β the conversion increased with increase in reaction temperature and decrease in WHSV. Whereas, in USY zeolite with pore opening of 8Å (12 ring pore system), the conversion increased more sharply within the temperature range of 400 – 450°C. The conversion and the product distribution were found to vary with the change in pore size of different zeolites as shown in Table 1.

Based on the palm oil feed, the gasoline yields were between 17 – 28 wt. % over HZSM-5, 15 - 26 wt. % for zeolite β and 4 – 17 wt. % for USY zeolite catalyst. The gasoline yields decreased sharply with increase of WHSV and decrease of reaction temperature for zeolite β and USY zeolite catalysts. The maximum gasoline yield was obtained with HZSM-5 indicating the high shape selectivity of a catalyst. The shape selectivity played an important role in a secondary cracking resulting in a high yield of organic liquid product. USY and zeolite β gave high selectivity for kerosene and diesel range hydrocarbons. The selectivity for aromatic hydrocarbons was higher with HZSM-5 catalyst, 40 – 60 wt. % based on OLP as compared to the other zeolite catalysts, 8 – 23 wt. % in zeolite β and 7 – 20 wt. % in USY zeolite. The gas yield increased with temperature and decreased with increase of WHSV. Ethylene, ethane, propylene, propane and butane were some of the major components of the gaseous products.

The performance of each catalyst in terms of the coke formation was also examined. Different zeolite spent catalysts gave different wt. % coke showing their shape selectivity in the coke formation. At 400°C and WHSV of 2.5 hr^{-1}, HZSM-5 gave 10 wt. % coke, whereas zeolite β and USY gave 30 wt. % and 15 wt. % coke, respectively based on the weight of spent catalyst.

Design of Experiment (DOE) was used to optimize the conversion and gasoline yield. Table 2 shows the analysis of variance (ANOVA) for the quadratic model that was used with 3^3 full factorial design. DESIGN-EXPERT software was used to analyze the response data. Five repeated cracking experiments were used at conditions of (400°C, 2.5 hr^{-1} and zeolite β catalyst) to obtain the estimate of experimental error. Equations 1 and 2 show the statistical quadratic model estimated from the significant effects and its interactions. The response surface methodology was used to obtain the optimum values of the two responses. The three zeolites gave different contours of conversion and gasoline yields against reaction temperature and weight hourly space velocity (WHSV) as shown in Figure 1.

Conversion = $396.98 - 1.50 \times T + 0.85 \times WHSV - 25.66 \times CPS + 1.252E\text{-}03 \times T^2 - 0.47 \times WHSV^2$
$- 337.03 \times CPS^2 + 0.014 \times T \times WHSV + 0.98 \times T \times CPS - 17.72 \times WHSV \times CPS$ ....(1)

Gasoline = $215.97 - 0.86 \times T + 2.06 \times WHSV - 42.64 \times CPS + 7.136E\text{-}04 \times T^2 - 0.75 \times WHSV^2$
$- 129.84 \times CPS^2 + 7.833E\text{-}03 \times T \times WHSV + 0.44 \times T \times CPS - 4.39 \times WHSV \times CPS$ .... (2)

Where: T: temperature; WHSV: weight hourly space velocity; CPS: catalyst pore size.

Temperature, °C	350			400			450		
WHSV, hr^{-1}	1	2.5	4	1	2.5	4	1	2.5	4
HZSM-5									
Conversion	99.0	80.3	77.3	96.9	89.5	78.7	97.2	91.6	86.0
Gas	42.9	19.9	17.3	47.9	32.4	36.3	46.8	44.1	34.3
Coke	5.0	1.4	7.2	5.1	5.1	2.2	4.6	1.3	2.0
OLP	43.6	53.0	47.2	40.2	49.0	31.1	40.1	43.5	46.2
Gasoline	28.3	26.6	23.5	20.4	25.8	17.28	24.1	22.9	24.4
Zeolite β									
Conversion	82.2	65.0	51.2	86.0	76.5	64.0	95.7	88.2	77.1
Gas	14.0	9.9	6.6	20.2	18.6	12	46.9	29.0	31.0
Coke	10.4	6.2	6.7	6.8	6.0	7.4	6.3	3.8	5.0
OLP	48.1	42.9	36.6	53.7	47.8	39.8	38.2	52.9	39.4
Gasoline	22.0	19.1	15.6	22.1	19.6	16.4	18.4	26.3	16.3
USY									
Conversion	53.2	45.3	28.3	61.5	51.1	29.6	93.2	70.7	57.6
Gas	6.4	8.4	2.0	16.1	15.5	9.7	39.8	38.3	28.9
Coke	12.2	11.0	2.2	7.5	4.7	3.8	3.7	3.8	2.4
OLP	29.2	24.3	20.5	30.2	27.8	11.7	45.1	25.1	26.3
Gasoline	7.3	5.2	3.8	13.0	13.0	4.3	17.5	14.3	11.7

Table 1. Conversion and major products obtained from catalytic cracking of palm oil over HZSM-5, zeolite β and USY catalysts at different cracking conditions

Source	Sum of Squares	Degree of Freedom	Main Square	F value
Conversion				
Model	10382.03	9	1153.56	65.02[a]
Residual	390.34	22	17.74	-
Lack of Fit	353.21	17	7.43	-
Gasoline				
Model	1147.69	9	127.52	22.86[a]
Residual	122.71	22	5.58	-
Lack of Fit	115.87	17	6.82	-

a: significant at 1 % level.

Table 2. Analysis of variance (ANOVA) for the conversion and the gasoline for quadratic model in Response Surface Methodology

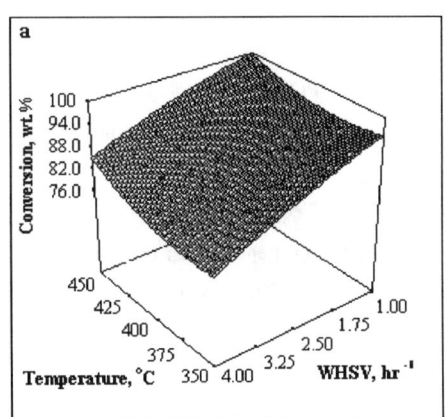
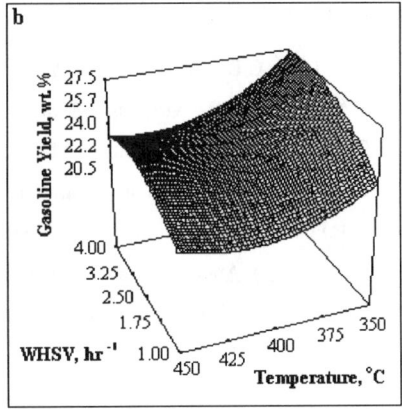

Figure 1. (a) Response surface plot for conversion of palm oil over HZSM-5 catalyst, (b) Response surface plot for gasoline yield obtained from palm oil cracking over HZSM-5 catalyst

The optimum values of gasoline yield and palm oil conversion obtained from the proposed model were 27.5 wt. % of gasoline based on palm oil feed at conversion of 92 % at WHSV of 1.7 hr^{-1} and temperature of 350°C for HZSM-5 zeolite catalyst. For zeolite β, the optimum value was 22.7 wt. % of gasoline yield at conversion 95 wt. % at 1.7 hr^{-1} WHSV and temperature of 450°C. For USY zeolite, the gasoline yield was 16.9 wt. % at conversion of 85.7 wt. % at 1 hr^{-1} WHSV and temperature of 450°C.

CONCLUSIONS

1. HZSM-5 zeolite catalyst gave the best performance among the three zeolite catalysts tested in terms of conversion, hydrocarbons in gasoline range, higher selectivity for aromatics, and lower coke formation.
2. USY and zeolite β catalysts gave higher selectivity for hydrocarbons in diesel range and lower production of gaseous products.
3. The optimum values of reaction parameters were obtained using Design Of Experiment (DOE) and Response Surface Methodology.

REFERENCES

1. T.S. Tang, C.L. Chong, M.S.A. Yusoff and M.T. Ab Gapor, PORIM technology, 1995, 17.
2. T.Y. Leng, A.R, Mohammed and S. Bhatia, Proceedings of Regional Symposium of Chemical Engineering, Johor - Malaysia Oct. 1997, **1**, pp. 75-81.
3. R. O. Idem, S.P.R. Katikaneni and N.N. Bakhshi, Fuel Proc. Tech., **51**, 1997, pp. 101-125.
4. G.E.P. Box, W.G. Hunter and J.S. Hunter, Statistics for Experimenters: An Introduction to Design, Data Analysis, and Model Building, John Wiley & Sons, New York, 1978.
5. J. D., Adjaye; S. P. R., Katikaneni and N. N., Bakhshi, Fuel Proc. Tech., **48**, 1996, pp. 115-143.

CARCINOGENICITY OF MINERAL ERIONITE FIBERS: MEASUREMENTS AND HYPOTHESIS OF ACTIVITY

B. D. HOGG, P. K. DUTTA, J.F. LONG[†] AND A. VAIDYALINGAM

Departments of Chemistry and [†]Veterinary Biosciences, The Ohio State University
Columbus, OH 43210; dutta.1@osu.edu

ABSTRACT

Inhalation of certain fibers, such as asbestos can lead to lung diseases, including mesothelioma, a type of malignant tumor around the lining of the lung. There is considerable effort in understanding the cause of asbestos toxicity, so that appropriate therapies can be designed. One of the most carcinogenic materials is the zeolite mineral erionite, considerably more so than asbestos. The extreme toxicity of erionite is puzzling, especially considering that other mineral zeolitic fibers, such as mordenite are relatively benign. In this paper, we report a technique which measures the toxic reactive oxygen metabolites (ROM) created upon interaction of erionite with rat lung macrophage cells. A peroxide marker dye [5-(and-6)-carboxy-2',7'-dichlorodihydrofluorescein diacetate] is used to determine intracellular levels of ROM after exposure to erionite mineral fibers via confocal fluorescence microscopy. Current theories of mineral toxicity not only require the creation of ROM, but also the presence of iron to turn the ROM into extremely toxic hydroxyl radicals. In this paper, we propose a hypothesis which states that the siting of iron in erionite alters its reduction potential allowing for the facile production of toxic hydroxyl radicals.

INTRODUCTION

Mineral-induced lung diseases pose a considerable threat to society. Alveolar macrophages (AM) are the first line of immunologic defense against dust deposition in the deep lung and accumulate in the alveoli ("air sacs") after fiber deposition [1]. These defense cells are capable of engulfing foreign matter and transporting the material out from the deep lung. In addition, during interaction with foreign matter, AM may release any of a variety of molecular species, including reactive oxygen metabolites (ROM), which include superoxide (O_2^-) and hydrogen peroxide (H_2O_2) and are thought to initiate chemical changes that lead to lung diseases.

Erionite is a naturally occurring fibrous zeolite and, if inhaled, is one of the most potent carcinogenic fibers known. In fact, erionite carcinogenic potency towards humans has been described as being significantly higher than asbestos [2]. Mordenite, on the other hand, though fibrous in nature, is relatively benign [3]. Thus, erionite and mordenite represent excellent models for use in understanding of fiber toxicity and is the goal of our research program.

Reactive oxygen metabolite (ROM) -specific assays provide an important method by which fiber characteristics may be directly associated with possible pathogenesis [4]. In these assays, where a large number of cells are exposed to a large number of fibers, firm control over fiber characterization and dosage is difficult. Flow cytometric procedures provide information on internal release of ROM in single cells, but temporal and spatial evolution within a given cell cannot be followed with these "one shot" flow methods [5]. This report describes a fluorescence technique which provides an approach to examining the link between the surface chemistry and morphology of a fiber and the reactive oxygen species produced upon phagocytosis of the fibers by lung macrophages.

EXPERIMENTAL

The mineral examined here is erionite (Minerals Research, Clarkson, NY). The cell line of macrophages that we have chosen (NR8383; obtained through the cooperation of R. J. Helmke at The University of Texas, Health Science Center, San Antonio) is derived from rat pulmonary alveolar macrophages and has been extensively characterized by Helmke and coworkers [6]. The following protocol, which has been described previously [7,8] was followed to establish cell-fiber contact. The erionite crystallites were primarily fibrous, with aspect ratios >>3. Cells (~750,000/ml) were washed, allowed to equilibrate with buffer, and were then exposed to dye $H_2DCF-DA$ (final concentration = 60µM) in an incubator (37 °C and 5% CO_2) for 40 minutes. Cells were then subjected to 1 ml buffer (control) or 1 ml fiber suspension (~12 x 10^6 particles/ml), vortexed briefly (10 sec.), and then centrifuged (600 rpm for 5 minutes) at 4 °C in order to promote contact of the fibers with the cell surface. The supernatant containing debris and excess dye was removed and discarded. The pelleted cells (and cells with fibers) were immediately resuspended in chilled buffer. This cell-fiber mixture was then taken to the confocal microscope (Meridian Instruments' ACAS 570; Okemos, MI) for fluorescence imaging [7].

RESULTS AND DISCUSSION

Procedure for measuring reactive oxygen metabolites (ROM): The role of the lung macrophages as defense cells involves engulfing or phagocytizing foreign matter. Figure 1 is a visualization of the fiber-cell interaction process examined by optical spectroscopy using natural erionite fibers. The cell pictured has clearly contacted one long (~10μm) fiber, and, more than likely, has completely taken this fiber within itself. The phagocytosis process results in release of ROM and a measure of this is important in defining the toxicity of the engulfed foreign matter.

The procedure we have developed for measuring ROM generation involves using a dye [5-(and-6)-carboxy-2',7'-dichlorodihydrofluorescein diacetate (H_2DCF-DA)] to determine intracellular levels of ROM after exposure to erionite mineral fibers. The nonfluorescent form of the dye is loaded into the macrophage. The reaction of the cellular dye with ROM produced within the cell results in oxidation of the dye and production of a highly fluorescent species which is detected via confocal fluorescence microscopy. Figure 2 is a schematic of the dye chemistry. The sub-micron spatial resolution of the technique allows dye concentrations to be followed sub-cellularly. The imaging process is rapid and may be followed on a 1-minute time scale.

Figure 3 illustrates the imaging aspect of the method and compares individual confocal slices from an erionite-exposed and a control cell. Increasing fluorescence is indicated by a darker shade. The erionite-exposed cell shows fluorescence considerably higher than the control, indicative of formation of ROM. Quantitation of the fluorescence signal provides for the amount of ROM. The total cellular fluorescence for both erionite-exposed and control cells over a 100-minute period was investigated for a number of cells. A gradual increase in fluorescence levels of control cells was observed and attributed to mitochondria-induced generation of ROM during normal respiration. However, for cells exposed to erionite, within 35 minutes after contact, the mean cellular fluorescence was more than three times that of control cells. These higher levels are expected and are consistent with erionite uptake and creation of ROM. This method can be used to compare the ROM creation for a

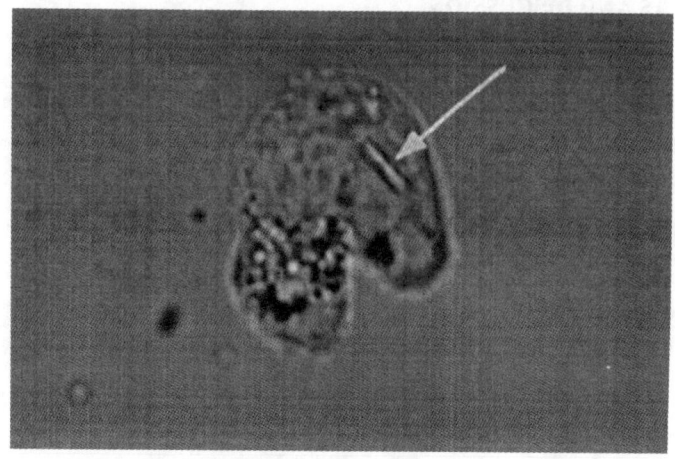

Figure 1. Macrophage, with arrow indicating internalized erionite fiber.

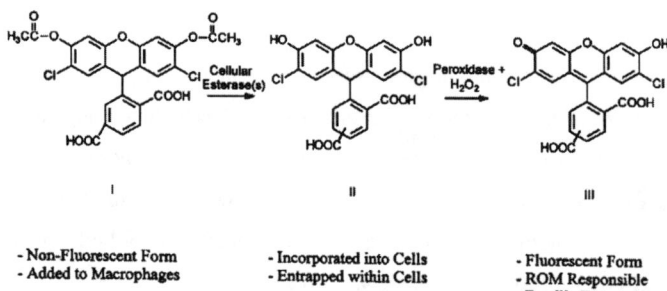

- Non-Fluorescent Form
- Added to Macrophages

- Incorporated into Cells
- Entrapped within Cells

- Fluorescent Form
- ROM Responsible
- Readily Detected

Figure 2. Schematic of Dye-Cell Interaction and Reactivity

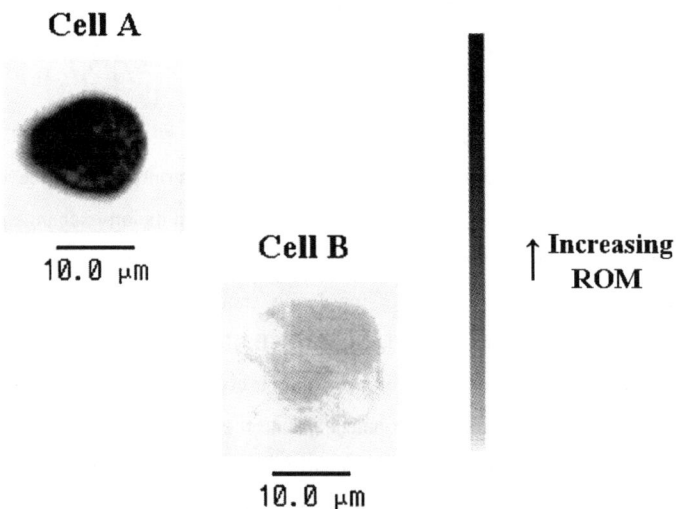

Figure 3. Internal ROM Levels for Erionite-exposed cell (A, Time = 28 minutes) vs. Control (B, Time = 26 minutes).

series of zeolitic fibers on a single cell basis and is currently under investigation.

Model of Toxicity:. The considerable toxicity of erionite as compared to other mineral fibers has been an area of considerable discussion [9]. Most models for fiber toxicity acknowledge the role of morphology so that the fibers can reach the lungs, as well as durability for long-term toxicity. Lung (alveolar) macrophages and neutrophils release hydrogen peroxide (H_2O_2) and superoxide ion (O_2^-) as they phagocytize free fibers. These are typically not considered to be toxic species in biological systems. Recent models of toxicity suggest that along with reactive oxygen metabolites, iron is also required in order to generate the toxicity [9]. In the presence of iron, the following iron-catalyzed Haber-Weiss reactions (reactions 1-4) can lead to the creation of hydroxyl radicals. Hydroxyl radicals react with biomolecules with diffusion controlled second-order rate constants ranging from 2×10^8 to 3.6×10^{10} M^{-1} s^{-1}. The damage to genetic material caused by hydroxyl radicals can lead to initiation of the diseased state.

$$\text{Fe (III) + reductant} \longrightarrow \text{oxidant + Fe(II)}. \quad (1)$$
$$\text{Fe(II)} + H_2O_2 \longrightarrow \text{Fe(III)} + OH^- + \cdot OH \quad (2)$$
$$\text{Fe(II)} + O_2 \longrightarrow \text{Fe(III)} + O_2^{-\cdot} \quad (3)$$
$$HO_2^{\cdot} + O_2^{-\cdot} + H^+ \longrightarrow O_2 + H_2O_2 \quad (4)$$

Erionite, as compared to amphibole forms of asbestos only has trace levels of iron. Toxicity for erionite is proposed in the literature to arise from its ion-exchange properties, which makes it possible for it to acquire Fe(II), (III) [9]. However, this model would then predict that all ion-exchangeable fibers, e.g. mordenite, should also exhibit the potent toxicity of erionite. This is not the case [3]. It is likely that Fe(II,III) does get exchanged into erionite and mordenite. But because of the differences in their crystal structure, the siting of the Fe ions and therefore the coordination environment around the metal ion is different for erionite and mordenite. The erionite structure is characterized by the presence of double six-membered rings, cancrinite cages and a large erionite cage accessible through eight-membered rings of dimensions of 3.6 x 5.2 Å. Mordenite is composed of five and four membered rings leading to channels of twelve-membered rings of dimensions of 6.7 x 7.0 Å. There is also a eight-ring channel parallel to the twelve-membered rings of dimensions of 2.9 x 5.7 Å Most importantly, the surface of the mordenite fiber has small, four membered rings, whereas, erionite has network of eight membered rings.[10].

It is clear from reactions (1) through (3) that the reduction potentials of Fe(III) will have a major effect on the creation of hydroxyl radicals. The reduction potential of Fe(III) is controlled by the coordination around the iron [11]. For example, the reduction potential of Fe(III) is 1.12V. 0.77V and -0.99V, depending on the ligand being phenanthroline, water and enterobactin, respectively [12]. Our hypothesis is that the surface of the erionite fiber is more effective in exchanging iron because of the larger eight membered rings and that the reduction potential of bound Fe(III) - erionite is more negative than that of Fe(III) mordenite, which facilitates the oxidation of Fe(II) in erionite promoting reactions (2) and (3) and leading to the formation of ·OH from ROM's generated by phagocytosis. Cellular reductants such as ascorbate, cysteine, glutathione are capable of reducing Fe(III) in reaction (1). There are a few reports on the reduction potential of iron in minerals. As compared to the reduction potential of Fe(III) of 0.77 V in water, in hornblende amphibole, magnetite and biotite, the reduction potential values are in the range of 0.27 to 0.52 V [13]. The mineral environments are destabilizing Fe(III) as compared to coordination by water, a proposal we are making for erionite. Cyclic voltammetric experiments are being planned to examine the reduction potentials of Fe in different zeolitic fibers to provide confirmation of the proposed hypothesis.

CONCLUSIONS

A procedure is reported for estimation of reactive oxygen metabolites produced in lung macrophage cells upon phagocytosis of carcinogenic erionite fibers. A model for hydroxyl radical toxicity based on the siting of iron in erionite is proposed.

REFERENCES

1. W.N. Rom, W.D.Travis and A.R.Brody, *Ann. Rev. Resp. Dis.*, **142**, 408-422 (1991).
2. K. Hansen and B.T. Mossman, *Cancer Res.*, **47**, 1681-1686 (1987).
3. Y. Suzuki, N. Kohyama, *Environ. Res.*, **35**, 277-292 (1984).
4. H.J. Cohen and M.E. Chovaniec, *J. Clin. Invest.*, **61**, 1081-1087 (1978).
5. P. Szeja, J.W. Parce, M.S. Seeds and D.A. Bass, *J. Immunol.*, **133**, 3303-3307, (1984).
6. (a) R.J. Helmke, V.F. German and J.A. Mangos, *In Vitro Cell Develop. Biol.* **25**, 44-48 (1989).
   (b) R.J. Helmke, R.L. Boyd, V.F. German, J.A. Mangos *In Vitro Cell Develop. Bio.*, **23**, 567-574 (1987).
7. B.D. Hogg, P.K. Dutta and J.F. Long, *Anal. Chem.*, **68**, 2309-2312 (1996).
8. J.F. Long, P.K. Dutta and B.D. Hogg, *Env. Health Perspect.*, **105**, 706-711 (1997).
9. J.A. Hardy and A.E. Aust, *Chem. Rev.*, **95**, 97-118 (1995).
10. D.W. Breck, Zeolite Molecular Sieves, (Robert E. Krieger, Malabar, Florida, 1974).
11. A. Jammal and O.M. Templeton, *Inorg. Chim. Acta.*, **245**, 199-207 (1996).
12. K.N. Raymond, T.J. McMurry and T.F. Garrett, *Pure and Appl. Chem.*, **60**, 545 (1988).
13. A.F. White, M.L. Peterson and M.F. Hochella, Jr., *Geochim. Cosmochim. Acta.* **58**, 1859 (1994).

INFLUENCE OF THE RARE EARTH CONTENT ON THE AMOUNT AND ON THE NATURE OF THE COKE FORMED FROM n-HEPTANE OVER Y ZEOLITES

C. A. HENRIQUES[*] and J. L F. MONTEIRO[†]

[*] Institute of Chemistry, State University of Rio de Janeiro, RJ, Brazil.

[†] Nucleus of Catalysis/COPPE, Federal University of Rio de Janeiro, CP: 68502, CEP:21945-970, Rio de Janeiro, RJ, Brazil, Fax: 55 21 590-7135, email:monteiro@peq.coppe.ufrj.br

ABSTRACT

The deactivation of USY zeolites with different rare earth contents by coking during n-heptane cracking at 723K was studied and the amount and the nature of the coke were determined. The results confirm that exchanging RE cations for protons leads to samples with larger hydrothermal stability and lower acidity. Also, decreasing coking and cracking activities were observed for increasing RE contents. RE cations catalyze the transformation of CH_2Cl_2-soluble coke into -insoluble coke but do not affect the composition of the soluble fraction. A single correlation between residual activity and total coke content was obtained, despite the variation of coke solubility among the samples. This suggests that deactivation is caused by site poisoning and not by pore blockage, for the relatively low coke contents obtained in this work. The tridimensional pore system of faujasites makes them not much sensitive to pore blockage, as compared to, for example, mordenite. Burning of the coke deposited on the samples indicated that the RE cations catalyze the coke oxidation reaction.

INTRODUCTION

Coke formation and its retention inside the pores is the main cause of deactivation of zeolitic catalysts in hydrocarbon processing [1]. Rare earth (RE) exchanged Y zeolites are widely used as the active component in cracking catalysts to improve their activity, their thermal and hydrothermal stability, the gasoline yield, and to lower gas production and coke formation. This work reports on the influence of the rare earth content on the amount and on the nature of the coke formed from n-heptane cracking over Y zeolite and on coke reactivity during its oxidation with air.

EXPERIMENTAL

Three samples of Y zeolite with similar sodium contents (Na$_2$O ≈ 4%) were obtained from the parent sodium form (NaY - Si/Al = 2.7) by ion-exchange with NH$_4$Cl, with both NH$_4$Cl and RECl$_3$, and with RECl$_3$. Steaming of these samples at 923 K gave, respectively, samples USY, CREY-1 and CREY-2. They were characterized by chemical analysis (XRF), FTIR spectroscopy, TPD of NH$_3$ and N$_2$ physissorption. Details on their preparation and characterization are presented by Henriques et al [2]. n-Heptane cracking was carried out in a glass fixed bed gas phase reactor at 723K and 1 atm with N$_2$ as carrier gas (N$_2$/nC7=7/3). The reaction products were analyzed on line by gas chromatography using an Al$_2$O$_3$/KCl PLOT column. WHSV was adjusted to allow a significant coke formation corresponding to initial conversions in the range of 50-60%. The experimental method for recovering the coke from the samples was that of Magnoux et al [3]. For TGA/DSC analyses, the coked samples were burnt in a stream of N$_2$ + O$_2$ (10% O$_2$) at a 10K/min heating rate up to 1023K in a Rigaku TS-1000 Thermoanalyser. The data so obtained allowed the calculation of total coke content and the inspection of the TPO/DSC profiles provided information on coke reactivity.

RESULTS AND DISCUSSION

Table 1 shows the main characteristics of the samples. Increasing the rare earth content increases the hydrothermal stability as can be observed from the framework Si/Al ratios values for the various samples. It also reduces the total acidity of the samples, with no significant modification of their microporosity. The beneficial effect of rare earth on zeolite stability is related to the formation of RE-O-RE bonds inside the zeolite cavities [4]. These bonds are highly resistant to hydrothermal treatment at high temperatures so explaining, according to Ward [5], the increased hydrothermal stability of REY zeolites.

	%Na$_2$O	%RE$_2$O$_3$	Si/Al global (XRF)	Si/Al framework (FTIR)	Total acidity (μmolNH$_3$/gcat)	Vmicro (cm^3/g)	Smeso (m^2/g)	S BET (m^2/g)
USY	3.9	0.0	2.7	5.3	650	0.262	33.9	623
CREY-1	3.5	2.83	2.7	4.9	420	0.273	27.1	649
CREY-2	4.6	12.3	2.7	2.9	310	0.275	20.2	645

Table 1: Main physicochemical characteristics of the samples.

Over the samples used in this work, n-heptane cracking proceeded predominantly through the classic bimolecular mechanism (involving the formation of carbenium ions and β-scission), the main reaction products being C3 and C4 (C3 + C4 ≥ 85%).

Figure 1 shows the activity (defined as moles of n-C7 reacted per hour per gram of catalyst) as a function of time on stream (TOS). The points were fitted to the Voorhies equation and the deactivation coefficients are 0.18 for USY and CREY-1 and 0.10 for CREY-2.

Comparing the results for CREY-1 with those for USY, one can see that a low RE content (± 12% degree of exchange) has little influence on the activity and on the stability of the samples, despite the large difference in their total acidity. Acid sites characterization by FTIR with pyridine adsorption, showed in Table 2, indicate that the concentration of strong Brönsted acid sites, taken as those retaining pyridine at 623K, is almost the same for those two samples, in accordance with the activity results observed.

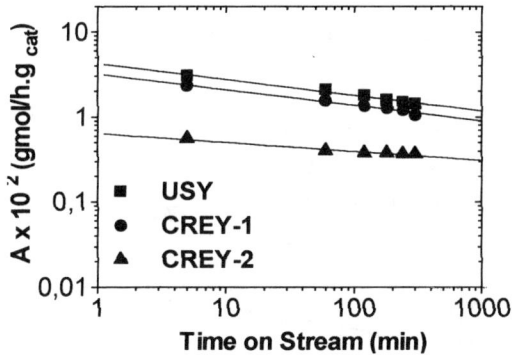

Figure 1. Mean activity (gmol/h.gcat) vs. TOS (nC7 cracking at 723 K).

On the other hand, as can be seen from the results for sample CREY-2, a large degree of RE exchange (± 60%) makes the cracking activity to decrease sharply and the catalytic stability to increase, so indicating that the sites active for both cracking and coking reactions were affected. A corresponding reduction in strong Brönsted acid sites concentration was observed.

While there is no doubt that the initiation step in the protolytic cracking of paraffins is the formation of a carbocation on a Brönsted acid site of the zeolite, it is still under discussion if paraffin cracking initiation via carbenium ions also occur and, if so, on what zeolite sites are the carbenium ions formed [6]. The results presented here are in accordance with the claim that for skeletal transformations of hydrocarbons the rate depends essentially on the protonic acidity of the catalysts [7]. No correlation was found in this work between the concentration of Lewis acid sites and activity, confirming that Lewis acid sites themselves do not seem to be active. However, when located in the vicinity of protonic sites they can increase their strength and consequently their activity [8].

Sample	USY	CREY-1	CREY-2
strong Brönsted acid sites (µmol/gcat)	0.20	0.20	0.13

Table 2: Concentration of strong Brönsted acid sites (those retaining pyridine at 623 K).

Figure 2 shows the coke content as a function of TOS. The greater the RE cations content, the lower the coking rate. This may be ascribed to both a decrease in acid sites density, which hinders bi and polymolecular reactions involved in coking, and to a reduced mesoporosity which limits the room available for voluminous coke precursor molecules.

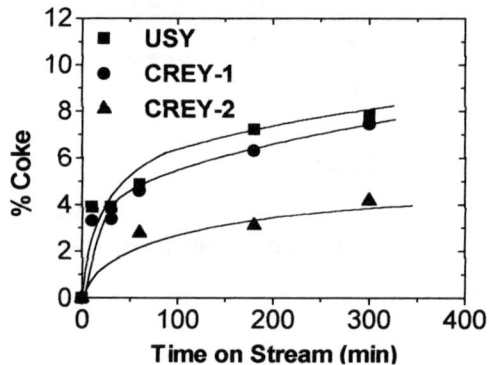

Figure 2. Coke build-up vs. TOS (nC7 cracking at 723K).

Figure 3 shows the amounts of both CH_2Cl_2-soluble and -insoluble coke for samples USY and CREY-2. For both samples, most of the coke was insoluble, within the range studied. Also, even for the lowest level of total coke content, the limiting amount of soluble coke was already attained. From then on, coke build up was due to insoluble coke formation, as previously observed on mordenites [9].

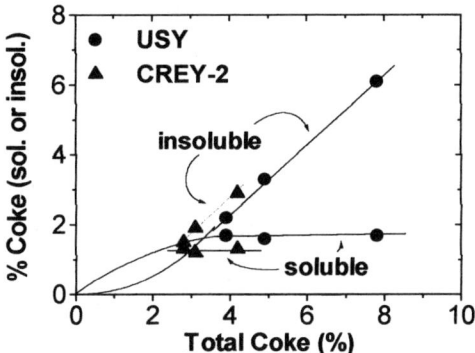

Figure 3. Nature of coke formed over USY and CREY-2 zeolites at 723K

The limiting amount of soluble coke was smaller for CREY-2 than for USY, while, for a given coke content, the amount of insoluble coke was larger for the former. This suggests that the RE cations catalyze at least some of the steps that transform soluble coke into insoluble one.

Figure 4 shows that, within the range studied, the residual activity (defined as the ratio between activity at a given time and that at 1 min) for all samples can be correlated with total coke content by means of a single curve, independently of the relative proportion soluble/insoluble coke. This suggests that deactivation is mainly due to site poisoning and not to pore blockage, which is consistent with the tridimensional nature of the pore structure of faujasites. We have previously shown that this does not hold for the monodimensional pore system of mordenite [9].

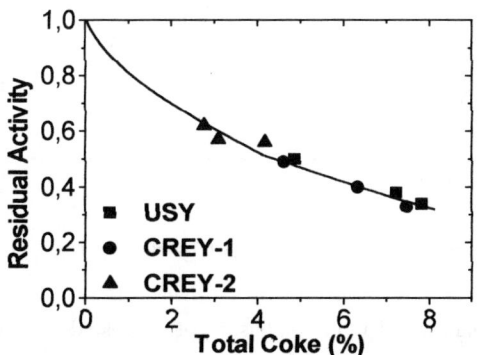

Figure 4. Residual activity as a function of coke content

Analysis of the CH$_2$Cl$_2$-soluble fraction by MS showed that its composition was not significantly affected by the presence of RE cations. The main components were families of polyaromatics with general formula C$_n$H$_{2n-26}$ (alkylcyclopentapyrenes) and C$_n$H$_{2n-32}$ (alkylindenepyrenes) while the alkylpyrenes, C$_n$H$_{2n-22}$, and the indeneanthracene/indenephenantrenes, C$_n$H$_{2n-28}$, appeared as minor components. The most probable structural formula of the main families are presented in Table 2. The formation of these families results from oligomerization of the olefins formed from n-heptane cracking, followed by cyclization, aromatization through hydrogen transfer reactions and alkylation of the aromatic compounds produced. These coke molecules, whose dimensions are close to that of zeolite cavities, remain trapped in the porous system due to their low volatility and/or by steric blockage.

Family Structural Formula	Molecular Formula
[alkylpyrene structure]–Cx	C$_n$H$_{2n-22}$
[alkylcyclopentapyrene structure]–Cx	C$_n$H$_{2n-26}$
[indeneanthracene]–Cx or [indenephenanthrene]–Cx	C$_n$H$_{2n-28}$
[alkylindenepyrene structure]–Cx or [alkylindenepyrene structure]–Cx	C$_n$H$_{2n-32}$

Cx ⇒ alkyl radical with x carbon atoms

Table 2: Main Components of the CH$_2$Cl$_2$ - Soluble Coke Fraction

TPO/DSC analyses indicated that the combustion of the coke deposited on USY was easier the lower the total coke content, i.e., the lower the insoluble fraction. This can be inferred from the profiles shown in Figures 5-a and 5-b which show that the relative contribution of the low temperature region is larger for the sample with lower coke content. Also, it can be seen from Figures 5-b and 5-c that for samples with similar total coke contents and similar fractions of insoluble coke (≈ 3%), the presence of RE cations favored coke oxidation. Similar trends were observed for the coke formed from tri-isopropylbenzene over the same samples [2].

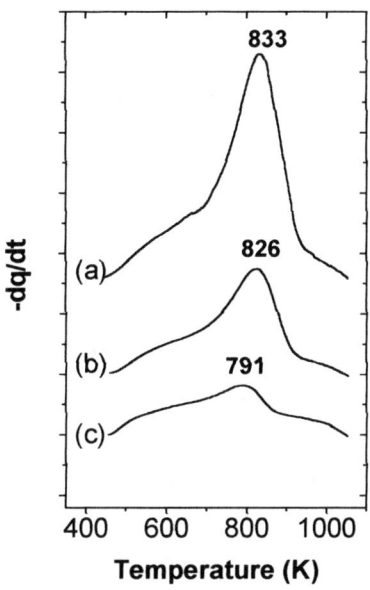

Figure 5. TPO/DSC profiles: (a) USY, 7.8% C, (b) USY, 4.9% C, (c) CREY-2, 4.2% C

CONCLUSIONS

Exchanging RE cations for protons in USY zeolites leads to samples with larger hydrothermal stability, lower acidity and reduced activity for both nC7 cracking and coke formation. The concentration of Brönsted sites that retain pyridine at 623K correlates well with the activity for nC7 cracking. A degree of exchange of about 60% causes a sharp decline in both activity and coking rate and this may be ascribed to a reduction in the concentration of both the Brönsted sites originally present and those resulting from hydrolysis of the trivalent cations.

Limiting amounts of soluble coke were reached on all samples, but the presence of RE cations seems to catalyze the transformation of soluble into insoluble coke. The composition of the soluble fraction was the same for all samples. A single correlation between residual activity and coke content holds for all samples, showing that deactivation is mainly due to site poisoning and not to pore blockage which is consistent with the porous network of faujasite.

TPO of the coke deposited on the samples indicated that the RE cations catalyze the oxidation reaction and that the soluble coke is more readily burned.

ACKNOWLEDGEMENTS

The authors thank Drs. E. F. Sousa-Aguiar and M. L. Murta Valle for supplying the zeolite samples. C. A. Henriques expresses her gratitude to Fabrica Carioca de Catalisadores for its financial support.

REFERENCES

1. M. Guisnet and P. Magnoux, Appl. Catal., **54**, 1 (1989).
2. C. A. Henriques, E. F. Sousa-Aguiar, M. L. Murta Valle, S. Varela and J. L. F. Monteiro, Studies in Surface Science and Catalysis, **111**, 427 (1997).
3. P. Magnoux, P. Roger, C. Canaff, V. Fouche, N. S. Gnep, and M. Guisnet; Studies in Surface Science and Catalysis, **34**, 317 (1987).
4. A. Rabo, C. L. Angel and B. Schomaker, in Proceedings of. 4th International Congress on Catalysis, Vol 2, Moscow, USSR, p.96 (1971).
5. J. W. Ward, Zeolite Chemistry and Catalysis, ACS Monograph 171 (J. Rabo, ed) Washington D.C., p.118 (1976).
6. A. Corma, Chem. Reviews, **95** (3), 559 (1995).
7. M. Guisnet and P.. Magnoux, Zeolite Microporous Solids: Synthesis, Structure and Reactivity, (E. Derouane et al, eds) NATO ASI Series, p. 457 (1991).
8. C. Mirodatos and D. Barthomeuf, J. Chem. Soc., Chem. Commun., 39 (1981).
9. C. A. Henriques, A. M. Bentes Jr, P. Magnoux, M. Guisnet and J. L. F. Monteiro, Appl. Catal., **166**, 301 (1998).

Part III
NMR Characterization

ENHANCED SURFACE NMR OF ZEOLITES AND RELATED MATERIALS USING LASER-POLARIZED XENON

E. BRUNNER*, M. HAAKE[§], A. PINES*, J. REIMER[§], AND R. SEYDOUX*

Materials Sciences Division, Lawrence Berkeley National Laboratory and the Departments of Chemistry* and Chemical Engineering[§], University of California, Berkeley, CA 94720, USA

ABSTRACT

A limitation of NMR is its low sensitivity caused by the relatively low equilibrium nuclear spin polarization in standard experiments. Increasing interest has therefore developed in the enhancement of NMR using laser-polarized noble gas nuclei such as ^{3}He or ^{129}Xe. In the present contribution it is shown that a continuously flowing stream of laser-polarized ^{129}Xe can be used to study zeolites and related materials. The sensitivity of ^{129}Xe NMR of adsorbed xenon can be enhanced by more than three orders of magnitude. The concentration of xenon atoms to be adsorbed on a sample is reduced correspondingly which leads to a minimization of the influence of Xe-Xe interactions upon the ^{129}Xe NMR chemical shift. The exchange of xenon atoms between different zeolite crystallites could be studied quantitatively by 2D exchange NMR spectroscopy of laser-polarized ^{129}Xe under continuous flow conditions. Finally, it is shown that spin polarization may be transferred from laser-polarized ^{129}Xe to surface nuclei such as ^{1}H.

INTRODUCTION

Noble gas nuclei, especially ^{129}Xe, have found numerous applications as probe molecules in surface NMR spectroscopy (see, e.g., [1-5]). Stimulated by the pioneering work of Ito and Fraissard [1] ^{129}Xe NMR spectroscopy has become one of the standard techniques for the characterization of zeolites and related materials. A quantity of particular interest is the chemical shift of adsorbed ^{129}Xe since it is correlated with characteristic structural and compositional parameters of the material such as the mean pore diameter and the type and concentration of the cations present. On the other hand, the ^{129}Xe NMR chemical shift is strongly influenced by the interactions between neighboring Xe atoms, i.e., by the xenon concentration. In order to minimize this influence it is advantageous to carry out ^{129}Xe NMR spectroscopic investigations at low xenon coverages. However, the sensitivity of NMR spectroscopy is relatively low in general because of the low equilibrium nuclear spin polarization in standard experiments. In practice, samples are normally loaded with amounts of xenon corresponding to partial pressures of the order of 10 kPa or more in order to avoid excessively long signal acquisition times. Under such circumstances, the influence of Xe-Xe interactions upon the chemical shift cannot be neglected in most cases. Typically, a series of experiments is carried out at different xenon concentrations and the chemical shift is extrapolated to zero xenon concentration; a procedure which finally yields a chemical shift value δ_0 determined only by the structural and compositional parameters of the material.

It has been demonstrated, however, that the sensitivity of ^{129}Xe NMR spectroscopy may be enhanced by more than three orders of magnitude using laser-polarized ^{129}Xe. Adsorbing a single portion of laser-polarized ^{129}Xe on a zeolite Na-Y, Raftery et al. [6] demonstrated the feasibility of using laser-polarized xenon to study zeolites. Under the conditions of such "single batch" experiments, however, enhanced spin polarization of the adsorbed ^{129}Xe nuclei decays with a rate $\rho_A = T_{1A}^{-1}$ where T_{1A} denotes the longitudinal relaxation time. For commercial zeolites T_{1A} is often governed by the influence of paramagnetic impurities. Characteristic values of 0.1 - 1 s were found for Na-Y [7]; for ultrapure zeolites values of a few seconds could be observed (see below). The short lifetime of the enhanced spin polarization of adsorbed laser-polarized ^{129}Xe prevents the application of more time-consuming techniques, such as multidimensional NMR spectroscopy, in single-batch experiments. Such considerations make highly desirable the use of a continuous flow of laser-polarized ^{129}Xe.

EXPERIMENTAL

The optical pumping process is carried out using a homebuilt apparatus similar to the one designed by Driehuys et al. [8] providing a continuously flowing helium gas stream carrying the laser-polarized ^{129}Xe gas. The pressurized gas mixture (up to 1 MPa), typically containing 0.5 - 40 kPa Xe (natural isotopic distribution, i.e., 26.44 % ^{129}Xe), enters the heated pump cell (420 - 470 K) containing a small amount of rubidium metal. The Rb atoms are optically pumped at the wavelength of their D$_1$ transition (794.8 nm) with circularly polarized light from a diode laser (Optopower Corp., Model OPC-A150-795-RPCZ) delivering about 90 W light power at 795 ± 1 nm. The enhanced ^{129}Xe nuclear spin polarization results from spin exchange with the Rb electron spins.

Static NMR experiments were carried out at a magnetic field of 4.2 T using the closed gas recirculation sytem described in [9]. The gas mixture is inserted into the probe via copper tubing and passes with a mean velocity v of 1 - 10 cm/s through the sample of thickness d placed in a glass tube. Steady-state spin polarizations for the ^{129}Xe gas P_G of 1 - 3 % are obtained so far which is 2500 - 7500 times the equilibrium spin polarization of ^{129}Xe of 0.4×10^{-3} % at room temperature. For magic angle spinning (MAS) experiments the apparatus described in [10] was used.

RESULTS AND DISCUSSION

Assuming that the concentration of xenon adsorbed on a surface c_A is equal to Kc_G (c_G: concentration of xenon in the gas phase, K: equilibrium constant), the spin polarization P_A of ^{129}Xe adsorbed on a surface under the conditions of continuous flow is given by

$$P_A = \frac{\frac{v}{d}}{\rho_A K + \left(\frac{v}{d} + \rho_G\right)\left(1 + \frac{\rho_A}{k_{AG}}\right)} P_G \approx \frac{\frac{v}{d}}{\rho_A K + \frac{v}{d}} P_G \quad . \quad (1)$$

The longitudinal relaxation rate ρ_G of gaseous ^{129}Xe is negligibly small compared to v/d. k_{AG} is the desorption rate corresponding to the inverse mean residence time τ_A of a xenon atom in a zeolite crystallite. Assuming that the crystallites do not exhibit diffusion barriers, τ_A follows from the diffusion coefficient D according to

$$\tau_A = \frac{\lambda^2}{D} \quad , \quad (2)$$

where λ is the characteristic diffusion length which is of the order of the mean crystallite diameter. For mean crystallite diameters of 1 - 5 μm and diffusion coefficients of the order of 10^{-9} m^{2}s^{-1} characteristic for adsorbed ^{129}Xe at room temperature [11] one expects mean residence times of 1 - 25 ms. Therefore, ρ_A/k_{AG} can be neglected for $T_{1A} > 100$ ms which leads to the approximation given in Eq. (1). For the experiment shown in Figure 1 a highly siliceous ZSM-11 with $T_{1A} = 6$ s was used.

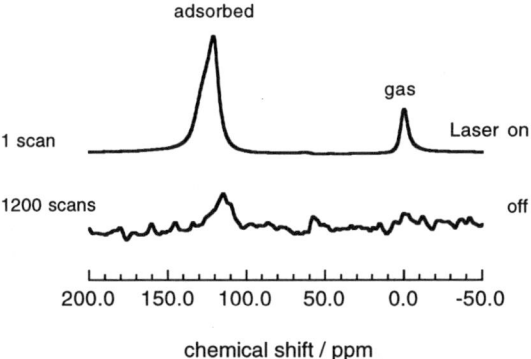

Figure 1. ^{129}Xe NMR spectra of xenon adsorbed on a highly siliceous ZSM-11 measured at room temperature when ^{129}Xe was laser polarized (Laser on, top) and in thermal equilibrium (Laser off, bottom). The sample was provided by Lucy M. Bull.

Assuming that the equilibrium constant K is approximately the same as observed for the adsorption of pure xenon (ca. 200 - 400 at room temperature) and with $P_G = 1.7$ % and $v/d \approx 50$ s^{-1} for this experiment, Eq. (1) predicts 0.7 % < P_A < 1.0 % in agreement with the observed value of $P_A = 1.0$ % corresponding to a signal enhancement factor of ca. 2500. It should be noted that a polarization P_A of 0.1 - 0.2 %, i.e., a signal enhancement factor of 250 - 500 is predicted by Eq. (1) even for a relatively short T_{1A} of 0.5 s under the same experimental conditions.

Two-dimensional (2D) exchange spectroscopy (EXSY [12]) has previously been used for the study of intracrystalline exchange processes especially in zeolite A at very high Xe pressures [4,13]. Intercrystalline exchange processes of xenon atoms adsorbed on mixtures of different zeolites and other composite catalysts were studied by Ryoo et al. [14,15] and by Chen and Fraissard [16]. However, 2D EXSY experiments have not been performed on such systems so far and the measurements reported in [14-16] were carried out at xenon partial pressures of 50 - 80 kPa. Since it is now possible to maintain large steady-state non-equilibrium ^{129}Xe spin polarizations P_A over sufficiently long times, 2D experiments may be performed even at low xenon pressures.

Figure 2 shows the 2D EXSY NMR spectrum (contour plot) of laser-polarized ^{129}Xe adsorbed on a mixture of highly siliceous zeolites Y and ZSM-11 recorded with a xenon partial pressure of 1.3 kPa and a mixing time of 25 ms for the EXSY experiment. The measurements were carried out at 220 K since the signal due to xenon atoms adsorbed on zeolite Y is strongly broadened and relatively weak at room temperature, in agreement with earlier observations [14,16]. The signal due to xenon adsorbed on zeolite Y is observed at 91 ppm and the signal due to xenon adsorbed on ZSM-11 occurs at 150 ppm at this temperature although the xenon partial pressure was only 1.3 kPa.

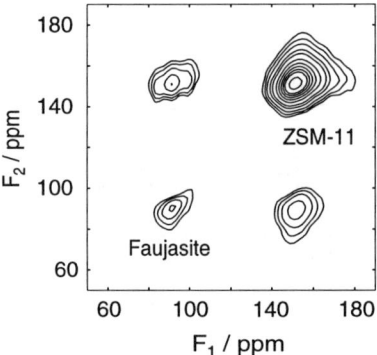

Figure 2. 2D EXSY NMR spectrum (contour plot) of laser-polarized ^{129}Xe adsorbed on a mixture of highly siliceous zeolites Y and ZSM-11 recorded at 220 K with a xenon partial pressure of 1.3 kPa and a mixing time for the EXSY experiment of 25 ms. The samples were provided by Lucy M. Bull.

The shift of these signals with respect to their room-temperature δ_0-values of ca. 60 ppm and 100 ppm, respectively [3,17-19], must be ascribed to the decreased temperature which results in a higher number of adsorbed xenon atoms and therefore in an increasing influence of Xe-Xe interactions upon the chemical shift. Furthermore, δ_0 itself is also temperature-dependent, especially for zeolite Y. The appearance of cross-peaks between the diagonal peaks at 91 and 150 ppm shows the occurrence of xenon exchange between the different crystallites within the selected mixing period. The mean correlation time between succeeding exchanges of 15 ± 5 ms could be determined by the analysis of the dependence of the cross-peak intensities on the mixing time [12].

The influence of Xe-Xe interactions upon the chemical shift of ^{129}Xe adsorbed on zeolites can be estimated according to the following formula [19]

$$\delta(n) = \delta_0 + n\delta_1 + n^2\delta_2 + ... \qquad (3)$$

where n denotes the number of xenon atoms adsorbed per gram sample. At room temperature the parameters δ_1 and δ_2 amount to 1.5×10^{-20} ppm·g and 1.2×10^{-41} ppm·g^2, respectively, for zeolites of pentasil type (ZSM-5, ZSM-11) [19]. Thus, the chemical shift contribution due to Xe-Xe interactions is smaller than 1.6 ppm for $n < 10^{20}$ g^{-1}; a condition which is fulfilled for siliceous pentasils at xenon partial pressures lower than ca. 1.3 kPa [19,20]. Using laser-polarized ^{129}Xe it is possible to carry out experiments at pressures even lower than this value. However, the gas mixture used in our continuous-flow experiments contains an excess of helium which might influence the ^{129}Xe chemical shift by collisions as has been shown for other noble gases [21]. A corresponding experiment indicated that the ^{129}Xe chemical shift of gaseous xenon remains constant within the experimental error of $\pm$ 0.1 ppm for He partial pressures up to 1.4 MPa confirming that the influence of He upon the ^{129}Xe chemical shift can be neglected in our experiments.

A further consideration of relevance for these studies is that the ^{129}Xe NMR signals of adsorbed xenon may be broadened, e.g., by susceptibility effects [22]. In order to achieve superior resolution of the ^{129}Xe spectra, magic angle spinning can be applied [23]. It was shown recently that laser-polarized ^{129}Xe may in fact be injected into a spinning rotor [10,24]. Figure 3 exhibits the ^{129}Xe MAS NMR spectrum of laser-polarized ^{129}Xe adsorbed on a zeolite Na-ZSM-5 measured at room temperature and a xenon partial pressure of 0.7 kPa. The signal at 108 ppm due to xenon adsorbed on the zeolite has a full-width-at-half-maximum of 2 ppm. It is interesting to note the report [19] of a discontinuity in δ_0 for Na-ZSM-5 at a framework Al concentration of 2 Al per unit cell close to the value of 1.9 Al per unit cell of our sample. For lower Al concentrations the observed values were $\delta_0 \approx$ 102 - 105 ppm and for higher concentrations $\delta_0 \approx$ 110 - 112 ppm. As expected, the chemical shift observed for our sample is between these values. A chemical shift of 61 ppm is observed for Na-Y

(Union Carbide, LZY-52, Si/Al = 2.5) at room temperature for a xenon partial pressure of 0.6 kPa in agreement with the δ_0-values of 58 - 61.5 ppm reported in [1,3,17,18]. These results indicate that (i) the influence of Xe-Xe interactions upon the ^{129}Xe NMR chemical shift is negligible at room temperature for xenon partial pressures of 0.7 kPa or less and (ii) interactions between helium atoms and the *adsorbed* xenon atoms do not result in a significant contribution to the ^{129}Xe chemical shift as already demonstrated for *gaseous* xenon above.

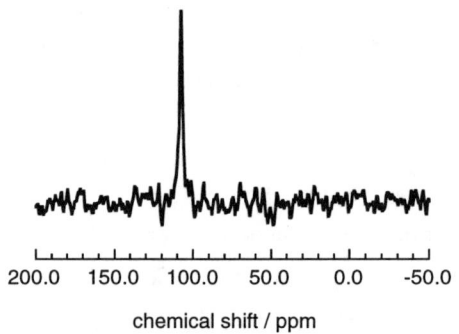

Figure 3. ^{129}Xe MAS NMR spectrum of laser-polarized xenon adsorbed on Na-ZSM-5 (provided by Mobil, Si/Al = 50) measured at room temperature with a xenon partial pressure of 0.7 kPa.

Finally, spin polarization can be transferred from laser-polarized noble gas nuclei to other nuclei S, e.g., protons, via the so-called Spin Polarization Induced Nuclear Overhauser Effect (SPINOE [25]). This effect provides an interesting tool for the selective enhancement of surface NMR signals since it requires a direct magnetic dipole interaction between the ^{129}Xe spins and the nuclei to be polarized. Transient signal enhancements up to a factor of 20 could be observed for the ^{1}H NMR signals of surface OH groups in SiO_2 using a single portion of laser-polarized ^{129}Xe with a spin polarization of 10 % [26]. Under continuous-flow conditions the nuclear spin polarization of the surface nuclei with $S = 1/2$ is given, according to the Solomon equations [27], by

$$\left| P_S - P_S^0 \right| = \left| \frac{\sigma_{IS}}{\rho_S} \left(P_A - P_A^0 \right) \right| \left\{ 1 - e^{-\rho_S t} \right\} \qquad (4)$$

provided that the longitudinal relaxation rate ρ_S of the S spins is higher than that of the adsorbed ^{129}Xe. P_A^0 and P_S^0 denote the equilibrium polarizations of ^{129}Xe and the S spins, respectively, and σ_{IS} is the cross relaxation rate. Figure 4 demonstrates the experimentally observed non-equilibrium polarization of ^{1}H spins located at the outer surface of SiO_2. The measurement was carried out at 150 K, corresponding to the maximum SPINOE enhancement. The observed deviation from the initial behavior predicted by Eq. (4) occurs because the non-equilibrium ^{129}Xe spin polarization of the adsorbed phase cannot be switched on instantaneously, an assumption that underlies Eq. (4). It is encouraging that a non-equilibrium ^{1}H spin polarization of 2.4 times the equilibrium spin polarization could be obtained with the relatively low spin polarization of adsorbed ^{129}Xe of 0.2 % and it should, furthermore, be mentioned that such studies can also be carried out under conditions of magic angle spinning [10,24].

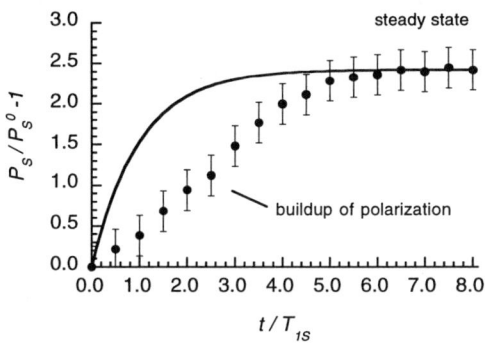

Figure 4. Non-equilibrium spin polarization of ^{1}H nuclei located at the outer surface of SiO_2 (AEROSIL300 provided by DEGUSSA AG). The measurements were carried out at 150 K with a spin polarization of ca. 0.2 % for the adsorbed ^{129}Xe (1.8 % gas phase ^{129}Xe spin polarization). At room temperature, the xenon partial pressure was 40 kPa. The longitudinal relaxation time of the ^{1}H nuclei under our experimental conditions was $T_{1S} = 8$ s. Filled circles: experimental data; solid line: result of Eq. (4) with $\sigma_{IS} = 2 \times 10^{-3}$ s^{-1}.

ACKNOWLEDGEMENTS

E.B. gratefully acknowledges a grant (Heisenberg fellowship) awarded by the Deutsche Forschungsgemeinschaft. M.H. thanks the Studienstiftung des deutschen Volkes and the BASF AG for a Post-doctoral fellowship. This work was supported by the Director, Office of Energy Research, Office of Basic Energy Sciences, Materials Sciences Division of the U.S. Department of Energy under contract no. DE-AC03-76SF00098.

REFERENCES

[1] T. Ito and J. Fraissard, Proc. 5th Int. Conf. on Zeolites, Naples, Ed.: L. Rees, Heyden, London 1980, p. 510; J. Chem. Phys. 76 (1982) 5225.
[2] J.A. Ripmeester and C.I. Ratcliffe, J. Phys. Chem. 94 (1990) 7652.
[3] B. Boddenberg and M. Hartmann, Chem. Phys. Lett. 203 (1993) 243.
[4] D. Raftery and B.F. Chmelka, NMR Basic Principles and Progress 30 (1994) 111.

[5] G. Cho, L.B. Moran, and J.P. Yesinowski, Appl. Magn. Reson. 8 (1995) 549.
[6] D. Raftery, H. Long, T. Meersmann, P.J. Grandinetti, L. Reven, and A. Pines, Phys. Rev. Lett. 66 (1991) 584.
[7] M.L. Smith and C. Dybowski, J. Phys. Chem. 95 (1991) 4942.
[8] B. Driehuys, G.D. Cates, E. Miron, K. Sauer, D.K. Walther, and W. Happer, Appl. Phys. Lett. 69 (1996) 1668.
[9] M. Haake, A. Pines, J.A. Reimer, and R. Seydoux, J. Am. Chem. Soc. 119 (1997) 11711.
[10] E. Brunner, R. Seydoux, M. Haake, A. Pines, and J.A. Reimer, J. Magn. Reson. 130 (1998) 145.
[11] J. Kärger, H. Pfeifer, F. Stallmach, N.N. Feoktistova, and S.P. Zhdanov, Zeolites 13 (1993) 50.
[12] R.R. Ernst, G. Bodenhausen, and A. Wokaun, Principles of Nuclear Magnetic Resonance in One and Two Dimensions, Clarendon Press, Oxford 1987.
[13] I. Moudrakovski, C.I. Ratcliffe, and J.A. Ripmeester, J. Am. Chem. Soc. 120 (1998) 3123.
[14] R. Ryoo, C. Pak, D.H. Ahn, L.-C. de Menorval, and F. Figueras, Catal. Lett. 7 (1990) 417.
[15] O.B. Yang, S.I. Woo, and R. Ryoo, J. Catal. 123 (1990) 375.
[16] Q.J. Chen and J. Fraissard, J. Phys. Chem. 96 (1992) 1814.
[17] L.C. de Menorval, D. Raftery, S.-B. Liu, K. Takegoshi, R. Ryoo, and A. Pines, J. Phys. Chem. 94 (1990) 27.
[18] V. Gupta, H.T. Davis, and A.V. McCormick, J. Phys. Chem. 100 (1996) 9824.
[19] Q. Chen, M.A. Springuel-Huet, J. Fraissard, M.L. Smith, D.R. Corbin, and C. Dybowski, J. Phys. Chem. 96 (1992) 10914.
[20] C. Tsiao, D.R. Corbin, V. Durante, D. Walker, and C. Dybowski, J. Phys. Chem. 94 (1990) 4195.
[21] A.K. Jameson, C. Jameson, and H.S. Gutowsky, J. Chem. Phys. 53 (1970) 2310.
[22] J. A. Ripmeester and C.I. Ratcliffe, Analytica Chimica Acta 283 (1993) 1103.
[23] A.K. Jameson, C.J. Jameson, A.C. de Dios, E. Oldfield, R.E. Gerald II, and G.L. Turner, Solid State Nuclear Magnetic Resonance 4 (1995) 1.
[24] D. Raftery, E. MacNamara, G. Fisher, C.V. Rice, and J. Smith, J. Am. Chem. Soc. 119 (1997) 8746.
[25] G. Navon, Y.-Q. Song, T. Rõõm, S. Appelt, R.E. Taylor, and A. Pines, Science 271 (1996) 1848.
[26] T. Rõõm, S. Appelt, R. Seydoux, E.L. Hahn, and A. Pines, Phys. Rev. B 55 (1997) 11604.
[27] I. Solomon, Phys. Rev. 99 (1955) 559.

^{19}F AND ^{29}Si SOLID-STATE NMR SPECTROSCOPY ON FIVE-COORDINATE SILICON SITES, $(SiO)_4SiF^-$, IN ZEOLITES

H. KOLLER[a], A. WÖLKER[a], S. VALENCIA[b], L.A. VILLAESCUSA[b], M.J. DÍAZ-CABAÑAS[b], M.A. CAMBLOR[b]

(a) Institute of Physical Chemistry, University of Münster, Schloßplatz 4/7, 48149 Münster, Germany, FAX:+49-251 8329159, e-mail: hkoller@uni-muenster.de
(b) Instituto de Tecnologia Quimica, Universidad Politecnica, Valencia, Spain

ABSTRACT

Five-coordinate silicon was found in $(SiO)_4SiF^-$ sites of high-silica zeolites (MFI, ITE, IFR, BEA, MTW) using ^{19}F and ^{29}Si solid-state NMR spectroscopy. The ^{29}Si NMR signals of these sites (chemical shifts between -140 and -148 ppm) are enhanced by ^{29}Si{^{19}F} cross-polarization. The ^{19}F chemical shifts are between -59 and -78 ppm. F^- ions are mobile at room temperature in MFI and IFR.

INTRODUCTION

High-silica zeolites which are made in basic medium with OH^- ions as mineralizing agent have many framework defects [1], whereas zeolites are very poor in defect sites when the mineralizing species are F^- ions [2]. The defects (SiO^- groups) are necessary to balance the charge of the structure directing agent (SDA), and they form hydrogen bonds with SiOH groups [3].
This contribution addresses the charge balance for high-silica zeolites made with F^- ions as mineralizing agent. The structure of [Cocp$_2$]F[Si-NON] contains five-coordinate silicon in $(SiO)_4SiF^-$ sites (^{29}Si NMR chemical shift of -145 ppm) to balance the charge of Cocp$_2^+$ cations [4,5]. At room temperature the fluoride ions are mobile in [TPA]F[Si-MFI], resulting in an averaged ^{29}Si NMR signal at -125 ppm [4]. At a temperature of 140 K the motion is frozen out, and two lines for $(SiO)_4SiF^-$ sites are observed at -144.1 and -147.0 ppm in the ^{29}Si{^{1}H} CPMAS NMR spectrum. These data show that five-coordinate silicon is the charge-balancing group in these zeolites. Here, these observations will be generalized on a larger variety of zeolites. We will characterize the $(SiO)_4SiF^-$ sites and the F^- motion in [TPA]F[Si-MFI] by ^{19}F and ^{29}Si solid state NMR spectroscopy.

EXPERIMENTAL SECTION

Silicalite-1 (MFI) was made at 448 K with NH_4F, Cabosil-M5, TPABr and distilled water [2]. Zeolite Beta (BEA) was crystallized according to a published procedure [6]. All other pure silica zeolites used in this study were synthesized in fluoride medium using tetraethylorthosilicate, deionized water, HF (40% aqueous solution) or NH_4F, and the appropriate SDA in hydroxide form. For the zeolite syntheses tetraethylorthosilicate was hydrolyzed in an aqueous solution of the SDA, and the mixture was stirred until complete evaporation of the ethanol produced. Then, HF was added and the mixture was stirred by hand or mechanically and transferred to teflon-lined stainless steel autoclaves which were kept under rotation (60 rpm) for the required time. Crystallization temperature and time and final chemical composition of each material are listed below (SDAs T_A to T_C are shown in Scheme 1):

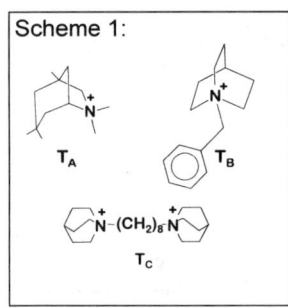

Scheme 1:

ITQ-3 (ITE): 28 days, 423 K; SiO_2 : 0.5 T_AOH : 0.50 HF : 7.7 H_2O
ITQ-4 (IFR): 13 days, 423 K; SiO_2 : 0.5 T_BOH : 0.50 NH_4F : 14 H_2O
ZSM-12 (MTW): 14 days, 448 K; SiO_2 : 0.25 $T_C(OH)_2$: 0.50 HF : 15 H_2O

The $^{29}Si\{^1H\}$ CPMAS NMR experiments were measured on a Bruker CXP-300 spectrometer. The contact times varied between 5 and 15 ms. The ^{19}F and $^{29}Si\{^{19}F\}$ solid state NMR experiments were carried out on a Bruker CXP-200 spectrometer. The $^{29}Si\{^{19}F\}$ cross-polarization conditions were optimized on Na_2SiF_6. The $^{29}Si\{^{19}F\}$ CPMAS NMR spectra of the zeolites were acquired without ^{19}F decoupling and contact times were 10 ms.

RESULTS

Figure 1 shows the $^{29}Si\{^1H\}$ and $^{29}Si\{^{19}F\}$ CPMAS NMR spectra of various zeolites. The signals of Q^4 sites are in the usual range between -105 and -118 ppm for all samples. In addition, the zeolites ITE, BEA and MTW show signals between -140 and -148 ppm which are assigned to $(SiO)_4SiF^-$ groups. These lines are relatively strong in the $^{29}Si\{^{19}F\}$ CPMAS NMR spectra which proves that the corresponding silicon sites are close to fluoride ions.

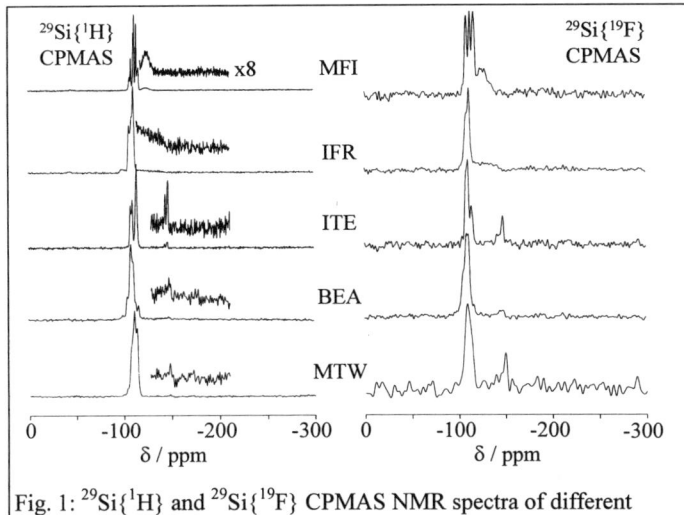

Fig. 1: ^{29}Si{^{1}H} and ^{29}Si{^{19}F} CPMAS NMR spectra of different zeolites

The ^{29}Si NMR spectra of the two zeolites, MFI and IFR, show broad lines between -120 and -140 ppm and no signal is observed in the range for five-coordinate silicon between -140 and -150 ppm. This is rationalized by a dynamic motion of the fluoride ions between different Si sites, resulting in an averaged broad ^{29}Si NMR signal between -120 and -140 ppm. The motion of F$^-$ in MFI is frozen out at low temperatures as shown by variable temperature NMR spectroscopy (Fig. 2). The ^{29}Si{^{1}H} CPMAS NMR spectrum at 140 K shows the two lines at -144.1 and -147.0 ppm for the (SiO)$_4$SiF$^-$ groups. These signals get very broad when the temperature is increased, and the broad signal at -125 ppm becomes apparent at 280 K. This observation indicates that the changes of F$^-$ mobility on the NMR time scale cover a large temperature range. The gradual changes in fluoride motion can also be followed by the ^{19}F MAS NMR spectra (Fig. 2). The spinning sidebands are caused by ^{19}F chemical shift anisotropy, CSA (tensor components δ_{11}, δ_{22}, δ_{33}), and the ^{19}F MAS NMR spectra are analyzed in order to obtain the isotropic chemical shift, $\delta_{iso} = 1/3(\delta_{11}+\delta_{22}+\delta_{33})$, the span, $\Omega = \delta_{11} - \delta_{33}$, and the asymmetry parameter, $\eta = (\delta_{22}-\delta_{11})/(\delta_{33}-\delta_{iso})$. These parameters are sensitive to dynamic motion and they change gradually over a large temperature range. The CSA parameters were obtained for all the zeolites studied, and the results show that rigid fluoride ions in (SiO)$_4$SiF$^-$ groups are characterized by a typical span value of $\Omega = 83\pm3$ ppm.

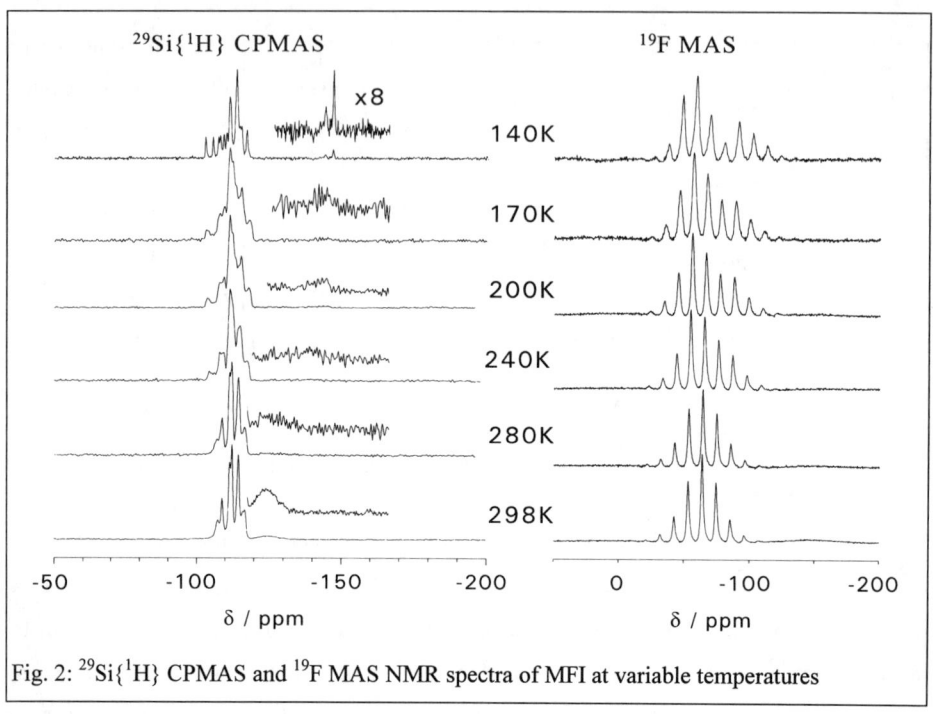

Fig. 2: ^{29}Si{^{1}H} CPMAS and ^{19}F MAS NMR spectra of MFI at variable temperatures

CONCLUSIONS

Our results show that fluoride ions are located in $(SiO)_4SiF^-$ groups in zeolites MFI, NON, BEA, ITE, IFR, and MTW. The fluoride ions undergo dynamic motion in MFI and IFR at room temperature.

ACKNOWLEDGEMENT

This work was funded by the Deutsche Forschungsgemeinschaft, the Fonds der Chemischen Industrie and the Bundesministerium für Bildung, Wissenschaft, Forschung und Technologie.

REFERENCES
[1] R.L. Lobo, S.I. Zones, M.E. Davis, *J. Incl. Phenom. Molec. Recogn. Chem.* **21**, 47 (1995).
[2] J.M. Chézeau, L. Delmotte, J.L. Guth, M. Soulard, *Zeolites* **9**, 78 (1989).
[3] H. Koller, R.F. Lobo, S.L. Burkett, M.E. Davis, *J. Phys. Chem.* **99**, 12588 (1995).
[4] H. Koller, A. Wölker, H. Eckert, C. Panz, P. Behrens, *Angew. Chem.* **109** (1997), 2939; *Angew. Chem. Int. Ed. Engl.* **36**, 2823 (1997).
[5] G. van de Goor, C.C. Freyhardt, P. Behrens, *Z. anorg. allg. Chem.* **621**, 311 (1995).
[6] M.A. Camblor, A. Corma, S. Valencia, *Chem. Commun.* 2365 (1996).

HIGH TEMPERATURE ¹H MAS NMR STUDIES OF THE PROTON MOBILITY IN ZEOLITES

H. ERNST, D. FREUDE, T. MILDNER and H. PFEIFER

Abteilung Grenzflächenphysik, Universität Leipzig, Germany; freude@physik.uni-leipzig.de

ABSTRACT

Proton mobility in hydrogen zeolites was observed up to 790 K by means of the laser heating of dehydrated and fused samples. ^{1}H MAS NMR spectra of the bare Brønsted sites of zeolites were monitored in the temperature range from 160 to 790 K. The full width at half maximum of the ^{1}H MAS NMR spectrum narrows by a factor of 24 for zeolite H-ZSM-5 and a factor of 55 for zeolite 85 H-Y. For the latter an activation energy of 78 kJ mol^{-1} has been determined. In situ ^{1}H MAS NMR spectroscopy of the proton transfer between bridging hydroxyl groups and benzene molecules yields temperature dependent exchange rates over more than five orders of magnitude. H-D exchange and NOESY MAS NMR experiments were performed by both conventional and laser heating up to 600 K. It is shown that the dealumination of the zeolite has a strong influence on this exchange rate, which represents a dynamic measure of Brønsted acidity. Activation energies of the proton transfer of 102 and 93 kJ mol^{-1} have been obtained for zeolites 85 H-Y and 92 H-Y, respectively.

INTRODUCTION

Bridging hydroxyl groups capable of donating protons to molecules in the cages of zeolites are Brønsted-acid sites in heterogeneous catalysis. ^{1}H NMR studies of the acid protons in dehydrated zeolites gave some insight into the proton mobility [1-3]. P. Sarv et al. [1] found by means of a fit of the temperature dependence of the second moment of the NMR line shape activation energies of 45, 54 and 61 kJmol^{-1} for zeolites H-ZSM-5, H-mordenite and H-Y, respectively, and explained the corresponding mobility with proton jumps in the first coordination sphere of the aluminium atom in the Brønsted site. The equation for the fit was taken from Abragam [4], who discussed an *ad hoc* introduced formula

$$M_2 = M_2'' + M_2' \frac{2}{\pi} \arctan[\alpha\, \delta\omega\tau_c], \quad (1)$$

where M''_2 denotes the second moment (unit s^{-2}) of a line shape which is narrowed under the influence of a fast internal motion. The sum $M'_2 + M''_2$ is the second moment of the rigid lattice, $\delta\omega$ is the line witdh, τ_c denotes the correlation time and α is a rather ill-defined factor of order unity. This equation describes the line narrowing only qualitatively. It should not be used for a fit of experimental data near to the onset of the line narrowing, since "the second moment of the absorption curve is unaffected by the narrowing motion" (Abragam [4]). A detailed consideration [5] shows that the narrowing onset of the line shape of bridging hydroxyl groups can be described quantitatively under the assumption of special jump models. From the temperature dependence of the ^{1}H MAS NMR central line $\delta\nu_{1/2}^{MAS}$ T. Baba et al. [2] obtained an activation energy of 17-20 kJmol^{-1} for the thermal motion of the protons in the zeolite H-ZSM-5. They used an equation given in Ref. [6]:

$$\delta v_{1/2}^{MAS} = \frac{M_2}{\pi}\left[\frac{2\tau_c}{1+(\omega_{rot}\tau_c)^2} + \frac{\tau_c}{1+(2\omega_{rot}\tau_c)^2}\right], \tag{2}$$

where M_2 denotes the second moment of the heteronuclear dipolar interaction and ω_{rot} is the MAS frequency. The maximum measuring temperature 473 K has been an important limitation in the study of Baba et al. [2] but also in other previous studies (660 K in [1,3]).

Proton transfer rates (from a Brønsted site to an adsorbed molecule) which were determined by NMR relaxation has been used as a dynamic measure of the Brønsted acidity more than 20 years ago, cf. Ref. [3]. Special attention was given to the H-D exchange between bridging hydroxyl groups and deuterated benzene molecules in zeolites [3,7,8]. The H-D exchange of deuterated benzene in zeolite H-ZSM-5 could be observed even at room temperature [7-8]. Beck et al. [9] found in the temperature range up to 393 K for the benzene exchange reaction in the zeolites H-ZSM-5, USY and H,Na-Y activation energies of 60, 85 and 107 kJ mol^{-1}, respectively, and carried out quantum chemical calculations of the proton transfer [9,10].

Several groups have shown that the application of new MAS techniques facilitates *in situ* NMR studies of catalytic reactions in zeolites [11-13]. In the present study four high temperature MAS NMR techniques were applied: (i) conventional heating of the MAS bearing gas is used up to 520 K yielding H-D exchange reaction rates up to about 0.1 per minute, (ii) the temperature jump method by means of laser heating [14] could be used for monitoring H-D exchange reaction rates up to about 10 min^{-1}, (iii) a two-dimensional ^{1}H NOESY MAS NMR exchange spectrum, which has been measured at 520 K, gives exchange rates around 100 min^{-1} for the H-H exchange, (iv) a continuous laser heating up to 790 K has been used for the ^{1}H MAS NMR studies of the bare Brønsted sites.

EXPERIMENTAL

Zeolite 92 H-Y (Si/Al = 3.1) containing 8% of the original content of Na$^+$ cations and some non-framework aluminium species was prepared by Dr. J. Meusinger by a repeated exchange of zeolite Na-Y in an aqueous solution of NH_4NO_3 followed by calcination. The defect-free zeolite 85 H-Y (85% NH_4^+ and 15% Na$^+$, Si/Al = 2.4) was donated by UOP. Zeolite H-ZSM-5 (Si/Al=35) was synthesized by Dr. W. Schwieger using mono-n-propylamine as the template, then calcined at 600 °C for 5 h and finally exchanged with 0.5M HCl. All samples were pretreated by heating 8 mm deep layers of zeolite in glass tubes 3 or 5 mm outer diameter at a rate of 10 K h^{-1} under vacuum. After maintaining the samples at 673 K (or 723 K) and less than 10^{-2} Pa for 24 h, some samples were loaded under vacuum at room temperature with two molecules benzene per supercage (1/8 unit cell) and then sealed. Benzene has been 99.5% enriched in ^{2}H for the H-D exchange experiments. The benzene loaded samples were kept frozen until the start of the NMR measurement. The high temperature ^{1}H MAS NMR spectra of the sealed samples were recorded at 300 MHz and spinning rates of 1-4 kHz. The laser probe, the temperature jump technique, the calibration of the temperature and the measurement of the temperature gradient are described in Refs. [14-15]. Details of the H-D exchange experiments and the NOESY experiment are given in Ref. [16]. Room and low temperature ^{1}H MAS NMR measurements were performed at 500 MHz. The C_6H_6 loaded samples do not show a variation of the signal before and after heating at 600 K. Thus, benzene molecules were not converted at temperatures up to 600 K.

RESULTS AND DISCUSSION

Hydrogen exchange between Brønsted sites and benzene molecules

^{1}H, ^{27}Al and ^{29}Si MAS NMR spectra of the dehydrated (^{1}H) and hydrated (^{27}Al and ^{29}Si) zeolites were already published [17]. Figure 1 gives the results of a H-D exchange experiment at 400 K. The time dependent ^{1}H NMR intensity can be described by

$$I(t) = I(\infty)\big(1 - b\exp(-k\,t)\big), \tag{3}$$

where k is the exchange rate and $I(\infty)$ denotes the intensity after a full exchange. The value b describes the exchange at $t = 0$. The initial state of the H-D exchange experiment is characterized by fully deuterated molecules and non-deuterated bridging hydroxyl groups. Therefore, if we denote by $n_B = n_B^H + n_B^D$ the total number of hydrogen positions in the benzene molecules per cavity (1/8 unit cell), it holds $n_B = n_B^D(0)$ at $t = 0$. Similarly, if we write for the number of bridging hydroxyl groups per unit cell which take part in the exchange $n_A = n_A^H + n_A^D$ it holds $n_A = n_A^H(0)$. For the equilibrium numbers after the exchange $n_A^H(\infty)$ and $n_B^H(\infty)$ it follows

$$n_A^H(\infty) = \frac{n_A^2}{n_A + n_B}, \quad n_B^H(\infty) = \frac{n_A n_B}{n_A + n_B}, \quad \text{and} \quad \frac{n_A^H(\infty)}{n_B^H(\infty)} = \frac{n_A}{n_B} = \frac{\tau_A}{\tau_B}. \tag{4}$$

with τ_A and τ_B as the mean life times of the protons in the bridging hydroxyl groups and benzene molecules, respectively [3]. The measured reaction rate k is described by the equation

$$k = \left(\frac{1}{\tau_A} + \frac{1}{\tau_B}\right) = \frac{1}{\tau_A}\left(1 + \frac{n_A}{n_B}\right) = \frac{1}{\tau_A n_B}(n_A + n_B). \tag{5}$$

A narrow signal of desorbed benzene molecules in the gas phase at 7.27 ppm could not be observed even at the maximum temperature of measurement. Therefore, $n_A + n_B$ can be taken as temperature independent. The temperature dependence of the reaction rate k is caused by that changes of the mean residence time τ_A of the hydroxyl proton before the transfer to the base molecule. Thus, the experimentally obtained reaction rate k represents a dynamic measure of Brønsted acidity.

With increasing temperatures the reciprocal value of the exchange rate becomes as short as the longitudinal relaxation time. Then the exchange reaction can be monitored by two-dimensional ^{1}H NOESY MAS NMR [17]. The cross peaks in the spectrum obtained at 520 K, cf. Figure 2, are caused by the magnetization transfer between the relatively broad signal of the bridging hydroxyl groups at ca. 4 ppm and the benzene signal at ca. 7.5 ppm. The signal at ca. 2 ppm due to non-acidic hydroxyl groups on framework defects [18] is not affected by a magnetization transfer, i.e., only the diagonal peak appears. It was shown [16], that at this temperature the pure chemical exchange is responsible for the magnetization transfer. Therefore, the exchange rates can be extracted from the mixing time dependent intensities, cf. Fig. 3.

The results of the H-H exchange experiments can be combined with those of the H-D exchange experiments. Figure 4 shows the Arrhenius plot of $k = k_0 \exp\{-E_A/RT\}$ for the benzene exchange reaction in the zeolites 85 H-Y and 92 H-Y. The values with $k > 10$ min^{-1} were determined by means of two-dimensional NOESY MAS NMR, whereas the other values were obtained by the above described time resolved H-D exchange experiments. The best fit is given by the solid lines in Figure 4 corresponding to activation energies E_A of 102 ± 5 kJ mol^{-1} and 100 ± 7 kJ mol^{-1} for the zeolites 85 H-Y and 92 H-Y, respectively.

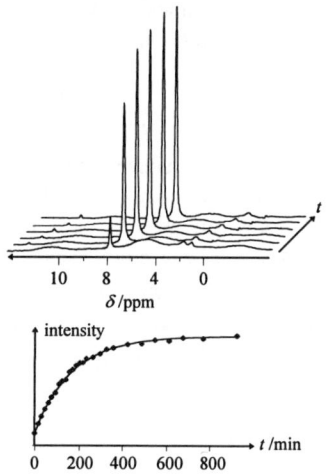

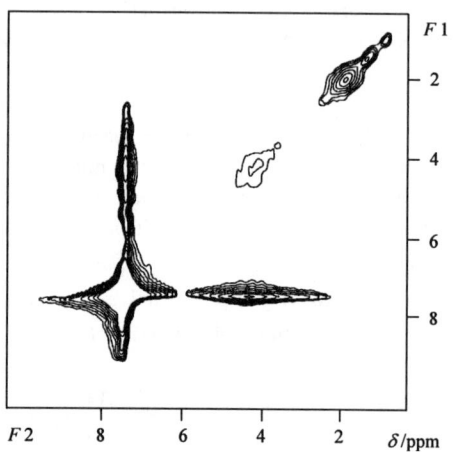

Figure 1. Time dependence of the ^{1}H MAS NMR spectra of the zeolite 85 H-Y loaded with fully deuterated benzene. The spectra were measured at 400 K.

Figure 2. 2D ^{1}H NOESY MAS NMR spectrum measured at 520 K with a MAS frequency of 4 kHz and a mixing period of 500 ms. Two benzene molecules per cavity were adsorbed on zeolite 92 H-Y.

It is well-known that the gas-phase acidity of bridging hydroxyl groups increases, if the aluminium content in the framework decreases and that the corresponding ^{1}H NMR shifts of hydroxyl groups pointing into the large cavities increases as well, so that a direct relation exists between both quantities [19], cf. also [16]. The zeolites 85 H-Y and 92 H-Y have Si/Al ratios of 2.4 and 3.1, respectively, which corresponds to a variation of the ^{1}H NMR shift of the bridging hydroxyl groups of only 0.1 ppm for the dehydrated zeolites [19]. On the other hand, in the present study an increase of the rate of the proton transfer between bridging hydroxyl groups and benzene molecules by a factor of ten going from zeolite 85 H-Y to 92 H-Y has been found, cf. Figure 4. Thus, the proton transfer rate is a much more sensitive measure of the strength of Brønsted acidity of the bridging hydroxyl groups than their ^{1}H NMR chemical shift in dehydrated samples (gas phase acidity). This is in accordance with the classic definition by J.N. Brønsted: "... that an acid is defined as a substance, which is able to split off H$^+$-ions simultaneously forming a base ..." [20].

In order to find out the elementary process which effects the temperature dependence of the proton transfer rate, Figure 4 can be reconsidered. The dotted line for zeolite 92 H-Y in Figure 4 is also within the limit of the experimental error and corresponds to a value of 93 kJ mol^{-1}. The extrapolation of the lines to $T = \infty$ gives the pre-exponential factor k_0 of the Arrhenius plot. A difference of about one order of magnitude is obtained, if the solid lines are used for the determination of the pre-exponential factors of both zeolites. But only one value (3×10^9 s^{-1}) is obtained for both zeolites, if the dotted line is used instead of the solid line for the zeolite 92 H-Y. Therefore, the experimental data do not allow distinction between, whether the activation energy or the pre-exponential factor gives rise to the factor of about ten between the rates for the two zeolites. Beck et al. [9] found activation energies for the benzene exchange reaction of 107 kJ mol^{-1} and 85 kJ mol^{-1} in a zeolite H-Y (which is similar to our zeolite 85 H-Y) and in an ultra-stable zeolite USY, respectively. The exchange rate at 370 K is higher by more than one order of magnitude for the zeolite USY, which has a Si/Al ratio of 5.4 and contains non-framework aluminium species. This result is similar to our measurement, but the small temperature range in Ref. [9] does not allow an extrapolation to $T = \infty$.

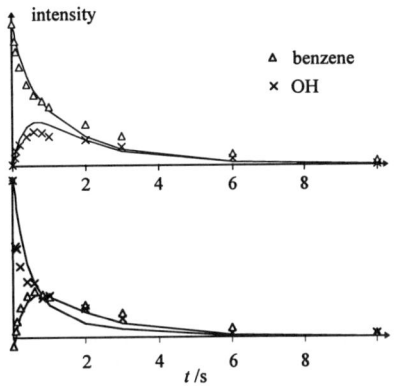

Figure 3. ^{1}H MAS NMR intensities of the benzene molecules and the bridging hydroxyl groups of zeolite 92 H-Y at 520 K (extracted from two mixing time dependent NOESY experiments). The best fit includes the values of the relaxation times of the two species and their exchange rate. Details are given in Ref. [16].

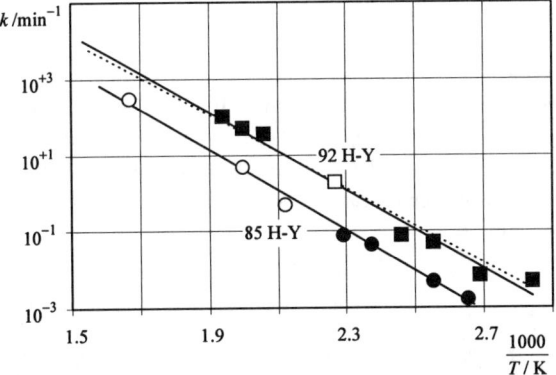

Figure 4. Arrhenius plot of the H-D and H-H exchange rates for benzene molecules in the zeolites 85 H-Y and 92 H-Y. The values which are marked by open or full circles and squares were measured by laser heating or conventional heating.

The pre-exponential factor k_0 is the reaction rate for infinite temperature or zero activation energy and can be discussed in terms of partition functions of the adsorbate complex for the H-D exchange in zeolites, as shown by Kramer and van Santen [21]. It is remarkable that the pre-exponential factor of 3×10^9 s^{-1} is nearly equal to the rotational constant 2.8×10^9 s^{-1} for the benzene rotation (about the C_6 axis) [22], whereas the stretching vibration of the bridging hydroxyl groups is around 10^{14} Hz. This hints that the rotation of the molecule may play a role for the proton transfer.

Ab initio quantum chemical calculations [23] yield a linear correlation between the ^{1}H NMR chemical shift δ of isolated surface hydroxyl groups and their deprotonation energy with a slope of (-84 ± 12) kJ mol^{-1} ppm^{-1}. The difference of the ^{1}H NMR shifts between the two zeolites H-Y under study of 0.1 ppm is related to a difference of their deprotonation energies of 8.4 kJ mol^{-1}. This agrees well with the difference of the activation energies, which were obtained under the assumption of a constant value of k_0: 102 kJ mol^{-1} for 85 H-Y (solid line in Figure 4) and 93 kJ mol^{-1} for 92 H-Y (dotted line in Figure 4). Thus, the variation of the Si/Al ratio, which causes a change of the deprotonation energy, can explain the differences of the exchange rate of one order of magnitude in the temperature region of 350-600 K. However, our experimental results are not sufficient to exclude that a variation of the pre-exponential factor caused by steric effects like the existence of non-framework aluminium species is the origin of the different rates of the proton transfer.

Proton mobility of the bare Brønsted sites

Figure 5 shows the ^{1}H MAS NMR spectra of the dehydrated zeolite 85 H-Y with a rotation frequency of 1 kHz. There are two reasons for choosing such a low rotation frequency. First, the envelope of many sidebands gives a good picture of the static (without sample rotation) line width. Second, the condition $\omega_{rot}\tau_c \ll 1$ giving the line widths as a linear function of the correlation time, cf. eq. (2), is fulfilled at lower temperature for a lower rotation frequency. A highly resolved spectrum of the same sample ($\nu_{rot} = 12$ kHz, $\nu_L = 500$ MHz, cf. [18]) shows that the ^{1}H MAS NMR spectrum consists only of the signals (b) and (c) of the bridging hydroxyl groups at 3.9 and 4.8 ppm, respectively. The two lines cannot be resolved at the rotation frequency of 1 kHz. Low temperature measurements ($1000/T > 3$/K in Figure 6) show that there is no significant line broadening with decreasing temperatures below room temperature. Thus, the narrowing onset takes place between 298 and 420 K, see Figure 5. The τ_c-dependence of the static line width is often described by the Anderson-Weiss model based on a static Markov process, cf. Ref. [6]. For bridging hydroxyl groups in zeolites it should be described by another model [5], which is based on a Gauss-Markov process. Independent of the special model it can be concluded that the correlation time τ_c which is given by the mean residence time τ of a proton at an oxygen atom is in order of magnitude of the reciprocal low temperature line width (about 100 µs) at the temperature of the narrowing onset.

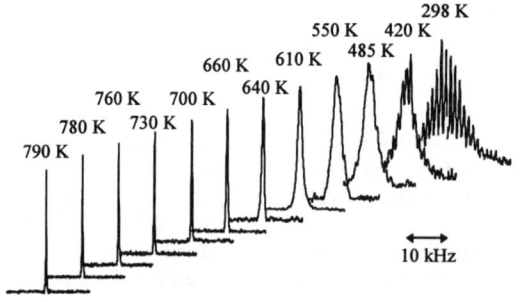

Figure 5. ^{1}H MAS NMR spectra of the dehydrated zeolite 85 H-Y for measuring temperatures from 298 K to 790 K. The rotation frequency is 1 kHz. No spinning sidebands can be observed at temperatures above 500 K, whereas at room temperature they are well-resolved.

The activation energy E_A of the mean residence time τ of a proton at one oxygen atom, which can be obtained by an Arrhenius plot $\tau = \tau_0 \exp(E_A/RT)$ of the temperature dependence, is more interesting for the modelling of the proton mobility than an absolute value for the proton jump frequency. It is most likely that the lowest energy barriers exist for the motion of the bridging hydroxyl protons around the four oxygen positions in the nearest neighbourhood of one aluminium atom. If these positions are of regular tetrahedral symmetry and the occupation probability for all positions would be identical, the dipolar proton aluminium interaction of the ^1H-^{27}Al pair must be averaged to zero. Since these conditions are probably not fulfilled and inter-pair dipolar interactions cannot be averaged out by this motion, a residual temperature independent line width should be observed at high temperatures, cf. [1]. Such a temperature dependence can be qualitatively described by eq. (1).

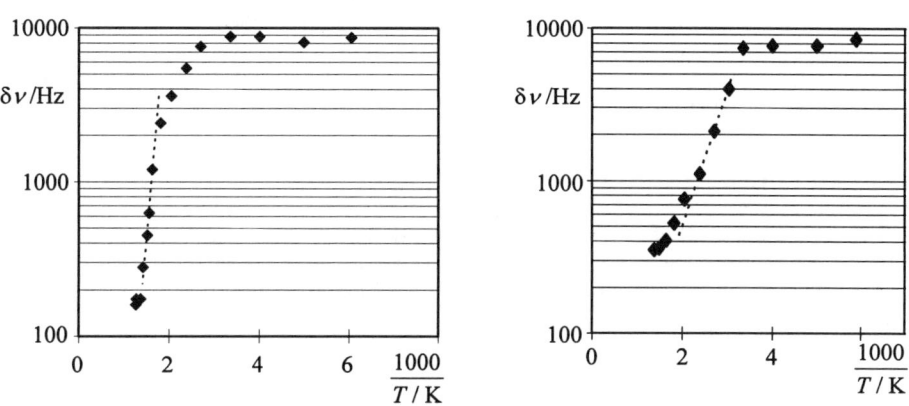

Figure 6. Temperature dependence of the full width at half maximum (*fwhm*) of the ^1H MAS NMR spectrum for the zeolites 85 H-Y and H-ZSM-5 at the left and right hand side, respectively.

The high temperature values in Figure 6 are also a hint for a step behaviour in our temperature dependencies, which could be described by eq. (1), if we take into consideration that an experimentally obtained second moment is proportional to the square of the *fwhm*. The *fwhm* narrows by a factor of 24 for the zeolite H-ZSM-5 and a factor of 55 for the zeolite 85 H-Y, respectively, see Figure 6, whereas the ratio of the values for the low and high temperature second moment in the study of P. Sarv et al. [1] was about 20 corresponding to a *fwhm* ratio of 4-5. We found such a small step in the temperature dependent fwhm for the zeolite 92 H-Y. In our opinion it is not possible to determine an activation energy from such a small step without an appropriate model. The situation is different for the *fwhm* ratio of 55 in the case of zeolite 85 H-Y. The *fwhm* narrowing from 8700 Hz, to 1300 Hz, to 630 Hz and then to 160 Hz for 298 K, 610 K, 640 K and 790 K, respectively, fulfils much better the condition that the line width range, in which a linear relation between line width and correlation time is expected, should be far from both, the low and the high temperature line width. Consequently, we determined an activation energy of 78 kJ mol^{-1} from the *fwhm* at 640 and 610 K for the zeolite 85 H-Y. In the same way the *fwhm* at 370 and 420 K for the zeolite H-ZSM-5 give an activation energy of 18 kJ mol^{-1}. The value for the zeolite H-ZSM-5 is in good agreement with that given by T. Baba et al. [2]. The accuracy of our activation energies are limited by the accuracy of the determination of the temperature in the sample, cf. [14]. The error of the activation energies is ±10%.

In the present study we have extended the temperature for which the ^{1}H MAS NMR signal of the bare Brønsted sites can be observed up to 790 K, which exceeds already the pre-treatment temperature of 723 K. A further increase of the measuring temperature above the pre-treatment temperature of the sample is not useful, since dehydroxylation and formation of water may take place. An irreversible change in the sample could be excluded in this study from the identity of the room temperature ^{1}H MAS NMR spectrum before and after the heating procedure, which was limited to 10 minutes at each temperature. After heating for one hour at the maximum temperature the number of the Brønsted sites detected has decreased. It cannot be excluded that at higher temperature there is a superposition of the proton mobility around an aluminium atom with a proton diffusion through the zeolite framework by means of proton vehicles like residual ammonia or water molecules. The values of the activation energy for the proton mobility around an aluminium atom are certainly useful for the evaluation of quantum chemical models, but, in contrast to the activation energies of the proton transfer between OH groups and benzene molecules, they cannot characterize the acidity of the Brønsted sites.

ACKNOWLEDGMENTS

We are grateful to Prof. Dr. Jörg Kärger, Dr. Mark E. Smith, Mrs. Dagmar Prager and Mr. Bernd Knorr for advice and valuable help. This work was supported by the *Deutsche Forschungsgemeinschaft*, (SFB 294/D2 and Fr 902/1) and the *Graduiertenkolleg Physikalische Chemie der Grenzflächen*.

REFERENCES

1. P. Sarv, T. Tuherm, E. Lippmaa, K. Keskinen, A. Root, J. Phys. Chem. **99**, 13763 (1995)
2. T. Baba, N. Komatsu, Y. Ono, H. Sugisawa, J. Phys. Chem. B **102**, 804 (1998)
3. D. Freude, W. Oehme, H. Schmiedel and B. Staudte, J. Catal. **49**,123 (1977), ibid. **32** 137 (1974).
4. A. Abragam, The Principles of Nuclear Magnetism, (Clarendon Press, Oxford, 1989) p. 456 and p. 433.
5. W. Ploß, D. Freude, H. Pfeifer und H. Schmiedel, Annalen der Physik **39**, 1 (1982).
6. D. Fenzke, B.C. Gerstein and H. Pfeifer, J. Magn. Reson. **98**, 469 (1992).
7. V.M. Mastikhin, K I. Zamaraev,.Zeitschrift f. Phys. Chem. **152**, 59 (1987).
8. J.L. White, L.W. Beck, J.F. Haw,.J. Am. Chem. Soc. **114**, 6182 (1992).
9. L.W. Beck, T. Xu, J.B. Nicholas and J.F. Haw, J. Am. Chem. Soc. **117**, 11594 (1995)
10. J. B. Nicholas, Topics in Catalysis **4**, 157 (1997).
11. D.B. Ferguson, T.R. Krawietz, J.F. Haw, J. Magn. Reson. A **109**, 273 (1994).
12. H. Ernst, D. Freude, T. Mildner, Chem. Phys. Letters **229**, 291 (1994).
13. M. Hunger, T.J. Horvath, Chem. Soc., Chem. Comm., 1423 (1995).
14. H. Ernst, D. Freude, T. Mildner, I. Wolf, Solid State NMR **6**, 147 (1996).
15. T. Mildner, H. Ernst, D. Freude, Solid State NMR **5**, 269 (1995).
16. T. Mildner and D. Freude, J. Catal. in press.
17. R.R.Ernst, G. Bodenhausen, A. Wokaun,. Principles of Nuclear Magnetic Resonance in One and Two Dimensions, (Clarendon Press, Oxford, 1990).
18. D. Freude, H. Ernst, I. Wolf, Solid State NMR **3**, 271 (1994).
19. H. Pfeifer, H. Ernst, Ann. Reports NMR Spectr., **91**, 28 (1994).
20. J.N. Brönsted, J. Phys. Chem., **30**, 777 (1926).
21. G. J. Kramer, R. A. van Santen, J. Am. Chem. Soc. **117**, 1766 (1995).
22. Landolt-Börnstein, New Series. Vol II/19a, p. 121, Springer, Berlin, 1992.
23. U. Eichler, M. Brändle, J. Sauer, J. Phys. Chem. B **101**, 10035 (1997).

INTERACTION OF CHLOROFLUOROCARBONS WITH ZEOLITES STUDIED BY IN SITU IR AND MULTINUCLEAR NMR SPECTROSCOPY

I. HANNUS[1], Z. KÓNYA[1], P. LENTZ[2], J. B.NAGY[2] and I. KIRICSI[1]

[1]Applied Chemistry Department, József Attila University,
H-6720 Szeged, Rerrich Béla tér 1, Hungary, e-mail:hannus@chem.u-szeged.hu
[2]Laboratoire de RMN, Facultés Universitaires Notre-Dame de la Paix,
B-5000 Namur, Rue de Bruxelles 61, Belgium

ABSTRACT

Adsorption and decomposition of CFC-12 (CCl_2F_2) and HCFC-22 ($CHClF_2$) on zeolites with different Si/Al ratios were investigated by IR and multinuclear NMR spectroscopy. The main intermediate products were carbon tetrachloride (CCl_4) and phosgene ($COCl_2$) in the CCl_2F_2, and chloroform ($CHCl_3$) in the $CHClF_2$ decomposition. When no oxygen was added to the reacting systems, the oxygen appearing in the intermediates has to stem from the framework of the zeolite. Consequently, continuous destruction of the framework took place during the CFCs decomposition reaction. This resulted in the dealumination of the zeolite. Oxygen present in the reacting mixture delayed the severe dealumination of zeolites. Both IR and NMR spectroscopy confirmed these findings.

INTRODUCTION

It has been established that zeolites are appropriate catalysts in the decomposition of chlorofluorocarbons (CFCs) [1]. Their acidity plays a dominant role in both the adsorption and the decomposition of CFCs. Furthermore transition metal ion-exchanged zeolites enhance the formation and decomposition of phosgene as intermediate product [2]. With increasing number of fluorine atoms in the molecule, decomposition of CFCs becomes more difficult and deactivation of catalysts is the consequence of acidity decrease and the dealumination of zeolites [3-5]. As far as the products of decomposition of CFCs over zeolite are concerned, continuous consumption of the framework constituents of zeolite takes place [6]. Therefore, the zeolite catalyst also becomes a true reactant.

On the other hand, one of the promising alternatives of the ozone depleting CFCs is the use of hydrochlorofluorocarbons (HCFCs) [7].

We report here on the results obtained for the investigation on the acidity, the structure of zeolites, the mechanism of deactivation and the influence of admixed oxygen to CFC-12 and HCFC-22 compounds during their decomposition.

EXPERIMENTAL

NaY-FAU (Union Carbide product) and its H-form (HY-FAU), HM-MOR (Norton product) and HZSM-5-MFI (made in Hamburg, $SiO_2/Al_2O_3=50$) were used in the experiments. CFC-12 (CCl_2F_2) and HCFC-22 ($CHClF_2$) reactants were Aldrich products.

For IR spectroscopic measurements self-supported wafers with a thickness of 10 mg/cm^2 were pressed from the zeolites, placed into the sample holder of the cell and outgassed at 723 K for 2 h in vacuum.

In situ MAS ^{13}C NMR studies were performed on an MSL-400 BRUKER spectrometer operating at 100.6 MHz. The zeolite samples were packed into special NMR tubes and evacuated at 723 K for 2 h. After CCl_2F_2 or $CHClF_2$ adsorption (0.01 g/0.05 g dry zeolite) the tubes were carefully sealed to achieve proper balance and high spinning rate.

^{29}Si, ^{27}Al and ^{23}Na NMR spectra of the zeolite samples were recorded before and after the surface reaction. For ^{29}Si (79.4 MHz) a 4 μs ($\Theta=\pi/6$) pulse was used with a repetition time of 6.0 s. For ^{27}Al (104.3 MHz) a 1.0 μs ($\Theta=\pi/12$) pulse was used with a repetition time of 0.1 s, and for ^{23}Na (105.8 MHz) a 1.0 μs ($\Theta=\pi/12$) pulse with a repetition time of 0.1 s was applied. All chemical shifts are referenced to external standards: aqueous NaCl and $AlCl_3$ solutions for ^{23}Na and ^{27}Al and tetramethylsilane for ^{29}Si NMR. The number of accumulations varied between 5000 and 20 000 to obtain a good signal to noise ratio.

RESULTS AND DISCUSSION

In the samples possessing Brönsted acidity even in very low concentration (for instance NaY-FAU), the decomposition of CFC-12 takes place via formation of phosgene as intermediate product. This was proved earlier for CCl_4 (CFC-10) [8]. The Fermi doublet in the IR spectrum (at 1720 and 1800 cm^{-1}) indicates the existence of adsorbed phosgene on the surface [9]. After high temperature treatment of the zeolite in the presence of CFC-12, the OH IR band in the spectrum could not be restored by evacuation proving the irreversible consumption of Brönsted acidity. Figure 1 shows the variation of the IR spectrum during the adsorption and reaction of CFC-12 on HY-FAU zeolite in the presence and the absence of O_2. It is clear from the spectra, that the doublet appears at higher temperature in the presence of O_2. The life-time of the phosgene intermediate is shorter without O_2, as it desorbs easily from the surface (see Figure 1/A at 1827 cm^{-1}). The appearance of phosgene as gas phase product indicates that the CFC-12 molecules have been already decomposed.

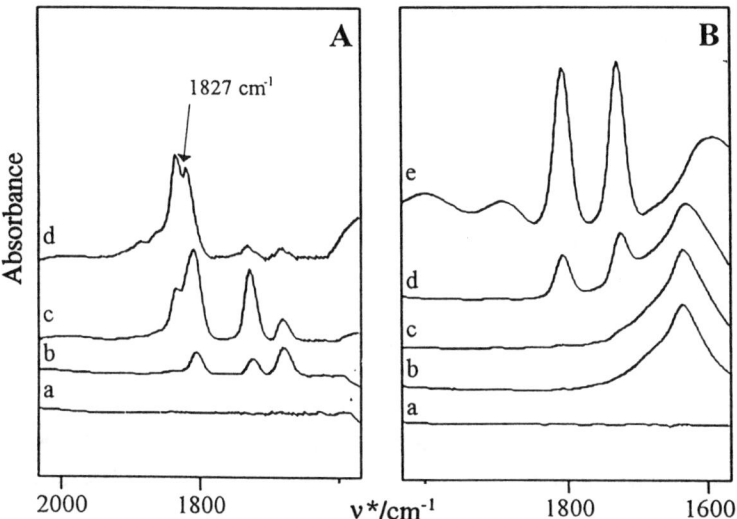

Figure 1. IR spectra of adsorbed CFC-12 on HY-FAU zeolite in absence (A) and in presence (B) of O_2; at room temperature (a), at 400 K (b), at 500 K (c), at 600 K (d) and at 700 K (e). (Differential spectra).

We found that Na-MOR is inactive, while the H-MOR is active in the decomposition of CFC-12. The investigations on the influence of oxygen show that in the absence of O_2 the Fermi doublet of adsorbed phosgene appears, but does not in the presence of O_2. It can be seen on Figure 2/A, that the amount of defect sites of the zeolite framework (characterized by the band at 930 cm^{-1} [10]) is smaller in the presence than in the absence of oxygen. This result was confirmed by the ^{27}Al NMR spectra (Figure 2/B), where practically no NMR line of octahedral Al at 0 ppm was found in presence of O_2.

Almost identical IR and ^{27}Al NMR results were found in the case of the decomposition of CFC-12 on ZSM-5 zeolites.

The adsorption and reaction of HCFC-22 (CHClF$_2$) seem to be more complex. In IR spectra taken at different stages of interaction of HCFC with NaY-FAU new bonds

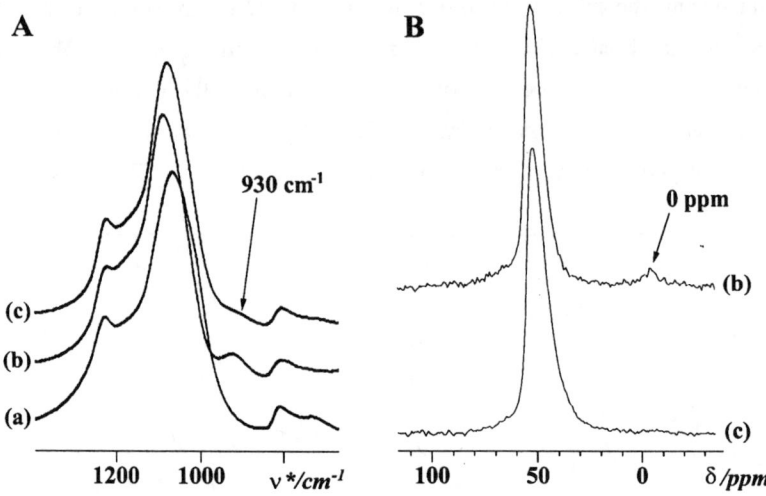

Figure 2. IR (A) and ^{27}Al NMR (B) spectra of adsorbed CFC-12 on H-MOR zeolite; activated zeolite (a), in absence (b) and in presence (c) of O_2.

appears in the OH-vibrations region. This result could only be explained by the formation of OH bonds. As can be seen in Figure 3 the frequencies of OH band(s) are somewhat higher than those obtained in the case of HNa-Y zeolite. If OH groups are formed in the

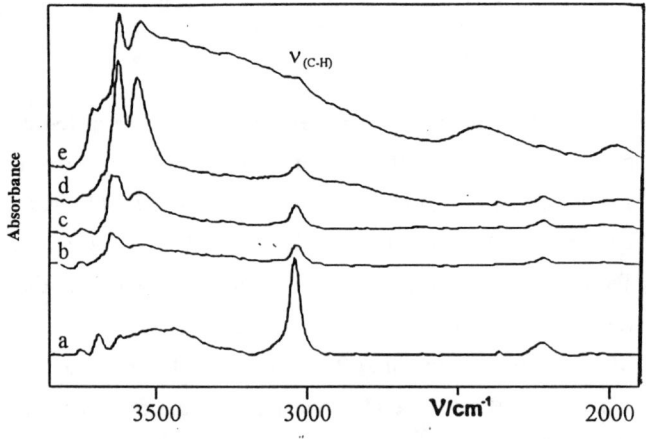

Figure 3. IR spectra of adsorbed HCFC-22 on NaY zeolite at room temperature (a), 400 K (b), 500 K (c), 600 K (d) and 700 K (e). (Differential spectra).

neighborhood of the halogen atoms upon halogenation of the surface, then their acidity should increase. This enhanced acid strength is proved by the development of new OH groups.

Figure 4 shows the ^{13}C NMR spectra of CFC-12 adsorbed on HZSM-5 zeolite. After adsorption at room temperature the chemical shift of the expected triplet is equal to 126.4 ppm vs TMS, which is almost identical to the chemical shift of the pure liquid (126.2 ppm). From this it follows that no transformation took place at room temperature. The same result was found at 100 °C. The spectra taken after treatment at 200 °C show a very clear sequence of surface reaction. After 1h of reaction, the triplet structure of CCl_2F_2 the

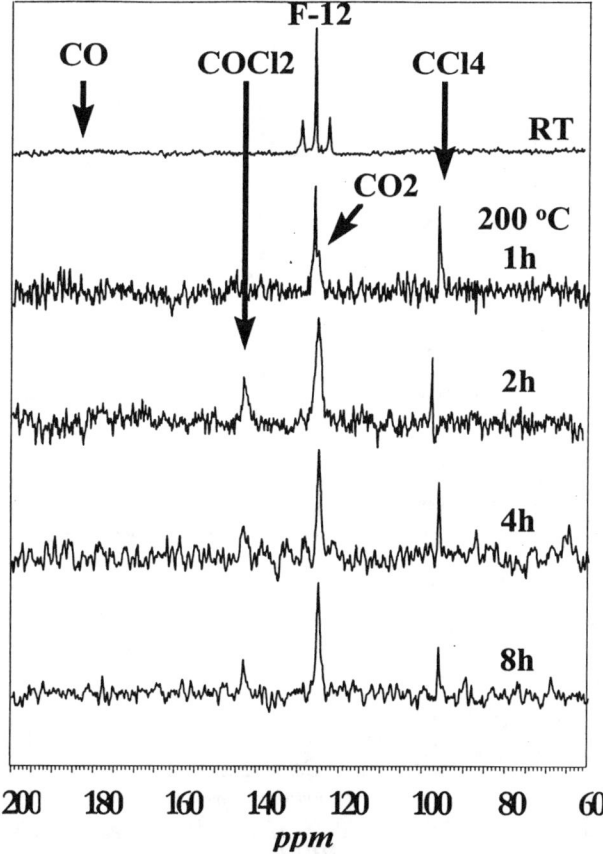

Figure 4. ^{13}C NMR spectra of CFC-12 on HZSM-5, adsorbed at room temperature (a), after reaction at 200 °C for two hours (b).

^{13}C spectrum has already changed. Carbon tetrachloride (at 96.4 ppm) and phosgene (at 143.5 ppm) are identified as intermediate products (their signal intensities pass through a maximum), and carbon dioxide (at 125.6 ppm), appears as final product (its signal intensity increases with time).

Figure 5 shows the ^{13}C NMR spectra of HCFC-22 adsorbed on HZSM-5 zeolite. A completely different product distribution was identified in this case. Instead of CCl_4, $COCl_2$ and CO_2, $CHCl_3$ and CO are the intermediate and the final products, respectively. Identical differences were observed for dealumination of zeolites with CCl_4 and $CHCl_3$ as reactants [11].

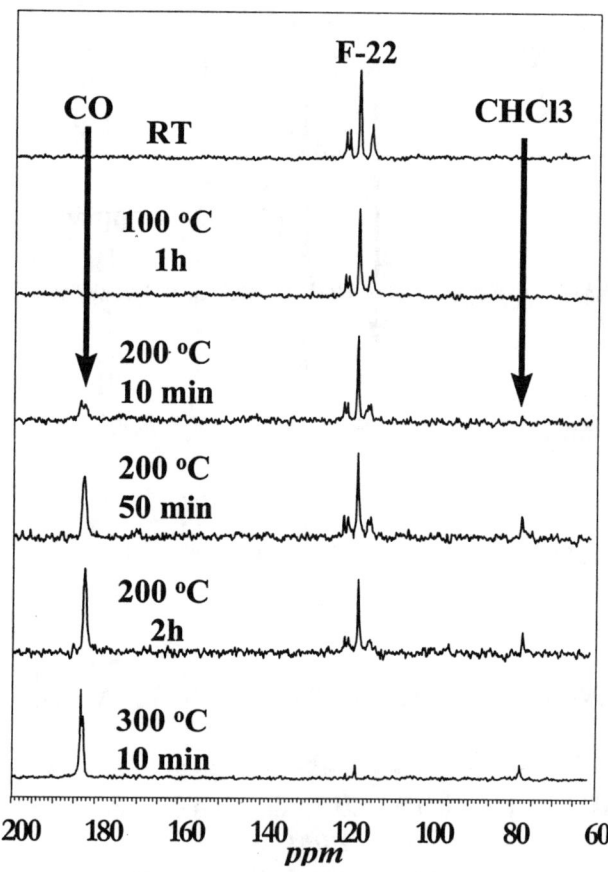

Figure 5. ^{13}C NMR spectra of HCFC-22 on HZSM-5, adsorbed at room temperature (a), after reaction at 200 °C for two hours (b).

The in situ ^{13}C NMR and IR proved that the decomposition of HCFC-22 follows a mechanism dissimilar to the CFC-12 one. On the basis of these observations a quite detailed reaction mechanism is proposed for the surface reactions;

CFC-12:

```
       |                                              |
     - Si -                                         - Si -
       |                                              |
       O                                              F
  |    |   M⁺   |                            |              |              MCl
- Si - O - Al⁻ - O - Si -  +  3 CCl₂F₂ -->  - Si - F    Cl - Si -    +    AlF₃
  |    |        |                            |              |              2 CO₂
       O        O                             F              O              CCl₄
       |        |  M⁺                         |              |  M⁺
     - Si - O - Al⁻ -                       - Si - O - Al⁻ -
       |        |                             |              |
```

$[AlO_2]^-M^+ + CCl_4 \quad \longrightarrow \quad AlCl_3 + MCl + CO_2 + \{...\}$

HCFC-22:

```
       |                                              |
     - Si -                                         - Si -
       |                                              |
       O                                              F              HCl
  |    |   M⁺   |                            |              |        2 MF
- Si - O - Al⁻ - O - Si -  +  4 CHClF₂ -->  - Si - F    HO - Si -  + AlF₃
  |    |        |                            |              |        3 CO
       O        O                             F              O        CHCl₃
       |        |  M⁺                         |              | H⁺
     - Si - O - Al⁻ -                       - Si - O - Al⁻ -
       |        |                             |              |
```

$[AlO_2]^-M^+ + 2\ CHCl_3 \quad \longrightarrow \quad AlCl_3 + MCl + 2\ HCl + 2CO + \{...\}$

where {...} denotes the vacancy formed upon aluminium removal.

CONCLUSIONS

The acidity plays a dominant role in the adsorption and decomposition of CFC-12 and HCFC-22. For H-forms of zeolites the decomposition reaction of CFC-12 occurs easily, while their Na-forms react only under severe conditions.

In the case of HCFC-22, surface OH groups are generated during the reaction. This may be the consequence of the complex halogenation, ion exchange and dealumination reactions occurring simultaneously.

Oxygen present in the reacting mixture delays the dealumination of zeolites. Significant differences are observed in the structure of zeolites after treatment with CFCs in the presence or in the absence of oxygen (depending on the Si/Al ratio of the zeolites). The reactivity order of zeolites showes the reverse order of their thermal stability, i.e. Y-FAU>MOR>ZSM-5.

^{13}C NMR spectroscopic studies reveals that decomposition of CFC-12 takes place via surface formation of CCl_4, while the decomposition of HCFC-22 leads to $CHCl_3$ as intermediate products.

As the consequence of these transformations, the crystal structure of the zeolite collapsed, and the zeolite should be therefore regarded as reaction partner and not as catalyst.

ACKNOWLEDGEMENTS

This work was performed with the help of grants FKFP 0992/1997 and OTKA T 025248, Hungary, and with financial help from CGRI, Belgium. P. Lentz acknowledges financial support from FRIA, Belgium.

REFERENCES

1. S. Karmakar, H.L. Greene, J. Catal., **138**, 364 (1992), **148**, 524 (1994)
2. Z. Kónya, I. Hannus, I. Kiricsi, Stud. Surf. Sci. Catal., **105**, 1509 (1997)
3. H. Nagata, T. Takakura, S. Tashiro, M. Kishida, K. Mizuno, I. Tamori, K. Wakabayashi, Applied Catal. B. Environmental, **5**, 23 (1994)
4. D.J. Wiley, R.P. Cooney, J.M. Seakins, G.J. Miller, Vibr. Spect., **9**, 245 (1995)
5. M. Tajima, M. Niwa, Y. Fujii, Y. Koinuma, R. Aizawa, S. Kushiyama, S. Kobayashi, K. Mizuno, H. Ohuchi, Applied Catal. B. Environmental, **9**, 167 (1994)
6. I. Hannus, Z. Kónya, J. B.Nagy, P. Lentz, I. Kiricsi, Applied Catal. B. Environmental, **9**, 167 (1997)
7. P.M. Midgley, in "Issues in Environmental Science and Technology 4, Volatile Organic Compounds in the Atmosphere" (Eds. R.E. Hester, R.M. Harrison) The Royal Society of Chemistry, 1995, p. 91.
8. I. Hannus, I.I. Ivanova, Gy. Tasi, I. Kiricsi, J. B.Nagy, ColloidsSurfaces A: Physicochem. Eng. Aspects **101**, 199 (1995)
9. I. Hannus, Z. Kónya, J. B.Nagy, I. Kiricsi, J. Mol. Struct., **410-411**, 89 (1997)
10. P. Fejes, I. Hannus, I. Kiricsi, Zeolites, **4**, 73 (1984)
11. P. Fejes, I. Kiricsi, I. Hannus, Gy. Schőbel, Magy. Kém. Foly., **89**, 264 (1983)

IN SITU AND EX SITU NMR METHODOLOGY TO STUDY MICROPOROUS PHASE CRYSTALLIZATION

C. GERARDIN [¥]*, M. HAOUAS [¥], F. TAULELLE [¥], C. ESTOURNES [§],
T. LOISEAU [£], G. FEREY [£]

[¥]RMN et Chimie du Solide, UMR 7510, Université L. Pasteur-Bruker-CNRS, 67070 Strasbourg, France, * present address : Geosciences Department and Princeton Materials Institute, Princeton University, NJ, USA, gerardin@chimie.u-strasbg.fr.
[§]IPCMS, UMR 46, 23, rue du Loess 67000 Strasbourg, France.
[£]Institut Lavoisier, IREM UMR CNRS 173, Université de Versailles-St Quentin, 78035 Versailles Cedex, France.

ABSTRACT

In situ solution NMR measurements were run under hydrothermal conditions up to 200°C in order to follow the formation of crystalline microporous phases. They lead to a direct real-time observation of the reactions occurring in solution while the crystal grows. The results presented for $AlPO_4$-CJ2 crystallization show the importance of *in situ* NMR to understand the formation mechanisms. First, some pentacoordinated Al species are shown to be directly responsible for the solid formation: they are revealed by *in situ* NMR, only in the presence of the solid; they disappear in the separated synthesis liquor. These species are believed to correspond to the secondary building units in solution. Knowing that $AlPO_4$-CJ2 contains pentacoordinated sites, it is also noticeable that the six-coordinated aluminum species seen at room temperature change to pentacoordinated sites on heating to 180°C. The pH conditions are also evaluated under hydrothermal conditions using a pH determination method by NMR.

Finally, an *ex situ* solid-state NMR study brings a complementary description of the hydrothermal mixture in identifying the main synthesis stages: dissolution of the initially formed amorphous network through a composition change - formation of $AlPO_4$-CJ2 - crystal growth from solution.

INTRODUCTION

The number of *in situ* hydrothermal studies of microporous materials crystallization has recently increased because it is now recognized that such investigations will bring a better understanding of how microporous crystals form [1]. *In situ* studies avoid invasive procedures and directly probe the reactions during the solid phase formation. In order to investigate the formation mechanisms of microporous phases, we developed *in situ* hydrothermal NMR techniques [2, 3] and applied our methodology to first study the crystallization at 180°C of an oxyfluorinated aluminophosphate: AlPO$_4$-CJ2 [4] is templated with ammonium ions. NMR presents the advantage of allowing the investigation of liquid or solid phases, and also liquid phases in the presence of solids, which can lead to a complete picture of the hydrothermal mixture throughout the synthesis.

NMR tubes were designed for use at high temperatures and pressures, up to 200°C, in corrosive conditions (various pH, presence of fluoride ions). *In situ* solution NMR studies of the synthesis mixture (liquid in the presence of solid) and of the homogeneous synthesis liquors (after separation of solid and liquid) were performed under hydrothermal synthesis conditions. It is shown how informative it is to compare both series of spectra characterizing the solution in the presence or absence of the solid being formed. The ^{14}N, ^{19}F, ^{27}Al, and ^{31}P NMR spectra were studied as a function of time and temperature so the dynamic processes could be investigated. The results lead to a description of the nature of the species in solution and a characterization of the exchange reactions. A pH determination method by *in situ* NMR was developed in order to follow the pH during hydrothermal syntheses. pH probing of the solution under hydrothermal conditions could be achieved using the ^{14}N NMR chemical shifts of well-chosen amines as pH indicators.

Ex situ solid-state NMR studies of intermediate synthesis products were also carried out in order to identify and quantitatively follow the solid phases present: the amorphous frameworks, which form and dissolve, and the competitive crystalline phases, which grow. Quantitative *in situ* and *ex situ* NMR characterizations are both necessary to follow the mass transfers from solution to solid and *vice-versa*.

EXPERIMENTAL

Crystalline microporous aluminophosphate AlPO$_4$-CJ2, whose formula is (NH$_4$)$_{0.88}$(H$_3$O)$_{0.12}$AlPO$_4$(OH)$_{0.33}$F$_{0.67}$, is obtained by mixing aluminum isopropoxyde with an aqueous solution containing orthophosphoric acid, ammonium fluoride and 1,4 DiAzaBiCyclo-2,2,2-Octane (DABCO) in the following molar ratios: 1 Al$_2$O$_3$ / 1 P$_2$O$_5$ / 2 NH$_4$F / 1 DABCO / 80

H$_2$O. The mixture is then heated at 180°C under autogenous pressure for 24 hours. The synthesis is carried out in a high pressure Torlon NMR tube protected with a Teflon insert; the device is described in a previous paper [1]. The synthesis takes place in the NMR spectrometer. The hydrothermal treatment yields a heterogeneous mixture of precipitate and liquid phase. After each *in situ* NMR measurement, the solid is isolated and X-ray diffraction and solid state NMR confirm AlPO$_4$-CJ2 formation. ^{14}N, ^{19}F, ^{27}Al and ^{31}P solution and solid-state NMR spectra were obtained using a DSX500 and a MSL300 Bruker spectrometer. *In situ* solution NMR spectra with a good signal-to-noise ratio could be acquired in 5 minutes up to 210°C.

RESULTS AND DISCUSSION

In situ NMR study of the solution phase

The methodology for *in situ* pH determination by ^{14}N NMR at high temperatures will be describ in another paper [5]. Imidazole was the amine probe that was added in small amounts to follow the p evolution during the synthesis. The results shown on fig 1 indicate a pH increase from about 4.9 to 5. These values indicate neutral pH conditions since pKa is 11 at 150°C.

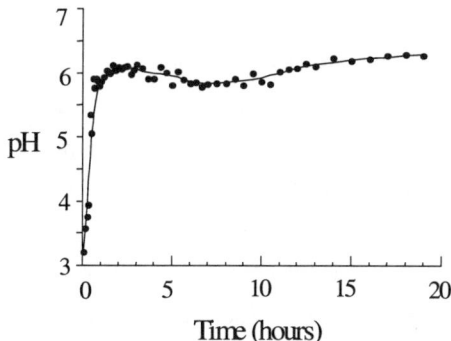

Figure 1: pH evolution at 150°C from ^{14}N NMR

Figure 2 shows *in situ* ^{19}F, ^{27}Al, and ^{31}P NMR spectra recorded above 150°C during the hydrothermal synthesis, after heating for 17 hours. The spectra were also recorded at different temperatures for the homogeneous liquors after separation from the solid.

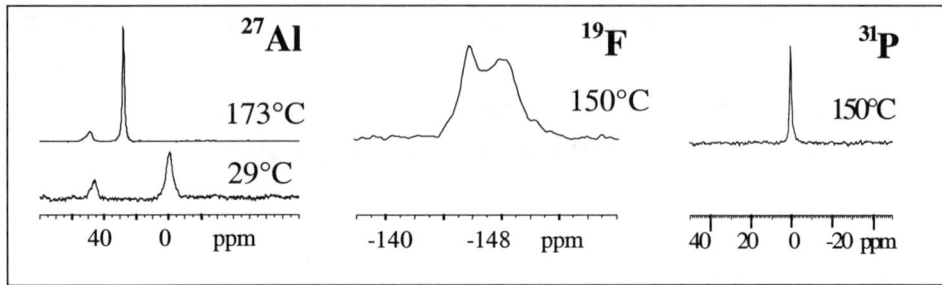

Figure 2: *In situ* hydrothermal NMR spectra of the reaction mixture

The *in situ* ^{27}Al NMR kinetic study shows the presence of two main signals at about 50 ppm and 30 ppm at 173°C. The aluminum species in solution under hydrothermal conditions are five-coordinated sites, which is consistent with the neutral pH conditions [6]. We should note that AlPO$_4$-CJ2 secondary building units contain five-coordinated Al species. A crucial observation is that the signal at higher field (and only that one) significantly changes in chemical shift with temperature. It moves from 30 ppm at 173°C to –1 ppm at room temperature, indicating a coordination change from five to six when cooling. This change is also observed in the homogeneous synthesis liquor. This result reflects the importance of *in situ* measurements in identifying the true species at high temperatures. It may also suggest the need to perform the synthesis at high temperatures in order to produce the pentacoordinated species necessary for the microporous phase formation. The comparison of *in situ* NMR spectra of the heterogeneous mixture with *ex situ* NMR spectra of the homogeneous liquor reveals the existence of species in solution (peak at 50 ppm) only when the solid phase is present, ie. when supersaturation occurs. These species are due to the solid framework formation and may be the evidence in the synthesis mixture of secondary building units.

The *in situ* ^{19}F NMR kinetic study shows that a single signal is present in the first hours of the synthesis; later, several signals appear in the same chemical shift range

(-145 to –150 ppm). The spectrum of the synthesis liquor obtained in hydrothermal conditions shows only one peak (-149 ppm). Again, the *in situ* NMR spectrum reveals new components due to the presence of the solid phase.

The *in situ* ^{31}P NMR kinetic observation is also shown. Only one peak, at about 1 ppm, is observed during the synthesis. The single signal is certainly the result of a rapid exchange between free phosphates and aluminum-complexing phosphates. An acquisition at room temperature for the synthesis liquor showed the presence of two signals due to free phosphates (3 ppm) and Al-complexing phosphates (-2 ppm).

Ex situ NMR characterization of the solid phases

^{27}Al, ^{19}F and ^{31}P MAS NMR spectra were obtained for solid samples separated after synthesis at 180°C for 1, 6, 12 and 24 hours. A multinuclear quantitative analysis allows us to estimate, first, the amount of F, P and Al atoms in the solid fraction relative to the total initial amount introduced (Figure 3), and second, the identity of the solid phases, both amorphous and crystalline phases in conjunction with X-ray diffraction powder analysis. The combined results show the following synthesis steps: first, the formation of an amorphous Al hydroxide-based phase rich in fluorine atoms; then, enrichment of this phase in phosphates at the expense of fluoride ions leading to the fast dissolution of the amorphous network and the simultaneous slower formation of AlPO$_4$-CJ2. This difference in dissolution and crystallization rates is reflected by the observed decrease in the solid fraction (Figure 3). Finally, the dissolution of the amorphous phase finishes while AlPO$_4$-CJ2 growth from solution becomes predominant leading to an increase in the solid fraction.

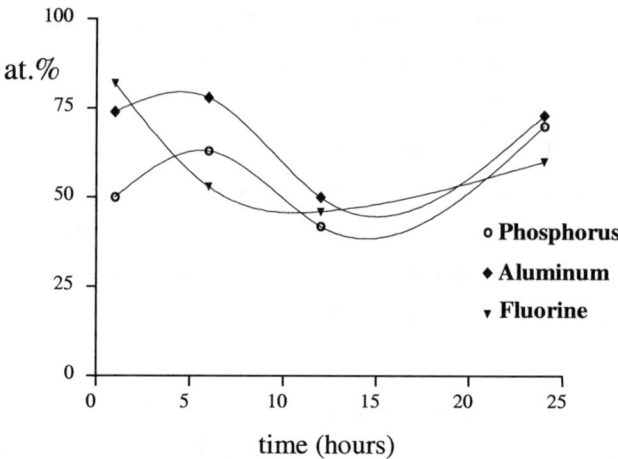

Figure 3: Atomic fractions of Al, P and F atoms in the solid part of the synthesis mixture.

CONCLUSIONS

High-resolution liquid-state *in situ* NMR measurements were obtained under hydrothermal conditions to investigate the formation of crystalline microporous phases. The medium is heterogeneous and, for the first time, results on the supernatant liquid are reported up to 200°C. They bring real-time information on the reactions occurring in solution while the crystal grows. These achievements are due to specific NMR tube developments for hydrothermal and corrosive conditions.

The results presented for $AlPO_4$-CJ2 crystallization show the importance of *in situ* NMR observations in order to understand the formation mechanisms of this type of compounds: Some Al species responsible for the solid formation are detected by *in situ* NMR only in the heterogeneous mixture in the presence of the solid, they are absent in the separated synthesis liquor. They are believed to correspond to the SBUs in solution. It is also noticeable that some Al species, which are hexacoordinated at room temperature, transform into five-coordinated species at 180°C. Finally, the *ex situ* solid-state NMR study yields essential results in describing the main synthesis stages, namely: dissolution of the amorphous Al hydroxide-based network via a composition change – crystallization of AlPO$_4$-CJ2 - crystal growth from solution.

REFERENCES

1. A.K. Cheetham and C.F. Mellot, Chem. Mater. **9** *(11)* 2269 (1997).
2. C. In-Gérardin, M. In, F. Taulelle, J. Chim. Phys. **92,** 1877 (1995).
3. M. Haouas, C. Gérardin-In, F. Taulelle, C. Estournes, T. Loiseau, G. Férey, J. Chim. Phys. **95,** 302 (1998).
4. G. Férey, T. Loiseau, P. Lacorre, F. Taulelle, J. Solid State Chem. **105,** 179 (1993).
5. Gérardin, C.; In, M.; Allouche, L.; Haouas, M.; Taulelle, F. J. Phys.Chem. B (1998), submitted.
6. R. F. Mortlock, A. T. Bell, C. J. Radke, J. Phys. Chem. **97,** 775 (1993).

SOLID STATE NMR INVESTIGATION OF CATION SITING IN LiX ZEOLITES

M. FEUERSTEIN, A. BURTON, R.F. LOBO*

*corresponding author:lobo@che.udel.edu
Center for Catalytic Science and Technology, Department of Chemical Engineering,
Colburn Laboratory, University of Delaware, Newark, DE 19716, USA

ABSTRACT

The lithium cations in the dehydrated zeolites LiX-1.0 [$(SiAlO_4)_{96}Li_{96}$] and LiX-1.25 [$(Si_{106}Al_{86}O_{384})Li_{86}$] are characterized by Magic-Angle Spinning (MAS) NMR spectroscopy. Both samples show in the ^6Li and the ^7Li MAS NMR spectra three lines assigned to Li cations in three different crystallographic sites (SI',SII, and SIII). The low field component in ^6Li and ^7Li MAS NMR spectra recorded at a temperature of 293 K belongs to mobile Li cations as proven by variable temperature ^7Li MAS NMR spectra. Adsorption of O_2 causes a strong shift of the SIII line, while the components belonging to SI' and SII cations are not changed.

INTRODUCTION

For over 12 years zeolites have been used industrially for air separations using the PSA (pressure swing adsorption) process. The nitrogen adsorption capacity of LiX zeolites is strongly correlated with the amount of charge balancing Li cations [1]. To understand the properties of zeolites it is necessary to determine the location of Li cations. In zeolites classical diffraction techniques have been used for this purpose, especially if single crystals are available. However, since most synthetic zeolites are not available in sufficiently large single crystals, powder diffraction techniques must be applied. The refinement of powder diffraction data is very time-consuming, and disorder and other factors can cause complications in these investigations. One typical problem is the dynamical disorder present when there are mobile cations or molecules in the zeolite. Because of the aforementioned difficulties, there is a need to use another technique in combination with the neutron diffraction for the characterization of Li cations. Solid State NMR is a very useful tool in this respect. Neutron diffraction data of LiX-1.0 were recently refined. Li cations are found at sites SI' (in front of six-ring windows inside the β-cages), SII (in front of six-ring windows inside the supercages), as well as SIII and SIII'(near the four-ring windows inside the supercages) [2]. We here present an investigation of the location of the Li cations in LiX zeolites using MAS NMR spectroscopy.

EXPERIMENTAL

MAS NMR spectra were carried out on a Bruker MSL 300 spectrometer operating at Larmor frequencies of 116.6 MHz for ^{7}Li and 44.2 MHz for ^{6}Li. All the MAS NMR spectra were recorded with a 4 mm double bearing MAS probe (Bruker) at spinning frequencies in the range of 5-10 kHz. Soft pulses with pulse lengths not longer than $\pi/8$ in case of ^{7}Li (spin 3/2) were applied for the single pulse experiments. For the variable ^{7}Li MAS NMR experiments, a Bruker B-VT1000 unit was used. Samples were carefully dehydrated at T = 673 K and were prepared in 4 mm MAS NMR glass inserts (Wilmad Glass) and sealed while under vacuum. One sample of LiX-1.0 was sealed after the adsorption of O_2 (99.997 %).

RESULTS AND DISCUSSION

The ^{6}Li and spectra of the dehydrated zeolites LiX-1.0 and LiX-1.25 are shown in Figure 1. The spectra contain three components attributed to Li cations at the sites SI', SII and SIII (see Table 1). The SI' and SII components show Gaussian lines in the ^{6}Li MAS NMR spectrum, in contrast to the SIII line, which was deconvoluted with a Lorentzian line shape. This difference in the line shape is due to the mobility of SIII cations at T = 293 K. Site populations were obtained from line intensities in the ^{6}Li MAS NMR spectra and within experimental errors (± 2 Li$^+$/u.c.) correspond very well with the SI' and SII cation populations from recent refinements of neutron diffraction data [2]. The determination of the site population of SIII by the Rietveld refinement of neutron diffraction data is critical because of the mobility of Li$^+$ in this site (see below). Therefore the SIII site population obtained by the simulation of the ^{6}Li MAS NMR spectra seems to be more reliable.

The SI' and SII components are much broader in the ^{7}Li MAS NMR (see Figure 2) spectrum of zeolite LiX-1.0 than the SIII line because the cations are more strongly influenced by the quadrupolar interaction. The second-order quadrupolar interaction can only be partly reduced by the use of the MAS NMR technique. The line width and shape of the SIII line are nearly the same in the ^{6}Li and ^{7}Li MAS NMR spectrum. Therefore, the influence of the quadrupole interaction on the Li cations at SIII is generally very small. ^{7}Li MAS NMR spectra of LiX-1.0

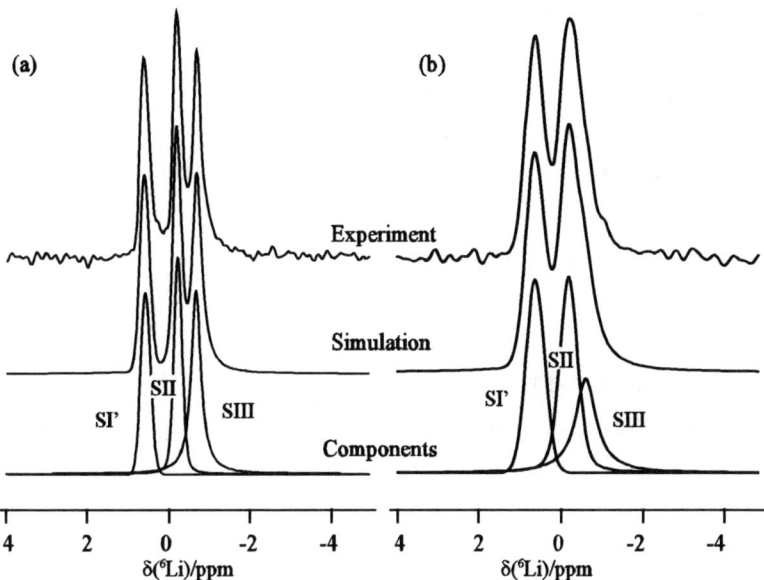

Figure 1: ^{6}Li MAS NMR spectra of (a) LiX-1.0 and (b) LiX-1.25

Table 1: Simulation of ^{6}Li MAS NMR spectra of LiX-1.0 and LiX-1.25

	SI'	SII	SIII		SI'	SII	SIII
LiX-1.0				LiX-1.25			
δ_{iso}/ppm	0.4	-0.3	-0.7	δ_{iso}/ppm	0.5	-0.3	-0.7
Li$^+$/u.c	29	33	34	Li$^+$/u.c	29	32	24

recorded at variable temperature are depicted in Figure 2. The spectra recorded at T = 273 K and T = 293 K show a narrow Lorentzian line at low field belonging to the Li cations at SIII. As the temperature is lowered (T = 253 K - T = 233 K), this line broadens dramatically. At temperatures below T = 213 K the SIII line becomes narrow again and has a more Gaussian line shape than at room temperature. For the SIII lines, at temperatures higher than T = 273 K a narrow Lorentzian line is observed because the motion of the cations is characterized by a correlation time τ_c which is much shorter than the reciprocal of the rotation frequency ω_{rot} [3]. In this limit, the motion of the SIII cations causes line narrowing. On the other hand, at temperatures below T = 213 K, τ_c is larger than $1/\omega_{rot}$ and the SIII-line is narrowed by MAS. The strong broadening in the middle temperature range (T = 253 - 233 K) is due to the fact that MAS NMR does not cause line narrowing when the motion of the cations is of the order of $1/\omega_{rot}$.

^{7}Li and ^{6}Li MAS NMR spectra are very sensitive to oxygen and water. Figure 3 shows the ^{7}Li MAS NMR spectrum of sample LiX-1.0 after adsorption of oxygen. Positions of SI' and SII lines are the same as seen in spectra of LiX-1.0 studied at vacuum in a sealed MAS NMR glass insert. However, the SIII component is shifted to higher field. This shift is caused by the paramagnetism of the O_2

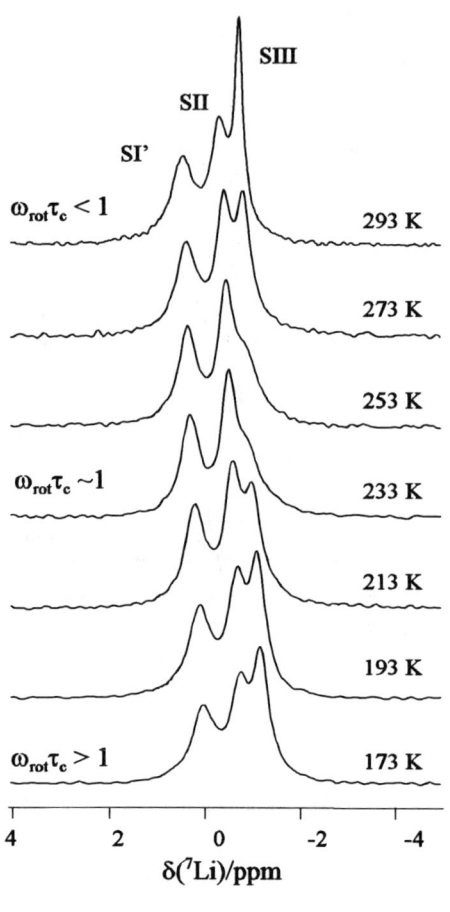

Figure 2: Variable temperature ^{7}Li MAS NMR spectra of LiX-1.25

molecules. A similar observation was published earlier from Plevert and coworkers [4], but the authors could not separate the single components of the Li cations in SI' and SII.

As known from refinement of the neutron diffraction data, the SII cations are very close inside the six-ring windows and sterically shielded by the framework oxygens. Hence these cations do not interact with the O_2 molecules. Cations inside the β-cages are not accessible for O_2 and therefore the position of the SI' line does not change due to O_2 adsorption. The line observed at δ_{iso} = -0.7 ppm in the ^{7}Li MAS NMR spectrum of the sample studied in vacuum and attributed to the SIII cations, shifts slightly to a lower frequency after the adsorption of nitrogen. This observation is explained by the preferred interaction of the nitrogen molecules with the Li cations at SIII and the inaccessibility of the SII and SI'cations.

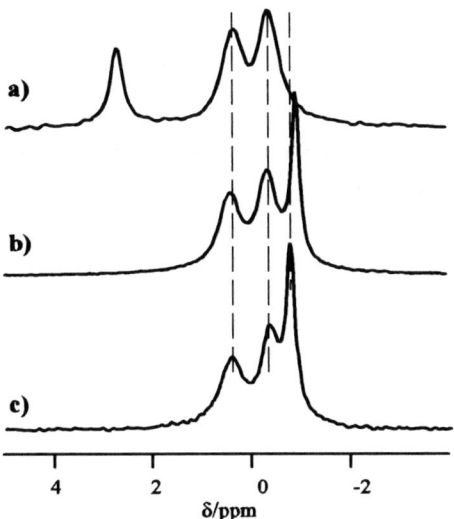

Figure 3: ^{7}Li MAS NMR spectra of LiX-1.0 (a) in the presence of O_2, (b) N_2, and (c) in vacuum.

CONCLUSIONS

^{6}Li and ^{7}Li MAS NMR spectra of LiX-1.0 and LiX-1.25 were simulated with three components belonging to cation at sites SI', SII and SIII. Variable temperature ^{7}Li MAS NMR spectra have shown that the SIII cations are mobile at temperatures larger than 253 K. SIII line position is strongly changed in the ^{7}Li MAS NMR spectra of LiX-1.0 after the adsorption of O_2 due to the paramagnetic shift. The presence of N_2 causes a small shift of the SIII line to lower frequencies and reflects the interaction of the N_2 molecules with the SIII Li cations.

ACKNOWLEDGMENT

The authors thank the Department of Chemistry and Biochemistry of the University of Delaware for access to the NMR spectrometer. This work was supported by the DRP program of the state of Delaware and PRAXAIR.

REFERENCES

[1] T.R. Gaffney, Solid State & Mater. Science, 1 (1996) 75.
[2] (a) M. Feuerstein, R.F. Lobo, Chem. Mater., in press.
    (b) J.Plevert, R. Di Renzo, F. Fajula and G. Chiari, J. Phys. Chem. B, 101 (1997) 10340.
[3] E.R. Andrew and A. Jasinski, J. Phys. C, 4 (1971) 391.
[4] J.Plevert, L.C. de Menorval, F. Di Renzo, F. Fajula, J. Phys .Chem B., 102 (1998) 3412 .

CHARACTERIZATION OF EXTRA FRAMEWORK CATIONS IN ETS-10 STUDIED THROUGH MQ-MAS NMR

L. DELEVOYE[1,†], S. GANAPATHY[2], T. KUMAR[2], C. FERNANDEZ[1] and J-P. AMOUREUX[1]

1 Laboratoire de Dynamique et Structure des Matériaux Moléculaires, CNRS URA 801 Université des Sciences et Techniques de Lille, F-59655 Villeneuve d'Ascq, France.
2 National Chemical Laboratory, Pune - 411 008, India.

ABSTRACT

Pure and highly crystalline ETS-10 and ETAS-10 titano-silicate molecular sieves, devoid of the impurity ETS-4 phase, were synthesized and fully studied by MQ-MAS NMR. Triple quantum ^{23}Na MAS experiments point to the presence of three distinct cation environments in ETS-10. Similar experiments in ETAS-10 additionally confirm that sodium environments are intact in the aluminum-substituted material, indicating an isomorphic aluminum substitution in the ETS-10 lattice.

INTRODUCTION

ETS-10 is a prominent member of a microporous titanosilicate family whose basic structural characteristics comprise corner sharing SiO_4 tetrahedra with TiO_6 octahedra, linked through bridging oxygens. This sample is titanium rich (Si/Ti=5) and is also the first microporous material wherein the location and coordination of metal atom, namely titanium, has been established from crystallographic and related techniques [1]. In order to utilize new materials in catalytic applications, it is essential to understand the features such as T-site ordering. Our paper focuses on these aspects with application to ETS-10 using MAS and MQ-MAS NMR.

[†] Corresponding author : E-mail: delevoye@lip5rx.univ-lille1.fr, Fax: +33-3 20 43 40 84

EXPERIMENTAL

MQ-MAS NMR experiments were performed on a Bruker ASX-400 FT-NMR spectrometer equipped with a specially-made MAS probehead capable of generating high speed spinning frequencies(15kHz). A three-pulse sequence was used and pulse lengths were experimentally optimized to obtain the best efficiency for the multiple quantum coherence transfers. Rotor synchronization along t_1 was used to eliminate the intense spinning side bands, which generally appear along the MQ dimension. Typical accumulation involved 2048 (t_2) x 128 (t_1) values. For each t_1 increment, the number of scans was typically 240 and 2400 for 3Q MAS on ^{23}Na and ^{27}Al, respectively. Pure-absorption mode 2D spectra are obtained by 2D Fourier transform with respect to t_1 and t_2.

RESULTS AND DISCUSSION

Recently, a new two-dimensional multiple quantum MAS (MQ-MAS) experiment that allows for a complete removal of the second-order quadrupolar broadening, was proposed by Frydman and Harwood [2]. Since this method only requires a MAS probehead to additionally average out first-order interactions such as dipolar and chemical shieldings, several researchers [3-6] have given the optimal experimental conditions to render the experiment accessible as a routine. Consequently, chemists have now access to a powerful tool for the study of half integer quadrupolar nuclei, such as the sodium cations and the aluminum sites in molecular sieves. In the MQ-MAS experiment, an isotropic spectrum of the quadrupolar nucleus can be obtained from the two dimensional correlation of triple quantum and single quantum transitions. Thus, our application of this method to ETS-10 and ETAS-10 allows for the characterization of cation environment in molecular sieves using MQ-MAS NMR.

In ETS-10 and ETAS-10, sodium cations counterbalance the negative framework charge induced by octahedrally coordinated titanium. Simple ^{23}Na MAS spectra (not shown) are featureless and consequently no information about the cation-environments can be deduced. In ETAS-10 samples, there is an increasing line broadening with increasing aluminum substitution. We performed triple quantum (3Q-MAS) correlation experiments on ETS-10 and ETAS-10 samples. These results are shown in Figure 1 as a two dimensional contour plot for the 3Q MAS experiment conducted on ETS-10. However, further explanation is necessary to fully understand this 2D spectrum. After a shearing transformation, a projection of the spectrum parallel to the vertical axis gives rise to the isotropic (δ_{ISO}) dimension and yields a high-resolution spectrum of sodium without second order quadrupole broadening. Moreover, the chemical shift and the relative magnitude of the quadrupolar interaction can be determined by projecting the center of gravity of a site onto the CS and QIS axes, which have a slope of 1 and –10/17, respectively. The isotropic spectra obtained on ETS-10 and ETAS-10 are compared in Figure 2. Several important aspects of MQ-MAS results are discussed in the following section.

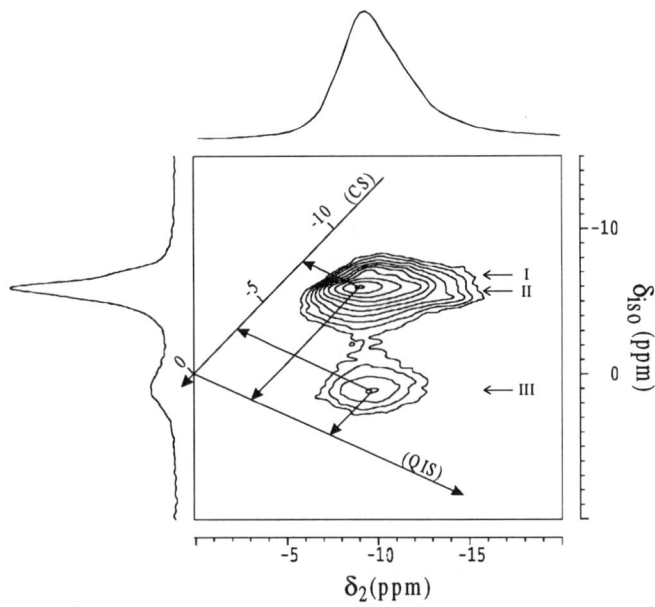

Figure 1. 105.8 MHz sheared ^{23}Na 3QMAS spectrum of ETS-10. The isotropic projection reveals three sodium cations sites labelled I, II and III as discussed in the text.

A careful inspection of the isotropic spectrum (Figure 2a) reveals three nonequivalent sodium sites in the structure of ETS-10. One of these sites is well resolved, whereas two other sites have considerable 3Q spectral overlap. The three sites are marked I, II and III in the figure. Recourse to ETS-10 structure shows that the asymmetric unit contains three crystallographically nonequivalent titaniums, of which Ti_1 and Ti_2 are similar due to the equivalent silicon and titanium connectivity to their immediate neighbors, while site Ti_3 is distinct. The charge-balancing role of the sodium cations present at these distinct titanium sites is therefore confirmed by our 3Q-MAS studies. Although for a complete charge balancing, we require two sodiums for each nonequivalent titanium, the observation of only three sodium sites from 3Q MAS experiments suggests that sodium cations occupy symmetrical related positions so that the two sodiums located on each titanium become pairwise equivalents. Such a picture does in fact emerge from cation modeling of sodium positions in ETS-10 [7]. Sodium ions are coordinated to titanium atoms in such a way that they are located on either side of the titanium atoms and form a chain of sodium atoms, which runs almost parallel to Ti-Ti chains.

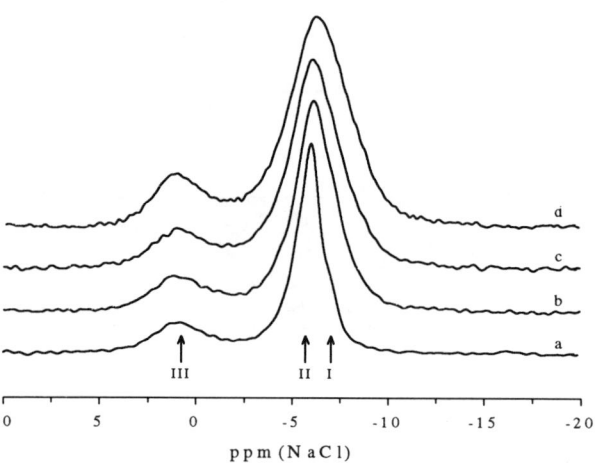

Figure 2. The isotropic projection of ^{23}Na 3Q MAS spectra of ETS-10 (a) and ETAS-10, Si/Al=63 (b), 43 (c) and 22 (d).

The expected 2:1 relative intensity of site (I,II) and (III) is not readily apparent in the isotropic spectrum of Figure 1. This is due to the fact that sites (I,II) and (III) which experience quadrupolar interactions of different magnitudes are not excited with the same efficiency with respect to the generation of multiple quantum coherences. Consequently, the relative population of certain sites is likely to be underestimated if one considers only the isotropic projection. In particular, site III, which experiences a much stronger quadrupolar interaction, develops weaker coherence compared to (I, II). However, quantitative information can be retrieved by a calculation taking into account the experimental parameters. In short, it is possible to simulate the pulse response of the sites to a given experimental condition by the knowledge of the rf field, pulse lengths as well as a first approximate of the quadrupolar constant C_q deduced from the 2D dataset. This procedure had previously been introduced for the quantification of 2D DAS spectra [8]. The result of this calculation is the mapping shown in Figure 3, representing the quadrupolar constant C_q versus the chemical shift δ_{CS}. The latter parameter is found identical to the one previously determined by projection onto the CS axis (see Figure 1). What is important here to underling is that the relative intensity given by this method is in remarkably good agreement with 2:1 ratio expected in the sodium site population.

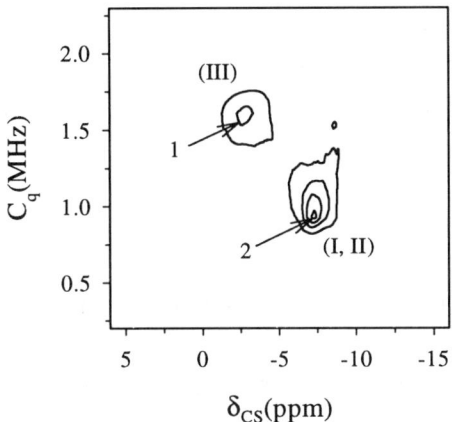

Figure 3. Quantification of the experimentally determined sodium sites (I,II) and (III) in ETS-10.

Figure 2 (b-d) exhibits the ^{23}Na 3Q MAS isotropic projection spectra for ETAS-10 with increasing aluminum substitution. When compared with the result obtained on ETS-10 (Figure 2-a), it is clear that the sodium environments are intact in the aluminum-substituted materials, except for the additional broadening. This broadening increases with increasing aluminum content and the distinction between I- and II-type environments is rendered quickly difficult. It may also be noted that the position of site III is unchanged. The ^{29}Si MAS NMR results discussed elsewhere [9] show that the framework structure of the aluminum-substituted material conforms to parent ETS-10 structure. In addition, the present 3Q MAS observations in ETAS-10 show that the cation environment remain also unchanged. Globally, our results provide support to the view that the overall structure of ETAS-10 is the same as that of ETS-10 and that the heteroatom substitution occurs isomorphically. We believe that the observed broadening arises mainly from a distribution of isotropic chemical shifts, and is only slightly due to a distribution caused by quadrupolar effects, since the 2D contour plots show distorted patterns with ridges that extend along the diagonal of slope 1.

CONCLUSION

In this paper we have presented a study of ETS-10 and ETAS-10 by MQ-MAS NMR. We focused on the sodium atom ^{23}Na and its cationic role in the framework. The counterbalancing role of ^{23}Na was confirmed and the aluminum substitution seemed to keep the framework unchanged, compared to parent ETS-10 structure. The capability of MQ-MAS to resolve the different sodium environments demonstrated that this recently discovered experiment is a powerful technique for the study of such materials.

REFERENCES

1. M.W. Anderson, O. Terasaki, T. Ohsuna, A. Philippou, S.P. Mackay, A. Ferreira, J. Rocha and S. Lidin, Nature **367**, 347 (1994)
2. L. Frydman and J.S. Harwood, J. Am. Chem. Soc. **117**, 5367 (1995)
3. C. Fernandez and J.P. Amoureux, Chem. Phys. Lett. **242**, 449 (1995)
4. D. Massiot, B. Touzo, D. Trumeau, J. Coutures, J. Virlet, P. Florian and P. Grandinetti, Solid-State Nuclear Magnetic Resonance **6**, 73 (1996)
5. G. Wu, D. Rovnyak, B. Sun and R.G. Griffin, Chem. Phys. Lett. **249**, 210 (1996)
6. J.P. Amoureux, C. Fernandez and L. Frydman, Chem. Phys. Lett. **259**, 347 (1996)
7. X. Yang and P.W. Blosser, Zeolites **17**, 237 (1996)
8. J.W. Zwanziger, Solid State Nuclear Magnetic Resonance **3**, 219 (1994)
9. S. Ganapathy, T.K. Das, R. Vetrivel, S.S. Ray, T. Sen, S. Sivasanker, L. Delevoye, C. Fernandez and J.P. Amoureux, J. Am. Chem. Soc. **120**, 4752 (1998)

APPLICATIONS OF ^{1}H NMR IMAGING AND ^{129}Xe NMR TO THE STUDY OF HYDROCARBON DIFFUSION IN ZEOLITES AND COKE DISTRIBUTION

T. DOMENICONI, P. N'GOKOLI-KEKELE, J.-L. BONARDET, M.-A. SPRINGUEL-HUET and J. FRAISSARD

Laboratoire de Chimie des Surfaces, CNRS-UPESA 7069, Université P. et M. Curie, 4 Place Jussieu, 75252 PARIS CEDEX 05, France. Fax: 33 1 44 27 55 36. E-mail: bonardet@ccr.jussieu.fr

ABSTRACT

Diffusion of benzene in H-ZSM-5 zeolite in powder or pellet form and of 2,3-dimethylpentane (DMP) in H-Y zeolite pellets, coked at different levels, was studied by ^{1}H NMR imaging and ^{129}Xe NMR of adsorbed xenon used as a probe. The shape and the evolution of the signal intensity of the 1D image, representing the spin distribution along the axis of the zeolite-containing cylindrical tube, make it possible to establish that the rate-limiting process is intracrystalline molecular diffusion when benzene is being adsorbed. Simulation of the variation of benzene concentration with time, obtained from the variation of the NMR signal intensity, gives a diffusion coefficient of about 1×10^{-14} m^{2}s^{-1}. Comparison of the ^{129}Xe NMR spectra, recorded during benzene adsorption, with those simulated assuming intracrystalline diffusion, as suggested by the ^{1}H imaging experiments, gives a similar value of the diffusion coefficient. For DMP in H-Y coked pellets, the NMR profile shows a maximum, which moves during DMP adsorption. This was interpreted as the superposition of two competing processes: intercrystalline diffusion and intracrystalline diffusion of the probe from uncoked to highly coked zones of the crystallites. 1D and 2D images at equilibrium adsorption of the probe clearly show that the coke is heterogeneously distributed at the macroscopic scale.

INTRODUCTION

Diffusion of reactants and products plays an important role in gas-solid reactions, such as reactions in the gas phase catalyzed by microporous solids.

^{1}H-NMR Imaging has been successfully used to study the solvent penetration into polymers [1] and coals [2], water permeation resistance of cement [3] or drying kinetics of porous catalyst pellets [4]. We report here for the first time a comparative study of hydrocarbon (benzene and 2,3-dimethylpentane (DMP)) diffusion in zeolites (H-ZSM-5 and H-Y) by ^{129}Xe NMR of xenon gas used as a probe [5] and by ^{1}H NMR imaging,. The coke distribution in more or less deactivated catalysts was also studied by these two techniques.

EXPERIMENTAL

Materials

H-ZSM-5 (Süd Chemie AG, Si/Al=130, mean crystallite size about 40 µm) and H-Y (LZ 54 from UOP, mean crystallite size about 1 µm) were used in powder or pellet form. In order to increase the detection sensitivity and to "mimic" industrial catalysts, most of the samples were compressed at a pressure of 1.5×10^3 Kg/cm^2 into pellets in the form of cylinders 10-12 mm long and 6-7 mm in diameter. Before adsorption, samples were outgassed overnight under vacuum at 673 K. H-Y pellets were deactivated by cracking n-heptane at 673 K for periods of 2-10 hours to obtain various coke contents.

Experimental procedure

^{1}H Imaging experiments were performed on a Bruker MSL 300 spectrometer. Images were obtained by superimposition of a pulse field gradient in one (1D imaging) or two (2D imaging) directions to the Hahn echo detection sequence. The 1D image along the z direction (vertical axis of the pellet, or tube axis) corresponds to the concentration profile of the resonant spins (^{1}H of the hydrocarbon) in this direction.

Adsorption experiments were performed by placing the sample in contact with hydrocarbon at its saturation vapor pressure either out of equilibrium, by recording spectra at different times (during the adsorption process) with a given time delay, τ, in the pulse sequence or at equilibrium for different τ.

Benzene and 2,3 DMP were chosen as NMR probes because they have magnetically equivalent protons, given the signal width, and because their kinetic diameters are close to the size of the micropore openings

^{129}Xe NMR experiments were performed on a Bruker MSL 400 spectrometer at 110.69 MHz. For diffusion experiments xenon gas was adsorbed before the hydrocarbon adsorption.

RESULTS AND DISCUSSION

Diffusion of benzene in H-ZSM-5

The NMR profiles, along the z direction, of the powdered and compressed zeolite (samples A and B, respectively) are horizontal (Fig. 1). The signal intensity increases during the diffusion-adsorption process. The rectangular shape of the profiles at any time proves that diffusion of benzene in the sample is controlled by micropore diffusion, as discussed by Heink et al. [7]. Moreover, equilibrium is reached within roughly the same time, about 8 hours for the two samples.

However, we observe a difference between these two samples when the relative amount of adsorbed benzene, $Q(t)/Q_\infty$, is plotted against the time and the square root of time (Fig. 2). The quantity $Q(t)$, concentration of adsorbed benzene at time t, is directly proportional to the NMR signal intensity, since a rapid calculation of the amount of benzene in the gas phase shows that it is negligible compared to the amount adsorbed, and we may assume that there is no variation of the transversal relaxation time T_2. Q_∞ is the adsorbate concentration at the adsorption equilibrium.

Initially the curves, $Q(t)/Q_\infty = f(\sqrt{t})$, are sigmoidal. This shape is attributed to the external surface barrier of the crystallites. The non-superposition of the curves can be explained by the difference in the bed porosity, ε, defined as the ratio of the macro- and mesopore volume by the total sample volume. Compression reduces the size of the intercrystallite spaces and, consequently, the bed porosity. The decrease in bed porosity should lead to a decrease in the mobility of benzene molecules which in turn reduces the transfer rate from the gas phase to the crystallite surface.

To obtain quantitative information from these NMR results, we attempt to simulate the kinetics curves.

Assuming that the benzene concentration in the gas phase is constant and equal to the saturation vapor presure ($\approx 5.35 \times 10^{-6}$ molecm^{-3}), Q(t) is obtained by solving the Fick equation, expressed in spherical coordinates:

$$\frac{\partial Q}{\partial t} = D\left(\frac{\partial^2 Q}{\partial r^2} + \frac{2}{r}\frac{\partial Q}{\partial r}\right)$$

where D and r are the diffusion coefficient and the radial coordinate, respectively.

The initial and boundary conditions are:
- $t = 0, Q = 0$;
- $t > 0, Q = Q^*$ for $r = R$
  $\delta Q/\delta r = 0$ for $r = 0$

where R is the mean radius of the crystallites assumed to be spherical, and Q* is the adsorbate concentration at the external surface of the crystallites, given by:

$Q^*(t) = Q_\infty [1 - \exp(-\beta t)]$

where β the transfer coefficient of molecules from the gas phase to the surface of crystallites. Under these conditions, Q(t) is given by the relation of Crank [7]:

$$\frac{Q(t)}{Q_\infty} = 1 - \frac{3}{\lambda}\exp(-\frac{\lambda t}{\theta})\left(1 - \sqrt{\lambda}\tan^{-1}\sqrt{\lambda}\right) + \frac{6\lambda}{\pi^2}\sum_{n=1}^{\infty}\frac{1}{n^2\left(n^2\pi^2 - \lambda\right)}\exp\left(\frac{-n^2\pi^2 t}{\theta}\right)$$

where $\lambda = \beta R^2/D$ and $\theta = R^2/D$.

Starting from an approximate value of θ, estimated as the time constant of the equilibrium adsorption, then using λ and θ as adjustable parameters, we simulated the experimental curves and obtained an identical benzene diffusivity, D, of about 1×10^{-14} m^{2}s^{-1} for A and B, which is in good agreement with the literature [8]. Conversely, the values of β are different, 1.18×10^{-3} s^{-1} for the powder and 0.78×10^{-3} s^{-1} for the pellet. This shows that compression has reduced the transfer rate of molecules to the surface.

The ^{129}Xe NMR spectra recorded during benzene adsorption under the same conditions as the ^{1}H imaging experiments are shown in Figure 3. The narrow high-field signal corresponds to Xe not interacting with benzene, the broad one to Xe interacting with benzene, the chemical shift and the linewidth depending on the benzene concentration. Xenon was adsorbed at a pressure low enough (26 KPa, 6×10^{20} atoms g^{-1}) for it to be assumed that benzene diffusion is not altered. Equilibrium is achieved after about 8 hours. We can then assume that the rate-limiting process of diffusion is intracrystalline diffusion, as suggested by ^{1}H NMR Imaging. ^{129}Xe spectra were simulated by taking Lorentzian signals and using the benzene concentration profile in zeolite crystallites, given by Crank for intracrystalline diffusion [7], which depends on D, t and the

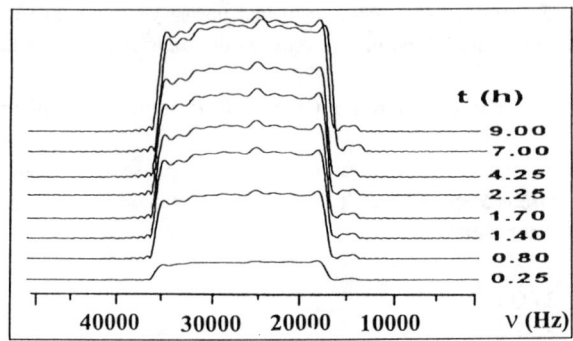

Figure 1. Evolution with time of the spin profiles during the adsorption of benzene on H-ZSM-5 (loose powder). Similar spin profiles are obtained with compressed H-ZSM-5 powder.

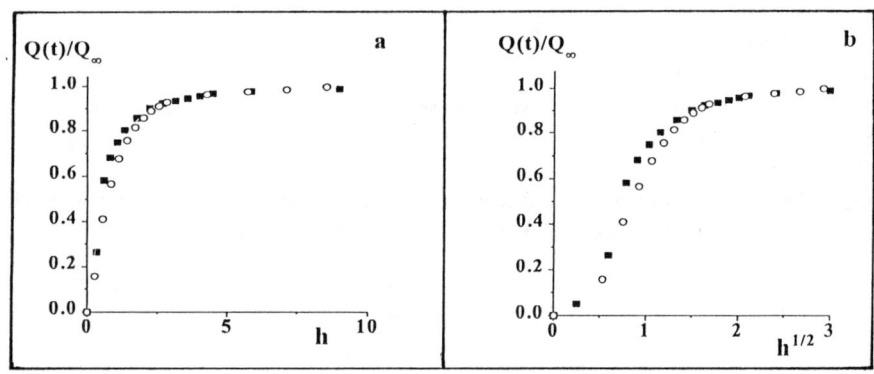

Figure 2. Variation of the relative amount of adsorbed benzene on H-ZSM-5, $Q(t)/Q_\infty$, in function of time (a) and of the square root of time (b), loose (■) and (o)compressed powder.

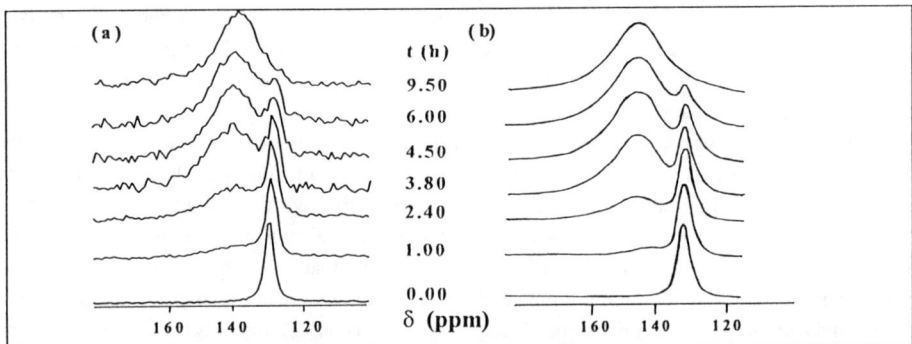

Figure 3. Evolution with time of experimental (a) and simulated (b) ^{129}Xe NMR spectra of adsorbed xenon during the adsorption of benzene on H-ZSM-5 (compressed powder).

crystallite radius R. Comparison of experimental and simulated spectra leads to a value of D of about 10^{-15} m^{2}s^{-1}, which is in good agreement with the above value. The method has been described elsewhere [8].

Diffusion of DMP in coked H-Y zeolites

Similar diffusion experiments were performed with DMP in a 10% coked H-Y zeolite pellet. In contrast with the rectangular benzene distribution profiles observed in a fresh catalyst, the shape of the profiles in a coked pellet is time dependent (Fig. 4). It can be seen that from t = 1 min to t = 30 min, the signal intensity increases, the maximum moving from the edge of the pellet towards the center. Up to 70 min this maximum continues to shift towards the lower face of the pellet; finally the signal intensity decreases over the whole pellet and remains constant for times greater than 100 min. This evolution in the shape and the intensity of the profiles can be related to two competing diffusion processes, one being intercrystalline and the other, slower, being attributed to intracrystalline diffusion of the probe from uncoked towards highly coked zones.

DMP imaging of coked H-Y zeolite pellets

Figure 5-a displays the spin distribution profile (1D imaging) of the proton probe, at adsorption equilibrium, along the vertical axis z of a 7.5% coked pellet saturated with DMP. Due to the short relaxation time T_2, the signal intensity and the shape of the profile depend markedly on the echo time, τ, imposed on the pulse sequence. As the echo time is decreased the signal intensity increases, regardless of z, but for a given τ there is a practically monotonic increase in the signal from the beginning of the pellet (z_0) to the end (z_f). The observed discontinuities can be attributed to structural heterogeneities due to pelletting. The real profile is determined by extrapolation to zero echo time. It indicates that the coke distribution, which is the "negative" of the spin profiles, is very heterogeneous at the macroscopic level and that the most active part in the reaction is that which is first exposed to the reagent flow.

The relaxation time T_2 is very short but increases linearly from z_0 to z_f. It should be remembered that the more aromatic the coke, the shorter T_2 (9). These results are consistent with the fact that, at high coke content, the carbonaceous residues are essentially polyaromatic (graphitic) and that the coke is the most graphitic in the most coked zones (T_2 shortest).

In contrast to the previous sample, and down to the lowest echo time, the profiles obtained for a more coked pellet (10% w/w) show a certain homogeneity in the distribution of the resonant spins and therefore of the coke (Fig. 5-b). This result is in agreement with those obtained by Barrage et al. [10] using ^{129}Xe NMR adsorbed in coked HY powders. The profile extrapolated to zero echo time displays marked discontinuities attributable partly to structural heterogeneities but also to the imprecision of the measurements. The very short (220 µs) but constant relaxation time shows that the coke is very graphitic but chemically more homogeneous than in a less coked sample.

Finally, figure 6 shows an attempt to obtain 2D images of a lightly coked pellet (2.5%). The darker zones are the more coked regions of the pellet. The transversal sections, xy (Fig. 6-a) for

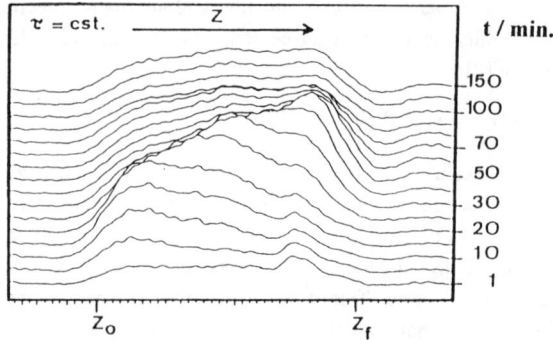

Figure 4. Evolution with time of the spin profiles during the DMP adsorption on 10% coked HY zeolite pellet.

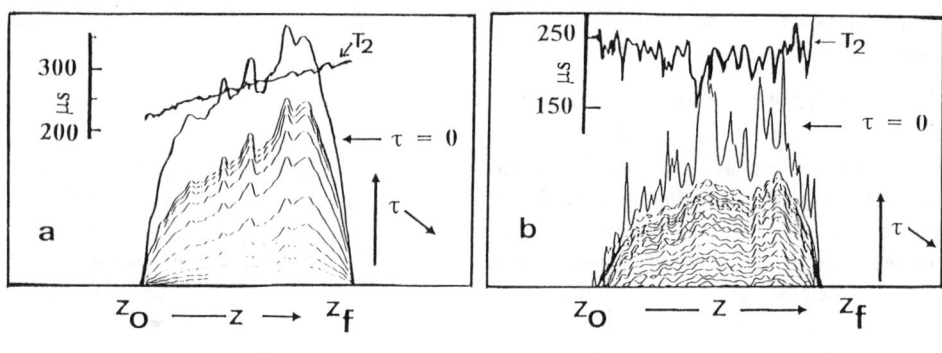

Figure 5. Spin distribution profiles along the z axis of 7.5% (a) and 10% (b) coked HY pellets.

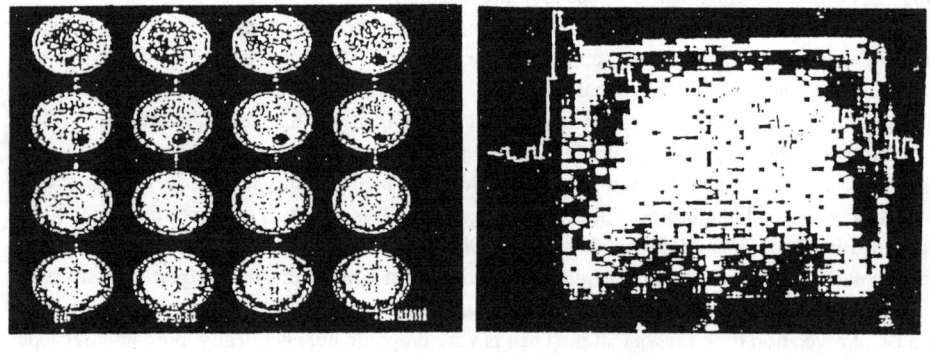

Figure 6. 2D images of coke in a 2.5 % coked pellet (7 x 12 mm). a) transversal xy planes for different values of z. b) longitudinal zx plane.

different values of z, or the yz (Fig. 6-b) planes of the pellet, show a very heterogeneous coke distribution. The coke is located essentially on the face attacked by the reagent flow and on the external surface of the pellet; these images show that a large part of the catalyst is inactive in the cracking reaction. Moreover, these images are in agreement with the observation of two ^{129}Xe NMR signals for adsorbed xenon (Fig. 7) whose chemical shift variations depend linearly on the xenon pressure. The δ-variation of the upfield signal (lower δ) is practically the same as that for a non-coked sample which proves that this signal is due to xenon adsorbed in non-coked regions of the pellet. The downfield signal (higher δ) is broad (chemical shift distribution) and comes from xenon adsorbed in the more or less coked zones. At the NMR time scale, xenon atoms do not exchange between the coked and non-coked zones.

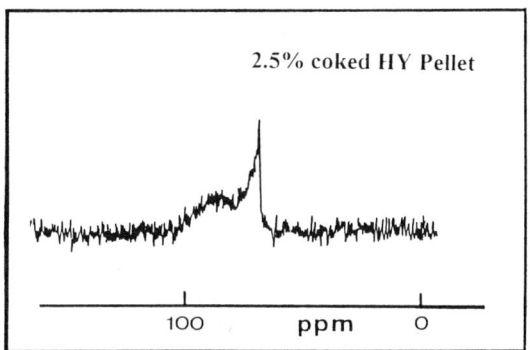

Figure 7. ^{129}Xe NMR spectrum of xenon adsorbed in a 2.5% coked pellet.

CONCLUSION

^{1}H NMR Imaging is a useful technique for studying molecular diffusion on the nanometer scale and the distribution of carbonaceous residues at the macroscopic level. Moreover, it should be possible to use this technique to optimize the shape and the size of industrial pelleted catalysts. The ^{129}Xe NMR of adsorbed xenon, which is much easier to use, gives similar results for molecular diffusion as well as for the coke distribution on the nanometer scale.

ACKNOWLEDGMENTS

^{1}H NMR Imaging experiments were performed in the Laboratory of Physical Chemistry and Crystallography in Louvain la Neuve, Belgium. We thank Professor J.M. Dereppe, head of this laboratory, for his material and technical help. His judicious advice and our fruitful discussions were very useful for the interpretation of the experimental results.

REFERENCES

1. S. Blackband and P. Mansfield, J. Phys. C: Solid State Phys. **19**, L49 (1986)
2. G. D. Cody and R. E. Botto, Macromolecules **27**, 2607 (1994)
3. G. Papavassiliou, F. Milia, M. Fardis, R. Rumm, E. Laganas, O. Jarh, A. Sepe, R. Blinc and M. M. Pintar, J. Am. Ceram. Soc. **76**, 2109 (1993)
4. I. V. Koptyug, V. B. Fenelonov, L. Y. Khitrina, R. Z. Sagdeev and V. N. Parmon, J. Phys. Chem. B **102**, 3090 (1998)
5. M.-A. Springuel-Huet, J.-L. Bonardet and J. Fraissard, Appl. Magn. Reson. **8**, 427 (1995)
6. W. Heink, J. Kärger and H. Pfeifer, Chem. Eng. Sci. **33**, 1019 (1978)
7. J. Crank, The Mathematics of Diffusion, (Clarendon Press, Oxford, U. K., 1956)
8. J. Kärger, H. Pfeifer, T. Wutscherk, S. Ernst, J. Weitkamp and J. Fraissard, J. Phys. Chem. **96**, 5059 (1992) and references therein
9. R.H. Meinhold and D.M. Bibby, Zeolites **10**, 2532 (1990)
10. M.C. Barrage, J.L. Bonardet and J. Fraissard, Catal. Lett. **5**, 733 (1990)

FAULTED ZEOLITE FRAMEWORK STRUCTURES

H. GIES*, R. KIRCHNER@, H. VAN KONINGSVELD$ and M. M. J. TREACY#

* Institut für Mineralogie, Ruhr-Universität Bochum, Universitätsstrasse 150, D-44780 Bochum 1, Germany
@ Department of Chemistry, Manhattan College, Riverdale, NY 10471, U.S.A.
$ Department of Organic Chemistry and Catalysis, Delft University of Technology, Julianalaan 136, 2628 BL Delft, The Netherlands; havank@cad4sun.tn.tudelft.nl
# NEC Research Institute Inc., 4 Independence Way, Princeton, NJ 08540, U.S.A.

ABSTRACT

An outline is proposed to describe families of faulted zeolite framework structures. Several examples are given.

INTRODUCTION

Since the description of the structure of zeolite beta, some years ago, a debate was started on the question whether a zeolite which does not show 3-dimensional periodicity should be assigned a structure type code. As a compromise a code for zeolite beta was included in the "Atlas of Zeolite Structure Types" [W. M. Meier, D. H. Olson and Ch. Baerlocher. Atlas of Zeolite Structure Types, 4th rev. ed., Elsevier, London, 1996] preceeded by an asterisk indicating that the code stands for a particular ordered stacking of a well-defined subunit for which pure end-member materials were not yet obtained synthetically. The subunit, periodic in two dimensions, but differently stacked in the third dimension, produces a whole family of zeolite structures which are all related to zeolite beta.

In the Atlas, the structure types have been arranged in alphabetic order according to their structure type code. A number of zeolites, included in the Atlas, can be considered as periodically ordered end-members of a family of zeolites that is built from one type of well-defined subunits and that consists of an infinite number of intermediate structures, e.g. the ABC-6 family of zeolites.

The Structure Commission of the International Zeolite Association has appointed the "Disordered Structures Committee" to formulate a first approach to a more general description of faulted frameworks exhibiting statistical stacking disorder and to present these structural information in a comprehensive way to the zeolite community. This communication reports on the proposed outline of a "Record of Families of faulted Zeolite Framework Structures". Readers are kindly requested to react and to send their suggestions, comments and criticisms to Henk van Koningsveld, member of the "Disordered Structures Committee", at his e-mail address.

CONCEPT

Regular crystal structures are periodically ordered in three dimensions, both chemically and structurally, whereas faulted structures are periodic in two or lower dimensions. Faulted structures can be thought of as exhibiting a statistical stacking disorder of structurally invariant Periodic Building Units (PBUs).

In zeolites, we are concerned with stacking disorder of the PBUs that define the framework structure. We exclude chemical disorder, e.g. different T atoms or cations on a particular site, and dynamic disorder, e.g. rotational disorder of template molecules. We also exclude structural disorder of any guest molecules within the cavities of zeolite frameworks.

Consequently, we consider a faulted zeolite framework structure to be built from PBUs with stacking disorder. The PBU constitutes of an array of (not necessarily connected) smaller units such as T atoms. The smaller units are related by simple symmetry operations, e.g. translations and rotations.

The relative orientation of neighboring PBUs can be described by connection modes between the parallel alligned PBUs. The connection mode contains the symmetry elements which relate the PBUs to each other, including lateral translation components given as a fraction of the basis vectors of the invariant PBU.

Regular framework structures built from PBUs are called end-member structures if periodic ordering is achieved in all three dimensions. Faulted framework structures are those where the stacking sequence of PBUs is irregular along one or more directions. All structures built from the same PBU, ordered end-members as well as well as the faulted intermediates, belong to the same family of structures.

DESCRIPTION

- Give a skeletal drawing of the PBU in top and side view; highlight the smaller unit and analyse whether different side projections can be related by rotation and/or mirroring.
- Give a drawing of the possible connection modes between neighboring PBUs. Analyse whether the connection modes introduce new symmetry elements between neighboring PBUs.
- Give a drawing of an intermediate structure (when feasible) and of the simplest ordered end-members illustrating the stacking sequence. Add a Table of the sequence of the connection modes and space group of each end-member, if clarity demands.

EXAMPLES

See next pages.

The members of the "Disordered Structures Committee"
    Hermann Gies
    Rich Kirchner
    Henk van Koningsveld
    Mike Treacy

Building units of faulted zeolite framework structures and
their simplest ordered end-member structures

The pentasil family

1. The Periodic Building Unit (PBU) equals the YZ layer:

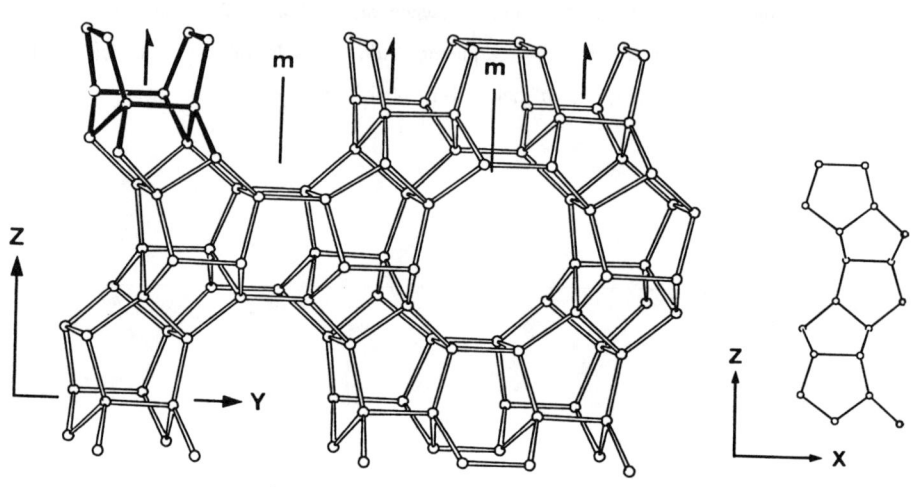

The PBU of the pentasil family of zeolite structures, the pentasil layer (at left), is composed of T12-units (in bold) forming left- and right-handed chains along Z. The chains, related by a mirror plane perpendicular to Y, are connected along Y to give the characteristic YZ pentasil layer. A parallel projection of the pentasil layer along Y is shown to the right of the Figure.

2. Type of faulting: 1-dimensional stacking disorder of the PBUs along X.

3. The symmetry of the pentasil layer is indicated in the Figure of the PBU.

4. Connectivity pattern of the PBU.
Neighbouring PBU's are connected via O-bridges along X through

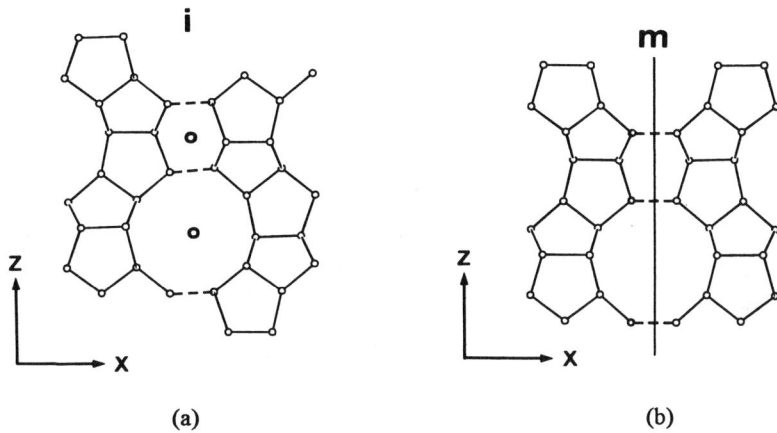

a) inversion (i); successive pentasil layers are connected after a 180° rotation about X (or Y) with respect to each other. The resulting connectivity exhibits inversion symmetry (symbol: o).

b) reflection (m); successive pentasil layers are connected after a 180° rotation about Z. The connectivity now shows mirror symmetry (symbol: |) between successive layers.

Once the distribution of the symmetry elements i and m along X is known the 3-dimensional structure is defined.

Example of an intermediate structure in the pentasil family of zeolites:

5. The simplest ordered end-members in the pentasil family:

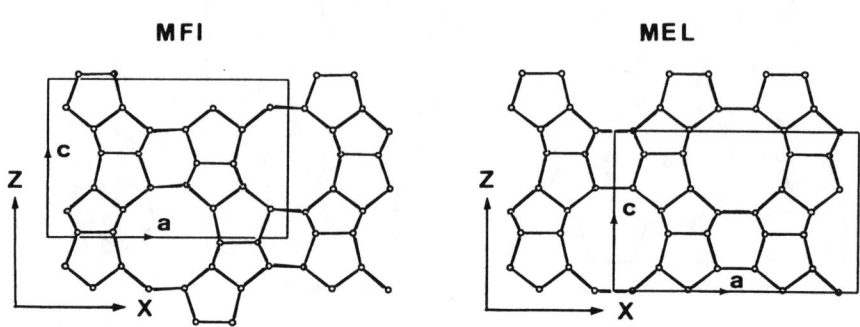

Pure MFI and MEL are obtained when neighboring pentasil layers along X are exclusively related by inversion or reflection, respectively.

6. Faulted materials synthesized and/or characterized so far:
1. Bor-D

7. References
1. G.Perego and M.Cesari, J. Appl. Cryst. **17**, 403 (1984).

Building units of faulted zeolite framework structures and
their simplest ordered end-member structures

The ITE/RTH family

1. The Periodic Building Unit (PBU) equals the XY layer:

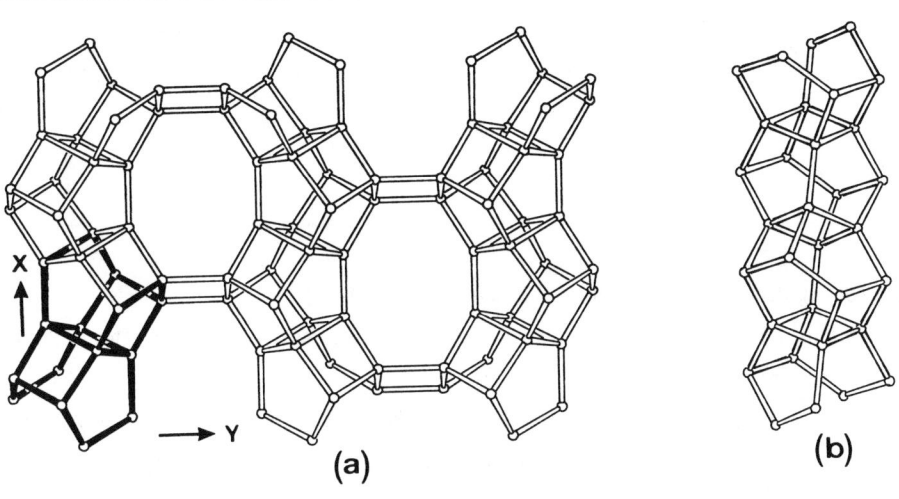

T16-units (in bold), consisting of three 4-rings and four 5-rings, are connected into chains after pure translations (a) along X. Chains, related by a shift vector of ½a (or by a mirror plane perpendicular to Y), are connected along Y to form the PBU of the ITE/RTH family of zeolite structures. A top view of the PBU (a) and a parallel view down Y (b) is shown.

2. Type of faulting: 1-dimensional stacking disorder of the PBUs along Z.

3. The point group symmetry of the PBU is c1m1.

4. Connectivity pattern of the PBU.

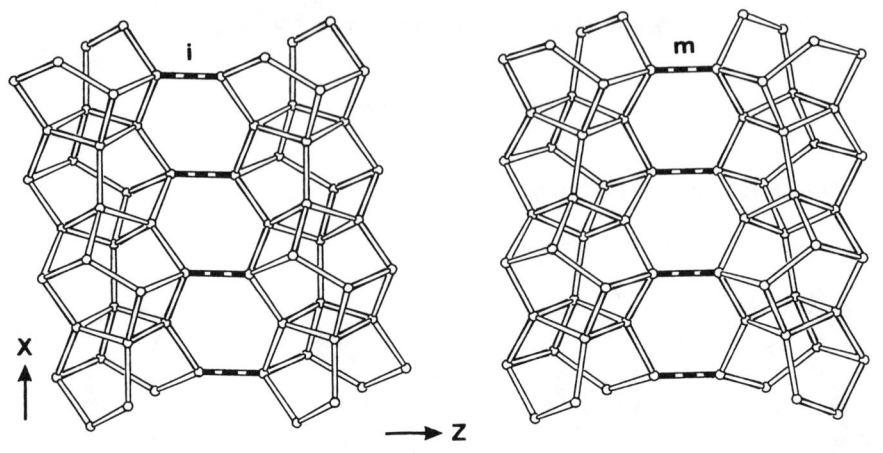

Neighboring PBU's are connected via O-bridges along Z through

i) inversion (i); successive layers are connected after a pure translation along Z. The resulting connectivity exhibits inversion symmetry.

ii) reflection (m); successive layers are connected after a 180° rotation about X (or Z). The connectivity now shows mirror symmetry (symbol: m) between successive layers.

Once the distribution of the symmetry elements i and m along Z is known the 3-dimensional structure is defined.

Example of an intermediate structure in the ITE/RTH family of zeolites:

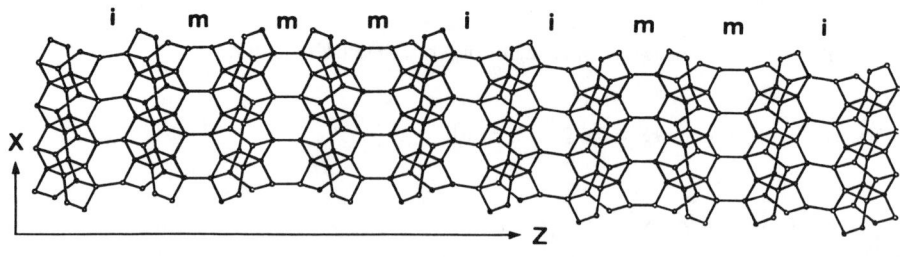

5. The simplest ordered end-members in the ITE/RTH family:

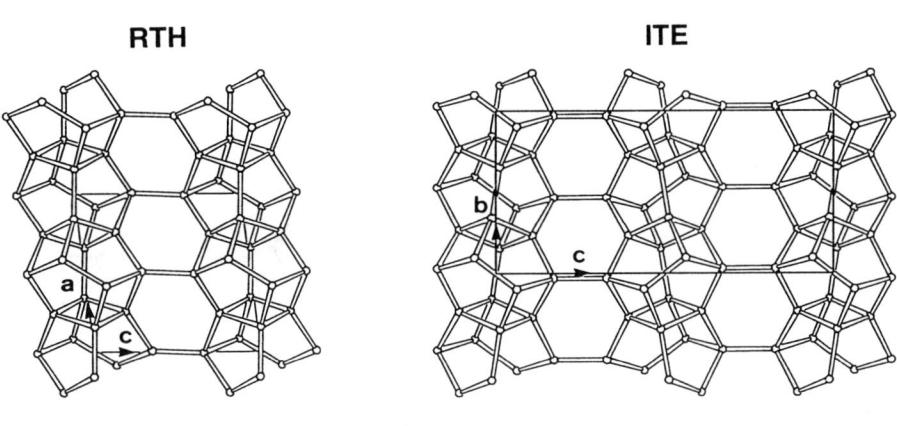

Pure RTH and ITE are obtained when neighboring layers along Z are exclusively related by inversion or reflection, respectively.

6. Faulted materials synthesized and/or characterized so far:
none

7. References

Building units of faulted zeolite framework structures and
their simplest ordered end-member structures

The faujasite family

1. The Periodic Building Unit (PBU) equals the faujasite layer:

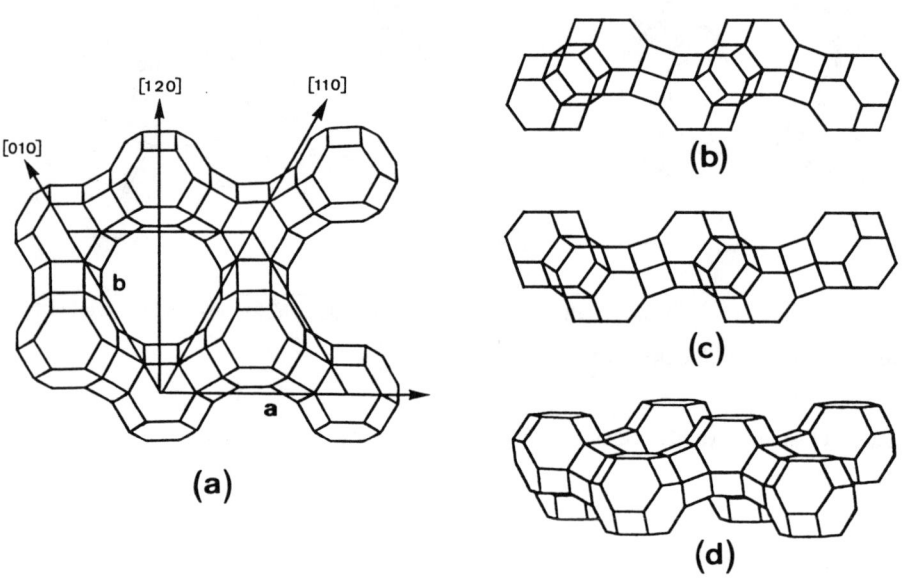

The PBU of the faujasite family of structures, the hexagonal faujasite layer, is composed of sodalite cages which are linked through double 6-rings. The faujasite layer corresponds to the (111) layer in cubic FAU and to the (001) layer in hexagonal EMT. Hexagonal axes are given. Views along [001] (a), [110] (b), [010] (c) and ~[120] (d) are shown. The layers, depicted in (b) and (c) are identical and related by a 60° rotation about (the hexagonal) [001] or by a mirror operation perpendicular to [001].

2. Type of faulting: 1-dimensional stacking disorder of the PBUs along [001].

3. The point group symmetry of the PBU is $\bar{3}m(1)$.

4. Connectivity pattern of the PBU.

Neighboring PBUs can be connected via O-bridges along [001] in two different ways:
a) the top layer is shifted over ⅓(-**a** + **b**) before connecting it to the bottom layer. The resulting connectivity exhibits inversion (i) symmetry (symbol: o) between successive layers.
b) the top layer is rotated over 60° about [001] (followed by the shift vector ⅓(-**a** + **b**)) before connecting it to the bottom layer. The connectivity now shows mirror (m) symmetry (symbol: |) between successive layers (See also 1(b,c)).

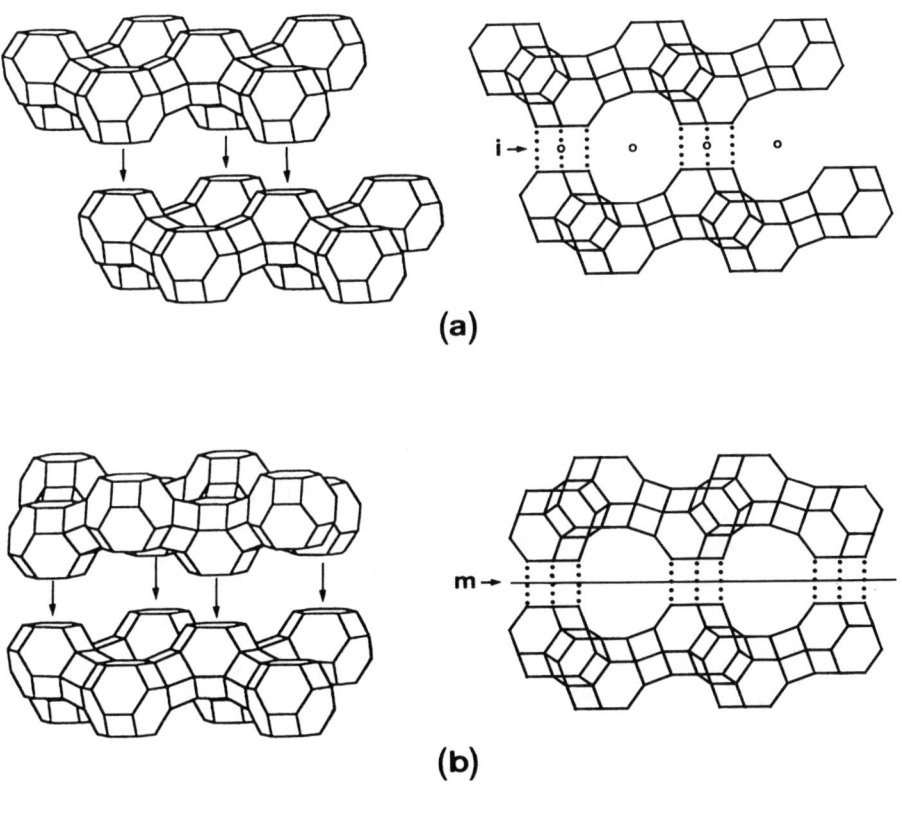

(a)

(b)

Once the distribution of the symmetry elements i and m along [001] is known the 3-dimensional structure is defined.

Example of an intermediate structure in the faujasite family of zeolites:

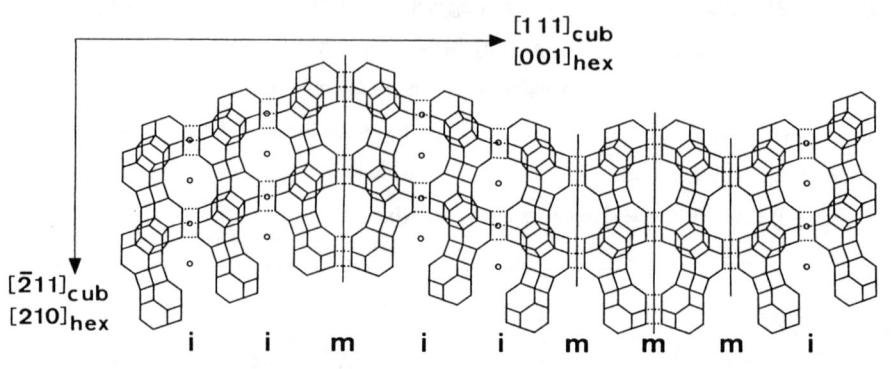

5. The simplest ordered end-members in the faujasite family:

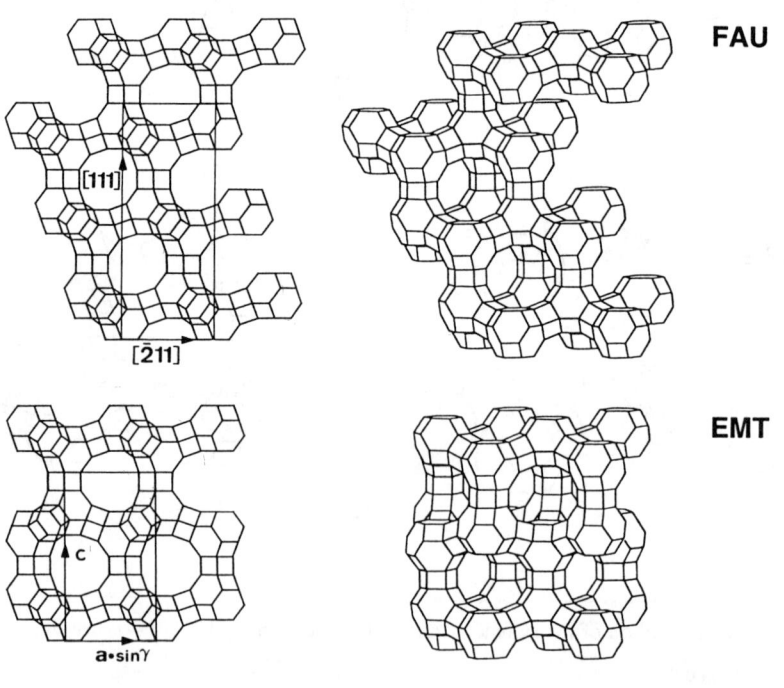

Pure FAU and EMT are obtained when neighboring faujasite layers along [001] are exclusively related by inversion or reflection, respectively.

6. Faulted materials synthesized and/or characterized so far:
CSZ-1(1,2)   CSZ-3(1,3)   ZSM-3(1,4)   ZSM-20(1,5)   ECR-30(1,6)

7. References
(1) M. M. J. Treacy, D. E. W. Vaughan, K. G. Strohmaier and J. M. Newsam, Proc. R. Soc. Lond. A **452**, 813 (1996).
(2) D. E. W. Vaughan and M. G. Barret, US Patent 4,309,313(1982).
(3) D. E. W. Vaughan and M. G. Barret, US Patent 4,333,859(1982).
(4) G. T. Kokotailo and J. Ciric, Adv. in Chem. Series, **101**, 109 (1971).
(5) J. Ciric, US Patent 3,972,983(1976).
   E. W. Valyocsik, Eur. Pat. Appl. 12572(1983).
   J. M. Newsam, M. M. J. Treacy, D. E. W. Vaughan, K. G. Strohmaier and W. J. Mortier, J. Chem. Soc., Chem. Commun. **1989**, 493 (1989).
(6) D. E. W. Vaughan, E Patent 0,351,461(1989).

Building units of faulted zeolite framework structures and
their simplest ordered end-member structures

The beta family

1. The Periodic Building Unit (PBU) equals the layer:

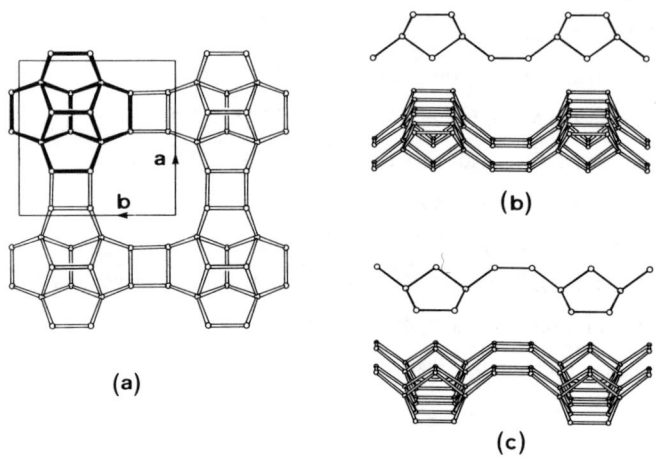

The PBU of the beta family of structure types, the tetragonal beta layer (a), is composed of T16 units (in bold) related by pure translations along **a** and **b**. Views along [001] (a), [100] (b) and [010] (c) are shown. The layers, depicted in (b) and (c) are identical and related by a 90° rotation about the plane normal or by a mirror operation perpendicular to the plane normal.

2. Type of faulting: 1-dimensional stacking disorder of the PBUs along [001].

3. The point group symmetry of the PBU is ($\bar{4}$)m2.

4. Connectivity pattern of the PBU.
Neighboring PBUs, related by a mirror operation, can be connected along [001] via O-bridges in three different ways:

a) the lateral shift of the top layer along **a** or **b** is zero, (this connection mode has not been observed yet)
b) the lateral shift of the top layer is ⅓a or ⅓b,
c) or the lateral shift of the top layer is -⅓a or -⅓b,

denoted as a):(0,0); b):(+⅓,0) or (0,+⅓); c):(-⅓,0) or (0,-⅓).

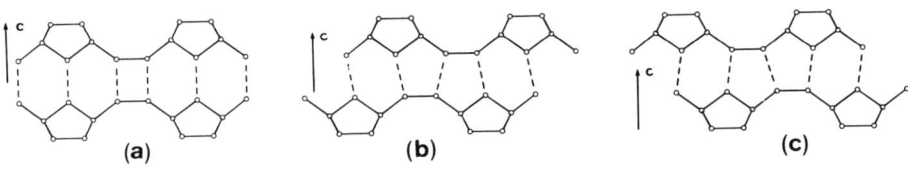

Once the distribution of the lateral shifts between the layers stacked along [001] is known, the 3-dimensional structure is defined.

5. The simplest ordered end-members in the beta family are given below.
None of them has been observed yet as pure single crystal material.

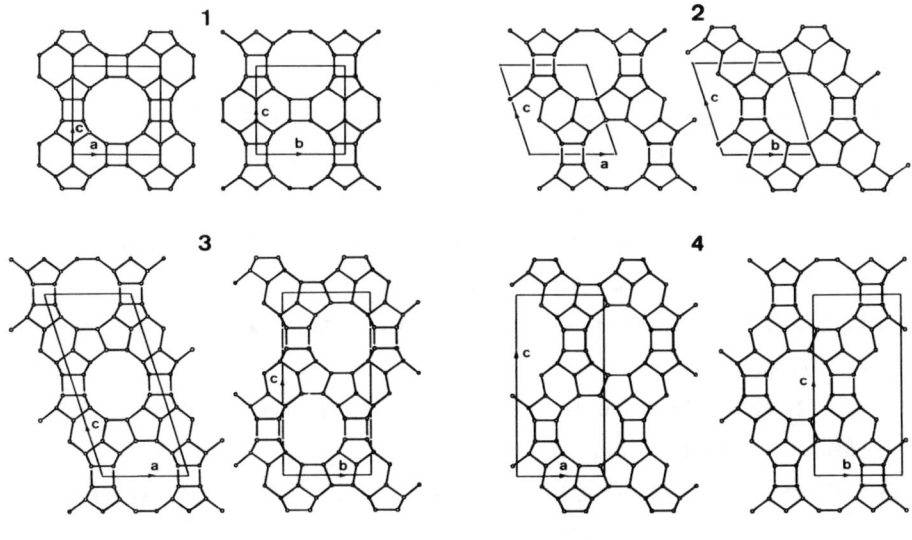

end-member	lateral shifts between subsequent PBUs along [001]; shifts are in fractions of (a, and b).					space group
1	(0,0);	(0,0);	(0,0);........			P4$_2$/mmc
2	(0,-⅓);	(-⅓,0);	(0,-⅓);.........			P$\bar{1}$@*
3	(0,-⅓);	(-⅓,0);	(0,+⅓);	(-⅓,0);	(0,-⅓);....	P2/c@
4	(-⅓,0);	(0,-⅓);	(+⅓,0);	(0,+⅓);	(-⅓,0);....	P4$_1$22**
5	(+⅓,0);	(0,-⅓);	(-⅓,0);	(0,+⅓);	(+⅓,0);....	P4$_3$22$

@ Space group is centrosymmetric and the same structure is obtained by reversing the signs of all lateral shifts.

* For comparison reasons the maximum topological symmetry of end-member number 2 has been transformed from C2/c to P$\bar{1}$.

** This is the end-member with structure type code *BEA; 4 and 5 are enantiomorphs.

$ In P4$_3$22 the coordinates in P4$_1$22 are transformed to xy$\bar{z}$.

6. Faulted materials synthesized and/or characterized so far:

Beta(1,2,3)   Borosilicate   *BEA(4,5)   Gallosilicate   *BEA(5)
Tschernichite(6)

7. References

(1) R. L. Wadlinger, G. T. Kerr and E. J. Rosinski, US Patent 3,308,069(1967).

(2) J. M. Newsam, M. M. J. Treacy, W.T Koetsier and C. B. de Gruyter, Proc. R. Soc. Lond. A **420**, 375 (1988).

(3) J. B. Higgins, R. B. LaPierre, J. L. Schlenker, A. C. Rohrman, J. D. Wood, G. T. Kerr and W. J. Rohrbaugh, Zeolites, **8**, 446 (1988).

(4) M. Marler R. Boehme and H. Gies, Proc. 9th IZC, Montreal, Butterworth-Heinemann (1993) p.425.

(5) K. S. N. Reddy, M. J. Eapen, P. N. Joshi, S. P. Mirajkar and V. P. Shiralkar, J. Incl. Phenom. Mol. Recogn. Chem. **20**, 197 (1994).

(6) R. C. Boggs, D. G. Howard, J. V. Smith and G. L. Klein, Am. Mineral. **78**, 822 (1993).

Building units of faulted zeolite framework structures and
their simplest ordered end-member structures

The SSZ-33 family

1. The Periodic Building Unit (PBU) equals the layer depicted in (b):

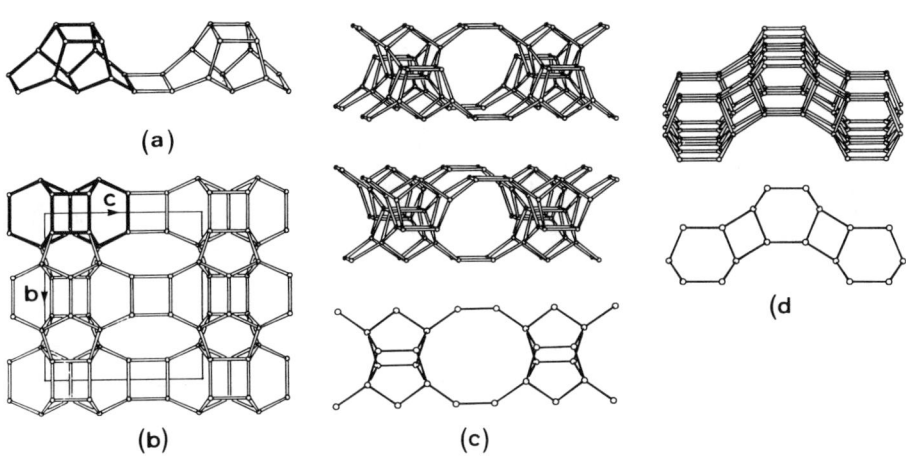

The PBU of the SSZ-33 family of structures is composed of parrallel chains (a) formed by connecting T14 units (in bold) related by pure translations along **c**. In the PBU two neighboring chains are related by a 180° rotation about **b** (or **c**). Chains are connected along **b** to form a puckered basic layer (d). Views along the plane normal (b), [010] (c) and [001] (d) are shown. The PBUs, depicted in (c) are identical and related by a 180° rotation about [010] or by a mirror operation perpendicular to the plane normal.

2. Type of faulting: 1-dimensional stacking disorder of the PBUs along [100].

3. The point group symmetry of the PBU is (2)mm.

4. Connectivity pattern of the PBU.

Neighboring PBUs, related by a mirror operation, can be connected along [100] via O-bridges in three different ways:
a) the lateral shift of the top layer along **c** is zero, (this mode has not been observed yet)
b) the lateral shift of the top layer is $+\frac{1}{3}$**c**,
c) or the lateral shift of the top layer is $-\frac{1}{3}$**c**,
denoted as a):(0); b):($+\frac{1}{3}$); c):($-\frac{1}{3}$).

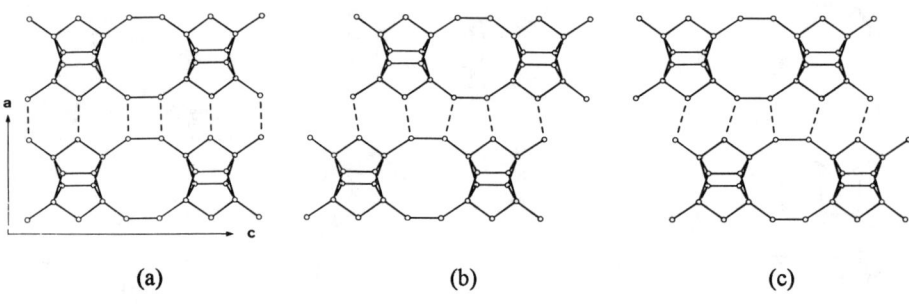

Once the distribution of the lateral shifts between the layers stacked along [100] is known, the 3-dimensional structure is defined.

5. The simplest ordered end-members in the SSZ-33 family are given below.
Only end-member 2 has been observed as pure single crystal material and represents the structure with type code CON. There is no difference in the projection of the structure of the end-members along [001].

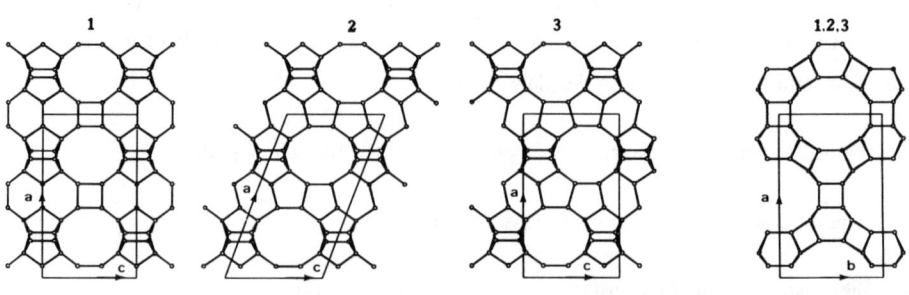

end-member	lateral shifts between subsequent PBUs along [001]; shifts are in fractions of (c).			space group
1	(0);	(0);	(0);........	Pmmm
2	(+⅓);	(+⅓);	(+⅓);.........	C2/m#*
3	(+⅓);	(-⅓);	(+⅓);..........	Pncm#**

\# Space group is centrosymmetric; the same structure is obtained by reversing the signs of all lateral shifts.

* This is the end-member with structure type code CON.

** For comparison reasons the standard space group Pmna (with a, b and c) has been transformed to the non-standard space group Pncm (with b, c, and a).

6. Faulted materials synthesized and/or characterized so far:
SSZ-26(1,2)   SSZ-33(2,3)

7. References

(1) S. I. Zones, D. S. Santilli, J. N. Holtermann, T. A. Pecoraro, R. A. Innes, US Patent 4,910,006(1990).

(2) R. F. Lobo, M. Pan, I. Chan, R. C. Medrud, S. I. Zones, P. A. Crozier and M. E. Davis, J. Phys. Chem. **98**, 12040 (1994).

(3) S. I. Zones, US Patent 4,963,337(1990).

Building units of faulted zeolite framework structures and
their simplest ordered end-member structures

The lovdarite family

1. The Periodic Building Unit (PBU) equals the layer depicted in (a):

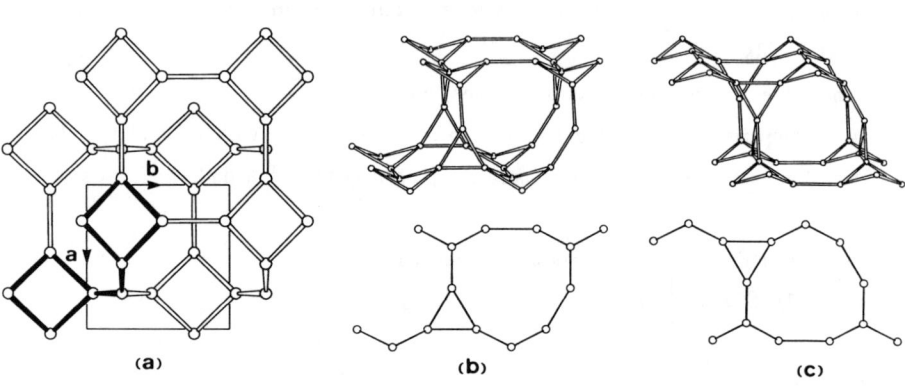

The PBU of the lovdarite family of structures is composed of T9 units (in bold: two 4-rings connected through a single T atom) related by pure translations along **a** and **b**. Views down [001] (a), down [100] (b) and along [010] (c) are shown. The PBUs, depicted in (b) and (c), are identical and related by a 90° rotation about the plane normal or by a mirror operation perpendicular to the plane normal.

2. Type of faulting: 1-dimensional stacking disorder of the PBUs along [001].

3. The point group symmetry of the PBU is mm(2).

4. Connectivity pattern of the PBU.
Neighboring PBU's, related by a mirror operation, can be connected along [001] via O-bridges in two different ways:
a) the lateral shift of the top layer along **a** or **b** is zero,
b) the lateral shift of the top layer is (plus or minus) ½**a** or ½**b**,

leading to the following connectivity codes: a): (0,0); b): (½,0) or (0,½).

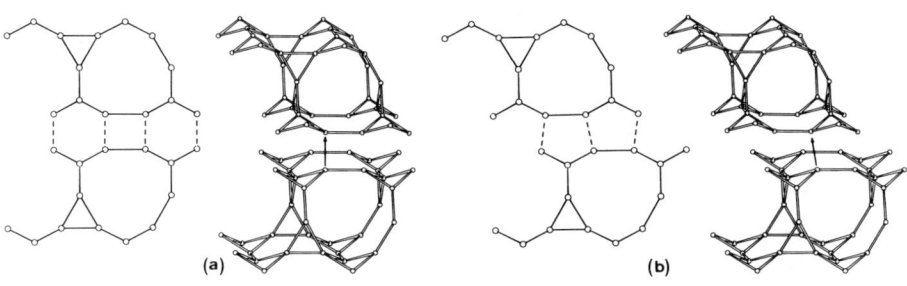

Once the distribution of the lateral shifts between the layers stacked along [001] is known, the 3-dimensional structure is defined.

5. The simplest ordered end-members in the lovdarite family are given below.
All three members have been observed as pure single crystal material.

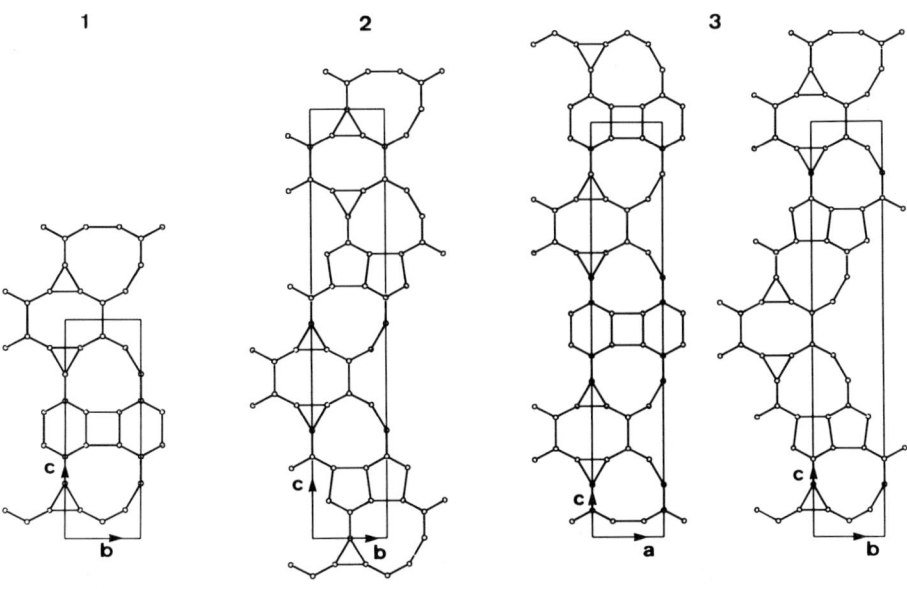

end-member	lateral shifts between subsequent mirrored basic layers along [001]; shifts are in fractions of (**a**, and **b**).					space group
1*	(0,0);	(0,0);	(0,0);.....			$P4_2/mmc$
2**	(0,½);	(½,0);	(0,½);	(½,0);	(0,½);	$I4_1/amd$
3***	(0,0);	(0,½);	(0,0);	(0,½);	(0,0);.....	$A2/m^{\#}$

*   This is the end-member with structure type code LOV.
**   This is the end-member with structure type code VSV.
***   This is the end-member with structure type code RSN.
\#   For comparison reasons the maximum topological symmetry of end-member number 3 has been transformed from C2/m to A2/m.

6. Faulted materials synthesized and/or characterized so far:
none

7. References

Building units of faulted zeolite framework structures and their simplest ordered end-member structures

The montesommaite family

1. The Periodic Building Unit (PBU) equals the 4-ring layer depicted in (a):

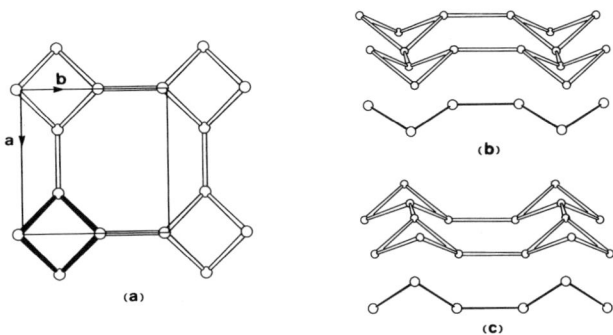

The PBU of the montesommaite family of structures is composed of T4 rings (in bold) related by pure translations along **a** and **b**. Views along [001] (a), [100] (b) and [010] (c) are shown. The PBUs, depicted in (b) and (c) are identical and related by a 90° rotation about the plane normal or by a mirror operation perpendicular to the plane normal.

2. Type of faulting: 1-dimensional stacking disorder of the PBUs along [001].

3. The point group symmetry of the PBU is $(\bar{4})m2$.

4. Connectivity pattern of the PBU.

Neighboring PBUs, related by a mirror operation, can be connected along [001] via O-bridges in two different ways:

a) the lateral shift of the top layer is (plus or minus) ½**a** or ½**b**,
b) the lateral shift of the top layer along **a** or **b** is zero, (this connection mode has not been observed yet)

denoted as the following connectivity codes: a): (½,0) or (0,½); b): (0,0).

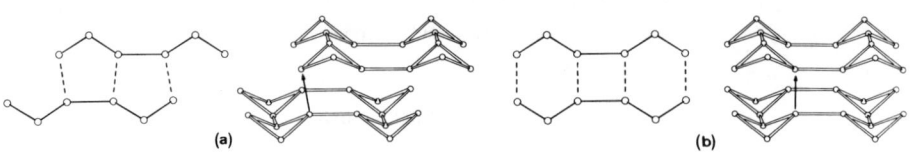

Once the distribution of the lateral shifts between the layers stacked along [001] is known, the 3-dimensional structure is defined.

5. The simplest ordered end-member in the montesommaite family, given below, has been observed as pure single crystal material.

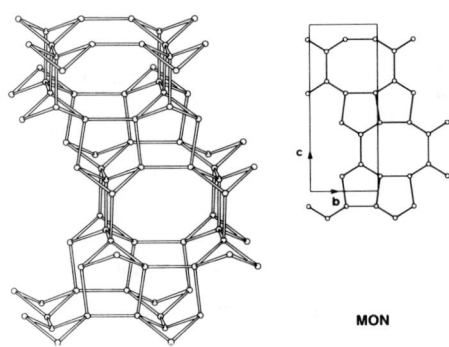

MON

end-member	lateral shifts between subsequent mirrored PBUs along [001]; shifts are in fractions of (**a**, and **b**).					space group
MO1*	(0,0);	(0,0);	(0,0);.....			$P4_2/mmc$
MON	(0,½);	(½,0);	(0,½);	(½,0);	(0,½);	$I4_1/amd$

* Hypotetical structure (not shown in drawing).

6. Faulted materials synthesized and/or characterized so far:
none

7. References

Building units of faulted zeolite framework structures and
their simplest ordered end-member structures

The ABC-6 family

1. The Periodic Building Unit (PBU) equals the 6-ring layer shown in Figure 1:

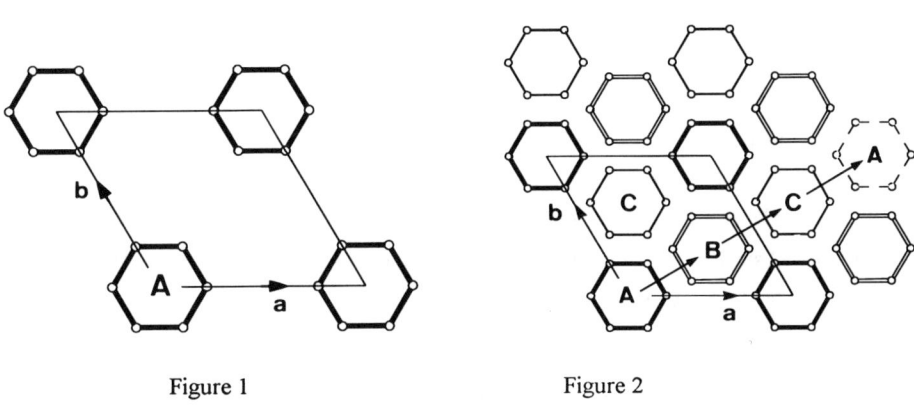

Figure 1   Figure 2

The PBU of the ABC-6 family of structure types consists of an hexagonal array of non-connected planar T6-rings (in bold), which are related by pure translations along **a** and **b**. The 6-rings are centered at (0,0) in the ab layer. This position is usually called the A position (Figure 1).

2. Type of faulting: 1-dimensional stacking disorder of the PBUs along [001].

3. The point group symmetry of the PBU is 6mm.

4. Connectivity pattern of the PBU.
Neighboring PBUs can be connected through tilted 4-rings along [001] in three different ways:
a) the next layer (second layer) is shifted by $+(\frac{2}{3}\mathbf{a} + \frac{1}{3}\mathbf{b})$ before connecting it to the first layer. The 6-rings in the second layer are centered at $(\frac{2}{3}, \frac{1}{3})$. This position is usually denoted as

the B position as illustrated in Fig. 2. The same connection mode can be repeated: a third PBU is shifted with respect to the second layer by +(⅔a + ⅓b). The 6-rings are now centered at ($^4/_3$, ⅔) [or at (⅓, ⅔)]. This position is called the C position (See Fig. 2). Adding a fourth layer with the same connection mode gives a shift with respect to the first layer of to (2a + b) [or zero] and an A position of the 6-rings is again obtained. The resulting stacking sequences, exhibiting the same connection mode, are denoted as AB, BC and CA, respectively.

The connection mode is viewed down [001] (top), nearly along [010] (middle), and along [010] (bottom).

b) the added layers are shifted by -(⅔a + ⅓b) before connecting them to the previous layer. The resulting stacking sequences AC, CB and BA are obtained.

The ensuing connection modes are equal and viewed as in a).

c) the added layer has a zero lateral shift along **a** and **b**. This connection mode leads to an AA, BB or CC stacking sequence depending on whether the added layer is connected to a layer with 6-rings in the A, B or C position, respectively.

The connection mode is viewed as in a).

(a)   (b)   (c)

Once the stacking sequence along [001] is known the 3-dimensional structure is defined.

Examples of faulted structures in the ABC-6 family of zeolites:

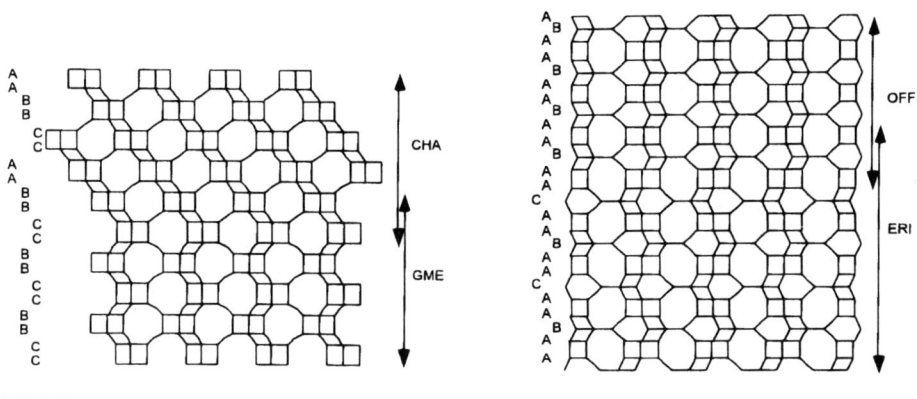

5. The simplest ordered end-members in the ABC-6 family:

Name	Code	Repeat Layers	Sequence
Cancrinite	CAN	2	AB(A)............
Sodalite	SOD	3	ABC(A)...........
Losod	LOS	4	ABAC(A)..........
Liottite	LIO	6	ABABAC(A)........
Afganite	AFG	8	ABABACAC(A)......
Offretite	OFF	3	AAB(A)...........
Erionite	ERI	6	AABAAC(A)........
TMA-E(AB)	EAB	6	ABBACC(A)........
Levyne	LEV	9	AABCCABBC(A).....
STA-2	SAT	12	ABAACACCBCBB(A)...
Gmelinite	GME	4	AABB(A)..........
Chabazite	CHA	6	AABBCC(A)........
SAPO-56	AFX	8	AABBCCBB(A)......
AlPO-52	AFT	12	AABBCCAACCBB(A)...

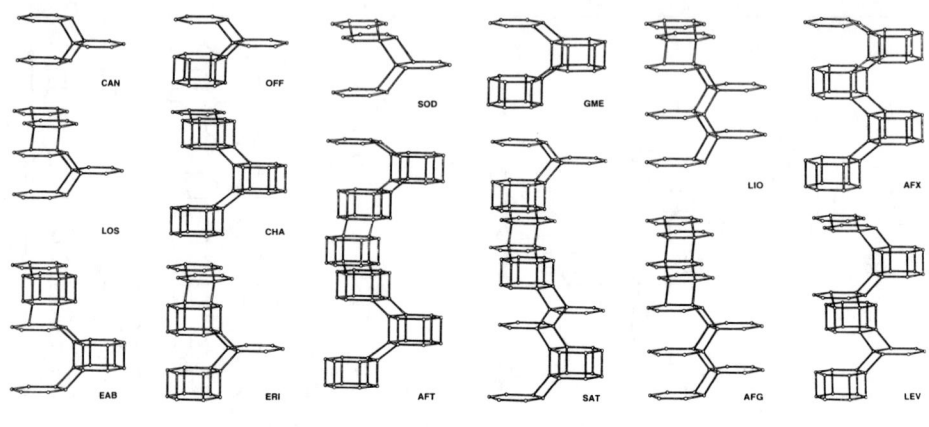

6. Faulted materials synthesized and/or characterized so far:

Linde T;   Babelite;   Linde D;   Phi;   ZK-14;   LZ-276;   LZ-277;

7. References

to be added

Building units of faulted zeolite framework structures and
their simplest ordered end-member structures

The decasil family

1. The Periodic Building Unit (PBU) equals the chain depicted in (a):

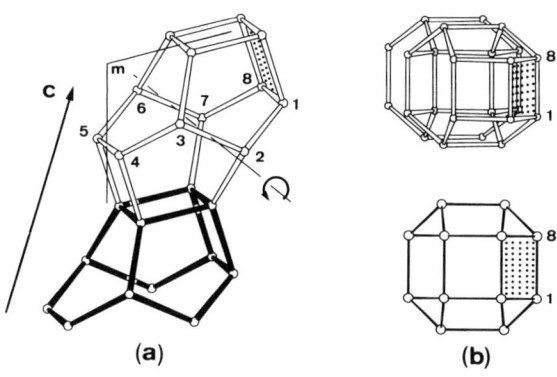

The PBU of the decasil family of structures, is formed by connecting T12-units (in bold) related by pure translations along **c**. As orientation sensitive indicator one of the T4-rings is shaded. The numbered T atoms are used in describing the connection modes. A perspective and parallel view down the chain axis is shown in (b).

2. Type of faulting: 2-dimensional stacking disorder of the PBUs along [100] and [010].

3. The rod symmetry (2/m) is indicated in the Figure.

4. Connectivity pattern of the PBU.
Neighboring PBUs can be connected via O-bridges in several ways:
a) the chains are connected after pure translations. The resulting connection modes are illustrated in 1, 4, 5 and 7. Connection modes 5 and 7 are related by a mirror operation parallel to the plane of the connected chains.

b) the chains are connected after translation accompanied by a 180° rotation about the chain axis. The resulting connection modes are shown in 2 and 3.

c) the chains are connected after translation followed by a +90° or -90° rotation about the chain axis. The resulting connection modes are given in 6 and 8.

The connection modes 2 and 3, 5 and 7, and 6 and 8, are pairwise identical. The modes in each pair are related by a 180° rotation about an axis parallel to the connecting TOT bridges.

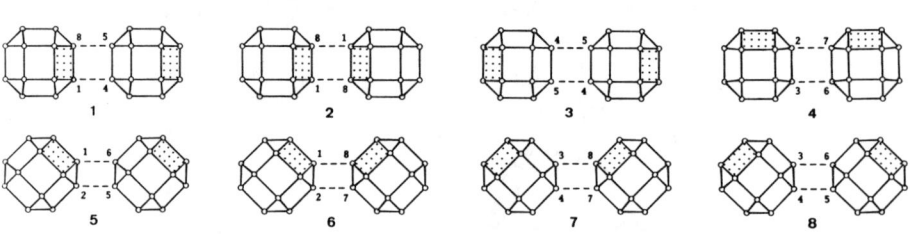

Once the distribution of the connection modes in two dimensions is known, the 3-dimensional framework is defined.

5. The simplest ordered end-members in the decasil family are given below. Only end-member 1 has been observed as single crystal material and represents the structure with type code RTE.

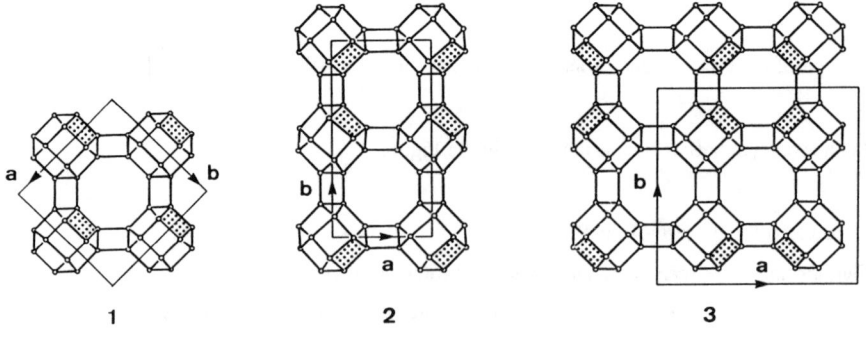

end-member	sequence of the connection modes along **a** and **b**: (along **a**,........;along **b**,......)	space group
1	(5,5,......;7,7,......)	C2/m*
2	(7,7,......;8,6,8,....)	P2/m
3	(8,6,8,....;8,6,8,....)	P4/nmm

* This is the end-member with structure type code RTE; the sequence of the conection modes is along (-**a** + **b**) and (**a** + **b**), respectively.

6. Faulted materials synthesized and/or characterized so far:
RUB-4(1,2)

7. References

(1) A. Grünewald-Lücke and H. Gies, Microporous Mater. **3**, 159 (1994).

(2) B. Marler, A. Grünewald-Lücke and H. Gies, Zeolites, **15**, 388 (1995).

AUTHOR INDEX

I, II, III and IV refer to the volume number

Abbenhuis, H.C.L., I-645
Abraham, M., III-2111
Ackermann, J., IV-2727
Adrián, R., II-801
Agger, J.R., III-1487, IV-2457
Aguado, J., II-1441
Aguiar, S., II-1463
Aiello, R., III-1619
Akolekar, D.B., IV-2751
Alberti, A., IV-2345, IV-2361
Al-Shamali, M., II-1129
Amarilli, S., I-575
Amoureux, J-P., IV-2985
Anderson, M.W., I-697, II-733, III-1487, IV-2457, IV-2465
Ando, K., I-445
André, G., I-659
Andrews, S.J., IV-2355
Andy, P., II-1433
Anpo, M., I-473, I-667
Antonic, T., III-2049, III-2057
Antunes, A.P., III-1771
Aoyagi, K., IV-2863
Ariyuki, M., I-667
Asaftei, I., IV-2759
Ashtekar, S., II-1003
Astrelin, I.M., III-2129
Auerbach, S.M., I-357, I-509
Ayala, A., III-1779
Azimova, Yu.I., IV-2651

Babitz, S.M., I-465
Babouchkina, E., III-1857
Baerlocher, Ch., I-533, I-537, IV-2519
Baggaley, A.K., IV-2457
Bagshaw, S.A., II-741, II-1337
Baker, M.D., III-2073, III-2137
Bakhranov, K.N., I-481
Balkus, Jr., K.J., II-1403, III-1779, III-1931
Bao, X., II-1283
Bär, N-K., I-77, I-149
Barger, P.T., I-567
Barrault, J., I-651
Barrett, P.A., III-1495
Barrie, P.J., II-1003
Basaldella, E.I., III-1663
Batamack, P., IV-2279, IV-2595
Bee, M., I-59
Beelen, T.P.M., III-1529
Bein, T., III-1787, III-2209
Béland, F., III-1877
Bell, A.T., I-393
Bell, R.G., II-839
Bellocq, N., I-675
Bellussi, G., I-541, I-575, II-1169
Benazzi, E., II-1011, II-1433, IV-2735

Bengueddach, A., III-1803
Berreghis, A., IV-2719
Beyer, H.K., IV-2875
Bhalla, A.S., III-2179
Bhargava, S.K., IV-2751
Bhat, S.G.T., I-127
Bhatia, S., IV-2921
Bhaumik, A., II-1227, III-1611
Biaglow, A.I., IV-2535
Bilba, N., IV-2759
Bischof, C., II-809
Blasco, T., IV-2473
Bockstette, M., III-2233
Boddenberg, B., I-481, IV-2589
Bogdanchikova, N.E., II-1165, III-2143
Böhlmann, W., II-795
Böhringer, W., I-591
Boix, E.T., III-1495
Bonardet, J-L., IV-2991
Bonneviot, L., III-1877
Bonnin, D., III-2079
Bordiga, S., I-317, IV-2571
Borgmann, C., III-2241
Borovkov, V.Yu., I-269, III-2079
Böttcher, R., II-795
Bouchard, C., II-1399
Boudreau, L.C., I-621
Bousmina, M., I-263
Braun, I., III-2233
Brenn, U., II-847
Brenner, S., I-533
Breuninger, M., IV-2697
Briend, M., IV-2681
Broach, R.W., III-1715, IV-2445
Brodkorb, A., II-1213
Bronic, J., III-2057
Brouwer, D.H., III-2137
Brueva, T.R., IV-2629, IV-2637
Brunel, D., I-675
Brunner, E., IV-2943
Bug, A.L.R., I-119
Bülow, M., I-111
Burger, B., IV-2503
Burton, A., IV-2979

Cahill, C.L., IV-2401
Cai, Q., II-923
Cairon, O., II-1025
Callanan, L.H., II-1065
Camblor, M.A., III-1495, IV-2951
Campbell, B.J., I-541
Cao, J., II-863
Cappellazzo, O., I-575
Caputo, D., III-1663
Cardoso, D., II-1019, II-1463, III-1885, IV-2481
Carluccio, L., I-541
Caro, J., III-2217
Carr, S.W., II-733, IV-2401
Carrazza, J., II-801

Casci, J.L., IV-2355
Castagnola, M.J., III-1627
Castillo, E., III-1917
Catlow, C.R.A., I-457, III-1671, III-1723
Caullet, P., III-1731
Cavers, M., IV-2615
Cejka, J., II-1419
Centi, G., II-1359
Ceulemans, A., I-387
Chakraborty, A.K., I-393
Chambellan, A., II-1025
Chang, J-S., I-437, I-667, II-1253, III-2187
Chao, K.J., II-857
Chatterjee, A., I-489
Che, M., III-2171
Cheetham, A.K., I-541, III-1487, IV-2453, IV-2767
Chen, C.Y., III-1945
Chen, J., III-1937, III-2147
Chen, L-W., II-885
Chen, T-H., III-1655, IV-2497
Chernenko, J.V., II-767
Chevreau, T., II-1025
Chia, L.S., II-787
Chiari, G., I-135
Chihara, K., I-203
Chmelka, B.F., I-43
Chon, H., IV-2259
Chou, C-Y., II-885
Chun, Y., II-989
Chung, S.Y., II-1149
Ciambelli, P., IV-2345, IV-2361
Ciraolo, M.F., IV-2295, IV-2301
Clacens, J.M., I-651
Clark, L.A., I-251
Clark, S.M., III-1671
Colella, C., III-1663
Coluccia, S., III-1569, IV-2775
Concepcion, P., IV-2473
Conner, W.C., I-97, I-409
Cook, M., I-409
Corbin, D.R., III-2095, IV-2295
Corker, J.M., III-2143
Corma, A., II-817, III-1495
Cortés-Jácome, M.A., II-1105
Costa, C.S., III-1771
Cottin, X., I-91
Coulomb, J.P., I-51, I-189, I-197, I-659
Cox, D.E., I-541
Cox, P.A., III-1603, IV-2355
Crea, F., III-1619
Cruciani, G., IV-2361
Cundy, C.S., III-1591
Curtiss, L.A., I-333, I-415

Dai, L-X., II-1455
Dann, S.E., IV-2379
da Silva, F.A., I-243
Datka, J., IV-2595, IV-2601
Davidova, N.P., II-981
Davidson, A., III-2171
Davidson, J.M., IV-2615

Davies, A.T., III-1671
Davis, M.E., III-1479, III-2065
Decyk, P., II-833
Dedecek, J., I-401, II-941, II-1193
de Haan, J.W., I-91
Delabie, A., I-387
Delevoye, L., IV-2985
de Medeiros, G.G., II-1019
de Ménorval, L.C., I-85
de Moor, P-P.E.A., III-1529
Demortier, G., IV-2395
den Exter, M.J., IV-2419
Denoyel, R., II-779
Derewinski, M., I-209, IV-2781
Derouault, A., I-651
Deson, J., IV-2425
De Vos, D.E., II-1213, III-1825
Díaz-Cabañas, M.J., III-1495, IV-2951
Di Renzo, F., I-85, I-135, I-675, III-1803, IV-2705
Djaouadi, D., II-801
Domen, K., IV-2577
Domeniconi, T., IV-2991
Dong, W-Y., II-909
Dry, M.E., IV-2557
Du, H., II-863, II-869, II-1277, IV-2665
Dubkov, K.A., II-1245
Dubsky, J., I-215
Dutartre, R., I-85, III-1803
Dutta, P.K., III-1627, IV-2927
Dwyer, J., I-183, II-1081, III-1591
Dyhr, K., II-1185
Dzwigaj, S., III-2171

Earl, W.L., III-1953, IV-2265, IV-2489
Ebina, T., I-489
Echchahed, B., III-1877
Eckert, J., I-119
Eic, M., I-215
Eldewik, A., III-1507
Elings, J.A., III-1975
El Malki, E.M., III-2171
Elomari, S.A., III-1945
Emig, G., I-235, III-1969, IV-2727, IV-2847
Engler, C., I-223
Enhbold, T., III-2129
Erdem-Senatalar, A., I-257, I-583
Ernst, H., IV-2955
Ernst, S., II-749, II-1409, III-2155
Escola, J.M., II-1441
Esteves, P.M., I-429
Estournes, C., IV-2971

Fajula, F., I-85, I-135, I-675, III-1803, IV-2273
Falabella, E., II-1463
Fan, B., II-901
Fan, W., II-863, II-869, II-901
Favre, D.E., I-43
Feast, S., IV-2889
Férey, G., III-1649, III-1737, IV-2409, IV-2453, IV-2971
Fernandez, C., IV-2985

Fernandez, L., IV-2473
Feuerstein, M., IV-2979
Filippov, A.P., II-767
Fischer, R.X., III-2209
Floquet, N., I-197, I-659
Flytzani-Stephanopoulos, M., IV-2787
Fonseca, A., II-917
Frache, A., III-1569
Fraissard, J.P., II-1049, II-1239, III-2079,
    IV-2279, IV-2425, IV-2595, IV-2991
Freude, D., III-1849, IV-2955
Frick, B., I-59
Fritzsche, S., I-149, I-371
Fu, G., III-1561
Fu, Q., IV-2585
Fudala, A., II-917, III-1685
Fuentes, S., II-1165, III-2143
Fukuya, S., I-301, III-2041
Fyfe, C.A., III-1561

Gabelica, Z., I-651, II-1207, III-1633,
    III-1771, III-1871, III-1925, IV-2395,
    IV-2711
Gaffney, T.R., I-505
Gaillard, L., II-1121
Galarneau, A., I-675
Galli, A., II-1359
Ganapathy, S., IV-2985
Gao, F., III-1863
Garces, J.M., I-457, I-551, II-1331, III-1535
Garforth, A.A., II-1081
Garrone, E., IV-2705, IV-2711
Gates, B.C., IV-2265
Gaub, M., I-371
Gedeon, A., IV-2425
Geidel, E., I-223, I-423, IV-2609
Geobaldo, F., I-317
Gérardin, C., III-1737, IV-2971
Geus, J.W., I-637
Gianotti, E., IV-2775
Gies, H., IV-2999
Gil, B., IV-2595
Gimon-Kinsel, M.E., III-1779
Giotto, M.V., IV-2481
Girotti, G., I-575
Gladden, L.F., II-1003
Gläser, R., II-1447
Gnep, N.S., II-1093, II-1433
Goh, N.K., II-787
Gola, A., II-1011, IV-2735
Gonzales, N.O., I-393
González, F., II-801
Gore, B.E., II-733
Gorte, R.J., I-103, IV-2287
Goryashenko, S.S., II-1157
Gougeon, R.D.M., II-779
Gouzinis, A., I-621
Graafsma, H., IV-2419
Graff, J.M.v.d., I-603
Grey, C.P., IV-2295, IV-2301, IV-2317
Grillet, Y., I-51, I-189, I-197, I-659, I-1707
Grobet, P.J., III-1825, IV-2325, IV-2497

Grubert, G., II-825, III-2249
Grün, M., II-757
Grünert, W., II-825, IV-2795
Guan, N.J., II-1313, IV-2585, IV-2803
Guczi, L., IV-2809
Guillemot, D., III-2079
Guiru, W., I-141, III-1795
Guisnet, M., II-1033, II-1093, II-1433,
    II-1463, IV-2719
Guo, K., II-1177
Guo, X., II-1283
Gupta, A., I-251
Gutjahr, M., II-795
Gutsze, A., IV-2589
Guzmán, M.L., II-1105

Haake, M., IV-2943
Haberlandt, R., I-149, I-371
Hahn, K., I-153
Halász, J., IV-2809
Hamidi, F., III-1803
Han, X., II-1283
Han, Y.W., IV-2839
Hanaoka, T., II-1165
Hannus, I., IV-2963
Hanson, J.C., IV-2295, IV-2401
Hanssen, R.W.J.M., I-645
Haouas, M., IV-2971
Hari Prasad Rao, P.R., III-1515, III-2009
Harkness, I.R., IV-2615
Harlick, P., I-309
Harris, R.K., II-779
Harris, T.V., III-1743, III-1945, III-2023
Hartmann, M., II-795, II-809, II-1409
Hay, P.J., I-379
Hayashi, A., III-1841
Hayashi, M., IV-2833
Hayashi, S., IV-2863
He, H-Y., III-1983
Healey, A.M., IV-2379
Hedlund, J., III-1809, III-1857
Hegde, S.G., III-1909
Heinichen, H.K., II-1085
Henriques, C., III-1771
Henriques, C.A., IV-2935
Henson, N.J., I-379
Hernadi, K., II-917
Hernández, F., II-1105
Hernández-Del Angel, M., II-1105
Heuchel, M., I-223
Higashimoto, S., I-667
Hiraga, K., III-1693, III-1937
Hirose, C., IV-2577
Hoefelmeyer, J., II-1403
Hoffmann, K., III-2121
Hofland, G.W., III-1553
Hogg, B.D., IV-2927
Hölderich, W.F., II-1085, II-1351, III-1893
Holsenburger, R.S., II-1081
Hongchen, G., I-141, III-1795
Howe, R.F., III-1507
Hu, R., IV-2413

Huang, L., I-707
Huang, Y., IV-2431
Hughes, C.D., III-1953
Hughes, E.M., III-2087
Huguenard, C., IV-2409
Hui, T.W., III-2073
Hunger, B., I-223
Hunger, M., IV-2503, IV-2697
Hurgobin, S., II-1073
Hursthouse, M.B., III-1937

Ichihashi, Y., I-667
Igi, H., IV-2643
Ihlein, G., III-2111
Ikemoto, Y., III-2103
Ikeya, H., I-293
Il'in, V.G., II-767, III-1679
Imbert, F.E., II-1093
Inui, T., II-1345, III-1871
Iofcea, Gh., IV-2759
Ione, K.G., III-1743, III-2023
Ito, S., IV-2863
Iton, L.E., I-333, I-415
Ivanov, A.V., IV-2541
Ivanov, D.P., II-1245
Ivanova, I.I., II-1207, IV-2273
Iwamoto, S., III-1871
Iwasaki, A., III-1817
Iwasaki, T., I-489

Jackson, C.L., I-697
Jacob, B., II-1317
Jacob, C.R., III-2195
Jacobs, M., IV-2395
Jacobs, P.A., II-1213, II-1233, II-1293, II-1299, II-1331, III-1825, III-2163, IV-2897, IV-2905
Jaeger, N.I., III-2249
Jahn, S.L., III-1885
Jama, M., I-215
Jänchen, J., IV-2623
Jankowska, A., I-277, IV-2847
Janowski, F., III-1849
Jansen, J.C., I-603, II-1073, III-1553, III-2117
Jappar, N., II-1391
Jaroniec, C.P., I-725
Jaroniec, M., I-681, I-725, II-761
Jasra, R.V., I-127
Jaumain, D., IV-2681, IV-2689
Jayat, F., II-1463
Jentys, A., IV-2817
Jialiang, W., II-1323
Jiang, H-W., III-1983, III-1991
Jiao, X.L., III-1641, IV-2387
Jirka, I., IV-2439
Jobic, H., I-59, I-77, IV-2609
Jones, C.W., III-1479
Jönsson, B.J., IV-2875
Jorda, E., II-1269
Jorda, J.L., II-817
Jost, S., I-149
Jousse, F., I-357, I-509

Joy, A., III-2095
Joyner, R.W., II-1367
Jung, K.T., III-1523
Jüstel, T., III-2241

Kahn, R., I-51
Kaliaguine, S., I-263, II-1221, II-1399
Kamalakar, G., II-897
Kamath, V., I-127
Kang, M., II-1345
Kannan, S., II-877
Kanougi, T., I-489
Kao, H-M., IV-2317
Kapoor, M.P., II-1221
Kapustin, G.I., IV-2629, IV-2637
Karetina, I.V., III-1961
Karge, H.G., I-269, II-847, IV-2697
Kärger, J., I-35, I-77, I-149, I-153, I-231
Karlsson, A., I-713
Kasture, M., IV-2781
Katada, N., IV-2549, IV-2643
Kato, M., III-2009
Kaucic, V., III-1585, III-1765
Kaucky, D., II-1193
Kazakov, A.V., II-1049
Kazansky, V.B., I-269
Keene, M.T.J., II-779
Kemmitt, T., II-741
Kentgens, A.P.M., IV-2515
Kervennal, J., II-1269
Kessler, H., II-1033, IV-2525
Kevan, L., IV-2825
Khanmamedova, A.K., II-1403
Kharitonov, A.S., II-1245
Khatyr, A., III-1757
Khelkovskaya-Sergeeva, E.G., IV-2541
Khvoshchev, S.S., III-1961
Kibby, C.L., III-1743, III-2023
Kikhtyanin, O.V., III-1743
Kikuchi, E., III-1515, III-2009, IV-2833
Kim, A.N., IV-2839
Kim, D.S., II-1157
Kim, J.M., I-689
Kim, J.S., I-437
Kim, J.W., II-1199
Kim, M.H., II-1149
Kim, W.Y., III-2187
Kim, Y., IV-2839
Kim, Y.G., II-1149
Kimbara, H., I-203
Kimura, A., I-719
Kimura, T., II-771
Kinger, G., IV-2817
Kirchner, R.M., III-1715, IV-2445, IV-2999
Kiricsi, I., II-917, III-1685, IV-2809, IV-2963
Kirsch-De Mesmaeker, A., II-1213
Kirschhock, C., III-1825
Kishida, M., II-1141
Kiyozumi, Y., III-1685, III-1833
Klap, G.J., III-2117
Klemm, E., I-235, III-1969, IV-2727, IV-2847
Klemt, R., IV-2665

Klepel, O., I-223
Klier, K., I-401
Knops-Gerrits, P.P.H.J.M., II-1299, III-2163
Knudsen, K.D., IV-2361
Ko, A-N., II-885
Kocirik, M., I-209, I-215
Koegler, J.H., IV-2419
Kogelbauer, A., II-1471
Kohtoku, Y., III-2041
Kokotailo, G.T., III-1561
Kolboe, S., II-1057
Koller, H., IV-2951
Kolobov, A.V., III-2217
Komarneni, S., III-2179
Kondo, J.N., IV-2577
König, T., III-1561
Koningsberger, D.C., I-637, IV-2515
Kónya, Z., IV-2809, IV-2963
Kornatowski, J., I-285, III-1577
Koros, W.J., I-25
Kosanovic, C., IV-2309
Koudriachova, M.V., I-423, I-497
Kovalakova, M., IV-2325
Kovalenko, A.S., II-767
Kowalak, S., I-277, IV-2847
Krauss, O., III-2111
Krijnen, S., I-645
Krishna, R., I-325
Kruk, M., I-681, I-725, II-761
Kruszona, K., I-277
Krysciak, J., I-209
Kubo, M., I-489
Kubota, Y., III-1693
Kucherov, A.V., II-1373, IV-2855
Kulkarni, S.J., II-893, II-897, II-1101
Kumar, R., II-1227, III-1611
Kumar, T., IV-2985
Kung, H.H., I-465
Kunieda, T., IV-2549
Kunimori, K., IV-2863
Kunkeler, P.J., III-1975, IV-2565
Kuno, M., I-349
Kuperman, A., I-457
Kuroda, K., I-719, II-771
Kurten, D.M., III-2087
Kustov, L.M., II-1049, IV-2541
Kynast, U., III-2241

Labouriau, A., III-1953, IV-2265, IV-2489
Lacombe, S., IV-2735
Laeri, F., III-2111
Lalo, C., IV-2425
Laman, C.D., IV-2419
Lamberti, C., I-317, IV-2571
Landry, C.C., II-801
Larach, F., IV-2869
Laspéras, M., I-675, IV-2705
Laufer, W., II-1351
Laurence, R.L., I-97
Lavalley, J.C., II-833
Lázár, K., IV-2875
Lee, C.W., I-437, II-1253, III-2187

Lee, J.F., II-857
Lee, J.H., II-1041
Lee, J.M., I-667
Lee, S-I., IV-2259
Lee, W.J., II-1253
Lee, Y., IV-2401
Lentz, P., IV-2963
Lercher, J.A., II-1425, IV-2881, IV-2889
Le Van Mao, R., III-1543
Lévesque, S., IV-2869
Levinbuk, M.I., II-1049, IV-2651
Lewis, D.W., I-341, III-1723
Li, C., II-1177
Li, G., II-1283, III-2147
Li, H-X., III-1655
Li, L., II-1323
Li, Q., I-707
Li, R., II-863, II-869, II-901
Li, T., II-1307
Li, W., II-1313, IV-2585
Li, X., III-1701
Li, X-W., IV-2659
Li, X-Y., IV-2897, IV-2905
Li, Z., IV-2787
Licea-Claverie, A., II-1165
Liese, T., IV-2795
Light, M., III-1937
Lim, K.H., IV-2301
Limburg, B., III-2111
Limtrakul, J., I-349
Lin, H-P., II-931
Lin, L., II-1283
Lin, W., II-923
Lin, Y-H., II-1081
Ling, T-R., II-931
Liu, B., II-923
Liu, C., III-2137
Liu, H., IV-2317
Liu, X., II-1307
Liu, X-Y., IV-2659
Llewellyn, P.L., I-189, I-197, I-659, II-779, III-1707
Lobo, R.F., I-119, III-1503, IV-2979
Loiseau, T., III-1737, IV-2409, IV-2453, IV-2971
Long, J.F., IV-2927
Long, Y-C., II-909, III-1983, III-1991
Long, Z-B., III-1655
López Nieto, J.M., IV-2473
López-Salinas, E., II-1105
Lourenço, J.P., III-1771
Lu, J.J., II-787
Luca, V., III-1507
Lugstein, A., IV-2817
Luigi, D.P., I-697, IV-2457, IV-2465
Lujano, J., II-801
Luna, F.J., II-1233
Lunsford, J.H., II-1381
Luo, X.M., III-2001
Lynch, J., II-1011, IV-2735

Ma, M-H., III-1991
MacDougall, J., I-505

Macedonia, M.D., I-363
Machado, M. da S., IV-2481
Machej, T., IV-2781
Madier, D., II-1033
Maeda, K., III-1833
Maginn, E.J., I-363
Magnoux, P., IV-2719
Maisuls, S.E., IV-2889
Mali, G., III-1585
Malossa, D., II-1471
Marchese, L., III-1569, IV-2775
Marcilly, C., II-1011, IV-2735
Marler, B., III-1731, III-1757
Marlow, F., III-2121
Martens, J.A., III-1825
Martin, C., I-51, I-189, I-197, I-659
Martin, D., II-1033
Martinez-Arias, A., IV-2473
Martra, G., IV-2775
Martucci, A., IV-2345, IV-2361
Marturano, P., IV-2705
Masloboishchikova, O.V., IV-2541
Massiani, P., III-2171, IV-2595
Masuda, T., II-1113
Matachowski, L., IV-2781
Matsukata, M., I-613, III-1515, III-2009
Matsumura, Y., I-667
Matsuoka, F., III-2041
Matsuoka, M., I-473
Matusek, K., IV-2875
McBrien, M., III-2137
McCormick, A.V., III-1595
McCusker, L.B., I-533, I-537, IV-2519
McDougall, G.S., IV-2615
Meden, A., III-1585, III-1765
Meiers, R., II-1351
Mello, M., I-215
Mellot, C.F., IV-2767
Melnikov, M.Ya., IV-2651
Meloni, D., II-1033
Menéndez, J.M., II-1441
Mériaudeau, P., II-1239, IV-2913
Methivier, A., I-59
Meyer zu Altenschildesche, H., III-1561
Michaels, A.S., I-25
Micke, A., I-215
Mikkelsen, Ø., II-1057
Mildner, T., IV-2955
Milestone, N.B., II-741, II-1337, III-1833, III-1901
Millar, D.M., III-1535
Miller, J.T., I-465
Millini, R., I-541
Mintova, S., III-1787, III-1809, IV-2525
Mirajkar, S.P., III-2203
Mishima, H., I-473
Mishin, I.V., II-1299, IV-2629, IV-2637
Mitchell, M., III-1931
Miyake, M., I-445
Miyamoto, A., I-489
Mizukami, F., III-1685, III-1833
Mo, S., III-1787

Mogica, E., II-1105
Moller, K., III-2209
Möller, K.P., I-591, III-2015
Monteiro, J.L.F., IV-2935
Montes de Correa, C., II-1331
Moore, J.G., II-801
Moreau, P., I-675
Moriyama, Y., III-1693
Morris, R.E., IV-2453
Mota, C.J.A., I-429
Mou, C-Y., II-931
Mougenel, J.C., III-1731
Mueller, K.T., IV-2331
Mukherjee, P., II-1227, III-1611
Mukhopadhyay, K., II-917
Munch, V., IV-2409, IV-2453
Munoz, Jr., T., III-1779
Murthy, K.V.V.S.B.S.R., II-893
Myers, A.L., I-103
Myrick, M.L., IV-2401

Naccache, C., II-1239, IV-2913
Naderi, M., IV-2457
Nagata, H., II-1141
Nagy, J.B., II-917, III-1619, III-1685, IV-2963
Nakano, T., III-2103
Nakata, S., III-1937
Nam, I-S., II-1149
Nascimento, M.A.C., I-429
Nash, R.J., IV-2557
Naum, N., IV-2759
Nava, N., II-1105
Navarro, M.T., II-817
Navrotsky, A., I-505, III-1737
Neugebauer Crawford, S., III-1953, IV-2265, IV-2489
Newalkar, B.L., I-127
Ngalula, K., IV-2681
Nghiem, Vu.T., II-1239, IV-2913
N'Gokoli-Kekele, P., IV-2991
Niederer, J.P.M., III-1893
Nishiyama, N., I-613
Nivarthy, G.S., II-1425
Niwa, M., IV-2549, IV-2643
Niwa, S-I., III-1685, III-1833
Njo, S.L., IV-2419, IV-2519
Noble, G.W., III-1603
Nomura, M., I-613
Nomura, H., II-1345
Norberg, V., I-167
Norby, P., IV-2295, IV-2309
Novak Tusar, N., III-1585
Nowak, I., II-833, II-997
Nowinska, K., IV-2847
Nozue, Y., III-2103, III-2225

O'Connor, C.T., I-591, II-1065, II-1073, III-2015, IV-2557
Ogden, J.S., III-2143
Ogura, M., IV-2833
Oh, S-H., II-1149
Oh, Y.S., II-1199

Ohayon, D., III-1543
Ohgushi, T., III-2179
Ohsuna, T., III-2225
Ohtsuka, H., II-1169
Olson, D.H., IV-2413
On, D.T., III-1877
Onida, B., IV-2705, IV-2711
Ono, T., I-667
Onyestyák, Gy., I-159, IV-2673, IV-2875
O'Rear, D.J., III-1743, III-2023
Otero Areán, C., I-317
Ott, K.C., IV-2489
Oumi, Y., I-489
Oyanagi, H., III-2217

Paillaud, J-L., IV-2525
Pandey, R.K., III-1611
Pang, W.Q., II-923, II-1277, III-1641, III-1751, IV-2387
Panjabi, G., IV-2265
Panov, G.I., II-1245
Parise, J.B., IV-2401
Park, S-E., I-437, I-667, II-1157, II-1253, III-2187
Park, S.H., II-1041
Park, Y-K., I-437, II-1157
Parker, S.F., IV-2609
Patarin, J., III-1707, III-1757
Patinec, V., III-1603
Paukshtis, E., II-1165
Pavlov, M.L., II-1049
Paweewan, B., II-1003
Pazé, C., IV-2571
Perego, C., I-575
Perego, G., I-541
Pérez de Vargas, S.L., III-1663
Pérez-Pariente, J., II-817, IV-2481
Pervaiz, N., III-1487
Peters, A.W., IV-2331, IV-2413
Petranovskii, V.P., II-1165, III-2143
Petrik, L.F., I-257, II-1073
Pfeifer, H., IV-2955
Philippou, A., IV-2465
Phillips, J.C., IV-2401
Pierard, F., II-1213
Pierloot, K., I-387
Pietrass, T., IV-2265
Pines, A., IV-2943
Pingel, U., III-1849
Pirngruber, G.D., IV-2881
Pirutko, L.V., II-1245
Plaisted, R.J., III-1591
Plévert, J., I-85, I-135, IV-2445
Poborchi, V.V., III-2217
Polisset-Thfoin, M., III-2079
Pöppl, A., II-795
Prakash, A.M., IV-2825
Preckwinkel, U.W., IV-2401
Prins, R., II-1471
Puil, N.v.d., I-603
Purnell, S., IV-2413

Qiu, P., IV-2431
Qiu, S., II-1277, III-1701, III-1863, III-1937
Quartieri, S., IV-2345
Quirin, J., III-1953

Raghavan, K.V., II-893, II-897, II-1101
Rajic, N., III-1765
Rakhmatkariev, G.U., I-481
Rakiewicz, E.F., IV-2331
Rakoczy, J., IV-2601
Rakshe, B., III-1909
Ramakrishna Prasad, M., II-897
Ramamurthy, V., III-2095
Ramaswamy, A.V., III-1909, III-2033
Ramaswamy, V., III-1909, III-2033
Ramsaran, A., III-1931
Randall, K.H., I-517
Rao, S., I-517
Rapacciuolo, M., IV-2345, IV-2361
Rathousky, J., II-825
Ratnasamy, P., III-2195
Rauscher, M., III-1849
Ravishankar, R., III-1825
Reddy, K.M., II-1133
Redondo, A., I-379
Rees, L.V.C., I-13, I-67, I-159, IV-2615, IV-2673
Reimer, J., IV-2943
Reinert, P., III-1757
Reitzmann, A., IV-2847
Renaudin, J., IV-2409
Rey, F., II-817
Rhee, H-K., II-1041
Ribeiro, F.R., III-1771
Ribeiro, M.F., III-1771
Rice, M.J., I-393
Rios, S.P.O., IV-2481
Riou, D., III-1649
Ristic, A., III-1585
Ritsch, S., III-1693
Rocha, J., IV-2457
Rödenbeck, C., I-153
Rodrigues, A.E., I-243
Roduner, E., II-863, II-869, IV-2665
Roest, A.L., IV-2515
Röger, H.P., III-2015
Romanovsky, B.V., II-1207
Romero, R., III-1953
Rønning, P.O., I-1057
Rosynek, M.P., II-1381
Rouquerol, F., III-1707
Rouquerol, J., II-779
Rouxhet, P., III-2163
Rozwadowski, M., I-285
Rutenbeck, D., IV-2795
Ryoo, R., I-689, II-1199

Sabatino, L.M.F., II-1169
Sacerdoti, M., IV-2345
Sahasrabudhe, N.S., III-1901
Sakamoto, T., III-1863

Sakamoto, Y., III-1937, III-2225
Salehirad, F., IV-2465
Samoson, A., IV-2409
Sankar, G., III-1671, III-1723
Sano, T., I-293, I-301, II-1113, III-1817, III-1841, III-2041, IV-2743
Santos, R.F., II-975
Sanz, R., III-1917
Sapaly, G., IV-2913
Saravanan, C., I-357, I-509
Sasaki, Y., III-2009
Sastre, G., I-341
Sato, M., I-445
Sato, T., IV-2863
Sauer, J., III-2241
Sauerbeck, S., III-2155
Sauerland, C., III-1707
Saur, O., II-833
Sayari, A., I-681, I-725, II-761, IV-2869
Scarano, D., I-317, IV-2571
Schaefer, D.J., I-43
Schay, Z., IV-2809
Scheffler, F., III-1849
Schell, F., IV-2665
Schemmert, U., I-231
Schenk, U., IV-2503
Schifter, I., II-1105
Schmidt, R., I-713
Schmitz, A.D., II-1133
Schnitzler, A.E., I-591
Schoeman, B.J., III-1825
Schomburg, C., III-2233
Schoonheydt, R.A., I-387
Schreyeck, L., III-1731
Schuchardt, U., II-1233
Schulz-Ekloff, G., II-825, III-2233
Schüth, F., III-2111, III-2241
Schwieger, W., II-847, III-1561, III-1849
Scott, A., II-1403
Sefcík, J., III-1595
Seff, K., IV-2839
Seidel, A., IV-2589
Seijger, S.B.G., I-603
Seitz, M., III-1969
Sekine, T., II-1113
Selle, M., II-749
Selvaraj, K., III-2033
Sepa, J., IV-2287
Seriu, A., III-1693
Serrano, D.P., II-1441, III-1917
Seshan, K., II-1425, IV-2881, IV-2889
Seydoux, R., IV-2943
Shahriari, D.Y., II-801
Shan, X.L., IV-2803
Shannon, M.D., IV-2355
Shantz, D.F., III-1503
Shao, C., III-1701
Shapoval, Yu.V., III-1961
Sharma, S., III-2203
Sharratt, P.N., II-1081
Shashkov, A.A., II-1207
Sheldon, R.A., III-1975

Shelef, M., II-1373
Shen, D., I-111, IV-2673
Shi, Q., II-1177
Shibata, M., III-1925
Shih, S., I-451
Shim, S-H., I-505
Shimada, T., II-1113
Shio, S., I-719
Shubaeva, M.A., III-1961
Shul, Y.G., III-1523
Siffert, S., II-1121
Silva, J.A.C., I-243
Simon, N., IV-2409
Singh, A.P., II-1317, III-2203
Singh, N.K., III-1507
Siperstein, F., I-103
Sirkecioglu, A., I-257, I-309
Sivasanker, S., II-877
Slangen, P.M., III-1553
Slinkin, A.A., II-1373, IV-2855
Smirnov, A.V., II-1207
Smirnov, V.K., II-1049
Smit, B., I-91, I-325
Smith, K., II-1129
Smith, S.P.J., I-603
Smorodinskaya, J.Ya., IV-2651
Snurr, R.Q., I-251, I-465
Sobalík, Z., II-941, IV-2339
Sobrinho, E.V., II-1463
Soga, K., I-293, I-301, II-1113, III-1841, III-2041, IV-2743
Sokol, A.A., I-457
Song, C., II-1133
Song, L., I-67
Sopa, M., IV-2847
Sotelo, J.L., II-1441
Soufek, M., IV-2309
Sponer, J.E., II-1419
Spoto, G., I-317, IV-2571
Springuel-Huet, M-A., IV-2991
Srinivas, N., II-893, II-1101
Stach, H., IV-2623
Steel, A., II-733
Sterte, J., I-629, II-1185, III-1809, III-1857
Stockenhuber, M., II-1367
Stöcker, M., I-713
Stojakovic, D., III-1765
Stucky, G.D., IV-2453
Stuglik, Z., I-277
Su, B-L., I-167, II-1121, IV-2681, IV-2689
Su, X., IV-2659
Subotic, B., III-2049, III-2057, IV-2309
Sugahara, Y., II-771
Sugi, Y., II-1165, III-1693
Sugiyama, K., III-1937
Suits, B.H., IV-2287
Sun, D.K., II-989
Sun, Y-J., II-909, III-1983, III-1991
Sung-Suh, H.M., III-2187
Sutovich, K.J., IV-2331
Sutra, P., I-675
Suzuki, A., III-2041

Suzuki, T., III-2009
Swaminathan, V.S., IV-2287
Swiecicka, M., IV-2847
Sychev, M., II-1261, III-2129
Szostak, R., III-1931

Tabata, T., II-1169
Tadenuma, R., IV-2743
Takahashi, I., II-1113
Takeuchi, Y., I-203
Takiyama, Y., II-1141
Tamási, A., IV-2809
Tanaka, K., III-2217
Tanaka, Y., III-1937
Tang, Y., I-175
Tao, K., II-1313, IV-2585
Tashiro, S., II-1141
Tatlier, M., I-583
Tatro, M.E., IV-2511
Tatsumi, T., II-1391, II-1455
Taulelle, F., IV-2409, IV-2453, IV-2971
Teissier, R., II-1269
Terasaki, O., III-1693, III-1863, III-1937, III-2225
Terskikh, V.V., II-1245
Testa, F., III-1619
Tezel, F.H., I-309
Theodorou, D.N., I-371
Thomas, J.M., III-1569, III-1723, III-1937
Tiddy, G.J.T., II-733
Toba, M., III-1685
Toby, B., IV-2413
Togashi, N., III-1937, III-2225
Toledo, J.A., II-1105
Torres, J.C., IV-2371
Torres Sánchez, R., III-1663
Tosheva, L., I-629
Totktarev, A.V., III-2023
Toufar, H., III-1561
Travers, C., II-1433
Treacy, M.M.J., I-517, IV-2999
Trouw, F., I-119
Tsapatsis, M., I-621
Tsuji, K., III-1479, III-2065
Tsuruya, H., I-489
Tuan, Vu.A., II-1239, IV-2913
Tuel, A., II-1269
Turner, M.D., I-97
Turnes Palomino, G., I-317
Turutina, N.V., III-1679
Tvaruzková, Z., IV-2339
Twaiq, F., IV-2921

Uehara, H., I-445
Ueyama, K., I-613, III-1515, III-2009
Uguina, M.A., III-1917
Unger, K.K., II-757
Uno, Y., IV-2743
Urquieta-González, E.A., II-975

Vaidyalingam, A., IV-2927
Valange, S., I-651, IV-2395, IV-2711

Valcheva-Traykova, M.L., II-981
Valencia, S., III-1495, IV-2951
Valtchev, V., I-629, IV-2525
Valyon, J., I-159
van Bekkum, H., III-1553, III-1975, III-2117, IV-2565
van Bokhoven, J.A., IV-2515
van de Graaf, B., IV-2519
van der Eerden, M.J., I-637
van der Ham, F., III-1553
van Grieken, R., II-1441
van Hooff, J.H.C., I-91, I-645
van Klooster, S.M., III-2117
van Koningsveld, H., IV-2419, IV-2519, IV-2999
van Laar, F.M.P.R., II-1213
van Ommen, J.G., IV-2889
Vanoppen, D.L., II-1213
Van Rhijn, W.M., II-1293
van Santen, R.A., I-91, I-645, III-1529
van Steen, E., I-591, II-1065
van Turnhout, J., III-2117
van Well, W.J.M., I-91, IV-2623
van Wolput, J.H.M.C., IV-2623
Varkey, S.R., III-2195
Vasina, T.V., II-1049, IV-2541
Vassena, D., II-1471
Vaughan, D.E.W., I-3
Verberckmoes, A.A., I-387
Verrelst, W., II-1293
Vezzalini, G., IV-2345
Vidal, L., III-1633
Viets, J., I-481
Vietze, U., III-2111
Villaescusa, L.A., III-1495, IV-2951
Vinek, H., IV-2817
Vlugt, T.J.H., I-325
Vogel, R.F., III-1743

Wakabayashi, F., IV-2577
Wakabayashi, K., II-1141
Wallau, M., II-1233
Wan, B-Z., II-931
Wang, C-B., I-175
Wang, D., II-1381
Wang, D.S., IV-2803
Wang, J., II-1313
Wang, J.D., I-613
Wang, J-Z., III-1655
Wang, X., II-1283
Wang, Y., II-989
Wang, Z.B., I-293, I-301, II-1113, III-1841, III-2041, IV-2743
Wark, M., II-825, III-2249
Warnken, M., III-2249
Washmon, L., III-1779
Weber, R.W., III-2015
Weber, W.A., IV-2265
Weckhuysen, B.M., II-1381
Wei, A.C., II-857
Weigel, S.J., III-1487, IV-2453

Weitkamp, J., II-1409, II-1447, IV-2503,
    IV-2665, IV-2697
Wellach, S., II-1409
Weller, M.T., III-2087, IV-2379
Wessels, T., I-537
White, D., IV-2287
Wichterlová, B., I-401, II-941, II-1193,
    II-1419, IV-2339
Williams, B.A., I-465
Wilson, S.T., I-567, III-1715
Witkamp, G.J., III-1553
Wöhrle, D., III-2233
Wölker, A., IV-2951
Wormsbecher, R.F., IV-2331
Wouters, B.H., IV-2325, IV-2497
Wright, P.A., III-1603
Wu, A-M., III-1983
Wu, H.C., II-857
Wu, S.Q., II-1399
Wübbenhorst, M., III-2117
Wyles, J.K., III-1723

Xia, Q., II-1391
Xiang, S.H., IV-2803
Xiangsheng, W., I-141
Xiao, F-S., II-1277, III-1701, III-1863
Xie, J., I-263
Xie, Y., I-175
Xomeritakis, G., I-621
Xu, G., I-263
Xu, Q.H., II-989
Xu, R., II-1277, III-1641, III-1751, III-1863,
    III-1937, III-2147
Xu, T-M., II-909
Xu, Y., II-787, II-923, II-1177, III-1751,
    IV-2387

Yamada, K., III-1841
Yamaguchi, M., I-719
Yamashita, H., I-473, I-667
Yan, Y., III-1983, III-1991
Yanagishita, H., III-1841
Yang, H., II-901
Yang, X., III-2155
Yang, Y., IV-2869

Yao, Y., III-1751
Yaremov, P.S., III-1679
Yarimizu, T., IV-2863
Yie, J.E., II-1199
Yngvesson, S., I-97
Yoo, J.W., II-1253
Yoshida, K., I-719
Yu, J., III-1863, III-1937
Yu, R., II-1277
Yuan, Z-Y., III-1655
Yuen, L.T., III-1945
Yushan, L., III-1795
Yuzaki, K., IV-2863

Zabidi, N.A.M., IV-2921
Zabukovec Logar, N., III-1585
Zadrozna, G., III-1577, IV-2601
Zahedi-Niaki, M.H., II-1221
Zecchina, A., I-317, IV-2571
Zeuthen, P., IV-2541
Zhai, Q., III-1701
Zhan, B-Z., IV-2897, 2905
Zhang, K., IV-2803
Zhang, S.G., I-667
Zhang, Z., I-1701
Zhanpeisov, N.U., I-473
Zhao, J.P., III-1591
Zhao, L., III-1991
Zhao, Q., II-1283
Zhao, X., II-1177
Zheng, S., II-1313, III-1701, III-1937, IV-2585
Zholobenko, V.L., I-183
Zhong, B., II-869, II-901
Zhou, F., III-1641
Zhou, L., IV-2585
Zhu, G., III-1863
Zhu, J.H., II-989
Zhuravlev, V.V., III-2217
Zikánová, A., I-209, I-215
Zilková, N., II-1419
Ziolek, M., II-833, II-997
Zones, S.I., III-1945
Zukal, A., II-825
Zuurdeeg, B.J., IV-2565
Zygmunt, S.A., I-333, I-415

SUBJECT INDEX

I, II, III and IV refer to the volume number

A, zeolite, I-119, I-159, II-1041, III-1595,
    III-1671, III-2087, III-2179
  ESCA of, IV-2439
  kinetics of crystal growth, III-2049
  PbI_2 clusters in, III-2225
  synthesis, III-1863
  thin films, III-1857
a-alanine, use in dosimetry, I-277
ABC-6 materials, IV-2445
ab-initio calculations, I-387, II-1419
  adsorption in zeolites, I-415
  molecular orbit theory, I-333
  pyrrole in Na-Y, IV-2609
absorption bands, III-2147
acetic acid, wet oxidation, IV-2869
acetonitrile, adsorption in H-ZSM-5, IV-2287
acetylation of methylfuran, II-1293
acetylene pyrolysis, preparation of
    nanotubes, II-917
acid
  sites in mordenite, II-1165
    distribution of, IV-2639
  strength, IV-2571
    in Y, QM study, I-429
  treatment, of H-ZSM-5, IV-2743
acidic
  centers, IV-2651
  media, Ti-MCM-41, II-817
  property, IV-2643
  sites, in ZSM-5, IV-2413
acidity, II-1141, IV-2963
  in MCM-41 and MCM-48, II-801
  in MCM-48, II-801
  of OH groups in H-Y, IV-2595
acrylonitrile, II-1337
activation
  barrier, I-333
  energy, III-2057
  of n-hexane consumption, IV-2651
activity
  coefficient, I-103
  of ZSM-5, I-591
acylation
  gas phase of phenol, II-1463
  of aromatics over beta, II-1085
adsorbed water, in A, III-2179
adsorption, I-91, I-97, I-103, I-363,
    III-2117, IV-2589, IV-2609, IV-2651
  ab-initio and DFT calculations in
    zeolites, I-415
  calculations in zeolites, I-415
  competitive, I-119
  complexes, IV-2535
  diatomic gases on ETS-10, I-317

equilibria, binary, I-257
heat pumps, I-583
isotherm of water
  in AlPO-5, I-285
  in ZSM-5, I-301
  in ZSM-12, I-293
N_2 and CO_2 in ZSM-5, I-309
NOx in Co-ZSM-5, I-437
of CD_3OH in H-Y and H-ZSM-5, IV-2279
of chloromethane on beta, IV-2681
of CO and NO on SAPO, IV-2751
of n-alkanes in ZK-5, IV-2623
of pyridine, IV-2697
properties of Al and Nb MCM-41,
    II-833
sites, of benzene in Na-beta, I-167
state, of phenol on H-Y, IV-2659
AEL, I-51
AFI, I-51, I-341
Ag particles, II-1337
alkali
  aqua complexes, II-981
  cations, III-1619
    affect on zeolite basicity, II-981
alkane
  activation on H-ZSM-5, IV-2273
  branched, I-325
  cracking, IV-2535
  sorption, in $AlPO_4$-5, I-127
alkene cation radicals, I-451
alkoxyl groups, IV-2601
alkyl chain length, II-771
alkylation, II-897, II-1425, IV-2601
  of benzene, I-575
    over beta, II-1121
    with propylene, II-1207
  of furan, II-1293
  of naphthalene, II-1307
alkyltrimethylammonium, II-771
allylic alcohols, III-2163
$AlPO_4$, LTA-type, III-1731
$AlPO_4$-5, I-59, I-127, I-197, I-341, III-1671,
    III-1765, III-2111, III-2117, III-2147,
    III-2217
  dynamics of neopentane in, I-189
  encapsulation of dyes, III-2233
  films of, III-1787
  water sorption in, I-285
$AlPO_4$-11, I-51, I-197
  substituted, II-1239
$AlPO_4$-14, III-1715
$AlPO_4$-31, III-1743
$AlPO_4$-34, I-341
$AlPO_4$-CJ2, IV-2971
$AlPO_4$-HDA, synthesis and structure,
    III-1937
$AlPO_4$-SOD, III-1633
alumina, II-1433

lxv

aluminophosphate, III-1585, III-1633
　chabazitic, III-1723
　Co, I-379
　mesoporous, II-771
aluminum, extra-framework, IV-2643
ammination of methanol, II-1065
ammonia, II-1323, IV-2673
　heats of adsorption, IV-2637
　in Na-beta, I-167
ammonium
　hexafluoro metal complexes, III-1901
　hexafluorosilicate, IV-2735
angular overlap modelling, I-387
anisotropic diffusivities, I-67
AOM, I-387
argon, sorption, I-111
aromatics, II-1381, II-1471
aromatization
　of n-heptane on ZSM-5, IV-2759
　on basic zeolites, IV-2727
　using Pt/KL, IV-2557
asbestos, IV-2927
a-SnP, III-1641
assembly routes, II-741
ATO, III-1743
atomic force microscopy, III-1487
atomistic forcefield, I-379
Auger, A 1 KLL lines, IV-2439
autocatalytic reaction, II-1433
automobile exhaust gas converters, IV-2803
autoreduction, II-1351
a,w-diamino templates, III-1561
azobenzene, III-2121

ball milling, IV-2309
Barrer, R.M., I-3, I-13, I-25
basic
　properties of zeolites, II-981
　zeolites, as aromatization catalysts, IV-2727
basicity
　of framework O in Na-beta, I-167
　use
　　of methylacetylene, IV-2705
　　of pyrrole, IV-2705
BEA, II-941, III-1707
Beckmann rearrangement, II-1455
benzene, I-43, II-1337, IV-2273, IV-2955
　alkylation, I-525, II-1121, II-1207, II-1419
　diffusion, IV-2991
　in Y, I-509
　hydroxylation with H_2 and O_2, II-1253
　in Na-beta, I-167
　sorption complex, IV-2839
benzyltrimethylammonium chloride (BTMACl), III-1655
beta, I-523, II-941, II-1463, IV-2565, IV-2681, IV-2803, IV-2817, IV-2855
　acylation of aromatics, II-1085
　Al-free Ti-BEA, III-1515
　alkylation of benzene over, II-1121

benzene
　hydroconversion over, II-1025
　in, I-167
　catalytic cracking of palm oil, IV-2921
　Co-beta, IV-2339
　crystallization, III-1515
　cumene synthesis, II-1073
　ethane conversion and coking, II-1003
　Fe-beta, III-1707
　furan alkylation, II-1293
　Ga-beta, III-1707
　H-beta, II-1337, II-1419, II-1425, II-1455, II-1471, III-2203, IV-2689
　hydration of propylene, II-1313
　Mo beta, III-1893
　Na-beta, IV-2681
　(Na)Fe-beta, III-1707
　post synthesis modification, III-1893
　propionylation of toluene over, II-1317
　pure silica, III-1515
　selective catalytic reduction, II-1169
　surface passivation, III-1975
　synthesis of Ti/Al, III-1885
　Ti-, II-1391
　transition metal containing, III-1893
　V containing for ethanol oxidation, II-877
　Zr containing, III-1909
beta-naphthyl methyl ether, synthesis with MCM-41, II-885
bifunctional catalysis, IV-2817, IV-2913
　Pt-ZSM-5, IV-2881
bimetallic catalysts, IV-2809
　PdAu/HY, III-2079
bimolecular mechanism, II-1433
binary adsorption, I-251
binderless zeolite, III-2001
biomimetic V oxidation catalysis, III-2171
biphasic silicates, III-1849
bond angles, I-341
borosilicate, ZSM-12, III-1655
branched alkanes, I-325
　sorption in $AlPO_4$-5, I-127
Breck-Flanigen correlation, II-1049
bridging hydroxyl groups, IV-2259
broad-line ^1H NMR at 4 K, IV-2279
bromination, liquid phase, III-2203
Brönsted
　acid, I-341, IV-2689, IV-2711, IV-2913
　site, I-333, I-349, III-1771, IV-2535, IV-2549, IV-2571, IV-2697, IV-2809, IV-2955
　in USY, IV-2331
　modelling of, I-457
acidity,
　in stilbite, IV-2345
　of CoAPO-18, IV-2775
　of NaH-X, NaH-Y, IV-2601
B-spodumene, III-2033
build-bottle-around-ship
　CrP@NaY, IV-2905
　NiTMPyP@NaY, IV-2897

butane, IV-2273
butanol, II-1177

calcination, III-1707, IV-2565
calorimetry, III-1737
canonical Monte Carlo simulations, IV-2767
capillary hydrodynamic fractionation, III-1523
carbon
  dioxide
    dynamics in A, X and Y, I-159
    sorption in
      SAPO-5 and SAPO-11, I-257
      ZSM-5, I-309
  monoxide, II-1409
  nanotubes, preparation on Co and Fe impregnated magadiite, II-917
carbonylation
  of formaldehyde, II-1113
  over Cu-ZSM-5, II-1177
carcinogenicity, of mineral erionite, IV-2927
catalase, III-2195
catalysis
  in 1D channel systems, I-153
  with organically functionalized zeolites, III-1479
catalytic
  conversion, of chloromethane on beta, IV-2681
  cracking, of palm oil, IV-2921
  degradation of polyolefins, II-1081
  distillation, I-603
  hydration, of propylene to isopropanol, II-1313
  membrane, II-1399
  oxidation, II-1337, II-1403
  properties, of MCM-41 and MCM-48, II-801
  reduction, of nitric oxide with methane, IV-2787
cation positions
  in dehydrated Li,Na-LSX, I-135
  in ETS-4 and ETS-10, IV-2457
  predicting, IV-2767
CD_3OH, in H-Y and H-ZSM-5, IV-2279
CeO_2-H-ZSM-5, IV-2795
ceramic support, ZSM-5 on, III-1833
cerium-promotion, IV-2787
CHA, I-341
chabazite, III-1723
  enthalpy of formation, I-505
CHA-GME Group, IV-2445
chain(-)
  branching, I-325
  length, I-325
  type species, II-1359
chelate, III-1701
chemical
  modification, of MCM-41, I-725
  transport reaction, IV-2855
  vapor deposition, II-825

chemisorption, of pyridine, IV-2673
chiral
  aminoalcohols, I-675
  derivatives of the Schiff base salen, III-2155
  induction, III-2095
  manganese-salen-complexes, II-749
chlorobenzene bromination, III-2203
chlorofluorocarbons, decomposition of, IV-2963
chloromethane
  adsorption and conversion on beta, IV-2681
  as probe molecule, IV-2689
chromium, I-667, III-1577
  (III) porphyrin, nitrile hydratase activity in Na-Y, IV-2905
chromophore, inclusion in $AlPO_4$-5, III-2233
cinnamaldehyde, I-651
clays, III-1663
clinoptilolite, IV-2371
  Co and Cu in, III-2129
cluster(s)
  model, I-333
  Rh_6 in Na-Y, IV-2265
Co, I-379
  in Na-A, I-387
  in Na-Y, IV-2615
  ions, II-941
  sites in dehydrated ZSM-5, mordenite, and ferrierite, II-1193
CO
  adsorption, I-175
  isotherm in H-Y, IV-2589
  on SAPO, IV-2751
  detection, III-2249
  hydrogenation with metal substituted ZSM-48, II-901
CO_2, photoinduced activation and reduction, III-2187
coadsorption, I-349
$CoAlPO_4$-5, III-1671
Co-$AlPO_4$-5, spin-echo mapping NMR, IV-2489
CoAPO-18, NO activation, IV-2775
CoAPO-37, III-1771
coatings, structure and synthesis, I-603
cobalt, I-379, I-387, II-941, II-1193, III-1671, III-1751, IV-2339, IV-2489, IV-2615, IV-2833, IV-2889
  phosphates, IV-2387
Co-beta, IV-2339
Co-ferrierite, IV-2339
Co-$GaPO_4$-LTA, III-1751
coke
  distribution, IV-2991
  during n-heptane cracking, IV-2935
  formation in ZSM-22, II-1033
  in beta during ethene conversion, II-1003
  oxidation, IV-2935
combinatorial crystallography, I-517

lxvii

Co-mordenite, IV-2339
competitive adsorption, I-97
completely dissolved solutions, I-719
composite
    membrane reactor, III-1795
    microporous compounds, I-707, III-1649
computer simulation, MCM-41 structure, II-839
concentration pulse chromatography, I-309
configurational-bias, I-363
    Monte Carlo, I-91, I-325
confined molecules, I-51
constant acid strength, IV-2643
continuous flow conditions, IV-2945
control, of pore diameter, II-771
coordination, IV-2565
CoO_x/ZSM-5, in SCR of NO, II-1157
Co/Pd/H-ZSM-5 catalysts, IV-2833
$CoPO_4$, IV-2387
copper-oxide, II-1359
Co-Pt/ZSM-5, IV-2889
cordierite, III-2033
    zeolite synthesis on, IV-2803
Co-ZSM-5, IV-2339
Cr^{3+} ions, III-1577
cracking, II-1049
    alkane, IV-2535
    enhanced in Y zeolite, I-465
    ethane, I-333
    in ZSM-5, DFT calculation, I-489
    n-heptane
        in Y zeolite, II-975
        over ZSM-22, II-1033
    of n-hexane, IV-2651
    of octanes, IV-2629
    of polyolefins, II-1441
    reaction mechanism, I-465
CrAPO-5, III-1577
CrAPO-14, III-1715
cresol transformation over Y and ZSM-5, II-1093
CREY-1, IV-2935
CREY-2, IV-2935
Cr-HMS, I-667
crystal
    growth, III-1487, III-1671, III-2057
    of mordenite, III-1803
    morphology, I-689
    size, III-1877
    of ZSM-5, I-591
    structure, of zincophosphate, III-1757
crystallization, IV-2971
    diagram, III-1595
    kinetics of ZSM-5, III-1983
    mechanisms, III-1633
Cs,Na-X, IV-2705
Cs,Na-Y, IV-2705
Cs-X, IV-2317
Cs-Y, IV-2317
    sorption of hydrofluorocarbons, IV-2301
CTABr surfactant, II-779

Cu ions, II-941
    Cu^{2+}, II-1359
Cu,Co,Pd and Ga-exchanged ZSM-5, I-393
Cu-exchange, of SAPO, IV-2751
cumene, I-575
    isomerization, II-1207
    synthesis over beta, II-1073
CuNaA, II-1409 Cu-SAPO-34, IV-2751
Cu-Y, IV-2425
Cu-ZSM-5, II-1359
CVD
    modification of ZSM-5 and Y, III-1969
    of TEOS on ZSM-5, III-2015
cyclic voltammetry, III-2073, III-2137
cyclization, of C_4 alcohols, II-1101
cyclohexane, I-97
    oxidation
        over Cr-VPI-5, II-1233
        of, II-749
cyclohexanone oxime, III-1655

DAF-5, III-1723
DCM-2, I-551
deactivation, II-1447, II-1471
    by coking, of USY, IV-2935
    in beta, II-1003
    in ZSM-22, II-1033
    of FCC catalysts, II-1299
    of ZSM-5, I-591
dealumination, IV-2955, IV-2963
    beta, III-2171
    H-Y, IV-2735
    mordenite, II-1133
    Y zeolite, II-1011
    zeolite omega, II-1041
    ZSM-5, III-1953
    ZSM-12, I-293
decomposition of N_2O, on Fe-ZSM-22, IV-2781
defect-free silica zeolite, III-1495
deformation shift, IV-2339
dehydration, of diethylene glycol, III-1795
dehydroisomerization of n-butane, IV-2881
demetallation, I-393
de novo design, III-1723
de-NOx
    activity, IV-2795, IV-2803, IV-2809
    catalyst, II-1199
        Co-Pt/ZSM-5, IV-2889
density functional theory, I-333, I-349, I-393
    adsorption calculations, I-415
depth profiling by RBS, IV-2395
desorption energy distribution functions, I-223
dialkylformamide media, III-1633
diaminoalkanes, III-1757
diatomic gases, sorption on ETS-10, I-317
diaza-polyoxa-macrocycles, III-1731
dichloromethane, gravimetric measurement in USY, I-203

dielectric properties of A-type zeolites, III-2179
diethylamine, oxidation, II-1269
diethylene glycol, dehydration of, III-1795
difference adsorption isotherm, I-301
diffuse reflectance
  infra-red spectroscopy, DRIFTS, IV-2615
    control, I-269
    of adsorbed CO, III-2079
  spectroscopy, I-387
  UV-vis spectroscopy, IV-2775
    of {M}-MSU-X, II-741
  - see also UV-visible spectroscopy
diffusion, I-35, I-67
  coefficient, I-235
  in 1D channels, I-153
  of alcohols in erionite/offretite, III-1961
  of benzene, I-357
  of hydrocarbons in
    13X and ZSM-5, I-243
    silicalite I, I-183
  of propane in silicalite I, IV-2615
  of pyridine in mordenite, IV-2673
  properties of H-ZSM-5, I-141
diffusivity, I-59, I-77
diisopropylbenzenes, I-575
dimer, phenol, IV-2659
dimerization, II-1425
dimethyl ether, IV-2577
dimethylamine, III-1757
di-n-pentylamine, III-1743
disproportionation of ethylbenzene, II-1019, IV-2629
D_2O-MCM-41, I-659
DOR NMR, of fluorinated $AlPO_4$, IV-2409
dosimetry, I-277
DQ NMR, of fluorinated $AlPO_4$, IV-2409
DRIFTS
  - see diffuse reflectance infra-red spectroscopy
dry gel conversion technique, II-1391, III-1515, III-2009
dye(-)
  alignment in dipolar $AlPO_4$-5, III-2233
  cell, IV-2927
dynamic(s)
  of neopentane in $AlPO_4$-5, I-189
  structure function, I-371

8-ring pores, III-1715
electric field gradient tensors, ab initio calculation, IV-2457
electrochemistry, III-2079, III-2137
electrocyclic reactions, III-2095
electron
  microscopies, III-1685
    - see also HREM and SEM
  spin
    density equation, I-451
    echo modulation (ESEM), IV-2825
    resonance (ESR), II-1373, III-2121, IV-2665, IV-2825, IV-2855

HYSCORE, II-795
  transfer, III-2187
electroneutrality, III-1679
electronic ceramics, III-2033
electrostatic potential, I-341
Eley-Rideal mechanism, I-349
EMC-2, Ga substituted, III-1871
EMT, palladium-salen-complexes in, III-2155
encapsulated molecular sieves, II-893
encapsulation of chalcogens in $AlPO_4$-5 and SAPO-44, III-2147
end-capping, I-675
enentioselective alkylation, I-675
energy dispersive x-ray diffraction (EDXRD), III-1671
enhanced mechanical stability, III-1685
enthalpy of formation, chabazite, I-505
entropy of adsorption, IV-2623
enumeration, of zeolite frameworks, I-517
enzyme mimics, III-2195
epoxidation, I-645, III-2163
  of propylene over TS-1, II-1283
  of styrene, II-749
EPR, II-1381, III-2163
  of NO in H-Y, IV-2589
erionite(-),
  Ag in, III-2143
  fibers, IV-2927
  offretite, synthesis, III-1961
ERS-7, structure solution, I-541
ESCA, of zeolites, IV-2439
ESR - see electron spin resonance
ETAS-10, IV-2985
ethane
  conversion in beta, II-1003
  cracking, I-333
ethene, II-1337, II-1425
ethylation of ethylbenzene, in H-ZSM-5, I-141
ethylbenzene, I-141, II-1337, IV-2727
  disproportionation, in zeolite Y, III-1969
ethylene, I-567
  from methanol, II-1345
ethylenediaminetetraacetic acid, IV-2735
ETS-4, IV-2825
  cation siting, IV-2457
ETS-10, III-1507, IV-2825, IV-2985
  cation siting, IV-2457
  sorption of diatomic gases on, I-317
Eu doping of zeolite X, III-2241
EU-1, location of template in, IV-2355
EXAFS, II-1367, III-2079
  of Ag clusters, III-2143
  of dispersed species, I-175
  of ETS-10, III-1507
exchangeable cations, in A, III-2179
exhaust gas converters, IV-2803
external
  acidity, of H-ZSM-5, I-141
  surface acidity of mordenite and beta, III-1975

extra-framework
   aluminium, IV-2643
      effect on selectivity and
         deactivation, II-975
      in Y zeolite, II-1011
   chromium, III-1577
   species, I-401, IV-2735

faujasite, I-43, I-111, IV-2705
faulting, in CHA-GME, IV-2445
FAU-type zeolites, I-43, II-941
   palladium-salen-complexes, III-2155
   sorbate immobilization, I-215
   structure, III-1771
FCC catalyst, deactivation and post-mortem analysis, II-1299
Fe
   metallic, IV-2875
   modification, IV-2847
   nanoparticles, IV-2875
Fe^{3+} in ETS10, III-1507
Fe-beta, III-1707
FER, II-941, IV-2697
ferrierite, I-91, I-323, IV-2855
   dehydrated, Co ions in, II-923
   Pd loaded, II-1331
   refinement, IV-2361
   synthesis, III-1991
Fe-SAPO, II-1345
Fe-X, IV-2875
Fe-Y, IV-2875
Fe-ZSM-5, II-1367, III-1707
Fe-ZSM-22, IV-2781
fibers, erionite, IV-2927
Ficks' second law, I-231
field emission SEM, I-689
fine mesoporous powders, I-719
finite loadings, I-357
Fischer-Tropsch hexane, IV-2559
5A, zeolite, III-2179
fluoride media, III-1619
   synthesis of ZSM-5 in, III-1901
fluorinated
   alumino-phosphates, IV-2409
   $GaPO_4$, IV-2453
fly-ash, zeolitized, II-1105
FOCUS, I-533
force field, I-423
formaldehyde, II-1337
4A, zeolite, I-583, III-1795, III-2087, III-2179
framework
   aluminum, IV-2565
   relaxation, IV-2767
   substitution of Cr, III-1577
frequency response (FR) technique, I-67, I-159, IV-2615, IV-2673
Fries rearrangement, II-1463
FTIR
   microscopy, I-183

   spectroscopy, I-209, I-675, II-1381, III-1771, IV-2577, IV-2689, IV-2697, IV-2711, IV-2775
   of acetonitrile adsorption in H-ZSM-5, IV-2287
   of beta, IV-2681
FT-Raman Spectroscopy, of ZSM-5, IV-2431
furan, alkylation, II-1293

Ga substitution in EMT/FAU intergrowths, III-1871
Ga-beta, III-1707
gallophosphates, ULM-n, III-1737
gallosilicate, ZSM-5, IV-2395
gamma-alumina, from catapal A, IV-2331
$GaPO_4$-14, III-1715
$GaPO_4$-LTA, III-1751
gas(-)
   mixtures, I-103
   separation, I-25
   solid chromatography, use of HMS mesoporous adsorbent, II-1323
Ga-ZSM-5, III-1619, IV-2395
Gd^{3+}, II-1373
gel
   dissolution, III-2057
   sedimentation, III-1817
generation and distribution of new cages in CHA-GME, IV-2445
geometry optimization, I-379
global softness of different molecules, I-223
gold nanoparticles, III-2079
grand canonical Monte Carlo simulations, I-251, I-363, IV-2767
gravimetric analysis, I-67, I-243
greenhouse gas, IV-2863
growth, III-1529
   mechanism, III-1487
   rate constant, zeolite A, III-2049
guest assisted laser ablation (GALA), III-1779

H_2 chemisorption, III-2079
^1H NOESY, 2D, IV-2259, IV-2955
Hahn echo NMR, I-43
H-(Al)ZSM-5, II-1419
H-(Al)ZSM-11, II-1419
H-(Al)ZSM-23, II-1419
Hartree-Fock, I-349
H-beta, I-1337, II-1419, II-1425, II-1455, II-1471, IV-2689
H-CHA, II-1455
HCl, IV-2571
H-D exchange, IV-2955
H_2-D_2 exchange, I-393
heats of adsorption, I-103
   of ammonia, IV-2637
   of methanol, III-1969
heavy metals, IV-2371
heterogeneous catalyst, MCM-41, I-591, III-2163

H-EU-1, II-1433
hexafluorotitanic acid, III-1885
H-FER, II-1433
H-(Fe)ZSM-5, II-1419
H-(Fe)ZSM-11, II-1419
high
    resolution
        TEM, I-713, III-1693, III-2009, IV-2787
        x-ray photoelectron spectroscopy, I-401, II-1367
    silica
        USY, I-203
        zeolites, II-1373, IV-2331
        5-coordinated Si in, IV-2951
H-LTL, II-1455
H-MFI, II-1433
H-MOR, II-1455
H-mordenite, II-1419, II-1447, IV-2689, IV-2673
HMS mesoporous sieve, adsorbent in gas-solid chromatography, II-1323
H-MTW, II-1455
H,Na-EMT, IV-2689
H,NA-Y, IV-2643
H_2O_2, II-1185, IV-2869
    catalyzed disproportionation, II-1213
    decomposition, II-1331
    oxidation, II-1221, II-1391
H-OFF-ERI, II-1455
hopping dynamics, I-43
host-guest
    interaction, III-1765
    species, III-2147
    systems, III-2111
HREM, I-713, III-2009
    of Ce,Ag-ZSM-5, IV-2787
    of ZSM-12, III-1693
H-TON, II-1433
H-USY, II-1455
H-Y, II-1419, IV-2317, IV-2589, IV-2659, IV-2673, IV-2689, IV-2735, IV-2991
    CD_3OH, IV-2279
    distribution of acid strength, IV-2535
hybrid organic-inorganic catalysts, I-675
hydrocarbon diffusion, IV-2991
    in silicalite I, I-183
hydroconversion
    of benzene over beta and mordenite, II-1025
    of light naphtha, IV-2817
hydrocracking
    of n-decane in Y zeolite, II-1011
    of n-heptane, IV-2817
hydrofluorocarbons, sorption of, IV-2301
hydrogen, I-51
    adsorption
        in NaX and NaY, I-269
        on Ru-MCM-41/48, II-807
    bonding, IV-2689
    peroxide
        Mn catalyzed decomposition, II-1185

over Na-Y, IV-2869
oxidation of hydrocarbons and alcohols, II-1221
role in benzene hydroconversion, II-1025
hydrogenation, I-651, IV-2897
    of cinnamaldehyde, III-2155
hydroisomerization
    of C8 paraffins, IV-2913
    of n-heptane, IV-2817
hydrothermal
    aging, III-1877
    stability of metallosilicates, II-1149
    synthesis, III-1649, III-1765
    treatment, effect on ammination, II-1065
hydroxyl
    nest transformation, I-457
    radicals, IV-2927
hydroxylation, of benzene with NO, IV-2847
hydroxysodalite, III-1595
HYSCORE, hyperfine sublevel correlation ESR, II-795
H-ZSM-5, I-209, I-333, II-1471, IV-2259, IV-2273, IV-2317, IV-2577, IV-2689, IV-2955, IV-2991
    adsorption of acetonitrile, IV-2287
    catalytic cracking of palm oil, IV-2921
    CD_3OH, IV-2279
    nanocrystalline, I-141
    realumination of, IV-2743
H-ZSM-12, II-1419
H-ZSM-35, IV-2697

ICP analysis, for Si/Al ratios, IV-2511
ideal adsorbed solution (IAS) theory, I-363
immobilization, I-645
immobilized species, I-209
impact grinding on kaolinite, III-1663
impregnation
    by Fe salt, IV-2847
    of V in beta, III-2111
    V in Si-MCM-41, II-825
incorporation, I-621
    of cobalt and copper cations in clinoptilolite, III-2129
induction period of crystallization, III-2057
inelastic neutron scattering, pyrrole in Na-Y, IV-2609
infrared
    spectroscopy, II-1367, IV-2659
        pyrrole in Na-Y, IV-2609
    transmission window, IV-2339
in situ
    ^1H MAS NMR, IV-2955
    inclusion, III-2111
    NMR, IV-2971
    studies, III-1671
interference microscopy, I-35, I-231
interphase contactor, II-1399
intracrystalline diffusion, I-67, I-231, I-591
introduction of Mo^{5+} ions, IV-2855
inverse gas chromatography, I-263

ion
    exchange, I-13
        Cd^{2+}, IV-2371
        in Zeolite P, IV-2379
        scattering spectroscopy, of Ti-MCM-41, II-825
ionizing radiation, dosimetry, I-277
IR spectroscopy, II-1049, III-1685, IV-2517, IV-2963
    of adsorbed CO_2 in Cs,Na-X, IV-2705
    of alkoxyl groups, IV-2601
    of steamed H-Y, IV-2595
I^2R windows, of Na-beta, I-167
iron(-), III-1877
    in erionite, IV-2927
    oxygen nanoclusters, II-1367
Ising model for diffusion in Na-Y, I-509
isobutane, II-1425
iso-butanol, II-897
isomer selectivity, II-1425
isomerization
    of butane in ferrierite, III-1991
    of butene(s) over
        metal substituted $AlPO_4$-11, II-1239
        metallosilicate ZSM-5, II-1141
    of cis-2-butene, I-667
    of n-butene, II-1433
    p- to m-ethyltoluene, I-209
isomorphous substitution of beta, III-1893
isopropanol, decomposition, II-989
isopropylation of naphthalene, II-1133
isopropylbenzene, II-1419
isosteric method, I-111
isotopic labeling in ZSM-5, II-1057
ITQ-1, as-synthesized, crystal structure, IV-2519

JBW, ion exchange in, IV-2379
jetloop recycle reactor, I-591

K clusters, III-2103
kanemite, II-847
kaolinite, impact grinding, III-1663
Kelvin equation, for nitrogen adsorption, II-761
ketone reduction, IV-2565
K-FAU, III-2103
kinetic(s)
    of nucleation/growth, III-1671
    of zeolite A crystal growth, III-2049
    of ZSM-5 dealumination, III-1953
    permeation-reaction model, II-1399
KIT-1, Pt-containing for catalytic reduction of NO, II-1199
kryptofix, III-1731
K-X, III-2103

La-modified Fe-Y, IV-2875
La,Na-Y, II-1447
lanthanide additives, II-1373

laser(-)
    dye molecules, III-2111
    polarized noble gas nuclei, IV-2945
    pulsed laser deposition, III-1779
lattice
    deformation, IV-2339
    energy, ERS-7, I-541
    structure, III-2117
LAY, low-alumina Y, IV-2651
layer
    growth, III-1487
    silicates, III-1685
layered titanosilicate, II-1277
lead titanate, microporous, II-1277
Lewis
    acid sites, IV-2589
        in H-Y, IV-2425
        in USY, IV-2331
    acidity, IV-2535, IV-2549, IV-2565
        in stilbite, IV-2345
        in titanosilicates, IV-2825
Li-ABW, IV-2309
Li-LSX, I-43, I-85, I-135, IV-2767, IV-2979
Li,Na-LTA, IV-2309
liquid
    hydrocarbon fuels, IV-2921
    ion exchange, IV-2847
    phase
        bromination, III-2203
        oxidation, II-893, II-1391
Li-X, IV-2979
long-term sorption, I-209
low(-)
    alumina Y (LAY), II-1049, IV-2651
    valent cations, in P and JBW, IV-2379
LSX, I-51, I-85, IV-2767, IV-2979
    cation positions in, I-135
LTA, I-371, I-387, II-941, III-2103
lube oil, II-1441
luminescence, III-2241
lung cancer, IV-2927
LZ-210, IV-2651

macrophage, with erionite fiber, IV-2927
macropore diffusion, I-159
magadiite
    Fe and Co impregnated, II-917
    pillaring of, III-1685
magic-angle-turning NMR, IV-2465
magnesioaluminophosphates, III-1603
Malaysian palm oil triglyceride, IV-2921
MAPO-5, synthesis and NMR characterization, IV-2481
MAPO-31, III-1603, III-1743
MAPO-36, synthesis and NMR characterization, IV-2481
MAPO-39, I-675, III-1779
MAS NMR, IV-2259
    of fluorinated $AlPO_4$, IV-2409

MCM-22, I-575
  HMI configuration in, I-445
MCM-41, I-645, I-659, I-713, I-725, II-1419, III-1685, III-1779
  catalytic properties, II-801
  hydrothermal restructuring, I-681
  in ether synthesis, II-885
  KNO_3, II-989
  MCM-41/ZSM-5 material, I-707
  mesopore analysis, II-761
  Mn-Salen, II-749
  modelling of structure, II-839
  monodisperse, II-757
  morphologically well-defined, II-757
  Nb and Al, sorption properties, II-833
  oriented films, II-923
  sample controlled thermal analysis, II-779
  surface modified, II-909
  tetrapyridine copper(II) complex in, II-795
  tubular shaped, II-931
  V containing, redox behavior, II-857
MCM-48, I-697, II-733
  catalytic properties, II-801
  synthesis of, I-689
MCM-50, II-733
MeAlPO$_4$-34, III-1723
MeAPO, IV-2751
medium pore size zeolites, II-1433
MEL, II-941, IV-2419
membrane, II-1403
  asymmetric, I-25
  reactor, III-1795
    parallel passage, I-603
  structure and synthesis, I-603
meso-micro-porous, I-707
mesophase, silica, II-779
mesopores, IV-2735
mesoporous
  aluminophosphates, II-771
  (Cu,Zn,Al) mixed oxide catalysts, I-651
  silicoaluminophosphate, II-771
metal(-)
  catalyst, II-1337
  insulator transition, III-2103
  ion complexes, II-795
  nitrides, III-1535
  supports, I-583
  surfaces, synthesis of zeolites on, I-637
metallocene templates, III-1931
metallophosphonates, III-1649
metalloporphyrin, in Na-Y, IV-2905
metallosilicate(s), III-1619, IV-2803
  (M)-MSU-X, II-741
methane, I-51
  CD$_4$ in
    mordenite, I-197
    silicalite, I-197
  conversion, II-1381
  selective catalytic reduction of NO, IV-2787, IV-2795
  sorption in SAPO-5 and SAPO-11, I-257
methanol, IV-2577
  conversion over ZSM-5, II-1057
  to olefins, I-567
methyl viologen, III-2073
methyloxy species reactivity, IV-2577
methylamine formation, I-349
methylation of toluene, IV-2585
MFI, I-67, I-393, II-941, III-1619, III-1707
  sorbate immobilization in, I-215
MgAPO-20, magic-angle-turning NMR, IV-2465
MgVAPO-5, NMR and ESR, IV-2473
microcalorimetry, I-103
microelectronic packaging, III-2023
microlasers, III-2111
microwave heating, I-97, III-1591, III-1787, III-1917
  assisted crystallization, III-2233
  of A-type zeolites, III-2179
mild steaming, IV-2565
MIL-n, III-1649
mixed gas isotherms, I-363
M-MSU-X metallosilicates, II-741
Mn-cancrinite, III-2087
Mn(II)-exchanged zeolite X, IV-2893
MnSalen*MCM-41, II-749
Mo beta, III-1893
Mobil-badger alkylation process, I-551
modelling, I-349
  benzene diffusion, I-357
modified(-)
  electrode, III-2073
  zeolites, III-2095
    by Mo, IV-2855
molecular
  dynamics, I-371
    HMI in MCM-22, I-445
    simulations
      of diffusion, I-153
      of self diffusion in silicalite, I-149
    vibrational spectra of Y, I-423
  modelling, IV-2355
  packing, I-127
  segregation effects, I-251
  sieve, III-1765
    inclusion pigments, III-2233
molybdenum (VI), templated polyoxoanions, II-787
Monte Carlo, grand cononical simulations, I-251, I-363, IV-2767
mordenite, I-197, II-944, III-1803, IV-2673
  Ag clusters in, II-1165, III-2143
  alkylation of naphthalene, II-1307
  benzene hydroconversion over, II-1025
  dealumination, II-1133
  dehydrated, Co ions in, II-1193
  fibers, IV-2927
  methane and ethane in, I-363
  methanol ammination, II-1065
  morphology of, III-1803

surface passivation, III-1975
uses, I-551
morphological control, I-719, III-1627
Mössbauer, ETS10, III-1507
Mott transition, III-2103
MQ MAS NMR, IV-2985
    of fluorinated $AlPO_4$, IV-2409
M41S, II-733
MTG process, IV-2577
MTO conversion, I-567, I-591
MTW, III-1655
Mu-4, III-1633
Mu-7, III-1633
multinuclear NMR spectroscopy, IV-2963
m-xylene isomerization, III-1771

Na-A, I-77, III-1663
    ceramic precursor, III-2033
NaA, synthesis, III-1553
Na-beta, IV-2681
    benzene in, I-167
Na,Ca-A, methanol in, I-231
NaCl titration, IV-2833
NaH-X, alkylation, IV-2601
NaH-Y, alkylation, IV-2601
Na,K-LSX, IV-2401
n-alkanes, I-67
    adsorption, IV-2623
Na-mordenite, IV-2673
nanoclusters, II-1367
nanocomposites, III-2209
nanocrystals, ZSM-5, II-1441
nano-oxide particles, III-2249
nanopowders, silicalite-1 and TS-1, III-1825
nanosized crystals, III-1787
nano-ZSM-5, I-141
naphthalene, II-897
    alkylation over mordenite, II-1307
    isopropylation, II-1447
narsarsukite, transformation from ETS-4, IV-2457
Na-X, I-59, I-167, I-357, III-1663, IV-2839
    desorption from, I-223
    hydrogen adsorption in, I-269
    tetrapyridine copper(II) complex in, II-795
    Zn^{2+}-exchanged, IV-2295
Na-Y, I-175, I-357, II-1337, III-2095, IV-2265, IV-2673
    "build-bottle-around-ship", IV-2897
    ceramic precursor, III-2033
    CO in, IV-2615
    hydrogen adsorption in, I-269
    laser-induced luminescence, IV-2425
    Na-Y/KNO_3, II-989
    nitrile hydratase activity, IV-2905
    ship-in-the-bottle catalysis, III-2163
    sorption of hydrofluorocarbons, IV-2301
    synthesis, III-1553
    tetrapyridine copper(II) complex in, II-795
    transition metal-exchanged, IV-2869

Na-ZSM-5, IV-2945
$NbCl_5$, exchange in zeolite Y, II-997
Nb_2O_5, exchange in zeolite Y, II-997
Nb-substituted ZSM-5, III-1901
n-butane adsorption, I-91
n-butanol, II-897
n-butene, II-1425
n-decane, II-1359
neopentane, I-51
    dynamics in $AlPO_4$-5, I-189
neutral media synthesis, III-1925
neutron
    diffraction, I-197, I-659
        of H in ZSM-5, IV-2413
        of LSX, I-135
    scattering(/)
        incoherent quasi-elastic, I-35, I-51, I-59, I-77
        inelastic, I-119
NH_3, thermodesorption, IV-2651
n-heptane
    aromatization, IV-2759
    hydrocracking, IV-2719, IV-2817, IV-2935
    hydroisomerization, IV-2817
n-hexane aromatization, IV-2557
    cracking, III-1849, III-1969
NH_4-stilbite, IV-2345
Ni(II) chelate complex, III-1765
Ni(II)-tetrakis(n-methyl-4-pyridyl)-porphyrin, IV-2897
nitric acid, IV-2735
nitrile hydratase, IV-2905
nitrogen adsorption, I-675, I-725
    in MCM-41, II-761
    in ZSM-5, I-309
nitrous oxide, oxidation of benzene to phenol, II-1245
Ni-ZSM-5, IV-2817
NMR, III-1503, III-1685, IV-2295, IV-2945
    ^{27}Al MAS NMR, IV-2863, IV-2971
    Al MQ MAS of La-exchanged Y, IV-2515
    ^{13}C EIS NMR, I-43
    ^{13}C MAS NMR, I-91, I-681, IV-2273, IV-2325, IV-2535
    ^{133}Cs MAS NMR, IV-2301
    DOR NMR, of fluorinated $AlPO_4$, IV-2409
    DQ NMR, of fluorinated $AlPO_4$, IV-2409
    ^{19}F MAS NMR, IV-2971
    ^{19}F RFDR NMR, of $GaPO_4$, IV-2453
    ^{19}F solid-state NMR spectroscopy, IV-2951
    ^1H MAS NMR, II-779
    ^1H NMR imaging, IV-2991
    Hahn echo NMR, I-43
    ^3He NMR, IV-2945
    invisible aluminum in Y, IV-2497
    ^6Li MAS NMR, I-85, IV-2979
    ^7Li MAS NMR, IV-2979
    magic-angle-turning, IV-2465
    MAS NMR, IV-2259, IV-2409
    ^{14}N MAS NMR, of M41S, II-733

^{23}Na MAS NMR, I-85, IV-2301, IV-2985
NOESY, IV-2259, IV-2955
  of acetonitrile adsorption in H-ZSM-5, IV-2287
  of Co-AlPO$_4$-5 and TS-1, IV-2489
  of fluorinated AlPO$_4$, IV-2409
  of H-ZSM-35, IV-2697
  of MAPO-5 and MAPO-36, IV-2481
  of MCM-41, II-749, II-801
  of MCM-48, II-801
  of M41S, II-733
  of steamed H-Y, IV-2595
  of VAPO-5 and MgVAPO-5, IV-2473
^{31}P MAS NMR, IV-2971
pulsed field gradient NMR, I-35, I-77
RFDR NMR, of fluorinated AlPO$_4$, IV-2409
^{29}Si MAS NMR, IV-2317, IV-2863, IV-2951
2D EXSY NMR, IV-2945
2D 3Q MAS of titano-silicates, IV-2457
variable temperature, IV-2265
^{129}Xe NMR, III-2079, IV-2589, IV-2945, IV-2991
NO
  activation on CoAPO-18, IV-2775
  adsorption
    isotherms in H-Y, IV-2589
    on SAPO, IV-2751
  decomposition, I-473, IV-2803, IV-2809
  hydroxylation of benzene, IV-2847
  reduction, II-1359
    by hydrocarbons, II-1149
    with methane, IV-2833
  SCR over ZSM-5, II-1157
  selective catalytic reduction of, IV-2795
N$_2$O decomposition, IV-2863
NOESY
  MAS-NMR, IV-2259, IV-2955
  nuclear Overhauser effect spectroscopy MAS-NMR, IV-2259, IV-2955
non-aqueous
  media, III-1633
  synthesis, of CoPO$_4$, IV-2387
nonframework
  aluminum, IV-2549, IV-2743
  iron, Fe-ZSM-5, IV-2781
nonlinear optical properties, III-2121
nonporous silica particles, II-757
normal
  alkanes, I-67
  analysis, I-423
n-pentane, I-325
  adsorption, I-91
n-propylbenzene, I-575, II-1419
nuclear Overhauser effect spectroscopy, NOESY, IV-2259, IV-2955
nucleation, III-1529, III-1627
  of silver, III-2137
  of zeolite X, III-2001

O$_2$ sorption, IV-2317

OH groups, IV-2595
o-hydroxyacetophenone, II-1463
olefins, III-2163
1,3-butadiene, aromatization of, IV-2727
1-butanol, in synthesis of theta-1, III-2041
1-butene isomerization, III-1771
1-dimensional Se chains, III-2217
one-dimensional phase, I-189
1-hexanol, as diluent, II-1455
optical sensors, III-2249
ordered mesoporous materials, I-725, II-761
organic(-)
  functionalized molecular sieves, III-1479
  halides, chemo-selective oxidation, II-1227
  silica interaction, III-1693
organosilanes, I-725
organosiloxanes, II-767
orientated films, III-1779
OU-1, III-2009
oxidation
  liquid phase, II-893
  of alcohols and hydrocarbons, II-1221
  of alkenes, II-1391
  of benzene with NO, II-1245
  of cyclohexane, II-749, II-1233
  of ethanol, II-877
  of olefins, II-1213
  of organic halides, II-1227
oxyfunctionalization of n-hexane by aqueous H$_2$O$_2$, II-1399
oxygen, sorption, I-111
oxygenated compounds, IV-2557
oxygenates, IV-2557
o-xylene, desorption from Na-X, I-215

P, ion-exchange in, IV-2379
palladium(-)
  complexes, III-2155
  nanoparticles, III-2079
  salen-complexes, III-2155
palm oil, catalytic cracking of, IV-2921
paraffin isomerization, II-1049, IV-2541
parallel passage membrane reactor, I-603
paramagnetic O$_2$, I-85
para-nitration, of toluene, II-1471
particle
  growth mechanism, TS-1, III-1523
  size distribution, TS-1, III-1523
PBE-1, I-575
PbI$_2$ clusters in zeolite A, III-2225
Pd clusters, II-1337, IV-2665
Pd^{2+}, IV-2833
Pd-H-ferrierite, II-1331
Pd-H-ZSM-5, II-1331
Pd/Na-Y, IV-2665
Pd-Pt-containing TS-1, II-1351
PEM-FC, II-1409
periodic nodal surfaces, I-533
permanganate, III-2087
permselectivity, I-613

pervaporation, III-1841
p-ethyltoluene, I-209
PFG NMR, of silicalite, I-149
phenol
    acylation, gas phase, II-1463
    adsorption on H-Y, IV-2659
    conversion, II-1277
    dimer, IV-2659
    trimer, IV-2659
photocatalytic reactivity, I-667
photochemical activation, III-2187
photochemistry, III-2095
photoelectrocyclization, III-2095
photoluminescence, of mesoporous molecular sieves, I-667
photosensitive refractive indices, III-2121
pillared material, III-1641
pillaring of Al-magadiite, III-1685
pinholes, in membranes, I-603
platelet faujasite, III-1627
PMMA, III-2209
polar molecules, III-2117
polarization dependent absorption, III-2121
pollutants, separation of, I-97
polydimethylsiloxane, II-1399
polymeric templates, III-1603
polymerization, III-2209
polyolefin
    acid catalyzed degradation, II-1081
    cracking, II-1441
polysulphide, III-2087
population balance method, III-2057
pore
    diameter, control of, II-771
    mouth catalysis, IV-2913
    size distribution, I-409
        of MCM-41, II-761
    walls, I-713
porous
    glass, III-1849
    silica, I-689
powder diffraction, structure solution, I-533, I-537
precalcination effect, IV-2863
precursor layer, I-621
preferred orientation, III-1809
pressure swing adsorption (PSA), I-203
pressurized sol-gel technique, III-1841
probe molecules, I-223
    benzene, IV-2259
    chloromethane, IV-2689
    trimethylphosphine oxide, IV-2331
propane, II-1359, IV-2273
    adsorption, I-91
    diffusion in silicalite I, IV-2615
propene, II-1425
propionylation of toluene over beta, II-1317
propylene, I-567
    oxide synthesis, II-1351
    to isopropanol, II-1313
protolytic cracking, I-333
protonated zeolites, IV-2571, IV-2689

Pt particles, II-1337
Pt/BaKL, IV-2557
Pt,Co-ZSM-5, IV-2809
Pt,Cu-ZSM-5, IV-2809
Pt/H-EMT catalyst, IV-2719
Pt/H-FAU catalysts, IV-2719
Pt/H-MOR, IV-2541
Pt/H-ZSM-5, IV-2541
Pt/KL, IV-2557
Pt/Na-Y, IV-2665
Pt/rare-earth-FAU, IV-2541
Pt-ZSM5, IV-2881
pulp bleaching, II-1185
pulsed
    field gradient NMR, I-35, I-77
    laser deposition, III-1779
pyridine
    adsorption of, IV-2697
    in Y and mordenite, IV-2673
pyrocatechol, III-1701

QENS, - see quasi-elastic neutron scattering
QM calculations, NO on Ti-silicalite, I-473
quantum chemical
    modelling, I-379
    techniques, I-333
quasi-elastic neutron scattering (QENS), I-35, I-51, I-59, I-77

Raman spectroscopy, III-2217
    of $AlPO_4$-5 and SAPO-44, III-2147
    of ZSM-5, IV-2431
rare earth
    in USY, IV-2935
    sodalite, III-2241
rate spectrum, of sorption in A, X and Y, I-159
rational
    design, III-1723
    synthesis, zeolite A, III-1863
reaction pathways, I-349
realumination, of H-ZSM-5, IV-2743
$Re(Co)_3(bpy)Cl$ in zeolite Y, III-2187
recycle reactor, I-235
redox
    behavior
        of T-MCM-41, II-825
        of V-MCM-41, II-857
    catalysis, II-941, III-2129
reduction
    of Fe ions, IV-2875
    of NO, II-1359, IV-2889
reinsertion, of aluminum, IV-2743
relaxation shift, IV-2339
renucleation, I-621
RFDR NMR, of fluorinated $AlPO_4$, IV-2409
Rh particles, II-1337, IV-2265, IV-2863
Rh_6 clusters, in Na-Y, IV-2265
Rh/NaY catalyst, IV-2863
RHO, methanol ammination, II-1065
Rh/USY catalyst, IV-2863

Rietveld refinement, III-1585, III-1715, III-1877
 of Na,K-LSX, IV-2401
 of NH_4- and H-ferrierite, IV-2361
rod-shaped mesoporous powders, I-719
role of metal and Brönsted acid sites, IV-2809
rotational dynamics of methane in ZSM-5, I-497
Ru-MCM-41, II-809
Ru-MCM-48, II-809
Ru metal cluster, II-809
Rutherford backscattering spectroscopy, of metallosilicates, IV-2395

salen complexes, I-675, III-2195
salt(s)(-)
 and oxides, I-175
 and polyfluorinated hydrocarbons addition, II-1351
sample controlled thermal analysis (SCTA), III-1707
 MCM-41, II-779
SAPO-5, IV-2913
 binary adsorption of CO_2 and CH_4, I-257
 2D 1-H NOESY, IV-2259
SAPO-11, IV-2913
 binary adsorption of CO_2 and CH_4, I-257
 disproportionation of ethylbenzene over, II-1019
 synthesis, II-1019
 2D ^{1}H NOESY, IV-2259
SAPO-31, III-1743
SAPO-34, I-567, III-1723
 magic-angle-turning NMR, IV-2465
SAPO-37, III-1771
 templates in, IV-2325
SAPO-41, IV-2913
SAPO-44, III-2147
saponite, II-1261
SAPOs, I-341
scanning pyroelectric microscopy (SPEM), III-2117
scattering function, I-371
SCR of NO, II-1157
SCTA, of MCM-41, II-779
SDAs, III-1693
Se in $AlPO_4$-5, III-2217
secondary growth, I-621
seed film method, III-1857
seeding,
 in synthesis of high silica zeolites, III-2065
 of silicalite-1 thin films, III-1809
selected area electron diffraction, of mordenite, III-1803
selective
 catalytic reduction
  of NO, IV-2795
  over Co-beta, II-1169
  with Pt/KIT-1, II-1199

oxidation, III-2195
 Ti-MCM-41 catalyst, II-817
selectivity, I-103
 in ZSM-5, I-591
selenium, in $AlPO_4$-5 and SAPO-44, III-2147
self-assembled catalyst, I-645
self-diffusion, I-371
 of benzene in Na-X, Na-Y, I-357
 of methane and xenon in silicalite, I-149
self-reduction, III-2079
SEM, I-713, I-719, III-1627, III-1877
 field emission, I-689
 of B-ZSM-12, III-1655
 of ceramic membrane, III-1795
 of CoAPO-37, III-1771
 of $GaPO_4$-LTA, III-1751
 of MAPO-39, III-1779
 of mordenite, III-1803
 of pillared stannic phosphate, III-1641
 of silica sodalite, III-1701
 of silicalite-1 thin film, III-1809
 of SiO_2/magadiite, III-1685
 of spherical monodisperse MCM-41, II-757
 of zincophosphates, III-1757
 of ZSM-12, III-1693
separation, I-111
 of N_2/O_2, I-119
shape selectivity, II-1447, II-1471, IV-2549
ship-in-the-bottle catalysts, III-2155, III-2163
Si beta, III-1893
silanation of ZSM-5 and Y, III-1969
silanol groups, III-2171
silica(-)
 aerosil300, IV-2945
 mesophase, II-779
 porous, I-689
 sodalite, III-1701
 xerogel, III-2001
silicalite, I-67, I-97, I-149, I-183, I-197, I-235, I-363, II-1455, III-1701, III-1877, IV-2571, IV-2615
 membrane, III-1841
 methane and ethane in, I-363
 nanopowders, III-1825
 synthesis, III-1553, III-1983
 thin films, III-1809
 Ti, interaction with NO, I-473
silicoaluminophosphate, mesoporous, II-771
silicoaluminophosphate-5, IV-2259
silicoaluminophosphate-11, IV-2259
silicon, 5-coordinated, IV-2951
siloxy, I-457
silsesquioxane, I-645
silver
 clusters, in mordenite and erionite, III-2143
 nucleation, III-2137
simulated annealing
 of frameworks, I-517
 structure solution of ERS-7, I-541

simulation, of solvent recovery, I-203
single-file systems, I-153
singlet molecular oxygen, II-1213
skeletal butene isomerization, II-1433
small-angle x-ray scattering, III-1529
Sn particles, II-1337
Sn-zeolites, III-2249
sodium
   azide, IV-2875
   zinc
      germanates, IV-2379
      silicates, IV-2379
solid(-)
   state
      ion exchange, I-175
         Cu/Co in clinoptilolite, III-2129
      reaction, III-2171
      superacids, IV-2541
solubility, III-1595
solvent recovery, I-203
solvothermal synthesis, III-1641
sorbate(-)
   concentration, I-231
   immobilization, in MFI, I-215
   induced phase transitions, IV-2431
sorbic acid, III-2155
sorption, I-13, I-111
   energetics, in $AlPO_4$-5, I-127
   in A, X and Y, I-159
   of hydrocarbons in 13X and ZSM-5, I-243
   simulations, I-251
spectrasol Z-B, IV-2511
spectroscopic characterization, III-1569
spherical monodisperse porous MCM-41, II-757
spin-echo mapping NMR, Co-$AlPO_4$-5 and TS-1, IV-2489
spontaneous thermal dispersion, I-175
SSZ-13, I-341
SSZ-24, I-341
SSZ-42, synthesis and characterization, III-1945
stability, III-1737
stacking sequences, III-1693
stannic phosphate, III-1641
statistical
   Design of Experiment (DOE), IV-2921
   thermodynamics of cations in ZSM-5, I-481
steam reforming, methanol, II-1409
steaming, IV-2595
   of ZSM-12, I-293
stilbite
   crystal structure, IV-2345
   NH_4-exchanged, IV-2345
stoichiometric generation of acid sites, IV-2643
Stop-Go method, I-203
strong acid site, IV-2549, IV-2629
structural lability, III-1679

structure(-)
   AlPO-HAD, III-1937
   control, II-1367
   directed transition state selectivity, II-1419
   directing agents, III-1693, III-2065
      dynamics of, III-1503
      importance of OH, III-2065
   envelope mask, I-533
   ITQ-1, IV-2519
   OU-1, III-2009
   solution
      use of structure envelope, I-533
      with textured samples, I-537
   STS titanosilicate, IV-2525
structured catalysts, I-603
structuring, I-713
STS titanosilicate, structure, IV-2525
styrene
   epoxidation of, II-749
   from C_4, I-551
   oxide, rearrangement over ZSM-5, II-1129
sulfated zirconia, IV-2541
sulfonic acid sites, III-1479
sulfur
   in $AlPO_4$-5 and SAPO-44, III-2147
   tolerance, of Co-Pt/ZSM-5, IV-2889
super acid sites, IV-2549, IV-2629
supercritical fluids, II-1447
superlattice reflections in zeolite A, III-2225
superoxide dismutase, III-2195
surface
   hydroxyls, I-349
   modification, of MCM-41, I-725
surfactant templating, II-741
switchable optical properties, III-2121
synchrotron diffraction, I-537, I-541, IV-2413, IV-2295
synthesis,
   AlPO-HAD, III-1937
   aluminophosphate, III-1561
   aluminosilicates in liquid ammonia, III-1535
   B/Ti ZSM-5, III-1925
   continuous, III-1553
   CrAPO-5, III-1577
   defect-free high silica zeolites, III-1495
   erionite/offretite, III-1969
   ERS-7, I-541
   ferrierite, III-1991
   fluoride ion, III-1535
   high silica zeolites, III-1503
   methylacrylate polymer, III-2209
   mixture dilution, III-1495
   OU-1, III-2009
   promoter(s)
      assisted, III-1611
      oxy-anion, III-1611

SAPO-11, II-1019
SSZ-24, III-1945
TAPO-34, III-1569
TAPSO-34, III-1569
theta-1, III-2041
Ti/Al beta, III-1885
TS-1, III-1917
tubular reactor, III-1553
UDT1, UDT12, UDT-18, III-1931
zincophosphates, III-1757
Zr containing beta, III-1909
ZSM-5, I-637
    membrane, I-613
ZSM-25, III-2033

TAPO-34, III-1569
TAPSO-34, III-1569
TEM, I-689, I-719, III-2079
    of CeO_2-H-ZSM-5, IV-2795
    - see also HREM
temperature programmed desorption, I-91, I-223
    of ammonia, IV-2549, IV-2643
templated molybdenum (VI) polyoxoanions, II-787
templates, III-1723, IV-2355
    in ZSM-5 and SAPO-37, IV-2325
templating effect, III-1731
TEOTi, III-1885
tert-butanol, II-897
tert-butyl ether, I-325
tertiary amines, for Si/Al analysis, IV-2511
tetraethylammonium, III-1707
tetrahedral metal oxide moieties, I-667
tetrapropylammonium, III-1707
tetrapyridine copper(II) complex, II-795
textured samples, structure solution, I-537
thermal transformations
    of Li-ABW, IV-2309
    of Li,Na-LTA, IV-2309
thermodesorption, of NH_3, IV-2651
thermodynamic models, IV-2371
thermogravimetry, I-675
theta-1, synthesis and characterization, III-2041
3,5-diethyl pyridine, II-1101, III-1585
3A, zeolite, III-2179
three-dimensional channel system, III-1715
Ti-beta, II-1391
Ti-containing molecular sieves, II-1213
    oxidation of diethylamine, II-1269
Ti-MCM-41, IV-2825
    from acidic media, II-817
time-resolved synchrotron x-ray powder diffraction, IV-2401
Ti-silicalite, II-1351
titanium(-), I-645, III-1877
    containing aluminophosphate molecular sieves, II-1221
titanosilicate molecular sieves, II-1403, III-1507, IV-2825, IV-2985
    layered, II-1277

Ti-UTD-1, II-1403
Ti-zeolites, III-2249
toluene, II-1337, II-1471
    bromination, III-2203
    disproportionation over ZSM-5, III-2015
    methylation, I-175, IV-2585
TPD-mass spectroscopy, I-215, IV-2295, IV-2681
    of alkoxyl groups in NaH-X,Y, IV-2601
TPO, I-651
TPR, I-651, II-809, II-1409
transalkylation, I-575
    over zeolite omega, II-1041
transient experiments, I-235
transition(-)metal(-)(/)
    complex encapsulation, III-1765
    exchanged Na-Y, IV-2869
    ligand, I-675
    ZSM-5, II-1381
translational mobility, I-51
transmission FT-IR spectroscopy,
    - see FTIR spectroscopy
transport diffusivities, I-371
trimer, phenol, IV-2651
trivalent (T) atoms, incorporation of, III-1619
tropolone ethers, III-2095
TS-1, II-1351, III-1523, IV-2825
    and catalytic properties, II-1283
    nanopowders, III-1825
    oxidation of
        diethylamine, II-1269
        organic halides, II-1227
    spin-echo mapping NMR, IV-2489
    synthesis, III-1917
twinning, of FAU-like zincophosphate, III-1627
2D EXSY NMR, IV-2945
2D, ^1H NOESY, IV-2259, IV-2955
2-methyl-3-butyn-2-ol decomposition, III-2129
2-methyl naphthalene, II-893
2-methyl-1,4, naphthaquinone, II-893
2-methylpentane, I-325
two-step crystallization, I-707
2,3,5-collidine, II-1101

ULM-3/4/5, III-1737, IV-2409
uninodal zeolites, I-517
US patents, statistics, I-551
USY, I-575, II-1463, IV-2331, IV-2549, IV-2735, IV-2855
    catalytic cracking of palm oil, IV-2929
    coking, IV-2935
    solvent recovery, I-203
UTD-1, II-1403, III-1779
    synthesis, III-1931
UTD-12/18, synthesis, III-1931

UV-visible spectroscopy, I-675, II-817,
  II-825, III-1877, III-2249, IV-2775
  beta, III-1515
  of Pd complexes of salen, III-2155

vanadium
  insertion, III-2171
  supported complexes, III-2163
vanadodiphosphates, III-1649
van Hove correlation function, I-371
VAPO-5, NMR and ESR, IV-2473
vapor-phase transport, I-613
V-beta, III-1893, III-2171
V-HMS, I-667
vibrational spectroscopy, diatomic gases on ETS-10, I-317
vicinal silanol pair defect, I-457
V-MCM-41, II-825
  redox behavior, II-857
volatile organic compounds, I-97, II-1337
VPI-5, Cr-containing, cyclohexane oxidation, II-1233

washing conditions, of Ti-MCM-41, II-817
water
  confinement, I-659
  sorption in AlPO-5, I-285
  tolerance, IV-2889
  vapor treatment method, IV-2643
wet oxidation, of acetic acid, IV-2869
whispering gallery mode, III-2111

X, zeolite, I-111, I-119, I-159, II-1941, III-1595
  base strength of cesium containing, IV-2503
  BaX, I-59
  Cs,Na-X, IV-2705
  Cs-X, IV-2317
  doping with Eu, III-2241
  Fe-X, IV-2875
  inverse gas chromatography study of, I-263
  K-X, III-2103
  Li-X, IV-2979
  NaH-X, alkylation, IV-2601
  Na-X, I-77, I-357, III-1663, IV-2839
    desorption from, I-223
    tetrapyridine copper(II) complex in, II-795
    Zn^{2+}-exchanged, IV-2295
  Zn-X, IV-2295
XANES, III-1877, III-2217
XPS, II-1381, III-1877, III-2163
x-ray
  diffraction, I-675, III-1685, III-1877, IV-2301
  photoelectron spectroscopy, I-401, II-1367
  powder diffraction
    ITQ-1, IV-2519
    EU-1, IV-2355

reflection partitioning, I-537
xylene, I-59
  bromination, III-2203
  isomerization with Zr beta, III-1909

Y, zeolite, I-159, II-941, III-2073, IV-2945
  acid strength, QM study, I-429
  activity and selectivity, II-1011
  base strength of cesium-containing, IV-2503
  cracking of 2-methylpentane, I-465
  cresol transformation over, II-1093
  dye synthesis in, III-2233
  encapsulation of Re complex, III-2187
  ESCA of, IV-2439
  exchange with Nb_2O_5 and $NbCl_5$, II-997
  molecular dynamics simulation, I-423
  n-heptane cracking over, II-975
  Pd containing, in benzene oxidation, II-1253
  polarization of La cations in, IV-2515
  reversible Al coordination, IV-2497
  vibrational spectra, I-423
    Cs,Na-Y, IV-2705
    Cs-Y, IV-2319
      sorption of hydrofluorocarbons, IV-2301
    Cu-Y, IV-2425
    Fe-Y, IV-2875
    H,Na-Y, IV-2643
      alkylation, IV-2601
    H-Y, II-1419, IV-2317, IV-2589, IV-2659, IV-2673, IV-2689, IV-2735, IV-2991
      CD_3OH, IV-2279
      distribution of acid strength, IV-2595
    La-modified Fe-Y, IV-2875
    La,Na-Y, II-1447
    La,NaY-73, II-1447
    LAY, low-alumina Y, IV-2651
    Na-Y, I-175, I-357, II-1337, III-2095, IV-2265, IV-2678
      build-bottle-around-ship, IV-2897
      ceramic precursor, III-2033
      CO in, IV-2615
      laser-induced luminescence, IV-2425
    Na-Y/KNO_3, II-989
    nitrile hydratase activity, IV-2905
    ship-in-the-bottle catalysis, III-2163
    sorption of hydrofluorocarbons, IV-2301
    tetrapyridine copper(II) complex in, II-795
    transition metal-exchanged, IV-2869

zeolite
　catalysis, III-2195
　coatings, I-583
　dye composites, III-2121
　films, I-621
　layer thickness, I-583
　membranes, I-621
　modification, I-175
　nucleation, III-2057
　pigments, III-2087
　surface, I-451
　　structure, III-1487
　synthesis, IV-2355
　　modelling, III-1595
　tubes, hollow, bilayered, I-629
zeolitized fly-ash, catalytic activity of, II-1105
zinc, in ZSM-5, IV-2711
zincophosphate-X, III-1627
ZK-5, n-alkane adsorption, IV-2623
ZLC chromatography, I-215, I-243
ZnAPO-50, III-1585
Zn-X, IV-2295
Zn-ZSM-5, IV-2711
Zr containing beta, III-1909
ZSM-5, I-67, I-77, I-235, I-393, I-437,
　　I-481, I-489, I-575, I-591, I-613,
　　I-629, I-637, I-713, II-941, II-1093,
　　II-1113, II-1141, II-1177, II-1193,
　　II-1207, II-1245, II-1337, II-1409,
　　II-1463, III-1529, III-1543, III-1619,
　　III-1707, III-1817, III-1833, III-1953,
　　IV-2259, IV-2413, IV-2431, IV-2549,
　　IV-2803, IV-2855
　activation energy of dealumination, III-1953
　activity, I-591
　　of C-H and C-C bonds in cracking, I-489
　Ag-, IV-2787
　Al-, III-1619
　alkali exchanged, use in benzene oxidation, II-1245
　ammonia-alkali cations in, I-481
　aspect ratio, III-1817
　B-ZSM-5, III-1619
　carbonylation of formaldehyde over, II-1113
　Co-, IV-2339
　Co, NOx adsorption, I-437
　coating stainless steel, I-637
　Co/Pd/H- catalysts, IV-2833
　Co-Pt, IV-2889
　cresol transformation over, II-1093
　crystallization on ceramic support, III-1833
　Cu(-), II-1359
　　complex as carbonylation catalyst, II-1177
　dealumination kinetics, III-1953
　dehydrated, Co ions in, II-1193
　desilication, III-1543
　Fe-, II-1367, III-1707
　FT Raman spectroscopy, IV-2431
　Ga-, III-1619, IV-2395
　Ga containing for alkylation and isomerization, II-1207
　H-(Al), II-1419
　H-(Fe), II-1419
　hollow, bilayered tubes, I-629
　H-ZSM-5, I-209, I-303, II-1419, II-1433,
　　II-1471, IV-2259, IV-2273, IV-2317,
　　IV-2577, IV-2689, IV-2955, IV-2991
　　adsorption of acetonitrile, IV-2287
　　catalytic cracking of palm oil, IV-2921
　　nanocrystalline, I-141
　　realumination of, IV-2743
　isomerization of butane, II-1141
　location of protons in, IV-2413
　magic-angle-turning NMR, IV-2465
　membranes, synthesis, I-613
　metal substituted, III-1901
　methanol conversion over, II-1057
　modified, IV-2759
　morphology, III-1817
　Na-, IV-2945
　nanocrystalline, II-1441
　Pd loaded, II-1331
　pore size control, III-2015
　precursors, III-1529
　promoter assisted synthesis, III-1611
　rearrangement of styrene oxide, II-1129
　rotational dynamics of CH_4 in, I-497
　silanisation, III-2015
　sorbate immobilization, I-215
　sorption of N_2 and CO_2, I-309
　surface modification, III-2015
　synthesis, III-1849
　　in alkali free media, III-1925
　　of Ti/B substituted, III-1925
　　with microwave heating, III-1591
　templates in, IV-2325
　thermal stability, III-1543
　2D ^1H NOESY, IV-2259
　V containing for ethanol oxidation, II-877
　water sorption in, I-301
　Zn-, IV-2711
ZSM-11, II-1419, IV-2803, IV-2945
　structure analysis, IV-2419
　use in benzene oxidation, II-1245
　V-containing for ethanol oxidation, II-877
ZSM-12, I-525, III-1663, III-1693
　B-, III-1655
　dealumination by steaming, I-293
　morphology, III-1817
　promoter assisted synthesis, III-1611
　water sorption in, I-293
ZSM-22, IV-2913
　n-heptane cracking over, II-1033
ZSM-23, IV-2913

ZSM-25, synthesis and characterization, III-2023
ZSM-35, synthesis, III-1849
ZSM-48
    Fe-
        and Cr-containing, II-869
        containing, II-863, II-869

Mn, Al, and Co, catalytic performance, II-901
promoter-assisted synthesis, III-1611